physica status solidi (b)

SUBJECT AND AUTHOR INDEX (b) VOLUMES 73 TO 78 · 1976

SUBJECT AND AUTHOR INDEX (a) VOLUMES 33 TO 38 · 1976

physica status solidi (a)

Classification Scheme

1. Structure of Crystalline Solids
   1.1 Perfectly Periodic Structure
   1.2 Solid-State Phase Transformations
   1.3 Alloys. Metallurgy
   1.4 Microstructure (Magnetic Domains See 18; Ferroelectric Domains See 14.4.1)
   1.5 Films
   1.6 Surfaces
2. Non-Crystalline State
3. Crystal Growth
4. Bonding Properties
5. Mössbauer Spectroscopy
6. Lattice Dynamics. Phonons
7. Acoustic Properties
8. Thermal Properties
9. Diffusion
10. Defect Properties (Irradiation Defects See 11)
    10.1 Metals
    10.2 Non-Metals
11. Irradiation Effects (X-Ray Diffraction Investigations See 1 and 10)
12. Mechanical Properties (Plastic Deformations See 10)
    12.1 Metals
    12.2 Non-Metals
13. Electron States
    13.1 Band Structure
    13.2 Fermi Surfaces
    13.3 Surface and Interface States
    13.4 Impurity and Defect States
    13.5 Elementary Excitations (Phonons See 6)
        13.5.1 Excitons
        13.5.2 Plasmons
        13.5.3 Polarons
        13.5.4 Magnons
14. Electrical Properties. Transport Phenomena
    14.1 Metals. Semi-Metals
    14.2 Superconductivity. Superconducting Materials and Devices
    14.3 Semiconductors
        14.3.1 Films
        14.3.2 Surfaces and Interfaces
        14.3.3 Devices. Junctions (Contact Problems See 14.3.4)
        14.3.4 High-Field Phenomena, Space-Charge Effects, Inhomogeneities, Injected Carriers (Electroluminescence See 20.3; Junctions See 14.3.3)
    14.4 Dielectrics
        14.4.1 Ferroelectrics
15. Thermoelectric and Thermomagnetic Properties
16. Photoconductivity. Photovoltaic Effects
17. Emission of Electrons and Ions
    17.1 Field Emission Microscope Investigations
18. Magnetic Properties
    18.1 Paramagnetic Properties
    18.2 Ferromagnetic Properties
        18.2.1 Ferromagnetic Films
    18.3 Ferrimagnetic Properties
    18.4 Antiferromagnetic Properties

(Continued on cover three)

physica status solidi (b)
basic research

physica status solidi (a)
applied research

SUBJECT AND AUTHOR INDEX

Volumes 73 to 78, 1976 — **Volumes 33 to 38, 1976**

Edited by

E. BORCHARDT

AKADEMIE-VERLAG · BERLIN
and
ACADEMIC PRESS · LONDON-NEW YORK

Subscriptions and orders for this Subject and Author Index
or single copies of physica status solidi (b) and physica status solidi (a)
for the British Commonwealth and USA should be directed to:

ACADEMIC PRESS INC. (London) Limited
24 – 28 Oval Road
London N. W. 1

or

ACADEMIC PRESS INC., 111, Fifth Avenue, New York 10003, USA
whichever address is the most convenient.

For all other countries should be directed to:
AKADEMIE-VERLAG, DDR-108 Berlin, Leipziger Straße 3 – 4.

Distributed in Great Britain by Academic Press, London.

Application to mail at second-class postage rates is pending at Jamaica, New York 11431, USA.

Herausgeber und verantwortlich für den Inhalt des Registers:
Dipl.-Phys. E. Borchardt, DDR-102 Berlin, Neue Schönhauser Straße 20.
Verlag: Akademie-Verlag, DDR-108 Berlin, Leipziger Straße 3—4; Fernruf 22 36 221 und 22 36 229; Telex-Nr. 114420;
Postscheckkonto: Berlin 350 21; Bank: Staatsbank der DDR, Berlin, Kto.-Nr.: 6836-26-20712.
Veröffentlicht unter der Lizenznummer 1310 und 1620
des Presseamtes beim Vorsitzenden des Ministerrates der Deutschen Demokratischen Republik.
Gesamtherstellung: VEB Druckerei „Thomas Müntzer", DDR-582 Bad Langensalza.
Bestellnummer des Registers 1068/85 R 9.

AN(EDV) 20831/20735

Preface

The ninth index volume contains volumes 73 to 78 of "physica status solidi (b) — basic research" as well as volumes 33 to 38 of "physica status solidi (a) — applied research", i.e. all volumes having appeared in 1976 containing 1360 papers.

The general scheme of the last index volume was maintained. The basis for quoting is the subject classification given for every paper in the respective issue. Nevertheless, before taking over these subject classifications into the index volume they were once more carefully revised and in some cases improved and completed in order to get a most uniform and clear classification for all papers. The average number of quotations per paper in this index volume is nearly the same as in the last one, namely 3.2 instead of 3.

Again I have to express my thanks to the readers for their interest, and I will always readily take up any critical remarks and proposals for improvements. I am especially obliged to Prof. Dr. P. Görlich, Prof. Dr. E. Gutsche, Dr. H.-J. Hänsch, Dr. H. Lange, Dr. S. Oberländer, and Dipl. Phys. K. Wannewitz for giving me advice and unselfish help. Furthermore I have to thank Mrs. E. Schwan for her exact and reliable secretary's work at the manuscript.

Berlin, January 1977

Eva Borchardt

Classification Scheme

1. Structure of Crystalline Solids
   1.1 Perfectly Periodic Structure
   1.2 Solid-State Phase Transformations
   1.3 Alloys. Metallurgy
   1.4 Microstructure (Magnetic Domains See 18; Ferroelectric Domains See 14.4.1)
   1.5 Films
   1.6 Surfaces
2. Non-Crystalline State
3. Crystal Growth
4. Bonding Properties
5. Mössbauer Spectroscopy
6. Lattice Dynamics. Phonons
7. Acoustic Properties
8. Thermal Properties
9. Diffusion
10. Defect Properties (Irradiation Defects See 11)
    10.1 Metals
    10.2 Non-Metals
11. Irradiation Effects (X-Ray Diffraction Investigations See 1 and 10)
12. Mechanical Properties (Plastic Deformations See 10)
    12.1 Metals
    12.2 Non-Metals
13. Electron States
    13.1 Band Structure
    13.2 Fermi Surfaces
    13.3 Surface and Interface States
    13.4 Impurity and Defect States
    13.5 Elementary Excitations (Phonons See 6)
        13.5.1 Excitons
        13.5.2 Plasmons
        13.5.3 Polarons
        13.5.4 Magnons
14. Electrical Properties. Transport Phenomena
    14.1 Metals. Semi-Metals
    14.2 Superconductivity. Superconducting Materials and Devices
    14.3 Semiconductors
        14.3.1 Films
        14.3.2 Surfaces and Interfaces
        14.3.3 Devices. Junctions (Contact Problems See 14.3.4)
        14.3.4 High-Field Phenomena, Space-Charge Effects, Inhomogeneities, Injected Carriers (Electroluminescence See 20.3; Junctions See 14.3.3)
    14.4 Dielectrics
        14.4.1 Ferroelectrics
15. Thermoelectric and Thermomagnetic Properties
16. Photoconductivity. Photovoltaic Effects
17. Emission of Electrons and Ions
    17.1 Field Emission Microscope Investigations

18. Magnetic Properties
    18.1 Paramagnetic Properties
    18.2 Ferromagnetic Properties
        18.2.1 Ferromagnetic Films
    18.3 Ferrimagnetic Properties
    18.4 Antiferromagnetic Properties

19. Magnetic Resonance

20. Optical Properties
    20.1 Spectra. Optical Constants (X-Ray Spectra See 20)
    20.2 Lasers
    20.3 Luminescence (X-Ray Spectra See 20)

21. Metals. Alloys
    21.1 Transition Metals (Group Sc to Zn) and Their Alloys
        21.1.1 Fe and Fe Alloys
    21.2 Alkali Metals
    21.3 Alkali-Earth Metals
    21.4 Rare-Earth Metals
    21.5 Actinides
    21.6 Noble Metals
    21.7 Semi-Metals (Semiconducting Alloys See 22.7)

22. Semiconductors and Ionic Crystals
    22.1 Elements
        22.1.1 Germanium
        22.1.2 Silicon
        22.1.3 Group VI Elements
    22.2 III–V Compounds
        22.2.1 Arsenides
        22.2.2 Phosphides
        22.2.3 Antimonides
        22.2.4 Mixed Crystals
    22.3 IV–IV, II–V, and III–VI Compounds
        22.3.1 Mixed Crystals
    22.4 II–VI Compounds
        22.4.1 Sulphides
        22.4.2 Selenides
        22.4.3 Tellurides
        22.4.4 Mixed Crystals
    22.5 Halides
        22.5.1 Silver Halides
        22.5.2 Alkali Halides
        22.5.3 Alkali-Earth Halides
        22.5.4 Mixed Crystals
    22.6 Simple Oxides (CdO, ZnO See 22.4)
        22.6.1 Mixed Crystals
    22.7 Semiconducting Intermetallic Compounds
    22.8 Tertiary and Higher Compounds
        22.8.1 Salts of Oxy-Acids (Carbonates, Phosphates, Silicates Including Mica etc.)
        22.8.2 Spinels, Garnets, Ferrites
    22.9 Organic Semiconductors

23. Solidified Gases

For comments on the use of the classification scheme see p. 6.

Some Comments on the Use of the Subject and Author Index

The classification system[1]) comprises 20 broad sections each devoted to a major field of solid state physics and 3 broad sections devoted to the main groups of substances. A section may contain a number of subsections dealing with aspects of this field or special substance, respectively, which tend to receive particular attention. A paper dealing with this field will either be classified under one or more of the subsections or, if none of these are relevant, it will just appear under the general section heading (i.e. sections and subsections are equivalent in this sense). For example, a paper dealing with problems of paramagnetism would appear under 18.1 but not 18, whereas a paper on diamagnetism would be quoted under 18; a paper dealing with GaP would appear under 22.2.2, not under 22 or 22.2, whereas a paper on AlN would be quoted under 22.2, and a paper on a II–IV or a IV–VI compound would be quoted under 22.

All papers will be quoted under each relevant section. There is made a distinction between one or two prinicipalentries(refering to the main subject matter of the paper) and secondary entries (refering to subsidiary subject matter and substance, respectively). In the prinicipal entries the title of the paper is given in its original language and in English. In addition, every principal entry is marked by * in front of the title of the paper. In the secondary entries the title is only given in English.

Each entry gives the volume, the number of the issue, the initial and final page numbers, and the year in which the paper appeared. A paper of physica status solidi (b) or physica status solidi (a) is marked by (b) or (a) in front of the line, respectively. The letter "K" before the page numbers indicates a short note published in the offset part of the journal. Each entry is followed by references to the other entries. These references give the subject classifications and the pages in the index where the other entries appear. The references to the principal entries are given in italics.

Review articles appear at the head of the list of entries under a given subject and are given the subtitle "Review Articles". These articles are listed in the chronological order of their publication, irrespective of being a paper of physica status solidi (b) or physica status solidi (a). The subject classification for a review article consisting of several parts is based on the article as a whole rather than on the individual parts.

The entries for original papers, short notes, and errata are made together in a single list for each subject and this follows the entries for review articles. These are again ordered according to data of publication, irrespective of being a paper or physica status solidi (b) or physica status solidi (a). Original papers are sometimes published in several parts under the same principal title (in general with special subtitles). The various parts of papers of this type are treated separately and each part is classified according to its particular contents.

In quoting errata the following rule is obeyed. If the erratum refers to a paper quoted in the preceding index, it is listed in the chronological order of its publication, otherwise the erratum is quoted together with the paper to which it refers.

---

[1]) See cover two and three, and pages 4 and 5. For the case of losing the cover by bookbinding we repeat the classification scheme on pages 4 and 5.

In using the subject index attention should be given to the following details:

subject	quoted in	remarks
representations of space groups	1 and 13.1	
Alfvén waves, helicon waves	13 and 20	
hot electrons	13 and 14.3.4	
transport phenomena	14 or subsection of 14	if not clearly to be assigned to 8 or 15
galvanomagnetic effects	14 or subsection of 14	not in 18
electro-optical effects	14 and 20 (or subsection of 14 and 20)	
piezoelectricity, electrostriction	14 or subsection of 14	not in 12 or subsection of 12; different electromechanical effects are quoted in 12 or subsection of 12
ionic conduction	14.4	not in 14.3
magneto-optical effects	18 and 20 (or subsection of 18 and 20)	possibly in additional sections, e.g. Faraday effect in 13
ferromagnetic films	18.2.1	not additionally in 1.5
X-ray spectra	20	but not X-ray studies of structure
MIS and MOS structures	the section assigned to the semiconductor	not in the section assigned to the oxide or the metal
MIM structures	the section assigned to the insulator	generally not in the section assigned to the metal
epitaxial layers	the section assigned to the layer	not in the section assigned to the substrate

The author index gives the names of the authors in alphabetical order. Each name is followed by the bibliographic details of the paper and the numbers of the pages on which the paper appears in the subject index. A paper of physica status solidi (b) or physica status solidi (a) is again marked by (b) or (a) in front of the number of the volume, respectively. The principal entry is again given in italics. When several papers have been published by the same author they are arranged according to the dates of publication, irrespective of being a paper of physica status solidi (b) or physica status solidi (a).

Subject Index

1. Structure of Crystalline Solids

Original Papers and Short Notes

A. C. McLaren and W. S. MacKenzie
*The Spatial Coherence of X-Ray and Electron Beams and Its Influence on the Diffraction Patterns from Materials with Long-Period Superlattices
(a) Vol. **33**, No. 2, p. 491—495 (1976).
See also *1.4*, p. *25*; 22.6.1, p. 409.

R. E. Vandenberghe, G. G. Robbrecht, and V. A. M. Brabers
*Structure and Ionic Configuration of Oxidic Copper–Manganese Spinels ($Cu_xMn_{3-x}O_4$)
(a) Vol. **34**, No. 2, p. 583—592 (1976).
See also 1.2, p. 15; *18.3*, p. *240*; 22.8.2, p. 428.

B. N. Oshcherin
*On Surface Energies of A^NB^{8-N} Semiconducting Compounds
(a) Vol. **34**, No. 2, p. K181—K186 (1976).
See also *4*, p. *45*; 22, p. 336.

B. N. Oshcherin
*Debye Temperature and Melting Criterion of Ternary Chalcopyrite $A^1B^3C_2^6$- and $A^2B^4C_2^5$-Type Semiconductors
(a) Vol. **35**, No. 1, p. K35—K39 (1976).
See also *8*, p. *68*; 22.2, p. 361; 22.4, p. 376; 22.8, p. 415.

A. K. Rai, R. S. Tiwari, and O. N. Srivastava
Intrapolytypic and Interpolytypic Transformation in Cadmium Iodide Crystals
(a) Vol. **35**, No. 2, p. 719—724 (1976).
See also 22.5, p. 389.

J. C. Phillips
*Structure and Conductivity of Bilinear Organic Compounds
(b) Vol. **77**, No. 1, p. 259—264 (1976).
See also *14*, p. *163*; 22.9, p. 437.

D. W. Bullett
Non-Empirical Calculation of Structural Properties and Valence Force Fields. Application to Se and Te
(b) Vol. **77**, No. 2, p. 645—650 (1976).
See also *4*, p. *46*; 6, p. 61; 22.1.3, p. 360.

N. K. Radhakrishnan and A. B. Biswas
*A Neutron Diffraction Study of the Cation Migration in $MgMn_2O_4$
(a) Vol. **37**, No. 2, p. 719—722 (1976).
See also 10.2, p. 96; 22.8.2, p. 431.

J. C. Phillips
*Phase Transitions in the Bilinear Organic Compounds (TTF, HMTSF)-TCNQ
(b) Vol. **78**, No. 1, p. 371—379 (1976).
See also *14*, p. *163*; 22.9, p. 438.

1.1 Perfectly Periodic Structure

Review Article

R. PRASAD
*Present State of Polytypism in Cadmium Iodide Crystals
(a) Vol. **38**, No. 1, p. 11–44 (1976).
See also *1.2*, p. *14*; 10.2, p. 87; 22.3, p. 372; 22.4.1, p. 379; 22.5, p. 388; 22.5.3, p. 399; 22.8, p. 413.

Original Papers and Short Notes

H. H. SCHMIDT and P. DEIMEL
*Dynamical Theory of Neutron Diffraction by Magnetic Crystals for Laue Geometry
(b) Vol. **73**, No. 1, p. 87–93 (1976).

J. HANSEN-SCHMIDT
*Untersuchungen zur anomalen Absorption von Elektronen in Si in der Nähe von Zonenachsen
(Investigations of the Anomalous Absorption of Electrons in Si near the Zone Axes)
(b) Vol. **73**, No. 1, p. 117–130 (1976).
See also 22.1.2, p. 351.

V. L. INDENBOM and I. SH. SLOBODETSKII
*On the Theoretical Interpretation of the Hart-Milne Experiment
(b) Vol. **73**, No. 1, p. K9–K12 (1976).

E. N. KISLOVSKII, L. I. DATSENKO, and A. S. VASILKOVSKII
*Effect of Crystal Net Plane Curvature on X-Ray Intensity in the Case of Laue Diffraction
(a) Vol. **33**, No. 1, p. 275–280 (1976).
See also *10.2*, p. *87*; 22.1.2, p. 351.

R. DE RIDDER, G. VAN TENDELOO, and S. AMELINCKX
*Electron Microscopic Study of the Chimney Ladder Structures $MnSi_{2-x}$ and $MoGe_{2-x}$
(a) Vol. **33**, No. 1, p. 383–393 (1976).
See also 1.4, p. 25; 22, p. 334.

E. CABRERA and T. GORDILLO
*On the Unique Solution of Kossel Patterns
(a) Vol. **33**, No. 1, p. K19–K22 (1976).

S. POPOVIĆ, B. ČELUSTKA, and D. BIDJIN
*A Remark on the Paper "Phase Transitions in In_2Se_3 as Studied by Electron Microscopy and Electron Diffraction"
(a) Vol. **33**, No. 1, p. K23–K24 (1976).
See also 1.2, p. 14; 22.3, p. 372.

J. W. ANDREW and S. S. SHEININ
*Approximate Methods for Taking Absorption into Account in the Dynamical Theory of Electron Diffraction
(b) Vol. **73**, No. 2, p. 431–442 (1976).
See also 21, p. 290; 21.6, p. 329.

J. HANSEN SCHMIDT
*Anwendung gruppentheoretischer Methoden zur Lösung des Grundgleichungssystems der elastischen Elektronenstreuung in Einkristallen (Si)
(Application of Group-Theoretical Methods to the Solution of the Equations of Electron Diffraction in Single Crystals (Si))
(b) Vol. **73**, No. 2, p. 461–467 (1976).
See also 22.1.2, p. 351.

G. Arancia, M. Grandolfo, C. Manfredotti, and A. Rizzo
*Electron Diffraction Study of Melt- and Vapour-Grown $GaSe_{1-x}S_x$ Single Crystals
(a) Vol. **33**, No. 2, p. 563–571 (1976).
See also *3*, p. *42*; 22.3.1, p. 375.

O. Brümmer, H. R. Höche und J. Nieber
*Einfluß der Totalreflexion auf die Röntgenstrahlbeugung im stark asymmetrischen Bragg-Fall
(Influence of the Total Reflection on the X-Ray Diffraction in the Strongly Asymmetric Bragg-Case)
(a) Vol. **33**, No. 2, p. 587–593 (1976).
See also 22.1.2, p. 352.

M. David, R. Gevers, and R. Serneels
*The Critical Voltage Effect in Transmission Electron Microscopy (VI): Renormalized Perturbation Theory for the Treatment of Absorption
(b) Vol. **74**, No. 1, p. 359–373 (1976).

T. Soma
Elastic Constant of Hypothetical Face-Centered Cubic In
(b) Vol. **74**, No. 1, p. K67–K70 (1976).
See also *12.1*, p. *107*; 21, p. 291.

J. B. A. A. Elemans and K. H. J. Buschow
Comment on the Crystal and Magnetic Structures of $CaCu_5$-Type Compounds of Th with Fe, Co, and Ni
(a) Vol. **34**, No. 1, p. 355–359 (1976).
See also *18.2*, p. *230*; *18.4*, p. *243*; 21.1, p. 305; 21.1.1, p. 317; 21.5, p. 328.

S. Popović and M. Šljukić
*X-Ray Diffraction Measurements of Unit Cell Parameters of Differently Crystallized Samples
(a) Vol. **34**, No. 1, p. K91–K95 (1976).
See also 22.1.1, p. 346; 22.8, p. 414.

Y. Uesu and J. Kobayashi
*Crystal Structure and Ferroelectricity of Cesium Dihydrogen Phosphate CsH_2PO_4
(a) Vol. **34**, No. 2, p. 475–481 (1976).
See also *14.4.1*, p. *206*; 22.8.1, p. 421.

K. Godwod, A. T. Nagy, and Z. Rek
*The Application of the X-Ray Triple-Crystal Spectrometer for Measuring the Radius of Curvature of Bent Single Crystals
(a) Vol. **34**, No. 2, p. 705–710 (1976).
See also 10.2, p. 91; 22.1.2, p. 353.

H.-G. Brühl, L. Hildisch, H. Morwinski, and W. Schmidt
*X-Ray Study of Order in the Solid Solution $Ga_xIn_{1-x}P$
(a) Vol. **34**, No. 2, p. K147–K148 (1976).
See also 22.2.4, p. 370.

A. Mazel et F. Ajustron
*Influence des interactions systématiques sur les profils d'intensité des lignes de Kikuchi obtenues sous très haute tension
(Influence of Systematic Interactions on the Intensity Profiles of Kikuchi Lines Obtained at Very High Voltage)
(a) Vol. **35**, No. 1, p. 197–211 (1976).
See also 22.6, p. 404.

B. PARDO et D. RENARD
*Etude des contrastes obtenus en microscopie électronique, a partir de monocristaux d'épaisseur irrégulière
(Electron Microscopic Study on Single Crystals of Non-Uniform Thickness)
(a) Vol. **35**, No. 1, p. 329—338 (1976).
See also 22.6, p. 404.

N. WAGNER and O. BRÜMMER
*Remarks on the Relation between Crystal Structure and Electronic Band Structure in Bismuth-Antimony Alloys
(b) Vol. **75**, No. 2, p. K157—K161 (1976).
See also *13.1*, p. *124*; 22.7, p. 411.

C. W. ALLEN and P. R. OKAMOTO
*An Imaging Technique in Transmission Electron Microscopy Utilizing Diffraction Fine Structure
(a) Vol. **36**, No. 1, p. 107—116 (1976).
See also *1.4*, p. *27*; 21.1.1, p. 319.

D. PETRASCHECK und R. FOLK
*Theory of a Symmetric LLL Interferometer with Arbitrary Absorption
(a) Vol. **36**, No. 1, p. 147—159 (1976).

A. KUHN, A. CHEVY, and R. CHEVALIER
*Interatomic Distances in $GaSe_xS_{1-x}$
(a) Vol. **36**, No. 1, p. 181—187 (1976).
See also 22.3.1, p. 375.

W. H. HUBBIG, J. AULEYTNER, and M. MACIASZEK
*Changes of X-Ray Topographic Contrast Due to Annealing of Boron-Implanted Silicon
(a) Vol. **36**, No. 1, p. 209—215 (1976).
See also 10.2, p. 93; 22.1.2, p. 355.

D. W. FIELD
*An X-Ray Diffraction Study of White Tin
(a) Vol. **36**, No. 2, p. 597—601 (1976).
See also 6, p. 59; 21.7, p. 332.

J. VAN LANDUYT, G. VAN TENDELOO, and S. AMELINCKX
Electron Diffraction Study of Inter- and Intrapolytypic Phase Transitions in Transition Metal Dichalcogenides (III): Complementary Diffraction Studies and Lattice Imaging of the Deformation Waves
(a) Vol. **36**, No. 2, p. 757—777 (1976).
See also *1.2*, p. *18*; 22, p. 339.

D. VAN DYCK
*The Importance of Backscattering in High-Energy Electron Diffraction Calculations
(b) Vol. **77**, No. 1, p. 301—308 (1976).
See also *13*, p. *118*.

S. R. SINGH, R. S. RAI, and G. SINGH
**K*-Surfaces of Cubic Silicon Carbide (β-SiC)
(a) Vol. **37**, No. 1, p. 139—142 (1976).
See also *12.2*, p. *111*; 22.3, p. 373.

Z. JIRÁK, S. VRATISLAV, and J. ZAJÍČEK
*Oxygen Parameters and Debye-Waller Factors in $Mn_xCr_{3-x}O_4$ Spinels
(a) Vol. **37**, No. 1, p. K47—K51 (1976).
See also 22.8.2, p. 430.

K. Hübner and V. K. Bashenov
Ionicity and Structure of Solids
(b) Vol. 77, No. 2, p. 473–482 (1976).
See also 2, p. 40; *4*, p. *46*; *13.1*, p. *127*; 22.2, p. 361; 22.2.4, p. 371.

J. Härtwig
*X-Ray Diffraction in an Extremely Asymmetric Laue Case
(a) Vol. **37**, No. 2, p. 417–425 (1976).
See also 22.1.2, p. 356.

D. Bloor and F. H. Preston
Optical Properties of Bis(p-Toluene Sulphonate)Diacetylene (I): Crystal Optics of Partially and Fully Polymerized Crystals
(a) Vol. **37**, No. 2, p. 427–438 (1976).
See also *20.1*, p. *274*; 22.9, p. 438.

O. Brümmer, H. R. Höche, and J. Nieber
*Study of Extremely Asymmetric X-Ray Diffraction in the Bragg Case of Grazing Incidence
(a) Vol. **37**, No. 2, p. 529–532 (1976).
See also 22.1.2, p. 357.

M. Umeno
*Six-Beam Anomalous Transmission of Ge Crystals (I): Anomalous Transmission Pattern
(a) Vol. **37**, No. 2, p. 561–570 (1976).
See also 22.1.1, p. 349.

T. Soma and T. Umenai
*Madelung Energies of Tetragonal and Orthorhombic Systems
(b) Vol. 78, No. 1, p. 229–240 (1976).
See also 1.2, p. 19; 22.1.2, p. 357.

T. Shishido and M. Tanaka
*Determination of Structure Factors of Germanium by the Critical-Voltage and Convergent-Beam Diffraction Methods
(a) Vol. **38**, No. 2, p. 453–461 (1976).
See also 1.4, p. 29; 22.1.1, p. 350.

M. Umeno
*Six-Beam Anomalous Transmission of Ge Crystals (II): Interpretations from Eigensolutions
(a) Vol. **38**, No. 2, p. 701–711 (1976).
See also 22.1.1, p. 350.

1.2 Solid-State Phase Transformations

Review Article

R. Prasad
*Present State of Polytypism in Cadmium Iodide Crystals
(a) Vol. **38**, No. 1, p. 11—44 (1976).
See also *1.1*, p. *10*; 10.2, p. 87; 22.3, p. 372; 22.4.1, p. 379; 22.5, p. 388; 22.5.3, p. 399; 22.8, p. 413.

Original Papers and Short Notes

M. Ausloos
*On a Structural Phase Transition in Ytterbium
(b) Vol. **73**, No. 1, p. 211—221 (1976).
See also 13.1, p. 120; *14.1*, p. *164*; 21.4, p. 326.

M. Natarajan and E. A. Secco
*Electrical Conductivity and Phase Transformation Studies on the Mixed Metal Halides $RbCdX_3$ (X = Cl, Br, I) and Cs_2CuBr_4
(a) Vol. **33**, No. 1, p. 427—433 (1976).
See also 8, p. 67; 14.4, p. 202; 22.5.4, p. 401.

S. Popović, B. Čelustka, and D. Bidjin
A Remark on the Paper "Phase Transitions in In_2Se_3 as Studied by Electron Microscopy and Electron Diffraction"
(a) Vol. **33**, No. 1, p. K23—K24 (1976).
See also *1.1*, p. *10*; 22.3, p. 372.

J. J. Rousseau, M. Rousseau, and J. C. Fayet
*EPR Investigations of a Structural Phase Change in $RbCaF_3$, $RbCdF_3$, and $TlCdF_3$
(b) Vol. **73**, No. 2, p. 625—631 (1976).
See also *19*, p. *248*; 22.8, p. 413.

K. Gesi
*Dielectric Study on the Phase Transitions in Triammonium Hydrogen Disulfate $(NH_4)_3H(SO_4)_2$
(a) Vol. **33**, No. 2, p. 479—482 (1976).
See also *14.4*, p. *202*; 22.8.1, p. 419.

J. Vicens and P. Delavignette
*A Particular Domain Configuration Observed in a New Phase of the Ta–N System
(a) Vol. **33**, No. 2, p. 497—509 (1976).
See also *1.4*, p. *25*; 22, p. 334.

T. Schober and H. Wenzl
*β-Phase Melting and Solidification Phenomena in the Niobium–Hydrogen System
(a) Vol. **33**, No. 2, p. 673—681 (1976).
See also *1.4*, p. *25*; 10.1, p. 79; 21, p. 291.

G. van Tendeloo, J. van Landuyt, and S. Amelinckx
*The $\alpha \rightarrow \beta$ Phase Transition in Quartz and $AlPO_4$ as Studied by Electron Microscopy and Diffraction
(a) Vol. **33**, No. 2, p. 723—735 (1976).
See also *1.4*, p. *26*; 22.6, p. 403; 2.2.8.1, p. 419.

V. A. Povitskii, E. F. Makarov, N. V. Murashko, and A. N. Salugin
Mössbauer Study of Superparamagnetic δ-FeOOH and Its Transformation into α-Fe_2O_3
(a) Vol. **33**, No. 2, p. 783—787 (1976).
See also *5*, p. *48*; 22.8.2, p. 427.

J. Winter, K. Rössler, J. Bolz, and J. Pelzl
*Mössbauer Effect and Raman Studies of the Structural Phase Transitions in $K_2[SnCl_6]$
(b) Vol. **74**, No. 1, p. 193—198 (1976).
See also 5, p. 48; 20.1, p. 263; 22.8, p. 414.

P. Bassoul, A. Dubon, A. Lefebvre et J. C. Gilles
*Structure à antiphases périodiques de la phase non-stoechiométrique ε_{Mg} du système $MgGa_2O_4$–Ga_2O_3
(Periodic Antiphase Structures in the Non-Stoichiometric Mg Phase of the System $MgGa_2O_4$–Ga_2O_3
(a) Vol. **34**, No. 1, p. 125—134 (1976).
See also *1.4*, p. *26*; 22.8.2, p. 427.

E. Bernard and P. Duval
*On the Period of Antiphase Domain Structures in X- and α_2-Phases of Copper–Aluminium Alloys
(a) Vol. **34**, No. 1, p. 135—144 (1976).
See also *1.3*, p. *22*; 1.4, p. 26; 21, p. 291; 21.1, p. 304.

K. I. Arai, K. Ohmori, N. Tsuya, and S. Iida
Magnetostriction of Magnetite in the Vicinity of the Low-Temperature Transition
(a) Vol. **34**, No. 1, p. 325—330 (1976).
See also *18.3*, p. *240*; 22.8.2, p. 427.

A. Fernandez, L. Tejedor, and L. Bru
*Electron Microscopy Study of the Phase Change $DO_{19} \rightarrow L1_2$ in the Fe_3Ge Compound
(a) Vol. **34**, No. 1, p. K17—K19 (1976).
See also 21.1.1, p. 317.

Č. Bárta, J. P. Chapelle, G. Hauret, Cao-Xuân An, A. Fousková, and Č. Koňák
*Brillouin Scattering in Hg_2Cl_2
(a) Vol. **34**, No. 1, p. K51—K54 (1976).
See also 22.5, p. 388.

H. Göbel, E. Wolfgang, and H. Harms
*Properties of MnBi Compounds Partially Substituted with Cu, Zn, Ti, Sb, and Te (I): Formation of Mixed Phases and Crystal Structures
(a) Vol. **34**, No. 2, p. 553—564 (1976).
See also *1.3*, p. *22*; 22.7, p. 410.

W. Moldenhauer and W. Brückner
*Physical Properties of Nonstoichiometric Iron Sulfide $Fe_{1-x}S$ near the α-Phase Transition
(a) Vol. **34**, No. 2, p. 565—571 (1976).
See also *14.3*, p. *175*; 15, p. 209; 18.4, p. 243; 22, p. 336.

R. E. Vandenberghe, G. G. Robbrecht, and V. A. M. Brabers
Structure and Ionic Configuration of Oxidic Copper–Manganese Spinels ($Cu_xMn_{3-x}O_4$)
(a) Vol. **34**, No. 2, p. 583—592 (1976).
See also *1*, p. *9*; *18.3*, p. *240*; 22.8.2, p. 428.

J. Klimowski, W. Wanarski, and D. Ożgo
*Effect of Hydrostatic Pressure on the Phase Transition Temperatures and Spontaneous Polarization of Thiourea Monocrystals
(a) Vol. **34**, No. 2, p. 697—704 (1976).
See also **14.4,1**, p. 206; 22.9, p. 435.

V. G. Vaks and V. E. Schneider
*On the Theory of Phase Transitions in the Ammonium Halides
(a) Vol. **35**, No. 1, p. 61—72 (1976).
See also 22.5, p. 389; 22.5.4, p. 401.

H. Göbel, E. Wolfgang, and H. Harms
*Properties of MnBi Compounds Partially Substituted with Cu, Zn, Ti, Sb, and Te (II): Stability and Magnetooptic Properties of Thin Films
(a) Vol. **35**, No. 1, p. 89—95 (1976).
See also 1.3, p. 22; *18.2.1*, p. *237*; 22.7, p. 411.

K.-D. Toepfer and H. W. Helberg
Dielectric Dispersion of Thiourea Single Crystals at Microwave Frequencies between 8.25 and 75.0 GHz
(a) Vol. **35**, No. 1, p. 131—136 (1976).
See also *14.4.1*, p. *207*; 22.9, p. 436.

M. Ali, S. J. H. Bukhari, K. A. Shoaib, and F. H. Hashmi
*Monoclinic ⇌ Tetragonal Phase Transformation in ZrO_2 Single Crystals
(a) Vol. **35**, No. 1, p. 181—184 (1976).
See also 22.6, p. 404.

P. Duhaj, J. Sitek, M. Prejsa, and P. Butvin
*Electron Microscopy and Mössbauer Study of the Amorphous Metallic Alloys $Pd_{80-x}Fe_xSi_{20}$
(a) Vol. **35**, No. 1, p. 223—233 (1976).
See also *2*, p. *37*; 5, p. 49; 21.1.1, p. 318.

P. Ramasamy and T. Nagarajan
*Positron Angular Correlation Studies of Phase Transitions in Ammonium Halides
(a) Vol. **35**, No. 1, p. 379—386 (1976).
See also 4, p. 45; 22.5, p. 389.

W. Brückner, W. Moldenhauer, B. Thuss, and G. Försterling
*Crystallographic Changes at the Conductivity Transition of V_3O_5
(a) Vol. **35**, No. 1, p. K49—K51 (1976).
See also 15, p. 210; 22.6, p. 404.

I. G. Ratishvili
*On a Peculiarity of Ordering in F.C.C. Binary Alloys with a Distribution Function Containing Two Long-Range Order Parameters
(b) Vol. **75**, No. 2, p. 401—406 (1976).
See also *1.3*, p. *22*; 21, p. 294.

A. M. Ozorio de Almeida and S. Caticha Ellis
*X-Ray Scattering from Crystals with Intracellular Disorder
(b) Vol. **75**, No. 2, p. 657—668 (1976).
See also 14.4.1, p. 207; 22.8.1, p. 421.

H. P. Aubauer
*Elastic Energy of Isotropic Solid Solutions
(a) Vol. **35**, No. 2, p. 487—497 (1976).
See also 4, p. 45; 21, p. 295.

F. W. Boswell, A. Prodan, and J. M. Corbett
Electron Diffraction Study of the Structures of NbS_2 and Order–Disorder in Cu-Intercalated NbS_2
(a) Vol. **35**, No. 2, p. 591—604 (1976).
See also *1.4*, p. *27*; 22, p. 337.

G. HEGER, D. MULLEN, and K. KNORR
*On the Importance of Hydrogen Bonding for the Structural Phase Transitions in $(CH_3NH_3)_2MnCl_4$. A Single-Crystal Neutron Diffraction Study
(a) Vol. **35**, No. 2, p. 627—637 (1976).
See also **4**, p. 46; 22.9, p. 436.

M. GOETTEL, J. M. JANIK, J. A. JANIK, and M. RACHWALSKA
*Phase Transitions and NH_3 Rotations in $[Co(NH_3)_6]I_3$ Studied by Calorimetry and IR Profile Methods
(a) Vol. **35**, No. 2, p. 675—680 (1976).
See also **8**, p. 69; 22.8, p. 416.

A. L. ROITBURD and N. S. KOSENKO
*Orientational Dependence of the Elastic Energy of a Plane Interlayer in a System of Coherent Phases
(a) Vol. **35**, No. 2, p. 735—746 (1976).
See also *1.4*, p. *27*.

V. N. BYKOV, V. A. TROYAN, G. G. ZDOROVTSEVA, and V. S. KHAIMOVICH
*Erratum to: Phase Transformations at Bombardment of Thin Films with Ions
(a) Vol. **35**, No. 2, p. 779 (1976).
See also *1.5*, p. *31*; 11, p. 102; 21, p. 295; 21.1, p. 308; 21.1.1, p. 319; 22.1.1, p. 347.

J.-P. RIVERA, H. BILL, and R. LACROIX
EPR of Powder Mg-OH:Mn^{2+} Boracite Suggesting a Rhombohedral Phase
(a) Vol. **35**, No. 2, p. K105—K108 (1976).
See also *19*, p. *252*; 22.8, p. 416.

K. H. GERMANN, W. MÜLLER-LIERHEIM, H. H. OTTO, and T. SUSKI
*High Temperature Phase Transition in $Pb_5Ge_3O_{11}$
(a) Vol. **35**, No. 2, p. K165—K167 (1976).
See also 14.4.1, p. 207; 20.1, p. 270; 22.8.1, p. 422.

C. N. W. DARLINGTON
*Phase Transitions in Perovskite Crystals
(b) Vol. **76**, No. 1, p. 231—239 (1976).
See also 14.4.1, p. 207; 22.8, p. 416.

E. SCHUBERTH, M. CREUTZBURG, and W. MÜLLER-LIERHEIM
*Evidence for a Partial F.C.C.–H.C.P. Phase Transition in Solid Neon from Raman Scattering
(b) Vol. **76**, No. 1, p. 301—306 (1976).
See also **6**, p. 58; 20.3, p. 284; 23, p. 440.

D. B. LITVIN
*$\boldsymbol{k}$–$\boldsymbol{q}$ Functions and the Landau Theory of Second-Order Phase Transitions
(b) Vol. **76**, No. 1, p. 343—349 (1976).

H. BEIGE, C. BIRKHOLZ, E. CIESLA, and G. SCHMIDT
Elastic and Electromechanical Properties of Some Crystals near Structural Phase Transitions
(b) Vol. **76**, No. 1, p. K47—K50 (1976).
See also *12.2*, p. *111*; *14.4*, p. *203*; 22.8.1, p. 422.

G. CHAPIUS, R. KIND, and H. AREND
*X-Ray Study of Structural Phase Transitions in the Perovskite-Type Layer Compound $(CH_3NH_3)_2CdCl_4$
(a) Vol. **36**, No. 1, p. 285—295 (1976).
See also 22.9, p. 437.

M. GUYMONT and D. GRATIAS
On the Stability of Periodically Antiphased Alloys
(a) Vol. **36**, No. 1, p. 329–334 (1976).
See also *1.3*, p. *23*; *1.4*, p. *28*; 21.1, p. 309; 21.6, p. 330.

L. N. ALEKSANDROV and B. I. KIDYAROV
Kinetic Theory of Three-Barrier Processes of Non-Steady-State Nucleation at Phase Transformation
(a) Vol. **36**, No. 1, p. 403–413 (1976).
See also *3*, p. *43*.

Y. KASHIWADA and Y. FURUHATA
*Domains in Magnesium Pyroborate
(a) Vol. **36**, No. 1, p. K29–K31 (1976).
See also *3*, p. *43*; 22.8, p. 416.

T. V. QUYNH, G. SCHMIDT N. R. IVANOV, and L. A. SHUVALOV
*Determination of the Temperature Dependence of the Stiffness Coefficient c_{55} in $KH_3(SeO_3)_2$ by Ultrasonic Velocity Measurements
(a) Vol. **36**, No. 1, p. K85–K88 (1976).
See also *7*, p. *65*; 22.8, p. 416 .

Y. LUSPIN and G. HAURET
*Study of the Velocity and Damping of Acoustic Waves Obtained by Brillouin Scattering in the Paraelastic Phase in GMO
(b) Vol. **76**, No. 2, p. 551–558 (1976).
See also 7, p. 65; 22.8.1, p. 423.

K. D. KUNDRA and S. Z. ALI
*X-Ray Study of Thermal Expansion and Phase Transformation in β-In_2S_3
(a) Vol. **36**, No. 2, p. 517–525 (1976).
See also 8, p. 70; 22.3, p. 373.

J. VAN LANDUYT, G. VAN TENDELOO, and S. AMELINCKX
Electron Diffraction Study of Inter- and Intrapolytypic Phase Transitions in Transition Metal Dichalcogenides (III): Complementary Diffraction Studies and Lattice Imaging of the Deformation Waves
(a) Vol. **36**, No. 2, p. 757–777 (1976).
See also *1.1*, p. *12*; 22, p. 339.

R. HIRSCH and G. STÜHMER
*A Simple Technique to Measure Optical SHG in Powder Samples by Means of Electronic Integration
(a) Vol. **37**, No. 1, p. 105–108 (1976).
See also *20*, p. *258*; 22.5.4, p. 401.

D. KUCHARCZYK, A. PIETRASZKO, and K. ŁUKASZEWICZ
*Temperature Dependence of Lattice Parameters of $NaNO_2$ Single Crystals
(a) Vol. **37**, No. 1, p. 287–294 (1976).
See also *8*, p. *70*; 14.4.1, p. 208; 22.8, p. 417.

H. S. RANDHAWA, L. K. MALHOTRA, H. K. SEHGAL, and K. L. CHOPRA
*Structural Transformations in a-Ge Alloy Films
(a) Vol. **37**, No. 1, p. 313–320 (1976).
See also 1.5, p. 32; *2*, p. *40*; 22.1.1, p. 348.

H. ZABEL and H. PEISL
*X-Ray Study of the Phase Diagram of Hydrogen in Niobium
(a) Vol. **37**, No. 1, p. K67–K70 (1976).
See also 1.3, p. 23; 10.1, p. 84; 21, p. 298.

S. Franck, A. Kühnel, T. Nattermann, and S. Wendt
*Noncritical Fluctuations and the Temperature Dependence of the Order Parameter
(b) Vol. 77, No. 2, p. 631—637 (1976).
See also *14.4.1*, p. *208*.

J. C. Castro, H. C. Basso, and M. de Souza
Local Mode of H^- Substitutional Ion in KCN
(b) Vol. 77, No. 2, p. 685—692 (1976).
See also *6*, p. *61*; 10.2, p. 95; *20.1*, p. *274*; 22, p. 340.

R. de Ridder, G. van Tendeloo, J. van Landuyt, D. van Dyck, and S. Amelinckx
*Microstructural Study of Copper-Intercalated Niobium Disulphide and Tantalum Disulphide Cu_xNbS_2 and Cu_xTaS_2 (I): Phase Transitions in $Cu_{0.5}NbS_2$ and $Cu_{0.5}TaS_2$
(a) Vol. 37, No. 2, p. 591—606 (1976).
See also 1.4, p. 28; 22, p. 340.

M. I. Zakharova and H. Fuentes
*Structural Phase Transformations in the Alloys Ti-25%Nb-15%V and Ti-40%Nb
(a) Vol. 37, No. 2, p. K121—K125 (1976).
See also *1.3*, p. *24*; 21.1, p. 313.

E. M. Savitskii, M. I. Bychkova, V. V. Baron, B. E. Dzevitskii, and N. I. Savateev
*Lattice Instability of Compound Nb_3Sn in Strained State
(a) Vol. 37, No. 2, p. K165—K167 (1976).
See also 5, p. 50; 21, p. 299.

T. Soma and T. Umenai
Madelung Energies of Tetragonal and Orthorhombic Systems
(b) Vol. 78, No. 1, p. 229—240 (1976).
See also *1.1*, p. *13*; 22.1.2, p. 357.

W. Brückner, U. Gerlach, W. Moldenhauer, H. P. Brückner, N. Mattern, H. Oppermann and E. Wolf
*Phase Transitions and Semiconductor-Metal Transition in $V_{1-x}Ga_xO_2$ Single Crystals
(a) Vol. 38, No. 1, p. 93—102 (1976).
See also *3*, p. *44*; 14.3, p. 181; 15, p. 211; 18.1, p. 227; 22.6.1, p. 409.

A. Golański, A. Fiderkiewicz, H. Rzewuski, M. Lefeld-Sosnowska, J. Gronkowski, R. Grötzschel, U. Kreissig, and H. Bartsch
*Particularities of Crystalline to Amorphous State Conversion in Silicon Heavily Damaged by 140 keV Si^{++} Ions
(a) Vol. 38, No. 1, p. 139—149 (1976).
See also *11*, p. *104*; 22.1.2, p. 357.

G. van Tendeloo, R. de Ridder, D. van Dyck, J. van Landuyt, and S. Amelinckx
Microstructural Study of Copper-Intercalated Niobium Disulphide and Tantalum Disulphide Cu_xNbS_2 and Cu_xTaS_2 (II): Order–Disorder Transitions and Incommensurate Superstructures in Cu_xNbS_2 $(0.8 < x < 1)$
(a) Vol. 38, No. 1, p. 185—198 (1976).
See also 1.4, p. 29; 22, p. 341.

W. Brückner, H.-P. Brückner, U. Gerlach, B. Thuss, and G. Försterling
*The Phase Transition $M1 \rightarrow T$ in $V_{1-x}Ga_xO_2$
(a) Vol. 38, No. 1, p. K13—K16 (1976).
See also *18.1*, p. *227*; 22.6.1, p. 409.

W. Brückner and W. Moldenhauer
Anisotropy of Thermoelectric Power in $V_{1-x}Me_xO_2$(Me = Al, Ga, Fe)
(a) Vol. **38**, No. 1, p. K17–K21 (1976).
See also *15*, p. *211*; 22.6.1, p. 409.

K. H. Germann
Optical Anisotropy Measurements in Multidomain Samples: Application to Phase Transitions in NiO and $Pb_5Ge_3O_{11}$
(a) Vol. **38**, No. 1, p. K81–K84 (1976).
See also 14.4.1, p. 208; 18.4, p. 246; *20.1*, p. *277*; 22.6, p. 408; 22.8.1, p. 424.

V. I. Zinenko
*On the Theory of the Structural Phase Transition in Squaric Acid
(b) Vol. 78, No. 2, p. 721–727 (1976).
See also *14.4*, p. *205*; 22.9, p. 439.

A. M. Ziatdinov, M. M. Zaripov, Yu. V. Yablokov, and R. L. Davidovich
The Depression of the Dynamic Jahn-Teller Effect
(b) Vol. 78, No. 2, p. K69–K71 (1976).
See also *6*, p. *64*; *13.4*, p. *145*; 19, p. 256; 22.8, p. 418.

M. Guymont, D. Gratias, R. Portier, and M. Fayard
Space Group Theoretical Determination of Translation, Twin, and Translation-Twin Boundaries in Cell-Preserving Phase Transitions
(a) Vol. **38**, No. 2, p. 629–636 (1976).
See also *1.4*, p. *29*.

R. Moret, M. Huber, and R. Comès
*Diffuse Scattering and Titanium Short-Range Order in $Ti_{1+x}S_2$
(a) Vol. **38**, No. 2, p. 695–700 (1976).
See also 1.3, p. 24; 22, p. 342.

1.3 Alloys. Metallurgy

Review Article

E. Vogt
The Origin of Ferromagnetism and Antiferromagnetism in Alloys of Transition Metals (II): Alloys of Transition Metals with Other Transition Metals and with Polyvalent Non-Transition Elements
(a) Vol. **34**, No. 1, p. 11–67 (1976).
See also 18.1, p. 225; *18.2*, p. *228*; *18.4*, p. *242*; 21, p. 289; 21.1, p. 302; 21.6, p. 329.

Original Papers and Short Notes

V. A. Goltsov and L. I. Smirnov
*Diffusion of Interstitial Atoms in Ordered Pt_2Mo-Type Alloys (II)
(a) Vol. **33**, No. 1, p. 67–73 (1976).
See also *9*, p. *72*; 21, p. 289; 21.6, p. 329.

R. Wirths, P. Runow, and H. W. Schöpgens
Magnetic Study of the Ordered and Quenched Equiatomic FePt Alloy
(a) Vol. **33**, No. 1, p. 135–142 (1976).
See also *18.1*, p. *225*; *18.2*, p. *229*; 21.1.1, p. 316.

G. J. Davies and C. M. Sargent
*Preferred Orientation and Elastic Anisotropy in Sheets of Copper–Nickel Alloys
(a) Vol. **33**, No. 1, p. 281–288 (1976).
See also *12.1*, p. *107*; 21.1, p. 303.

H. Löffler and H.-G. Fabian
*Application of the Jump Model to the Decomposition of Some Binary and Ternary Aluminium Alloys
(a) Vol. **33**, No. 1, p. 345–353 (1976).
See also *10.1*, p. *78*; 14.1, p. 164; 21, p. 290; 21.1, p. 303.

A. J. Morton
Inversion Anti-Phase Domains in Cu-Rich γ-Brasses (II): Defects in the Domain Structure
(a) Vol. **33**, No. 1, p. 395–403 (1976).
See also *1.4*, p. *25*; 21.1, p. 303.

S. Tanigawa and M. Doyama
The Formulation of the Formation Energy of a Vacancy as an Alloy Problem
(b) Vol. **73**, No. 2, p. 517–526 (1976).
See also *10.1*, p. *78*; 21, p. 290.

Y. Aoki and M. Yamamoto
X-Ray and Magnetic Investigations of the High-Temperature Phase in the Co-Rich Co–V Alloy System
(a) Vol. **33**, No. 2, p. 625–632 (1976).
See also *18.1*, p. *225*; *18.2*, p. *229*; 21.1, p. 304.

B. K. Ray and G. D. Nigam
Variation of Elastic Constants of Partially Ordered Cu_3Au Alloy with Temperature
(a) Vol. **33**, No. 2, p. K101–K105 (1976).
See also *12.1*, p. *107*; 21.1, p. 304; 21.6, p. 329.

S. Kiss, I. Z. Harangozó, and F. J. Kedves
*Solubility Study of Cu in Al Using Zener Relaxation
(a) Vol. **33**, No. 2, p. K107–K109 (1976).
See also 21, p. 291; 21.1, p. 304.

A. W. Smith and R. D. Rawlings
*A Neutron Diffraction Study of Ordering in an Iron–Cobalt–1.8% Vanadium Alloy
(a) Vol. **34**, No. 1, p. 117–123 (1976).
See also 21.1, p. 304; 21.1.1, p. 317.

E. Bernhard and P. Duval
*On the Period of Antiphase Domain Structures in X- and α_2-Phases of Copper–Aluminium Alloys
(a) Vol. **34**, No. 1, p. 135–144 (1976).
See also *1.2*, p. *15*; 1.4, p. 26; 21, p. 291; 21.1, p. 304.

D. Preininger
Work Hardening of F.C.C. Metals Subjected to Internal Stresses. Dispersion and Irradiation Hardening
(a) Vol. **34**, No. 1, p. 227–242 (1976).
See also *10.1*, p. *80*; 21, p. 292.

G. L. Bleris, T. Karakostas, J. Stoemenos, and N. A. Economou
*On the Short-Range Order Arrangement at the Transition State of β-In_2Te_3
(a) Vol. **34**, No. 1, p. 243–254 (1976).
See also 10.2, p. 90; 22.7, p. 410.

C. Ghezzi
*Short-Range Order and Diffuse X-Ray Scattering in Single Crystal $Pb_xSn_{1-x}Te$ Alloys
(a) Vol. **34**, No. 2, p. 525–531 (1976).
See also 22.7, p. 410.

H. Göbel, E. Wolfgang, and H. Harms
*Properties of MnBi Compounds Partially Substituted with Cu, Zn, Ti, Sb, and Te (I): Formation of Mixed Phases and Crystal Structures
(a) Vol. **34**, No. 2, p. 553–564 (1976).
See also *1.2*, p. *15*; 22.7, p. 410.

H. Göpel, E. Wolfgang, and H. Harms
*Properties of MnBi Compunds Partially Substituted with Cu, Zn, Ti, Sb, and Te (II): Stability and Magnetooptic Properties of Thin Films
(a) Vol. **35**, No. 1, p. 89–95 (1976).
See also *1.2*, p. *16*; *18.2.1*, p. *237*; 22.7, p. 411.

I. G. Ratishvili
*On a Peculiarity of Ordering in F.C.C. Binary Alloys with a Distribution Function Containing Two Long-Range Order Parameters
(b) Vol. **75**, No. 2, p. 401–406 (1976).
See also *1.2*, p. *16*; 21, p. 294.

J. F. Graczyk and P. Chaudhari
*Scattering Properties of Dense Random Packings of Hard Spheres
(b) Vol. **75**, No. 2, p. 593–606 (1976).
See also *2*, p. *38*; 21, p. 294.

J. P. Muscat, M. Lannoo, and G. Allan
*Study of Growth of Guinier-Preston Zones in Al–Ag Compounds
(a) Vol. **35**, No. 2, p. 499–509 (1976).
See also 10.1, p. 82; 21, p. 295; 21.6, p. 330.

D. Fargeot, J. C. Glandus et P. Boch
*Constantes élastiques des alliages cuivre–aluminium riches en cuivre entre 77 et 293 K
(Elastic Constants of Copper-Rich Copper–Aluminium Alloys between 77 and 293 K)
(a) Vol. **35**, No. 2, p. 687–695 (1976).
See also *12.1*, p. *108*; 21, p. 295; 21.1, p. 308.

A. J. Morton
Erratum to: Inversion Anti-Phase Domains in Cu-Rich γ-Brasses (I)
(a) Vol. **35**, No. 2, p. 777 (1976).
See also *1.4*, p. *27*; 21.1, p. 308.

A. I. Olemskoi
*On Stability Conditions of Disperse Order
(a) Vol. **35**, No. 2, p. K101–K104 (1976).

V. Seetharaman, S. Banerjee, and R. Krishnan
*The Precipitation of $Ta_{64}C$ in Dilute Tantalum–Carbon Alloys
(a) Vol. **36**, No. 1, p. 39–52 (1976).
See also *1.4*, p. *27*; 21, p. 296.

H. D. Wiesinger
*Zustandsdiagramm und Sprungtemperatur zur Supraleitung von A15-Phasen
(Phase Diagram and Transition Temperature to Superconductivity of A15 Phases)
(a) Vol. **36**, No. 1, p. 61–72 (1976).
See also *14.2*, p. *170*; 21, p. 296.

H.-G. Fabian and H. Löffler
*On the Interpretation of the Decomposition Kinetics of Some Al–Zn(6 at%)–Mg(0.1 to 3.5 at%) Alloys by Means of the Enlarged Jump Model
(a) Vol. **36**, No. 1, p. 127–132 (1976).
See also *10.1*, p. *83*; 14.1, p. 165; 21, p. 296; 21.1, p. 309.

M. Guymont and D. Gratias
*On the Stability of Periodically Antiphased Alloys
(a) Vol. **36**, No. 1, p. 329–334 (1976).
See also 1.2, p. 18; *1.4*, p. *28*; 21.1, p. 309; 21.6, p. 330.

J. Schneider, H. Wiesner, and R. Gemperle
*Annealing Effects on the Magnetic Properties of Rapidly Quenched Transition Metal Alloys
(a) Vol. **36**, No. 1, p. K59–K64 (1976).
See also *18.2*, p. *233*; 21.1, p. 309; 21.1.1, p. 320.

J. Schneider and J. Konetzky
*Magnetization Studies of Rapidly Quenched Fe–Ni Alloys
(a) Vol. **36**, No. 1, p. K77–K80 (1976).
See also *18.2*, p. *233*; 21.1, p. 310; 21.1.1, p. 320.

Y. Aoki, Y. Gotoh, and Y. Obi
*On the Phase Diagram of the Co-Rich Co–V Alloy System
(a) Vol. **36**, No. 2, p. K149–K152 (1976).
See also 14.1, p. 166; 18.2, p. 234; 21.1, p. 310.

H. Zabel and H. Peisl
X-Ray Study of the Phase Diagram of Hydrogen in Niobium
(a) Vol. **37**, No. 1, p. K67–K70 (1976).
See also *1.2*, p. *18*; 10.1, p. 84; 21, p. 298.

P. Rennert and A. M. Radwan
*Calculation of the Electrostatic Energy of the Laves Phases C14, C15, and C36
(b) Vol. **77**, No. 2, p. 615–621 (1976).
See also 21, p. 298.

L. E. Popov, Yu. P. Sharkeev, N. A. Koneva, N. A. Enshina, and G. A. Perov
*Conservative Jog Slip on Superdislocations and Work-Hardening of Ordered Alloys
(a) Vol. 37, No. 2, p. 701—708 (1976).
See also *10.1*, p. *84*; 21.1, p. 312.

M. I. Zakharova and H. Fuentes
*Structural Phase Transformations in the Alloys Ti-25%Nb-15%V and Ti-40%Nb
(a) Vol. 37, No. 2, p. K121—K125 (1976).
See also *1.2*, p. *19*; 21.1, p. 313.

R. Moret, M. Huber, and R. Comès
Diffuse Scattering and Titanium Short-Range Order in $Ti_{1+x}S_2$
(a) Vol. 38, No. 2, p. 695—700 (1976).
See also *1.2*, p. *20*; 22, p. 342.

1.4 Microstructure (Magnetic Domains See 18; Ferroelectric Domains See 14.4.1)

Original Papers and Short Notes

A. R. Jones, P. R. Howell, and B. Ralph
*Weak Beam Microscopy of Grain Boundaries
(a) Vol. **33**, No. 1, p. 107–117 (1976).
See also 21.1.1, p. 316.

A. Melander
*Strong- and Weak-Beam Contrast from Cuboidal Inclusions
(a) Vol. **33**, No. 1, p. 255–264 (1976).
See also 10.1, p. 78; 21, p. 290; 21.1, p. 303.

L. M. Clarebrough and C. T. Forwood
*Diffraction from Single and Overlapping Stacking Faults in F.C.C. Crystals
(a) Vol. **33**, No. 1, p. 355–366 (1976).
See also *10.1*, p. *78*; 21.1, p. 303.

R. de Ridder, G. van Tendeloo, and S. Amelinckx
Electron Microscopic Study of the Chimney Ladder Structures $MnSi_{2-x}$ and $MoGe_{2-x}$
(a) Vol. **33**, No. 1, p. 383–393 (1976).
See also *1.1*, p. *10*; 22, p. 334.

A. J. Morton
*Inversion Anti-Phase Domains in Cu-Rich γ-Brasses (II): Defects in the Domain Structure
(a) Vol. **33**, No. 1, p. 395–403 (1976).
See also 1.3, p. 21; 21.1, p. 303.

A. C. McLaren and W. S. MacKenzie
*The Spatial Coherence of X-Ray and Electron Beams and Its Influence on the Diffraction Patterns from Materials with Long-Period Superlattices
(a) Vol. **33**, No. 2, p. 491–495 (1976).
See also *1*, p. *9*; 22.6.1, p. 409.

J. Vicens and P. Delavignette
*A Particular Domain Configuration Observed in a New Phase of the Ta–N System
(a) Vol. 33, No. 2, p. 497–509 (1976).
See also *1.2*, p. *14*; 22, p. 334.

C. Boulesteix, J. van Landuyt, and S. Amelinckx
*Identification of Rotation and Reflection Twins by Diffraction and Contrast Experiments in the Electron Microscope
(a) Vol. **33**, No. 2, p. 595–606 (1976).
See also *10.2*, p. *89*; 22.6, p. 403.

T. Schober and H. Wenzl
*β-Phase Melting and Solidification Phenomena in the Niobium–Hydrogen System
(a) Vol. **33**, No. 2, p. 673–681 (1976).
See also *1.2*, p. *14*; 10.1, p. 79; 21, p. 291.

L. B. Harris and J. Fiasson
*Cation Diffusion along Boundaries in Sodium Chloride
(a) Vol. **33**, No. 2, p. 697–706 (1976).
See also *9*, p. *72*; 22.5.2, p. 393.

G. van Tendeloo, J. van Landuyt, and S. Amelinckx
*The $\alpha \to \beta$ Phase Transition in Quartz and $AlPO_5$ as Studied by Electron Microscopy and Diffraction
(a) Vol. **33**, No. 2, p. 723–735 (1976).
See also *1.2*, p. *14*; 22.6, p. 403; 22.8.1, p. 419.

C. Manolikas, P. Delavignette, and S. Amelinckx
*Electron Microscopy of Phases in the $Cu_{2-x}S$ System
(a) Vol. **33**, No. 2, p. K77–K79 (1976).
See also 22, p. 335.

F. W. Schapink
*The Interaction of Thickness Contours and Dislocations under Dynamical Diffraction Conditions
(a) Vol. **33**, No. 2, p. K121–K123 (1976).
See also *10*, p. *75*.

P. Bassoul, A. Dubon, A. Lefebvre et J. C. Gilles
*Structure à antiphases périodiques de la phase non-stoechiométrique ε_{Mg} du système $MgGa_2O_4$–Ga_2O_3
(Periodic Antiphase Structures in the Non-Stoichiometric Mg Phase of the System $MgGa_2O_4$–Ga_2O_3)
(a) Vol. **34**, No. 1, p. 125–134 (1976).
See also *1.2*, p. *15*; 22.8.2, p. 427.

E. Bernard and P. Duval
On the Period of Antiphase Domain Structures in X- and α_2-Phases of Copper–Aluminium Alloys
(a) Vol. **34**, No. 1, p. 135–144 (1976).
See also *1.2*, p. *15*; *1.3*, p. *22*; 21, p. 291; 21.1, p. 304.

J. Woltersdorf
*Interface Structure in Systems of Small Misfit Studied in the Case of Gold on Silver
(a) Vol. **34**, No. 1, p. 183–194 (1976).
See also *1.5*, p. *31*; 21.6, p. 330.

G. Nihoul-Boutang and C. Boulesteix
*Observation of Lattice-Conservating Twins by Electron Micrscopy. Study of the Image Contrasts
(a) Vol. **34**, No. 1, p. 331–338 (1976).
See also *10.2*, p. *90*; 22.6, p. 403.

D. E. Temkin
*On the Theory of Impurity Effect in Grain Boundary Motion
(a) Vol. **34**, No. 1, p. 413–425 (1976).
See also *10*, p. *75*.

R. A. Varin, J. A. Kozubowski, M. W. Grabski, and R. Buczkowski
*Analysis of Transmission Electron Microscopy Contrast at Extrinsic Grain Boundary Dislocations
(a) Vol. **34**, No. 2, p. 711–722 (1976).
See also *10.1*, p. *81*; 21.1.1, p. 318.

H. S. Chen
*Positron Lifetime Study on the Structure of an Electron Irradiated Metallic Glass
(a) Vol. **34**, No. 2, p. K127–K129 (1976).
See also 2, p. 37; 21, p. 293; 21.1, p. 306.

J. RABIER, P. VEYSSIÈRE, and J. GRILHÉ
*Possibility of Stacking Faults and Dissociation of Dislocations in the Garnet Structure
(a) Vol. **35**, No. 1, p. 259—268 (1976).
See also *10.2*, p. *92*; 22.8.2, p. 428.

L. B. HARRIS, D. A. MONCRIEFF, and V. N. E. ROBINSON
*High-Resolution Examination of Uncoated Insulators by SEM Applied to Grain Boundaries in Sodium Chloride
(a) Vol. **35**, No. 1, p. 371—377 (1976).
See also *10.2*, p. *92*; 22.5.2, p. 394.

M. LEFELD-SOSNOWSKA and E. ZIELIŃSKA-ROHOZIŃSKA
*Symmetrical and Asymmetrical X-Ray Section Topographs of Ion-Implanted Silicon
(a) Vol. **35**, No. 1, p. K1—K3 (1976).
See also 10.2, p. 92; 22.1.2, p. 354.

F. W. BOSWELL, A. PRODAN, and J. M. CORBETT
*Electron Diffraction Study of the Structures of NbS_2 and Order–Disorder in Cu-Intercalated NbS_2
(a) Vol. **35**, No. 2, p. 591—604 (1976).
See also 1.2, p. 16; 22, p. 337.

H. MENNIGER, H. RAIDT, and G. VOIGT
*On the Influence of Microdefects on Charge Carriers in Silicon
(a) Vol. **35**, No. 2, p. 639—643 (1976).
See also 14.3, p. 177; 22.1.2, p. 354.

A. L. ROITBURD and N. S. KOSENKO
*Orientational Dependence of the Elastic Energy of a Plane Interlayer in a System of Coherent Phases
(a) Vol. **35**, No. 2, p. 735—746 (1976).
See also *1.2*, p. *17*.

A. J. MORTON
*Erratum to: Inversion Anti-Phase Domains in Cu-Rich γ-Brasses (I)
(a) Vol. **35**, No. 2, p. 777 (1976).
See also 1.3, p. 23; 21.1, p. 308.

W. HAUFFE
*Faceting Mechanism in the Sputtering Process
(a) Vol. **35**, No. 2, p. K93—K96 (1976).
See also 11, p. 102; 21.1.1, p. 319; 21.6, p. 330.

V. SEETHARAMAN, S. BANERJEE, and R. KRISHNAN
*The Precipitation of $Ta_{64}C$ in Dilute Tantalum–Carbon Alloys
(a) Vol. **36**, No. 1, p. 39—52 (1976).
See also *1.3*, p. *23*; 21, p. 296 .

C. W. ALLEN and P. R. OKAMOTO
*An Imaging Technique in Transmission Electron Microscopy Utilizing Diffraction Fine Structure
(a) Vol. **36**, No. 1, p. 107—116 (1976).
See also *1.1*, p. *12*; 21.1.1, p. 319.

G. VOIGT, R. RAIDT, H. PEIBST, H. MENNIGER, and L. HILDISCH
*Investigations of Growth Defects in Solution-Grown $Ga_xIn_{1-x}P$ with a SEM Operating in the Cathodoluminescence Mode
(a) Vol. **36**, No. 1, p. 173—179 (1976).
See also *3*, p. *43*; 22.2.4, p. 370.

S. IKENO
*Direct Observation of Cell Formation in Niobium by a High-Voltage Electron Microscope
(a) Vol. **36**, No. 1, p. 317–328 (1976).
See also *10.1*, p. *83*; 21, p. 296.

M. GUYMONT and D. GRATIAS
*On the Stability of Periodically Antiphased Alloys
(a) Vol. **36**, No. 1, p. 329–334 (1976).
See also 1.2, p. 18; *1.3*, p. *23*; 21.1, p. 309; 21.6, p. 330.

A. PRODAN, F. W. BOSWELL, and J. M. CORBETT
*An Electron Microscopic Investigation of $CoTe_2$ Single Crystals
(a) Vol. **36**, No. 1, p. K21–K23 (1976).
See also 10.1, p. 83; 21.1, p. 309.

P. R. W. HUDSON and P. P. PHAKEY
*Observations of Planar Faults or Giant Platelets in Natural Gem Diamond
(a) Vol. **36**, No. 2, p. 627–636 (1976).
See also 22.1, p. 344.

R. S. RAI, S. R. SINGH, and G. SINGH
*On the Origin of Large Spiral Steps on SiC Growth Faces
(a) Vol. **36**, No. 2, p. 699–703 (1976).
See also *3*, p. *43*; 22.3, p. 373.

T. L. ANDERSON, R. E. NEWNHAM, L. E. CROSS, and J. W. LAUGHNER
Laser-Induced Twinning in Quartz
(a) Vol. **37**, No. 1, p. 235–245 (1976).
See also *10.2*, p. *95*; *12.2*, p. *112*; 22.6, p. 407.

A. L. ROITBURD
Equilibrium Structure of Epitaxial Layers
(a) Vol. **37**, No. 1, p. 329–339 (1976).
See also *1.5*, p. *32*.

H. BETHGE, E. PIPPEL, and J. WOLTERSDORF
Problems of Epitaxial Growth in Systems of Weak Interaction Observed in the Case of Al on MgO
(a) Vol. **37**, No. 2, p. 457–466 (1976).
See also *1.5*, p. *32*; *3*, p. *44*; 21, p. 298; 22.6, p. 407.

R. DE RIDDER, G. VAN TENDELOO, J. VAN LANDUYT, D. VAN DYCK, and S. AMELINCKX
Microstructural Study of Copper-Intercalated Niobium Disulphide and Tantalum Disulphide Cu_xNbS_2 and Cu_xTaS_2 (I): Phase Transitions in $Cu_{0.5}NbS_2$ and $Cu_{0.5}TaS_2$
(a) Vol. **37**, No. 2,p. 591–606 (1976).
See also *1.2*, p. *19* ; 22, p. 340.

J. KIES, R. KÖHLER, and W. MÖHLING
*Double Crystal X-Ray Topographs of Microdefects in Silicon
(a) Vol. **37**, No. 2, p. K113–K115 (1976).
See also 10.2, p. 97; 22.1.2, p. 357.

J. M. CORBETT and S. S. SHEININ
*Electron Microscope Images of Twin Boundaries (I): Comparison of Experiment with Theory
(a) Vol. **38**, No. 1, p. 151–159 (1976).
See also 21.1, p. 313.

G. van Tendeloo, R. de Ridder, D. van Dyck, J. van Landuyt, and S. Amelinckx
Microstructural Study of Copper-Intercalated Niobium Disulphide and Tantalum Disulphide Cu_xNbS_2 and Cu_xTaS_2 (II): Order–Disorder Trantitions and Incommensurate Superstructures in Cu_xNbS_2 $(0.8 < x < 1)$
(a) Vol. **38**, No. 1, p. 185–198 (1976).
See also *1.2*, p. *19*; 22, p. 341.

M. J. Marcinkowski
*The Differential Geometry of Grain Boundaries: Twist Boundaries
(a) Vol. **38**, No. 1, p. 223–235 (1976).
See also 10, p. 76.

W. Polanschütz und E. Krautz
*Zur feldionenmikroskopischen Untersuchung von Korngrenzen bei Metallen
(Field Ion Microscopic Study of Grain Boundaries in Metals)
(a) Vol. **38**, No. 1, p. 237–242 (1976).
See also 10.1, p. 85; *17.1*, p. *221*; 21, p. 300.

O. V. Presnyakova and V. I. Zaitsev
*Effect of High Pressure on Nickel Polygonization
(a) Vol. **38**, No. 2, p. 447–452 (1976).
See also *10.1*, p. *86*; 21.1, p. 314.

T. Shishido and M. Tanaka
Determination of Structure Factors of Germanium by the Critical-Voltage and Convergent-Beam Diffraction Methods
(a) Vol. **38**, No. 2, p. 453–461 (1976).
See also *1.1*, p. *13*; 22.1.1, p. 350.

S. M. Ohr
*Direction of the Black–White Contrast of Dislocation Loops
(a) Vol. **38**, No. 2, p. 553–557 (1976).
See also *10.1*, p. *86*; 21, p. 300.

D. Gratias, J. P. Boilot, Y. Le Cars, and J. Thery
*Syntaxy between β and β″ Structures in Alumina/Sodium Oxide and Gallia/Sodium Oxide Systems
(a) Vol. **38**, No. 2, p. 595–600 (1976).
See also 22.8.2, p. 431.

M. Gyumont, D. Gratias, R. Portier, and M. Fayard
*Space Group Theoretical Determination of Translation, Twin, and Translation-Twin Boundaries in Cell-Preserving Phase Transitions
(a) Vol. **38**, No. 2, p. 629–636 (1976).
See also 1.2, p. 20.

R. de Ridder, G. van Tendeloo, D. van Dyck, and S. Amelinckx
*A Cluster Model for the Transition State and Its Study by Means of Electron Diffraction (I): Theoretical Model
(a) Vol. **38**, No. 2, p. 663–674 (1976).
See also 10.2, p. 99; 22, p. 342.

S. S. Sheinin and J. M. Corbett
*Application of the Multi-Beam Dynamical Theory to Crystals Containing Twins
(a) Vol. **38**, No. 2, p. 675–684 (1976).
See also 21.1, p. 314.

K. Borggreen and A. R. Thölén
*Diffraction Contrast from Misfitting Spherical Particles Precipitated Close Together on a Dislocation (I): Individuel Spheres Close Together on a Straight Line
(a) Vol. 38, No. 2, p. 739–752 (1976).
See also 10.2, p. 99; 22, p. 342.

F. W. Schapink
*Weak-Beam Electron Microscopy of Nearly Transverse Grain Boundaries
(a) Vol. 38, No. 2, p. K107–K109 (1976).
See also 21, p. 301.

B. E. Smith
*(101) Order-Twinning Mode in a CuAuI Superlattice
(a) Vol. 38, No. 2, p. K167–K170 (1976).
See also 12.1, p. 109; 21, p. 301.

1.5 Films

Original Papers and Short Notes

E. M. TRUKHANOV, E. B. GOROKHOV, and S. I. STENIN
*Specific Features of the Dislocation Structure of Germanium in the System Germanium-Pyrolitic Silica Film
(a) Vol. **33**, No. 1, p. 435–442 (1976).
See also 2, p. 36; 10.2, p. 88; 22.1.1, p. 345; 22.6, p. 402.

Y. SAGARA, O. UEMURA, Y. SUZUKI, and T. SATOW
The Amorphous Structure of the Sb-Se System
(a) Vol. **33**, No. 2, p. 691–696 (1976).
See also *2*, p. *36*; 22.7, p. 410.

G. JUNGK
Determination of Optical Constants. A Null-Method for Non-Absorbing Surface Films
(a) Vol. **34**, No. 1, p. 69–72 (1976).
See also *20.1*, p. *264*; 22.6, p. 403.

J. WOLTERSDORF
*Interface Structure in Systems of Small Misfit Studied in the Case of Gold on Silver
(a) Vol. **34**, No. 1, p. 183–194 (1976).
See also *1.4*, p. *26*; 21.6, p. 330.

W. HOFBERGER
The Birefringence of Crystalline Tetracene Derived from Diffraction Properties of Polycrystalline Films
(a) Vol. **34**, No. 1, p. K55–K58 (1976).
See also *20.1*, p. *265*; 22.9, p. 434.

R. DRAGILA, L. BÁRDOŠ, and G. LONČAR
*Model of SiO_2 Film Growth on Si in Oxygen Microwave Discharge
(a) Vol. **35**, No. 1, p. 291–301 (1976).
See also *3*, p. *42*; 22.6, p. 404.

M. JANDA and A. KUBOVÝ
*Growth of Vacuum-Deposited Tellurium Films on Glass Substrates and Some of Their Transport Properties
(a) Vol. **35**, No. 1, p. 391–402 (1976).
See also *3*, p. *42*; 14.3.1, p. 183; 22.1.3, p. 359.

E. RUSKE
*Photoexpansion of Amorphous As_2S_3 Films
(a) Vol. **35**, No. 2, p. 477–480 (1976).
See also 2, p. 38; *20.1*, p. *269*; 22, p. 337.

V. N. BYKOV, V. A. TROYAN, G. G. ZDOROVTSEVA, and V. S. KHAIMOVICH
*Erratum to: Phase Transformations at Bombardment of Thin Films with Ions
(a) Vol. **35**, No. 2, p. 779 (1976).
See also *1.2*, p. *17*; 11, p. 102; 21, p. 295; 21.1, p. 308; 21.1.1, p. 319; 22.1.1, p. 347.

P. THOMAS
Addendum to: On Anisotropic Hopping Conduction in Thin Amorphous Germanium Films
(b) Vol. **76**, No. 1, p. 439 (1976).
See also *14.3.1*, p. *184*; 22.1.1, p. 347.

P. V. PAVLOV, T. A. KRUZE, D. I. TETELBAUM, E. I. ZORIN, E. V. SHITOVA, and N. V. GUDKOVA
*Electron Microscopic Studies of Silicon Layers Irradiated with High Doses of Nitrogen Ions
(a) Vol. **36**, No. 1, p. 81–88 (1976).
See also 2, p. 39; *11*, p. *102*; 22.1.2, p. 355.

S. PETERSSON, E. MGBENU, and P. A. TOVE
Silicide Formation Correlated with Surface Resistivity Measurements
(a) Vol. **36**, No. 1, p. 217–225 (1976).
See also *14.3.1*, p. *184*; 22, p. 338.

M. WENDT, TH. KRAJEWSKI, and R. BIMBERG
*Detection of Thin Surface Films by Electron Beam Microanalysis. A Comparison between Wavelength-Dispersive and Energy-Dispersive Microprobes
(a) Vol. **36**, No. 1, p. 253–261 (1976).
See also 21.1, p. 309; 21.1.1, p. 319; 22.1.1, p. 348.

V. KOOS and H.-G. NEUMANN
*Effect of Heat Treatment on Stress in Amorphous Silicon Films
(a) Vol. **36**, No. 1, p. K47–K49 (1976).
See also *12.2*, p. *111*; 22.1.2, p. 355.

S. K. LILOV, YU. M. TAIROV, V. F. TSVETKOV, and M. A. CHERNOV
*Structural and Morphological Peculiarities of the Epitaxial Layers and Monocrystals of Silicon Carbide Highly Doped by Nitrogen
(a) Vol. **37**, No. 1, p. 143–150 (1976).
See also 22.3, p. 374.

V. KOOS, U. BECK, and H.-G. NEUMANN
*Structure and Intrinsic Stress of Nickel Silicide Films
(a) Vol. **37**, No. 1, p. 193–198 (1976).
See also *12.2*, p. *111*; 22, p. 339.

R. H. WADE
*Tilted-Beam Electron Microscopy of Amorphous Films
(a) Vol. **37**, No. 1, p. 247–256 (1976).
See also *2*, p. *40*; 22.1.1, p. 348.

H. S. RANDHAWA, L. K. MALHOTRA, H. K. SEHGAL, and K. L. CHOPRA
Structural Transformations in a-Ge Alloy Films
(a) Vol. **37**, No. 1, p. 313–320 (1976).
See also *1.2*, p. *18*; *2*, p. *40*; 22.1.1, p. 348.

A. L. ROITBURD
*Equilibrium Structure of Epitaxial Layers
(a) Vol. **37**, No. 1, p. 329–339 (1976).
See also 1.4, p. 28.

L. N. ALEKSANDROV and R. N. LOVYAGIN
*Epitaxial Germanium–Silicon Structures Obtained in Ultrahigh Vacuum
(a) Vol. **37**, No. 1, p. 341–352 (1976).
See also *3*, p. *32*; 22.1.1, p. 348; 22.1.2, p. 356.

H. BETHGE, E. PIPPEL, and J. WOLTERSDORF
*Problems of Epitaxial Growth in Systems of Weak Interaction Observed in the Case of Al on MgO
(a) Vol. **37**, No. 2, p. 457–466 (1976).
See also 1.4, p. 28; *3*, p. *44*; 21, p. 298; 22.6, p. 407.

R. I. BOLKHOVITYANOVA and YU. B. BOLKHOVITYANOV
*The Influence of Indium on the p–n Junction Formation in Si-Doped LPE Layers of Gallium Arsenide
(a) Vol. **37**, No. 2, p. K193–K195 (1976).
See also 3, p. 44; *14.3.1*, p. *185*; 22.2.4, p. 371.

A. E. GERSHINSKII, L. V. MIRONOVA, and E. I. CHEREPOV
*SiO_2–Si (n-Type) Interface Investigation with the Electrochemical Method
(a) Vol. **38**, No. 1, p. 369–**373** (1976).
See also *14.3.2*, p. *188*; 22.1.2, p. 358; 22.6, p. 408.

B. BARANOV and L. DÄWERITZ
*Vapour-Phase Epitaxial Growth of $Ga_{1-x}Al_xN$ on Sapphire
(a) Vol. **38**, No. 2, p. K111–K113 (1976).
See also *3*, p. *44*; 22.2.4, p. 371.

1.6 Surfaces

Original Papers and Short Notes

R. Mehrotra and M. M. Pant
*Static and Dynamic Displacements near a Crystal Surface
(b) Vol. **73**, No. 1, p. K27–K30 (1976).
See also *6*, p. *53*; 8, p. 67; 21.1.1, p. 316.

G. Filscher
*On the Theoretical Calculation of the Intensity Distribution in X-Ray Section Topographs of Crystals with Superposed Films
(a) Vol. **33**, No. 2, p. K81–K82 (1976).
See also 10.2, p. 89; 22.1.2, p. 352.

M. Euler and A. Scharmann
Thermally Stimulated Exoelectron Emission (TSEE) of BeO Layers. Auger Electron Spectroscopy Investigations
(a) Vol. **34**, No. 1, p. 297–306 (1976).
See also 13.3, p. 132; *17*, p. *218*; 22.6, p. 403.

N. Tajabor and D. R. Lovett
*Etching and Chemical Polishing of {112} Surfaces of Cd_3As_2 and Zn_3As_2 Crystals
(a) Vol. **34**, No. 2, p. K191–K194 (1976).
See also 22.3, p. 372.

V. Siegel, W. Rasp, and H.-H. Kirchner
The Influence of Alkyl Groups on Exoelectron Emission from BeO
(a) Vol. **34**, No. 2, p. K203–K204 (1976).
See also *17*, p. *218*; 22.6, p. 404.

W. Schulz and W. Gründler
*Quantum Mechanical ab initio Calculations of the Surface Structure of Crystalline Lithium Hydride
(b) Vol. **78**, No. 1, p. 183–189 (1976).
See also 22.5.2, p. 397.

J. G. Antonopoulos
*Contrast from Atomic Surface Steps Using TEM
(a) Vol. **38**, No. 2, p. K115–K117 (1976).
See also 21.1, p. 315; 21.6, p. 331.

2. Non-Crystalline State

Review Articles

V. L. BONCH-BRUEVICH
Problems of the Electron Theory of Small-Gap Disordered Semiconductors
(b) Vol. 74, No. 1, p. 11–22 (1976).
See also *13.1*, p. *120*; 22, p. 333.

H. BÖTTGER and V. V. BRYKSIN
Hopping Conductivity in Ordered and Disordered Solids (I)
(b) Vol. 78, No. 1, p. 9–56 (1976).
See also 6, p. 52; 13, p. 113; 13.5.3, p. 159; *14.3*, p. *173*; 22, p. 333.

H. BÖTTGER and V. V. BRYKSIN
Hopping Conductivity in Ordered and Disordered Solids (II)
(b) Vol. 78, No. 2, p. 415–451 (1976).
See also 6, p. 52; 13, p. 113; 13.5.3, p. 159; *14.3*, p. *173*; 22, p. 333.

Original Papers and Short Notes

I. V. CHEPELEVA, E. A. ZHILINSKAYA, V. N. LAZUKIN, and A. P. CHERNOV
Electron Paramagnetic Resonance of Cr^{3+} in Chalcogenide Glasses
(b) Vol. 73, No. 1, p. 65–72 (1976).
See also 18, p. 222; *19*, p. *247*; 22.8, p. 413.

W. LÖSER
*A Self-Consistent Cluster Approximation for the Disordered Hubbard Model
(b) Vol. 73, No. 1, p. 179–185 (1976).
See also 13, p. 113; 14, p. 162; *18*, p. *222*.

H. BÖTTGER and V. V. BRYKSIN
*Theoretical Investigation of the Hopping Conductivity in Disordered Systems in a Large Frequency Range
(b) Vol. 73, No. 1, p. 229–237 (1976).
See also 6, p. 52; 13, p. 113; *14.3*, p. *173*.

B. ESSER and P. KLEINERT
Erratum to: On the Theory of Interband $\varepsilon(\omega)$ and External Electric Field Modulated $\Delta\varepsilon(\omega)$ in Disordered Semiconductors
(b) Vol. 73, No. 1, p. 379 (1976).
See also *14.3.4*, p. *195*; *20.1*, p. *260*; 22, p. 333.

M. KUMEDA, N. KOBAYASHI, E. MARUYAMA, and T. SHIMIZU
Electron Spin Resonance and Optical Absorption in Amorphous Se–As
(b) Vol. 73, No. 1, p. K19–K22 (1976).
See also *19*, p. *247*; *20.1*, p. *261*; 22, p. 333.

S. KOBE and K. HANDRICH
*Correlation Function and Misfit in a Computer-Simulated Two-Dimensional Amorphous Ising Antiferromagnet
(b) Vol. 73, No. 1, p. K65–K67 (1976).
See also *18.4*, p. *242*.

NGUYEN VAN DONG, TRAN HUU DANH, B. AUGUIN, and A. DEFRESNE
*Transport Properties of Glassy $CdGe_{1-x}Sb_xAs_2$
(a) Vol. 33, No. 1, p. 295–299 (1976).
See also *14.3*, p. *173*; 15, p. 209; 22.8, p. 413.

K. L. CHOPRA and P. NATH
*Amorphous Ge Alloy Films (II): Transport Properties
(a) Vol. **33**, No. 1, p. 333–344 (1976).
See also *14.3.1*, p. *183*; 15, p. 209; 22.1.1, p. 345.

E. M. TRUKHANOV, E. B. GOROKHOV, and S. I. STENIN
Specific Features of the Dislocation Structure of Germanium in the System Germanium–Pyrolitic Silica Film
(a) Vol. **33**, No. 1, p. 435–442 (1976).
See also *1.5*, p. *31*; 10.2, p. 88; 22.1.1, p. 345; 22.6, p. 402.

I. M. BELOVA
On the Theory of Tunnelling in Disordered Semiconductors
(b) Vol. **73**, No. 2, p. 409–414 (1976).
See also *14.3.4*, p. *195*; 22, p. **334**.

B. ESSER
X-Ray Scattering from Electrons in Random Semiclassical Potential Wells
(b) Vol. **73**, No. 2, p. 499–502 (1976).
See also *13.4*, p. *135*; 22, p. **334**.

Y. SAGARA, O. UEMURA, Y. SUZUKI, and T. SATOW
*The Amorphous Structure of the Sb–Se System
(a) Vol. **33**, No. 2, p. 691–696 (1976).
See also 1.5, p. 31; 22.7, p. 410.

N. L. KRAMARENKO, I. V. KURNOSOV, and YU. V. NABOIKIN
*Light Scattering by Liquid Crystals in an Electric Field
(a) Vol. **33**, No. 2, p. 773–781 (1976).
See also *20*, p. *257*; 22.9, p. **434**.

T. SHIMIZU and N. ISHII
*Electronic Structure of As_2S_3, As_2Se_3, and As_2Te_3 Glasses by the Extended Hückel Theory
(b) Vol. **74**, No. 1, p. K39–K42 (1976).
See also *13*, p. *115*; 22.3, p. 372.

E. DONTH
*Universal Thermodynamic and Kinetic Functions for a Classical Freezing-in Process in Noncrystalline Bodies
(b) Vol. **74**, No. 1, p. K49–K52 (1976).
See also *8*, p. *67*.

V. V. VOSKOBOYNIKOV, V. A. GRITSENKO, V. M. EFIMOV, V. E. LESNIKOVSKAYA, and F. L. EDELMAN
*Structure and Electrophysical Properties of Boron Nitride Thin Films
(a) Vol. **34**, No. 1, p. 85–94 (1976).
See also *14.3.1*, p. *183*; 14.3.3, p. 190; 22, p. 335.

P. NATH, D. K. PANDYA, and K. L. CHOPRA
*Amorphous Ge Alloy Films (III): Optical Properties
(a) Vol. **34**, No. 1, p. 405–411 (1976).
See also *20.1*, p. *265*; 22.1.1, p. **346**.

O. BOSTANJOGLO and W. GIESE
*Magnetization and Crystallization of Stress-Stabilized Amorphous Iron Films
(a) Vol. **34**, No. 1, p. K1–K4 (1976).
See also *18.2.1*, p. *237*; 21.1.1, p. **317**.

J. Krempaský and D. Barančok
*The Use of Modified Relaxation Time in the Transport Theory of Non-Crystalline Semiconductors
(b) Vol. 74, No. 2, p. 741—751 (1976).
See also 13.4, p. 137; *14.3*, p. *175*; 15, p. 209; 22, p. 335.

J. D. Patterson and R. C. Weger
The Specific Heat of Small Magnetic Clusters with Simulated Amorphousness
(b) Vol. 74, No. 2, p. K115—K118 (1976).
See also *8*, p. *68*; *18.2*, p. *230*.

J. Richter, J. Schreiber, and K. Handrich
*Influence of Structure Fluctuations on Ferromagnetism in the Band Model
(b) Vol. 74, No. 2, p. K125—K128 (1976).
See also *18.2*, p. *231*.

H. Böttger and V. V. Bryksin
*On the Equivalence of Different Theoretical Descriptions of the Hopping Conductivity in Disordered Systems
(b) Vol. 74, No. 2, p. K151—K156 (1976).
See also 6, p. 55; 13, p. 115; *14.3*, p. *175*.

V. Šmíd
*A Contribution to the Interpretation of the High-Field Electrical Conductivity of Amorphous Semiconductors
(b) Vol. 74, No. 2, p. K169—K171 (1976).
See also *14.3.4* p. *196*.

H. S. Chen
Positron Lifetime Study on the Structure of an Electron Irradiated Metallic Glass
(a) Vol. 34, No. 2, p. K127—K129 (1976).
See also *1.4*, p. *26*; 21, p. 293; 21.1, p. 306.

P. Duhaj, J. Sitek, M. Prejsa, and P. Butvin
*Electron Microscopy and Mössbauer Study of the Amorphous Metallic Alloys $Pd_{80-x}Fe_xSi_{20}$
(a) Vol. 35, No. 1, p. 223—233 (1976).
See also *1.2*, p. *16*; 5, p. 49; 21.1.1, p. 318.

M. Popescu
*Relaxed Random Network Model of the $GeAs_2Te_7$ Glass
(a) Vol. 35, No. 1, p. 355—360 (1976).
See also *4*, p. *45*; 22.7, p. 411.

B. Bahadur, S. Chandra, and N. K. Sanyal
*Phase Transition Studies of the Liquid Crystal n-(p-Octyloxybenzylidene)-p-Toluidine
(a) Vol. 35, No. 1, p. 387—390 (1976).
See also *18*, p. *223*; 22.9, p. 436.

R. M. Hill
*On the Observation of Variable Range Hopping
(a) Vol. 35, No. 1, p. K29—K34 (1976).
See also *14.3*, p. *176*.

S. Balevičius, A. Česnys, and A. Deksnys
*Nanosecond Switching in Amorphous In_2Te_3 Films
(a) Vol. 35, No. 1, p. K41—K43 (1976).
See also *14.3.1*, p. *184*; 22.3, p. 372.

W. HENRION, M. ZÁVĚTOVÁ, T. MAZETS
*Infrared Optical Properties of Pseudobinary Chalcogenide Glasses
(a) Vol. **35**, No. 1, p. K77–K79 (1976).
See also *20.1*, p. *268*; 22.8, p. 415.

F. ITOH, K. KAI, M. KUROHA, and S. TAKEUCHI
*Positron Annihilation in Liquid Copper–Bismuth and Copper–Tin Alloys
(b) Vol. **75**, No. 2, p. 559–565 (1976).
See also 13, p. 116; 21.1, p. 307; 21.7, p. 332.

J. F. GRACZYK and P. CHAUDHARI
*Scattering Properties of Dense Random Packings of Hard Spheres
(b) Vol. **75**, No. 2, p. 593–606 (1976).
See also *1.3*, p. *22*; 21, p. 294.

E. KOLLEY and W. KOLLEY
*Effect of Thermal Fluctuations on the Electronic Densities of States in Disordered Binary Alloys. Numerical Results of CPA Model Calculations
(b) Vol. **75**, No. 2, p. 721–730 (1976).
See also 6, p. 57; *13*, p. *116*; 21, p. 294.

G. CUBIOTTI, E. DONATO, E. S. GIULIANO, R. RUGGERI, and A. STANCANELLI
Transport Properties of Liquid Cadmium
(b) Vol. **75**, No. 2, p. K93–K96 (1976).
See also *14.1*, p. *165*; *15*, p. *210*; 21, p. 295.

R. VACHER and J. PELOUS
*Measurement of the Intensity of the Brillouin Lines in Amorphous Media at Very Low Temperature
(b) Vol. **75**, No. 2, p. K101–K103 (1976).
See also *6*, p. *57*; 20.1, p. 269; 22.9, p. 436.

E. RUSKE
Photoexpansion of Amorphous As_2S_3 Films
(a) Vol. **35**, No. 2, p. 477–480 (1976).
See also *1.5*, p. *31*; *20.1*, p. *269*; 22, p. 337.

D. K. PANDYA and K. L. CHOPRA
*Obliquely Deposited Amorphous Ge Films (I): Optical Properties
(a) Vol. **35**, No. 2, p. 725–734 (1976).
See also *20.1*, p. *269*; 22.1.1, p. 347.

K. L. CHOPRA, S. K. BARTHWAL, and D. K. PANDYA
Structural Sensitivity of Transport and Optical Properties of Amorphous GeTe
(a) Vol. **35**, No. 2, p. 761–768 (1976).
See also *14.3*, p. *178*; *15*, p. *210*; 20.1, p. 270; 22.7, p. 411.

E. RUSKE
Reversibility of Photoinduced Edge Shift in Evaporated Amorphous As_2S_3 Films
(a) Vol. **35**, No. 2, p. K113–K114 (1976).
See also *20.1*, p. *270*; 22, p. 337.

H.-J. UNGER
*An Analytically Tractable Structure Model for Disordered Systems (I): Systems Based on a Primitive Network
(b) Vol. **76**, No. 1, p. 207–215 (1976).

W. BRAUN and M. CARDONA
Electronic Density of States of Amorphous $CuGaS_2$ and $ZnGeAs_2$
(b) Vol. **76**, No. 1, p. 251–257 (1976).
See also *13.1*, p. *125*; 17, p. 219; 22.8, p. 416.

P. V. Pavlov, T. A. Kruze, D. I. Tetelbaum, E. I. Zorin, E. V. Shitova, and N. V. Gudkova
Electron Microscopic Studies of Silicon Layers Irradiated with High Doses of Nitrogen Ions
(a) Vol. **36**, No. 1, p. 81—88 (1976).
See also *1.5*, p. *31*; *11*, p. *102*; 22.1.2, p. 355.

K. L. Chopra and D. K. Pandya
*Obliquely Deposited Amorphous Ge Films (II): Electrical Properties
(a) Vol. **36**, No. 1, p. 89—100 (1976).
See also *14.3.1*, p. *184*; 15, p. 210; 22.1.1, p. 348.

C. Birac and D. Lesueur
Lithium Diffusion in an Amorphous Metallic Pd–Si Alloy
(a) Vol. **36**, No. 1, p. 247—251 (1976).
See also *9*, p. *73*; 21, p. 296.

S. K. Barthwal and K. L. Chopra
*Transport Properties of Amorphous InSb and GaAs
(a) Vol. **36**, No. 1, p. 345—355 (1976).
See also *14.3.1*, p. *184*; 15, p. 210; 22.2.1, p. 363; 22.2.3, p. 368.

V. Bezák and J. Banský
A Generalization of the Sondheimer-Wilson Formula for the Avaraged Density Matrix and Magnetic Susceptibility of Conduction Electrons in a Model of Disordered Semiconductors
(b) Vol. **76**, No. 2, p. 569—580 (1976).
See also *13*, p. *117*; *18*, p. *224*; 22, p. 338.

H.-J. Unger
*An Analytically Tractable Structure Model for Disordered Systems (II): Complicated Structures and Application to Amorphous Selenium
(b) Vol. **76**, No. 2, p. 841—850 (1976).
See also 22.1.3, p. 359.

R. V. Sharma and K. C. Sharma
*A Possible Closed Form Expression for the Static Structure Factor for Liquid Metals
(b) Vol. **76**, No. 2, p. K75—K79 (1976).
See also *13*, p. *118*; 21.2, p. 323.

S. K. Barthwal and K. L. Chopra
Thermoelectric Power of Amorphous Germanium Films
(a) Vol. **36**, No. 2, p. 533—549 (1976).
See also *14.3.1*, p. *184*; *15*, p. *210*; 22.1.1, p. 348.

J. E. Lowther
*Mechanisms of Energy Transfer between Rare Earth Ions
(b) Vol. 77, No. 1, p. 359—366 (1976).
See also 6, p. 60; 13.4, p. 142; 21.4, p. 327.

N. G. Cherenda and D. M. Yudin
*Additives and Cage Effect in Glasses
(b) Vol. 77, No. 1, p. K27—K30 (1976).
See also 11, p. 103; *19*, p. *254*; 22.8.1, p. 423.

J. Richter
*Influence of Structure Fluctuations on Ferromagnetism of Strongly Correlated Electrons
(b) Vol. 77, No. 1, p. K43—K47 (1976).
See also *18.2*, p. *234*.

V. A. Vassilev, B. T. Kolomiets, T. N. Mamontova, and E. M. Raspopova
Photoluminescence and Photoconductivity in Densified Vitreous As_2Se_3
(a) Vol. 37, No. 1, p. 45—49 (1976).
See also *16*, p. *216*; *20.3*, p. *286*; 22, p. 339.

R. H. Wade
*Tilted-Beam Electron Microscopy of Amorphous Films
(a) Vol. 37, No. 1, p. 247—256 (1976).
See also *1.5*, p. *32*; 22.1.1, p. 348.

H. S. Randhawa, L. K. Malhotra, H. K. Sehgal, and K. L. Chopra
*Structural Transformations in a-Ge Alloy Films
(a) Vol. 37, No. 1, p. 313—320 (1976).
See also *1.2*, p. *18*; 1.5, p. 32; 22.1.1, p. 348.

K. Hübner and V. K. Bashenov
Ionicity and Structure of Solids
(b) Vol. 77, No. 2, p. 473—482 (1976).
See also 1.1, p. 13; *4*, p. *46*; *13.1*, p. *127*; 22.2, p. 361; 22.2.4, p. 371.

P. Blaudeck and R. Lenk
*Multiple s-Scattering in a Correlated Medium in the Low-Density Limit
(b) Vol. 77, No. 2, p. 601—605 (1976).
See also 11, p. 103.

L. I. Horváth and T. Szörényi
EPR Studies on Vanadium Phosphate Glasses
(a) Vol. 37, No. 2, p. K169—K171 (1976).
See also *19*, p. *255*; 22.8.1, p. 424.

L. Kubler, G. Gewinner, J. J. Koulmann, and A. Jaéglé
EPR Study on Amorphous Germanium. Oxygen, Deposition Angle, Annealing, and Substrate Temperature Effects
(b) Vol. 78, No. 1, p. 149—158 (1976).
See also *19*, p. *255*; 22.1.1, p. 349.

J. Schreiber and W. John
*On the Theory of Electron Localization in Amorphous Systems
(b) Vol. 78, No. 1, p. 199—210 (1976).
See also *13*, p. *119*.

R. N. Singh and P. L. Srivastava
Liquid Electrical Resistivity of Alkali Metals: Exchange and Correlation Effects
(b) Vol. 78, No. 1, p. K45—K49 (1976).
See also *14.1*, p. *167*; 21.2, p. 323.

J. Salik and S. Nadiv
*Structural Studies of the Photodissolution of Ag in Amorphous As_2S_3 Films
(a) Vol. 38, No. 1, p. 177—184 (1976).
See also 11, p. 104; 22, p. 341.

S. R. Herd
*Lorentz Microscopy on Amorphous Films with Perpendicular Anisotropy as a Function of Temperature
(a) Vol. 38, No. 1, p. 305—314 (1976).
See also *18.2.1*, p. *238*; 21.1, p. 314; 21.4, p. 327.

T. Boţilă and H. K. Henisch
*Study of Trapping in Amorphous Materials
(a) Vol. 38, No. 1, p. 331—337 (1976).
See also 13.4, p. 145; 14.3, p. 181; 16, p. 217; 22.8, p. 418.

A. Mansingh, J. K. Vaid, and R. P. Tandon
*Switching and Memory Behaviour of Vanadate Glasses
(a) Vol. **38**, No. 1, p. K1—K3 (1976).
See also *14.3.3*, p. *194*; 22.8.1, p. 424.

L. Moseley, R. S. Tripathi, and T. Lukes
*Theory of the Hopping Conductivity in Non-Crystalline Systems
(b) Vol. 78, No. 2, p. 563—567 (1976).
See also 13.4, p. 145; *14*, p. *163*.

W. John and J. Schreiber
*Localization Theory for the Random Weaire Model
(b) Vol. 78, No. 2, p. 577—583 (1976).
See also *13*, p. *119*; 22, p. 341.

W. Grevendonk, G. Ruymbeek, P. van den Keybus, and A. de Bock
*Faraday Rotation in Non-Crystalline As_2S_3
(b) Vol. 78, No. 2, p. K89—K91 (1976).
See also 18, p. 224; 20, p. 259; 22, p. 342.

R. V. Sharma and K. C. Sharma
*The Structure of Molten Cuprous Chloride
(b) Vol. 78, No. 2, p. K93—K95 (1976).
See also 22.5, p. 390.

J. Škácha
*Electrical Structure and Transport in Amorphous GeS Layers
(a) Vol. **38**, No. 2, p. K143—K146 (1976).
See also *14.3.1*, p. *186*; 22, p. 342.

3. Crystal Growth

Original Papers and Short Notes

G. Arancia, M. Grandolfo, C. Manfredotti, and A. Rizzo
*Electron Diffraction Study of Melt- and Vapour-Grown $GaSe_{1-x}S_x$ Single Crystals
(a) Vol. **33**, No. 2, p. 563–571 (1976).
See also *1.1*, p. *11*; 22.3.1, p. 375.

J. M. Galligan
Growth Characteristics of Crystals Containing Disclinations
(a) Vol. **34**, No. 2, p. K101–K103 (1976).
See also *10.1*, p. *81*; 21.6, p. 330.

Yu. A. Vodakov, G. A. Lomakina, E. N. Mokhov, E. I. Radovanova, V. I. Sokolov, M. M. Usmanova, G. F. Yuldashev, and B. S. Machmudov
Silicon Carbide Doped with Gallium
(a) Vol. **35**, No. 1, p. 37–42 (1976).
See also *13.4*, p. *138*; 14.3, p. 176; 20.1, p. 268; 20.3, p. 283; 22.3, p. 372.

A. Prasad, R. Beserman, P. Germain, and J. C. Bourgoin
*Kinetics of Recrystallization in Ion-Implanted Silicon Layers
(a) Vol. **35**, No. 1, p. 109–112 (1976).
See also 22.1.2, p. 353.

R. Dragila, L. Bárdoš, and G. Lončar
*Model of SiO_2 Film Growth on Si in Oxygen Microwave Discharge
(a) Vol. **35**, No. 1, p. 291–301 (1976).
See also *1.5*, p. *31*; 22.6, p. 404.

R. Hergt, M. Wendt, P. Görnert, and S. Bornmann
*Investigation of the Growth Mechanism in High-Temperature Solutions by the Induced Striation Method (I): Dissolution Striations
(a) Vol. **35**, No. 1, p. 347–353 (1976).
See also 22.8.2, p. 429.

M. Janda and A. Kubový
*Growth of Vacuum-Deposited Tellurium Films on Glass Substrates and Some of Their Transport Properties
(a) Vol. **35**, No. 1, p. 391–402 (1976).
See also *1.5*, p. *31*; 14.3.1, p. 183; 22.1.3, p. 359.

P. Görnert, S. Bornmann, and R. Hergt
*Investigation of the Growth Mechanism in High-Temperature Solutions by the Induced Striation Method (II): Driving Forces in Slow-Cooling Experiments
(a) Vol. **35**, No. 2, p. 583–590 (1976).
See also 22.8.2, p. 429.

P. Matthys, M. Vanhaelst, and E. Boesman
*A Modified Doping Procedure for High O_2^- Concentrations in Potassium and Rubidium Halides
(a) Vol. **35**, No. 2, p. K137–K139 (1976).
See also *19*, p. *252*; 22.5.2, p. 395.

O. A. Troitskii
*Isotope Enrichment of Metal in the Process of Zone Melting
(a) Vol. **35**, No. 2, p. K151–K155 (1976).
See also 21, p. 295; 21.1, p. 308.

G. Voigt, R. Raidt, H. Peibst, H. Menniger, and L. Hildisch
*Investigations of Growth Defects in Solution-Grown $Ga_xIn_{1-x}P$ with a SEM Operating in the Cathodoluminescence Mode
(a) Vol. **36**, No. 1, p. 173—179 (1976).
See also *1.4*, p. *27*; 22.2.4, p. 370.

D. Ball and B. M. Wanklyn
*Coloured Synthetic Zircon Crystals. Flux Growth and EPR of V^{4+} Impurity Centres
(a) Vol. **36**, No. 1, p. 307—316 (1976).
See also *19*, p. *253*; 22.8.1, p. 422.

L. N. Aleksandrov and B. I. Kidyarov
*Kinetic Theory of Three-Barrier Processes of Non-Steady-State Nucleation at Phase Transformation
(a) Vol. **36**, No. 1, p. 403—413 (1976).
See also 1.2, p. 18.

H. Rogass, J. Lauckner, and J. Finke
*On the Influence of an Electric Field on the Condensation of Selenium Layers
(a) Vol. **36**, No. 1, p. K9—K11 (1976).
See also 22.1.3, p. 359.

Y. Kashiwada and Y. Furuhata
*Domains in Magnesium Pyroborate
(a) Vol. **36**, No. 1, p. K29—K31 (1976).
See also *1.2*, p. *18*; 22.8, p. 416.

B. I. Fomin, A. E. Gershinskii, E. I. Cherepov, and F. L. Edelman
*Investigation of Phase Growth Kinetics in the System of Si Single Crystals and V Thin Films
(a) Vol. **36**, No. 1, p. K89—K91 (1976).
See also 21.1, p. 310; 22.1.2, p. 355.

R. S. Rai, S. R. Singh, and G. Singh
*On the Origin of Large Spiral Steps on SiC Growth Faces
(a) Vol. **36**, No. 2, p. 699—703 (1976).
See also *1.4*, p. *28*; 22.3, p. 373.

I. V. Bodnar, E. E. Matyas, and L. A. Makovetskaya
*Phase Diagram of $Ga_xIn_{1-x}P$ Solid Solution
(a) Vol. **36**, No. 2, p. K141—K144 (1976).
See also 22.2.4, p. 371.

L. N. Aleksandrov and R. N. Lovyagin
Epitaxial Germanium–Silicon Structures Obtained in Ultrahigh Vacuum
(a) Vol. **37**, No. 1, p. 341—352 (1976).
See also *1.5*, p. *32*; 22.1.1, p. 348; 22.1.2, p. 356.

Burma Bansaragtschin, G. Lehmann, and R. Link
*Application of the Travelling Solvent Method to Growing of $Pb_{1-x}Sn_xTe$ Single Crystals
(a) Vol. **37**, No. 1, p. K13—K15 (1976).
See also 14.3, p. 179; 22, p. 339.

L. G. Sakvarelidze, A. A. Mirtskhulava, I. A. Mirtskhulava, and G. G. Gegiadze
*Production of Bulk Single Crystals of $Ga_{1-x}In_xAs$ Solid Solutions
(a) Vol. **37**, No. 1, p. K71—K73 (1976).
See also 22.2.4, p. 371.

H. Bethge, E. Pippel, and J. Woltersdorf
*Problems of Epitaxial Growth in Systems of Weak Interaction Observed in the Case of Al on MgO
(a) Vol. **37**, No. 2, p. 457–466 (1976).
See also 1.4, p. 28; *1.5*, p. *32*; 21, p. 298; 22.6, p. 407.

P. Görnert and G. Wende
*Study of the Mass Transport during Crystal Growth from High-Temperature Solutions by the "High-Resolution Induced Striation Method"
(a) Vol. **37**, No. 2, p. 505–513 (1976).
See also 22.8.2, p. 430.

R. I. Bolkhovityanova and Yu. B. Bolkhovityanov
The Influence of Indium on the p–n Junction Formation in Si-Doped LPE Layers of Gallium Arsenide
(a) Vol. **37**, No. 1, p. K193–K195 (1976).
See also *1.5*, p. *32*; *14.3.1*, p. *185*; 22.2.4, p. 371.

W. Brückner, U. Gerlach, W. Moldenhauer, H. P. Brückner, N. Mattern, H. Oppermann, and E. Wolf
Phase Transitions and Semiconductor–Metal Transition in $V_{1-x}Ga_xO_2$ Single Crystals
(a) Vol. **38**, No. 1, p. 93–102 (1976).
See also *1.2*, p. *19*; 14.3, p. 181; 15, p. 211; 18.1, p. 227; 22.6.1, p. 409.

F. V. Wald
*Self-Compensation in CdTe and ZnTe Crystals Grown from Indium Solvents
(a) Vol. **38**, No. 1, p. 253–259 (1976).
See also 10.2, p. 97; 14.3, p. 181; 20.1, p. 277; 22.4.3, p. 386.

R. S. Momtaz and L. N. Rashkovich
*The Distribution of Deuterium between Solid and Liquid Phases during the Crystallization of Crystals of the KDP Group from Aqueous Solutions
(a) Vol. **38**, No. 1, p. 401–408 (1976).
See also 14.4.1, p. 208; 22.8.1, p. 424.

B. Baranov and L. Däweritz
*Vapour-Phase Epitaxial Growth of $Ga_{1-x}Al_xN$ on Sapphire
(a) Vol. **38**, No. 2, p. K111–K113 (1976).
See also *1.5*, p. *33*; 22.2.4, p. 371.

4. Bonding Properties

Original Papers and Short Notes

S. Benckert
*Erratum to: Third-Order Contribution to the Energy and the Elastic Constants of Simple Metals
(b) Vol. **73**, No. 2, p. 753 (1976).
See also *12.1*, p. *107*; 21, p. 291; 21.2, p. 323.

K. Hübner
*Strain Dependence of the Baldereschi Point and of the Bond Charge
(b) Vol. **73**, No. 2, p. K141—K144 (1976).
See also *13.1*, p. *122*; 22.1, p. 343.

K. Kumazaki
*Elastic Properties and Ionicity of Zero-Gap Semiconductors
(a) Vol. **33**, No. 2, p. 615—623 (1976).
See also 7, p. 65; *12.2*, p. *110*; 22.4, p. 376.

R. Blinc, H. Arend, and A. Kandušer
Vibrational Spectra, Hydrogen Bonding, and Ferroelectricity in the $PbHPO_4$ Family
(b) Vol. **74**, No. 2, p. 425—435 (1976).
See also *6*, p. *55*; 14.4.1, p. 206; *20.1*, p. *265*; 22.8.1, p. 420.

B. N. Oshcherin
*On Surface Energies of A^NB^{8-N} Semiconducting Compounds
(a) Vol. **34**, No. 2, p. K181—K186 (1976).
See also *1*, p. *9*; 22, p. 336.

A. P. Rusakov and M. V. Anfimov
*Non-Central Forces and Stability of the A^NB^{8-N} Tetrahedral Compounds (I): Bond Charges and the Valence-Force-Field Model
(b) Vol. **75**, No. 1, p. 73—77 (1976).
See also 12.2, p. 110; 22.1, p. **343**; 22.2, p. 361; 22.4, p. 376; 22.5, p. 389.

V. Kondawar and C. Mande
Chemical Shifts in the K X-Ray Absorption Discontinuity of Arsenic
(b) Vol. **75**, No. 1, p. 79—83 (1976).
See also 20, p. 258; 22, p. 336.

M. Popescu
*Relaxed Random Network Model of the $GeAs_2Te_7$ Glass
(a) Vol. **35**, No. 1, p. 355—360 (1976).
See also *2*, p. *37*; 22.7, p. 411.

P. Ramasamy and T. Nagarajan
Positron Angular Correlation Studies of Phase Transitions in Ammonium Halides
(a) Vol. **35**, No. 1, p. 379—386 (1976).
See also *1.2*, p. *16*; 22.5, p. 389.

H. P. Aubauer
Elastic Energy of Isotropic Solid Solutions
(a) Vol. **35**, No. 2, p. 487—497 (1976).
See also *1.2*, p. *16*; 21, p. 295.

G. Heger, D. Mullen, and K. Knorr
On the Importance of Hydrogen Bonding for the Structural Phase Transitions in $(CH_3NH_3)_2MnCl_4$. A Single-Crystal Neutron Diffraction Study
(a) Vol. **35**, No. 2, p. 627–637 (1976).
See also *1.2*, p. *17*; 22.9, p. 436.

T. Wasiutyński
Self-Consistent Phonon Calculation for Hexamethylenetetramine Molecular Crystals
(b) Vol. **76**, No. 1, p. 175–181 (1976).
See also *6*, p. *57*; 12.2, p. 110; 22.9, p. 437.

K. Hübner and M. Schäfer
*Bond Charge Model for the Chemical Shift of the Photoemission in Semiconductors
(b) Vol. **76**, No. 1, p. K63–K67 (1976).
See also *20.1*, p. *271*; 22.2, p. 361.

A. Batana and E. R. Gonzalez
*On the Pressure Dependence of the Effective Ionic Charge of Lithium, Sodium, and Rubidium Halides
(b) Vol. **76**, No. 2, p. K105–K108 (1976).
See also *12.2*, p. *111*; 22.5.2, p. 396.

J. J. Rousseau and J. C. Fayet
*Influence of the Crystal Field on Bent Configuration of V_k Centres in AMF_3 Compounds
(b) Vol. 77, No. 1, p. 195–201 (1976).
See also *13.4*, p. *142*; 19, p. 253; 22.8, p. 417.

S. K. Chakrabarti, S. K. Sarkar, and S. Sengupta
*On the Change of Electronic Polarisability of an Ion in a Crystal
(b) Vol. 77, No. 1, p. 329–340 (1976).
See also 22.5.2, p. 396.

K. Hübner and V. K. Bashenov
*Ionicity and Structure of Solids
(b) Vol. 77, No. 2, p. 473–482 (1976).
See also 1.1, p. 13; 2, p. 40; *13.1*, p. *127*; 22.2, p. 361; 22.2.4, p. 371.

L. K. Aminov, R. Kirmse, and B. V. Solovev
*Electron Spin–Lattice Relaxation of Cu^{2+} Ions in Planar Coordination Geometries. Calculation and Experimental Data of T_1 for Cu^{2+} in Ni(II)bis(diethyl-dithiocarbamate) Crystals
(b) Vol. 77, No. 2, p. 505–515 (1976).
See also 6, p. 61; 18.1, p. 226; 19, p. 254; 22.9, p. 438.

R. Pickenhain and A. Milchev
*The Quadratic Response of a Fermi Gas. Electronic Charge Density of Valence Electrons in Zincblende Semiconductors
(b) Vol. 77, No. 2, p. 571–579 (1976).
See also 13, p. 119; 22.1.1, p. 348.

D. W. Bullett
*Non-Empirical Calculation of Structural Properties and Valence Force Fields. Application to Se and Te
(b) Vol. 77, No. 2, p. 645–650 (1976).
See also 1, p. 9; 6, p. 61; 22.1.3, p. 360.

H. Ö. Pamuk and T. Halicioğlu
*Evaluation of Morse Parameters for Metals
(a) Vol. 37, No. 2, p. 695–699 (1976).
See also 12.1, p. 108; 21, p. 299.

H. J. Stolz, H. Wendel, A. Otto, L. Pintschovius, and H. Kahlert
Acoustic and Optical Phonons in $(SN)_x$
(b) Vol. 78, No. 1, p. 277–286 (1976).
See also *6*, p. *63*; *20.1*, p. *276*; 22, p. 341.

H. P. Niedermann and M. Wagner
On the Quantum Mechanical Foundation of Classical Shell Models in Lattice Dynamics
(b) Vol. 78, No. 2, p. 615–624 (1976).
See also *6*, p. *64*; 22, p. 342.

M. Schlüter, M. L. Cohen, S. E. Kohn, and C. Y. Fong
Electronic Structure of BiI_3
(b) Vol. 78, No. 2, p. 737–747 (1976).
See also *13.1*, p. *129*; 20.1, p. 278; 22.5, p. 390.

5. Mössbauer Spectroscopy

Review Article

H. W. DE WIJN, A. M. VAN DIEPEN, and K. H. J. BUSCHOW
Samarium in Crystal Fields, a Case of Strong J-Mixing
(b) Vol. **76**, No. 1, p. 11–30 (1976).
See also *13*, p. *113*; *18*, p. *222*; 19, p. 247; 21, p. 289; 21.1, p. 302; 21.1.1, p. 316; 21.4, p. 326.

Original Papers and Short Notes

A. M. AFANASEV and V. D. GOROBCHENKO
* On the Theory of Mössbauer Emission Relaxation Spectra
(b) Vol. **73**, No. 1, p. 73–79 (1976).

S. K. ARIF and D. S. P. BUNBURY
*Mössbauer and X-Ray Studies of $Dy(Fe_{1-x}Ni_x)_3$ and $Y(Fe_{1-x}Ni_x)_3$ Intermetallic Compounds
(a) Vol. **33**, No. 1, p. 91–95 (1976).
See also 18.2, p. 228; 18.3, p. 239; 21.1.1, p. 316.

V. A. TSURIN, A. E. ERMAKOV, YU. G. LEBEDEV, and B. N. FILIPPOV
*A Mössbauer Study of the Structural Characteristics of Equiatomic FePd and FePt Alloys
(a) Vol. **33**, No. 1, p. 325–332 (1976).
See also 18.2, p. 229; 21.1.1, p. 316.

E. DE GRAVE, C. DAUWE, A. GOVAERT, and J. DE SITTER
*Cation Distribution and Magnetic Exchange Interactions in $MgFe_2O_4$ Studied by the Mössbauer Effect
(b) Vol. **73**, No. 2, p. 527–532 (1976).
See also *18.3*, p. *239*; 22.8.2, p. 426.

R. G. GABRIELYAN, L. A. KOCHARYAN, A. R. MKRTCHYAN, and S. S. TUMANYAN
*Double γ-Acoustic Resonance in Crystals
(b) Vol. **73**, No. 2, p. 681–687 (1976).

A. ANDREEFF, H.-J. HUNGER, and S. UNTERRICKER
*The Magnetic Hyperfine Interaction at Cd^{111} in Iron–Aluminium Alloys
(b) Vol. **73**, No. 2, p. K89–K92 (1976).
See also 18.2, p. 229; 21, p. 291; 21.1.1, p. 317.

V. A. POVITSKII, E. F. MAKAROV, N. V. MURASHKO, and A. N. SALUGIN
*Mössbauer Study of Superparamagnetic δ-FeOOH and Its Transformation into α-Fe_2O_3
(a) Vol. **33**, No. 2, p. 783–787 (1976).
See also 1.2, p. 15; 22.8.2, p. 427.

M. J. TRICKER, R. K. THORPE, J. H. FREEMAN, and G. A. GARD
*Conversion Electron Mössbauer Spectroscopy Study of ^{57}Fe Ion Implanted into Single Crystals of Graphite
(a) Vol. **33**, No. 2, p. K97–K100 (1976).
See also 22.1, p. 343.

A. G. MADDOCK, A. F. WILLIAMS, K. E. SIEKIERSKA, and J. FENGER
*The Mössbauer Emission Spectra of Impurity ^{57}Cobalt in a Halide Matrix
(b) Vol. **74**, No. 1, p. 183–192 (1976).
See also 22.5.2, p. 393; 22.5.3, p. 399.

J. WINTER, K. RÖSSLER, J. BOLZ, and J. PELZL
*Mössbauer Effect and Raman Studies of the Structural Phase Transitions in $K_2[SnCl_6]$
(b) Vol. **74**, No. 1, p. 193–198 (1976).
See also *1.2*, p. *15*; 20.1, p. 263; 22.8, p. 414.

H. Pollak
*Charge Transfer in Cordierite
(b) Vol. 74, No. 1, p. K31—K34 (1976).
See also *20.1*, p. *264*; 22.8.1, p. 420.

K. Ruebenbauer and W. Żurek
*Mössbauer Line-Shape after Nuclear Resonant Scattering
(b) Vol. 74, No. 2, p. K165—K167 (1976).

P. C. M. Gubbens and K. H. J. Buschow
*Mössbauer Effect Study of $Tm_2Fe_{17-x}Co_x$ and $Tm_2Fe_{17-x}Ni_x$ Compounds
(a) Vol. 34, No. 2, p. 729—735 (1976).
See also 18.2, p. 231; 21.1.1, p. 318; 21.4, p. 326.

J. P. Mathieu and J. Chappert
*Electronic Properties and Relaxation Processes of Fe^{2+} in Iron Bisacetylacetonate Dihydrate from Mössbauer Spectroscopy
(b) Vol. 75, No. 1, p. 163—171 (1976).
See also 22.9, p. 436.

G. Ya. Selyutin
*Mössbauer Spectrum near the Critical Point
(b) Vol. 75, No. 1, p. 233—238 (1976).
See also 18, p. 223.

S. Ligenza
A Study of the $^6S_{5/2}$-Term Splitting of an Fe^{3+} Ion in Zinc Ferrite by Neutron Spectroscopy
(b) Vol. 75, No. 1, p. 315—326 (1976).
See also *13.4*, p. *138*; 18.3, p. 240; 22.8.2, p. 428.

G. Chandra, C. Bansal, and J. Ray
*Temperature-Dependent Mössbauer Study of Equiatomic Disordered CoFe and NiFe Alloys
(a) Vol. 35, No. 1, p. 73—77 (1976).
See also 21.1, p. 307; 21.1.1, p. 318.

A. M. van der Kraan, J. N. J. van der Velden, J. H. F. van Apeldoorn, P. C. M. Gubbens, and K. H. J. Buschow
*Mössbauer Effect Study and Magnetization Measurements of RFe_3 Intermetallic Compounds with R = Sm, Gd, Tb, Dy, Ho, and Th
(a) Vol. 35, No. 1, p. 137—151 (1976).
See also 18.2, p. 232; 21.1.1, p. 318; 21.4, p. 326.

F. Basile et P. Poix
*Etude par effet Mössbauer et measures magnétiques de la solution solide $Fe_{2-2x}Mg_{1+x}Sn_xO_4$ (Mössbauer Effect and Magnetic Measurement on the Solid Solution $Fe_{2-2x}Mg_{1+x}Sn_xO_4$)
(a) Vol. 35, No. 1, p. 153—164 (1976).
See also 18.3, p. 240; 22.8.2, p. 428.

P. Duhaj, J. Sitek, M. Prejsa, and P. Butvin
Electron Microscopy and Mössbauer Study of the Amorphous Metallic Alloys $Pd_{80-x}Fe_xSi_{20}$
(a) Vol. 35, No. 1, p. 223—233 (1976).
See also *1.2*, p. *16*; *2*, p. *37*; 21.1.1, p. 318.

R. Lindsey
*Measurement of Mössbauer Diffusion Broadening in α-Iron Stabilized with Vanadium
(b) Vol. 75, No. 2, p. 583—592 (1976).
See also *9*, p. *73*; 21.1.1, p. 319.

A. M. Afanasev and V. D. Gorobchenko
*Theory of Time-Dependent Mössbauer Emission Spectra under Relaxation Conditions
(b) Vol. 76, No. 2, p. 465–473 (1976).

Debadideb Bhattacharya
*The Study of Dislocations by the Mössbauer Effect
(b) Vol. 76, No. 2, p. K81–K84 (1976).
See also *10.2*, p. *94*.

J. Sawicki, B. Sawicka, J. Stanek, and J. Kowalski
*Mössbauer Study of Heat Treated ^{57}Fe Implanted Silicon
(b) Vol. 77, No. 1, p. K1–K4 (1976).
See also 22.1.2, p. 356.

F. P. Livi, J. D. Rogers, and P. J. Viccaro
*Experimental Studies of Hyperfine Interactions and Magnetization in $HfFe_2$
(a) Vol. 37, No. 1, p. 133–138 (1976).
See also *18.2*, p. *235*; 21.1.1, p. 320.

T. Ruskov, T. Tomov, and S. Georgiev
*Mössbauer Investigation of the Morin Transition in Hematite
(a) Vol. 37, No. 1, p. 295–302 (1976).
See also 18.4, p. 245; 22.6, p. 407.

M. Shiga and Y. Nakamura
*Mössbauer Study of B.C.C. Fe–Cr Alloys near the Critical Concentration
(a) Vol. 37, No. 1, p. K89–K92 (1976).
See also 18.2, p. 235; 21.1, p. 311; 21.1.1, p. 321.

Reva Garg and V. K. Garg
*On the Quadrupole Splitting–Isomer Shift Linear Correlationship and the Sign of Electric Field Gradient
(b) Vol. 77, No. 2, p. K165–K170 (1976).
See also 22.8, p. 417.

K. V. Reddy and S. C. Chetty
*Mössbauer Studies on the Fe–Te System
(a) Vol. 37, No. 2, p. 687–694 (1976).
See also 18.4, p. 245; 21.1.1, p. 321.

K. Knoch and H. Dannheim
*Temperature Dependence of the Cation Distribution in Magnesium Ferrite
(a) Vol. 37, No. 2, p. K135–K137 (1976).
See also 22.8.2, p. 431.

E. M. Savitzkii, M. I. Bychkova, V. V. Baron, B. E. Dzevitskii, and N. I. Savateev
Lattice Instability of Compound Nb_3Sn in Strained State
(a) Vol. 37, No. 2, p. K165–K167 (1976).
See also *1.2*, p. *19*; 21, p. 299.

S. K. Godovikov, M. G. Kozin, V. V. Turovtsev, and V. S. Shpinel
*Hyperfine Fields Acting on Diamagnetic Impurities and the Anisotropic Exchange via the s- and d-Electrons in the Rare Earth Metals
(b) Vol. 78, No. 1, p. 103–111 (1976).
See also 18.2, p. 235; 18.4, p. 245; 21.4, p. 327.

S. Ligenza
A Study of Spin Interaction in a Manganese-Substituted Zinc Ferrite by Neutron Spectroscopy
(b) Vol. 78, No. 1, p. 301–310 (1976).
See also *18.4*, p. *245*; 22.8.2, p. 431.

V. S. Zasimov, N. N. Lobanov, J. Rüdiger und R. N. Kusmin
*Untersuchung der thermischen Diffusionsstreuung im Graphitkristall mit Hilfe des Mössbauer-Effektes
(Study of Thermal Diffusion Scattering in Graphite Crystal by the Mössbauer Effect)
(a) Vol. **38**, No. 1, p. K45–K47 (1976).
See also 22.1, p. 344.

D. C. Gupta and N. D. Sharma
*Pressure Dependence of the Mössbauer Fraction for ^{57}Fe in Ti, V, and Cu
(b) Vol. 78, No. 2, p. K125–K129 (1976).
See also 6, p. 64; 21.1, p. 314.

S. C. Bhargava and P. K. Iyengar
*Study of Hyperfine Fields at Iron Nuclei in $(Co, Fe)_5Ge_3$ Using the Mössbauer Effect
(a) Vol. **38**, No. 2, p. 601–609 (1976).
See also 18.1, p. 227; 18.2, p. 236; 21.1.1, p. 322.

6. Lattice Dynamics. Phonons

Review Articles

W. Haubenreisser, Th. Klupsch, and F. Voigt
Quantum Statistical Theory of Parametric Spin-Wave Excitations in Magnetoelastic Ferromagnets under Parallel Pumping (II): Calculation of Macroscopic Observables
(b) Vol. **73**, No. 1, p. 11–64 (1976).
See also *13*, p. *113*; 13.5.4, p. 160; *18.2*, p. *228*.

M. N. Sapozhnikov
*Zero-Phonon Transitions in the Optical Spectra of Impurity Molecular Crystals
(b) Vol. **75**, No. 1, p. 11–51 (1976).
See also 13.4, p. 134; 20.1, p. 260; *20.3*, p. *280*; 22.9, p. 433.

H. Böttger and V. V. Bryksin
Hopping Conductivity in Ordered and Disordered Solids (I)
(b) Vol. **78**, No. 1, p. 9–56 (1976).
See also 2, p. 35; 13, p. 113; 13.5.3, p. 159; *14.3*, p. *173*; 22, p. 333.

H. Böttger and V. V. Bryksin
Hopping Conductivity in Ordered and Disordered Solids (II)
(b) Vol. **78**, No. 2, p. 415–451 (1976).
See also 2, p. 35; 13, p. 113; 13.5.3, p. 159; *14.3*, p. *173*; 22, p. 333.

Original Papers and Short Notes

F. Bechstedt, R. Enderlein, A. Kolpakov, and K. Peuker
*Theory of Resonance Raman Scattering near Critical Points
(b) Vol. **73**, No. 1, p. 141–149 (1976).
See also 13.1, p. 120; 13.5.1, p. 149; *20.1*, p. *260*.

M. Taut and H. Eschrig
*Electron and Phonon Spectra of Calcium
(b) Vol. **73**, No. 1, p. 151–154 (1976).
See also *13.1*, p. *120*; 13.2, p. 130; 21.3, p. 324.

N. J. Doran and E. Gerlach
*A Theory for Electron Transport Due to Phonon Scattering Including Screening
(b) Vol. **73**, No. 1, p. 203–209 (1976).
See also *14.3*, p. *173*; 22.2.1, p. 362.

H. Böttger and V. V. Bryksin
Theoretical Investigation of the Hopping Conductivity in Disordered Systems in a Large Frequency Range
(b) Vol. **73**, No. 1, p. 229–237 (1976).
See also *2*, p. *35*; 13, p. 113; *14.3*, p. *173*.

J. Höner zu Siederdissen
*Impurity-Induced Raman Spectra of Potassium Halides
(b) Vol. **73**, No. 1, p. 239–244 (1976).
See also 13.4, p. 134; *20.1*, p. *260*; 22.5.2, p. 392.

P. Plumelle and M. Vandevyver
*Lattice Dynamics of ZnTe and CdTe
(b) Vol. **73**, No. 1, p. 271–281 (1976).
See also 22.4.3, p. 385.

J. W. J. Timans and H. W. den Hartog
*Local-Mode Absorption of H^- and D^- in $SrCl_2$:RE and SrF_2:RE Single Crystals
(b) Vol. 73, No. 1, p. 283—292 (1976).
See also *20.1*, p. *260*; 22.5.3, p. 399.

L. M. Falicov and W. Y. Hsu
*Phonons and Quasi-Periodic Static Fluctuations
(b) Vol. 73, No. 1, p. 301—306 (1976).
See also *13.2*, p. *130*; 14.1, p. 164; 21.1, p. 302.

T. Fukumoto, S. Nakashima, K. Tabuchi, and A. Mitsuishi
*Temperature Dependence of Raman Spectra of Cuprous Halides
(b) Vol. 73, No. 1, p. 341—351 (1976).
See also *20.1*, p. *260*; 22.5, p. 388.

C. Uher
Erratum to: Low-Field Thermomagnetic Tensor for the Bi-Like Band Structure
(b) Vol. 73, No. 1, p. 379 (1976).
See also 13.1, p. 121; *15*, p. *209*; 21.7, p. 332.

P. K. Basu and B. R. Nag
Iterative Calculation of the Polar Optical Mobility in Semiconductors in the Extreme Quantum Limit
(b) Vol. 73, No. 1, p. K15—K17 (1976).
See also *14.3*, p. *173*; 22.2.3, p. 367.

W. Bausch
*Accuracy of the Boundary Relaxation Time Approximation
(b) Vol. 73, No. 1, p. K23—K25 (1976).
See also 8, p. 67; 22.5.2, p. 392.

R. Mehrotra and M. M. Pant
*Static and Dynamic Displacements near a Crystal Surface
(b) Vol. 73, No. 1, p. K27—K30 (1976).
See also *1.6*, p. *34*; 8, p. 67; 21.1.1, p. 316.

A. V. Alekseev and U. Kh. Kopvillem
*The Collapse of Radiation Stimulated by the Dipole–Dipole Temperature
(b) Vol. 73, No. 1, p. K41—K46 (1976).
See also 19, p. 247; 22.8, p. 413.

M. Leppihalme and T. Tuomi
Photovoltage Spectra of Metal–Silicon (p-Type) Diodes near the Indirect Absorption Edge of Silicon
(a) Vol. 33, No. 1, p. 125—133 (1976).
See also 13.5.1, p. 149; *14.3.3*, p. *190*; *16*, p. *213*; 20.1, p. 261; 22.1.2, p. 351.

M. L. Swanson, L. M. Howe, and A. F. Quenneville
The Combined Effects of Lattice Vibrations and Irradiation-Induced Defects on Dechanneling in Copper
(a) Vol. 33, No. 1, p. 265—274 (1976).
See also *11*, p. *100*; 21.1, p. 303.

S. I. Masharov
*Anharmonic Effects in Heat Capacity and Saturation Vapour Pressure of Dilute Substitutional Alloys
(b) Vol. 73, No. 2, p. 401—408 (1976).
See also *8*, p. *67*; 21, p. 290.

E. M. LEDOVSKAYA and M. I. PETRASHEN
The Determination of the Symmetry Group of Molecular Centres in Crystals
(b) Vol. **73**, No. 2, p. 495–498 (1976).
See also *10.2*, p. *89*; 22.5.2, p. 393.

M. GIEHLER and E. JAHNE
*Effect of Damping on the Plasmon–Phonon Coupling in CdS and GaP
(b) Vol. **73**, No. 2, p. 503–516 (1976).
See also *13.5.2*, p. *157*; 20.1, p. 261; 22.2.2, p. 365; 22.4.1, p. 379.

P. PETELENZ
Effective Electron–Hole Interaction in Molecular Crystals (I)
(b) Vol. **73**, No. 2, p. 565–573 (1976).
See also *13*, p. *114*; *13.5.1*, p. *150*; 14.3, p. 174; 22, p. 334.

A. GRISEL and PH. SCHMID
*Polytypism and Lattice Vibrations of PbI_2
(b) Vol. **73**, No. 2, p. 587–591 (1976).
See also 20.1, p. 262; 22.5, p. 388.

J. VON BORZESZKOWSKI
On the Electric Field Dependence of the Conductivity of Holes in Si at Low Temperatures (I): The Drift Velocity under Consideration of the Non-Equipartition of Acoustic Phonon Energy
(b) Vol. **73**, No. 2, p. 607–614 (1976).
See also 13.1, p. 121; *14.3.4*, p. *195*; 22.1.2, p. 351.

B. DORNER, R. E. GHOSH, and G. HARBEKE
*Phonon Dispersion in the Layered Compound PbI_2
(b) Vol. **73**, No. 2, p. 655–659 (1976).
See also 7, p. 65; 12.2, p. 110; 22.5, p. 388.

V. K. ARORA
Effect of Inelasticity on Magnetoresistance
(b) Vol. **73**, No. 2, p. K93–K94 (1976).
See also *14.3*, p. *174*; 22.2.3, p. 367.

L. HULDT
*Phonon-Assisted Auger Recombination in Germanium
(a) Vol. **33**, No. 2, p. 607–614 (1976).
See also *13.1*, p. *122*; 20, p. 257; 22.1.1, p. 345.

B. V. NOVIKOV, N. S. SOKOLOV, and S. V. GASTEV
Free-Carrier Generation via Exciton–Phonon and Exciton–Impurity Interaction in Ge Crystals
(b) Vol. **74**, No. 1, p. 81–89 (1976).
See also *13.4*, p. *136*; *13.5.1*, p. *150*; 16, p. 214; 22.1.1, p. 345.

T. IIDA and R. MONNIER
Static Properties of the Self-Trapped Hole in Simple Cubic Cesium Halides
(b) Vol. **74**, No. 1, p. 91–102 (1976).
See also *13.4*, p. *136*; 20.1, p. 263; 22.5.2, p. 393.

D. V. KAPOR, S. D. STOJANOVIĆ, M. J. ŠKRINJAR, and B. S. TOŠIĆ
*A New Approach to Exciton–Phonon Interaction and Some Consequences
(b) Vol. **74**, No. 1, p. 103–110 (1976).
See also *13.5.1*, p. *150*.

J. KURITTU and M. MERISALO
*Mean-Square Displacements for NH_4Cl and ND_4Cl Crystals
(b) Vol. **74**, No. 1, p. 177–181 (1976).
See also 22, p. 335.

C. A. Bates and H. Szymczak
*The Phonon-Induced Temperature Dependences of Spin-Hamiltonian Parameters for S-State Ions
(b) Vol. 74, No. 1, p. 225–233 (1976).
See also *19*, p. *249*; 22.4.1, p. 379; 22.6, p. 403.

E. Jahne
*Long-Wavelength Optical Phonons in Mixed Crystals (I): A System of Two Coupled Modes
(b) Vol. 74, No. 1, p. 275–283 (1976).
See also *20.1*, p. *263*; 22.2.4, p. 370; 22.3.1, p. 375.

Shi-yu Wu
*The Resonance Modes in the Vibrational Spectrum of a Chain with Light Impurities
(b) Vol. 74, No. 1, p. 349–357 (1976).
See also *10*, p. *75*.

J. K. Pribram and C. M. Penchina
Phonon-Assisted Electroabsorption by Contact Excitons
(b) Vol. 74, No. 1, p. 375–381 (1976).
See also *13.5.1*, p. *151*; 14.3.4, p. 196; 20.1, p. 264.

V. I. Alshits and Yu. M. Sandler
Erratum to: Flutter Mechanism of Dislocation Drag
(b) Vol. 74, No. 1, p. K85 (1976).
See also *10*, p. *75*.

R. Blinc, H. Arend, and A. Kandušer
*Vibrational Spectra, Hydrogen Bonding, and Ferroelectricity in the $PbHPO_4$ Family
(b) Vol. 74, No. 2, p. 425–435 (1976).
See also 4, p. 45; 14.4.1, p. 206; *20.1*, p. *265*; 22.8.1, p. 420.

M. I. Dykman
*Theory of Nondegenerate Two-Dimensional Electrons Interacting with Phonons in Quantizing Magnetic Fields
(b) Vol. 74, No. 2, p. 547–559 (1976).
See also *13.1*, p. *122*; 20.1, p. 265.

B. Ulrici and E. Jahne
*Phonon Frequencies of InP Derived from Two-Phonon Absorption
(b) Vol. 74, No. 2, p. 601–607 (1976).
See also *20.1*, p. *265*; 22.2.2, p. 365.

P. Petelenz
*Effective Electron–Hole Interaction in Molecular Crystals (II): Phonon Mechanism
(b) Vol. 74, No. 2, p. 655–658 (1976).
See also *13*, p. *115*; 13.5.1, p. 152; 14.3, p. 175; 22, p. 335.

J. W. Martin and R. Paetsch
Generalised Viscosity and Dislocation Damping Constant in Diatomic Crystalline Solids
(b) Vol. 74, No. 2, p. 761–772 (1976).
See also *10*, p. *75*; 12, p. 106.

J. Niewodniczańska-Zawadzka and J. Raułuszkiewicz
*The Observation of LO Phonons in HgSe and HgTe by Electron Tunneling
(b) Vol. 74, No. 2, p. K93–K95 (1976).
See also *14.3.3*, p. *191*; 22.4.2, p. 383; 22.4.3, p. 385.

H. Böttger and V. V. Bryksin
On the Equivalence of Different Theoretical Descriptions of the Hopping Conductivity in Disordered Systems
(b) Vol. 74, No. 2, p. K151–K156 (1976).
See also *2*, p. *37*; 13, p. 115; *14.3*, p. *175*.

P. Charsley and M. H. S. Rahim
The Effect of Ageing on the Lattice Thermal Conductivity of Copper Alloys Deformed at Liquid Helium Temperatures
(a) Vol. **34**, No. 2, p. 533–539 (1976).
See also *8*, p. *68*; *10.1*, p. *80*; 21.1, p. 306.

G. B. Abdullaev, K. R. Allakhverdiev, R. Kh. Nani, E. Yu. Salaev, and R. M. Sardarly
*Far-Infrared Reflection Spectra of $TlGaSe_2$
(a) Vol. **34**, No. 2, p. K115–K117 (1976).
See also *20.1*, p. *266*; 22.8, p. 414.

S. Duraiswamy and T. M. Haridasan
*Moments of the Phonon Spectrum in Lithium Chloride
(b) Vol. **75**, No. 1, p. 53–59 (1976).
See also *8*, p. *68*; 22.5.2, p. 394.

O. P. Bansal
On the Pressure Dependence of Electron Transport in Noble Metals
(b) Vol. **75**, No. 1, p. 85–90 (1976).
See also 12.1, p. 107; *13.2*, p. *130*; 21.6, p. 330.

R. P. Goyal, K. K. Sarkar, and S. C. Goyal
*Dispersion of Phonons in Transition Metal Oxides (MnO)
(b) Vol. **75**, No. 1, p. 101–104 (1976).
See also 22.6, p. 404.

E. L. Bokhenkov, I. Natkaniec, and E. F. Sheka
*Neutron Spectroscopy of Naphthalene Crystal Internal Phonon Modes
(b) Vol. **75**, No. 1, p. 105–116 (1976).
See also 20.1, p. 267; 22.9, p. 435.

D. Billard, F. Gervais, and B. Piriou
*Analysis of Multiphonon Absorption in Corundum
(b) Vol. **75**, No. 1, p. 117–126 (1976).
See also *20.1*, p. *267*; 22.6, p. 404.

E. Jahne
*Long-Wavelength Optical Phonons in Mixed Crystals (II): The Persistence of Common Gaps
(b) Vol. **75**, No. 1, p. 221–232 (1976).
See also *20.1*, p. *267*; 22.2.4, p. 370.

R. Mazumdar and L. S. Kothari
*Anharmonic Vibrations of a Small Linear Chain
(b) Vol. **75**, No. 1, p. 381–390 (1976).
Erratum
(b) Vol. **78**, No. 2, p. 843 (1976).

T. R. Rao, M. L. Bansal, V. C. Sahni, and A. P. Roy
*Study of Internal Modes of SO_4^{2-} in $LiKSO_4$
(b) Vol. **75**, No. 1, p. K31–K34 (1976).
See also 20.1, p. 267; 22.8.1, p. 421.

W. Siegel, A. Heinrich, and E. Ziegler
Electron and Hole Mobility in $ZnSiP_2$
(a) Vol. **35**, No. 1, p. 269–279 (1976).
See also 13.4, p. 139; *14.3*, p. *176*; 22.8, p. 415.

S. V. Gantsevich, V. L. Gurevich, V. D. Kagan, and R. Katilius
Transformation of Incoherent Phonon Flux into a Coherent Acoustical Signal as the Result of the Nonlinear Amplification of Acoustical Phonons in Semiconductors
(b) Vol. **75**, No. 2, p. 407–422 (1976).
See also *7*, p. *65*; *14.3.4*, p. *197*; 22, p. 336.

H. UEBA and S. ICHIMURA
Plasmon-Phonon-Coupled Mode Sideband of Excitons
(b) Vol. 75, No. 2, p. 501—506 (1976)
See also *13.5.1*, p. *153*; 13.5.2, p. 157; *20.1*, p. *268*; 22, p. 336.

H. MICHEL and M. WAGNER
*Non-Thermal Stimulation of Three-Level Impurities by Phonons
(b) Vol. 75, No. 2, p. 507—517 (1976).
See also 8, p. 68; *13.4*, p. *139*; 22.5.2, p. 395.

H. BILL and W. VON DER OSTEN
Raman and Vibronic Spectra of a New Oxygen Molecule Ion in CaF_2
(b) Vol. 75, No. 2, p. 613—619 (1976).
See also *10.2*, p. *93*; 20.3, p. 284; 22.5.3, p. 399.

C. M. GOEL and B. DAYAL
*Second-Order Raman Spectra of RbCl, RbBr, and RbI Crystals
(b) Vol. 75, No. 2, p. 697—703 (1976).
See also *20.1*, p. *268*; 22.5.2, p. 395.

E. KOLLEY and W. KOLLEY
Effect of Thermal Fluctuations on the Electronic Densities of States in Disordered Binary Alloys. Numerical Results of CPA Model Calculations
(b) Vol. 75, No. 2, p. 721—730 (1976).
See also *2*, p. *38*; *13*, p. *116*; 21, p. 294.

R. VACHER and J. PELOUS
*Measurement of the Intensity of the Brillouin Lines in Amorphous Media at Very Low Temperature
(b) Vol. 75, No. 2, p. K101—K103 (1976).
See also *2*, p. *38*; 20.1, p. 269; 22.9, p. 436.

R. PAETSCH
*Dislocation Damping Constant in Dispersive Monatomic Crystals
(b) Vol. 75, No. 2, p. K105—K109 (1976).
See also *10*, p. *76*.

C. UHER and H. J. GOLDSMID
Phonon-Drag Generated Umkehreffekt in Bismuth
(b) Vol. 75, No. 2, p. K137—K141 (1976).
See also *15*, p. *210*; 21.7, p. 332.

J. ZIRKE, G. FRAHM, A. TAUSEND, and D. WOBIG
*Infrared and Raman Studies of TlSe
(b) Vol. 75, No. 2, p. K149—K152 (1976).
See also *20.1*, p. *269*; 22, p. 337.

B. N. N. ACHAR and G. R. BARSCH
*Phonon Dispersion in V_3Si
(b) Vol. 76, No. 1, p. 133—143 (1976).
See also 22.7, p. 411.

A. K. GARG
*Pressed-Pellet Infrared Absorption Spectra of Molybdenum Dichalcogenides
(b) Vol. 76, No. 1, p. 171—174 (1976).
See also *20.1*, p. *270*; 22, p. 337.

T. WASIUTYŃSKI
Self-Consistent Phonon Calculation for Hexamethylenetetramine Molecular Crystals
(b) Vol. 76, No. 1, p. 175—181 (1976).
See also 4, p. 46; 12.2, p. 110; 22.9, p. 437.

E. Schuberth, M. Creutzburg, and W. Müller-Lierheim
Evidence for a Partial F.C.C.-H.C.P. Phase Transition in Solid Neon from Raman Scattering
(b) Vol. **76**, No. 1, p. 301—306 (1976).
See also *1.2*, p. *17*; 20.3, p. 284; 23, p. 440.

E. Sigmund and M. Wagner
An Operator Method for Optical Lineshape Calculations in Non-Adiabatic Electron–Phonon Systems
(b) Vol. **76**, No. 1, p. 325—336 (1976).
See also 13.4, p. 140; *20.1*, p. *270*.

R. Cavalheiro and M. M. Shukla
*Lattice Dynamics of Gold on an Extended de Launay Model
(b) Vol. **76**, No. 1, p. 371—375 (1976).
See also 21.6, p. 330.

F. M. Gashimzade and K. A. Rustamov
*Erratum to: Selection Rules for the Spinel Structure
(b) Vol. **76**, No. 1, p. 437 (1976).
See also *13.4*, p. *140*; 22.8.2, p. 429.

J. Szymański
Erratum to: Theory of Free Carrier Absorption in n-PbSe
(b) Vol. **76**, No. 1, p. 439 (1976).
See also 13.1, p. 125; *20.1*, p. *271*; 22, p. 338.

H. Köhler and C. R. Becker
*Infrared-Active Lattice Vibrations in $SnSe_2$
(b) Vol. **76**, No. 1, p. K15—K17 (1976).
See also *20.1*, p. *271*; 22, p. 338.

N. J. Doran and E. Gerlach
Frequency Dependent Conductivity Due to Deformation Potential and Piezoelectric Scattering by Acoustic Phonons in Semiconductors Including Screening
(b) Vol. **76**. No. 1, p. K43—K46 (1976).
See also *14.3*, p. *178*; 22.2.1, p. 363.

L. Holz, K. Kneipp, A. Lau, and W. Werncke
*Stimulated Raman Scattering of Picosecond Laser Pulses by Polaritons in $LiIO_3$ and $LiNbO_3$ Single Crystals
(a) Vol. **36**, No. 1, p. K5—K8 (1976).
See also 13.5, p. 147; 20.1, p. 271; 20.2, p. 279; 22.8.1, p. 422.

V. Yakhot
*Theory of Multiphonon Vibrational Relaxation of a Guest Molecule in a Host Crystal
(b) Vol. 76, No. 2, p. 451—456 (1976).
See also 23, p. 440.

S. K. S. Ma, G. P. Alldredge, and F. W. de Wette
Surface and Size Effects in the Specific Heat of Linear Chains
(b) Vol. **76**, No. 2, p. 457—464 (1976).
See also *8*, p. *69*.

P. Konsin
Ferroelectric Phase Transitions in Systems with Broad Electronic Bands Caused by Interband Electron–Phonon Interaction
(b) Vol. **76**, No. 2, p. 487—496 (1976).
See also 13.1, p. 125; *14.4.1*, p. *207*; 22, p. 338.

A. Loidl, H. Jex, J. Daubert, and M. Müllner
*Zero and First Sound in NaF and RbI
(b) Vol. **76**, No. 2, p. 581–592 (1976).
See also 7, p. 66; 8, p. 69; *12.2*, p. *111*; 22.5.2, p. 395.

L. G. Grechko, G. L. Falko, and A. M. Fedorchenko
Surface Specific Heat of Isotropic Solids
(b) Vol. **76**, No. 2, p. 593–598 (1976).
See also *8*, p. *69*.

R. Pässler
*Calculation of Nonradiative Multiphonon Capture Coefficients and Ionization Rates for Neutral Centres According to the Static Coupling Scheme (II): Alternative Trap Models
(b) Vol. **76**, No. 2, p. 647–659 (1976).
See also 13.1, p. 126; *13.4*, p. *141*; 22, p. 339; 22.4.1, p. 381.

B. N. N. Achar and G. R. Barsch
*Group Theoretical Analysis and Phonon Modes of V_3Si
(b) Vol. **76**, No. 2, p. 677–688 (1976).
See also 22.7, p. 411.

G. Röpke, V. Christoph, and G. Vojta
Influence of Phonon Damping on the Electrical Resisitivity of Open Electron–Phonon Model Systems
(b) Vol. **76**, No. 2, p. 695–706 (1976).
See also 8, p. 69; *14.1*, p. *165*.

P. Clauws and J. Vennik
*Lattice Vibrations of V_2O_5. Determination of TO and LO Frequencies from Infrared Reflection and Transmission
(b) Vol. **76**, No. 2, p. 707–713 (1976).
See also *20.1*, p. *272*; 22.6, p. 406.

S. V. Vlachos, A. P. Lambros, A. Thanailakis, and N. A. Economou
Anisotropic Indirect Absorption Edge in GeSe
(b) Vol. **76**, No. 2, p. 727–735 (1976).
See also *20.1*, p. *272*; 22.7, p. 411.

T. Soma
*Modified Perturbation Method for Diamond-Type Crystals
(b) Vol. **76**, No. 2, p. 753–758 (1976).
See also 12.2, p. 111; 13.1, p. 126; 22.1, p. 344; 22.1.1, p. 348; 22.1.2, p. 355.

F. F. Sokolov
*Memory Function of Frenkel Exciton in the Presence of Exciton–Phonon Interaction
(b) Vol. **76**, No. 2, p. K131–K135 (1976).
See also *13.5.1*, p. *154*.

E. Ziegler, W. Siegel, and A. Heinrich
*First-Order NPO Scattering in p-$ZnSiAs_2$ and p-$CdSiAs_2$
(a) Vol. **36**, No. 2, p. 491–494 (1976).
See also *14.3*, p. *179*; 22.8, p. 417.

D. W. Field
An X-Ray Diffraction Study of White Tin
(a) Vol. **36**, No. 2, p. 597–601 (1976).
See also *1.1*, p. *12*; 21.7, p. 332.

U. VOLAND, P. DEUS, and H. A. SCHNEIDER
*Debye Temperature of Polycrystalline Silicon
(a) Vol. **36**, No. 2, p. K165—K169 (1976).
See also *8*, p. *70*; 22.1.2, p. 356.

A. V. LELYAKOV
Bose-Einstein Condensation of Free Excitons in AgBr-Type Crystals at High Excitation
(b) Vol. 77, No. 1, p. 59—70 (1976).
See also 13.1, p. 127; *13.5.1*, p. *154*; 20.3, p. 285; 22.5.1, p. 391.

V. K. ARORA and E. QURESHI
*Quantum Theory of Microwave Conductivity
(b) Vol. 77, No. 1, p. 77—84 (1976).
See also *14*, p. *163*.

G. P. SRIVASTAVA
*Derivation and Calculation of a Sequence of Lower-Bound Results for Lattice Thermal Conductivity
(b) Vol. 77, No. 1, p. 131—140 (1976).
See also *8*, p. *70*.

R. S. SORBELLO
*Transport Theory of Ultrasonic Attenuation
(b) Vol. 77, No. 1, p. 141—152 (1976).
See also 7, p. 66; 14.1, p. 166.

P. BRÜESCH, M. BÖSCH, W. KÄNZIG, M. ZIEGLER, and W. BÜHRER
*Lattice Dynamics of the Phases I, II, and III of Sodium Hyperoxide NaO_2
(b) Vol. 77, No. 1, p. 153—164 (1976).
See also 20.1, p. 272; 22.6, p. 406.

L. MERTEN and G. BORSTEL
*Phenomenological Theory of Coupling of Debye Relaxations with IR-Active Phonons
(b) Vol. 77, No. 1, p. 221—227 (1976).
See also *14.4.1*, p. *207*.

F. SPIEGELBERG, E. GUTSCHE, and J. VOIGT
*Exciton–Phonon Interaction in CdS
(b) Vol. 77, No. 1, p. 233—242 (1976).
See also *13.5.1*, p. *154*; 20.1, p. 272; 22.4.1, p. 381.

J. SZYMAŃSKI
*Deformation Potential Theory for Optical and Acoustic Phonons in PbSe-Like Semiconductors
(b) Vol. 77, No. 1, p. 347—351 (1976).
See also 22, p. 339.

J. E. LOWTHER
Mechanisms of Energy Transfer between Rare Earth Ions
(b) Vol. 77, No. 1, p. 359—366 (1976).
See also *2*, p. *39*; 13.4, p. 142; 21.4, p. 327.

J. A. COCKS and SHI-YU WU
*The Dynamics of and the Effect of a Short Range Order on a Segregated Binary Chain
(b) Vol. 77, No. 1, p. K5—K9 (1976).
See also 10, p. 76.

V. G. KVACHADZE and Z. G. AKHVLEDIANI
*Influence of γ-Irradiation on the Thermal Conductivity of KCl:KOH Crystals
(b) Vol. 77, No. 1, p. K21—K25 (1976).
See also *8*, p. *70*; 11, p. 103; 20.1, p. 273; 22.5.2, p. 396.

R. Martens and F. Freund
*The Potential Energy Curve of the Proton and the Dissociation Energy of the OH^- Ion in $Mg(OH)_2$
(a) Vol. 37, No. 1, p. 97–104 (1976).
See also *20.1*, p. *273*; 22.8, p. 417.

A. Kiveris, Š. Kudžmauskas, and P. Pipinys
Release of Electrons from Traps by an Electric Field with Phonon Participation
(a) Vol. 37, No. 1, p. 321–327 (1976).
See also *13.4*, p. *142*; 14.3.4, p. 199; *20.3*, p. *286*; 22.4.1, p. 381.

L. A. Bertolo and M. M. Shukla
*A Calculation of Debye-Waller Factors for F.C.C. Metals
(b) Vol. 77, No. 2, p. 433–439 (1976).
See also 21, p. 298; 21.1, p. 312; 21.5, p. 328; 21.6, p. 331.

D. Klähn and K. Kilimann
The Damping of Excitons by Phonons
(b) Vol. 77, No. 2, p. 483–493 (1976).
See also *13.5.1*, p. *155*; 22.1.1, p. 348.

L. K. Aminov, R. Kirmse, and B. V. Solovev
Electron Spin–Lattice Relaxation of Cu^{2+} Ions in Planar Coordination Geometries. Calculation and Experimental Data of T_1 for Cu^{2+} in Ni(II)bis(diethyl-dithiocarbamate) Crystals
(b) Vol. 77, No. 2, p. 505–515 (1976).
See also *4*, p. *46*; 18.1, p. 226; 19, p. 254; p. 22.9, p. 438.

H. Böttger
*On the Theory of Lattice Vibrations in Mixed Crystals. The Connection between the REI Model, ATA, and CPA
(b) Vol. 77, No. 2, p. 561–570 (1976).
See also 22, p. 340.

S. C. Goyal and A. K. Goyal
*Dispersion of Phonons in Metal Carbides (UC)
(b) Vol. 77, No. 2, p. 639–643 (1976).
See also 12.2, p. 112; 22, p. 340.

D. W. Bullett
Non-Empirical Calculation of Structural Properties and Valence Force Fields. Application to Se and Te
(b) Vol. 77, No. 2, p. 645–650 (1976).
See also 1, p. 9; *4*, p. *46*; 22.1.3, p. 360.

M. Obiditsch and H. Kahlert
Electric Field Induced Changes of the Fundamental Absorption and Low-Temperature Energy Relaxation in n-Type GaSb
(b) Vol. 77, No. 2, p. 677–684 (1976).
See also *14.3*, p. *180*; *20.1*, p. *274*; 22.2.3, p. 368.

J. C. Castro, H. C. Basso, and M. de Souza
*Local Mode of H^- Substitutional Ion in KCN
(b) Vol. 77, No. 2, p. 685–692 (1976).
See also 1.2, p. 19; 10.2, p. 95; *20.1*, p. *274*; 22, p. 340.

J. Aslam, S. Rolandson, M. M. Beg, N. M. Butt, and Q. H. Khan
*Phonons in the Mixed Crystal $K_{0.5}Rb_{0.5}I$
(b) Vol. 77, No. 2, p. 693–700 (1976).
See also 12.2, p. 112; 22.5.4, p. 401.

A. Ramakanth and S. K. Ghatak
*Electron-Mediated Nuclear Spin–Phonon Interaction in Metals
(b) Vol. 77, No. 2, p. K123–K127 (1976)
See also *19*, p. *255*.

P. R. W. Hudson
*The Significance of the Frequency Dependence of Phonon Scattering by Platelets in Type I_a Diamond
(a) Vol. 37, No. 2, p. 645–651 (1976).
See also 8, p. 71; 22.1, p. 344.

A. M. Bon, C. Benoit, and O. Bernard
*Dynamical Properties of Crystals of $Sr(NO_3)_2$, $Ba(NO_3)_2$, and $Pb(NO_3)_2$ (I): Infrared Spectra and Structure
(b) Vol. 78, No. 1, p. 67–78 (1976).
See also *20.1*, p. *275*; 22.8.1, p. 424.

J. T. Devreese and R. Evrard
On the Momentum Distribution of Electrons in Polar Semiconductors for High Electric Field
(b) Vol. 78, No. 1, p. 85–92 (1976).
See also *14.3.4*, p. *200*; 22.2.3, p. 369.

H. J. Falge, E. Schuller, and G. Borstel
*Directional Dispersion of Polar Phonons Measured by Infrared Reflection
(b) Vol. 78, No. 1, p. 123–132 (1976).
See also 20.1, p. 275; 22.6, p. 408.

K. E. Gauss and H. Happ
Millimeter Wave Investigation of the Complex Dielectric Constant in DKDP
(b) Vol. 78, No. 1, p. 133–138 (1976).
See also *14.4.1*, p. *208*; 20.1, p. 275; 22.8.1, p. 424.

G. A. Gehring and J. Laugsch
Theory of Side-Bands in a Cooperative Jahn-Teller Material
(b) Vol. 78, No. 1, p. 211–218 (1976).
See also 13.5.1, p. 155; *20.1*, p. *275*; 22.8.1, p. 424.

S. K. S. Ma, F. W. de Wette, and G. P. Alldredge
Anharmonicity in the Surface and Size Effects in the Specific Heat of Linear Chains
(b) Vol. 78, No. 1, p. 219–228 (1976).
See also *8*, p. *71*.

A. K. Bohra and R. K. Gupta
*Lattice Dynamics of Rare-Gas Solids by Kihara's Potential (III): Green Functions
(b) Vol. 78, No. 1, p. 255–258 (1976).
See also 23, p. 440.

M. B. Haxton
*Parameter Calculation from Zinc Blende Zone Boundary Conditions of CuI
(b) Vol. 78, No. 1, p. 259–264 (1976).
See also 22.5, p. 390.

R. H. Prince
Low-Temperature Diffusion of Electrons in Ice. Evidence for Polaron Bands
(b) Vol. 78, No. 1, p. 271–276 (1976).
See also *13.5.3*, p. *159*; 22, p. 341.

H. J. Stolz, H. Wendel, A. Otto, L. Pintschovius, and H. Kahlert
*Acoustic and Optical Phonons in $(SN)_x$
(b) Vol. 78, No. 1, p. 277–286 (1976).
See also 4, p. 47; *20.1*, p. *276*; 22, p. 341.

V. L. Broude, L. M. Umarov, and E. F. Sheka
*Multiphonon Absorption in Molecular Crystals
(b) Vol. 78, No. 1, p. 325–333 (1976).
See also *20.1*, p. *276*; 22.9, p. 438.

E. V. Chisler, V. Yu. Davydov, I. N. Goncharuk, and E. A. Ivanova
*On Fermi Resonance Theory
(b) Vol. 78, No. 1, p. 359–370 (1976).
See also *20.1*, p. *276*.

V. P. Singh and M. P. Hemkar
*Dynamical Study of the Temperature Dependence of Grüneisen Parameters in Alkali Metals
(b) Vol. 78, No. 1, p. 381–387 (1976).
See also 8, p. 71; 12.1, p. 109; 21.2, p. 323.

S. Permogorov and V. Travnikov
*Resonant Scattering of Exciton-Polaritons in CdS Crystals
(b) Vol. 78, No. 1, p. 389–398 (1976).
See also *13.5*, p. *148*; 20.3, p. 287; 22.4.1, p. 381.

A. N. Pandey and K. S. Upadhyaya
*Lattice Dynamics of BaO
(b) Vol. 78, No. 1, p. 399–404 (1976).
See also 22.6, p. 408.

K. P. Meletov and E. F. Sheka
*Asymmetry of Phonon Side Bands of Optical Transitions in Impurity Molecular Crystals Caused by a Change in External Phonon Frequencies
(b) Vol. 78, No. 1, p. K1–K5 (1976).
See also *13.4*, p. *144*; 20.1, p. 276; 20.3, p. 287; 22.9, p. 439.

H. Maier
*Jahn-Teller Effect, Stacking Faults, and the Formation of Polytypes in ZnS
(b) Vol. 78, No. 1, p. K65–K68 (1976).
See also *13.4*, p. *144*; 20.1, p. 276; 20.3, p. 287; 22.4.1, p. 382.

A. M. Bon, C. Benoit, and J. Giordano
*Dynamical Properties of Crystals of $Sr(NO_3)_2$, $Ba(NO_3)_2$, and $Pb(NO_3)_2$ (II): Temperature Dependence of the Infrared Spectra
(b) Vol. 78, No. 2, p. 453–464 (1976).
See also 20.1, p. 277; 22.8.1, p. 424.

J. C. Phillips
High-Temperature Conductivity of Organic Bilinear Addition Compounds
(b) Vol. 78, No. 2, p. 471–476 (1976).
See also *14*, p. *163*; 22.9, p. 439.

A. Miller, G. D. Holah, W. D. Dunnett, and G. W. Iseler
*Optical Phonons in $AgGaSe_2$
(b) Vol. 78, No. 2, p. 569–576 (1976).
See also 20.1, p. 277; 22.8, p. 418.

H. P. NIEDERMANN and M. WAGNER
*On the Quantum Mechanical Foundation of Classical Shell Models in Lattice Dynamics
(b) Vol. 78, No. 2, p. 615–624 (1976).
See also 4, p. 47; 22, p. 342.

R. PÄSSLER
Relationships between the Nonradiative Multiphonon Carrier-Capture Properties of Deep Charged and Neutral Centres in Semiconductors
(b) Vol. 78, No. 2, p. 625–635 (1976).
See also 13, p. 119; *13.4*, p. *145*; 22, p. 342; 22.1, p. 344; 22.2, p. 361; 22.4, p. 378.

R. K. SINGH and K. P. SINGH
*Lattice Dynamics of AgBr at 295 K
(b) Vol. 78, No. 2, p. 677–680 (1976).
See also 20.1, p. 277; 22.5.1, p. 391.

J. WEBER
Indirect Near Edge Emission in Pure AgBr
(b) Vol. 78, No. 2, p. 699–709 (1976).
See also *13.5.1*, p. *156*; *20.3*, p. *288*; 22.5.1, p. 391.

F. BECHSTEDT, R. ENDERLEIN, and K. PEUKER
Resonance Raman Scattering in a Strong Electric Field
(b) Vol. 78, No. 2, p. 711–719 (1976).
See also *14.3.4*, p. *200*; *20.1*, p. *278*; 22, p. 342.

K. N. SHRIVASTAVA
*Thermal Effects in Magnetic Exchange Interaction
(b) Vol. 78, No. 2, p. 749–755 (1976).
See also *18.2*, p. *236*; 22.5.4, p. 401.

M. DURAND, B. AYRAULT, Y. MARQUETON, and E. A. DECAMPS
*Lattice Dynamics of Paratellurite
(b) Vol. 78, No. 2, p. 767–777 (1976).
See also *20.1*, p. *278*; 22.6, p. 408.

J. P. BUISSON, S. LEFRANT, A. SADOC, L. TAUREL, and M. BILLARDON
Raman Scattering by KI Containing F Centres
(b) Vol. 78, No. 2, p. 779–790 (1976).
See also *10.2*, p. *98*; 13.4, p. 145; 20.1, p. 278; 22.5.2, p. 398.

A. M. ZIATDINOV, M. M. ZARIPOV, YU, V. YABLOKOV, and R. L. DAVIDOVICH
*The Depression of the Dynamic Jahn-Teller Effect
(b) Vol. 78, No. 2, p. K69–K71 (1976).
See also 1.2, p. 20; *13.4*, p. *145*; 19, p. 256; 22.8, p. 418.

W. BAUHOFER, L. GENZEL, and W. KÖNIG
*Far Infrared Studies of the Impurity Mode $NH_4Br:Cl^-$
(b) Vol. 78, No. 2, p. K121–K123 (1976).
See also *20.1*, p. *278*; 22.5.4, p. 401.

D. C. GUPTA and N. D. SHARMA
Pressure Dependence of the Mössbauer Fraction for ^{57}Fe in Ti, V, and Cu
(b) Vol. 78, No. 2, p. K125–K129 (1976).
See also *5*, p. *51*; 21.1, p. 314.

I. VILFAN
*On the Electron-Intramolecular Vibration Coupling in Aromatic Molecular Crystals
(b) Vol. 78, No. 2, p. K131–K135 (1976).
See also 13.5.3, p. 159; *14.3*, p. *182*; 22.9, p.439.

7. Acoustic Properties

Original Papers and Short Notes

B. Dorner, R. E. Ghosh, and G. Harbeke
Phonon Dispersion in the Layered Compound PbI_2
(b) Vol. **73**, No. 2, p. 655–659 (1976).
See also *6*, p. *54*; 12.2, p. 110; 22.5, p. 388.

W. Fuhs, M. Milleville, and J. Niklas
*Acoustoelectric Current Saturation and Sound Velocity in Tellurium Crystals under Hydrostatic Pressure
(b) Vol. **73**, No. 2, p. K127–K130 (1976).
See also *14.3.4*, p. *195*; 22.1.3, p. 359.

K. Kumazaki
Elastic Properties and Ionicity of Zero-Gap Semiconductors
(a) Vol. **33**, No. 2, p. 615–623 (1976).
See also *4*, p. *45*; *12.2*, p. *110*; 22.4, p. 376.

T. P. Lepètre, J.-M. Dusseau, and J.-L. Robert
*Expression of Acoustoelectric Net Gain in p-Type GaSb
(a) Vol. **33**, No. 2, p. 715–722 (1976).
See also *14.3.4*, p. *196*; 22.2.3, p. 367.

E. Salje
*New Crystals $Pb_3(P_xV_{1-x}O_4)_2$ for Acousto-Optical Device Applications
(a) Vol. **33**, No. 2, p. K165–K167 (1976).
See also *20.1*, p. *263*; 22.8, p. 413.

H. Kuzmany
*Light Scattering Experiments and Nonlinear Behaviour of the Acoustoelectric Oscillator
(a) Vol. **34**, No. 1, p. 347–353 (1976).
See also *14.3*, p. *174*; 22.4.1, p. 379.

S. V. Gantsevich, V. L. Gurevich, V. D. Kagan, and R. Katilius
*Transformation of Incoherent Phonon Flux into a Coherent Acoustical Signal as the Result of the Nonlinear Amplification of Acoustical Phonons in Semiconductors
(b) Vol. **75**, No. 2, p. 407–422 (1976).
See also 6, p. 56; *14.3.4*, p. *197*; 22, p. 336.

L. Opilska and A. Opilski
*Relation between Energy Gap and Velocity of Ultrasonic Wave Propagation in $A^{II}B^{V}$ Group Semiconductors
(a) Vol. **35**, No. 2, p. K183–K187 (1976).
See also 22.3, p. 373; 22.3.1, p. 375.

T. V. Quynh, G. Schmidt, N. R. Ivanov, and L. A. Shuvalov
*Determination of the Temperature Dependence of the Stiffness Coefficient c_{55} in $KH_3(SeO_3)_2$ by Ultrasonic Measurements
(a) Vol. **36**, No. 1, p. K85–K88 (1976).
See also *1.2*, p. *18*; 22.8, p. 416.

Y. Luspin and G. Hauret
Study of the Velocity and Damping of Acoustic Waves Obtained by Brillouin Scattering in the Paraelastic Phase in GMO
(b) Vol. **76**, No. 2, p. 551–558 (1976).
See also *1.2*, p. *18*; 22.8.1, p. 423.

A. Loidl, H. Jex, J. Daubert, and M. Müllner
Zero and First Sound in NaF and RbI
(b) Vol. 76, No. 2, p. 581–592 (1976).
See also *6*, p. *59*; 8, p. 69; *12.2*, p. *111*; 22.5.2, p. 395.

R. S. Sorbello
Transport Theory of Ultrasonic Attenuation
(b) Vol. 77, No. 1, p. 141–152 (1976).
See also *6*, p. *60*; 14.1, p. 166.

J. C. Peuzin et M. Tasson
*Génération et détection d'hypersons. Transducteur feuilleté
(Generation and Detection of Hypersound. Domain Transducer)
(a) Vol. 37, No. 1, p. 119–125 (1976).
See also *14.4.1*, p. *207*; 22.8.1, p. 423.

H. Veeser, U. Bogner, and W. Eisenmenger
*Analysis of the Acoustic Transients in the Pulse Response of the Linear Electro-Optic Effect
(a) Vol. 37, No. 1, p. 161–170 (1976).
See also 12.2, p. 111; *14.4.1*, p. *208*; 20.1, p. 273; 22.8.1, p. 423.

A. Jucys and V. Kunigelis
*Influence of the Surface Potential upon the Acoustoelectric Interaction in CdS
(b) Vol. 77, No. 2, p. K113–K117 (1976).
See also *14.3.2*, p. *188*; 22.4.1, p. 381.

N. P. Kolmakova, D. I. Sirota, and M. Acquarone
*Anharmonicity of Sound Propagation near First-Order Orientational Phase Transitions in Antiferromagnets
(b) Vol. 77, No. 2, p. K143–K146 (1976).
See also *18.4*, p. *245*.

G. Splitt
*Magnetic Properties of Superconducting Lead Films Investigated with Hypersonic Waves
(a) Vol. 37, No. 2, p. K131–K134 (1976).
See also *14.2*, p. *171*; 21, p. 299.

G. Johri and H. N. Spector
*Second Harmonic Generation in an Electric Field
(a) Vol. 38, No. 1, p. K61–K64 (1976).

8. Thermal Properties

Original Papers and Short Notes

T. Romanovskis
*Konzentrationsabhängigkeit der Debye-Temperatur in idealen Mischkristallen (Concentration Dependence of the Debye Temperature in Ideal Solid Solutions)
(b) Vol. **73**, No. 1, p. K13—K14 (1976).
See also 22.5.4, p. 401.

W. Bausch
Accuracy of the Boundary Relaxation Time Approximation
(b) Vol. **73**, No. 1, p. K23—K25 (1976)
See also *6*, p. *53*; 22.5.2, p. 392.

R. Mehrotra and M. M. Pant
Static and Dynamic Displacements near a Crystal Surface
(b) Vol. **73**, No. 1, p. K27—K30 (1976).
See also *1.6*, p. *34*; *6*, p. *53*; 21.1.1, p. 316.

M. Natarajan and E. A. Secco
Electrical Conductivity and Phase Transformation Studies on the Mixed Metal Halides $RbCdX_3$ (X = Cl, Br, I) and Cs_2CuBr_4
(a) Vol. **33**, No. 2, p. 427—433 (1976).
See also *1.2*, p. *14*; 14.4, p. 202; 22.5.4, p. 401.

S. I. Masharov
Anharmonic Effects in Heat Capacity and Saturation Vapour Pressure of Dilute Substitutional Alloys
(b) Vol. **73**, No. 2, p. 401—408 (1976).
See also *6*, p. *53*; 21, p. 290.

A. N. Aleksandrovskii, V. I. Kuchnev, V. G. Manzhelii, and A. M. Tolkachev
*Studies on the Low Temperature Anomaly in the Thermal Expansion of Solid Methane
(b) Vol. **73**, No. 2, p. K111—K114 (1976).
See also 23, p. 440.

R. Heller and H. Wipf
Diffusion Coefficient, Heat of Transport, and Solubility Limit of H in V
(a) Vol. **33**, No. 2, p. 525—529 (1976).
See also *9*, p. *72*; 14.1, p. 164; 21.1, p. 304.

G. A. Gehring, H. G. Kahle, W. Nägele, A. Simon, and W. Wüchner
Two-Level Singlet Magnetism in $TbAsO_4$ and $TbVO_4$
(b) Vol. **74**, No. 1, p. 297—309 (1976).
See also *18.4*, p. *242*; 20.1, p. 264; 22.8.1, p. 420.

W. Löser and W. Tiersch
*Thermodynamics of Finite Non-Half-Filled-Band Hubbard Rings
(b) Vol. **74**, No. 1, p. K43—K47 (1976).
See also *18.4*, p. *243*.

E. Donth
*Universal Thermodynamic and Kinetic Functions for a Classical Freezing-in Process in Noncrystalline Bodies
(b) Vol. **74**, No. 1, p. K49—K52 (1976).
See also *2*, p. *36*.

H. E. Flotow, R. Kuentzler, and D. W. Osborne
*The Specific Heat of the Nearly Ferromagnetic Compound $(Mn_{0.865}Fe_{0.135})Be_2$ from 0.1 to 23 K
(a) Vol. **34**, No. 1, p. 291–295 (1976).
See also 18, p. 222; 22.8, p. 414.

E. Hegenbarth, S. V. Pirogova, E. M. Savitskii, and B. Göldner
*Thermal Conductivity of the Intermetallic Compound NiAl in the Temperature Range from 20 to 80 K
(a) Vol. **34**, No. 1, p. 319–323 (1976).
See also 10.1, p. 80; 21.1, p. 305.

K.-D. Ufert
*Stress Induced Switching in VO_2 Thin Films
(a) Vol. **34**, No. 1, p. K83–K86 (1976).
See also *12.2*, p. *110*; 14.3.1, p. 183; 22.6, p. 404.

J. D. Patterson and R. C. Weger
*The Specific Heat of Small Magnetic Clusters with Simulated Amorphousness
(b) Vol. **74**, No. 2, p. K115–K118 (1976).
See also 2, p. 37; *18.2*, p. *230*.

P. Charsley and M. H. S. Rahm
*The Effect of Ageing on the Lattice Thermal Conductivity of Copper Alloys Deformed at Liquid Helium Temperatures
(a) Vol. **34**, No. 2, p. 533–539 (1976).
See also 6, p. 56; *10.1*, p. *80*; 21.1, p. 306.

S. Duraiswamy and T. M. Haridasan
*Moments of the Phonon Spectrum in Lithium Chloride
(b) Vol. **75**, No. 1, p. 53–59 (1976).
See also *6*, p. *56*; 22.5.2, p. 394.

B. N. Oshcherin
*Debye Temperature and Meling Criterion of Ternary Chalcopyrite $A^1B^3C^6_2$- and $A^2B^4C^5_2$-Type Semiconductors
(a) Vol. **35**, No. 1, p. K35–K39 (1976).
See also *1*, p. *9*; 22.2, p. 361; 22.4, p. 376; 22.8, p. 415.

L. M. Valiev, I. G. Kerimov, N. G. Aliev, and A. A. Abdurragimov
The Thermal Expansion and Electrical Properties of $Cu_4Mn_2X_4$(X: S, Se, and Te)
(a) Vol. **35**, No. 1, p. K85–K88 (1976).
See also 14, p. 162; 15, p. 210; *18.1*, p. *226*; *18.4*, p. *244*; 22.8, p. 415.

C. Scherer and I. Aveline
*Reaction Field Approximation for Heisenberg Paramagnets and the Magnetic Specific Heat
(b) Vol. **75**, No. 2, p. 465–470 (1976).
See also *18.1*, p. *226*.

H. Michel and M. Wagner
Non-Thermal Stimulation of Three-Level Impurities by Phonons
(b) Vol. **75**, No. 2, p. 507–517 (1976).
See also *6*, p. *57*; *13.4*, p. *139*; 22.5.2, p. 395.

M. A. Boubel, R. P. Mainard, H. G. Fousse, A. J. Pointon, and F. R. Jeannot
The Specific Heat Anomaly of Solid Solutions of Isostructural Antiferromagnetic Oxides (*p*FeO–*q*CoO)
(a) Vol. **35**, No. 2, p. 459–464 (1976).
See also *18.4*, p. *244*; 22.6, p. 405.

F. GILLETTA, M. CHABIN, and CAO-XUAN AN
*Thermal Properties of Lead Phosphate
(a) Vol. 35, No. 2, p. 545–549 (1976).
See also 22.8.1, p. 421.

M. GOETTEL, J. M. JANIK, J. A. JANIK, and M. RACHWALSKA
Phase Transitions and NH_3 Rotations in $[Co(NH_3)_6]I_3$ Studied by Calorimetry and IR Profile Methods
(a) Vol. 35, No. 2, p. 675–680 (1976).
See also *1.2*, p. *17*; 22.8, p. 416.

Z. BIEGAŃSKI and M. DRULIS
*Low Temperature Phase Transition in Neodymium Trihydride. Specific Heat Measurements
(a) Vol. 35, No. 2, p. K127–K130 (1976).
See also *18*, p. *223*; 21.4, p. 326.

E. HEGENBARTH and B. SCHMIDT
*Heat Condu ᵒ ɛ vity Measurement at V_3Si in the Temperature Range 4.2 to 80 K
(b) Vol. 76, Noi. 1, p. 307–311 (1976).
See also 13.1, p. 125; 14.2, p. 170; 21, p. 296.

D. ROY CHAUDHURI, P. K. BASU, and A. N. CHAKRAVARTI
*Effect of the Surface Electric Field on the Specific Heat of Semiconductor Inversion Layers
(a) Vol. 36, No. 1, p. K65–K69 (1976).
See also 14.3.4, p. 198; 22.1.2, p. 355.

H.-J. ANKLAM and K.-D. UFERT
*Electrothermal Behaviour of VO_2 Thin Film Switching Devices
(a) Vol. 36, No. 1, p. K71–K75 (1976).
See also *14.3.3*, p. *192*; 14.3.4, p. 198; 22.6, p. 405.

G. WOLF and C. HOHLFELD
*The Molar Heat Capacity of Superconductive $(PdRh_{0.06}H_x$ in the Temperature Range from 2 to 7.5 K
(a) Vol. 36, No. 1, p. K99–K103 (1976).
See also 14.2, p. 170; 21, p. 296.

S. K. S. MA, G. P. ALLDREDGE, and F. W. DE WETTE
*Surface and Size Effects in the Specific Heat of Linear Chains
(b) Vol. 76, No. 2, p. 457–464 (1976).
See also 6, p. 58.

V. S. LUBIMOV, L. A. MANAKOVA, E. F. MAKAROV, and S. P. IONOV
*The Thermodynamic Behaviour of Charge-Ordered Crystals
(b) Vol. 76, No. 2, p. 505–514 (1976).
See also *13*, p. *117*.

A. LOIDL, H. JEX, J. DAUBERT, and M. MÜLLNER
Zero and First Sound in NaF and RbI
(b) Vol. 76, No. 2, p. 581–592 (1976).
See also *6*, p. *59*; 7, p. 66; *12.2*, p. *111*; 22.5.2, p. 395.

L. G. GRECHKO, G. L. FALKO, and A. M. FEDORCHENKO
*Surface Specific Heat of Isotropic Solids
(b) Vol. 76, No. 2, p. 593–598 (1976).
See also 6, p. 59.

G. RÖPKE, V. CHRISTOPH, and G. VOJTA
Influence of Phonon Damping on the Electrical Resistivity of Open Electron–Phonon Model Systems
(b) Vol. 76, No. 2, p. 695–706 (1976).
See also 6, p. 59; *14.1*, p. *165*.

K. D. KUNDRA and S. Z. ALI
X-Ray Study of Thermal Expansion and Phase Transformation in β-In_2S_3
(a) Vol. 36, No. 2, p. 517–525 (1976).
See also *1.2*, p. *18*; 22.3, p. 373.

G. WOLF, J. JAHNKE, and K. BOHMHAMMEL
*The Molar Heat Capacity of $(PdAg_{0.2})H_x$ in the Temperature Range from 2 to 6 K
(a) Vol. 36, No. 2, p. K125–K128 (1976).
See also 14.2, p. 170; 21, p. 297.

U. VOLAND, P. DEUS, and H. A. SCHNEIDER
*Debye Temperature of Polycrystalline Silicon
(a) Vol. 36, No. 2, p. K165–K169 (1976).
See also *6*, p. *60*; 22.1.2, p. 356.

A. SLEBARSKI and A. CHEŁKOWSKI
*Thermal Expansion and the Heat Capacity of $GdAl_2$
(a) Vol. 36, No. 2, p. K171–K174 (1976).
See also 18.2, p. 234; 21.4, p. 327.

G. P. SRIVASTAVA
*Derivation and Calculation of a Sequence of Lower-Bound Results for Lattice Thermal Conductivity
(b) Vol. 77, No. 1, p. 131–140 (1976).
See also *6*, p. *60*.

V. G. KVACHADZE and Z. G. AKHVLEDIANI
*Influence of γ-Irradiation on the Thermal Conductivity of KCl:KOH Crystals
(b) Vol. 77, No. 1, p. K21–K25 (1976).
See also *6*, p. *60*; 11, p. 103; 20.1, p. 273; 22.5.2, p. 396.

P. R. W. HUDSON
*Heat Flow Rectification
(a) Vol. 37, No. 1, p. 93–96 (1976).
See also 22.2.1, p. 364.

G. WOLF, M. ZIMMERMANN, and K. BOHMHAMMEL
*The Molar Heat Capacity of Non-Superconducting PdH_x in the Temperature Range from 2 to 7 K
(a) Vol. 37, No. 1, p. 179–182 (1976).
See also 13.1, p. 127; 21, p. 297.

D. KUCHARCZYK, A. PIETRASZKO, and K. ŁUKASZEWICZ
*Temperature Dependence of Lattice Parameters of $NaNO_2$ Single Crystals
(a) Vol. 37, No. 1, p. 287–294 (1976).
See also *1.2*, p. *18*; 14.4.1, p. 208; 22.8, p. 417.

Cz. MARUCHA, J. MUCHA, and J. RAFAŁOWICZ
*Phenomenological Interpretation of Heat Flux Volume Rectification in Non-Homogeneous Media
(a) Vol. 37, No. 1, p. K5–K7 (1976).
See also 22.2.1, p. 364.

G. WOLF and M. ZIMMERMANN
*The Molar Heat Capacity of Superconducting PdD_x in the Temperature Range from 2 to 12 K
(a) Vol. 37, No. 2, p. 485–489 (1976).
See also 14.2, p. 171; 21, p. 299.

P. R. W. Hudson
The Significance of the Frequency Dependence of Phonon Scattering by Platelets in Type I_a Diamond
(a) Vol. 37, No. 2, p. 645—651 (1976).
See also *6*, p. *62* ; 22.1, p. 344.

K.-H. Müller, M. Wolf, and F. Kammler
Thermal Switching and Current-Voltage Characteristics for Semiconducting Sandwich Samples
(a) Vol. 37, No. 2, p. K201—K203 (1976).
See also *14.3.3*, p. *194*; 14.3.4, p. 200; 22, p. 340.

E. Mrosan and G. Lehmann
Localized Non-Magnetic Impurities in Metals (II): Application to 3d-Transition Metal Impurities in Aluminium
(b) Vol. 78, No. 1, p. 159—167 (1976).
See also *13.4*, p. *144*; 14.1, p. 167; 14.2, p. 172; 15, p. 211; 21, p. 299.

S. K. S. Ma, F. W. de Wette, and G. P. Alldredge
*Anharmonicity in the Surface and Size Effects in the Specific Heat of Linear Chains
(b) Vol. 78, No. 1, p. 219—228 (1976).
See also 6, p. 62.

V. P. Singh and M. P. Hemkar
Dynamical Study of the Temperature Dependence of Grüneisen Parameters in Alkali Metals
(b) Vol. 78, No. 1, p. 381—387 (1976).
See also *6*, p. *63*; 12.1, p. 109; 21.2, p. 323.

G. I. Georgiev and S. G. Gocheva
Numerical Computations on a Rigorously Solvable Model
(b) Vol. 78, No. 1, p. K11—K13 (1976).
See also *18*, p. *224*.

Ah Mee Hor, P. W. M. Jacobs, and K. S. Moodie
*Thermal Depolarization in Crystals of Potassium Chloride Containing Strontium Chloride
(a) Vol. 38, No. 1, p. 293—298 (1976).
See also 10.2, p. 98; 22.5.2, p. 398.

N. P. Kekelidze, Z. V. Kvinikadze, Z. F. Davitaya, G. N. Eritsyan, V. A. Saakyan, and F. K. Karapetyan
*Studies of Thermoelectric Characteristics of InP_xAs_{1-x} Alloys
(a) Vol. 38, No. 1, p. K49—K52 (1976).
See also *15*, p. *211*; 22.2.4, p. 371.

K. Suzuki
*Piezo-Thermal Conductivity of n-Ge
(b) Vol. 78, No. 2, p. K77—K79 (1976).
See also 22.1.1, p. 349.

9. Diffusion

Original Papers and Short Notes

H. MEHRER, A. SEEGER, and E. STEINER
*The Relationship between Correlation Factor and Isotope Effect of Diffusion in Crystals
(b) Vol. **73**, No. 1, p. 131–140 (1976).
See also 10, p. 75.

V. A. GOLTSOV and L. I. SMIRNOV
*Diffusion of Interstitial Atoms in Ordered Pt_2Mo-Type Alloys (II)
(a) Vol. **33**, No. 1, p. 67–73 (1976).
See also *1.3*, p. *21*; 21, p. 289; 21.6, p. 329.

R. HELLER and H. WIPF
*Diffusion Coefficient, Heat ot Transport, and Solubility Limit of H in V
(a) Vol. **33**, No. 2, p. 525–529 (1976).
See also 8, p. 67; 14.1, p. 164; 21.1, p. 304.

L. B. HARRIS and J. FIASSON
*Cation Diffusion along Boudaries in Sodium Chloride
(a) Vol. **33**, No. 2, p. 697–706 (1976).
See also *1.4*, p. *25*; 22.5.2, p. 393.

K. G. BANERJEE and B. MILLS
*The Zener Relaxation and Diffusion in the Copper-Nickel-Zinc System
(a) Vol. **33**, No. 2, p. 707–714 (1976).
See also *10.1*, p. *79*; 21.1, p. 304.

D. I. PUSHKAROV and KH. I. PUSHKAROV
*Defecton Diffusion in the Very Quantum Region
(b) Vol. **74**, No. 2, p. K97–K100 (1976).
See also 13.5, p. 147.

V. B. VYKHODETS, V. B. DEMIN, and P. V. GELD
*Hydrogen Isotope Diffusivity and Solubility in Nickel, Palladium, and Copper
(a) Vol. **34**, No. 2, p. 787–795 (1976).
See also 10.1, p. 81; 21, p. 293; 21.1, p. 306.

G. E. PIKE and C. H. SEAGER
*Transport in Microscopically Inhomogeneous Materials
(b) Vol. **75**, No. 1, p. 289–297 (1976).
See also 14.3, p. 176; 22.1.1, p. 347; 22.1.2, p. 353.

E. STEINER, H. MEHRER, and A. SEEGER
*Correlation Factors and Isotope Effect for Self-Diffusion by Mono- and Divacancies in H.C.P. Crystals
(b) Vol. **75**, No. 1, p. 361–373 (1976).
See also 10, p. 76.

UDAYAN DE
*The Effect of Thermal Expansion on the Vacancy Mechanism of Diffusion of Impurities in Metals
(b) Vol. **75**, No. 1, p. K55–K60 (1976).

C. SCHAFFRIN
*Oxygen Diffusion in $BaTiO_3$ Ceramic
(a) Vol. **35**, No. 1, p. 79–88 (1976).
See also 14.4, p. 202; 22.8.1, p. 421.

S. J. ROTHMAN and NGHI Q. LAM
*Interstitial Retention into Stage III and Radiation-Enhanced-Diffusion in Concentrated Alloys
(a) Vol. **35**, No. 1, p. K13—K16 (1976).
See also *11*, p. *101*; 21.1, p. 307.

R. LINDSEY
*Measurement of Mössbauer Diffusion Broadening in α-Iron Stabilized with Vanadium
(b) Vol. 75, No. 2, p. 583—592 (1976).
See also *5*, p. *49*; 21.1.1, p. 319.

A. Z. TRIFONOV, M. D. MIKHAILOV, I. A. TSENOV, and B. K. SHISHEDIEV
*A Spectroscopic Method for Evaluating the Diffusion of a Liquid into Another One
(a) Vol. **35**, No. 2, p. 709—714 (1976).
See also 22.9, p. 436.

J. HAMERSKÝ
Electrical Properties of CdSe Films under the Influence of Oxygen
(a) Vol. **35**, No. 2, p. K97—K99 (1976).
See also *14.3.1*, p. *184*; 22.4.2, p. 383.

C. BIRAC et D. LESUEUR
*Diffusion atomique du lithium dans l'alliage métallique amorphe Pd–Si
(Lithium Diffusion in an Amorphous Metallic Pd–Si Alloy)
(a) Vol. **36**, No. 1, p. 247—251 (1976).
See also 2, p. 39; 21, p. 296.

M. D. MIKHAILOV, A. Z. TRIFONOV, I. A. TSENOV, and M. A. ALADZHEM
*A Spectroscopic Method for Determination of the Cross-Term Diffusion Coefficients in a Three-Component Liquid System
(a) Vol. **36**, No. 1, p. 391—396 (1976).
See also 22.9, p. 437.

V. S. VAVILOV, A. E. KIV, and O. R. NIYAZOVA
Erratum to: The Subthreshold Radiation Effects in Semiconductors
(a) Vol. **36**, No. 1, p. 427 (1976).
See also 10.2, p. 94; *11*, p. *102*; 14.3, p. 178; 22.1, p. 343; 22.2, p. 361; 22.3, p. 373; 22.4, p. 376.

Y. KAMIURA and F. HASHIMOTO
Annealing Stage above 400 °C in Quenched Gallium-Doped Germanium
(a) Vol. **36**, No. 1, p. K81—K84 (1976).
See also *10.2*, p. *94*; 14.3, p. 178; 22.1.1, p. 348.

T. D. DZHAFAROV and YU. P. DEMAKOV
*Diffusion of Group I Impurities in Epitaxial Layers of AlAs–GaAs Solid Solutions
(a) Vol. **36**, No. 2, p. 439—443 (1976).
See also 10.2, p. 94; 22.2.1, p. 363; 22.2.4, p. 370.

C. G. CHEN and H. K. BIRNBAUM
*Low-Temperature H–O and H–N Relaxations: Confirmation of the Flynn-Stoneham Tunneling Theory
(a) Vol. **36**, No. 2, p. 687—692 (1976).
See also 21, p. 297.

M. A. LITOVSKII and R. SH. MALKOVICH
*Method of Determining the Diffusion Profile for a Concentration Dependent Diffusion Coefficient
(a) Vol. **36**, No. 2, p. K145—K148 (1976).

M. H. Yoo and W. H. Butler
*Steady-State Diffusion of Point Defects in the Interaction Force Field
(b) Vol. 77, No. 1, p. 181–193 (1976).
See also *10.1*, p. *83*; 21, p. 297; 21.1, p. 310.

P. Spindler and K. Nachtrieb
*Diffusion of Phosphorus in Copper
(a) Vol. 37, No. 2, p. 449–456 (1976).
See also 21.1, p. 312.

W. K. Chen and H. Downing
*Cation Self-Diffusion in $CoFe_2O_4$
(a) Vol. 37, No. 2, p. 515–520 (1976).
See also 10.2, p. 96; 22.8.2, p. 430.

U. A. Shakirov and M. S. Yunusov
Influence of X-Irradiation on Silicon Schottky Diodes
(a) Vol. 37, No. 2, p. 681–686 (1976).
See also *11*, p. *104*; *14.3.2*, p. *188*; 22.1.2, p. 357.

M. N. Pletnev
*Description of Reactions between Defects in Solids
(b) Vol. 78, No. 1, p. 57–66 (1976).
See also *10*, p. *76*.

H. Müller and Ch. Opitz
*Quantum Chemical Investigations on Vacancy Diffusion and Vacancy Lattices in Crystallites
(b) Vol. 78, No. 1, p. 177–182 (1976).
See also *10.1*, p. *85*; 21.1, p. 313; 21.1.1, p. 321.

L. J. Cheng, M. Süli, J. P. Kowall, P. W. Wallace, and J. W. Corbett
*Effects of Radiation on Gold Diffusion in Silicon
(a) Vol. 38, No. 1, p. 321–330 (1976).
See also 11, p. 104; 22.1.2, p. 358.

K. Maier, H. Mehrer, E. Lessmann, and W. Schüle
*Self-Diffusion in Nickel at Low Temperatures
(b) Vol. 78, No. 2, p. 689–698 (1976).
See also 10.1, p. 86; 21.1, p. 314.

A. F. Lubchenko and V. N. Pavlovich
*Laser Stimulation of Light Interstitial Diffusion in Semiconductors and Insulators
(b) Vol. 78, No. 2, p. K97–K101 (1976).
See also 22, p. 342.

P. Fahrentholtz and J. Mimkes
*Diffusion of Sb along Dislocations in Si
(b) Vol. 78, No. 2, p. K137–K140 (1976).
See also 10.2, p. 98; 22.1.2, p. 358.

A. I. Koifman, A. N. Narkulov, B. L. Oksengendler, and M. S. Yunusov
*The Low-Temperature Doping of Silicon by Radiation-Induced Diffusion
(a) Vol. 38, No. 2, p. 439–446 (1976).
See also *11*, p. *105*; 14.3.2, p. 189; 22.1.2, p. 358.

10. Defect Properties
(Irradiation Defects See 11)

Original Papers and Short Notes

H. Mehrer, A. Seeger, and E. Steiner
The Relationship between Correlation Factor and Isotope Effect of Diffusion in Crystals
(b) Vol. 73, No. 1; p. 131—140 (1976).
See also *9*, p. *72*.

I. Jäger
*Verzerrungsinduziertes Potential eines Fremdatoms in der Nähe einer freien Oberfläche
(Strain-Induced Potential of an Impurity in the Vicinity of a Free Surface)
(a) Vol. 33, No. 1, p. 167—172 (1976).

F. W. Schapink
*The Interaction of Thickness Contours and Dislocations under Dynamical Diffraction Conditions
(a) Vol. 33, No. 2, p. K121—K123 (1976).
See also *1.4*, p. *26*.

H. O. K. Kirchner
*Dislocation Symmetries in the Integral Theory
(b) Vol. 74, No. 1, p. 255—260 (1976).

Shi-yu Wu
*The Resonance Modes in the Vibrational Spectrum of a Chain with Light Impurities
(b) Vol. 74, No. 1, p. 349—357 (1976).
See also *6*, p. *55*.

V. I. Alshits and Yu. M. Sandler
*Erratum to: Flutter Mechanism of Dislocation Drag
(b) Vol. 74, No. 1, p. K85 (1976).
See also 6, p. 55.

N. K. Gilra
*Behaviour of Like Dislocations near an Interface
(a) Vol. 34, No. 1, p. 79—83 (1976).

D. E. Temkin
*On the Theory of Impurity Effect in Grain Boundary Motion
(a) Vol. 34, No. 1, p. 413—425 (1976).
See also *1.4*, p. *26*.

J. W. Martin and R. Paetsch
*Generalised Viscosity and Dislocation Damping Constant in Diatomic Crystalline Solids
(b) Vol. 74, No. 2, p. 761—772 (1976).
See also 6, p. 55; 12, p. 106.

Y. T. Chou
*On Antiplane Line Force in a Two-Phase Anisotropic Medium
(a) Vol. 34, No. 2, p. 645—650 (1976).

B. M. Strunin and A. B. Popov
*Comparison of Crystal Hardening by Arrays of Independently and Correlatedly Spaced Point Obstacles
(a) Vol. 34, No. 2, p. 761—768 (1976).

E. STEINER, H. MEHRER, and A. SEEGER
Correlation Factors and Isotope Effect for Self-Diffusion by Mono- and Divacancies in H.C.P. Crystals
(b) Vol. 75, No. 1, p. 361–373 (1976).
See also *9*, p. *72*.

J. KUBÁT and M. RIGDAHL
*The Exponential and Power Laws of Stress Relaxation Kinetics and a General Relation between the Activation Volume and Effective Stress
(a) Vol. 35, No. 1, p. 173–180 (1976).

H. O. K. KIRCHNER and K. H. BLUEMEL
*Elastic Quasi-Isotropy Normal to the Basal Plane in the Hexagonal System
(b) Vol. 75, No. 2, p. 527–532 (1976).
See also 12, p. 106.

R. J. DEMELFI, W. D. NIX, D. M. BARNETT, J. H. HOLBROOK, and G. M. POUND
*An Analysis of the Entropy of Thermally Activated Dislocation Motion Based on the Theory of Thermoelasticity
(b) Vol. 75, No. 2, p. 573–582 (1976).

R. PAETSCH
*Dislocation Damping Constant in Dispersive Monatomic Crystals
(b) Vol. 75, No. 2, p. K105–K109 (1976).
See also *6*, p. *57*.

B. M. STRUNIN
*On the Statistical Theory of Hardening of Crystals by Point Obstacles
(a) Vol. 35, No. 2, p. 551–561 (1976).

M. SUEZAWA and K. SUMINO
*Lattice Disortion and the Effective Shear Modulus along a Coherent Twin Boundary
(a) Vol. 36, No. 1, p. 263–268 (1976).

J. A. COCKS and SHI-YU WU
The Dynamics of and the Effect of a Short Range Order on a Segregated Binary Chain
(b) Vol. 77, No. 1, p. K5–K9 (1976).
See also *6*, p. *60*.

W. G. WOLFER and L. K. MANSUR
*The Mechanical Interaction of Point Defects with Spherical Surface Layers
(a) Vol. 37, No. 1, p. 211–222 (1976).

G. J. C. CARPENTER
*Unsafe Orientations for the Characterisation of Dislocation Loops Using Transmission Electron Microscopy
(a) Vol. 37, No. 1, p. K61–K66 (1976).

M. N. PLETNEV
*Description of Reactions between Defects in Solids
(b) Vol. 78, No. 1, p. 57–66 (1976).
See also *9*, p. *74*.

M. J. MARCINKOWSKI
The Differential Geometry of Grain Boundaries: Twist Boundaries
(a) Vol. 38, No. 1, p. 223–235 (1976).
See also *1.4*, p. *29*.

K.-H. KATERBAU
*The Contrast of Dynamical Images of Small Lattice Defects in the Electron Microscope
(a) Vol. 38, No. 2, p. 463–476 (1976).

B. WIELKE and G. SCHOECK
*Transient Functions for Plastic Strain Rate Changes
(a) Vol. **38**, No. 2, p. 539–546 (1976).

D. M. BARNETT
*The Singular Nature of the Self-Stress Field of a Plane Dislocation Loop in an Anisotropic Elastic Medium
(a) Vol. **38**, No. 2, p. 637–646 (1976).

10.1 Metals

Review Article

S. V. Tsivinskii
*Dislocation Multiplication in Single and Polycrystals in the Initial Stage of Deformation
(a) Vol. **36**, No. 1, p. 11–31 (1976).
See also *10.2*, p. *87*; 21, p. 289 ; 21.1, p. 302; 21.1.1, p. 316; 21.6, p. 329; 22.1.1, p. 345; 22.2.3, p. 367; 22.5.2, p. 392.

Original Papers and Short Notes

D. H. Sastry, M. J. Luton, and J. J. Jonas
*Thermally Activated Flow of H.C.P. Metals at Low Temperatures
(a) Vol. **33**, No. 1, p. 173–187 (1976).
See also 21, p. 289; 21.1, p. 302; 21.3, p. 324.

B. Wielke
*Easy-Glide Hardening of Zn and Cd between 4.2 and 400 K
(a) Vol. **33**, No. 1, p. 241–246 (1976).
See also 21, p. 290; 21.1, p. 302.

A. Melander
Strong- and Weak-Beam Contrast from Cuboidal Inclusions
(a) Vol. **33**, No. 1, p. 255–264 (1976).
See also *1.4*, p. *25*; 21, p. 290; 21.1, p. 303.

I. V. Gektina, F. F. Lavrentev, and V. I. Startsev
*Effect of Structure Imperfections on the Drag of Dislocations in Case of Their Viscous Motion in Zn Crystals
(a) Vol. **33**, No. 1, p. 289–294 (1976).
See also 21.1, p. 303.

G. Haneczok, J. Ilczuk, R. Kuśka, and J. W. Moroń
*Migrational Relaxations in α-Fe–V–C in the Region of the Snoek Effect
(a) Vol. **33**, No. 1, p. 313–324 (1976).
See also *18.2*, p. *229*; 21.1.1, p. 316.

H. Löffler and H.-G. Fabian
*Application of the Jump Model to the Decomposition of Some Binary and Ternary Aluminium Alloys
(a) Vol. **33**, No. 1, p. 345–353 (1976).
See also *1.3*, p. *21*; 14.1, p. 164; 21, p. 290; 21.1, p. 303.

L. M. Clarebrough and C. T. Forwood
*Diffraction from Single and Overlapping Stacking Faults in F.C.C. Crystals
(a) Vol. **33**, No. 1, p. 355–366 (1976).
See also *1.4*, p. *25*; 21.1, p. 303.

F. A. Mohamed and T. G. Langdon
*The Determination of the Activation Energy for Superplastic Flow
(a) Vol. **33**, No. 1, p. 375–381 (1976).
See also 21, p. 290; 21.1, p. 303.

S. Tanigawa and M. Doyama
*The Formulation of the Formation Energy of a Vacancy as an Alloy Problem
(b) Vol. **73**, No. 2, p. 517–526 (1976).
See also 1.3, p. 21; 21, p. 290.

R. YAMAMOTO
*Comments on Faber's Expression for the Vacancy Formation Energy in Metals
(b) Vol. 73, No. 2, p. K103–K105 (1976).
See also 21, p. 291.

F. KROUPA and V. PAIDAR
*Structure of DO_3 Superlattice Dislocations in Fe–Si Alloys
(a) Vol. 33, No. 2, p. 555–561 (1976).
See also 21.1.1, p. 317.

P. GONDI, R. TOGNATO, and E. EVANGELISTA
*High-Temperature Relaxation Peaks in Dispersion-Hardened S.A.P. Alloys
(a) Vol. 33, No. 2, p. 579–586 (1976).
See also 12.1, p. 107; 21, p. 291.

M. S. BAPNA and V. R. PARAMESWARAN
*The Elastic Interaction of Divacancies and Extended Dislocations in F.C.C. Metals
(a) Vol. 33, No. 2, p. 639–650 (1976).
See also 21.6, p. 329.

T. SCHOBER and H. WENZL
β-Phase Melting and Solidification Phenomena in the Niobium–Hydrogen System
(a) Vol. 33, No. 2, p. 673–681 (1976).
See also *1.2*, p. *14*; *1.4*, p. *25*; 21, p. 291.

K. G. BANERJEE and B. MILLS
*The Zener Relaxation and Diffusion in the Copper–Nickel–Zinc System
(a) Vol. 33, No. 2, p. 707–714 (1976).
See also *9*, p. *72*; 21.1, p. 304.

B. HERTEL and J. DIEHL
*On the Formation of Vacancy Clusters in Copper Irradiated with 5 keV Argon Ions
(a) Vol. 33, No. 2, p. K73–K75 (1976).
See also *11*, p. *100*; 21.1, p. 304.

R. YAMAMOTO
*The Binding Energy of an Impurity–Vacancy Pair in Simple Metals
(b) Vol. 74, No. 1, p. 159–168 (1976).
See also 21, p. 291.

E. SCHNEIDER and H. KRONMÜLLER
*The Elementary Interaction between a Crystal Dislocation and the Flux Line Lattice of a Type II Superconductor
(b) Vol. 74, No. 1, p. 261–273 (1976).
See also *14.2*, p. *169*; 21, p. 291.

G. ALLAN and M. LANNOO
*Influence of Self-Consistency on the Relaxation near a Vacancy or a Surface in Transition Metals
(b) Vol. 74, No. 1, p. 409–414 (1976).
See also *13.4*, p. *136*; 21.6, p. 329.

V. I. BETEKHTIN, A. G. KADOMTSEV, A. I. PETROV, and V. I. VLADIMIROV
*Reversibility of the First Stage of Fracture in Metals
(a) Vol. 34, No. 1, p. 73–78 (1976).
See also 21, p. 291.

F. APPEL and U. MESSERSCHMIDT
*The Estimation of the Effective Stress in Aluminium Foils Deformed in the High-Voltage Electron Microscope (HVEM)
(a) Vol. 34, No. 1, p. 175–181 (1976).
See also 21, p. 292.

A. I. Bondarenko and Yu. V. Konobeev
*Void Growth Kinetics in Irradiated Metals
(a) Vol. **34**, No. 1, p. 195–205 (1976).
See also 11, p. 100; 21.1, p. 305.

D. Preininger
*Work Hardening of F.C.C. Metals Subjected to Internal Stresses. Dispersion and Irradiation Hardening
(a) Vol. **34**, No. 1, p. 227–242 (1976).
See also 1.3, p. 22; 21, p. 292.

R. H. J. Fastenau, L. M. Caspers, and A. van Veen
*Small Clusters of Vacancies and Helium in Molybdenum
(a) Vol. **34**, No. 1, p. 277–289 (1976).
See also 21, p. 292.

E. Hegenbarth, S. V. Pirogova, E. M. Savitskii, and B. Göldner
Thermal Conductivity of the Intermetallic Compound NiAl in the Temperature Range from 20 to 80 K
(a) Vol. **34**, No. 1, p. 319–323 (1976).
See also *8*, p. *68*; 21.1, p. 305.

R. Gonzalez, J. Piqueras, and L. Bru
*Annealing of Fatigued Copper in the Electron Microscope
(a) Vol. **34**, No. 1, p. K9–K12 (1976).
See also 21.1, p. 305.

F. Dobeš
*Stochastic Approach to the Interpretation of Measured Effective Stress
(a) Vol. **34**, No. 1, p. K27–K29 (1976).
See also 12.1, p. 107.

J. Ullrich and H. Müllner
*Temperature Dependence of Internal Friction in Copper Due to Oxygen Contamination
(a) Vol. **34**, No. 1, p. K31–K34 (1976).
See also 12.1, p. 107; 21.1, p. 305.

B. L. Mordike
*Anomalous Slip and Discontinuous Yielding
(a) Vol. **34**, No. 1, p. K59–K61 (1976).
See also 21, p. 292.

L. M. Schafe, P. L. Rossiter, and W. A. Rachinger
Field History Effects in Superconducting Pb–Sn Alloys
(b) Vol. **74**, No. 2, p. 785–794 (1976).
See also *14.2*, p. *169*; 21, p. 292.

F. Maury and P. Lucasson
*On the Configurational Anisotropy of Frenkel Pairs in B.C.C. Molybdenum
(a) Vol. **34**, No. 2, p. 513–518 (1976).
See also 21, p. 293.

P. Charsley and M. H. S. Rahim
*The Effect of Ageing on the Lattice Thermal Conductivity of Copper Alloys Deformed at Liquid Helium Temperatures
(a) Vol. **34**, No. 2, p. 533–539 (1976).
See also 6, p. 56; *8*, p. *68*; 21.1, p. 306.

G. Schroeder and H. Kronmüller
*Investigation of the Annealing of the Dislocations in Ni Crystals by Means of the Magneto-Optical Kerr Effect
(a) Vol. **34**, No. 2, p. 623—638 (1976).
See also 18.2, p. 231; 20, p. 258; 21.1, p. 306.

L. J. Chen and A. J. Ardell
*The Observation of Multiple-Layer Loops in Nickel Base Alloys under Ion Bombardment
(a) Vol. **34**, No. 2, p. 679—690 (1976).
See also *11*, p. *101*; 21.1, p. 306.

R. A. Varin, J. A. Kozubowski, M. W. Grabski, and R. Buczkowski
*Analysis of Transmission Electron Microscopy Contrast at Extrinsic Grain Boundary Dislocations
(a) Vol. **34**, No. 2, p. 711—722 (1976).
See also *1.4*, p. *26*; 21.1.1, p. 318.

G. Dlubek, O. Brümmer, and E. Hensel
*Positron Annihilation Investigation for an Estimation of the Dislocation Density and Vacancy Concentration of Plastically Deformed Polycrystalline Ni of Different Purity
(a) Vol. **34**, No. 2, p. 737—746 (1976).
See also 21.1, p. 306.

V. B. Vykhodets, V. B. Demin, and P. V. Geld
Hydrogen Isotope Diffusivity and Solubility in Nickel, Palladium, and Copper
(a) Vol. **34**, No. 2, p. 787—795 (1976).
See also *9*, p. *72*; 21, p. 293; 21.1, p. 306.

M. L. Grossbeck, P. Williams, C. A. Evans, Jr., and H. K. Birnbaum
*Application of Ion Probe Analysis to Studies of Hydrogen Behavior in Solids
(a) Vol. **34**, No. 2, p. K97—K99 (1976).
See also 21.1, p. 306.

J. M. Galligan
*Growth Characteristics of Crystals Containing Disclinations
(a) Vol. **34**, No. 2, p. K101—K103 (1976).
See also *3*, p. *42*; 21.6, p. 330.

M. Antonini and H. D. Carstanjen
*Location of Interstitial Deuterium in $TaD_{0.067}$ by Channeling
(a) Vol. **34**, No. 2, p. K153—K157 (1976).
See also 21.1, p. 306.

L. N. Bystrov and S. N. Shakhmaev
*The Effect of Thermal Treatment on Stress Relaxation in Metals
(a) Vol. **34**, No. 2, p. K187—K190 (1976).
See also 21, p. 293; 21.6, p. 330.

H. E. Schaefer and W. Dander
The Initial Susceptibility of Cobalt after Electron Irradiation at 4.2 K
(b) Vol. **75**, No. 1, p. 173—178 (1976).
See also *11*, p. *101*; 14.1, p. 165; *18.2*, p. *231*; 21.1, p. 307.

K. H. Westmacott
*The Origin and Nature of Multiple Loop Configurations in Quenched Aluminium and Aluminium Alloys
(a) Vol. **35**, No. 1, p. 185—196 (1976).
See also 21, p. 294; 21.3, p. 324.

P. Merle, F. Fouquet, J. Merlin et P. F. Gobin
*Relations, dans un alliage aluminium–cuivre, entre les variations des propriétés mécaniques et les caractéristiques morphologiques de la précipitation de la phase Θ'
(Relations between the Variations of the Mechanical Properties and the Morphologic Characteristics of Phase Precipitation in an Aluminium–Copper Alloy)
(a) Vol. 35, No. 1, p. 213–222 (1976).
See also 12.1, p. 108; 21, p. 294; 21.1, p. 307.

B. L. Mordike
*Athermal Solid Solution Hardening in Niobium Alloy Single Crystals
(a) Vol. 35, No. 1, p. 303–308 (1976).
See also 12.1, p. 108; 21, p. 294.

O. A. Kaibyshev, R. Z. Valiev, and V. V. Astanin
*On the Nature of Superplastic Deformation
(a) Vol. 35, No. 1, p. 403–413 (1976).
See also 21, p. 294; 21.1, p. 307.

J. Topolsky, S. Y. Chuang, and S. J. Tao
*Positron Lifetimes in Heat Treated and Plastically Deformed Iron
(a) Vol. 35, No. 1, p. K17–K22 (1976).
See also 21.1.1, p. 318.

E. Ya. Mikhlin and V. V. Nelaev
*On the Increase of the Frenkel Defect Recombination Zone in α-Iron Caused by Hydrostatic Compression
(a) Vol. 35, No. 1, p. K81–K84 (1976).
Erratum
(a) Vol. 38, No. 1, p. K101 (1976).
See also 21.1.1, p. 318.

P. Tzanetakis, J. Hillairet, and G. Revel
*The Formation Energy of Vacancies in Aluminium and Magnesium
(b) Vol. 75, No. 2, p. 433–439 (1976).
See also 14.1, p. 165; 21, p. 294; 21.3, p. 324.

M. Kulić and L. Dobrosavljević
Pinning in Superconducting Films with Modulated Concentration of Impurities
(b) Vol. 75, No. 2, p. 677–686 (1976).
See also *14.2*, p. *170*; 21, p. 294.

J. P. Muscat, M. Lannoo, and G. Allan
Study of Growth of Guinier-Preston Zones in Al–Ag Compounds
(a) Vol. 35, No. 2, p. 499–509 (1976).
See also *1.3*, p. *22*; 21, p. 295; 21.6, p. 330.

P. V. Ryabko and K. P. Ryaboshapka
*On the Energy Change at Dislocation Motion in a B2-Type Solid Solution
(a) Vol. 35, No. 2, p. 575–581 (1976).
See also 12.1, p. 108.

M. M. Savin, V. M. Chernov, and A. M. Strokova
*Energy Factor of Dislocations in Hexagonal Crystals
(a) Vol. 35, No. 2, p. 747–754 (1976).
See also 12.1, p. 108; 21.1, p. 308; 21.3, p. 324.

Kishore and K. I. Vasu
*On the Low-Temperature Deformation Behaviour of a Quenched Aluminium–2% Germanium Alloy
(a) Vol. 35, No. 2, p. 769–775 (1976).
See also 21, p. 295.

F. DOBEŠ and J. ČADEK
*Measured Internal Stress in Some Models of Creep Controlled by Dislocation Climb
(a) Vol. **35**, No. 2, p. K123–K125 (1976).
See also 12.1, p. 108.

R. J. SALZBRENNER and S. H. CARPENTER
*Anelastic Relaxations in the Fe-Mn-N System
(a) Vol. **36**, No. 1, p. 117–126 (1976).
See also 21.1.1, p. 319.

H.-G. FABIAN and H. LÖFFLER
*On the Interpretation of the Decomposition Kinetics of Some Al-Zn(6 at%)-Mg(0.1 to 3.5 at%) Alloys by Means of the Enlarged Jump Model
(a) Vol. **36**, No. 1, p. 127–132 (1976).
See also *1.3*, p. *23*; 14.1, p. 165; 21, p. 296; 21.1, p. 309.

S. IKENO
*Direct Observation of Cell Formation in Niobium by a High-Voltage Electron Microscope
(a) Vol. **36**, No. 1, p. 317–328 (1976).
See also *1.4*, p. *28*; 21, p. 296.

D. LEPSKI and R. MICHAEL
*Comparison of the Properties of Electron Microscope Images from Dislocations and Rows of Tetragonal Point Defects
(a) Vol. **36**, No. 1, p. K17–K20 (1976).
See also 21.1.1, p. 320.

A. PRODAN, F. W. BOSWELL, and J. M. CORBETT
An Electron Microscopic Investigation of $CoTe_2$ Single Crystals
(a) Vol. **36**, No. 1, p. K21–K23 (1976).
See also *1.4*, p. *28*; 21.1, p. 309.

J. F. DUFRESNE, A. SEEGER, P. GROH, and P. MOSER
*Hydrogen Relaxation in α-Iron
(a) Vol. **36**, No. 2, p. 579–589 (1976).
See also *18.2*, p. *234*; 21.1.1, p. 320.

K. ISHIDA
*Direct Estimation of Stacking Fault Energy by Thermodynamic Analysis
(a) Vol. **36**, No. 2, p. 717–728 (1976).
See also 21.1, p. 310; 21.1.1, p. 320; 21.6, p. 331.

J. M. ANDERSON and A. D. STEWART
*The Internal Friction of Electron Irradiated and Cold-Worked Chromium
(a) Vol. **36**, No. 2, p. K119–K123 (1976).
See also 11, p. 103; 12.1, p. 108; 21.1, p. 310.

M. H. YOO and W. H. BUTLER
*Steady-State Diffusion of Point Defects in the Interaction Force Field
(b) Vol. 77, No. 1, p. 181–193 (1976).
See also *9*, p. *74*; 21, p. 297; 21.1, p. 310.

P. VAŠEK
*Negative Deviations from Matthiessen's Rule in Quenched Platinum at Low Temperatures
(b) Vol. 77, No. 1, p. K67–K70 (1976).
See also *14.1*, p. *166*; 21.6, p. 331.

W. WYCISK and M. FELLER-KNIEPMEIER
*Quenching Experiments on High-Purity Nickel
(a) Vol. **37**, No. 1, p. 183–191 (1976).
See also 14.1, p. 166; 21.1, p. 311.

F. PRINZ, H. P. KARNTHALER, and H. O. K. KIRCHNER
*Asymmetric Dislocation Nodes in Cu-Al Alloys
(a) Vol. 37, No. 1, p. 223–233 (1976).
See also 12.1, p. 108; 21, p. 297; 21.1, p. 311.

J. PETIT et M. QUINTARD
*Frottement intérieur du titane écroui dans le domaine des basses températures
(Internal Friction of Cold-Worked Titanium at Low Temperatures)
(a) Vol. 37, No. 1, p. 257–262 (1976).
See also *12.1*, p. *108*; 21.1, p. 311.

S. TAKÁCS
The Volume Pinning Force in Superconductors with Statistically Distributed Point Defects
(a) Vol. 37, No. 1, p. 279–286 (1976).
See also *14.2*, p. *171*.

H. ZABEL and H. PEISL
X-Ray Study of the Phase Diagram of Hydrogen in Niobium
(a) Vol. 37, No. 1, p. K67–K70 (1976).
See also *1.2*, p. *18*; 1.3, p. 23; 21, p. 298.

R. KROGGEL, O. KABISCH, and H. LÖFFLER
*On the Connection between Microhardness and the Mean Radius of the Guinier-Preston Zones in Al-Zn (12, 15, 22 at%) Alloys
(a) Vol. 37, No. 1, p. K85–K88 (1976).
See also 21, p. 298; 21.1, p. 311.

R. RODLOFF and H. NEUHÄUSER
Application of an "Evolution Strategy" to Calculate Static and Dynamic Dislocation Group Configurations
(a) Vol. 37, No. 1, p. K93–K96 (1976).
See also 11, p. 103; 21.1, p. 312.

H. KRONMÜLLER and H. RIEDEL
*The Critical Field for the Nucleation of n-Domains in Type I Superconductors
(b) Vol. 77, No. 2, p. 581–590 (1976).
See also *14.2*, p. *171*; 21, p. 298.

L. M. CASPERS, A. VAN VEEN, A. A. VAN GORKUM, A. VAN DEN BEUKEL, and C. M. VAN BAAL
*Helium Desorption from a (110) Mo Crystal. Evidence for the Vacancy Model in Stage III Annealing
(a) Vol. 37, No. 2, p. 371–383 (1976).
See also *11*, p. *103*; 21, p. 298.

L. E. POPOV, YU. P. SHARKEEV, N. A. KONEVA, N. A. ENSHINA, and G. A. PEROV
*Conservative Jog Slip on Superdislocations and Work-Hardening of Ordered Alloys
(a) Vol. 37, No. 2, p. 701–708 (1976).
See also *1.3*, p. *24*; 21.1, p. 312.

A. I. LANDAU and V. I. DOTSENKO
*Power-Like Dependence of Effective Dislocation Velocity on Load Resulting from the Stochastic Character of Motion through a Random Array of Point Obstacles
(a) Vol. 37, No. 2, p. 709–718 (1976).
See also 21.1, p. 312.

H. J. BLYTHE and L. H. DUNLEAVY
*Low Temperature Magnetic Disaccommodation in Neutron Irradiated Iron
(a) Vol. 37, No. 2, p. K197–K200 (1976).
See also *18.2*, p. *235*; 21.1.1, p. 321.

G. A. HASSAN and F. H. HAMMAD
*A Resistivity Decrement in Deformed Aluminium
(a) Vol. 37, No. 2, p. K209–K211 (1976).
See also *14.1*, p. *167*; 21, p. 299.

G. A. HASSAN, K. H. GEORGY, F. H. HAMMAD, and J. CADEK
*Effect of Pre-Torsional Deformation on the Transient Creep of Aluminium
(a) Vol. 37, No. 2, p. K213–K216 (1976).
See also 21, p. 299.

H. MÜLLER and CH. OPITZ
*Quantum Chemical Investigations on Vacancy Diffusion and Vacancy Lattices in Crystallites
(b) Vol. 78, No. 1, p. 177–182 (1976).
See also *9*, p. *74*; 21.1, p. 313; 21.1.1, p. 321.

K. M. MILLER and P. T. HEALD
*The Lattice Distortion around Interstitials in F.C.C. Metals
(b) Vol. 78, No. 1, p. 341–351 (1976).
See also 12.1, p. 109; 21, p. 299; 21.1, p. 313; 21.6, p. 331.

YU. M. PLISHKIN and I. E. PODCHINENOV
*Vacancy Migration Energy Calculation in an F.C.C. Copper Lattice by Computer Simulation
(a) Vol. 38, No. 1, p. 51–55 (1976).
See also 21.1, p. 313.

H. KOUSEK, P. BAJONS, and H. MÜLLNER
*Analysis of Structural Changes in High-Frequency Fatigued Copper by Means of Damping and Modulus Studies
(a) Vol. 38, No. 1, p. 199–205 (1976).
See also *12.1*, p. *109*; 21.1, p. 313.

P. GUYOT and J. P. SIMON
Symmetrical High-Angle Tilt Boundary Energy Calculation in Aluminium and Lithium
(a) Vol. 38, No. 1, p. 207–216 (1976).
See also *13.4*, p. *144*; 21, p. 300; 21.2, p. 323 .

W. POLANSCHÜTZ and E. KRAUTZ
Field Ion Microscopic Study of Grain Boundaries in Metals
(a) Vol. 38, No. 1, p. 237–242 (1976).
See also *1.4*, p. *29*; *17.1*, p. *221*; 21, p. 300.

H. MUGHRABI, R. KÜTTERER, K. LUBITZ, and H. KRONMÜLLER
*Magnetic Studies on Cyclically Deformed α-Iron Single Crystals
(a) Vol. 38, No. 1, p. 261–270 (1976).
See also 18.2, p. 236; 21.1.1, p. 321.

P. P. PAL-VAL, V. YA. PLATKOV, and V. I. STARTSEV
*Temperature Dependence of the Dislocation Drag Constant in Antimony
(a) Vol. 38, No. 1, p. 383–391 (1976).
See also 21.7, p. 332.

S. D. MAMCHICH and V. N. DROBYASIN
*On the Analysis of the Creep Dynamics in the Stress Relaxation Regime
(a) Vol. 38, No. 1, p. K5–K7 (1976).
See also 21, p. 300; 21.1, p. 314.

S. KISS
Coherency Loss of θ' Phase Due to Deformation
(a) Vol. 38, No. 1, p. K9–K11 (1976).
See also 12.1, p. 109; 21, p. 300.

K. Maier, H. Mehrer, E. Lessmann, and W. Schüle
Self-Diffusion in Nickel at Low Temperatures
(b) Vol. 78, No. 2, p. 689–698 (1976).
See also *9*, p. *74*; 21.1, p. 314.

O. V. Presnyakova and V. I. Zaitsev
*Effect of High Pressure on Nickel Polygonization
(a) Vol. 38, No. 2, p. 447–452 (1976).
See also *1.4*, p. *29*; 21.1, p. 314.

K. W. Ingle and A. G. Crocker
*A Computer Simulation Study of the Effect of Applied Stress on Divacancy Binding Energies in Body-Centred Cubic Crystals
(a) Vol. 38, No. 2, p. 523–531 (1976).
See also 12.1, p. 109; 21.1.1, p. 321.

S. M. Ohr
*Direction of the Black–White Contrast of Dislocation Loops
(a) Vol. 38, No. 2, p. 553–557 (1976).
See also *1.4*, p. *29*; 21, p. 300.

N. Kiesewetter and P. Schiller
*The Acoustic Emission from Moving Dislocations in Aluminium
(a) Vol. 38, No. 2, p. 569–576 (1976).
See also 21, p. 301.

K. Aoki and O. Izumi
*The Relation between Defect Hardening and Substitutional Solid Solution Hardening in an Intermetallic Compound Ni_3Al
(a) Vol. 38, No. 2, p. 587–594 (1976).
See also 12.1, p. 109; 21.1, p. 314.

K. Rosan and H. Wipf
*Nitrogen Traps for Hydrogen in Tantalum
(a) Vol. 38, No. 2, p. 611–620 (1976).
See also 14.1, p. 167; 21, p. 301.

B. A. Greenberg, M. A. Ivanov, Yu. N. Gornostirev, and L. E. Karkina
*Some Aspects of Plastic Deformation Theory with an Account for Thermally Activated Dislocation Transformations
(a) Vol. 38, No. 2, p. 653–662 (1976).
See also 21.1, p. 314.

J. P. Rivière and J. Grilhé
Neutron Irradiation Damage in a Plastically Deformed Fe–40 at% Al Ordered Alloy
(a) Vol. 38, No. 2, p. 713–720 (1976).
See also *11*, p. *105*; 14.1, p. 168; 21.1.1, p. 322.

V. M. Chernov and M. M. Savin
*Elastic Interaction of Dislocations with Tetragonal Point Defects in Anisotropic H.C.P. Crystals
(a) Vol. 38, No. 2, p. 761–768 (1976).
See also 21, p. 301; 21.1, p. 315; 21.3, p. 325; 21.6, p. 331.

S. M. Ashimov, T. P. Batsankalashvili, N. L. Nedzelyak, and J. S. Tsakadze
Torque Anisotropy Due to Crystal Defect Structure of the Superconducting Niobium Single Crystal
(a) Vol. 38, No. 2, p. 769–779 (1976).
See also *14.2*, p. *172*; 21, p. 301.

F. J. Kedves, M. Hordós, and F. Schuszter
*Precipitation in Al–Mn and Al–Cr Alloys
(a) Vol. 38, No. 2, p. K123–K126 (1976).
See also *14.1*, p. *168*; 21, p. 301; 21.1, p. 315.

10.2 Non-Metals

Review Articles

S. V. TSIVINSKII
*Dislocation Multiplication in Single and Polycrystals in the Initial Stage of Deformation
(a) Vol. **36**, No. 1, p. 11–31 (1976).
See also *10.1*, p. *78*.; 21, p. 289; 21.1, p. 302; 21.1.1, p. 316; 21.6, p. 329; 22.1.1, p. 345; 22.2.3, p. 367; 22.5.2, p. 392.

R. PRASAD
Present State of Polytypism in Cadmium Iodide Crystals
(a) Vol. **38**, No. 1, p. 11–44 (1976).
See also *1.1*, p. *10*; *1.2*, p. *14*; 22.3, p. 372; 22.4.1, p. 379; 22.5, p. 388; 22.5.3, p. 399; 22.8, p. 413.

Original Papers and Short Notes

A. N. ERMOSHKIN, R. A. EVARESTOV, and E. A. KOTOMIN
*Molecular Cluster Approach to Magnesium and Calcium Oxide Crystals (II): F^+ and F Centres
(b) Vol. **73**, No. 1, p. 81–86 (1976).
See also *13.4*, p. *134*; 22.6, p. 402.

W. BAUMHOER, J. THIEMANN, and O. KANERT
Monte Carlo Simulation of NMR Line Shapes in Distorted Non-Metallic Solids
(b) Vol. **73**, No. 1, p. 155–164 (1976).
See also *19*, p. *247*; 22.5.2, p. 392.

R. SATI
*Calculations of Optical Transitions of the F_3^+ Center
(b) Vol. **73**, No. 1, p. 353–359 (1976).
See also 13.4, p. 134; 22.5.2, p. 392.

L. BOSI
*A Study of Electric Field Effect on the Radiative Lifetime of the M Center in NaCl and KCl
(b) Vol. **73**, No. 1, p. K61–K64 (1976).
See also 22.5.2, p. 392.

J. VARVAS and T. NIRK
*High-Temperatures Conductivity of Undoped and Copper-Doped Cadmium Selenide
(a) Vol. **33**, No. 1, p. 75–84 (1976).
See also 13.4, p. 135; *14.3*, p. *173*; 22.4.2, p. 383.

HJ. BERNHARDT
*Comparative Studies of the LNT Coloration of $Y_3Al_5O_{12}$, $YAlO_3$, and $LiNbO_3$ Crystals
(a) Vol. **33**, No. 1, p. 211–216 (1976).
See also *20.1*, p. *261*; 20.3, p. 280; 22.8.1, p. 419; 22.8.2, p. 426.

J. E. STRUTT and E. LILLEY
*Structural Aspects of Clustering Reactions in Alkali Halides Doped with Divalent Impurities
(a) Vol. **33**, No. 1, p. 229–239 (1976).
See also 22.5.2, p. 392.

E. N. KISLOVSKII, L. I. DATSENKO, and A. S. VASILKOVSKII
*Effect of Crystal Net Plane Curvature on X-Ray Intensity in the Case of Laue Diffraction
(a) Vol. **33**, No. 1, p. 275–280 (1976).
See also *1.1*, p. *10*; 22.1.2, p. 351.

S. V. TSIVINSKII
*Statistical Theory of Thermally Activated Dislocation Motion in Crystals in the Case of Zero Internal Stresses
(a) Vol. **33**, No. 1, p. 301–311 (1976).
See also 22.1.1, p. 345; 22.1.2, p. 351; 22.2.3, p. 367; 22.5.2, p. 392.

P. BRAILLON and J. SERUGHETTI
*Mechanical Twinning in Calcite. Observation of Twinning Dislocations and Planar Defects by Transmission Electron Microscopy
(a) Vol. **33**, No. 1, p. 405–413 (1976).
See also 22.8.1, p. 419.

T. J. MAGEE, N. M. JOHNSON, and J. PENG
*The Effect of Illumination on the 10.6 μm Absorption Coefficient of Electron-Irradiated Single Crystal KCl
(a) Vol. **33**, No. 1, p. 415–419 (1976).
See also *11*, p. *100*; 20.1, p. 261; 22.5.3, p. 399.

E. M. TRUKHANOV, E. B. GOROKHOV, and S. I. STENIN
Specific Features of the Dislocation Structure of Germanium in the System Germanium–Pyrolitic Silica Film
(a) Vol. **33**, No. 1, p. 435–442 (1976).
See also *1.5*, p. *31*; 2, p. 36; 22.1.1, p. 345; 22.6, p. 402.

V. Z. BENGUS and T. P. KOVALENKO
*Dependence of the Dislocation Distribution in the Glide Band on the Density of Immobile Dislocations
(a) Vol. **33**, No. 1, p. K1–K3 (1976).
See also 22.5.2, p. 392.

E. NES
*On the Nucleation of Precipitate Colonies from Swirl Defects in Silicon
(a) Vol. **33**, No. 1, p. K5–K8 (1976).
See also 22.1.2, p. 351.

G. SCHIFFMACHER, P. E. CARO et C. BOULESTEIX
*Dislocations de macles observées en microscopie électronique à haute résolution (Twin Dislocations Observed by Electron Microscopy at High Temperatures)
(a) Vol. **33**, No. 1, p. K9–K13 (1976).
See also 22.6, p. 402.

C. F. CONDE, A. DOMINGUEZ-RODRIGUEZ, A. CONDE, and R. MARQUEZ
*Microhardness Tests in Nickel Oxide Single Crystals
(a) Vol. **33**, No. 1, p. K25–K29 (1976).
See also 22.6, p. 402.

A. TÓTH
*On the Mechanism of Charge Transport Caused by Moving Dislocations in NaCl Crystals Doped with Ca
(a) Vol. **33**, No. 1, p. K47–K50 (1976).
See also *14.4*, p. *202*; 22.5.2, p. 392.

H.-J. MÖLLER and P. HAASEN
*Thermally Activated Cross Slip of Screw Dislocations in Ge
(a) Vol. **33**, No. 1, p. K59–K62 (1976).
See also 22.1.1, p. 345.

L. NAGLI
*$In^+v_a^+v_c^-$ Centres in a KCl–In Crystal
(b) Vol. **73**, No. 2, p. 427–430 (1976).
See also 20.3, p. 280; 22.5.2, p. 393.

R. A. Evarestov, A. N. Ermoshkin, and E. A. Kotomin
*Molecular Cluster Approach to Magnesium and Calcium Oxide Crystals (III): Charge Distribution Analysis of Some Hole Centres
(b) Vol. **73**, No. 2, p. 483—486 (1976).
See also *13.4*, p. *135*; 22.6, p. 402.

E. M. Ledovskaya and M. I. Petrashen
*The Determination of the Symmetry Group of Molecular Centres in Crystals
(b) Vol. **73**, No. 2, p. 495—498 (1976).
See also 6, p. 54; 22.5.2, p. 393.

C. M. Scala and D. R. Hutton
*Site Assignment of Fe^{3+} in α-Quartz
(b) Vol. **73**, No. 2, p. K115—K117 (1976).
See also 19, p. 248; 22.6, p. 403.

C. Boulesteix, J. van Landuyt, and S. Amelinckx
*Identification of Rotation and Reflection Twins by Diffraction and Contrast Experiments in the Electron Microscope
(a) Vol. **33**, No. 2, p. 595—606 (1976).
See also *1.4*, p. *25*; 22.6, p. 403.

V. D. Verner, S. K. Maksimov, and D. K. Nichugovskii
*The Nature of Defects of the Crystalline Structure in GaAs Heavily Doped with Te
(a) Vol. **33**, No. 2, p. 755—763 (1976).
See also 22.2.1, p. 362.

L. I. Khandros, G. S. Pekar, M. K. Sheinkman, and E. L. Shtrum
High-Temperature Investigations of CdS Single Crystals (I): Electron Concentration, Cd Solubility, and Charge of Intrinsic Defects
(a) Vol. **33**, No. 2, p. 765—771 (1976).
See also *14.3*, p. *174*; 22.4.1, p. 379.

S. Mader and A. E. Michel
*Dislocation Reactions in Arsenic-Implanted and Annealed Silicon
(a) Vol. **33**, No. 2, p. 793—805 (1976).
See also 11, p. 100; 22.1.2, p. 352.

G. Filscher
On the Theoretical Calculation of the Intensity Distribution in X-Ray Section Topographs of Crystals with Superposed Films
(a) Vol. **33**, No. 2, p. K81—K82 (1976).
See also *1.6*, p. *34*; 22.1.2, p. 352.

N. Takeuchi, K. Inabe, and H. Nanto
Kinetics Order of Thermoluminescent Red Emission from MgO Single Crystals
(a) Vol. **33**, No. 2, p. K125—K127 (1976).
See also *20.3*, p. *281*; 22.6, p. 403.

E. Hartmann and Á. Péter
*The Ionic Conductivity of Plastically Twisted NaCl Whiskers
(a) Vol. **33**, No. 2, p. K169—K171 (1976).
See also *14.4*, p. *202*; 22.5.2, p. 393.

N. Kristianpoller, Á. Rehavy, and M. Israeli
*Effects of Ultraviolet Irradiation on KBr:Gd
(b) Vol. **74**, No. 1, p. 245—248 (1976).
See also 13.5.1, p. 151; 20.1, p. 263; 20.3, p. 281; 22.5.2, p. 393.

B. D. BHATTACHARYYA
*Tunnel Splitting in a Trigonal O^- Center
(b) Vol. 74, No. 1, p. K53–K56 (1976).
See also *19*, p. *249*; 22.5.3, p. 399.

D. D. RICHARDSON
Effects of an Impurity on Magnon Sidebands in Insulators
(b) Vol. 74, No. 1, p. K61–K65 (1976).
See also *18.2*, p. *230*; *20.1*, p. *264*; 22, p. 335.

G. M. DE'MUNARI, L. GABBA, F. GIUSIANO, and G. MAMBRIANI
*Energy of Vacancy Formation in Semiconductors
(a) Vol. 34, No. 1, p. 95–105 (1976).
See also 22.1, p. 343; 22.1.1, p. 346; 22.1.2, p. 352.

F. FRÖHLICH, P. GRAU, and M. SUSZYNSKA
*The Simple Power Law of the Critical Flow Stress by Solution Hardening with Respect to the Aggregation of Ca Impurities in NaCl
(a) Vol. 34, No. 1, p. 165–173 (1976).
See also 22.5.2, p. 393.

G. L. BLERIS, T. KARAKOSTAS, J. STOEMENOS, and N. A. ECONOMOU
On the Short-Range Order Arrangement at the Transition State of β-In_2Te_3
(a) Vol. 34, No. 1, p. 243–254 (1976).
See also *1.3*, p. *22*; 22.7, p. 410.

G. NIHOUL-BOUTANG and C. BOULESTEIX
*Observation of Lattice-Conservating Twins by Electron Microscopy. Study of the Image Contrasts
(a) Vol. 34, No. 1, p. 331–338 (1976).
See also *1.4*, p. *26*; 22.6, p. 403.

J. N. BOLAND
*An Electron Microscope Study of Dislocation Image Contrast in Olivine $(Mg, Fe)_2SiO_4$
(a) Vol. 34, No. 1, p. 361–367 (1976).
See also 22.8.1, p. 420.

L. I. KHANDROS, G. S. PEKAR, M. K. SHEINKMAN, and É. L. SHTRUM
High-Temperature Investigations of CdS Single Crystals (II): Electron Mobility and Scattering Mechanism
(a) Vol. 34, No. 1, p. 397–403 (1976).
See also *14.3*, p. *174*; 22.4.1, p. 379.

Z. LILIENTAL and H. BARTSCH
*The Effect of Diffusion of Ag on the Real Structure of ZnTe Crystals
(a) Vol. 34, No. 1, p. K5–K6 (1976).
See also 22.4.3, p. 385.

I. A. MARKOVA-OSORGINA and S. Z. SHMURAK
Deformation-Induced Exciton Luminescence of Alkali Halide Crystals
(b) Vol. 74, No. 2, p. 753–760 (1976).
See also 13.5.1, p. 152; *20.3*, p. *282*; 22.5.2, p. 394.

R. BALZER and F. R. ENGLER
*Mechanical Anisotropy of H Centers in CsBr
(b) Vol. 74, No. 2 p. K135–K140 (1976).
See also 11, p. 101; 22.5.2, p. 394.

V. Z. BENGUS and T. P. KOVALENKO
*Mobile and Immobile Dislocations in LiF Crystals under Straining
(a) Vol. 34, No. 2, p. 445–454 (1976).
See also 22.5.2, p. 394.

G. M. DE'MUNARI, L. GABBA, F. GIUSIANO, and G. MAMBRIANI
*Migration Energy of Vacancies in Semiconductors
(a) Vol. 34, No. 2, p. 455–460 (1976).
See also 22.1, p. 343; 22.1.1, p. 346; 22.1.2, p. 353.

D. CASASENT and F. CAIMI
*Suppressive Writing in Lithium-Doped NaF
(a) Vol. 34, No. 2, p. 461–466 (1976).
See also 20.1, p. 266; 22.5.2, p. 394.

D. CHAKRABORTY and G. LEHMANN
*On the Structures and Orientations of Hydrogen Defects in Natural and Synthetic Quartz Crystals
(a) Vol. 34, No. 2, p. 467–474 (1976).
See also 20.1, p. 266; 22.6, p. 404; 22.8, p. 414.

H. I. BERNHARDT
*The Structure of Coloration Centres in $Y_3Al_5O_{12}$Crystals after X-Ray Irradiation at Room Temperature
(a) Vol. 34, No. 2, p. 483–487 (1976).
See also 20.3, p. 282; 22.8.2, p. 427.

C. BECKER, E. ZSOLDOS, and A. WEBER
*Dislocation Configurations at Inclusions in GGG ($Gd_3Ga_5O_{12}$)
(a) Vol. 34, No. 2, p. 519–523 (1976).
See also 22.8.2, p. 428.

Y. V. G. S. MURTI, G. SAMUEL, and S. SIVARAMAN
*Studies on Tin Impurity Centres in KCl Crystals
(a) Vol. 34, No. 2, p. 615–621 (1976).
See also 13.4, p. 137; 19, p. 250; 20.1, p. 266; 22.5.2, p. 394.

R. METSELAAR, J. P. M. DAMEN, P. K. LARSEN, and M. A. H. HUYBERTS
*Investigation of Colour Centres in Gadolinium Gallium Garnet Crystals
(a) Vol. 34, No. 2, p. 665–670 (1976).
See also 20.1, p. 266; 22.8.2, p. 428.

K. GODWOD, A. T. NAGY, and Z. REK
The Application of the X-Ray Triple-Crystal Spectrometer for Measuring the Radius of Curvature of Bent Single Crystals
(a) Vol. 34, No. 2, p. 705–710 (1976).
See also *1.1*, p. *11*; 22.1.2, p. 353.

E. MIZERA and M. KLIMKIEWICZ
*Some Remarks on the Defect Structure in HgTe Crystals Revealed by Transmission Electron Microscopy
(a) Vol. 34, No. 2, p. K107–K108 (1976).
See also 22.4.3, p. 385.

T. WARMINSKI
*On the Detectability of Impurity Atoms in a Thin Surface Layer by an Electron Microprobe Analyser
(a) Vol. 34, No. 2, p. K119–K123 (1976).
See also 22.1.2, p. 353.

P. J. P. DE MAAYER and J. D. MACKENZIE
*Effect of Defect Structure on the Seebeck Coefficient of Thin Films of TiN_x
(a) Vol. 34, No. 2, p. K143–K146 (1976).
See also *15*, p. *209*; 22, p. 336.

H. Schaumburg and F. Willmann
*Optical Absorption of Plastically Deformed Germanium
(a) Vol. **34**, No. 2, p. K173–K177 (1976).
See also 13.4, p. 138; *20.1*, p. *266*; 22.1.1, p. 346.

L. Bosi
*Lifetime Studies on Excited M and M^+ Centers in NaF
(b) Vol. **75**, No. 1, p. K27–K30 (1976).
See also 22.5.2, p. 394.

R. Bubáková and Z. Šourek
*Contrast in the Dislocation Images for the Bragg Case
(a) Vol. **35**, No. 1, p. 55–60 (1976).
See also 22.1.2, p. 353.

M. Roulet, H. Huber, and C. Jaccard
*Thermal Disorder of Silver Bromide Studied by 100 keV Proton Channeling
(a) Vol. **35**, No. 1, p. 97–108 (1976).
See also 11, p. 101; 22.5.1, p. 391.

R. Nink and H.-J. Kos
*On the Role of Z Centers in the Trapping Mechanism of Thermoluminescent Lithium Fluoride
(a) Vol. **35**, No. 1, p. 121–129 (1976).
See also 13.4, p. 138; *20.1*, p. *268*; 22.5.2, p. 394.

J. Rabier, P. Veyssière, and J. Grilhé
*Possibility of Stacking Faults and Dissociation of Dislocations in the Garnet Structure
(a) Vol. **35**, No. 1, p. 259–268 (1976).
See also *1.4*, p. *27*; 22.8.2, p. 428.

J. Bouwma, K. J. de Vries, and A. J. Burggraaf
*Non-Stoichiometry, Defect Structure, and Dielectric Relaxation in Lanthana-Substituted $SrTiO_3$
(a) Vol. **35**, No. 1, p. 281–290 (1976).
See also *14.4*, p. *203*; 22.8, p. 415.

R. Visocekas, T. Ceva, C. Marti, F. Lefaucheux, and M. C. Robert
Tunneling Processes in Afterglow of Calcite
(a) Vol. **35**, No. 1, p. 315–327 (1976).
See also *20.3*, p. *283*; 22.8.1, p. 421.

L. B. Harris, D. A. Moncrieff, and V. N. E. Robinson
*High-Resolution Examination of Uncoated Insulators by SEM Applied to Grain Boundaries in Sodium Chloride
(a) Vol. **35**, No. 1, p. 371–377 (1976).
See also *1.4*, p. *27*; 22.5.2, p. 394.

M. Lefeld-Sosnowska and E. Zielińska-Rohozińska
Symmetrical and Asymmetrical X-Ray Section Topographs of Ion-Implanted Silicon
(a) Vol. **35**, No. 1, p. K1–K3 (1976).
See also *1.4*, p. *27*; 22.1.2, p. 354.

T. Warminski, E. Mizera, M. Klimkiewicz, and M. Iiyama
*On the Energy Dependence of K_αSi X-Ray Production in Silicon by Electron Bombardment
(a) Vol. **35**, No. 1, p. K9–K11 (1976).
See also 22.1.2, p. 354.

F. Matsuura, S. Teratani, S. Kishida, and I. Tsurumi
Temperature Dependence of the Half-Width of the 2.3 eV Optical Absorption Band in Neutron-Irradiated ZnS
(a) Vol. **35**, No. 1, p. K57–K60 (1976).
See also *11*, p. *101*; *20.1*, p. *268*; 22.4.1, p. 380.

N. Matsunami, Y. Yokoyama, and N. Itoh
Displacement of Impurities through Interaction with Radiation-Induced Interstitials in Alkali Halides
(b) Vol. **75**, No. 2, p. 483–492 (1976).
See also *11*, p. *101*; *20.1*, p. *268*; 22.5.2, p. 395.

H. Bill and W. von der Osten
*Raman and Vibronic Spectra of a New Oxygen Molecule Ion in CaF_2
(b) Vol. **75**, No. 2, p. 613–619 (1976).
See also 6, p. 57; 20.3, p. 284; 22.5.3, p. 399.

E. P. Pokatilov and L. V. Tarakanova
*F-Centre Ground State at the Ellipsoidal Dispersion Law of Electrons
(b) Vol. **75**, No. 2, p. K143–K148 (1976).
See also *13.5.3*, p. *159*.

L. Bosi
*On the Luminescence and Photoconductivity of the F Centres in NaBr
(b) Vol. **75**, No. 2, p. K163–K165 (1976).
See also 16, p. 214; 20.3, p. 284; 22.5.2, p. 395.

N. Dupont-Pavlovsky, F. Caralp, P. Delhaes et J. Amiell
*Caractérisation chimique et étude des propriétés électroniques de l'oxyde de zinc non stoechiométrique. Nature et répartition des défauts
(Chemical Characterization and Study of the Electronic Properties of Non-Stoichoimetric Zinc Oxide. Nature and Repartition of Defects)
(a) Vol. **35**, No. 2, p. 615–625 (1976).
See also *13.4*, p. *139*; 14.3, p. 177; 18, p. 223; 22.4, p. 376.

E. F. Harris and J. H. Crawford, Jr.
*The Influence of OH^- Impurity on the Oxidation of Iron Impurity in MgO above 1000 °C
(a) Vol. **35**, No. 2, p. 667–673 (1976).
See also 20.1, p. 269; 22.6, p. 405.

G. Dokter and H. J. Hoffmann
*Transient Bleaching of the F-Band in Alkali Halides
(b) Vol. **76**, No. 1, p. 241–249 (1976).
See also *13.4*, p. *140*; 16, p. 215; 22.5.2, p. 395.

T. Goworek and C. Rybka
*Impurity Induced Positronium in Molecular Crystals
(b) Vol. **76**, No. 1, p. K35–K37 (1976).
See also 22.9, p. 437.

Nawab Singh and A. W. Joshi
*Pressure Dependence of the Peak F-Band Absorption Energy in Alkali Halides
(b) Vol. **76**, No. 1, p. K39–K42 (1976).
See also *20.1*, p. *271*; 22.5.2, p. 395.

W. H. Hubrig, J. Auleytner, and M. Maciaszek
Changes of X-Ray Topographic Contrast Due to Annealing of Boron-Implanted Silicon
(a) Vol. **36**, No. 1, p. 209–215 (1976).
See also *1.1*, p. *12*; 22.1.2, p. 355.

V. P. Kisel
*Mobility of Dislocations in NaCl Single Crystals at Low Temperatures (I): 77 to 298 K
(a) Vol. **36**, No. 1, p. 297–306 (1976).
See also 22.5.2, p. 395.

C. Escribe and E. van der Voort
*Lattice Energy Calculations of Complexes Generated by Dissolution of Gd and Fe in Monoclinic Zirconia
(a) Vol. **36**, No. 1, p. 375–383 (1976).
See also 22.6, p. 405.

A. A. Urusovskaya, G. G. Knab, and Yu. Z. Estrin
*Comparison of Dislocation Mobility in LiF Measured by Direct and Indirect Methods
(a) Vol. **36**, No. 1, p. 397–402 (1976).
See also 22.5.2, p. 395.

V. S. Vavilov, A. E. Kiv, and O. R. Niyazova
Erratum to: The Subthreshold Radiation Effects in Semiconductors
(a) Vol. **36**, No. 1, p. 427 (1976).
See also 9, p. 73; *11*, p. *102*; 14.3, p. 178; 22.1, p. 343; 22.2, p. 361; 22.3, p. 373; 22.4, p. 376.

Y. Kamiura and F. Hashimoto
*Annealing Stage above 400 °C in Quenched Gallium-Doped Germanium
(a) Vol. **36**, No. 1, p. K81–K84 (1976).
See also 9, p. 73; 14.3, p. 178; 22.1.1, p. 348.

Debadideb Bhattacharya
*The Study of Dislocations by the Mössbauer Effect
(b) Vol. **76**, No. 2, p. K81–K84 (1976).
See also *5*, p. *50*.

T. D. Dzhafarov and Yu. P. Demakov
Diffusion of Group I Impurities in Epitaxial Layers of AlAs–GaAs Solid Solutions
(a) Vol. **36**, No. 2, p. 439–443 (1976).
See also *9*, p. *73*; 22.2.1, p. 363; 22.2.4, p. 370.

J. L. Farvacque, J. C. Doukhan, B. Escaig, and T. Tuomi
*Plastic Deformation of Selenium Single Crystals in ***a*** Glide and X-Ray Observation of the Induced Substructure
(a) Vol. **36**, No. 2, p. 461–470 (1976).
See also 22.1.3, p. 360.

B. Speit and G. Lehmann
*Hole Centers in the Feldspar Sanidine
(a) Vol. **36**, No. 2, p. 471–481 (1976).
See also 16, p. 215; 20.1, p. 272; 20.3, p. 285; 22.8.1, p. 423.

T. Bedyńska, R. Bubáková, and Z. Šourek
*Comparison between Experimental and Theoretical Dislocation Image for the Bragg Case
(a) Vol. **36**, No. 2, p. 509–516 (1976).
See also 22.1.2, p. 355.

P. Braillon et J. Serughetti
*Déformation plastique de monocristaux de calcite en compression suivant ⟨001⟩
(Plastic Deformation of Calcite Single Crystals by 001 Compression)
(a) Vol. **36**, No. 2, p. 637–646 (1976).
See also 22.8.1, p. 423.

V. V. Ratnam and H. D. Banerjee
*On the Glow Curves and TL Centres in CaF_2:Ce, Mn Crystals Irradiated by Cathode Rays
(a) Vol. **36**, No. 2, p. 693–697 (1976).
See also 20.3, p. 285; 22.5.3, p. 400.

J. Sárközi and R. Voszka
*The Variation of the Surface Charge of NaCl:Pb Monocrystals Due to Doping and Annealing
(a) Vol. **36**, No. 2, p. K105–K107 (1976).
See also 14.4, p. 203; 22.5.2, p. 396.

S. Etienne, C. Mai, G. Fantozzi et P. F. Gobin
*Phénomènes anélastiques dans le silicium à basse température
(Anelastic Phenomena in Silicon at Low Temperatures)
(a) Vol. **36**, No. 2, p. K113–K118 (1976).
See also 22.1.2, p. 356.

U. M. Grassano, A. Tanga, and M. Valli
*Electric Field Effect on the Thallium Emission Bands in KBr and KI
(b) Vol. 77, No. 1, p. 341–346 (1976).
See also 20.1, p. 273; *20.3*, p. *286*; 22.5.2, p. 396.

L. Mărculescu
*Optical Spectra of Aggregate Centres in KCl:Pb Crystals
(b) Vol. 77, No. 1, p. 353–357 (1976).
See also *20.1*, p. *273*; 22.5.2, p. 396.

I. N. Magden, L. I. Katurina, B. L. Meiler, and P. V. Musienko
[100] Silicon Doped with Arsenic (I): Electrical Parameters and Their Thermal Stability
(a) Vol. 37, No. 1, p. 29–38 (1976).
See also *14.3*, p. *179*; 22.1.2, p. 356.

I. M. Boswarva
*The Lattice Energies of Precipitate Phases and the Heats of Solution of Divalent Cations in Alkali Halide Crystals
(a) Vol. 37, No. 1, p. 65–72 (1976).
See also 22.5.2, p. 397.

S. Govinda
*Coloration and Luminescence in Pure and Chromium-Doped Al_2O_3 Single Crystals Irradiated with X-Rays at Room Temperature
(a) Vol. 37, No. 1, p. 109–117 (1976).
See also 11, p. 103; *20.1*, p. *273*; 20.3, p. 286; 22.6, p. 406.

T. L. Anderson, R. E. Newnham, L. E. Cross, and J. W. Laughner
*Laser-Induced Twinning in Quartz
(a) Vol. 37, No. 1, p. 235–245 (1976).
See also 1.4, p. 28; *12.2*, p. *112*; 22.6, p. 407.

Y. Motohashi, P. Braillon, and J. Serughetti
*Elastic Energy, Stress Field of Dislocations, and Dislocation Parameters in Calcite Crystals
(a) Vol. 37, No. 1, p. 263–270 (1976).
See also 12.2, p. 112; 22.8.1, p. 423.

T. Wosiński, T. Figielski, and A. Mąkosa
*Spin-Dependent Photoconductivity Spectrum of Dislocated Silicon
(a) Vol. 37, No. 1, p. K57–K59 (1976).
See also *16*, p. *216*; 19, p. 254; 22.1.2, p. 356.

J. L. Farvacque and E. Gerlach
*Scattering of Free Carriers Due to the Strain Field of Screw Dislocations in Semiconductors
(b) Vol. 77, No. 2, p. 651–656 (1976).
See also *14.3*, p. *179*; 22.1.3, p. 360.

J. C. Castro, H. C. Basso, and M. de Souza
Local Mode of H^- Substitutional Ion in KCN
(b) Vol. 77, No. 2, p. 685–692 (1976).
See also 1.2, p. 19; *6*, p. *61*; *20.1*, p. *274*; 22, p. 340.

L. Marculescu, M. Popescu, I. Baltog, and C. Ghita
*The State of Dispersion of Pb^{2+} in Highly Doped KCl Crystals
(b) Vol. 77, No. 2, p. 701–706 (1976).
See also 20.1, p. 274; 20.3, p. 287; 22.5.2, p. 397.

N. BANAII and G. JACOBS
*The Parallel and Perpendicular Stark Effect of the F-Center in RbBr and RbI
(b) Vol. 77, No. 2, p. 707–715 (1976).
See also 20.1, p. 274; 22.5.2, p. 397.

J. T. GOURLEY and E. R. VANCE
*Sharp Infrared Absorption Lines in Luminescent CaO
(b) Vol. 77, No. 2, p. K85–K89 (1976).
See also 20.1, p. 274; 22.6, p. 407.

HJ. BERNHARDT
*The Manganese-Induced O^- Centres in Yttrium Aluminium Garnet
(a) Vol. 37, No. 2, p. 445–448 (1976).
See also 20.1, p. 275; 20.3, p. 287; 22.8.2, p. 430.

M. E. VAN HULLE and W. MAENHOUT-VAN DER VORST
*Ionic Conductivity and Dielectric Loss of AgCl Doped with Pb^{2+}
(a) Vol. 37, No. 2, p. 473–477 (1976).
See also *14.4*, p. *204*; 22.5.1, p. 391.

J. W. MATTHEWS and T. S. PLASKETT
*Optical Birefringence Images of Edge Dislocations
(a) Vol. 37, No. 2, p. 499–503 (1976).
See also 22.8.2, p. 430.

W. K. CHEN and H. DOWNING
Cation Self-Diffusion in $CoFe_2O_4$
(a) Vol. 37, No. 2, p. 515–520 (1976).
See also *9*, p. *74*; 22.8.2, p. 430.

J. DAUNAY, J. BAILLOU, M. LEMERCIER, JAC. DAUNAY, P. BUGNET, and G. BATAILLER
Dispersion of the Parasitic Birefringence Induced by Lattice Plane Defects in ZnSe Single Crystals
(a) Vol. 37, No. 2, p. 521–527 (1976).
See also *20*, p. *258*; 22.4.2, p. 383.

H. G. LIPSON, P. LIGOR, and J. J. MARTIN
*The Effect of Ionizing Radiation on the 10.6 μm Absorption of KCl and NaCl
(a) Vol. 37, No. 2, p. 547–552 (1976).
See also *11*, p. *104*; 20.1, p. 275; 22.5.2, p. 397.

R. K. GARTIA
*Thermoluminescence Studies of Z_1 Centers in X-Irradiated Ca-Doped NaCl Crystals
(a) Vol. 37, No. 2, p. 571–578 (1976).
See also 20.3, p. 287; 22.5.2, p. 397.

K. KUMAZAKI, E. MATSUSHIMA, and A. ODAJIMA
*Point Defects and Non-Stoichiometry in HgSe
(a) Vol. 37, No. 2, p. 579–589 (1976).
See also *14.3*, p. *180*; 22.4.2, p. 383.

F. M. CAIMI, J. O. ARTMAN, W. W. LIN, and D. P. CASASENT
*Interpretation of Direct Writing M Center H-D Curve Characteristics of Photodichroic NaF Crystals
(a) Vol. 37, No. 2, p. 617–624 (1976).
See also 13.4, p. 143; 20.1, p. 275; 22.5.2, p. 397.

N. K. RADHAKRISHNAN and A. B. BISWAS
A Neutron Diffraction Study of the Cation Migration in $MgMn_2O_4$
(a) Vol. 37, No. 2, p. 719–722 (1976).
See also *1*, p. *9*; 22.8.2, p. 431.

J. Kies, R. Köhler, and W. Möhling
Double Crystal X-Ray Topographs of Microdefects in Silicon
(a) Vol. 37, No. 2, p. K113—K115 (1976).
See also *1.4*, p. *28*; 22.1.2, p. 357.

N. Banaii and G. Jacobs
*Stark Effect of the F Center in NaBr
(b) Vol. 78, No. 1, p. K51—K53 (1976).
See also 22.5.2, p. 397.

I. N. Magden and V. A. Safonov
[100] Silicon Doped with Arsenic (II): The Connection of Electrical and Structural Parameters with Microplasmas
(a) Vol. 38, No. 1, p. 45—50 (1976).
See also *14.3*, p. *180*; 22.1.2, p. 357.

J. Montemayor, R. Gómez-Ramírez, and E. Carrillo
*High-Temperature Creep on KBr Single Crystals
(a) Vol. 38, No. 1, p. 67—72 (1976).
See also 22.5.2, p. 397.

S. S. Paliwal and O. P. Sinha
*Calculation of Activation Energy and Frequency Factor from Glow Curves of SrS:Hf Phosphors
(a) Vol. 38, No. 1, p. 73—76 (1976).
See also 13.4, p. 144; *20.3*, p. *287*; 22, p. 341.

F. Appel, H. Bethge, and U. Messerschmidt
*Distribution of Point-Obstacle Distances during the Motion of Screw Dislocations in MgO Single Crystals
(a) Vol. 38, No. 1, p. 103—108 (1976).
See also 22.6, p. 408.

U. del Pennino and S. Mantovani
*Width of the Dislocation Energy Band in Germanium
(a) Vol. 38, No. 1, p. 109—113 (1976).
See also *13.1*, p. *128*; 14.3.4, p. 200; 22.1.1, p. 349.

G. Berg and J. Poźniak
*Theoretical Analysis of the Dipole Polarization and Depolarization in Me^{++}-Doped NaCl-Type Alkali Halide Crystals (I): Nearest-Neighbour Approach
(a) Vol. 38, No. 1, p. 115—122 (1976).
See also 14.4, p. 204; 22.5.2, p. 398.

N. Starbov, A. Buroff, and J. Malinowski
Surface Ionic Conductivity, Lattice Disorder, and Space Charge Region in Thin Silver Bromide Layers
(a) Vol. 38, No. 1, p. 161—170 (1976).
See also *14.4*, p. *204*; 22.5.1, p. 391.

F. V. Wald
Self-Compensation in CdTe and ZnTe Crystals Grown from Indium Solvents
(a) Vol. 38, No. 1, p. 253—259 (1976).
See also *3*, p. *44*; 14.3, p. 181; 20.1, p. 277; 22.4.3, p. 386.

J. L. Farvacque, J. C. Doukhan, and B. Escaig
*The Asymmetry and Orientation Dependence of **c**-Slip in Tellurium
(a) Vol. 38, No. 1, p. 271—284 (1976).
See also 22.1.3, p. 360.

AH MEE HOR, P. W. M. JACOBS, and K. S. MOODIE
Thermal Depolarization in Crystals of Potassium Chloride Containing Strontium Chloride
(a) Vol. 38, No. 1, p. 293–298 (1976).
See also *8*, p. *71*; 22.5.2, p. 398.

R. W. WHITWORTH
*A Charged-Dislocation Model for the Effect of an Electric Field on the Flow Stress of an Ionic Crystal
(a) Vol. 38, No. 1, p. 299–304 (1976).
See also 22.5.2, p. 398.

J. S. AHEARN, JR., C. A. B. BALL, and C. LAIRD
*Stress-Induced Birefringence of Mismatching III–V Heterojunctions
(a) Vol. 38, No. 1, p. 315–320 (1976).
See also *20*, p. *259*; 22.2.1, p. 364; 22.2.2, p. 365.

V. K. JAIN
Effect of Annealing on Thermoluminescence of LiF (TLD-100)
(a) Vol. 38, No. 1, p. K65–K68 (1976).
See also *20.3*, p. *288*; 22.5.2, p. 398.

A. A. ELANGO and T. N. NURAKHMETOV
*Structure and Generation Mechanism of Br_3^- Centers in X-Rayed KBr
(b) Vol. 78, No. 2, p. 529–536 (1976).
See also 20.1, p. 277; 22.5.2, p. 398.

J. P. BUISSON, S. LEFRANT, A. SADOC, L. TAUREL, and M. BILLARDON
Raman Scattering by KI Containing F Centres
(b) Vol. 78, No. 2, p. 779–790 (1976).
See also 6, p. 64; 13.4, p. 145; 20.1, p. 278; 22.5.2, p. 398.

S. ZAZUBOVICH and V. OSMININ
* Hole Mechanisms of H and F Centre Production in KCl Crystals
(b) Vol. 78, No. 2, p. 813–819 (1976).
See also *20.3*, p. *288*; 22.5.2, p. 398.

F. J. LÓPEZ and F. AGULLÓ-LÓPEZ
ESR of Conduction Electrons in Electrolytically Coloured Cation-Doped NaCl
(b) Vol. 78, No. 2, p. K81–K83 (1976).
See also *19*, p. *256*; 22.5.2, p. 398.

P. FAHRENTHOLTZ and J. MIMKES
Diffusion of Sb along Dislocations in Si
(b) Vol. 78, No. 2, p. K137–K140 (1976).
See also *9*, p. *74*; 22.1.2, p. 358.

R. MORINEAU
Structural Defects and Oxidation–Reduction Equilibrium in Mn, Zn Ferrites
(a) Vol. 38, No. 2, p. 559–568 (1976).
See also *18.3*, p. *241*; 22.8.2, p. 431.

J. W. MATTHEWS and T. S. PLASKETT
*Formation of Large Climb Loops at Decorated Dislocations in Gadolinium Gallium Garnet
(a) Vol. 38, No. 2, p. 577–585 (1976).
See also 22.8.2, p. 431.

V. M. ORERA and R. ALCALÁ
*Formation and Size Evolution of Ca Colloids in Additively Colored CaF_2
(a) Vol. 38, No. 2, p. 621–627 (1976).
See also 20.1, p. 278; 22.5.3, p. 400.

R. DE RIDDER, G. VAN TENDELOO, D. VAN DYCK, and S. AMELINCKX
A Cluster Model for the Transition State and Its Study by Means of Electron Diffraction (I): Theoretical Model
(a) Vol. **38**, No. 2, p. 663–674 (1976).
See also *1.4*, p. *29*; 22, p. 342.

K. BORGGREEN and A. R. THÖLÉN
Diffraction Contrast from Misfitting Spherical Particles Precipitated Close Together on a Dislocation (I): Individual Spheres Close Together on a Straight Line
(a) Vol. **38**, No. 2, p. 739–752 (1976).
See also *1.4*, p. *30*; 22, p. 342.

H. IWANAGA, K. SUZUKI, and S. TAKEUCHI
*Direct Observation of Dislocation Motion in ZnO
(a) Vol. **38**, No. 2, p. K119–K122 (1976).
See also 22.4, p. 378.

11. Irradiation Effects (X-Ray Diffraction Investigations See 1 and 10)

Review Article

P. V. Pavlov, E. I. Zorin, D. I. Tetelbaum, and A. F. Khokhlov
Nitrogen as Dopant in Silicon and Germanium
(a) Vol. **35**, No. 1, p. 11–36 (1976).
See *13.4*, p. *134*; *14.3*, p. *173*; 22.1.1, p. 345; 22.1.2, p. 351.

Original Papers and Short Notes

M. L. Swanson, L. M. Howe, and A. F. Quenneville
*The Combined Effects of Lattice Vibrations and Irradiation-Induced Defects on Dechanneling in Copper
(a) Vol. **33**, No. 1, p. 265–274 (1976).
See also 6, p. 53; 21.1, p. 303.

T. J. Magee, N. M. Johnson, and J. Peng
*The Effect of Illumination on the 10.6 μm Absorption Coefficient of Electron-Irradiated Single Crystal KCl
(a) Vol. **33**, No. 1, p. 415–419 (1976).
See also *10.2*, p. *88*; 20.1, p. 261; 22.5.3, p. 399.

A. Tamura and T. Kawamura
*Quantum Theory of Rosette-Motion Channeling
(b) Vol. **73**, No. 2, p. 391–400 (1976).
See also 13, p. 114; 21, p. 290; 22.5.2, p. 393.

H. Nakata and H. Morishima
Dose Dependence of ESR Spectra for On-Channel-Irradiated Silicon
(a) Vol. **33**, No. 2, p. 633–638 (1976).
See also *19*, p. *248*; 22.1.2, p. 352.

E. V. K. Rao
Optical Properties of Oxygen-Implanted GaAs
(a) Vol. **33**, No. 2, p. 683–690 (1976).
See also *20.1*, p. *263*; 22.2.1, p. 362.

S. Mader and A. E. Michel
Dislocation Reactions in Arsenic-Implanted and Annealed Silicon
(a) Vol. **33**, No. 2, p. 793–805 (1976).
See also *10.2*, p. *89*; 22.1.2, p. 352.

B. Hertel and J. Diehl
*On the Formation of Vacancy Clusters in Copper Irradiated with 5 keV Argon Ions
(a) Vol. **33**, No. 2, p. K73–K75 (1976).
See also *10.1*, p. *79*; 21.1, p. 304.

P. Apel and U. Müller-Jahreis
*Critical Angles of Low-Energy Light Ions in Gold
(a) Vol. **33**, No. 2, p. K129–K132 (1976).
See also 21.6, p. 329.

A. I. Bondarenko and Yu. V. Konobeev
Void Growth Kinetics in Irradiated Metals
(a) Vol. **34**, No. 1, p. 195–205 (1976).
See also *10.1*, p. *80*; 21.1, p. 305.

V. A. Girii, A. V. Kondrachuk, S. I. Kornyushin, V. G. Litovchenko, R. O. Litvinov, and V. I. Shakhovtsov
Effect of Impurities on the Radiation Sensitivity of the Si–SiO_2 Interface in MIS Structures
(a) Vol. **34**, No. 1, p. K35—K38 (1976).
See also *13.3*, p. *132*; *14.3.3*, p. *191*; 22.1.2, p. 353.

B. M. Latta and P. J. Scanlon
*Average Electronic Energy Loss in the Modified Firsov Theory
(b) Vol. 74, No. 2, p. 711—719 (1976).
See also 21, p. 292.

R. Balzer and F. R. Engler
*Mechanical Anisotropy of H Centers in CsBr
(b) Vol. 74, No. 2, p. K135—K140 (1976).
See also *10.2*, p. *90*; 22.5.2, p. 394.

E. W. Kreutz
X-Radiation Effects in Bulk Si at Subthreshold Energies
(a) Vol. **34**, No. 2, p. 489—495 (1976).
See also *14.3*, p. *175*; 22.1.2, p. 353.

L. J. Chen and A. J. Ardell
*The Observation of Multiple-Layer Loops in Nickel Base Alloys under Ion Bombardment
(a) Vol. **34**, No. 2, p. 679—690 (1976).
See also *10.1*, p. *81*; 21.1, p. 306.

H. E. Schaefer and W. Dander
*The Initial Susceptibility of Cobalt after Electron Irradiation at 4.2 K
(b) Vol. 75, No. 1, p. 173—178 (1976).
See also 10.1, p. 81; 14.1, p. 165; *18.2*, p. *231*; 21.1, p. 307.

M. Roulet, H. Huber, and C. Jaccard
*Thermal Disorder of Silver Bromide Studied by 100 keV Proton Channeling
(a) Vol. **35**, No. 1, p. 97—108 (1976).
See also *10.2*, p. *92*; 22.5.1, p. 391.

S. J. Rothman and Nghi Q. Lam
*Interstitial Retention into Stage III and Radiation-Enhanced Diffusion in Concentrated Alloys
(a) Vol. **35**, No. 1, p. K13—K16 (1976).
See also *9*, p. *73*; 21.1, p. 307.

A. S. Rudnev, V. I. Roslyakov, E. I. Sirotinin, and A. F. Tulinov
*A Study of Relative Energy Losses of Channeled Protons in a Tungsten Single Crystal
(a) Vol. **35**, No. 1, p. K23—K27 (1976).
See also 21, p. 294.

F. Matsuura, S. Teratani, S. Kishida, and I. Tsurumi
*Temperature Dependence of the Half-Width of the 2.3 eV Optical Absorption Band in Neutron-Irradiated ZnS
(a) Vol. **35**, No. 1, p. K57—K60 (1976).
See also 10.2, p. 92; *20.1*, p. *268*; 22.4.1, p. 380.

N. Matsunami, Y. Yokoyama, and N. Itoh
*Displacement of Impurities through Interaction with Radiation-Induced Interstitials in Alkali Halides
(b) Vol. 75, No. 2, p. 483—492 (1976).
See also 10.2, p. 93; *20.1*, p. *268*; 22.5.2, p. 395.

W. DANDER, H. KRONMÜLLER, and H. E. SCHAEFER
*Magnetic Aftereffect Phenomena in Cobalt between 4.2 and 20 K after Low-Temperature Electron Irradiation
(b) Vol. 75, No. 2, p. 647–655 (1976).
See also *18.2*, p. *232*; 21.1, p. 308.

V. N. BRUDNYI, D. L. BUDNITSKII, M. A. KRIVOV, and V. G. MELEV
*p–n-Type Conversion and Optical Properties of 2.0 MeV Electron-Irradiated $ZnSnAs_2$
(a) Vol. 35, No. 2, p. 425–430 (1976).
See also 13.4, p. 139; *14.3*, p. *177*; *20.1*, p. *269*; 22.8, p. 415.

S. RADHAKRISHNA and T. BHASKAR RAO
*Radiation Damage of Chromium Doped Potassium Alum
(a) Vol. 35, No. 2, p. 715–718 (1976).
See also 19, p. 252; 20.1, p. 269; 22.8.1, p. 422.

V. N. BYKOV, V. A. TROYAN, G. G. ZDOROVTSEVA, and V. S. KHAIMOVICH
Erratum to: Phase Transformations at Bombardment of Thin Films with Ions
(a) Vol. 35, No. 2, p. 779 (1976).
See also *1.2*, p. *17*; *1.5*, p. *31*; 21, p. 295; 21.1, p. 308; 21.1.1, p. 319; 22.1.1, p. 347.

W. HAUFFE
Faceting Mechanism in the Sputtering Process
(a) Vol. 35, No. 2, p. K93–K96 (1976).
See also *1.4*, p. *27*; 21.1.1, p. 319; 21.6, p. 330.

M. S. YUNUSOV, M. A. ZAIKOVSKAYA, B. L. OKSENGENDLER, and K. R. TOKHIROV
*Subthreshold Defect Production in Silicon
(a) Vol. 35, No. 2, p. K145–K149 (1976).
See also 22.1.2, p. 354.

M. A. KUMAKHOV and R. WEDELL
*A Theory of Resonance Dechannelling
(b) Vol. 76, No. 1, p. 119–131 (1976).

H. NAKATA
*Notes on the Stopping Functions in the Reduced Velocity Region
(b) Vol. 76, No. 1, p. K57–K62 (1976).

P. V. PAVLOV, T. A. KRUZE, D. I. TETELBAUM, E. I. ZORIN, E. V. SHITOVA, and N. V. GUDKOVA
*Electron Microscopic Studies of Silicon Layers Irradiated with High Doses of Nitrogen Ions
(a) Vol. 36, No. 1, p. 81–88 (1976).
See also *1.5*, p. *31*; 2, p. 39; 22.1.2, p. 355.

K. MURATA
*Exit Angle Dependence of Penetration Depth of Backscattered Electrons in the Scanning Electron Microscope
(a) Vol. 36, No. 1, p. 197–208 (1976).
See also 21, p. 296; 21.1, p. 309; 21.6, p. 330.

V. S. VAVILOV, A. E. KIV, and O. R. NIYAZOVA
*Erratum to: The Subtreshold Radiation Effects in Semiconductors
(a) Vol. 36, No. 1, p. 427 (1976).
See also 9, p. 73; 10.2, p. 94; 14.3, p. 178; 22.1, p. 343; 22.2, p. 361; 22.3, p. 373; 22.4, p. 376.

H. DÜSTERHÖFT, R. MANNS, and D. HILDEBRANDT
*Experimental Proof of a Theoretical Relation for the Probability of Positive Ion Excitation by Bombardment of Solid Surfaces with 12 keV Ar^+ Ions
(a) Vol. 36, No. 1, p. K93–K97 (1976).
See also 21, p. 296; 21.1, p. 310; 21.1.1, p. 320; 22.1.1, p. 348; 22.1.2, p. 355; 22.6, p. 405.

V. V. KAPLIN, D. E. POPOV, S. A. VOROBEV
*Bound States of Fast Electrons in Planar Channeling
(b) Vol. **76**, No. 2, p. 779–786 (1976).
See also 13, p. 118; 22.1.2, p. 355.

K. MURATA
*Depth Resolution of the Low- and High-Deflection Backscattered Electron Images in the Scanning Electron Microscope
(a) Vol. **36**, No. 2, p. 527–532 (1976).
See also 21.1, p. 310.

G. QUELARD and D. LESUEUR
*Neutron Radiation Defects in Bismuth and Tellurium-Doped Bismuth
(a) Vol. **36**, No. 2, p. 729–734 (1976).
See also 14.1, p. 165; 21.7, p. 332.

J. M. ANDERSON and A. D. STEWART
The Internal Friction of Electron Irradiated and Cold-Worked Chromium
(a) Vol. **36**, No. 2, p. K119–K123 (1976).
See also *10.1*, p. *83*; 12.1, p. 108; 21.1, p. 310.

V. G. KVACHADZE and Z. G. AKHVLEDIANI
Influence of γ-Irradiation on the Thermal Conductivity of KCl:KOH Crystals
(b) Vol. **77**, No. 1, p. K21–K25 (1976).
See also *6*, p. *60*; *8*, p. *70*; 20.1, p. 273; 22.5.2, p. 396.

N. G. CHERENDA and D. N. YUDIN
Additives and Cage Effect in Glasses
(b) Vol. **77**, No. 1, p. K27–K30 (1976).
See also *2*, p. *39*; *19*, p. *254*; 22.8.1, p. 423.

N. P. MOROZOV, D. I. TETELBAUM, P. V. PAVLOV, and E. I. ZORIN
*The Calculation of Secondary Defect Formation at Ion Implantation of Silicon
(a) Vol. **37**, No. 1, p. 57–64 (1976).
See also 22.1.2, p. 356.

S. GOVINDA
Coloration and Luminescence in Pure and Chromium-Doped Al_2O_3 Single Crystals Irradiated with X-Rays at Room Temperature
(a) Vol. **37**, No. 1, p. 109–117 (1976).
See also *10.2*, p. *95*; *20.1*, p. *273*; 20.3, p. 286; 22.6, p. 406.

R. RODLOFF and H. NEUHÄUSER
Application of an "Evolution Strategy" to Calculate Static and Dynamic Dislocation Group Configurations
(a) Vol. **37**, No. 1, p. K93–K96 (1976).
See also *10.1*, p. *84*; 21.1, p. 312.

P. BLAUDECK and R. LENK
Multiple s-Scattering in a Correlated Medium in the Low-Density Limit
(b) Vol. **77**, No. 2, p. 601–605 (1976).
See also *2*, p. *40*.

L. M. CASPERS, A. VAN VEEN, A. A. VAN GORKUM, A. VAN DEN BEUKEL, and C. M. VAN BAAL
*Helium Desorption from a (110) Mo Crystal. Evidence for the Vacancy Model in Stage III Annealing
(a) Vol. **37**, No. 2, p. 371–383 (1976).
See also *10.1*, p. *84*; 21, p. 298.

PHAM MINH TAN
Determination of Ion Implantation Profiles from Measurements of the Differential Capacity at the Semiconductor-Electrolyte Contact
(a) Vol. 37, No. 2, p. 439—444 (1976).
See also *13.4*, p. *143*; *14.3.3*, p. *193*; 22.1.2, p. 356.

H. G. LIPSON, P. LIGOR, and J. J. MARTIN
*The Effect of Ionizing Radiation on the 10.6 μm Absorption of KCl and NaCl
(a) Vol. 37, No. 2, p. 547—552 (1976).
See also *10.2*, p. *96*; 20.1, p. 275; 22.5.2, p. 397.

R. F. EGERTON
*Measurement of Inelastic/Elastic Scattering Ratio for East Electrons and Its Use in the Study of Radiation Damage
(a) Vol. 37, No. 2, p. 663—668 (1976).
See also 13.5.2, p. 158; 21, p. 299; 21.3, p. 324; 21.7, p. 332; 22.9, p. 438.

U. A. SHAKIROV and M. S. YUNUSOV
* Influence of X-Irradiation on Silicon Schottky Diodes
(a) Vol. 37, No. 2, p. 681—686 (1976).
See also 9, p. 74; *14.3.2*, p. *188*; 22.1.2, p. 357.

H. E. SCHAEFER and W. DANDER
*The Magnetic After-Effect Spectrum in Cobalt between 20 and 350 K after Electron Irradiation at 4.2 K
(b) Vol. 78, No. 1, p. 139—147 (1976).
See also *18.2*, p. *235*; 21.1, p. 313.

V. D. PARKHOMENKO, S. F. DUBININ, B. N. GOSHCHITSKII, YU. G. CHUKALKIN, S. K. SIDOROV, V. G. VOLOGIN, and V. V. PETROV
Peculiarities of Radiation Damage in Ferrites with Spinel Structure
(a) Vol. 38, No. 1, p. 57—66 (1976).
See also *18.3*, p. *241*; 22.8.2, p. 431.

V. S. LYSENKO, A. V. SACHENKO, A. P. DUBCHAK, A. N. NAZAROV, V. T. ATAMANENKO, M. P. LOKSHIN, and YU. YA. POLISHUK
Properties of MOS Structures after B^+ Ion Bombardment
(a) Vol. 38, No. 1, p. 131—138 (1976).
See also *14.3.3*, p. *194*; 22.1.2, p. 357.

A. GOLAŃSKI, A. FIDERKIEWICZ, H. RZEWUSKI, M. LEFELD-SOSNOWSKA, J. GRONKOWSKI, R. GRÖTZSCHEL, U. KREISSIG, and H. BARTSCH
*Particularities of Crystalline to Amorphous State Conversion in Silicon Heavily Damaged by 140 keV Si^{++} Ions
(a) Vol. 38, No. 1, p. 139—149 (1976).
See also *1.2*, p. *19*; 22.1.2, p. 357.

J. SALIK and S. NADIV
Structural Studies of the Photodissolution of Ag in Amourphous As_2S_3 Films
(a) Vol. 38, No. 1, p. 177—184 (1976).
See also *2*, p. *40*; 22, p. 341.

L. J. CHENG, M. SÜLI, J. P. KOWALL, P. W. WALLACE, and J. W. CORBETT
Effects of Radiation on Gold Diffusion in Silicon
(a) Vol. 38, No. 1, p. 321—330 (1976).
See also *9*, p. *74*; 22.1.2, p. 358.

A. E. KARKIN, V. E. ARKHIPOV, B. N. GOSHCHITSKII, E. P. ROMANOV, and S. K. SIDOROV
*Radiation Effects in the Superconductor Nb_3Sn
(a) Vol. 38, No. 2, p. 433—438 (1976).
See also *14.2*, p. *172*; 21, p. 300.

A. I. Koifman, A. N. Narkulov, B. L. Oksengendler, and M. S. Yunusov
*The Low-Temperature Doping of Silicon by Radiation-Induced Diffusion
(a) Vol. **38**, No. 2, p. 439–446 (1976).
See also *9*, p. *74*; 14.3.2, p. 189; 22.1.2, p. 358.

J. P. Rivière and J. Grilhé
*Neutron Irradiation Damage in a Plastically Deformed Fe–40 at% Al Ordered Alloy
(a) Vol. **38**, No. 2, p. 713–720 (1976).
See also 10.1, p. 86; 14.1, p. 168; 21.1.1, p. 322.

V. G. Vologin, S. F. Dubinin, V. D. Parkhomenko, Yu. G. Chukalkin, B. N. Goshchitskii, S. K. Sidorov, and V. V. Petrov
Experimental Determination of the Preference Energy in Ferrites–Spinels
(a) Vol. **38**, No. 2, p. K147–K150 (1976).
See also *18.3*, p. *241*; 22.8.2, p. 432.

D. Hildebrandt and R. Manns
Reflection of Kinetic Energy during the Bombardment of Polycrystalline Solids with He and Ne Ions
(a) Vol. **38**, No. 2, p. K155–K157 (1976).

12. Mechanical Properties (Plastic Deformations See 10)

Original Papers and Short Notes

J. W. Martin and R. Paetsch
Generalised Viscosity and Dislocation Damping Constant in Diatomic Crystalline Solids
(b) Vol. **74**, No. 2, p. 761–772 (1976).
See also 6, p. 55; *10*, p. *75*.

H. O. K. Kirchner and K. H. Bluemel
Elastic Quasi-Isotropy Normal to the Basal Plane in the Hexagonal System
(b) Vol. **75**, No. 2, p. 527–532 (1976).
See also *10*, p. *76*.

T. D. Shermergor and S. A. Patlazhan
*Elastic Constants of Quasi-Isotropic Polycrystals
(a) Vol. **38**, No. 1, p. 375–381 (1976).

12.1 Metals

Original Papers and Short Notes

U. HOFMANN
Physical Properties of Nickel–Rhenium Alloys
(a) Vol. **33**, No. 1, p. 119–123 (1976).
See also **14.1**, p. 164; *18.2*, p. *228*; 21.1, p. 302.

T. SEIDENKRANZ and E. HEGENBARTH
*Single-Crystal Elastic Constants of $MgZn_2$ in the Temperature Range from 4.2 to 300 K
(a) Vol. **33**, No. 1, p. 205–210 (1976).
See also 21.1, p. 302; 21.3, p. 324.

G. J. DAVIES and C. M. SARGENT
*Preferred Orientation and Elastic Anisotropy in Sheets of Copper–Nickel Alloys
(a) Vol. **33**, No. 1, p. 281–288 (1976).
See also *1.3*, p. *21*; 21.1, p. 303.

S. BENCKERT
*Erratum to: Third-Order Contribution to the Energy and the Elastic Constants of Simple Metals
(b) Vol. **73**, No. 2, p. 753 (1976).
See also *4*, p. *45*; 21, p. 291; 21.2, p. 323.

P. GONDI, R. TOGNATO, and E. EVANGELISTA
High-Temperature Relaxation Peaks in Dispersion-Hardened S.A.P. Alloys
(a) Vol. **33**, No. 2, p. 579–586 (1976).
See also *10.1*, p. *79*; 21, p. 291.

B. K. RAY and G. D. NIGAM
*Variation of Elastic Constants of Partially Ordered Cu_3Au Alloy with Temperature
(a) Vol. **33**, No. 2, p. K101–K105 (1976).
See also 1.3, p. 21; 21.1, p. 304; 21.6, p. 329.

F. I. BEKKERMAN and D. V. GITSU
*The Nernst-Ettingshausen Effect in Bismuth under the Influence of Directed Deformation
(b) Vol. **74**, No. 1, p. K1–K4 (1976).
See also *15*, p. *209*; 21.7, p. 332.

T. SOMA
*Elastic Constant of Hypothetical Face-Centered Cubic In
(b) Vol. **74**, No. 1, p. K67–K70 (1976).
See also 1.1, p. 11; 21, p. 291.

F. DOBEŠ
Stochastic Approach to the Interpretation of Measured Effective Stress
(a) Vol. **34**, No. 1, p. K27–K29 (1976).
See also *10.1*, p. *80*.

J. ULLRICH and H. MÜLLNER
Temperature Dependence of Internal Friction in Copper Due to Oxygen Contamination
(a) Vol. **34**, No. 1, p. K31–K34 (1976).
See also *10.1*, p. *80*; 21.1, p. 305.

O. P. BANSAL
On the Pressure Dependence of Electron Transport in Noble Metals
(b) Vol. **75**, No. 1, p. 85–90 (1976).
See also 6, p. 56; *13.2*, p. *130*; 21.6, p. 330.

P. MERLE, F. FOUQUET, J. MERLIN and P. F. GOBIN
Relations between the Variations of the Mechanical Properties and the Morphologic Characteristics of Phase Precipitation in an Aluminium-Copper Alloy
(a) Vol. **35**, No. 1, p. 213—222 (1976).
See also *10.1*, p. *82*; 21, p. 294; 21.1, p. 307.

B. L. MORDIKE
Athermal Solid Solution Hardening in Niobium Alloy Single Crystals
(a) Vol. **35**, No. 1, p. 303—308 (1976).
See also *10.1*, p. *82*; 21, p. 294.

P. V. RYABKO and K. P. RYABOSHAPKA
On the Energy Change at Dislocation Motion in a B2-Type Solid Solution
(a) Vol. **35**, No. 2, p. 575—581 (1976).
See also *10.1*, p. *82*.

D. FARGEOT, J. C. GLANDUS et P. BOCH
*Constantes élastiques des alliages cuivre-aluminium riches en cuivre entre 77 et 293 K
(Elastic Constants of Copper-Rich Copper-Aluminium Alloys between 77 and 293 K)
(a) Vol. **35**, No. 2, p. 687—695 (1976).
See also *1.3*, p. *22*; 21, p. 295; 21.1, p. 308.

M. M. SAVIN, V. M. CHERNOV, and A. M. STROKOVA
Energy Factor of Dislocations in Hexagonal Crystals
(a) Vol. **35**, No. 2, p. 747—754 (1976).
See also *10.1*, p. *82*; 21.1, p. 308; 21.3, p. 324.

F. DOBEŠ and J. ČADEK
Measured Internal Stress in Some Models of Creep Controlled by Dislocation Climb
(a) Vol. **35**, No. 2, p. K123—K125 (1976).
See also *10.1*, p. *83*.

A. MAGERL, B. BERRE, and G. ALEFELD
*Changes of the Elastic Constants of V, Nb, and Ta by Hydrogen and Deuterium
(a) Vol. **36**, No. 1, p. 161—171 (1976).
See also 21, p. 296; 21.1, p. 309.

J. M. ANDERSON and A. D. STEWART
The Internal Friction of Electron Irradiated and Cold-Worked Chromium
(a) Vol. **36**, No. 2, p. K119—K123 (1976).
See also *10.1*, p. *83*; 11, p. 103; 21.1, p. 310.

B. L. MORDIKE
*Single Crystal Elastic Constants of Silver Base Alloys
(a) Vol. **36**, No. 2, p. K161—K163 (1976).
See also 21.6, p. 331.

F. PRINZ, H. P. KARNTHALER, and H. O. K. KIRCHNER
Asymmetric Dislocation Nodes in Cu-Al Alloys
(a) Vol. **37**, No. 1, 223—233 (1976).
See also *10.1*, p. *84*; 21, p. 297; 21.1, p. 311.

J. PETIT et M. QUINTARD
*Frottement intérieur du titane écroui dans le domaine des basses températures
(Internal Friction of Cold-Worked Titanium at Low Temperatures)
(a) Vol. **37**, No. 1, p. 257—262 (1976).
See also *10.1*, p. *84*; 21.1, p. 311.

H. Ö. PAMUK and T. HALICIOĞLU
Evaluation of Morse Parameters for Metals
(a) Vol. **37**, No. 2, p. 695—699 (1976).
See also *4*, p. *46*; 21, p. 299.

K. M. Miller and P. T. Heald
The Lattice Distortion around Interstitials in F.C.C. Metals
(b) Vol. 78, No. 1, p. 341–351 (1976).
See also *10.1*, p. *85*; 21, p. 299; 21.1, p. 313; 21.6, p. 331.

L. I. Vinokurova, A. V. Vlasov, and M. Pardavi-Horváth
Pressure Effects on Magnetic Phase Transitions in FeRh and FeRhIr Alloys
(b) Vol. 78, No. 1, p. 353–357 (1976).
See also 18.1, p. 227; *18.2*, p. *236*; *18.4*, p. *245*; 21.1.1, p. 321.

V. P. Singh and M. P. Hemkar
Dynamical Study of the Temperature Dependence of Grüneisen Parameters in Alkali Metals
(b) Vol. 78, No. 1, p. 381–387 (1976).
See also *6*, p. *63*; 8, p. 71; 21.2, p. 323.

Y. Li
*The Anisotropic Behavior of Poisson's Ratio, Young's Modulus, and Shear Modulus in Hexagonal Materials
(a) Vol. 38, No. 1, p. 171–175 (1976).
See also 21, p. 299; 21.3, p. 324.

H. Kousek, P. Bajons, and H. Müllner
*Analysis of Structural Changes in High-Frequency Fatigued Copper by Means of Damping and Modulus Studies
(a) Vol. 38, No. 1, p. 199–205 (1976).
See also *10.1*, p. *85*; 21.1, p. 313.

S. Kiss
Coherency Loss of θ' Phase Due to Deformation
(a) Vol. 38, No. 1, p. K9–K11 (1976).
See also *10.1*, p. *85*; 21, p. 300.

K. W. Ingle and A. G. Crocker
A Computer Simulation Study of the Effect of Apllied Stress on Divacancy Binding Energies in Body-Centred Cubic Crystals
(a) Vol. 38, No. 2, p. 523–531 (1976).
See also *10.1*, p. *86*; 21.1.1, p. 321.

K. Aoki and O. Izumi
The Relation between Defect Hardening and Substitutional Solid Solution Hardening in an Intermetallic Compound Ni_3Al
(a) Vol. 38, No. 2, p. 587–594 (1976).
See also *10.1*, p. *86*; 21.1, p. 314.

J. P. Day and A. L. Ruoff
*Erratum to: The Variation of the Elastic Constants of Lithium with Temperature and Pressure
(a) Vol. 38, No. 2, p. 781 (1976).
See also 21.2, p. 323.

B. E. Smith
(101) Order-Twinning Mode in a CuAuI Superlattice
(a) Vol. 38, No. 2, p. K167–K170 (1976).
See also *1.4*, p. *30*; 21, p. 301.

12.2 Non-Metals

Original Papers and Short Notes

R. I. Cottam and G. A. Saunders
*An Extension of Keyes' Correlation
(a) Vol. **33**, No. 1, p. 367–373 (1976).
See also 22.1, p. 343; 22.2, p. 361; 22.4, p. 376.

B. Dorner, R. E. Ghosh, and G. Harbeke
Phonon Dispersion in the Layered Compound PbI_2
(b) Vol. **73**, No. 2, p. 655–659 (1976).
See also *6*, p. *54*; 7, p. 65; 22.5, p. 388.

L. Dmowski, M. Kończykowski, R. Piotrzkowski, and S. Porowski
Pressure-Induced Slow Relaxation of the Free Electron Concentration in Undoped n-Type InSb
(b) Vol. **73**, No. 2, p. K131–K134 (1976).
See also *14.3*, p. *174*; 22.2.3, p. 367.

K. Kumazaki
*Elastic Properties and Ionicity of Zero-Gap Semiconductors
(a) Vol. **33**, No. 2, p. 615–623 (1976).
See also *4*, p. *45*; 7, p. 65; 22.4, p. 376.

A. M. Kapitonov and E. M. Smokotin
*Elastic Constants of Single-Crystal Lithium Ferrite
(a) Vol. **34**, No. 1, p. K47–K49 (1976).
See also 22.8.2, p. 427.

K.-D. Ufert
*Stress Induced Switching in VO_2 Thin Films
(a) Vol. **34**, No. 1, p. K83–K86 (1976).
See also *8*, p. *68*; 14.3.1, p. 183; 22.6, p. 404.

A. P. Rusakov and M. V. Anfimov
Non-Central Forces and Stability of the A^NB^{8-N} Tetrahedral Compounds (I): Bond Charges and the Valence-Force-Field Model
(b) Vol. **75**, No. 1, p. 73–77 (1976).
See also *4*, p. *45*; 22.1, p. 343; 22.2, p. 361; 22.4, p. 376; 22.5, p. 389.

W. Wardzyński and K. Pataj
Bound Exciton–Neutral Acceptor Complex in ZnTe
(b) Vol. **75**, No. 1, p. 341–346 (1976).
See also *13.5.1*, p. *152*; *20.3*, p. *283*; 22.4.3, p. 385.

G. I. Peresada, E. G. Ponyatovskii, and Zh. D. Sokolovskaya
*Pressure Dependence of the Elastic Constants of PbS
(a) Vol. **35**, No. 2, p. K177–K180 (1976).
See also 22, p. 337.

T. Wasiutyński
Self-Consistent Phonon Calculation for Hexamethylenetetramine Molecular Crystals
(b) Vol. **76**, No. 1, p. 175–181 (1976).
See also 4, p. 46; *6*, p. *57*; 22.9, p. 437.

W. Kühn
Stress Dependence of the $(E_1, E_2 + \Delta_1)$ Transition in Germanium
(b) Vol. **76**, No. 1, p. 289–299 (1976).
See also *13.1*, p. *125*; *20.1*, p. *270*; 22.1.1, p. 347.

H. BEIGE, C. BIRKHOLZ, E. CIESLA, and G. SCHMIDT
*Elastic and Electromechanical Properties of Some Crystals near Structural Phase Transitions
(b) Vol. 76, No. 1, p. K47—K50 (1976).
See also 1.2, p. 17; *14.4*, p. *203*; 22.8.1, p. 422.

J. ALBERS, R. W. SAILER, and H. E. MÜSER
*Dielectric, Elastic, and Piezoelectric Properties of Mg–Cl Boracite
(a) Vol. 36, No. 1, p. 189—195 (1976).
See also *14.4.1*, p. *207*; 22.8, p. 416.

K. P. BELOV, A. M. KADOMTSEVA, N. M. KOVTUN, V. N. DERKACHENKO, V. N. MELOV, and V. A. KHOKHLOV
On the Character of Phase Transitions in $ErFeO_3$
(a) Vol. 36, No. 1, p. 415—425 (1976).
See also *18.4*, p. *244*; 22.8.2, p. 429.

G. A. PETRAKOVSKII and A. I. PANKRATS
*The Magnetoelastic Coupling in $FeBO_3$
(a) Vol. 36, No. 1, p. K43—K46 (1976).
See also *18.4*, p. *244*; 22.8.1, p. 423.

V. KOOS and H.-G. NEUMANN
*Effect of Heat Treatment on Stress in Amorphous Silicon Films
(a) Vol. 36, No. 1, p. K47—K49 (1976).
See also *1.5*, p. *32*; 22.1.2, p. 355.

A. LOIDL, H. JEX, J. DAUBERT, and M. MÜLLNER
*Zero and First Sound in NaF and RbI
(b) Vol. 76, No. 2, p. 581—592 (1976).
See also *6*, p. *59*; 7, p. 66; 8, p. 69; 22.5.2, p. 395.

T. SOMA
Modified Pertubation Method for Diamond-Type Crystals
(b) Vol. 76, No. 2, p. 753—758 (1976).
See also *6*, p. *59*; 13.1, p. 126; 22.1, p. 344; 22.1.1, p. 348; 22.1.2, p. 355.

A. BATANA and E. R. GONZALEZ
*On the Pressure Dependence of the Effective Ionic Charge of Lithium, Sodium, and Rubidium Halides
(b) Vol. 76, No. 2, p. K105—K108 (1976).
See also *4*, p. *46*; 22.5.2, p. 396.

S. R. SINGH, R. S. RAI, and G. SINGH
**K*-Surfaces of Cubic Silicon Carbide (β-SiC)
(a) Vol. 37, No. 1, p. 139—142 (1976).
See also *1.1*, p. *12*; 22.3, p. 373.

H. VEESER, U. BOGNER, and W. EISENMENGER
Analysis of the Acoustic Transients in the Pulse Response of the Linear Electro-Optic Effect
(a) Vol. 37, No. 1, p. 161—170 (1976).
See also *7*, p. *66*; *14.4.1*, p. *208*; 20.1, p. 273; 22.8.1, p. 423.

V. KOOS, U. BECK, and H.-G. NEUMANN
Structure and Intrinsic Stress of Nickel Silicide Films
(a) Vol. 37, No. 1, p. 193—198 (1976).
See also *1.5*, p. *32*; 22, p. 339.

T. L. Anderson, R. E. Newnham, L. E. Cross, and J. W. Laughner
*Laser-Induced Twinning in Quartz
(a) Vol. 37, No. 1, p. 235—245 (1976).
See also 1.4, p. 28; *10.2*, p. *95*; 22.6, p. 407.

Y. Motohashi, P. Braillon, and J. Serughetti
Elastic Energy, Stress Field of Dislocations, and Dislocation Parameters in Calcite Crystals
(a) Vol. 37, No. 1, p. 263—270 (1976).
See also *10.2*, p. *95*; 22.8.1, p. 423.

F. Perrot
Bulk Properties of Lithium Hydride up to 100 Mbar
(b) Vol. 77, No. 2, p. 517—525 (1976).
See also *13.1*, p. *127*; 22, p. 340.

S. C. Goyal and A. K. Goyal
Dispersion of Phonons in Metal Carbides (UC)
(b) Vol. 77, No. 2, p. 639—643 (1976).
See also *6*, p. *61*; 22, p. 340.

J. Aslam, S. Rolandson, M. M. Beg, N. M. Butt, and Q. H. Khan
Phonons in the Mixed Crystal $K_{0,5}Rb_{0,5}I$
(b) Vol. 77, No. 2, p. 693—700 (1976).
See also *6*, p. *61*; 22.5.4, p. 401.

D. S. Puri, V. K. Garg, and M. P. Verma
*Many-Body Effects on the Third-Order Elastic Constants and Pressure Derivatives of the Second-Order Elastic Constants of CsCl Structure Solids
(b) Vol. 78, No. 1, p. 113—122 (1976).
See also 22.5.2, p. 397.

V. Dienys and Ž. Kancleris
Influence of Uniaxial Stress and Electron–Electron Scattering on the Phenomenological Conductivity Relaxation Time in n-Type Germanium
(b) Vol. 78, No. 2, p. 757—765 (1976).
See also *14.3.4*, p. *200*; 22.1.1, p. 349.

13. Electron States

Review Articles

W. Haubenreisser, Th. Klupsch, and F. Voigt
*Quantum Statistical Theory of Parametric Spin-Wave Excitations in Magnetoelastic Ferromagnets under Parallel Pumping (II): Calculation of Macroscopic Observables
(b) Vol. 73, No. 1, p. 11–64 (1976).
See also 6, p. 52; 13.5.4, p. 160; *18.2*, p. *228*.

H. W. de Wijn, A. M. van Diepen, and K. H. J. Buschow
*Samarium in Crystal Fields, a Case of Strong J-Mixing
(b) Vol. 76, No. 1, p. 11–30 (1976).
See also 5, p. 48; *18*, p. *222*; 19, p. 247; 21, p. 289; 21.1, p. 302; 21.1.1, p. 316; 21.4, p. 326.

S. V. Vonsovskii and M. S. Svirskii
*Multiplicity and Pseudomultiplicity in Electron–Ion Systems (I)
(b) Vol. 77, No. 1, p. 9–52 (1976).
See also 14.1, p. 164; 14.2, p. 169; 18.2, p. 228; 18.4, p. 242; 21, p. 289.

S. V. Vonsovskii and M. S. Svirskii
*Multiplicity and Pseudomultiplicity of Electron–Ion Systems (II)
(b) Vol. 77, No. 2, p. 403–432 (1976).
See also 14.1, p. 164; 14.2, p. 169; 18.2, p. 228; 18.4, p. 242; 21, p. 289.

H. Böttger and V. V. Bryksin
Hopping Conductivity in Ordered and Disordered Solids (I)
(b) Vol. 78, No. 1, p. 9–56 (1976).
See also 2, p. 35; 6, p. 52; 13.5.3, p. 159; *14.3*, p. *173*; 22, p. 333.

H. Böttger and V. V. Bryksin
Hopping Conductivity in Ordered and Disordered Solids (II)
(b) Vol. 78. No. 2, p. 415–451 (1976).
See also 2, p. 35; 6, p. 52; 13.5.3, p. 159; *14.3*, p. *173*; 22, p. 333.

Original Papers and Short Notes

W. Löser
A Self-Consistent Cluster Approximation for the Disordered Hubbard Model
(b) Vol. 73, No. 1, p. 179–185 (1976).
See also *2*, p. *35*; 14, p. 162; *18*, p. *222*.

H. Böttger and V. V. Bryksin
Theoretical Investigation of the Hopping Conductivity in Disordered Systems in a Large Frequency Range
(b) Vol. 73, No. 1, p. 229–237 (1976).
See also *2*, p. *35*; 6, p. 52; *14.3*, p. *173*.

P. Petelenz
*Effect of Polarization on Charge Carrier Transfer in Molecular Crystals
(b) Vol. 73, No. 1, p. 293–299 (1976).
See also 14.4, p. 202; 22.9, p. 433.

M. G. Kaplunov, Yu. M. Shulga, K. I. Pokhodnya, and Yu. G. Borodko
X-Ray Photoelectron Spectra of Quasi-One-Dimensional Conductors
(b) Vol. 73, No. 1, p. 335–339 (1976).
See also 17, p. 218; *20*, p. *257*; 22.9, p. 433.

G. GROSSO, G. PASTORI PARRAVICINI, and R. RESTA
*Isotropic Compton Profile of LiH in the Tight-Binding Approximation
(b) Vol. 73, No. 1, p. 371—378 (1976).
See also 22, p. 333.

A. ČÍŽEK, M. ŠOB, and K. KRÁSENSKÝ
*The Determination of Ionic Radii of Some Metals by the Positron Annihilation Method
(b) Vol. 73, No. 1, p. K1—K4 (1976).
See also 21, p. 289; 21.1, p. 302; 21.1.1, p. 316; 21.6, p. 329.

I. MOORE, M. KING, M. JONES, R. J. DECK, and A. M. HERMANN
Microwave Faraday Rotation in Molybdenum Disulfide
(b) Vol. 73, No. 1, p. K31—K34 (1976).
See also *18*, p. *222*; *20*, p. *257*; 22, p. 333.

S. ROBASZKIEWICZ and R. MICNAS
*Absence of Magnetic Ordering in One and Two Dimensions in a Model with Localized and Itinerant Electrons
(b) Vol. 73, No. 1, p. K35—K39 (1976).
See also *18.2*, p. *228*.

F. M. GALPERIN
*Hyperfine Field on the Nucleus of 3d Transition Metal Atoms
(b) Vol. 73, No. 1, p. K73—K77 (1976).
See also *18.2*, p. *228*; 21.1, p. 302; 21.1.1, p. 316.

A. TAMURA and T. KAWAMURA
Quantum Theory of Rosette-Motion Channeling
(b) Vol. 73, No. 2, p. 391—400 (1976).
See also *11*, p. *100*; 21, p. 290; 22.5.2, p. 393.

M. AUSLOOS
*Renormalized Density of States by Conduction Electron–Localized Spin Scattering
(b) Vol. 73, No. 2, p. 469—474 (1976).
See also 14.1, p. 164; 21, p. 290.

P. PETELENZ
*Effective Electron–Hole Interaction in Molecular Crystals (I)
(b) Vol. 73, No. 2, p. 565—573 (1976).
See also 6, p. 54; *13.5.1*, p. *150*; 14.3, p. 174; 22, p. 334.

P. NOVÁK and I. VELTRUSKÝ
*Overlap and Covalency Contribution to the Zero-Field Splitting of S-State Ions
(b) Vol. 73, No. 2, p. 575—586 (1976).
See also *18.3*, p. *239*; 22.8.2, p. 426.

J. R. L. DE ALMEIDA and S. M. REZENDE
Scattering of Light by Magnetoelastic Excitations in Ferromagnets
(b) Vol. 73, No. 2, p. 661—672 (1976).
See also 18, p. 222; *20*, p. *257*; 22.8.2, p. 427.

C. MANOHAR and V. K. KELKAR
*Effects of Three-Body Interactions on Hyperfine Fields
(b) Vol. 73, No. 2, p. 735—740 (1976).
See also *18*, p. *222*; 21.1, p. 303.

K. BREUER, W. TEWS, and E. STEINBEISS
*Energy Loss Measurements on Thin VO_2 Films
(b) Vol. 73, No. 2, p. K95—K97 (1976).
See also *20*, p. *257*; 22.6, p. 403.

G. Dlubek and O. Brümmer
*A Comparison of Positron Annihilation and Compton Effect Measurements on Si Single Crystals
(b) Vol. 73, No. 2, p. K107—K110 (1976).
See also *20*, p. *257*; 22.1.2, p. 352.

O. B. Sokolov, V. K. Finashkin, and E. A. Turov
*Spin Polarization of X-Ray Photoelectrons in Nickel
(b) Vol. 74, No. 1, p. 35—43 (1976).
See also 21.1, p. 304.

F. Brosens, L. F. Lemmens, and J. T. Devreese
*Frequency-Dependent Exchange Correction to the Dielectric Function of the Electron Gas
(b) Vol. 74, No. 1, p. 45—55 (1976).
See also *14.4*, p. *202*.

T. Shimizu and N. Ishii
*Electronic Structure of As_2S_3, As_2Se_3, and As_2Te_3 Glasses by the Extended Hückel Theory
(b) Vol. 74, No. 1, p. K39—K42 (1976).
See also *2*, p. *36*; 22.3, p. 372.

J. Monecke
*On the Phase Diagram of the Hubbard Model
(b) Vol. 74, No. 1, p. K57—K60 (1976).
See also *18.4*, p. *243*.

W.-H. Steeb
High Temperature Properties of a Doubly-Degenerate Hubbard Model
(b) Vol. 74, No. 1, p. K71—K74 (1976).
See also *18.2*, p. *230*; *18.4*, p. *243*.

J. Mizia
*CPA Treatment of the Hubbard Model with Intra-Atomic Exchange Interaction
(b) Vol. 74, No. 2, p. 461—470 (1976).
See also 18.1, p. 225; 18.2, p. 230; 21.1, p. 305.

P. Petelenz
*Effective Electron–Hole Interaction in Molecular Crystals (II): Phonon Mechanism
(b) Vol. 74, No. 2, p. 655—658 (1976).
See also *6*, p. *55*; 13.5.1, p. 152; 14.3, p. 175; 22, p. 335.

J. Zieliński
*On the Existence of Spin Waves in a Model Transition Metal with Finite Anderson Mixing Potential
(b) Vol. 74, No. 2, p. 779—783 (1976).
See also *18.2*, p. *230*; 21.1, p. 305.

H. Böttger and V. V. Bryksin
On the Equivalence of Different Theoretical Descriptions of the Hopping Conductivity in Disordered Systems
(b) Vol. 74, No. 2, p. K151—K156 (1976).
See also *2*, p. *37*; 6, p. 55; *14.3*, p. *175*.

L. Skála
*On the Electron Structure of Dilute Substitutional Alloys
(b) Vol. 74, No. 2, p. K177—K180 (1976).
See also 21, p. 293.

G. Debnath and P. Rudra
*Magnetization and Susceptibility in the Truncated Anderson Model
(b) Vol. 75, No. 1, p. 145—153 (1976).
See also *18*, p. *223*.

K. TAKANAKA and R. YAMAMOTO
*The Self-Damping of RKKY Interaction in Dilute Magnetic Alloys
(b) Vol. 75, No. 1, p. 279–288 (1976).
See also *18*, p. *223*.

A. MATULIONIS, J. POŽELA, and A. REKLAITIS
*Monte Carlo Calculations of Hot-Electron Transient Behaviour in CdTe and GaAs
(a) Vol. 35, No. 1, p. 43–48 (1976).
See also *14.3.4*, p. *197*; 22.2.1, p. 363; 22.4.3, p. 385.

F. ITOH, K. KAI, M. KUROHA, and S. TAKEUCHI
Positron Annihilation in Liquid Copper–Bismuth and Copper–Tin Alloys
(b) Vol. 75, No. 2, p. 559–565 (1976).
See also *2*, p. *38*; 21.1, p. 307; 21.7, p. 332.

P. M. BISCH and A. A. GOMES
*Formal Results on the Susceptibility of Dilute Alloys (I): Green's Function Approach
(b) Vol. 75, No. 2, p. 705–715 (1976).
See also *18*, p. *223*; 21.1, p. 308.

E. KOLLEY and W. KOLLEY
*Effect of Thermal Fluctuations on the Electronic Densities of States in Disordered Binary Alloys. Numerical Results of CPA Model Calculations
(b) Vol. 75, No. 2, p. 721–730 (1976).
See also *2*, p. *38*; 6, p. 57; 21, p. 294.

L. C. DAVIS
*A Reply to "Observations on Quasi-Classical Quantization Near Self-Intersecting Orbits"
(b) Vol. 75, No. 2, p. K81–K82 (1976).
See also 21, p. 295.

G. KAMIENIARZ
*On First Order Green Function Theory of $S = 1/2$ Transverse Ising Paramagnets
(b) Vol. 75, No. 2, p. K97–K99 (1976).
See also *18.1*, p. *226*.

A. ANDREEFF and G. BRAUER
*Positron Annihilation Study of the Metal–Insulator Transition in V_2O_3 and VO_2
(b) Vol. 75, No. 2, p. K115–K118 (1976).
See also 22.6, p. 405.

J. SIVARDIÈRE
*Magnetism in $S = \frac{3}{2}$ Models: Magnetization Jumps and Reorientations, Heat Magnetization, and Tricritical Points
(b) Vol. 76, No. 1, p. 77–84 (1976).
See also *18.2*, p. *233*.

A. M. BRODSKII and M. I. URBAKH
*Adsorption-Induced Changes of Electron Levels
(b) Vol. 76, No. 1, p. 93–104 (1976).
See also *13.1*, p. *124*; 21.2, p. 323.

V. S. KOMISSAROV
*Calculation of Hot Electron Mobility and Electron Temperature in n-Type InSb under the Assumption of a Maxwellian Distribution
(b) Vol. 76, No. 1, p. 105–112 (1976).
See also *14.3.4*, p. *197*; 22.2.3, p. 368.

V. P. KALANTARIAN and V. I. CHECHERNIKOV
Nuclear Magnetic Resonance and Susceptibility Measurements of Transition Metals and Alloys: V–Ta, V–Nb, and Nb–Ta Systems
(b) Vol. 76, No. 1, p. 113–118 (1976).
See also 18, p. 223; *19*, p. *252*; 21, p. 296; 21.1, p. 309.

F. Brouers
*Alloy Analogy Approximation of the Degenerate Hubbard Model
(b) Vol. 76, No. 1, p. 145–151 (1976).
See also *18.4*, p. *244*.

G. M. Obermair and G. H. Wannier
*Bloch Electrons in Magnetic Fields. Rationality, Irrationality, Degeneracy
(b) Vol. 76, No. 1, p. 217–222 (1976).

D. Barjon, A. Raymond, and J. L. Robert
*Low-Energy Gap Semiconductor Mobilities in Presence of Spatial Fluctuations of Carrier Density
(b) Vol. 76, No. 1, p. 223–229 (1976).
See also *14.3.4*, p. *197*; 22.2.3, p. 368.

K. I. Papunashvili and E. K. Frolova
*Longitudinal Anisotropy of Conductivity of n-Si in Heating Electric Fields at Low Temperatures
(b) Vol. 76, No. 1, p. 351–355 (1976).
See also *14.3.4*, p. *197*; 22.1.2, p. 354.

I. M. Dykman and P. M. Tomchuk
*Transport Phenomena and Eigenoscillations of Hot Electrons in Semiconductors with Periodic Structure
(b) Vol. 76, No. 1, p. 385–393 (1976).
Erratum
(b) Vol. 78, No. 2, p. 843 (1976).
See also *14.3.4*, p. *197*; 22, p. 338.

P. M. Bisch and A. A. Gomes
*Formal Results on the Susceptibility of Dilute Alloys (II): Mean-Field Approach and Applications
(b) Vol. 76, No. 1, p. 395–406 (1976).
See also *18*, p. *224*; 19, p. 252; 21.1, p. 309.

F. M. Galperin
*Hyperfine Fields of the Impurities in Ferromagnetic Iron
(a) Vol. 76, No. 1, p. K9–K13 (1976).
See also *18.2*, p. *233*; 21.1, p. 309; 21.1.1, p. 319.

V. S. Lubimov, L. A. Manakova, E. F. Makarov, and S. P. Ionov
*The Thermodynamic Behaviour of Charge-Ordered Crystals
(b) Vol. 76, No. 2, p. 505–514 (1976).
See also *8*, p. *69*.

V. Bezák and J. Banský
*A Generalization of the Sondheimer-Wilson Formula for the Averaged Density Matrix and Magnetic Susceptibility of Conduction Electrons in a Model of Disordered Semiconductors
(b) Vol. 76, No. 2, p. 569–580 (1976).
See also 2, p. 39; *18*, p. *224*; 22, p. 338.

S. Nagel, K. Maschke, and A. Baldereschi
*Pseudopotential Calculation of the Electronic Valence Charge Density and the Optical Dielectric Constant of NaCl and MgO
(b) Vol. 76, No. 2, p. 629–632 (1976).
See also *14.4*, p. *203*; 22.5.2, p. 396; 22.6, p. 406.

V. V. Kaplin, D. E. Popov, and S. A. Vorobev
Bound States of Fast Electrons in Planar Channeling
(b) Vol. 76, No. 2, p. 779–786 (1976).
See also *11*, p. *103*; 22.1.2, p. 355.

R. V. Sharma and K. C. Sharma
*A Possible Closed Form Expression for the Static Structure Factor for Liquid Metals
(b) Vol. 76, No. 2, p. K75–K79 (1976).
See also *2*, p. *39*; 21.2, p. 323.

Th. Eifrig
*Cluster-CPA without Breaking of Translational Symmetry
(b) Vol. 76, No. 2, p. K101–K104 (1976).
See also 21, p. 297.

S. Robaszkiewicz
*Magnetic, Charge and Orbital Orderings in a Two-Band Model of Transition Metal Compounds
(b) Vol. 76, No. 2, p. K127–K130 (1976).
See also *18*, p. *224*.

S. Guha and M. Sathe
*Nonlinear Collision-Induced Convective Instability in Solid-State Plasmas
(a) Vol. 36, No. 2, p. 453–459 (1976).
See also *14.3.4*, p. *198*; 22, p. 339.

E. Heiner
Roth's Method and Moment Conservation of the Spectral Weight Function for the Neutral Hubbard Model
(b) Vol. 77, No. 1, p. 93–103 (1976).
See also *18.1*, p. *226*; *18.4*, p. *245*.

D. O. Welch and K. G. Lynn
*Systematic Variation of the Mean Positron Lifetime and Gaussian Fraction in Annealed Metals and Semiconductors
(b) Vol. 77, No. 1, p. 277–286 (1976).
See also *14.3*, p. *179*; 22, p. 339.

B. Riccò
*Density of States of Layer Compounds: 2H-$NbSe_2$
(b) Vol. 77, No. 1, p. 287–293 (1976).
See also 21.1, p. 311.

D. van Dyck
*The Importance of Backscattering in High-Energy Electron Diffraction Calculations
(b) Vol. 77, No. 1, p. 301–308 (1976).
See also *1.1*, p. *12*.

R. Salchow, R. Liebmann und J. Appel
*Energiebetrachtung zum Peierls-Übergang
(On the Energy in the Peierls Transition)
(b) Vol. 77, No. 1, p. 373–379 (1976).

I. Nicoară
*Effect of the s–d Interaction on the Magnetoresistance of Magnetic Alloys
(b) Vol. 77, No. 1, p. K17–K20 (1976).
See also *14*, p. *163*.

I. M. Chapnik
*On Possible Formation of Closed Electron Shells with Coupling Spins outside the Inert-Gas Shells in Some Transition Metals and Their Alloys
(a) Vol. 37, No. 1, p. K29–K32 (1976).
See also 21.1, p. 311; 21.1.1, p. 321; 22.7, p. 412.

S. P. Ionov, I. I. Amelin, V. S. Lubimov, G. V. Ionova, and E. F. Makarov
*Theoretical Analysis of the Charge-Ordered State in a Quasi-One-Dimensional and a Finite Cyclic System with the Ab-Initio LCAO-MO Method
(b) Vol. 77, No. 2, p. 441–450 (1976).

R. Pickenhain and A. Milchev
The Quadratic Response of a Fermi Gas. Electronic Charge Density of Valence Electrons in Zincblende Semiconductors
(b) Vol. 77, No. 2, p. 571–579 (1976).
See also *4*, p. *46*; 22.1.1, p. 348.

G. M. Arutyunyan and S. M. Shahinyan
*Multiphoton Absorption of Intensive Electromagnetic Waves in Semiconductors
(b) Vol. 77, No. 2, p. K171–K176 (1976).
See also *20.1*, p. *274*; 22, p. 340.

J. Schreiber and W. John
*On the Theory of Electron Localization in Amorphous Systems
(b) Vol. 78, No. 1, p. 199–210 (1976).
See also *2*, p. *40*.

W.-H. Steeb and E. Marsch
Thermodynamics of a Two-Point Doubly Degenerate Hubbard Model in the Half-Filled Case
(b) Vol. 78, No. 1, p. K39–K44 (1976).
See also *18.2*, p. *236*; *18.4*, p. *245*.

W. John and J. Schreiber
*Localization Theory for the Random Weaire Model
(b) Vol. 78, No. 2, p. 577–583 (1976).
See also *2*, p. *41*; 22, p. 341.

R. Pässler
Relationships between the Nonradiative Multiphonon Carrier-Capture Properties of Deep Charged and Neutral Centres in Semiconductors
(b) Vol. 78, No. 2, p. 625–635 (1976).
See also 6, p. 64; *13.4*, p. *145*; 22, p. 342; 22.1, p. 344; 22.2, p. 361; 22.4, p. 378.

J. Th. M. de Hosson
*The Electronic States in Molybdenum
(b) Vol. 78, No. 2, p. 791–801 (1976).
See also 21, p. 300.

E. Yang
*Electrons in Finite One-Dimensional Crystals in a Uniform Electric Field
(b) Vol. 78, No. 2, p. 803–811 (1976).

13.1 Band Structure

Review Articles

V. L. Bonch-Bruevich
*Problems of the Electron Theory of Small-Gap Disordered Semiconductors
(b) Vol. 74, No. 1, p. 11–22 (1976).
See also 2, p. 35; 22, p. 333.

J. Kadlec and K. H. Gundlach
Results and Problems of Internal Photoemission in Sandwich Structures
(a) Vol. 37, No. 1, p. 11–28 (1976).
See also *14.3.3*, p. *190*; *16*, p. *213*; 22.6, p. 402.

Original Papers and Short Notes

H. Köhler
*Non-Parabolic $E(\boldsymbol{k})$ Relation of the Lowest Conduction Band in Bi_2Te_3
(b) Vol. 73, No. 1, p. 95–104 (1976).
See also *13.2*, p. *130*; 14.3, p. 173; 22.7, p. 410.

H. Meinhold and G. Weiser
Modulation Spectroscopy on MoS_2 and $MoSe_2$
(b) Vol. 73, No. 1, p. 105–115 (1976).
See also *13.5.1*, p. *149*; *20.1*, p. *260*; 22, p. 333.

F. Bechstedt, R. Enderlein, A. Kolpakov, and K. Peuker
Theory of Resonance Raman Scattering near Critical Points
(b) Vol. 73, No. 1, p. 141–149 (1976).
See also *6*, p. *52*; 13.5.1, p. 149; *20.1*, p. *260*.

M. Taut and H. Eschrig
*Electron and Phonon Spectra of Calcium
(b) Vol. 73, No. 1, p. 151–154 (1976).
See also *6*, p. *52*; 13.2, p. 130; 21.3, p. 324.

P. Leyral, D. Bois, and P. Pinard
*Strain-Associated Band Tailing Effect in Semiconductors
(b) Vol. 73, No. 1, p. 187–192 (1976).
See also *20.1*, p. *260*; 22.2.1, p. 362.

M. Ausloos
On a Structural Phase Transition in Ytterbium
(b) Vol. 73, No. 1, p. 211–221 (1976).
See also *1.2*, p. *14*; *14.1*, p. *164*; 21.4, p. 326.

H.-J. Wünsche and K. Henneberger
Interband-Current Theory for Esaki Diodes
(b) Vol. 73, No. 1, p. 245–254 (1976).
See also *14.3.3*, p. *190*; 14.3.4, p. 195; 22.2.1, p. 362.

R. H. Williams, I. T. McGovern, R. B. Murray, and M. Howells
*An Investigation of the Electronic Structure of GaSe and GaTe by Photoelectron Spectroscopy, Using a Synchrotron Source, and Electron Energy Loss Spectroscopy
(b) Vol. 73, No. 1, p. 307–316 (1976).
See also *17*, p. *218*; 20.1, p. 260; 22.3, p. 372.

A. A. Ignatov and Yu. A. Romanov
Nonlinear Electromagnetic Properties of Semiconductors with a Superlattice
(b) Vol. **73**, No. 1, p. 327—333 (1976).
See also *14.3.4*, p. *195*; 22, p. 333.

D. Fink and R. Braunstein
*Determination of the Conduction Band Effective Masses of the Germanium–Silicon Alloy by Infrared Cyclotron Resonance
(b) Vol. **73**, No. 1, p. 361—370 (1976).
See also 22.1.1, p. 345; 22.1.2, p. 351.

C. Uher
Erratum to: Low-Field Thermomagnetic Tensor for the Bi-Like Band Structure
(b) Vol. **73**, No. 1, p. 379 (1976).
See also 6, p. 53; *15*, p. *209*; 21.7, p. 332.

M. Fritsche, M. Pöhler, M. Schubert, and G. Staupendahl
*Investigation of the Absorption of Intense CO_2 Laser Radiation by the p-Band of Tellurium
(a) Vol. **33**, No. 1, p. 149—154 (1976).
See also *20.1*, p. *261*; 22.1.3, p. 359.

W. Ekardt and M. I. Sheboul
The Influence of the Electron–Hole Exchange Interaction on the Biexciton Ground State in CuCl and CuBr and Related Optical Transitions
(b) Vol. **73**, No. 2, p. 475—482 (1976).
See also *13.5.1*, p. *150*; 20.1, p. 261; 20.3, p. 280; 22.5, p. 388.

J. von Borzeszkowski
On the Electric Field Dependence of the Conductivity of Holes in Si at Low Temperatures (I): The Drift Velocity under Consideration of the Non-Equipartition of Acoustic Phonon Energy
(b) Vol. **73**, No. 2, p. 607—614 (1976).
See also 6, p. 54; *14.3.4*, p. *195*; 22.1.2, p. 351.

H. Lange, J. Donecker, and H. Friedrich
Electroreflectance and Wavelength Modulation Study of the Direct and Indirect Fundamental Transition Region of $In_{1-x}Ga_xP$
(b) Vol. **73**, No. 2, p. 633—639 (1976).
See also *20.1*, p. *262*; 22.2.4, p. 370.

J. M. Honig, A. P. B. Sinha, W. E. Wahnsiedler, and H. Kuwamoto
*Studies of the Band Structure of NbO by X-Ray Photoelectron Spectroscopy
(b) Vol. **73**, No. 2, p. 651—654 (1976).
See also 17, p. 218; 22.6, p. 403.

B. Reimer, H. Baessler, J. Hesse, and G. Weiser
Optical Properties of a Polydiacetylene Single Crystal — Evidence for van Hove Singularities
(b) Vol. **73**, No. 2, p. 709—716 (1976).
See also *20.1*, p. *262*; 22.9, p. 433.

R. S. Kim and S. Narita
*Far-Infrared Interband Magnetoabsorption and Band Structure of $Hg_{1-x}Cd_xTe$ Alloys
(b) Vol. **73**, No. 2, p. 741—752 (1976).
See also 18, p. 222; 20.1, p. 262; 22.4.4, p. 387.

N. N. Sirota, I. V. Bodnar, A. I. Lukomskii, G. F. Smirnova, and L. M. Finkelstein
*Indirect Absorption Edge of $(GaP)_x(InAs)_{1-x}$ Mixed Crystals
(b) Vol. **73**, No. 2, p. K135—K136 (1976).
See also *20.1*, p. *262*; 22.2.4, p. 370.

I. A. SMIRNOV, R. SURYANARAYANAN, and S. G. SHULMAN
Optical Absorption of Metallic SmS Films near the Interband Transitions
(b) Vol. **73**, No. 2, p. K137–K140 (1976).
See also *20.1*, p. *262*; 21.4, p. 326.

K. HÜBNER
*Strain Dependence of the Baldereschi Point and of the Bond Charge
(b) Vol. **73**, No. 2, p. K141–K144 (1976).
See also *4*, p. *45*; 22.1, p. 343.

K. LÖSCH and J. U. FISCHBACH
Measurement of High Near-Edge Absorption Coefficients in Surface Layers of InP
(a) Vol. **33**, No. 2, p. 473–478 (1976).
See also 13.5.1, p. 150; *20.1*, p. *262*; 22.2.2, p. 365.

L. HULDT
Phonon-Assisted Auger Recombination in Germanium
(a) Vol. **33**, No. 2, p. 607–614 (1976).
See also *6*, p. *54*; 20, p. 257; 22.1.1, p. 345.

J. C. JONES and P. R. WALLACE
On the Anomalous Magneto-Resistance of Gallium Arsenide in Intense Magnetic Fields
(b) Vol. **74**, No. 1, p. 23–33 (1976).
See also *14.3*, p. *174*; 22.2.1, p. 362.

M. B. KOZINTSEVA and V. E. ZILBERVARG
*The Electron and Magnon Spectra of Ferrimagnetic Semiconductors
(b) Vol. **74**, No. 1, p. 199–208 (1976).
See also *13.5.4*, p. *160*; 18.3, p. 239; 22.8.2, p. 427.

N. WAGNER and O. BRÜMMER
*XPS Investigations for the Purpose of the Research of Positron Annihilation Effects in Bi_xSb_{1-x} Alloys Concerning Band Structure Modifications
(b) Vol. **74**, No. 1, p. 249–254 (1976).
See also 17, p. 218; 22.7, p. 410.

U. BERG, G. DRÄGER, and O. BRÜMMER
*Combined Investigations of the Valence Band Structure of Graphite by K X-Ray Emission Spectroscopy and X-Ray Photoemission Spectroscopy
(b) Vol. **74**, No. 1, p. 341–348 (1976).
See also 17, p. 218; 20, p. 257; 22.1, p. 343.

S. S. LI, D. W. SCHOENFELD, and R. T. OWEN
*Optical Transmission Measurements and Electron Microprobe Analysis of a Bulk $InAs_{1-x}P_x$ Alloy System
(a) Vol. **34**, No. 1, p. 255–261 (1976).
See also *20.1*, p. *264*; 22.2.4, p. 370.

D. PUDEWILL, F.-J. HIMPSEL, V. SAILE, N. SCHWENTNER, M. SKIBOWSKI, and E. E. KOCH
*Excitons in Pure and Doped Solid Neon
(b) Vol. **74**, No. 2, p. 485–496 (1976).
See also *13.5.1*, p. *151*; 20.1, p. 265; 23, p. 440.

M. I. DYKMAN
*Theory of Nondegenerate Two-Dimensional Electrons Interacting with Phonons in Quantizing Magnetic Fields
(b) Vol. **74**, No. 2, p. 547–559 (1976).
See also *6*, p. *55*; 20.1, p. 265.

K. HENNEBERGER
On the Theory of Photon-Assisted Zener Tunneling
(b) Vol. **74**, No. 2, p. 567—577 (1976).
See also *14.3.4*, p. *196*; 22, p. 335.

H. KÖHLER
*Non-Parabolicity of the Highest Valence Band of Bi_2Te_3 from Shubnikov-de Haas Effect
(b) Vol. **74**, No. 2, p. 591—600 (1976).
See also *13.2*, p. *130*; 14.3, p. 175; 22.7, p. 410.

H. MALTER
*On the Electronic Molecular Structure of the Organic Semiconductor Copper Phthalocyanine
(b) Vol. **74**, No. 2, p. 627—633 (1976).
See also *20.1*, p. *265*; 22.9, p. 435.

R. JAGANNATHAN and N. R. RANGANATHAN
*A New Form of Magnetic Bloch Functions and Applications to Magnetic Band Structure Calculations
(b) Vol. **74**, No. 2, p. 667—672 (1976).
See also *18*, p. *222*.

E. V. BURSIAN, YA. G. GIRSHBERG, and A. V. RUZHNIKOV
The Correlation between Optical Absorption Spectra, Carrier Mobility, and Phase Transition Temperature in Some Ferroelectrics
(b) Vol. **74**, No. 2, p. 689—693 (1976).
See also *14.4.1*, p. *206*; *20.1*, p. *266*; 22.8.1, p. 420.

S. AULUCK
*On the Pseudopotential Coefficients for Zinc and Cadmium
(b) Vol. **74**, No. 2, p. K129—K133 (1976).
See also *13.2*, p. *130*; 20.1, p. 266; 21, p. 292; 21.1, p. 305.

V. V. BHOKARE and R. M. SINGRU
*Valence Electron Momentum Distribution in Ge
(b) Vol. **74**, No. 2, p. K161—K163 (1976).
See also 22.1.1, p. 346.

V. P. BHOLA
*Thermoreflectance of Cd_3As_2
(a) Vol. **34**, No. 2, p. K179—K180 (1976).
See also *20.1*, p. *267*; 22.3, p. 372.

H. KÖHLER
*Anisotropic *g*-Factor of the Conduction Electrons in Bi_2Te_3
(b) Vol. **75**, No. 1, p. 127—136 (1976).
See also *13.2*, p. *131*; 14.3, p. 176; 22.7, p. 410.

G. OELGART and R. HERRMANN
*Cyclotron Masses in Semiconducting $Bi_{1-x}Sb_x$ Alloys
(b) Vol. **75**, No. 1, p. 189—196 (1976).
See also 13.2, p. 131; 19, p. 250; 22.7, p. 411.

R. NILSSON, A. BERNDTSSON, N. MÅRTENSSON, R. NYHOLM, and J. HEDMAN
*The Valence Band of Os Studied by Electron Spectroscopy
(b) Vol. **75**, No. 1, p. 197—203 (1976).
See also 17, p. 219; 21, p. 293.

L. B. COLEMAN, C. R. FINCHER, JR., A. F. GARITO, and A. J. HEEGER
Far-Infrared Single Crystal Studies of TTF-TCNQ
(b) Vol. **75**, No. 1, p. 239—246 (1976).
See also *20.1*, p. *267*; 22.9, p. 436.

J. Camassel, M. Schlüter, S. Kohn, J. P. Voitchovsky, Y. R. Shen, and M. L. Cohen
*Wavelength Modulation Spectra and Electronic Band Structure of SnS_2 and $SnSe_2$
(b) Vol. 75, No. 1, p. 303–314 (1976).
See also *20.1*, p. *267*; 22.7, p. 411.

D. Mukherji and B. R. Nag
*Electron Effective Mass in Semiconductor Superlattices
(b) Vol. 75, No. 1, p. K35–K37 (1976).
See also 22.2.1, p. 363.

S. I. Radautsan, E. K. Arushanov, and A. N. Nateprov
*The Conduction Band of Cadmium Phosphide
(a) Vol. 35, No. 1, p. K53–K56 (1976).
See also 14.3, p. 177; 22.3, p. 372.

V. N. Holyavko, V. P. Dragunov, B. V. Morozov, E. M. Skok, and N. B. Velchev
Study of the Size Quantization in p-Type Silicon Inversion Layers by Means of Magnetoresistance
(b) Vol. 75, No. 2, p. 423–432 (1976).
See also *14.3*, p. *177*; 22.1.2, p. 354.

H. Köhler
*The g-Factor of the Holes in the Highest Valence Band of Bi_2Te_3
(b) Vol. 75, No. 2, p. 441–449 (1976).
See also *13.2*, p. *131*; 14.3, p. 177; 22.7, p. 411.

K.-R. Schulze, H. Neumann, and K. Unger
*Band Structure of $Ga_{1-x}In_xAs$
(b) Vol. 75, No. 2, p. 493–500 (1976).
See also 22.2.4, p. 370.

M. Brauwers and F. Brouers
*Optical Relaxation Time and Effective Mass in Noble Metal Alloys
(b) Vol. 75, No. 2, p. 519–525 (1976).
See also *20*, p. *258*; 21.6, p. 330.

Kwok Leung Yip, A. B. Kunz, and W. S. Williams
*Electronic Structure of TiC and VC from the Iterative Extended Hückel Theory
(b) Vol. 75, No. 2, p. 533–541 (1976).
See also 20, p. 258; 22.3, p. 373.

G. Bauer, M. Overhamm, P. Grosse, W. Müller, E. Gornick, and H. W. Pötzl
*Landau Emission in n-InSb
(b) Vol. 75, No. 2, p. 543–551 (1976).
See also *14.3.4*, p. *197*; 20.1, p. 268; 20.3, p. 283; 22.2.3, p. 368.

U. Berg, G. Dräger, K. Mosebach, and O. Brümmer
*Combined XS and XPS Investigations of the FeS_2 Valence Band
(b) Vol. 75, No. 2, p. K89–K92 (1976).
See also 17, p. 219; 20, p. 258; 22, p. 337.

N. Wagner and O. Brümmer
*Remarks on the Relation between Crystal Structure and Electronic Band Structure in Bismuth Antimony Alloys
(b) Vol. 75, No. 2, p. K157–K161 (1976).
See also *1.1*, p. *12*; 22.7, p. 411.

A. M. Brodskii and M. I. Urbakh
*Adsorption-Induced Changes of Electron Levels
(b) Vol. 76, No. 1, p. 93–104 (1976).
See also *13*, p. *116*; 21.2, p. 323.

W. BRAUN and M. CARDONA
*Electronic Density of States of Amorphous $CuGaS_2$ and $ZnGeAs_2$
(b) Vol. 76, No. 1, p. 251–257 (1976).
See also 2, p. 38; 17, p. 219; 22.8, p. 416.

H. BURKHARD, G. BAUER, P. GROSSE, and A. LOPEZ-OTERO
*Free Carrier Magneto-Optical Effects in the Reststrahlen Region of n-PbTe
(b) Vol. 76, No. 1, p. 259–266 (1976).
See also *18*, p. *223*; *20.1*, p. *270*; 22, p. 337.

W. KÜHN
*Stress Dependence of the (E_1, $E_1 + \Delta_1$) Transition in Germanium
(b) Vol. 76, No. 1, p. 289–299 (1976).
See also 12.2, p. 110; *20.1*, p. *270*; 22.1.1, p. 347.

E. HEGENBARTH and B. SCHMIDT
Heat Conductivity Measurement at V_3Si in the Temperature Range 4.2 to 80 K
(b) Vol. 76, No. 1, p. 307–311 (1976).
See also *8*, p. *69*; 14.2, p. 170; 21, p. 296.

N. E. ALEKSEEVSKII, E. P. KRASNOPEROV, and T. MYDLARZ
Susceptibility and Superconductivity of Compounds with A-15 Structure
(b) Vol. 76, No. 1, p. 357–362 (1976).
See also *14.2*, p. *170*; *18.1*, p. *226*; 22.8, p. 416.

R. A. EVARESTOV, M. I. PETRASHEN, and E. M. LEDOVSKAYA
*Comparison of the Large Unit Cell and Small Periodic Cluster Approaches to the Quasi-molecular Calculations of the Band Spectra of Crystals
(b) Vol. 76, No. 1, p. 377–383 (1976).
See also *13.4*, p. *140*; 22.5.2, p. 395.

V. V. MITIN and E. I. TOLPYGO
Piezoresistance of Uniaxially Deformed n-Si
(b) Vol. 76, No. 1, p. 407–411 (1976).
See also *14.3*, p. *178*; 22.1.2, p. 355.

D. I. SHEKA and A. N. KOROL
Dielectric Permittivity of Semiconductors with Narrow Forbidden Gap in the Long-Wave Limit
(b) Vol. 76, No. 1, p. 413–418 (1976).
See also *14.4*, p. *203*; 22.2, p. 361.

J. SZYMAŃSKI
Erratum to: Theory of Free Carrier Absorption in n-PbSe
(b) Vol. 76, No. 1, p. 439 (1976).
See also 6, p. 58; *20.1*, p. *271*; 22, p. 338.

L. MEYER, TH. EIFRIG, U. BEHN, and W. WELLER
*Side Bands in the Hubbard-III Solution near the Band Limit
(b) Vol. 76, No. 1, p. K23–K25 (1976).
Erratum
(b) Vol. 77, No. 1, p. K79 (1976).

P. KONSIN
Ferroelectric Phase Transitions in Systems with Broad Electronic Bands Caused by Interband Electron–Phonon Interaction
(b) Vol. 76, No. 2, p. 487–496 (1976).
See also 6, p. 58; *14.4.1*, p. *207*; 22, p. 338.

G. Jungk and R. Gründler
*Determination of Optical Constants: Au
(b) Vol. 76, No. 2, p. 541–546 (1976).
See also *20.1*, p. *271*; 21.6, p. 330.

H. Höchst, A. Goldmann, S. Hüfner, and H. Malter
*X-Ray Photoelectron Valence Band Studies on Phthalocyanine Compounds
(b) Vol. 76, No. 2, p. 559–568 (1976).
See also 17, p. 219; 22.9, p. 437.

A. R. Birks, S. P. Hind, and P. M. Lee
*Band Structure Changes in Intercalates of Niobium Diselenide
(b) Vol. 76, No. 2, p. 599–604 (1976).
See also 14.2, p. 170.

A. M. Radwan and M. Taut
*Band Structure of Al and the Laves Phase $CaAl_2$
(b) Vol. 76, No. 2, p. 605–612 (1976).
See also 20, p. 258; 21, p. 297; 22.7, p. 411.

R. Pässler
Calculation of Nonradiative Multiphonon Capture Coefficients and Ionization Rates for Neutral Centres According to the Static Coupling Scheme (II): Alternative Trap Models
(b) Vol. 76, No. 2, p. 647–659 (1976).
See also *6*, p. *59*; *13.4*, p. *141*; 22, p. 339; 22.4.1, p. 381.

A. E. Lee and J. Webster
*The Soft X-Ray Appearance Potential Spectra of Diamond and Graphite
(b) Vol. 76, No. 2, p. 661–667 (1976).
See also 13.3, p. 133; 20, p. 258; 22.1, p. 344.

A. Balzarotti and A. Bianconi
*Electronic Structure of Aluminium Oxide as Determined by X-Ray Photoemission
(b) Vol. 76, No. 2, p. 689–694 (1976).
See also 17, p. 219; 22.6, p. 406.

H. G. Maguire
*Investigation of the Band Structure of Diamond Using Low-Energy Electrons
(b) Vol. 76, No. 2, p. 715–726 (1976).
See also 17, p. 219; 22.1, p. 344.

A. Jędrzejczak and T. Dietl
*Thermomagnetic Properties of n-Type and p-Type HgTe
(b) Vol. 76, No. 2, p. 737–751 (1976).
See also 14.3, p. 179; *15*, p. *210*; 22.4.3, p. 386.

T. Soma
Modified Pertubation Method for Diamond-Type Crystals
(b) Vol. 76, No. 2, p. 753–758 (1976).
See also *6*, p. *59*; 12.2, p. 111; 22.1, p. 344; 22.1.1, p. 348; 22.1.2, p. 355.

A. J. H. Eijkelenkamp and K. Vos
Reflectance Measurements on Single Crystals of PbFCl, PbFBr, and $PbBr_2$
(b) Vol. 76, No. 2, p. 769–778 (1976).
See also 13.5.1, p. 154; *20.1*, p. *272*; 22.5, p. 389.

K. Suzuki, N. Miura, S. Uchida, and S. Tanaka
*Far-Infrared Magneto-Reflection in HgTe
(b) Vol. 76, No. 2, p. 787–796 (1976).
See also 20.1, p. 272; 22.4.3, p. 386.

A. S. Filipchenko and L. P. Bolshakov
*Mobility of Holes in p-InSb Crystals
(b) Vol. 77, No. 1, p. 53—58 (1976).
See also *14.3*, p. *179*; 22.2.3, p. 368.

A. V. Lelyakov
Bose-Einstein Condensation of Free Excitons in AgBr-Type Crystals at High Excitation
(b) Vol. 77, No. 1, p. 59—70 (1976).
See also 6, p. 60; *13.5.1*, p. *154*; 20.3, p. 285; 22.5.1, p. 391.

M. Taut
*Dielectric Function of Hexagonal Metals (I): General Considerations
(b) Vol. 77, No. 1, p. 165—170 (1976).
See also *14.1*, p. *166*; 21, p. 297.

J. Shy-Yih Wang, M. Schlüter, and M. L. Cohen
*The Electronic Structure of AgCl. Energy Bands, Densities of States, and Charge Densities
(b) Vol. 77, No. 1, p. 295—300 (1976).
See also 22.5.1, p. 391.

G. Wolf, M. Zimmermann, and K. Bohmhammel
The Molar Heat Capacity of Non-Superconducting PdH_x in the Temperature Range from 2 to 7 K
(a) Vol. 37, No. 1, p. 179—182 (1976).
See also *8*, p. *70*; 21, p. 297.

K. Hübner and V. K. Bashenov
*Ionicity and Structure of Solids
(b) Vol. 77, No. 2, p. 473—482 (1976).
See also 1.1, p. 13; 2, p. 40; *4*, p. *46*; 22.2, p. 361; 22.2.4, p. 371.

F. Perrot
*Bulk Properties of Lithium Hydride up to 100 Mbar
(b) Vol. 77, No. 2, p. 517—525 (1976).
See also 12.2, p. 112; 22, p. 340.

L. Pasemann, W. Cordts, A. Heinrich, and J. Monecke
*Band Structure Calculation of $ZnSiP_2$
(b) Vol. 77, No. 2, p. 527—533 (1976).
See also 22.8, p. 417.

M. Taut and W. Hanke
Dielectric Function of Hexagonal Metals (II): Beryllium
(b) Vol. 77, No. 2, p. 543—550 (1976).
See also 13.5.2, p. 157; *14.1*, p. *166*; 14.4, p. 204; 20.1, p. 274; 21.3, p. 324.

A. Faust, C. Merkel, and H. Bross
*Energy Band Structure and Fermi Surface of Potassium
(b) Vol. 77, No. 2, p. 591—600 (1976).
See also 13.2, p. 131; 21.2, p. 323.

J. Mertsching
*Static Approximation for the Hubbard Model: Magnetic Phase Transitions
(b) Vol. 77, No. 2, p. 717—724 (1976).
See also *18*, p. *224*.

J. Humlíček and F. Lukeš
Direct Identification of the $L_3' \rightarrow L_1$ Transition in Ge
(b) Vol. 77, No. 2, p. 731—738 (1976).
See also *14.3.4*, p. *199*; *20.1*, p. *274*; 22.1.1, p. 348.

J. H. YEE and G. MYERS
*Pseudopotential Calculation of the Acoustic Deformation Potential for Several Group IV, II–VI, and III–V Compounds
(b) Vol. 77, No. 2, p. K81–K84 (1976).
See also 22.1.1, p. 349; 22.1.2, p. 356; 22.2, p. 361; 22.4, p. 377.

K. SIERAŃSKI, J. SZATKOWSKI, and E. MAJCHROWSKA
*Effective Mass of Heavy Holes in $Cd_xHg_{1-x}Te$ at 4.2 to 300 K
(b) Vol. 77, No. 2, p. K91–K95 (1976).
See also 22.4.4, p. 387.

M. KERTÉSZ, J. KOLLER, and A. AŽMAN
*Ab Initio Treatment of a Polysulfur Nitride $(SN)_x$ Chain
(b) Vol. 77, No. 2, p. K157–K160 (1976).
See also 22, p. 340.

I. V. BODNAR, A. I. LUKOMSKII, and G. F. SMIRNOVA
*Absorption Edge of InP_xAs_{1-x} Mixed Crystals
(a) Vol. 37, No. 2, p. K173–K176 (1976).
See also 20.1, p. 275; 22.2.4, p. 371.

G. HOPPMANN and E. SALJE
Optical and Electrical Properties of H^+ Doped WO_3 Single Crystals
(a) Vol. 37, No. 2, p. K187–K191 (1976).
See also *14.3*, p. *180*; *20.1*, p. *275*; 22.6, p. 408.

M. TAUT and P. ZIESCHE
Dielectric Function of Hexagonal Metals (III): Magnesium
(b) Vol. 78, No. 1, p. 169–175 (1976).
See also 13.5.2, p. 158; *14.1*, p. *167*; 14.4, p. 204; 20.1, p. 275; 21.3, p. 324.

L. RUSINIAK
Electric Paramagnetic Resonance and Combined Resonance of Conductivity Electrons in InSb at 24 GHz
(b) Vol. 78, No. 1, p. 265–270 (1976).
See also *19*, p. *255*; 20.1, p. 276; 22.2.3, p. 369.

W. RÜHLE and E. GÖBEL
*New Aspects of the Magnetoluminescence of a Band to Acceptor Transition in GaAs
(b) Vol. 78, No. 1, p. 311–317 (1976).
See also 20.3, p. 287; 22.2.1, p. 364.

J. ZIELIŃSKI
*A Double d-Band Model of Transition Metals
(b) Vol. 78, No. 1, p. 319–324 (1976).
See also *18.2*, p. *236*; 21.1, p. 313.

I. NEIZVESTNY, N. OVSYUK, and M. SINYUKOV
Electro-Reflectance Spectra of E_1, $E_1 + \Delta_1$ Transitions in Germanium
(b) Vol. 78, No. 1, p. K19–K22 (1976).
See also *14.3.4*, p. *200*; *20.1*, p. *276*; 22.1.1, p. 349.

B. BRÜCKEL, U. BIRKHOLZ, and K. ZIEGLER
Fundamental Absorption and Franz-Keldysh Effect in Silicon Telluride
(b) Vol. 78, No. 1, p. K23–K25 (1976).
See also *20.1*, p. *276*; 22, p. 341.

U. DEL PENNINO and S. MANTOVANI
*Width of the Dislocation Energy Band in Germanium
(a) Vol. 38, No. 1, p. 109–113 (1976).
See also *10.2*, p. *97*; 14.3.4, p. 200; 22.1.1, p. 349.

L. P. Mokhracheva, V. A. Tskhai, and P. V. Geld
*One-Centre Exchange Potentials in a Crystal (TiC)
(b) Vol. 78, No. 2, p. 465–469 (1976).
See also 22.3, p. 374.

P. Byszewski, J. Kowalski, and R. R. Gałązka
*Shubnikov-de Haas Oscillations in HgSe at 77 K in Pulsed Magnetic Fields
(b) Vol. 78, No. 2, p. 477–481 (1976).
See also 14.3, p. 181; 22.4.2, p. 384.

K. M. van Vliet and A. H. Marshak
Conduction Current and Generalized Einstein Relations for Degenerate Semiconductors and Metals
(b) Vol. 78, No. 2, p. 501–517 (1976).
See also *14.1*, p. *167*; *14.3*, p. *181*; 15, p. 211; 21, p. 300; 22, p. 341.

W. Wegener and L. Fritsche
*A Study of the Band Structure of Graphite Based on the Rigorous Cellular Method
(b) Vol. 78, No. 2, p. 585–597 (1976).
See also 22.1, p. 344.

H. Neddermeyer
*X-Ray Spectroscopic Investigation of the Electronic Structure of Mg–Cu Alloys
(b) Vol. 78, No. 2, p. 609–614 (1976).
See also *20*, p. *259*; 21.1, p. 314; 21.3, p. 324.

H. Köhler, W. Haigis, and A. von Middendorff
*Shubnikov-de Haas Investigations on n-Type $Bi_2(Se_xTe_{1-x})_3$
(b) Vol. 78, No. 2, p. 637–642 (1976).
See also *13.2*, p. *131*; 14.3, p. 181; 22.7, p. 412.

A. Kosakov, K. Unger, and G. Leonhardt
*Calculation of Energy-Dependent Photoionization Probabilities for Valence Electrons in Diamond and Silicon Crystals
(b) Vol. 78, No. 2, p. 681–688 (1976).
See also 17, p. 220; 20, p. 259; 22.1, p. 344; 22.1.1, p. 349; 22.1.2, p. 358.

G. Dräger and O. Brümmer
*X-Ray Spectroscopic Investigation of Beryllium by Polarized K-Emission Valence Bands
(b) Vol. 78, No. 2, p. 729–735 (1976).
See also 20, p. 259; 21.2, p. 325.

M. Schlüter, M. L. Cohen, S. E. Kohn, and C. Y. Fong
*Electronic Structure of BiI_3
(b) Vol. 78, No. 2, p. 737–747 (1976).
See also 4, p. 47; 20.1, p. 278; 22.5, p. 390.

13.2 Fermi Surfaces

Original Papers and Short Notes

H. Köhler
*Non-Parabolic $E(k)$ Relation of the Lowest Conduction Band in Bi_2Te_3
(b) Vol. 73, No. 1, p. 95–104 (1976).
See also *13.1*, p. *120*; 14.3, p. 173; 22.7, p. 410.

M. Taut and H. Eschrig
Electron and Phonon Spectra of Calcium
(b) Vol. 73, No. 1, p. 151–154 (1976).
See also *6*, p. *52*; *13.1*, p. *120*; 21.3, p. 324.

L. M. Falicov and W. Y. Hsu
*Phonons and Quasi-Periodic Static Fluctuations
(b) Vol. 73, No. 1, p. 301–306 (1976).
See also *6*, p. *53*; 14.1, p. 164; 21.1, p. 302.

G. Oelgart, R. Herrmann, H. Krüger, and R. Stegmann
*Fermi Surface and Helicons in Semiconducting $Bi_{1-x}Sb_x$ Alloys
(b) Vol. 73, No. 2, p. 615–624 (1976).
See also 14.3, p. 174; 19, p. 248; 22.7, p. 410.

H. Sato and K. Yonemitsu
Anisotropic Size Effect in Thin Aluminium Films
(b) Vol. 73, No. 2, p. 723–733 (1976).
See also *14.1*, p. *164*; 21, p. 290.

G. Oelgart, G. Schneider, W. Kraak, and R. Herrmann
*The Semiconductor–Semimetal Transition in $Bi_{1-x}Sb_x$ Alloys
(b) Vol. 74, No. 1, p. K75–K78 (1976).
See also 14, p. 162; 19, p. 249; 22.7, p. 410.

I. Ya. Dekhtyar, V. I. Silantev, S. G. Sakharova, R. G. Fedchenko, V. I. Patoka, and V. N. Kolesnik
*Fermi Surface Anisotropy and Properties of Tungsten Crystals
(b) Vol. 74, No. 2, p. 471–476 (1976).
See also 21, p. 292.

H. Köhler
*Non-Parabolicity of the Highest Valence Band of Bi_2Te_3 from Shubnikov-de Haas Effect
(b) Vol. 74, No. 2, p. 591–600 (1976).
See also *13.1*, p. *123*; 14.3, p. 175; 22.7, p. 410.

V. A. Gasparov and M. H. Harutunian
*Precision Measurements of the Fermi Surface of Copper
(b) Vol. 74, No. 2, p. K107–K110 (1976).
See also 19, p. 250; 21.1, p. 305.

S. Auluck
*On the Pseudopotential Coefficients for Zinc and Cadmium
(b) Vol. 74, No. 2, p. K129–K133 (1976).
See also *13.1*, p. *123*; 20.1, p. 266; 21, p. 292; 21.1, p. 305.

O. P. Bansal
*On the Pressure Dependence of Electron Transport in Noble Metals
(b) Vol. 75, No. 1, p. 85–90 (1976).
See also 6, p. 56; 12.1, p. 107; 21.6, p. 330.

H. KÖHLER
*Anisotropic g-Factor of the Conduction Electrons in Bi_2Te_3
(b) Vol. 75, No. 1, p. 127–136 (1976).
See also *13.1*, p. *123*; 14.3, p. 176; 22.7, p. 410.

G. OELGART and R. HERRMANN
Cyclotron Masses in Semiconducting $Bi_{1-x}Sb_x$ Alloys
(b) Vol. 75, No. 1, p. 189–196 (1976).
See also *13.1*, p. *123*; 19, p. 250; 22.7, p. 411.

H. KÖHLER
*The g-Factor of the Holes in the Highest Valence Band of Bi_2Te_3
(b) Vol. 75, No. 1, p. 441–449 (1976).
See also *13.1*, p. *124*; 14.3, p. 177; 22.7, p. 411.

G. KUKA, W. BRAUNE, and R. HERRMANN
*Cyclotron Resonance with a Slighthly Anomalous Skin Effect
(b) Vol. 76, No. 2, p. 633–638 (1976).
See also 14.1, p. 165; 21, p. 297.

K. YONEMITSU and T. MATSUDA
Electrical Resistivity of Spherical Guinier-Preston Zone
(a) Vol. 36, No. 2, p. 791–798 (1976).
See also *14.1*, p. *166*; 21, p. 297; 21.6, p. 331.

B. BOSACCHI and P. FRANZOSI
*The Electron–Phonon Enhancement Factor on the Fermi Surface of Copper
(b) Vol. 77, No. 2, p. 457–464 (1976).
See also 21.1, p. 312.

A. FAUST, C. MERKEL, and H. BROSS
Energy Band Structure and Fermi Surface of Potassium
(b) Vol. 77, No. 2, p. 591–600 (1976).
See also *13.1*, p. *127*; 21.2, p. 323.

R. KUHL, W. KRAAK, H. HAEFNER, and R. HERRMANN
*The Semimetal–Semiconductor Transition in Bismuth–Antimony Alloys
(b) Vol. 77, No. 2, p. K109–K111 (1976).
See also *14.1*, p. *166*; 19, p. 254; 21.7, p. 332.

H. KÖHLER, W. HAIGIS, and A. VON MIDDENDORFF
*Shubnikov-de Hass Investigations on n-Type $Bi_2(Se_xTe_{1-x})_3$
(b) Vol. 78, No. 2, p. 637–642 (1976).
See also *13.1*, p. *129*; 14.3, p. 181; 22.7, p. 412.

K. BERNDT and O. BRÜMMER
*X-Ray Compton Scattering from LiMg Alloys
(b) Vol. 78, No. 2, p. 659–666 (1976).
See also 20, p. 259; 21.2, p. 323; 21.3, p. 325.

13.3 Surface and Interface States

Original Papers and Short Notes

F. Weber
*On the Dispersion Relation for Surface Density Waves in Semiconductors
(a) Vol. **33**, No. 2, p. 651–660 (1976).
See also *14.3.2*, p. *187*; 14.3.4, p. 196; 22.1.1, p. 345.

M. Euler and A. Scharmann
Thermally Stimulated Exoelectron Emission (TSEE) of BeO Layers. Auger Electron Spectroscopy Investigations
(a) Vol. **34**, No. 1, p. 297–306 (1976).
See also 1.6, p. 34; *17*, p. *218*; 22.6, p. 403.

V. A. Girii, A. V. Kondrachuk, S. I. Kornyushin, V. G. Litovchenko, R. O. Litvinov, and V. I. Shakhovtsov
*Effect of Impurities on the Radiation Sensitivity of the Si–SiO_2 Interface in MIS Structures
(a) Vol. **34**, No. 1, p. K35–K38 (1976).
See also 11, p. 101; *14.3.3*, p. *191*; 22.1.2, p. 353.

M. Lannoo and J. P. Muscat
*An Application of First-Order Perturbation Theory to Metal Surfaces
(b) Vol. **74**, No. 2, p. K141–K143 (1976).
See also 21, p. 293.

J. Jerhot and V. Šnejdar
*The Influence of Interface States on an Isotype Heterojunction between Two Wide-Gap Semiconductors
(a) Vol. **34**, No. 2, p. 505–511 (1976).
See also *14.3.3*, p. *191*; 22, p. 335.

I. S. Kaulach and E. A. Silinsh
*The Influence of the Hyperfine Range Magnetic Field on Charge Transfer States at the Metal–Organic Semiconductor Interface
(b) Vol. **75**, No. 1, p. 247–254 (1976).
See also 14.3.3, p. 191; 16, p. 214; 22.9, p. 436.

A. D. Boardman, B. V. Paranjape, and Y. O. Nakamura
*Surface Plasmon-Polaritons in a Spatially Dispersive Inhomogeneous Medium
(b) Vol. **75**, No. 1, p. 347–359 (1976).
See also *13.5.2*, p. *157*; 14.3.2, p. 188.

H. Kirihata and J. Akutsu
Exoelectron Emission by Slow Electron Impacts
(a) Vol. **35**, No. 1, p. K73–K75 (1976).
See also *17*, p. *219*; 22.5.2, p. 394; 22.6, p. 405.

S. F. Timashev and M. A. Krykin
Optical and Electrooptical Properties of the Interfaces of Media
(b) Vol. **76**, No. 1, p. 67–76 (1976).
See also *14.3.2*, p. *188*; *20.1*, p. *270*.

G. D. Bagratishvili, R. B. Dzhanelidze, N. I. Kurdiani, and O. V. Saksaganskii
MIS Structure GaAs–Ge_3N_4–Al
(a) Vol. **36**, No. 1, p. 73–79 (1976).
See also *14.3.2*, p. *188*; 20.3, p. 285; 22.2.1, p. 363.

R. K. Nahar and D. Nagchoudhuri
Depletion Layer Characteristics of Heretojunctions with Built-in Homojunction
(a) Vol. **36**, No. 1, p. 139–145 (1976).
See also *14.3.3*, p. *192*; 22, p. 338.

S. Martinuzzi, O. Mallem, and T. Cabot
Dark Currents in Cu_2S–CdS Single Crystal Heterojunctions. Consequences for Photocells
(a) Vol. **36**, No. 1, p. 227–234 (1976).
See also *14.3.3*, p. *192*; *16*, p. *215*; 22, p. 338; 22.4.1, p. 381.

E. Kröger and E. Kretschmann
*Surface Plasmon and Polariton Dispersion at Rough Boundaries
(b) Vol. **76**, No. 2, p. 515–523 (1976).
See also *13.5*, p. *147*; 21.6, p. 330.

A. E. Lee and J. Webster
The Soft X-Ray Appearance Potential Spectra of Diamond and Graphite
(b) Vol. **76**, No. 2, p. 661–667 (1976).
See also *13.1*, p. *126*; 20, p. 258; 22.1, p. 344.

W. Hanus and M. Oszwałdowski
The Dependence of Photoconductive and Photoelectromagnetic Effects on Slab Thickness in InSb at Room Temperature
(a) Vol. **36**, No. 2, p. 445–452 (1976).
See also *16*, p. *215*; 22.2.3, p. 368.

K. Saminadayar and J. C. Pfister
*Surface-State Generation on MOS Capacitors Irradiated with UV Light and Electrons
(a) Vol. **36**, No. 2, p. 679–686 (1976).
See also *14.3.3*, p. *192*; 22.1.2, p. 356.

T. N. Sitenko
Non-Equilibrium Low-Temperature Depletion
(a) Vol. **36**, No. 2, p. K157–K159 (1976).
See also *14.3.2*, p. *188*; 22.2.1, p. 363.

S. N. Karyagin, P. K. Kashkarov, V. F. Kiselev, and S. N. Kozlov
*On the Nature of Electron and Hole Traps in Germanium Dioxide Thermally Grown on Germanium
(a) Vol. **37**, No. 1, p. K17–K19 (1976).
See also *19*, p. *254*; 22.6, p. 407.

H. Flietner and Ngo Duong Sinh
*Character of Si–SiO_2 Interface States from Analysis of the CV Term Spectra
(a) Vol. **37**, No. 2, p. 533–539 (1976).
See also 14.3.2, p. 188; 22.1.2, p. 357.

P. Basu and H. Saha
Possible Performance of Back-Wall Schottky Barrier Solar Cells
(a) Vol. **37**, No. 2, p. 625–631 (1976).
See also *14.3.3*, p. *193*; *16*, p. *216*; 22.1.2, p. 357; 22.2.1, p. 364; 22.4.3, p. 386.

13.4 Impurity and Defect States

Review Articles

Z. S. Kachlishvili
Galvanomagnetic and Recombination Effects in Semiconductors in a Strong Electric Field
(a) Vol. **33**, No. 1, p. 15–51 (1976).
See also *14.3.4*, p. *195*; 22, p. 333.

M. N. Sapozhnikov
Zero-Phonon Transitions in the Optical Spectra of Impurity Molecular Crystals
(b) Vol. **75**, No. 1, p. 11–51 (1976).
See also *6*, p. *52*; 20.1, p. 260; *20.3*, p. *280*; 22.9, p. 433.

P. V. Pavlov, E. I. Zorin, D. I. Tetelbaum, and A. F. Khokhlov
*Nitrogen as Dopant in Silicon and Germanium
(a) Vol. **35**, No. 1, p. 11–36 (1976).
See also 11, p. 100; *14.3*, p. *173*; 22.1.1, p. 345: 22.1.2, p. 351.

Original Papers and Short Notes

A. N. Ermoshkin, R. A. Evarestov, and E. A. Kotomin
*Molecular Cluster Approach to Magnesium and Calcium Oxide Crystals (II): F^+ and F Centres
(b) Vol. **73**, No. 1, p. 81–86 (1976).
See also *10.2*, p. *87*; 22.6, p. 402.

D. J. Mickish and A. B. Kunz
*Self-Consistent Local Orbitals for Silver Halide Crystals
(b) Vol. **73**, No. 1, p. 193–202 (1976).
See also 22.5.1, p. 391.

J. Hörner zu Siederdissen
Impurity-Induced Raman Spectra of Potassium Halides
(b) Vol. **73**, No. 1, p. 239–244 (1976).
See also *6*, p. *52*; *20.1*, p. *260*; 22.5.2, p. 392.

W. Rühle, W. Jakowetz, C. Wölk, R. Linnebach, and M. Pilkuhn
Optical Studies of Free and Bound Excitonic States in GaSb. Evidence for Deep A^+-Complexes
(b) Vol. **73**, No. 1, p. 255–264 (1976).
See also *13.5.1*, p. *149*; 20.1, p. 260; *20.3*, p. *280*; 22.2.3, p. 367.

P. Loose and M. Wöhlecke
*A Simple Approach to the Energy of Bound Exciton Complexes
(b) Vol. **73**, No. 1, p. 265–270 (1976).
See also *13.5.1*, p. *149*; 22.4, p. 376.

R. Sati
Calculations of Optical Transitions of the F_3^+ Center
(b) Vol. **73**, No. 1, p. 353–359 (1976).
See also *10.2*, p. *87*; 22.5.2, p. 392.

R. Krishnan
Optical Absorption of Co^{3+} in YIG Crystals
(b) Vol. **73**, No. 1, p. K51–K56 (1976).
See also *20.1*, p. *261*; 22.8.2, p. 426.

J. Varvas and T. Nirk
High-Temperatures Conductivity of Undoped and Copper-Doped Cadmium Selenide
(a) Vol. **33**, No. 1, p. 75–84 (1976).
See also *10.2*, p. *87*; *14.3*, p. *173*; 22.4.2, p. 383.

S. H. Hagen and A. W. C. van Kemenade
On the Role of Boron in the Luminescence of Silicon Carbide Doped with Nitrogen and Boron
(a) Vol. **33**, No. 1, p. 97–105 (1976).
See also *20.3*, p. *280*; 22.3, p. 372.

M. Tapiero, J. P. Zielinger, and C. Noguet
Photomemory Effect in Cu_2O Single Crystals. Phenomenology and Interpretation
(a) Vol. **33**, No. 1, p. 155–166 (1976).
See also *16*, p. *213*; 22.6, p. 402.

M. Baj, L. Dmowski, M. Kończykowski, and S. Porowski
A New Explanation of the Cl Donor Deionization and Its Kinetics Observed in CdTe:Cl under Hydrostatic Pressure
(a) Vol. **33**, No. 1, p. 421–426 (1976).
See also *14.3*, p. *174*; 22.4.3, p. 385.

P. Hiesinger
*Hall-Effect Levels in Ag- and Au-Doped p-Type GaAs
(a) Vol. **33**, No. 1, p. K39–K41 (1976).
See also *14.3*, p. *174*; 22.2.1, p. 362.

R. A. Evarestov, A. N. Ermoshkin, and E. A. Kotomin
*Molecular Cluster Approach to Magnesium and Calcium Oxide Crystals (III): Charge Distribution Analysis of Some Hole Centres
(b) Vol. **73**, No. 2, p. 483–486 (1976).
See also *10.2*, p. *89*; 22.6, p. 402.

B. Esser
*X-Ray Scattering from Electrons in Random Semiclassical Potential Wells
(b) Vol. **73**, No. 2, p. 499–502 (1976).
See also 2, p. 36; 22, p. 334.

B. Andlauer, J. Schneider, and W. Tolksdorf
*Optical Absorption, Fluorescence, and Electron Spin Resonance of Ir^{4+} on Octahedral Sites in $Y_3Ga_5O_{12}$
(b) Vol. **73**, No. 2, p. 533–540 (1976).
See also 19, p. 248; 20.1, p. 261; 20.3, p. 280; 22.8.2, p. 426.

I. N. Douglas
Magnetic Circular Dichroism of Co^{2+}-Doped Cubic Crystals
(b) Vol. **73**, No. 2, p. 641–650 (1976).
See also *18*, p. *222*; *20.1*, p. *262*; 22.5.2, p. 393; 22.6, p. 403.

T. Markvart
*Crystal–Impurity Energy Transport by Molecular Excitons in the Wave-Like Limit (I): Capture Mechanism
(b) Vol. **73**, No. 2, p. 689–698 (1976).
See also *13.5.1*, p. *150*; 22.9, p. 433.

K. Hattori
*Polarizability of an Electron Trapped by an Interstitial Ion
(b) Vol. **73**, No. 2, p. 717–721 (1976).
See also 22.5.1, p. 391; 22.5.2, p. 393.

K. D. GLINCHUK, N. M. LITOVCHENKO, and R. MERKER
*The Role of Impurities in the Formation of Quenched-in Recombination Centres in Thermally Treated Silicon
(a) Vol. **33**, No. 2, p. K87—K90 (1976).
See also 16, p. 213; 22.1.2, p. 352.

E. M. ROCKAR and R. A. FORMAN
$Fe^{4+}(3d^4)$ Optical Absorption in Oxidized $SrTiO_3$
(a) Vol. **33**, No. 2, p. K91—K95 (1976).
See also *20.1*, p. *263*; 22.8, p. 413.

I. BIRCHAK, V. V. SERDYUK, and G. G. CHEMERESYUK
*Optical and Temperature Quenching of Photocurrent in Zinc Selenide Single Crystals
(a) Vol. **33**, No. 2, p. K145—K149 (1976).
See also *16*, p. *213*; 22.4.2, p. 383.

M. A. IVANOV, V. YA. MITROFANOV, L. D. FALKOVSKAYA, and A. YA. FISHMAN
The Properties of a Threefold Degenerate Orbital Magnetic Impurity in a Ferromagnetic Crystal
(b) Vol. **74**, No. 1, p. 57—67 (1976).
See also *18.2*, p. *230*; 22.1, p. 343.

B. V. NOVIKOV, N. S. SOKOLOV, and S. V. GASTEV
*Free-Carrier Generation via Exciton–Phonon and Exciton–Impurity Interaction in Ge Crystals
(b) Vol. **74**, No. 1, p. 81—89 (1976).
See also 6, p. 54; *13.5.1*, p. *150*; 16, p. 214; 22.1.1, p. 345.

T. IIDA and R. MONNIER
*Static Properties of the Self-Trapped Hole in Simple Cubic Cesium Halides
(b) Vol. **74**, No. 1, p. 91—102 (1976).
See also 6, p. 54; 20.1, p. 263; 22.5.2, p. 393.

T. MARKVART
*Crystal–Impurity Energy Transport by Molecular Excitons in the Wave-Like Limit (II): Diffusion in Crystals
(b) Vol. **74**, No. 1, p. 135—143 (1976).
Erratum
(b) Vol. **76**, No. 2, p. 853 (1976).
See also *13.5.1*, p. *151*; 22.9, p. 434.

U. KAUFMANN, A. RÄUBER, and J. SCHNEIDER
ESR Analysis of IIB–IV–V_2 Semiconductors with Mn^{2+} as a Paramagnetic Probe
(b) Vol. **74**, No. 1, p. 169—175 (1976).
See also *19*, p. *248*; 22.8, p. 414.

U. GIORGIANNI, G. MONDIO, G. SAITTA, and G. VERMIGLIO
*Role of the Jahn-Teller Effect in the Excited States of KBr:Tl-Type Crystals
(b) Vol. **74**, No. 1, p. 317—322 (1976).
See also *20.1*, p. *264*; 22.5.2, p. 393.

G. ALLAN and M. LANNOO
Influence of Self-Consistency on the Relaxation near a Vacancy or a Surface in Transition Metals
(b) Vol. **74**, No. 1, p. 409—414 (1976).
See also *10.1*, p. *79*; 21.6, p. 329.

A. M. WHITE, P. J. DEAN, B. DAY, and F. MANSFIELD
*Polaron, Zeeman, and Diamagnetic Effects in the Recombination Spectra of Donor Bound Excitons in Cadmium Telluride
(b) Vol. **74**, No. 1, p. K9—K13 (1976).
See also *13.5.1*, p. *151*; 13.5.3, p. 159; 20.3, p. 281; 22.4.3, p. 385.

W. Barth, K. Elsaesser, and W. Güth
*The Optical Absorption of 60° Dislocations in Germanium
(a) Vol. **34**, No. 1, p. 153–163 (1976).
See also *20.1*, p. *264*; 22.1.1, p. 346.

M. Manzel, W. Ecke, and H. Bruchlos
Negative Differential Resistance in Gold-Doped High-Resistivity Silicon
(a) Vol. **34**, No. 1, p. 219–226 (1976).
See also 14.3.3, p. 190; *14.3.4*, p. *196*; 22.1.2, p. 352.

P. M. Karageorgy-Alkalaev, I. Z. Karimova, P. I. Knigin, and A. Yu. Leiderman
Double-Injection Currents in Zn-Compensated Silicon p–i–n Structures
(a) Vol. **34**, No. 1, p. 391–395 (1976).
See also *14.3.4*, p. *196*; 22.1.2, p. 352.

S. G. Atarbekov, A. S. Filipchenko, and M. A. Rzaev
*Deep Level in Fused Tunnel Diodes Made from InSb
(a) Vol. **34**, No. 1, p. K7–K8 (1976).
See also *14.3.3*, p. *190*; 22.2.3, p. 367.

B. I. Sysoev, A. N. Likholet, and V. F. Synorov
*Study of the Effect of Electric Field Screening on Trap Level Filling by the Capacitor Thermostimulated Discharge Method
(a) Vol. **34**, No. 1, p. K71–K75 (1976).
See also *14.3.3*, p. *191*.

B. Kozarzewski, J. Nowak, and J. Zieliński
The Coherent Potential Approximation in the Theory of Paramagnetic Alloys
(b) Vol. 74, No. 2, p. 437–442 (1976).
See also *14.1*, p. *165*; 21, p. 292.

P. Koidl
*Jahn-Teller Effect in the $^4T_1(l)$ and $^4T_2(l)$ States of Tetrahedrally Coordinated Mn^{2+}
(b) Vol. 74, No. 2, p. 477–484 (1976).
See also 20.1, p. 265; 20.3, p. 282; 22.2.2, p. 365; 22.4.1, p. 380; 22.4.2, p. 383.

B. D. Bhattacharyya
*A New Aspect of Jahn-Teller Effect in the Low-Spin Ferrihaemoglobin Azide
(b) Vol. 74, No. 2, p. 695–700 (1976).
See also 19, p. 250; 20.1, p. 266; 22.9, p. 435.

J. Krempaský and D. Barančok
The Use of Modified Relaxation Time in the Transport Theory of Non-Crystalline Semiconductors
(b) Vol. 74, No. 2, p. 741–751 (1976).
See also *2*, p. *37*; *14.3*, p. *175*; 15, p. 209; 22, p. 335.

N. Karl and H. Feederle
*Absorption Spectrum of the Tetracene Monopositive Ion in the Anthracene Lattice
(a) Vol. **34**, No. 2, p. 497–503 (1976).
See also *20.1*, p. *266*; 22.9, p. 435.

R. M. Hill
*Variable-Range Hopping
(a) Vol. **34**, No. 2, p. 601–613 (1976).
See also *14.3.1*, p. *183*; 22.1, p. 343; 22.1.1, p. 346; 22.2.1, p. 362.

Y. V. G. S. Murti, G. Samuel, and S. Sivaraman
Studies on Tin Impurity Centres in KCl Crystals
(a) Vol. **34**, No. 2, p. 615–621 (1976).
See also *10.2*, p. *91*; 19, p. 250; 20.1, p. 266; 22.5.2, p. 394.

S. Nešpurek and E. A. Silinsh
Space-Charge-Limited Current Theory for Molecular Crystals with Gaussian Distribution of Local Trapping States
(a) Vol. **34**, No. 2, p. 747–759 (1976).
See also *14.3.4*, p. *197*; 22.9, p. 435.

K. D. Glinchuk, A. V. Prokhorovich, and V. I. Vovnenko
*The Internal Radiative Efficiency and the Mechanism of Temperature Quenching of the 1.03, 1.20, and 1.30 eV Emission Bands in n-GaAs
(a) Vol. **34**, No. 2, p. 777–786 (1976).
See also *20.3*, p. *282*; 22.2.1, p. 362.

H. Schaumburg and F. Willmann
Optical Absorption of Plastically Deformed Germanium
(a) Vol. **34**, No. 2, p. K173–K177 (1976).
See also *10.2*, p. *92*; *20.1*, p. *266*; 22.1.1, p. 346.

J. Emery and J. C. Fayet
*Computation of the Two-Centre Molecular Integrals Involved in the Exchange Ínteraction between Ligand and Metal Electrons for the $[VF_6]^{4-}$ Cluster in $KMgF_3$
(b) Vol. **75**, No. 1, p. 255–259 (1976).
See also 22.8, p. 414.

M. O. M. de Matos, H. S. Brandi, and B. Maffeo
Wave Function for the Interstitial Hydrogen Atom in SrF_2
(b) Vol. **75**, No. 1, p. 271–278 (1976).
See also *19*, p. *250*; 22.5.3, p. 399.

S. Ligenza
*A Study of the $^6S_{5/2}$-Term Splitting of an Fe^{3+} Ion in Zinc Ferrite by Neutron Spectroscopy
(b) Vol. **75**, No. 1, p. 315–326 (1976).
See also 5, p. 49; 18.3, p. 240; 22.8.2, p. 428.

V. Kumar and K. Chandra
*A Study of the Nature of the Crystalline Field of the Yb^{3+} Ion in C_{3h} and D_{2d} Site Symmetries
(b) Vol. **75**, No. 1, p. K1–K5 (1976).
See also 22.8, p. 414; 22.8.1, p. 421.

Yu. A. Vodakov, G. A. Lamokina, E. N. Mokhov, E. I. Radovanova, V. I. Sokolov, M. M. Usmanova, G. F. Yuldashev, and B. S. Machmudov
*Silicon Carbide Doped with Gallium
(a) Vol. **35**, No. 1, p. 37–42 (1976).
See also 3, p. 42; 14.3, p. 176; 20.1, p. 268; 20.3, p. 283; 22.3, p. 372.

R. Nink and H.-J. Kos
On the Role of Z Centers in the Trapping Mechanism of Thermoluminescent Lithium Fluoride
(a) Vol. **35**, No. 1, p. 121–129 (1976).
See also *10.2*, p. *92*; *20.1*, p. *268*; 22.5.2, p. 394.

A. V. Ferris-Prabhu
Tunnelling Theories of Non-Volatile Semiconductor Memories
(a) Vol. **35**, No. 1, p. 243–250 (1976).
See also *14.3.3*, p. *191*; 22.1.2, p. 353.

A. N. Georgobiani, Yu. V. Ozerov, E. E. Strumban, and V. V. Anisimov
*Investigation of Deep Centres of Gallium-Doped Zinc Sulfide Crystals
(a) Vol. **35**, No. 1, p. 251–257 (1976).
See also 16, p. 214; 20.1, p. 268; 20.3, p. 283; 22.4.1, p. 380.

W. Siegel, A. Heirich, and E. Ziegler
Electron and Hole Mobility in $ZnSiP_2$
(a) Vol. **35**, No. 1, p. 269–279 (1976).
See also 6, p. 56; *14.3*, p. *176*; 22.8, p. 415.

W. F. Braerman, K. W. Böer, and G. A. Dussel
*Observation of Deep-Trap Modulation with "Near" Band-Edge Light in Copper-Doped CdS Crystals
(a) Vol. **35**, No. 1, p. 309–313 (1976).
See also *20.1*, p. *268*; 22.4.1, p. 380.

M. Ėgée
Spectral Distribution of the Green Emission Band of ZnS:Cu in Dependence on the Localization Diversity of the Radiative Recombination Processes
(a) Vol. **35**, No. 1, p. 361–370 (1976).
See also *20.3*, p. *283*; 22.4.1, p. 380.

H. Michel and M. Wagner
*Non-Thermal Stimulation of Three-Level Impurities by Phonons
(b) Vol. **75**, No. 2, p. 507–517 (1976).
See also *6*, p. *57*; 8, p. 68; 22.5.2, p. 395.

F. F. Sokolov and V. V. Hizhnyakov
Fading Memory in the Theory of Electronic Excitation Transfer
(b) Vol. **75**, No. 2, p. 669–676 (1976).
See also *20.1*, p. *268*; 22, p. 337.

V. N. Brudnyi, D. L. Budnitskii, M. A. Krivov, and V. G. Melev
p–n-Type Conversion and Optical Properties of 2.0 MeV Electron-Irradiated $ZnSnAs_2$
(a) Vol. **35**, No. 2, p. 425–430 (1976).
See also 11, p. 102; *14.3*, p. *177*; *20.1*, p. *269*; 22.8, p. 415.

S. Mantovani, U. del Pennino, and E. Mazzega
*Capacitance Measurements as a New Tool to Investigate the Electronic States of Dislocations in Semiconductors
(a) Vol. **35**, No. 2, p. 451–457 (1976).
See also 14.3, p. 177; 22.1.1, p. 347.

M. Pugnet, J. Barbolla, J. C. Brabant, F. Saint-Yves, and M. Brousseau
*Energy Levels for Platinum and Palladium in Silicon Measured by the Dark Transient Capacitance Technique
(a) Vol. **35**, No. 2, p. 533–543 (1976).
See also *14.3.3*, p. *191*; 22.1.2, p. 354.

I. S. Gorban, G. A. Grishchenko, A. P. Sakalas, and I. I. Tychina
*Conducting and Semi-Insulating n- and p-Type $ZnSiP_2$ Single Crystals
(a) Vol. **35**, No. 2, p. 571–574 (1976).
See also 14.3, p. 177; 22.8, p. 415.

A. P. Levanyuk and V. V. Osipov
Optical Distortion in Crystals
(a) Vol. **35**, No. 2, p. 605–614 (1976).
See also *14.4.1*, p. *207*; *20.1*, p. *269*; 22.8.1, p. 422.

N. Dupont-Pavlovsky, F. Caralp, P. Delhaes et J. Amiell
*Caractérisation chimique et étude des propriétés électroniques de l'oxyde de zinc non stoechiométrique. Nature et répartition des défauts
(Chemical Characterization and Study of the Electronic Properties on Non-Stoichoimetric Zinc Oxide. Nature and Repartition of Defects)
(a) Vol. **35**, No. 2, p. 615–625 (1976).
See also *10.2*, p. *93*; 14.3, p. 177; 18, p. 223; 22.4, p. 376.

M. Kh. Ashurov, Yu. K. Voronko, V. V. Osiko, A. A. Sobol, B. P. Starikov, M. I. Timoschechkin, and A. Ya. Yablonskii
*Inequivalent Luminescence Centres of Er^{3+} in Gallium Garnet Single Crystals
(a) Vol. **35**, No. 2, p. 645–649 (1976).
See also 20.1, p. 269; 20.3, p. 284; 22.8.2, p. 429.

K. Somogyi
Relation between the Activation Energies ($\varepsilon_1 E_1$) and the Temperature of Carrier Mobility Maximum in Semiconductors. A New Method for Determination of E_1
(a) Vol. **35**, No. 2, p. 659–666 (1976).
See also *14.3*, p. *178*; 22.1.2, p. 354; 22.2.2, p. 365.

K. D. Glinchuk, N. M. Litovchenko, and R. Merker
*Effect of Quenching Rate and Annealing on the Concentration of Quenched-in Recombination Centres in Heat Treated Silicon
(a) Vol. **35**, No. 2, p. K157–K160 (1976).
See also 16, p. 214; 22.1.2, p. 354.

V. V. Samartsev, R. G. Usmanov, I. Kh. Khadiev, E. F. Kustov, and M. N. Baranov
Photon Echo in $CaWO_4:Nd^{3+}$ and $LiAl_5O_8:Cr^{3+}$
(b) Vol. **76**, No. 1, p. 55–66 (1976).
See also *20.1*, p. *270*; *20.3*, p. *284*; 22.8.1, p. 422.

F. El Akkad
*Free-to-Bound Transitions in Lithium-Doped Zinc Telluride
(b) Vol. **76**, No. 1, p. 85–92 (1976).
See also *20.3*, p. *284*; 22.4.3, p. 385.

G. Dokter and H. J. Hoffmann
*Transient Bleaching of the F-Band in Alkali Halides
(b) Vol. **76**, No. 1, p. 241–249 (1976).
See also *10.2*, p. *93*; 16, p. 215; 22.5.2, p. 395.

E. Sigmund and M. Wagner
An Operator Method for Optical Lineshape Calculations in Non-Adiabatic Electron–Phonon Systems
(b) Vol. **76**, No. 1, p. 325–336 (1976).
See also 6, p. 58; *20.1*, p. *270*.

R. A. Evarestov, M. I. Petrashen, and E. M. Ledovskaya
*Comparison of the Large Unit Cell and Small Periodic Cluster Approaches to the Quasimolecular Calculations of the Band Spectra of Crystals
(b) Vol. **76**, No. 1, p. 377–383 (1976).
See also *13.1*, p. *125*; 22.5.2, p. 395.

G. Ruszczynski and R. Boyn
Jahn-Teller Coupling Parameters from Stress- Dichroism Spectra for $CdS:Ti^{2+}$ and $CdS:Co^{2+}$
(b) Vol. **76**, No. 1, p. 427–436 (1976).
See also *20.1*, p. *271*; 22.4.1, p. 380.

F. M. Gashimzade and K. A. Rustamov
Erratum to: Selection Rules for the Spinel Structure
(b) Vol. **76**, No. 1, p. 437 (1976).
See also *6*, p. *58*; 22.8.2, p. 429.

V. V. Hizhnyakov
*Effect of Hot Transfer on Vibrational Relaxation of Electronic Excitation
(b) Vol. **76**, No. 1, p. K69–K72 (1976).
See also 20.3, p. 285; 22, p. 338.

L. Hernández, O. Vigil, and F. Gónzalez
*Optical Transitions and Distribution of Localized Levels in $ZnIn_2S_4$
(a) Vol. **36**, No. 1, p. 33—37 (1976).
See also 16, p. 215; *20.1*, p. *271*; 22.8, p. 416.

G. Müller
*Optical and Electrical Spectroscopy of Zinc Oxide Crystals Simultaneously Doped with Copper and Donors
(b) Vol. **76**, No. 2, p. 525—532 (1976).
See also 14.3, p. 178; 20.1, p. 271; 22.4, p. 377.

R. Pässler
*Calculation of Nonradiative Multiphonon Capture Coefficients and Ionization Rates for Neutral Centres According to the Static Coupling Scheme (II): Alternative Trap Models
(b) Vol. **76**, No. 2, p. 647—659 (1976).
See also *6*, p. *59*; 13.1, p. 126; 22, p. 339; 22.4.1, p. 381.

M. N. Botvinko and M. A. Ivanov
*Spectrum of Magnetic Excitations of a Ferromagnet with Weakly Bound Antiferromagnetic Interstitial Impurities
(b) Vol. **76**, No. 2, p. 797—809 (1976).
See also 13.5.4, p. 160; *18.2*, p. *234*.

R. Boyn
Tetrahedral Model for Stress Dichroism Due to Deep Impurities in Wurtzite Type Crystals
(b) Vol. **76**, No. 2, p. K119—K122 (1976).
See also *20.1*, p. *272*; 22.4, p. 377.

C. J. F. Graf, B. Maffeo, and H. S. Brandi
*Calculation of Magnetic Hyperfine Parameters for the H_i^0 Center in CaF_2, SrF_2 and BaF_2
(b) Vol. **76**, No. 2, p. K137—K140 (1976).
See also *19*, p. *253*; 22.5.3, p. 400.

F. Sprenger
Magnetoresistance of Frozen-in Photoconductivity in Semiconducting CdF_2:In
(a) Vol. **36**, No. 2, p. 483—490 (1976).
See also *14.3*, p. *179*; *16*, p. *215*; 22.5, p. 389.

J. Barbolia, M. Pugnet, J. C. Brabant, and M. Brousseau
*Thermal Capture Cross-Section of Free Electrons at Neutral Gold Centres in n-Type Silicon
(a) Vol. **36**, No. 2, p. 495—498 (1976).
See also 14.3.3, p. 192; 22.1.2, p. 355.

Ľ. Hrivnák
The Switching Effect in Crystalline Semiconductors Due to the Screening of Donors by Conduction Electrons
(a) Vol. **36**, No. 2, p. 551—556 (1976).
See also *14.3.4*, p. *198*; 22, p. 339.

D. Lapraz and P. Iacconi
On Some Luminescent and Optical Properties of Synthetic Calcite Single Crystals
(a) Vol. **36**, No. 2, p. 603—616 (1976).
See also *20.1*, p. *272*; *20.3*, p. *285*; 22.8.1, p. 423.

H. Krause
*Electron Injection and Trap Filling in Insulating Layers
(a) Vol. **36**, No. 2, p. 705—715 (1976).
See also *14.3.3*, p. *192*; 22.6, p. 406.

A. SAMOĆ, M. SAMOĆ, and J. SWORAKOWSKI
*Interpretation of One-Carrier Thermally Stimulated Currents and Isothermal Decay Currents (I): Basic Concepts
(a) Vol. **36**, No. 2, p. 735—745 (1976).
See also 14.3.4 p. 198; 14.4, p. 203.

V. S. GRUNIN, G. D. DAVTYAN, V. A. IOFFE, and I. B. PATRINA
*EPR of Cu^{2+} and Radiation Centres in Anatase (TiO_2)
(b) Vol. 77, No. 1, p. 85—92 (1976).
See also *19*, p. *253*; 22.6, p. 406.

J. J. ROUSSEAU and J. C. FAYET
*Influence of the Crystal Field on Bent Configuration of V_K Centres in AMF_3 Compounds
(b) Vol. 77, No. 1, p. 195—201 (1976).
See also *4*, p. *46*; 19, p. 253; 22.8, p. 417.

J. E. LOWTHER
Mechanisms of Energy Transfer between Rare Earth Ions
(b) Vol. 77, No. 1, p. 359—366 (1976).
See also *2*, p. *39*; 6, p. 60; 21.4, p. 327.

W. E. HAGSTON and H. H. SUTHERLAND
Properties of an Exciton Bound in an Asymmetric Electric Dipole Field
(b) Vol. 77, No. 1, p. K53—K56 (1976).
See also *13.5.1*, p. *155*; 22.4, p. 377.

M. RIGGIN, S. RADHAKRISHNA, and P. W. WHIPPEY
Thermoluminescence of X-Irradiated Alkali Fluoride Crystals
(a) Vol. 37, No. 1, p. 51—56 (1976).
See also *20.3*, p. *286*; 22.5.2, p. 396.

A. SAMOĆ, M. SAMOĆ, J. SWORAKOWSKI, J. M. THOMAS, and J. O. WILLIAMS
*Interpretation of One-Carrier Thermally Stimulated Currents and Isothermal Decay Currents (II): Application to the Evaluation of Trap Parameters in Anthracene
(a) Vol. 37, No. 1, p. 271—278 (1976).
See also 14.3.4, p. 199; 14.4, p. 204; 22.9, p. 438.

G. HOLZAPFEL and M. KRYSTEK
*Thermally Stimulated Luminescence and Exoelectron Emission from Barium and Strontium Sulphate Doped with Europium
(a) Vol. 37, No. 1, p. 303—312 (1976).
See also 17, p. 219; 20.3, p. 286; 22.8.1, p. 424.

A. KIVERIS, Š. KUDŽMAUSKAS, and P. PIPINYS
*Release of Electrons from Traps by an Electric Field with Phonon Participation
(a) Vol. 37, No. 1, p. 321—327 (1976).
See also 6, p. 61; 14.3.4, p. 199; *20.3*, p. *286*; 22.4.1, p. 381.

N. T. CHERPAK, V. V. MISHENKO, and T. A. SMIRNOVA
Inversion Ratio in Andalusite Crystals at 6 mm Wave Band
(a) Vol. 37, No. 1, p. K37—K38 (1976).
See also *19*, p. *254*; 22.8, p. 417.

E. MROSAN and G. LEHMANN
*Localized Non-Magnetic Impurities in Metals (I): General Theory
(b) Vol. 77, No. 2, p. 607—614 (1976).
See also 21, p. 298.

E. König and R. Schnaking
*Zero-Field Splitting of $^6S(d^5)$ Ions in Tetragonal and Rhombic Symmetry (I): Complete d-Electron Calculation of the Crystal-Field Dependence of D and E
(b) Vol. 77, No. 2, p. 657—666 (1976).
See also 18.1, p. 227; 19, p. 254.

V. Soriano and D. Galland
*Photosensitivity of the EPR Spectrum of the F^+ Center in ZnO
(b) Vol. 77, No. 2, p. 739—743 (1976).
See also *19*, p. *254*; 22.4, p. 377.

E. Mrosan and G. Lehmann
*Calculation of Residual Resistivity and Single Temperatures for Cu-Based 3d-Transition Metal Dilute Alloys
(b) Vol. 77, No. 2, p. K161—K164 (1976).
See also *14.1*, p. *167*; 21.1, p. 312.

V. A. Brodovoi, N. Z. Derikot, B. R. Kiyak, G. P. Peka, and A. I. Proskura
*Trigger Effect in Electroluminescent Ceramics ZnSe–Cu, I
(a) Vol. 37, No. 2, p. 363—369 (1976).
See also *14.3.4*, p. *199*; 20.3, p. 287; 22.4.2, p. 383.

Pham Minh Tan
*Bestimmung von Ionenimplantationsprofilen aus der Messung der differentiellen Kapazität am Halbleiter–Elektrolyt-Kontakt
(Determination of Ion Implantation Profiles from Mesurements of the Differential Capacity at the Semiconductor–Electrolyte Contact)
(a) Vol. 37, No. 2, p. 439—444 (1976).
See also 11, p. 104; *14.3.3*, p. *193*; 22.1.2, p. 356.

M. Manzel and W. Ecke
The Influence of Trapped Space Charge on the Double-Injection Switching
(a) Vol. 37, No. 2, p. 479—484 (1976).
See also *14.3.4*, p. *199*; 22.1.2, p. 356.

H. P. Braun
Current Oscillations in Photo-Oxidized Anthracene Crystals
(a) Vol. 37, No. 2, p. 553—559 (1976).
See also *14.3.4*, p. *199*; 16, p. 216; 22.9, p. 438.

F. M. Caimi, J. O. Artman, W. W. Lin, and D. P. Casasent
Interpretation of Direct Writing M Center H-D Curve Characteristics of Photodichroic NaF Crystals
(a) Vol. 37, No. 2, p. 617—624 (1976).
See also *10.2*, p. *96*; 20.1, p. 275; 22.5.2, p. 397.

R. Meaudre and M. Meaudre
*Capacitive Effects in Insulators or Semiconductors. Influence of the Density of States in the Gap
(a) Vol. 37, No. 2, p. 633—643 (1976).
See also *14.3.4*, p. *199*; 22.4, p. 377.

A. S. Filipchenko and A. Naurizbaev
*Moss-Burstein Effect in n-Type Crystals of Indium Antimonide
(a) Vol. 37, No. 2, p. K139—K142 (1976).
See also 20.1, p. 275; 22.2.3, p. 369.

K. Ziegler and U. Birkholz
Photoconductivity of Si_2Te_3 Single Crystals
(a) Vol. 37, No. 2, p. K147—K149 (1976).
See also *16*, p. *216*; 22, p. 340.

H. Kuwabara, K. Yamanaka, and S. Yamada
Donor–Acceptor Pair Emission from β-SiC Doped with Gallium
(a) Vol. **37**, No. 2, p. K157—K161 (1976).
See also *20.3*, p. *287*; 22.3, p. 374.

C. Hamann, Ch. Tantzscher, and G. Lehmann
A Model of the Electrical Transport Phenomena in Imperfect Crystals of Copper Phthalocyanine (IV): Concentration of Charge Carriers and Current–Voltage Characteristics
(b) Vol. **78**, No. 1, p. 79—84 (1976).
See also *14.3*, p. *180*; *14.3.1*, p. *185*; 22.9, p. 438.

V. P. Fedorov, V. V. Kind, and A. V. Korshunov
The Multiplet Structure of the Electron Absorption Spectra of Naphthalene–β-Naphthol Mixed Molecular Crystals
(b) Vol. **78**, No. 1, p. 93—101 (1976).
See also 13.5.1, p. 155; *20.1*, p. *275*; 22.9, p. 438.

E. Mrosan and G. Lehmann
*Localized Non-Magnetic Impurities in Metals (II): Application of 3d-Transition Metal Impurities in Aluminium
(b) Vol. **78**, No. 1, p. 159—167 (1976).
See also 8, p. 71; 14.1, p. 167; 14.2, p. 172; 15, p. 211; 21, p. 299.

K. P. Meletov and E. F. Sheka
*Asymmetry of Phonon Side Bands of Optical Transitions in Impurity Molecular Crystals Caused by a Change in External Phonon Frequencies
(b) Vol. **78**, No. 1, p. K1—K5 (1976).
See also *6*, p. *63*; 20.1, p. 276; 20.3, p. 287; 22.9, p. 439.

H. Maier
*Jahn-Teller Effect, Stacking Faults, and the Formation of Polytypes in ZnS
(b) Vol. **78**, No. 1, p. K65—K68 (1976).
See also *6*, p. *63*; 20.1, p. 276; 20.3, p. 287; 22.4.1, p. 382.

S. S. Paliwal and O. P. Sinha
Calculation of Activation Energy and Frequency Factor from Glow Curves of SrS:Hf Phosphors
(a) Vol. **38**, No. 1, p. 73—76 (1976).
See also *10.2*, p. *97*; *20.3*, p. *287*; 22, p. 341.

A. N. Georgobiani, R. G. Maev, Yu. V. Ozerov, and E. E. Strumban
*Investigation of Deep Centres of Chlorine-Doped Zinc Sulfide Crystals
(a) Vol. **38**, No. 1, p. 77—83 (1976).
See also 16, p. 217; 20.1, p. 276; 22.4.1, p. 382.

P. Guyot and J. P. Simon
*Symmetrical High-Angle Tilt Boundary Energy Calculation in Aluminium and Lithium
(a) Vol. **38**, No. 1, p. 207—216 (1976).
See also 10.1, p. 85; 21, p. 300; 21.2, p. 323.

Y. Ochiai and E. Matsuura
ESR in Heavily Doped n-Type Silicon near a Metal–Nonmetal Transition
(a) Vol. **38**, No. 1, p. 243—252 (1976).
See also *19*, p. *255*; 22.1.2, p. 357.

J. Świątek
*Electrical Conduction Mechanism in Thin Coronene Films
(a) Vol. **38**, No. 1, p. 285—291 (1976).
See also *14.3.1*, p. *185*; 22.9, p. 439.

T. Boţilă and H. K. Henisch
Study of Trapping in Amorphous Materials
(a) Vol. 38, No. 1, p. 331–337 (1976).
See also *2*, p. *40*; 14.3, p. 181; 16, p. 217; 22.8, p. 418.

H. Gollnast, G. Richter, and H. Herrmann
Nitrogen Concentration in GaP Diodes Determined by Photovoltage Measurements
(a) Vol. 38, No. 1, p. K89–K91 (1976).
See also *16*, p. *217*; 22.2.2, p. 366.

K. W. Blazey
Optical Absorption of $SrTiO_3:Fe^{4+}$
(a) Vol. 38, No. 1, p. K97–K99 (1976).
See also *20.1*, p. *277*; 22.8.1, p. 424.

V. E. Bozhevolnov, L. N. Ivanov, V. K. Kozlov, Yu. V. Voronov, Yu. P. Timofeev, and V. V. Karelin
Energy Transfer in CaF_2–Gd, Mn Crystals
(b) Vol. 78, No. 2, p. 483–487 (1976).
See also *20.3*, p. *288*; 22.5.4, p. 401.

J. Kossut
*Electron Transport Phenomena in Narrow- and Zero-Gap Semiconductors Containing Magnetic Impurities
(b) Vol. 78, No. 2, p. 537–542 (1976).
See also *14.3*, p. *181*; 15, p. 211; 22.4.3, p. 386.

E. König and R. Schnakig
*Zero-Field Splitting of $^6S(d^5)$ Ions in Tetragonal and Rhombic Symmetry (II): Application of Complete d-Electron Calculations to the Zero-Field Splitting in $Mn:MnF_2$ and $Mn:ZnF_2$
(b) Vol. 78, No. 2, p. 543–551 (1976).
See also 18.1, p. 227; 19, p. 255; 20.1, p. 277; 22.5, p. 390.

L. Moseley, R. S. Tripathi, and T. Lukes
*Theory of the Hopping Conductivity in Non-Crystalline Systems
(b) Vol. 78, No. 2; p. 563–567 (1976).
See also *2*, p. *41*; *14*, p. *163*.

R. Pässler
*Relationships between the Nonradiative Multiphonon Carrier-Capture Properties of Deep Charged and Neutral Centres in Semiconductors
(b) Vol. 78, No. 2, p. 625–635 (1976).
See also 6, p. 64; 13, p. 119; 22, p. 342; 22.1, p. 344; 22.2, p. 361; 22.4, p. 378.

J. P. Buisson, S. Lefrant, A. Sadoc, L. Taurel, and M. Billardon
Raman Scattering by KI Containing F Centres
(b) Vol. 78, No. 2, p. 779–790 (1976).
See also 6, p. 64; *10.2*, p. *98*; 20.1, p. 278; 22.5.2, p. 398.

A. M. Ziatdinov, M. M. Zaripov, Yu. V. Yablokov, and R. L. Davidovich
*The Depression of the Dynamic Jahn-Teller Effect
(b) Vol. 78, No. 2, p. K69–K71 (1976).
See also 1.2, p. 20; *6*, p. *64*; 19, p. 256; 22.8, p. 418.

C. Manfredotti, R. Murri, A. Quirini, and L. Vasanelli
Photoelectronic Properties of n-GaSe
(a) Vol. 38, No. 2, p. 685–693 (1976).
See also 14.3, p. 182; *16*, p. *217*; 22.3, p. 374.

G. Nimtz and E. Tyssen
Far Infrared Recombination Radiation from n-Type $Hg_{1-x}Cd_xTe$
(a) Vol. **38**, No. 2, p. K171–K173 (1976).
See also *20.1*, p. *278*; 22.4.4, p. 387.

M. Constantinescu, C. Ghita, L. Ghita, and I. Baltog
*Localized Levels in the Energy Gap of PbI_2 Crystals Evidenced by the TSC Method
(a) Vol. **38**, No. 2, p. K175–K178 (1976).
See also 22.5, p. 390.

13.5 Elementary Excitations (Phonons See 6)

Original Papers and Short Notes

W. D. KRAEFT and W. FENNEL
*Stability of an Electron-Hole Plasma
(b) Vol. **73**, No. 2, p. 487—494 (1976).

D. I. PUSHKAROV and KH. I. PUSHKAROV
Defecton Diffusion in the Very Quantum Region
(b) Vol. **74**, No. 2, p. K97—K100 (1976).
See also *9*, p. *72*.

V. S. LUBIMOV, G. V. IONOVA, O. M. PACHEV, L. A. MANAKOVA, E. F. MAKAROV, and S. P. IONOV
*Low-Lying Collective Electronic Excitations in Semiconducting Charge-Ordered Crystals
(b) Vol. **75**, No. 1, p. 91—99 (1976).
See also 22, p. 336.

R. ZIMMERMANN and M. RÖSLER
*Theory of Electron-Hole Plasma in CdS
(b) Vol. **75**, No. 2, p. 633—645 (1976).
See also 13.5.2, p. 157; *20.3*, p. *284*; 22.4.1, p. 380.

E. WIENER-AVNEAR
*Polaritons in an Ensemble of CuI Crystallites
(b) Vol. **75**, No. 2, p. 717—720 (1976).
See also 20.1, p. 269; 22.5, p. 389.

A. S. DAVYDOV and N. I. KISLUKHA
*Solitons in One-Dimensional Molecular Chains
(b) Vol. **75**, No. 2, p. 735—742 (1976).

F. HENNEBERGER and J. VOIGT
Polariton Effects in the Optical Spectra of Excitonic Molecules
(b) Vol. **76**, No. 1, p. 313—323 (1976).
See also *13.5.1*, p. *153*; *20.1*, p. *270*; 20.3, p. 284; 22, p. 337.

L. HOLZ, K. KNEIPP, A. LAU, and W. WERNCKE
Stimulated Raman Scattering of Picosecond Laser Pulses by Polaritons in $LiIO_3$ and $LiNbO_3$ Single Crystals
(a) Vol. **36**, No. 1, p. K5—K8 (1976).
See also *6*, p. *58*; 20.1, p. 271; 20.2, p. 279; 22.8.1, p. 422.

E. KRÖGER and E. KRETSCHMANN
*Surface Plasmon and Polariton Dispersion at Rough Boundaries
(b) Vol. **76**, No. 2, p. 515—523 (1976).
See also *13.3*, p. *133*; 21.6, p. 330.

G. BORSTEL, E. SCHULLER, and H. J. FALGE
*Surface Phonon Polaritons on Absorbing Crystals
(b) Vol. **76**, No. 2, p. 759—768 (1976).
See also *20.1*, p. *272*; 22.6, p. 406.

H. G. WESSJOHANN
*Relativistic Excitations of a Thin Graphite Film by Inelastic Electron Scattering in Theory and Experiment
(b) Vol. **77**, No. 2, p. 535—542 (1976).
See also 14.4, p. 204; 22.1, p. 344.

W. Ebeling, W. D. Kraeft, D. Kremp, and K. Kilimann
*Coexisting Phases in an Electron–Hole Plasma
(b) Vol. 78, No. 1, p. 241–253 (1976).
See also *13.5.1*, p. *155*; 22, p. 341.

S. Permogorov and V. Travnikov
*Resonant Scattering of Exciton-Polaritons in CdS Crystals
(b) Vol. 78, No. 1, p. 389–398 (1976).
See also *6*, p. *63*; 20.3, p. 287; 22.4.1, p. 381.

13.5.1 Excitons

Original Papers and Short Notes

H. Meinhold and G. Weiser
*Modulation Spectroscopy on MoS_2 and $MoSe_2$
(b) Vol. **73**, No. 1, p. 105–115 (1976).
See also 13.1, p. 120; *20.1*, p. *260*; 22, p. 333.

F. Bechstedt, R. Enderlein, A. Kolpakov, and K. Peuker
Theory of Resonance Raman Scattering near Critical Points
(b) Vol. **73**, No. 1, p. 141–149 (1976).
See also *6*, p. *52*; 13.1, p. 120; *20.1*, p. *260*.

M. I. Sheboul and W. Ekardt
*Theory of Biexcitons Using the Boson Formalism. The Ground State Binding Energy
(b) Vol. **73**, No. 1, p. 165–178 (1976).
See also 22.5, p. 388.

W. Rühle, W. Jakowetz, C. Wölk, R. Linnebach, and M. Pilkuhn
*Optical Studies of Free and Bound Excitonic States in GaSb. Evidence for Deep A^+-Complexes
(b) Vol. **73**, No. 1, p. 255–264 (1976).
See also 13.4, p. 134; 20.1, p. 260; *20.3*, p. *280*; 22.2.3, p. 367.

P. Loose and M. Wöhlecke
*A Simple Approach to the Energy of Bound Exciton Complexes
(b) Vol. **73**, No. 1, p. 265–270 (1976).
See also *13.4*, p. *134*; 22.4, p. 376.

G. B. Abdullaev, T. G. Kerimova, Sh. S. Mamedov, T. R. Mechtiev, R. Kh. Nani, and E. Y. Salaev
*Study of the Fundamental Absorption Edge of $CdInGaS_4$ by Wavelength Modulation
(b) Vol. **73**, No. 1, p. K69–K72 (1976).
See also *20.1*, p. *261*; 22.8, p. 413.

R. B. Hammond, T. C. McGill, and J. W. Mayer
Properties of the Electron–Hole Liquid Luminescence in Si Double Injection Diodes
(a) Vol. **33**, No. 1, p. 59–66 (1976).
See also 14.3.4, p. 195; *20.3*, p. *280*; 22.1.2, p. 351.

M. Leppihalme and T. Tuomi
Photovoltage Spectra of Metal–Silicon (p-Type) Diodes near the Indirect Absorption Edge of Silicon
(a) Vol. **33**, No. 1, p. 125–133 (1976).
See also 6, p. 53; *14.3.3*, p. *190*; *16*, p. *213*; 20.1, p. 261; 22.1.2, p. 351.

K. Marinova, A. Opanowicz, and W. Ruppel
Photoconductivity of Highly Excited CdSe and $CdS_{0.48}Se_{0.52}$ Crystals
(a) Vol. **33**, No. 1, p. 195–204 (1976).
See also *16*, p. *213*; 22.4.2, p. 383; 22.4.4, p. 387.

Yu. B. Gaididei, V. M. Loktev, A. F. Prikhotko, and L. I. Shanskii
*α-Oxygen Doublets (II): Experiment
(b) Vol. **73**, No. 2, p. 415–425 (1976).
See also *20.1*, p. *261*; 23, p. 440.

W. Ekardt and M. I. Sheboul
*The Influence of the Electron–Hole Exchange Interaction on the Biexciton Ground State in CuCl and CuBr and Related Optical Transitions
(b) Vol. **73**, No. 1, p. 475–482 (1976).
See also **13.1**, p. 121; 20.1, p. 261; 20.3, p. 280; 22.5, p. 388.

P. Petelenz
*Effective Electron–Hole Interaction in Molecular Crystals (I)
(b) Vol. **73**, No. 2, p. 565–573 (1976).
See also 6, p. 54; *13*, p. *114*; 14.3, p. 174; 22, p. 334.

R. A. Neville and B. L. Evans
*The Band Edge Excitons in 2H-MoS_2
(b) Vol. **73**, No. 2, p. 597–606 (1976).
See also 20.1, p. 262; 22, p. 334.

T. Markvart
*Crystal–Impurity Energy Transport by Molecular Excitons in the Wave-Like Limit (I): Capture Mechanism
(b) Vol. **73**, No. 2, p. 689–698 (1976).
See also *13.4*, p. *135*; 22.9, p. 433.

A. I. Onipko
*Annihilation of Incoherent Excitons in Molecular Crystals with Allowance for Their Interaction
(b) Vol. **73**, No. 2, p. 699–707 (1976).

R. Brodmann, G. Tolkiehn, and G. Zimmerer
*Luminescence Originating from Free and Self-Trapped Excitons in Solid Xenon
(b) Vol. **73**, No. 2, p. K99–K101 (1976).
See also *20.3*, p. *280*; 23, p. 440.

K. Lösch and J. U. Fischbach
Measurement of High Near-Edge Absorption Coefficients in Surface Layers of InP
(a) Vol. **33**, No. 2, p. 473–478 (1976).
See also 13.1, p. 122; *20.1*, p. *262*; 22.2.2, p. 365.

B. V. Novikov, N. S. Sokolov, and S. V. Gastev
*Free-Carrier Generation via Exciton–Phonon and Exciton–Impurity Interaction in Ge Crystals
(b) Vol. **74**, No. 1, p. 81–89 (1976).
See also 6, p. 54; *13.4*, p. *136*; 16, p. 214; 22.1.1, p. 345.

D. V. Kapor, S. D. Stojanović, M. J. Škrinjar, and B. S. Tošić
*A New Approach to Exciton–Phonon Interaction and Some Consequences
(b) Vol. **74**, No. 1, p. 103–110 (1976).
See also *6*, p. *54*.

T. Chong and N. Itoh
Interaction between Excitons and Radiation-Induced Radicals in Naphthalene Single Crystals
(b) Vol. **74**, No. 1, p. 111–120 (1976).
See also *20.3*, p. *281*; 22.9, p. 434.

P. Reineker
*To the Theory of the ESR Lineshape of Triplet Excitons in Molecular Crystals. Influence of the Quasi-Incoherent Exciton Motion and Comparison with Experimental Data
(b) Vol. **74**, No. 1, p. 121–134 (1976).
See also *19*, p. *248*; 22.9, p. 434.

T. MARKVART
*Crystal–Impurity Energy Transport by Molecular Excitons in the Wave-Like Limit (II): Diffusion in Crystals
(b) Vol. 74, No. 1, p. 135–143 (1976).
Erratum
(b) Vol. 76, No. 2, p. 853 (1976).
See also *13.4*, p. *136*; 22.9, p. 434.

J. ALIZON, G. BERTHET, J. P. BLANC, J. GALLICE, and H. ROBERT
*Exciton Dynamics Study by Magnetic Resonance in $TCNQ^- \cdot NH_4^+$
(b) Vol. 74, No 1, p. 217–224 (1976).
See also *19*, p. *249*; 22.9, p. 434.

N. KRISTIANPOLLER, A. REHAVY, and M. ISRAELI
Effects of Ultraviolet Irradiation on KBr:Gd
(b) Vol. 74, No. 1, p. 245–248 (1976).
See also *10.2*, p. *89*; 20.1, p. 263; 20.3, p. 281; 22.5.2, p. 393.

J. K. PRIBRAM and C. M. PENCHINA
*Phonon-Assisted Electroabsorption by Contact Excitons
(b) Vol. 74, No. 1, p. 375–381 (1976).
See also 6, p. 55; 14.3.4, p. 196; 20.1, p. 264.

W. B. WHITTEN and S. ARNOLD
*Pressure Modulation of Exciton Fission in Tetracene
(b) Vol. 74, No. 1, p. 401–407 (1976).
See also 20.3, p. 281; 22.9, p. 434.

A. M. WHITE, P. J. DEAN, B. DAY, and F. MANSFIELD
*Polaron, Zeeman, and Diamagnetic Effects in the Recombination Spectra of Donor Bound Excitons in Cadmium Telluride
(b) Vol. 74, No. 1, p. K9–K13 (1976).
See also *13.4*, p. *136*; 13.5.3, p. 159; 20.3, p. 281; 22.4.3, p. 385.

R. A. ELLIOTT
Erratum to: Comment on "Diffusion Theory of Luminescent Emmission from a Doped Organic Solid"
(b) Vol. 74, No. 1, p. K87–K88 (1976).
See also *20.3*, p. *281*; 22.9, p. 434.

M. S. BRODIN, M. A. DUDINSKII, S. V. MARISOVA, and E. N. MYASNIKOV
*Polariton Scattering and Formation of Exciton Luminescence Spectrum in Anthracene Crystals
(b) Vol. 74, No. 2, p. 453–460 (1976).
See also *20.3*, p. *282*; 22.9, p. 435.

D. PUDEWILL, F.-J. HIMPSEL, V. SAILE, N. SCHWENTNER, M. SKIBOWSKI, and E. E. KOCH
*Excitons in Pure and Doped Solid Neon
(b) Vol. 74, No. 2, p. 485–496 (1976).
See also *13.1*, p. *122*; 20.1, p. 265; 23, p. 440.

W. EKARDT and M. I. SHEBOUL
*On the First Excited State of the Biexciton in CuCl
(b) Vol. 74, No. 2, p. 523–529 (1976).
See also 20.3, p. 282; 22.5, p. 388.

CH. ACKERMANN, R. BRODMANN, U. HAHN, A. SUZUKI, and G. ZIMMERER
*Photoluminescence Excitation Spectra of Solid Krypton
(b) Vol. 74, No. 2, p. 579–590 (1976).
See also *20.3*, p. *282*; 23, p. 440.

W. DIETRICH and D. SCHMID
Electron Spin Resonance of the Triplet State of Pure and Doped $NaNO_2$ Single Crystals
(b) Vol. 74, No. 2, p. 609–616 (1976).
See also *19*, p. *250*; 22.8.1, p. 420.

P. PETELENZ
Effective Electron–Hole Interaction in Molecular Crystals (II): Phonon Mechanism
(b) Vol. 74, No. 2, p. 655–658 (1976).
See also *6*, p. *55*; *13*, p. *115*; 14.3, p. 175; 22, p. 335.

S. DEONARINE and S. J. JOSHUA
Symmetry Properties of the Planar Antiferromagnet K_2NiF_4
(b) Vol. 74, No. 2, p. 659–665 (1976).
See also *13.5.4*, p. *160*; 18.4, p. 243; 22.5.4, p. 401.

R. M. WESTERVELT
*Nucleation Phenomena in Electron–Hole Drop Formation in Ge and Si (I): Nucleation Rates
(b) Vol. 74, No. 2, p. 727–739 (1976).
See also 22.1.1, p. 346; 22.1.2, p. 353.

I. A. MARKOVA-OSORGINA and S. Z. SHMURAK
Deformation-Induced Exciton Luminescence of Alkali Halide Crystals
(b) Vol. 74, No. 2, p. 753–760 (1976).
See also 10.2, p. 90; *20.3*, p. *282*; 22.5.2, p. 394.

P. LOOSE, M. ROSENZWEIG, and M. WÖHLECKE
*Zeeman Effect of Bound Exciton Complexes in ZnO
(b) Vol. 75, No. 1, p. 137–144 (1976).
See also *20.1*, p. *267*; 20.3, p. 282; 22.4, p. 376.

J. VOIGT, M. SENONER, and I. RÜCKMANN
*Quantitative Analysis of Interference Structures in the Transmission Spectrum of Very Thin CdS Platelets
(b) Vol. 75, No. 1, p. 213–219 (1976).
See also 20.1, p. 267; 22.4.1, p. 380.

W. WARDZYŃSKI and K. PATAJ
*Bound Exciton–Neutral Acceptor Complex in ZnTe
(b) Vol. 75, No. 1, p. 341–346 (1976).
See also 12.2, p. 110; *20.3*, p. *283*; 22.4.3, p. 385.

E. SKAISTYS and L. VALKŪNAS
*Screening Effects in the Exciton–Electron Luminescence Spectra
(b) Vol. 75, No. 1, p. K23–K26 (1976).
See also *20.3*, p. *283*; 22.4.1, p. 380.

M. P. LISITSA, A. P. ZAKHARCHUK, S. F. TEREKHOVA, G. G. TSEBULYA, L. K. MLADOV, and S. M. TODOROV
Anisotropy of the Absorption Edge in Germanium Selenide Single Crystals
(b) Vol. 75, No. 1, p. K51–K54 (1976).
See also *20.1*, p. *267*; 22, p. 336.

J. L. STAEHLI
*Electron–Hole Droplet Formation Kinetics in Pure Germanium
(b) Vol. 75, No. 2, p. 451–463 (1976).
See also 20.3, p. 283; 22.1.1, p. 347.

A. OPANOWICZ, K. MARINOVA, H. LIEBING, and W. RUPPEL
*Luminescence of Highly Excited ZnTe Crystals
(b) Vol. 75, No. 2, p. 471–481 (1976).
See also *20.3*, p. *283*; 22.4.3, p. 385.

H. Ueba and S. Ichimura
*Plasmon–Phonon-Coupled Mode Sideband of Excitons
(b) Vol. 75, No. 2, p. 501–506 (1976).
See also 6, p. 57; 13.5.2, p. 157; *20.1*, p. *268*; 22, p. 336.

F. R. Kessler, J. Metzdorf, and R. H. Müller
Interband Faraday Rotation in Germanium. Screening of Excitons by Free Carriers
(b) Vol. 75, No. 2, p. 567–572 (1976).
See also *18*, p. *223*; *20*, p. *258*; 22.1.1, p. 347.

W. Arden, M. Kotani, and L. M. Peter
*Triplet Exciton Decay Processes in Crystalline Tetracene
(b) Vol. 75, No. 2, p. 621–631 (1976).
See also 20.3, p. 284; 22.9, p. 436.

U. Gösele
*Comment on "Diffusion Theory of Luminescent Emision from a Doped Organic Solid"
(b) Vol. 75, No. 2, p. K129–K132 (1976).
See also *20.3*, p. *284*; 22.9, p. 436.

R. M. Westervelt
*Nucleation Phenomena in Electron–Hole Drop Formation in Ge and Si (II): Application to Observable Phenomena
(b) Vol. 76, No. 1, p. 31–43 (1976).
See also 22.1.1, p. 347; 22.1.2, p. 354.

A. J. H. Eijkelenkamp
Powder Reflection Measurements of Some Divalent Lead Halides
(b) Vol. 76, No. 1, p. 153–160 (1976).
See also *20.1*, p. *270*; 22.5, p. 389.

R. Zimmermann
*Excitons and Electron–Hole Plasma: A Ground State Calculation
(b) Vol. 76, No. 1, p. 191–205 (1976).
See also 13.5.2, p. 157; 22, p. 337.

O. Goede, M. Blaschke, and K. H. Klohs
*Piezoemission Study of I_1- and I_2-Type Bound Excitons in CdS
(b) Vol. 76, No. 1, p. 267–280 (1976).
See also *20.3*, p. *284*; 22.4.1, p. 380.

K. Hattori
*Exciton Binding Energies in Polar Crystals
(b) Vol. 76, No. 1, p. 281–287 (1976).
See also 22.2.1, p. 363; 22.4, p. 376; 22.5, p. 389.

F. Henneberger and J. Voigt
*Polariton Effects in the Optical Spectra of Excitonic Molecules
(b) Vol. 76, No. 1, p. 313–323 (1976).
See also 13.5, p. 147; *20.1*, p. *270*; 20.3, p. 284; 22, p. 337.

A. I. Bobrysheva, S. A. Moskalenko, and V. I. Vybornov
*The Shift of the Exciton Level and the Biexciton Formation in Crystals under High Density Excitation
(b) Vol. 76, No. 1, p. K51–K56 (1976).
See also 22, p. 338.

A. Anedda and E. Fortin
*Temperature Shift of Excitons in CdS.
(a) Vol. 36, No. 1, p. 385–389 (1976).
See also 22.4.1, p. 381.

P. I. KHADZHI, K. G. PETRASHKU, and S. A. MOSKALENKO
*Biexciton Luminescence with Interserial Transition of the Formed Exciton
(b) Vol. 76, No. 2, p. 497—504 (1976).
See also *20.3*, p. *285*; 22.4.1, p. 381.

E. RIEFLIN, G. STÜHMER, and E. SCHWAB
*Exciton Band Width versus Peak Shift in NH_4Br in Phases II and III as Measured by Thermoreflectance
(b) Vol. 76, No. 2, p. 547—550 (1976).
See also 20.1, p. 271; 22.5, p. 389.

R. DINGES, D. FRÖHLICH, and CH. UIHLEIN
*Two-Photon Absorption of Anisotropic Excitons in β-AgI
(b) Vol. 76, No. 2, p. 613—622 (1976).
See also *20.1*, p. *271*; 22.5.1, p. 391.

H.-J. WÜNSCHE and F. HENNEBERGER
*Transmission and Reflectivity of Highly Excited Semiconductors Due to the Optical Conversion of Excitons into Excitonic Molecules
(b) Vol. 76, No. 2, p. 639—646 (1976).
See also *20.1*, p. *272*; 22, p. 339.

A. J. H. EIJKELENKAMP and K. VOS
Reflectance Measurements on Single Crystals of PbFCl, PbFBr, and $PbBr_2$
(b) Vol. 76, No. 2, p. 769—778 (1976).
See also 13.1, p. 126; *20.1*, p. *272*; 22.5, p. 389.

H. J. HESSE, W. FUHS, G. WEISER, and L. VON SZENTPALY
*Directional Dispersion of an Exciton Polariton in a Pentamethinium Cyanine Dye
(b) Vol. 76, No. 2, p. 817—826 (1976).
See also 20.1, p. 272; 22.9, p. 437.

G. O. BERIM and A. R. KESSEL
*The Model Hamiltonian and Some Properties of Magnetic Resonance of Triplet Excitons (I): Non-Localized Exciton Approximation
(b) Vol. 76, No. 2, p. 827—839 (1976).
See also *19*, p. *253*.

W. EKARDT and M. I. SHEBOUL
*Influence of the Electron–Hole Exchange Interaction on the Biexciton Ground State in CdS
(b) Vol. 76, No. 2, p. K89—K91 (1976).
See also 22.4.1, p. 381.

F. F. SOKOLOV
*Memory Function of Frenkel Exciton in the Presence of Exciton–Phonon Interaction
(b) Vol. 76, No. 2, p. K131—K135 (1976).
See also *6*, p. *59*.

A. V. LELYAKOV
*Bose-Einstein Condensation of Free Excitons in AgBr-Type Crystals at High Excitation
(b) Vol. 77, No. 1, p. 59—70 (1976).
See also 6, p. 60; 13.1, p. 127; 20.3, p. 285; 22.5.1, p. 391.

F. SPIEGELBERG, E. GUTSCHE, and J. VOIGT
*Exciton–Phonon Interaction in CdS
(b) Vol. 77, No. 1, p. 233—242 (1976).
See also *6*, p. *60*; 20.1, p. 272; 22.4.1, p. 381.

J. Donecker and J. Kluge
Magnetic Circular Dichroism of Indirect Excitonic Transitions in GaP and $In_{1-x}Ga_xP$
(b) Vol. 77, No. 1, p. 243–248 (1976).
See also *18*, p. *224*; *20.1*, p. *273*; 22.2.2, p. 365; 22.2.4, p. 371.

E. L. Frankevich, M. M. Tribel, and I. A. Sokolik
Photoconductivity of Sublimated Rubrene Films during Oxidation. The Magnetic Field Effect
(b) Vol. 77, No. 1, p. 265–276 (1976).
See also *14.3.1*, p. *185*; *16*, p. *215*; 20.3, p. 286; 22.9, p. 437.

R. Levy, C. Klingshirn, E. Ostertag, Vu Duy Phach, and J. B. Grun
*Luminescence of a "Cold" Gas of Biexcitons in CuCl
(b) Vol. 77, No. 1, p. 381–391 (1976).
See also *20.3*, p. *286*; 22.5, p. 390.

A. Opanowicz and K. Marinova
*Luminescence of ZnTe Single Crystals under High Two-Photon Excitation
(b) Vol. 77, No. 1, p. K35–K38 (1976).
See also *20.3*, p. *286*; 22.4.3, p. 386.

W. E. Hagston and H. H. Sutherland
*Properties of an Exciton Bound in an Asymmetric Electric Dipole Field
(b) Vol. 77, No. 1, p. K53–K56 (1976).
See also 13.4, p. 142; 22.4, p. 377.

V. G. Litovchenko
*On Superconducting State Peculiarities at the Surface of the Electron–Hole Droplets
(b) Vol. 77, No. 1, p. K71–K76 (1976).
See also *14.2*, p. *171*.

V. L. Alperovich, V. M. Zaletin, A. F. Kravchenko, and A. S. Terekhov
*The Influence of Phonons and Impurities on the Broadening of Excitonic Spectra in Gallium Arsenide
(b) Vol. 77, No. 2, p. 465–472 (1976).
See also *20.1*, p. *274*; 22.2.1, p. 364.

D. Klähn and K. Kilimann
*The Damping of Excitons by Phonons
(b) Vol. 77, No. 2, p. 483–493 (1976).
See also 6, p. 61; 22.1.1, p. 348.

V. P. Fedorov, V. V. Kind, and A. V. Korshunov
The Multiplet Structure of the Electron Absorption Spectra of Naphthalene-β-Naphthol Mixed Molecular Crystals
(b) Vol. 78, No. 1, p. 93–101 (1976).
See also 13.4, p. 144; *20.1*, p. *275*; 22.9, p. 438.

G. A. Gehring and J. Laugsch
Theory of Side-Bands in a Cooperative Jahn-Teller Material
(b) Vol. 78, No. 1, p. 211–218 (1976).
See also 6, p. 62; *20.1*, p. *275*; 22.8.1, p. 424.

W. Ebeling, W. D. Kraeft, D. Kremp, and K. Kilimann
*Coexisting Phases in an Electron–Hole Plasma
(b) Vol. 78, No. 1, p. 241–253 (1976).
See also *13.5*, p. *148*; 22, p. 341.

P. Petelenz
*Mixing of Frenkel Excitons and Ionic Excited States of a Linear Molecular Crystal with Two Molecules in the Unit Cell (I): General Model
(b) Vol. 78, No. 2, p. 489–500 (1976).
See also 20.1, p. 277; 22, p. 341.

S. Fujiwara, T. Nakayama, and N. Itoh
*Exciton Interactions in Naphthalene Single Crystals
(b) Vol. 78, No. 2, p. 519–528 (1976).
See also 20.3, p. 288; 22.9, p. 439.

B. Hönerlage, C. Klingshirn, and J. B. Grun
*Spontaneous Emission Due to Exciton-Electron Scattering in Semiconductors
(b) Vol. 78, No. 2, p. 599–608 (1976).
See also *20.3*, p. *288*; 22.4, p. 378; 22.4.1, p. 382; 22.4.2, p. 384.

J. Weber
*Indirect Near Edge Emission in Pure AgBr
(b) Vol. 78, No. 2, p. 699–709 (1976).
See also 6, p. 64; *20.3*, p. *288*; 22.5.1, p. 391.

13.5.2 Plasmons

Original Papers and Short Notes

M. Giehler and E. Jahne
*Effect of Damping on the Plasmon–Phonon Coupling in CdS and GaP
(b) Vol. 73, No. 2, p. 503–516 (1976).
See also *6*, p. *54*; 20.1, p. 261; 22.2.2, p. 365; 22.4.1, p. 379.

R. Ruppin and H. Yatom
*Size and Shape Effects on the Broadening of the Plasma Resonance Absorption in Metals
(b) Vol. 74, No. 2, p. 647–654 (1976).
See also *20.1*, p. *265*; 21, p. 292.

B. Rheinländer
Volume Plasma Faraday Effect in n-Type GaAs
(b) Vol. 75, No. 1, p. 261–270 (1976).
See also *18*, p. *223*; *20.1*, p. *267*; 22.2.1, p. 363.

A. D. Boardman, B. V. Paranjape, and Y. O. Nakamura
*Surface Plasmon-Polaritons in a Spatially Dispersive Inhomogeneous Medium
(b) Vol. 75, No. 1, p. 347–359 (1976).
See also *13.3*, p. *132*; 14.3.2, p. 188.

H. Ueba and S. Ichimura
Plasmon–Phonon-Coupled Mode Sideband of Excitons
(b) Vol. 75, No. 2, p. 501–506 (1976).
See also 6, p. 57; *13.5.1*, p. *153*; *20.1*, p. *268*; 22, p. 336.

E. Gerlach, P. Grosse, M. Rautenberg, and W. Senske
Dynamical Conductivity and Plasmon Excitation in Bi
(b) Vol. 75, No. 2, p. 553–558 (1976).
See also *14.1*, p. *165*; *20.1*, p. *268*; 21.7, p. 332.

R. Zimmermann and M. Rösler
Theory of Electron–Hole Plasma in CdS
(b) Vol. 75, No. 2, p. 633–645 (1976).
See also *13.5*, p. *147*; *20.3*, p. *284*; 22.4.1, p. 380.

R. Zimmermann
Excitons and Electron–Hole Plasma: A Ground State Calculation
(b) Vol. 76, No. 1, p. 191–205 (1976).
See also *13.5.1*, p. *153*; 22, p. 337.

S. Zawisz
*Cluster Approximation for Non-Homogeneous Disordered Systems
(b) Vol. 77, No. 1, p. 367–372 (1976).
See also *18.2*, p. *234*.

Th. Kokkinakis and G. C. Papavassiliou
*Surface Plasmons in Small Cu Particles
(b) Vol. 77, No. 1, p. K49–K51 (1976).
See also 20.1, p. 273; 21.1, p. 311.

M. Taut and W. Hanke
Dielectric Function of Hexagonal Metals (II): Beryllium
(b) Vol. 77, No. 2, p. 543–550 (1976).
See also *13.1*, p. *127*; *14.1*, p. *166*; 14.4, p. 204; 20.1, p. 274; 21.3, p. 324.

R. F. EGERTON

Measurement of Inelastic/Elastic Scattering Ratio for Fast Electrons and Its Use in the Study of Radiation Damage
(a) Vol. 37, No. 2, p. 663–668 (1976).
See also *11*, p. *104*; 21, p. 299; 21.3, p. 324; 21.7, p. 332; 22.9, p. 438.

M. TAUT and P. ZIESCHE

Dielectric Function of Hexagonal Metals (III): Magnesium
(b) Vol. 78, No. 1, p. 169–175 (1976).
See also *13.1*, p. *128*; *14.1*, p. *167*; 14.4, p. 204; 20.1, p. 275; 21.3, p. 324.

13.5.3 Polarons

Review Articles

H. BÖTTGER and V. V. BRYKSIN
Hopping Conductivity in Ordered and Disordered Solids (I)
(b) Vol. 78, No. 1, p. 9–56 (1976).
See also 2, p. 35; 6, p. 52; 13, p. 113; *14.3*, p. *173*; 22, p. 333.

H. BÖTTGER and V. V. BRYKSIN
Hopping Conductivity in Ordered and Disordered Solids (II)
(b) Vol. 78, No. 2, p. 415–451 (1976).
See also 2, p. 35; 6, p. 52; 13, p. 113; *14.3*, p. *173*; 22, p. 333.

Original Papers and Short Notes

J. RÖSELER, K. HENNEBERGER, and R. STREHLOW
*On the Theory of Electroabsorption in Polar Crystals
(b) Vol. 73, No. 2, p. 541–552 (1976).
See also *14.3.4*, p. *195*; *20.1*, p. *262*; 22, p. 334.

A. M. WHITE, P. J. DEAN, B. DAY, and F. MANSFIELD
Polaron, Zeeman, and Diamagnetic Effects in the Recombination Spectra of Donor Bound Excitons in Cadmium Telluride
(b) Vol. 74, No. 1, p. K9–K13 (1976).
See also *13.4*, p. *136*; *13.5.1*, p. *151*; 20.3, p. 281; 22.4.3, p. 385.

N. UHLE and J. WINDSCHEIF
Optical Absorption in V_2O_5 Doped with Li and Na
(b) Vol. 74, No. 1, p. K15–K18 (1976).
See also *20.1*, p. *264*; 22.6, p. 403.

E. P. POKATILOV and L. V. TARAKANOVA
*F-Centre Ground State at the Ellipsoidal Dispersion Law of Electrons
(b) Vol. 75, No. 2, p. K143–K148 (1976).
See also *10.2*, p. *93*.

R. H. PRINCE
*Low-Temperature Diffusion of Electrons in Ice. Evidence for Polaron Bands
(b) Vol. 78, No. 1, p. 271–276 (1976).
See also 6, p. 62; 22, p. 341.

H. FINKENRATH, K. KRUG, and N. UHLE
*Optical Absorption by Free Polarons in ZnO
(b) Vol. 78, No. 1, p. K27–K30 (1976).
See also *20.1*, p. *276*; 22.4, p. 377.

I. VILFAN
On the Electron-Intramolecular Vibration Coupling in Aromatic Molecular Crystals
(b) Vol. 78, No. 2, p. K131–K135 (1976).
See also *6*, p. *64*; *14.3*, p. *182*; 22.9, p. 439.

13.5.4 Magnons

Review Article

W. HAUBENREISSER, TH. KLUPSCH, and F. VOIGT
Quantum Statistical Theory of Parametric Spin-Wave Excitations in Magnetoelastic Ferromagnets under Parallel Pumping (II): Calculation of Macroscopic Observables
(b) Vol. **73**, No. 1, p. 11–64 (1976).
See also 6, p. 52; *13*, p. *113*; *18.2*, p. *228*.

Original Papers and Short Notes

M. B. KOZINTSEVA and V. E. ZILBERVARG
*The Electron and Magnon Spectra of Ferrimagnetic Semiconductors
(b) Vol. **74**, No. 1, p. 199–208 (1976).
See also *13.1*, p. *122*; 18.3, p. 239; 22.8.2, p. 427.

I. G. GOCHEV
*Dyson-Maleev Transformation and Three-Magnon Problem for an $s = \frac{1}{2}$ Ferromagnet
(b) Vol. **74**, No. 1, p. 311–316 (1976).

S. DEONARINE and S. J. JOSHUA
*Symmetry Properties of the Planar Antiferromagnet K_2NiF_4
(b) Vol. **74**, No. 2, p. 659–665 (1976).
See also 13.5.1, p. 152; 18.4, p. 243; 22.5.4, p. 401.

C. B. DE ARAUJO
*On the Coherence Properties of Parametric Magnon States
(b) Vol. **75**, No. 1, p. 327–332 (1976).
See also 18.2, p. 231.

A. I. BELYAEVA, M. M. KOTLYARSKII, E. A. POPOV, and B. V. BEZNOSIKOV
Spectroscopic Investigation of the Low Temperature Phase Transition in $NaMnCl_3$
(b) Vol. **75**, No. 2, p. K123–K127 (1976).
See also *20.1*, p. *269*; 22.8, p. 415.

A. I. BELYAEVA and M. M. KOTLYARSKII
*Spectroscopic Investigations of Magnons in Weak Ferromagnetic $RbMnCl_3$
(b) Vol. **76**, No. 1, p. 419–426 (1976).
See also *20.1*, p. *271*; 22.5, p. 389.

V. V. ZOLOTAREV
*Magnon Damping Caused by Spin–Spin Exchange Interaction
(b) Vol. **76**, No. 1, p. K31–K33 (1976).
See also 18.2, p. 233.

G. AERS, A. D. BOARDMAN, and E. D. ISAAC
*A Monte Carlo Study of the Gunn Effect Behaviour of a Ferromagnetic Semiconductor
(a) Vol. **36**, No. 1, p. 357–364 (1976).
See also *14.3.4*, p. *198*; 22.8.2, p. 429.

M. N. BOTVINKO and M. A. IVANOV
Spectrum of Magnetic Excitations of a Ferromagnet with Weakly Bound Antiferromagnetic Interstitial Impurities
(b) Vol. **76**, No. 2, p. 797–809 (1976).
See also *13.4*, p. *141*; *18.2*, p. *234*.

B. S. Tošić and M. M. Marinković
*On the Existence of Local Levels in a Heisenberg Ferromagnet
(b) Vol. 76, No. 2, p. K85—K88 (1976).
See also *18.2*, p. *234*.

G. S. Poo and D. M. Edwards
*Electron Interactions and the Spin-Wave Stiffness Constant in Ferromagnetic Two-Component Alloys
(b) Vol. 78, No. 1, p. 287—299 (1976).
See also *18.2*, p. *235*; 21.1, p. 313.

S. Kraśnicki
*Neutron Investigation of Magnons in the Cu_2MnAl Heusler Alloy
(b) Vol. 78, No. 1, p. 335—340 (1976).
See also 18.2, p. 236; 21.1, p. 313.

14. Electrical Properties. Transport Phenomena

Original Papers and Short Notes

W. Löser
A Self-Consistent Cluster Approximation for the Disordered Hubbard Model
(b) Vol. **73**, No. 1, p. 179–185 (1976).
See also *2*, p. *35*; 13, p. 113; *18*, p. *222*.

Ya. A. Iosilevskii
*Construction of the Reciprocal Fourth-Rank Tensors in Crystals
(b) Vol. **73**, No. 2, p. 673–680 (1976).
See also *20*, p. *257*.

S. Matthies
*Über die Berechnung des elektrostatischen Potentials in Kristallgittern (I): Grundlagen der Berechnungsmethode
(On the Calculation of the Electrostatic Potential in Crystal Lattices (I): Basic Formulae of the Method)
(b) Vol. **74**, No. 1, p. 69–80 (1976).
See also 22, p. 335.

G. Oelgart, G. Schneider, W. Kraak, and R. Herrmann
The Semiconductor–Semimetal Transition in $Bi_{1-x}Sb_x$Alloys
(b) Vol. **74**, No. 1, p. K75–K78 (1976).
See also *13.2*, p. *130*; 19, p. 249; 22.7, p. 410.

S. Matthies
*Über die Berechnung des elektrostatischen Potentials in Kristallgittern (II): Kubische Gitter – Berechnungsmethoden für die Summen $\sigma_{l,m}^{(n_0)}$
(On the Calculation of the Electrostatic Potential in Crystal Lattices (II): Cubic Lattice – Calculation Method for the Sum $\sigma_{l,m}^{(n)}$)
(b) Vol. **74**, No. 2, p. 531–546 (1976).
See also 22, p. 335.

G. Röpke, T. Frauenheim, and W. Schiller
*Electrical Conductivity in the Hubbard Model Using the Functional Integral Technique
(b) Vol. **74**, No. 2, p. K101–K105 (1976).

V. K. Arora and K. L. Gomber
*Quantum Transport Equation for Electric Conductivity
(b) Vol. **74**, No. 2, p. K111–K114 (1976).

L. M. Valiev, I. G. Kerimov, N. G. Aliev, and A. A. Abdurragimov
The Thermal Expansion and Electrical Properties of $Cu_4Mn_2X_4$(X:S, Se, and Te)
(a) Vol. **35**, No. 1, p. K85–K88 (1976).
See also 8, p. 68; 15, p. 210; *18.1*, p. *226*; *18.4*, p. *244*; 22.8, p. 415.

P. Ganguly, Om Parkash, and C. N. R. Rao
*Electron Transport and Magnetic Properties of Rare-Earth Ortho-Titanites and -Vanadites, $LnTiO_3$ and $LnVO_3$
(a) Vol. **36**, No. 2, p. 669–678 (1976).
See also 15, p. 210; *18*, p. *224*; 22.8.1, p. 423.

V. I. Litvinov·
*The Kondo Effect in a Magnetic Field
(b) Vol. **77**, No. 1, p. 71–76 (1976).
See also *18*, p. *224*; 21, p. 297.

V. K. Arora and E. Quershi
*Quantum Theory of Microwave Conductivity
(b) Vol. 77, No. 1, p. 77–84 (1976).
See also *6*, p. *60*.

J. C. Phillips
*Structure and Conductivity of Bilinear Organic Compounds
(b) Vol. 77, No. 1, p. 259–264 (1976).
See also *1*, p. *9*; 22.9, p. 437.

I. Nicoară
*Effect of the s–d Interaction on the Magnetoresistance of Magnetic Alloys
(b) Vol. 77, No. 1, p. K17–K20 (1976).
See also *13*, p. *118*.

W. Gasser and J. Schneider
*A Comment to Electrical Conduction in a Single Narrow Energy Band
(b) Vol. 77, No. 2, p. K103–K108 (1976).

J. C. Phillips
Phase Transitions in the Bilinear Organic Compounds (TTF, HMTSF)-TCNQ
(b) Vol. 78, No. 1, p. 371–379 (1976).
See also *1*, p. *9*; 22.9, p. 438.

J. C. Phillips
*High-Temperature Conductivity of Organic Bilinear Addition Compounds
(b) Vol. 78, No. 2, p. 471–476 (1976).
See also 6, p. 63; 22.9, p. 439.

L. Moseley, R. S. Tripathi, and T. Lukes
*Theory of the Hopping Conductivity in Non-Crystalline Systems
(b) Vol. 78, No. 1, p. 563–567 (1976).
See also *2*, p. *41*; 13.4, p. 145.

14.1 Metals. Semi-Metals

Review Articles

S. V. Vonsovskii and M. S. Svierskii
Multiplicity and Pseudomultiplicity in Electron–Ion Systems (I)
(b) Vol. 77, No. 1, p. 9–52 (1976).
See also *13*, p. *113*; 14.2, p. 169; 18.2, p. 228; 18.4, p. 242; 21, p. 289.

S. V. Vonsovskii and M. S. Svirskii
Multiplicity and Pseudomultiplicity of Electron–Ion Systems (II)
(b) Vol. 77, No. 2, p. 403–432 (1976).
See also *13*, p. *113*; 14.2, p. 169; 18.2, p. 228; 18.4, p. 242; 21, p. 289.

Original Papers and Short Notes

M. Ausloos
*On a Structural Phase Transition in Ytterbium
(b) Vol. **73**, No. 1, p. 211–221 (1976).
See also *1.2*, p. *14*; 13.1, p. 120; 21.4, p. 326.

L. M. Falicov and W. Y. Hsu
Phonons and Quasi-Periodic Static Fluctuations
(b) Vol. **73**, No. 1, p. 301–306 (1976).
See also *6*, p. *53*; *13.2*, p. *130*; 21.1, p. 302.

U. Hofmann
Physical Properties of Nickel–Rhenium Alloys
(a) Vol. **33**, No. 1, p. 119–123 (1976).
See also 12.1, p. 107; *18.2*, p. *228*; 21.1, p. 302.

H. Löffler and H.-G. Fabian
Application of the Jump Model to the Decomposition of Some Binary and Ternary Aluminium Alloys
(a) Vol. **33**, No. 1, p. 345–353 (1976).
See also *1.3*, p. *21*; *10.1*, p. *78*; 21, p. 290; 21.1, p. 303.

E. M. Gray and J. H. Smith
*The Effect of Heat Treatment on Micromagnetic Mn<u>Cu</u> Alloys
(a) Vol. **33**, No. 1, p. K43–K46 (1976).
See also *18.2*, p. *229*; 21.1, p. 303.

M. Ausloos
Renormalized Density of States by Conduction Electron–Localized Spin Scattering
(b) Vol. **73**, No. 2, p. 469–474 (1976).
See also *13*, p. *114*; 21, p. 290.

H. Sato and K. Yonemitsu
*Anisotropic Size Effect in Thin Aluminium Films
(b) Vol. **73**, No. 2, p. 723–733 (1976).
See also 13.2, p. 130; 21, p. 290.

R. Heller and H. Wipf
Diffusion Coefficient, Heat of Transport, and Solubility Limit of H in V
(a) Vol. **33**, No. 2, p. 525–529 (1976).
See also 8, p. 67; *9*, p. *72*; 21.1, p. 304.

B. Kozarewski, J. Nowak, and J. Zieliński
*The Coherent Potential Approximation in the Theory of Paramagnetic Alloys
(b) Vol. **74**, No. 2, p. 437—442 (1976).
See also 13.4, p. 137; 21, p. 292.

S. K. Bahl and H. D. Drew
Optical Properties of Mo–Re Alloys
(b) Vol. **74**, No. 2, p. 721—726 (1976).
See also *20.1*, p. *266*; 21, p. 292.

K. D. Schotte and D. Jacob
*Linear Magnetoresistance of a Metal with Holes
(a) Vol. **34**, No. 2, p. 593—600 (1976).
See also 21, p. 293.

H. E. Schaefer and W. Dander
The Initial Susceptibility of Cobalt after Electron Irradiation at 4.2 K
(b) Vol. **75**, No. 1, p. 173—178 (1976).
See also 10.1, p. 81; *11*, p. *101*; *18.2*, p. *231*; 21.1, p. 307.

P. Tzanetakis, J. Hillairet, and G. Revel
The Formation Energy of Vacancies in Aluminium and Magnesium
(b) Vol. **75**, No. 2, p. 433—439 (1976).
See also *10.1*, p. *82*; 21, p. 294; 21.3, p. 324.

E. Gerlach, P. Grosse, M. Rautenberg, and W. Senske
*Dynamical Conductivity and Plasmon Excitation in Bi
(b) Vol. **75**, No. 2, p. 553—558 (1976).
See also 13.5.2, p. 157; *20.1*, p. *268*; 21.7, p. 332.

G. Cubiotti, E. Donato, E. S. Giuliano, R. Ruggeri, and A. Stancanelli
*Transport Properties of Liquid Cadmium
(b) Vol. **75**, No. 2, p. K93—K96 (1976).
See also 2, p. 38; *15*, p. *210*; 21, p. 295.

H.-G. Fabian and H. Löffler
On the Interpretation of the Decomposition Kinetics of Some Al–Zn (6 at%)–Mg (0.1 to 3.5 at%) Alloys by Means of the Enlarged Jump Model
(a) Vol. **36**, No. 1, p. 127—132 (1976).
See also *1.3*, p. *23*; *10.1*, p. *83*; 21, p. 296; 21.1, p. 309.

A. L. Efros and B. I. Shklovskii
*Critical Behaviour of Conductivity and Dielectric Constant near the Metal–Non-Metal Transition Threshold
(b) Vol. **76**, No. 2, p. 475—485 (1976).
See also *14.4*, p. *203*.

G. Kuka, W. Braune, and R. Herrmann
Cyclotron Resonance with a Slightly Anomalous Skin Effect
(b) Vol. **76**, No. 2, p. 633—638 (1976).
See also *13.2*, p. *131*; 21, p. 297.

G. Röpke, V. Christoph, and G. Vojta
*Influence of Phonon Damping on the Electrical Resisitivity of Open Electron–Phonon Model Systems
(b) Vol. **76**, No. 2, p. 695—706 (1976).
See also 6, p. 59; 8, p. 69.

G. Quelard and D. Lesueur
Neutron Radiation Defects in Bismuth and Tellurium-Doped Bismuth
(a) Vol. **36**, No. 2, p. 729—734 (1976).
See also *11*, p. *103*; 21.7, p. 332.

K. Yonemitsu and T. Matsuda
*Electrical Resistivity of Spherical Guinier-Preston Zone
(a) Vol. **36**, No. 2, p. 791–798 (1976).
See also 13.2, p. 131; 21, p. 297; 21.6, p. 331.

Y. Aoki, Y. Gotoh, and Y. Obi
On the Phase Diagram of the Co-Rich Co–V Alloy System
(a) Vol. **36**, No. 2, p. K149–K152 (1976).
See also *1.3*, p. *23*; 18.2, p. 234; 21.1, p. 310.

R. S. Sorbello
Transport Theory of Ultrasonic Attenuation
(b) Vol. 77, No. 1, p. 141–152 (1976).
See also *6*, p. *60*; 7, p. 66.

M. Taut
*Dielectric Function of Hexagonal Metals (I): General Considerations
(b) Vol. 77, No. 1, p. 165–170 (1976).
See also *13.1*, p. *127*; 21, p. 297.

P. Vašek
*Negative Deviations from Matthiessen's Rule in Quenched Platinum at Low Temperatures
(b) Vol. 77, No. 1, p. K67–K70 (1976).
See also *10.1*, p. *83*; 21.6, p. 331.

W. Wycisk and M. Feller-Kniepmeier
Quenching Experiments on High-Purity Nickel
(a) Vol. **37**, No. 1, p. 183–191 (1976).
See also *10.1*, p. *83*; 21.1, p. 311.

P. G. Vassilev
Annealing Effects on the Thermoelectric Power and Visible Reflectance of Thin Copper Films
(a) Vol. **37**, No. 1, p. K25–K27 (1976).
See also *15*, p. *210*; *20.1*, p. *273*; 21.1, p. 311.

M. Taut and W. Hanke
*Dielectric Function of Hexagonal Metals (II): Beryllium
(b) Vol. 77, No. 2, p. 543–550 (1976).
See also *13.1*, p. *127*; 13.5.2, p. 157; 14.4, p. 204; 20.1, p. 274; 21.3, p. 324.

J. Krizek, K. N. R. Taylor, and H. Krizek
Room Temperature Optical Properties of Rare Earths
(b) Vol. 77, No. 2, p. 725–730 (1976).
See also *20.1*, p. *274*; 21.4, p. 327.

V. N. Andreev, T. V. Smirnova, and F. A. Chudnovskii
*Electrical Conductivity of Vanadium Dioxide Films in the Vicinity of the Metal–Semiconductor Phase Transition
(b) Vol. 77, No. 2, p. K97–K101 (1976).
See also *14.3.1*, p. *185*; 22.6, p. 407.

R. Kuhl, W. Kraak, H. Haefner, and R. Herrmann
*The Semimetal–Semiconductor Transition in Bismuth–Antimony Alloys
(b) Vol. 77, No. 2, p. K109–K111 (1976).
See also *13.2*, p. *131*; 19, p. 254; 21.7, p. 332.

C. M. Kachhava and D. Parihar
*Residual Resistivity of Dilute Binary Alloys
(b) Vol. 77, No. 2, p. K135—K137 (1976).
See also 21, p. 298; 21.1, p. 312; 21.2, p. 323.

K. I. Wysokiński
*Resistivity of Ternary Alloys: Weak-Scattering Limit
(b) Vol. 77, No. 2, p. K139—K142 (1976).

K. Elk and V. Christoph
*CPA Treatment of the Electrical Resistivity of Disordered Alloys
(b) Vol. 77, No. 2, p. K151—K155 (1976).
See also 21.1, p. 312.

E. Mrosan and G. Lehmann
*Calculation of Residual Resistivity and Single Temperatures for Cu-Based 3d-Transition Metal Dilute Alloys
(b) Vol. 77, No. 2, p. K161—K164 (1976).
See also *13.4*, p. *143*; 21.1, p. 312.

G. A. Hassan and F. H. Hammad
*A Resistivity Decrement in Deformed Aluminium
(a) Vol. 37, No. 2, p. K209—K211 (1976).
See also *10.1*, p. *85*; 21, p. 299.

E. Mrosan and G. Lehmann
Localized Non-Magnetic Impurities in Metals (II): Application to 3d-Transition Metal Impurities in Aluminium
(b) Vol. 78, No. 1, p. 159—167 (1976).
See also 8, p. 71; *13.4*, p. *144*; 14.2, p. 172; 15, p. 211; 21, p. 299.

M. Taut and P. Ziesche
*Dielectric Function of Hexagonal Metals (III): Magnesium
(b) Vol. 78, No. 1, p. 169—175 (1976).
See also *13.1*, p. *128*; 13.5.2, p. 158; 14.4, p. 204; 20.1, p. 275; 21.3, p. 324.

K. Sugawara and C. Y. Huang
*Microwave Resistivities of Some Rare Earth-Group VA Intermetallic Compounds
(b) Vol. 78, No. 1, p. K7—K9 (1976).
See also *18.4*, p. *245*; 21, p. 299.

R. N. Singh and P. L. Srivastava
*Liquid Electrical Resistivity of Alkali Metals: Exchange and Correlation Effects
(b) Vol. 78, No. 1, p. K45—K49 (1976).
See also 2, p. 40; 21.2, p. 323.

F. J. Kedves and L. Gergely
*Hall Coefficient and Electrical Resistivity of Dilute AlCr, AlMn. and AlCrMn Alloys
(a) Vol. 38, No. 1, p. K31—K35 (1976).
See also 21, p. 300; 21.1, p. 314.

K. M. van Vliet and A. H. Marshak
*Conduction Current and Generalized Einstein Relations for Degenerate Semiconductors and Metals
(b) Vol. 78, No. 2, p. 501—517 (1976).
See also 13.1, p. 129; *14.3*, p. *181*; 15, p. 211; 21, p. 300; 22, p. 341.

K. Rosan and H. Wipf
Nitrogen Traps for Hydrogen in Tantalum
(a) Vol. 38, No. 2, p. 611—620 (1976).
See also *10.1*, p. *86*; 21, p. 301.

J. P. Rivière and J. Grilhé
Neutron Irradiation Damage in a Plastically Deformed Fe–40 at% Al Ordered Alloy
(a) Vol. **38**, No. 2, p. 713–720 (1976).
See also 10.1, p. 86; *11*, p. *105*; 21.1.1, p. 322.

F. J. Kedves, M. Hordós, and F. Schusztek
*Precipitation in Al–Mn and Al–Cr Alloys
(a) Vol. **38**, No. 2, p. K123–K126 (1976).
See also *10.1*, p. *86*; 21, p. 301; 21.1, p. 315.

14.2 Superconductivity. Superconducting Materials and Devices

Review Articles

S. V. Vonsovskii and M. S. Svirskii
Multiplicity and Pseudomultiplicity in Electron–Ion Systems (I)
(b) Vol. 77, No. 1, p. 9–52 (1976).
See also *13*, p. *113*; 14.1, p. 164; 18.2, p. 228; 18.4, p. 242; 21, p. 289.

S. V. Vonsovskii and M. S. Svirskii
Multiplicity and Pseudomultiplicity of Electron–Ion Systems (II)
(b) Vol. 77, No. 2, p. 403–432 (1976).
See also *13*, p. *113*; 14.1, p. 164; 18.2, p. 228; 18.4, p. 242; 21, p. 289.

Original Papers and Short Notes

A. I. Akimenko, V. S. Solovev, and I. K. Yanson
*Harmonics Mixing in Superconducting Point Contacts
(a) Vol. **33**, No. 1, p. 53–58 (1976).
See also 21, p. 289.

M. Jergel, S. Takács, and V. Černuško
*Influence of Gas Impurities on the Angular Dependence of the Critical Currents in Nb_3Sn Vapour-Deposited Tapes
(a) Vol. **33**, No. 1, p. 85–90 (1976).
See also 21, p. 289.

E. Schneider and H. Kronmüller
*The Elementary Interaction between a Crystal Dislocation and the Flux Line Lattice of a Type II Superconductor
(b) Vol. **74**, No. 1, p. 261–273 (1976).
See also *10.1*, p. *79*; 21, p. 291.

A. J. Pindor
*On the Coulomb Pseudopotential in Superconducting PdH
(b) Vol. **74**, No. 1, p. K19–K22 (1976).
See also 21, p. 291.

T. Nattermann
*On the Ginzburg Criteria for Static and Dynamic Critical Phenomena and the Small Parameter of Perturbation Theory
(b) Vol. **74**, No. 2, p. 635–646 (1976).

L. M. Schafe, P. L. Rossiter, and W. A. Rachinger
*Field History Effects in Superconducting Pb–Sn Alloys
(b) Vol. **74**, No. 2, p. 785–794 (1976).
See also 10.1, p. 80; 21, p. 292.

N. E. Alekseevskii, V. M. Sakosarenko, K. Blüthner, and H.-J. Köhler
*Superconducting Properties of Vanadium Films
(a) Vol. **34**, No. 2, p. 541–546 (1976).
See also 21, p. 293.

H.-W. Pohl and H. Teichler
*Impurity Effects in the H_{c2} Anisotropy of Cubic Superconductors
(b) Vol. **75**, No. 1, p. 205–212 (1976).
See also 21, p. 294.

M. Kulić and L. Dobrosavljević
*Pinning in Superconducting Films with Modulated Concentration of Impurities
(b) Vol. **75**, No. 2, p. 677–686 (1976).
See also 10.1, p. 82; 21, p. 294.

W. A. Ortiz and F. P. Missell
*Superconducting Properties of Ta–19 at% Zr and Ta–29 at% Zr
(a) Vol. **35**, No. 2, p. K135–K136 (1976).
See also 21, p. 295.

E. P. Balsamo, G. Paternò, A. Barone, M. Russo, and R. Vaglio
*Temperature and Magnetic Field Dependence of the Critical Current in Sn-Sn_xO_y-In Josephson Junctions
(a) Vol. **35**, No. 2, p. K173–K175 (1976).
See also 21, p. 295; 22.6, p. 405.

E. Hegenbarth and B. Schmidt
Heat Conductivity Measurement at V_3Si in the Temperature Range 4.2 to 80 K
(b) Vol. **76**, No. 1, p. 307–311 (1976).
See also *8*, p. *69*; 13.1, p. 125; 21, p. 296.

N. E. Alekseevskii, E. P. Krasnoperov, and T. Mydlarz
*Susceptibility and Superconductivity of Compounds with A-15 Structure
(b) Vol. **76**, No. 1, p. 357–362 (1976).
See also 13.1, p. 125; *18.1*, p. *226*; 22.8, p. 416.

W. Zacharko, R. Kubiak, and J. Mazur
*The Anisotropy of Superconducting Parameters of $InSn_4$ Single Crystals
(b) Vol. **76**, No. 1, p. K1–K4 (1976).

H. D. Wiesinger
*Zustandsdiagramm und Sprungtemperatur zur Supraleitung von A15-Phasen
(Phase Diagram and Transition Temperature to Superconductivity of A15 Phases)
(a) Vol. **36**, No. 1, p. 61–72 (1976).
See also *1.3*, p. *23*; 21, p. 296.

G. Wolf and C. Hohlfeld
The Molar Heat Capacity of Superconductive $(PdRh_{0.06})H_x$ in the Temperature Range from 2 to 7.5 K
(a) Vol. **36**, No. 1, p. K99–K103 (1976).
See also *8*, p. *69*; 21, p. 296.

A. R. Birks, S. P. Hind, and P. M. Lee
Band Structure Changes in Intercalates of Niobium Diselenide
(b) Vol. **76**, No. 2, p. 599–604 (1976).
See also *13.1*, p. *126*.

G. Wolf, J. Jahnke, and K. Bohmhammel
The Molar Heat Capacity of $(PdAg_{0.2})H_x$ in the Temperature Range from 2 to 6 K
(a) Vol. **36**, No. 2, p. K125–K128 (1976).
See also *8*, p. *70*; 21, p. 297.

W. D. Schmidt and D. George
*Switching Behaviour of Nb–Nb_xO_y–Pb Josephson Tunnel Junctions
(a) Vol. **36**, No. 2, p. K153–K155 (1976).
See also 14.3.3, p. 193; 22.6, p. 406.

E. H. Brandt
*Microscopic Theory of Clean Type-II Superconductors in the Entire Field–Temperature Plane
(b) Vol. **77**, No. 1, p. 105–119 (1976).

V. Rudat and G. Kluge
*Strain Dependence of T_c and Discontinuity of Elastic Moduli in the Ginzburg-Landau Theory
(b) Vol. 77, No. 1, p. K63—K65 (1976).

V. G. Litovchenko
*On Superconducting State Peculiarities at the Surface of the Electron-Hole Droplets
(b) Vol. 77, No. 1, p. K71—K76 (1976).
See also *13.5.1*, p. *155*.

D. Eckert and A. Handstein
*Experimental Investigation of the Flux Density Profile in Hard Superconductors
(a) Vol. 37, No. 1, p. 171—178 (1976).
See also 21, p. 297; 21.1, p. 311.

S. Takács
*The Volume Pinning Force in Superconductors with Statistically Distributed Point Defects
(a) Vol. 37, No. 1, p. 279—286 (1976).
See also 10.1, p. 84.

N. E. Alekseevskii, C. Bazan, A. B. Mitin, T. Mydlarz, E. P. Krasnoperov, and B. Rączka
*Influence of Magnetic Impurities on Superconducting Properties of Some Compounds with A-15 Structure
(b) Vol. 77, No. 2, p. 451—455 (1976).
See also 21, p. 298.

E. H. Brandt
*On the Shear Modulus of the Flux Line Lattice
(b) Vol. 77, No. 2, p. 551—560 (1976).

H. Kronmüller and H. Riedel
*The Critical Field for the Nucleation of n-Domains in Type I Superconductors
(b) Vol. 77, No. 2, p. 581—590 (1976).
See also *10.1*, p. *84*; 21, p. 398.

H. Nowotny
*Approximative BCS-Like Solution of the Eliashberg Gap Equation
(b) Vol. 77, No. 2, p. 623—629 (1976).

N. V. Ageev, N. E. Alekseevskii, and V. F. Shamray
*Superconductivity of Nb_3Al Base Ternary Solid Solutions
(b) Vol. 77, No. 2, p. K129—K131 (1976).
See also 22.8, p. 417.

T. Nakajima
*On the Isotopic Volume Effect in Superconducting PdH(D) Systems
(b) Vol. 77, No. 2, p. K147—K150 (1976).
See also 21, p. 298.

G. Wolf and M. Zimmermann
The Molar Heat Capacity of Superconducting PdD_x in the Temperature Range from 2 to 12 K
(a) Vol. 37, No. 2, p. 485—489 (1976).
See also *8*, p. *70*; 21, p. 299.

G. Splitt
*Magnetic Properties of Superconducting Lead Films Investigated with Hypersonic Waves
(a) Vol. 37, No. 2, p. K131—K134 (1976).
See also *7*, p. *66*; 21, p. 299.

A. W. Szafrański, T. Skośkiewicz, and B. Baranowski
*Superconductivity in the $Pd_{1-x}M_xH_c$ (M = Noble Metal) and PdB_yH_c Alloy Systems
(a) Vol. 37, No. 2, p. K163–K164 (1976).
See also 21, p. 299.

E. Mrosan and G. Lehmann
Localized Non-Magnetic Impurities in Metals (II): Application to 3d-Transition Metal Impurities in Aluminium
(b) Vol. 78, No. 1, p. 159–167 (1976).
See also 8, p. 71; *13.4*, p. *144*; 14.1, p. 167; 15, p. 211; 21, p. 299.

G. Zerweck
*The Shear Modulus of the Flux Line Lattice in Low-$\varkappa$ Type-II Superconductors
(b) Vol. 78, No. 1, p. K15–K17 (1976).

H. R. Khan, K. Lüders, Ch. J. Raub, and Z. Szücs
*Knight-Shift and Superconductivity of Nb–Al σ- and A-15 Phases
(b) Vol. 78, No. 2, p. K107–K109 (1976).
See also *19*, p. *256*; 21, p. 300.

A. E. Karkin, V. E. Arkhipov, B. N. Goshchitskii, E. P. Romanov, and S. K. Sidorov
*Radiation Effects in the Superconductor Nb_3Sn
(a) Vol. 38, No. 2, p. 433–438 (1976).
See also *11*, p. *104*; 21, p. 300.

S. M. Ashimov, T. P. Batsankalashvili, N. L. Nedzelyak, and J. S. Tsakadze
*Torque Anisotropy Due to Crystal Defect Structure of the Superconducting Niobium Single Crystal
(a) Vol. 38, No. 2, p. 769–779 (1976).
See also 10.1, p. 86; 21, p. 301.

14.3 Semiconductors

Review Articles

P. V. Pavlov, E. I. Zorin, D. I. Tetelbaum, and A. F. Khokhlov
*Nitrogen as Dopant in Silicon and Germanium
(a) Vol. **35**, No. 1, p. 11–36 (1976).
See also 11, p. 100; *13.4*, p. *134*; 22.1.1, p. 345; 22.1.2, p. 351.

H. Böttger and V. V. Bryksin
*Hopping Conductivity in Ordered and Disordered Solids (I)
(b) Vol. **78**, No. 1, p. 9–56 (1976).
See also 2, p. 35; 6, p. 52; 13, p. 113; 13.5.3, p. 159; 22, p. 333.

H. Böttger and V. V. Bryksin
*Hopping Conductivity in Ordered and Disordered Solids (II)
(b) Vol. **78**, No. 2, p. 415–451 (1976).
See also 2, p. 35; 6, p. 52; 13, p. 113; 13.5.3, p. 159; 22, p. 333.

Original Papers and Short Notes

H. Köhler
Non-Parabolic $E(\boldsymbol{k})$ Relation of the Lowest Conduction Band in Bi_2Te_3
(b) Vol. **73**, No. 1, p. 95–104 (1976).
See also *13.1*, p. *120*; *13.2*, p. *130*; 22.7, p. 410.

N. J. Doran and E. Gerlach
*A Theory for Electron Transport Due to Phonon Scattering Including Screening
(b) Vol. **73**, No. 1, p. 203–209 (1976).
See also *6*, p. *52*; 22.2.1, p. 362.

H. Böttger and V. V. Bryksin
*Theoretical Investigation of the Hopping Conductivity in Disordered Systems in a Large Frequency Range
(b) Vol. **73**, No. 1, p. 229–237 (1976).
See also *2*, p. *35*; 6, p. 52; 13, p. 113.

P. K. Basu and B. R. Nag
*Iterative Calculation of the Polar Optical Mobility in Semiconductors in the Extreme Quantum Limit
(b) Vol. **73**, No. 1, p. K15–K17 (1976).
See also 6, p. 53; 22.2.3, p. 367.

J. Varvas and T. Nirk
*High-Temperature Conductivity of Undoped and Copper-Doped Cadmium Selenide
(a) Vol. **33**, No. 1, p. 75–84 (1976).
See also *10.2*, p. *87*; 13.4, p. 135; 22.4.2, p. 383.

R. Parker and C. J. Tinsley
*Electrical Conduction in Magnetite
(a) Vol. **33**, No. 1, p. 189–194 (1976).
See also 18.3, p. 239; 22.6, p. 402.

Nguyen van Dong, Tran Huu Danh, B. Auguin, and A. Defresne
*Transport Properties of Glassy $CdGe_{1-x}Sb_xAs_2$
(a) Vol. **33**, No. 1, p. 295–299 (1976).
See also *2*, p. *35*; 15, p. 209; 22.8, p. 413.

M. Baj, L. Dmowski, M. Kończykowski, and S. Porowski
*A New Explanation of the Cl Donor Deionization and Its Kinetics Observed in CdTe:Cl under Hydrostatic Pressure
(a) Vol. **33**, No. 1, p. 421—426 (1976).
See also 13.4, p. 135; 22.4.3, p. 385.

P. Hiesinger
*Hall-Effect Levels in Ag- and Au-Doped p-Type GaAs
(a) Vol. **33**, No. 1, p. K39—K41 (1976).
See also *13.4*, p. *135*; 22.2.1, p. 362.

P. Petelenz
Effective Electron–Hole Interaction in Molecular Crystals (I)
(b) Vol. **73**, No. 2, p. 565—573 (1976).
See also 6, p. 54; *13*, p. *114*; *13.5.1*, p. *150*; 22, p. 334.

G. Oelgart, R. Herrmann, H. Krüger, and R. Stegmann
*Fermi Surface and Helicons in Semiconducting $Bi_{1-x}Sb_x$ Alloys
(b) Vol. **73**, No. 2, p. 615—624 (1976).
See also *13.2*, p. *130*; 19, p. 248; 22.7, p. 410.

V. K. Arora
*Effect of Inelasticity on Magnetoresistance
(b) Vol. **73**, No. 2, p. K93—K94 (1976).
See also 6, p. 54; 22.2.3, p. 367.

L. Dmowski, M. Kończykowski, R. Piotrzyowski, and S. Porowski
*Pressure-Induced Slow Relaxation of the Free Electron Concentration in Undoped n-Type InSb
(b) Vol. **73**, No. 2, p. K131—K134 (1976).
See also 12.2, p. 110; 22.2.3, p. 367.

L. I. Khandros, G. S. Pekar, M. K. Sheinkman, and E. L. Shtrum
*High-Temperature Investigations of CdS Single Crystals (I): Electron Concentration, Cd Solubility, and Charge of Intrinsic Defects
(a) Vol. **33**, No. 2, p. 765—771 (1976).
See also 10.2, p. 89; 22.4.1, p. 379.

J. C. Jones and P. R. Wallace
*On the Anomalous Magneto-Resistance of Gallium Arsenide in Intense Magnetic Fields
(b) Vol. **74**, No. 1, p. 23—33 (1976).
See also 13.1, p. 122; 22.2.1, p. 362.

D. Lemoine, C. Pelletier, S. Rolland, and R. Granger
*Electronic Conduction in GaAs at Low Temperatures
(a) Vol. **34**, No. 1, p. 307—317 (1976).
See also 22.2.1, p. 362.

H. Kuzmany
*Light Scattering Experiments and Nonlinear Behaviour of the Acoustoelectric Oscillator
(a) Vol. **34**, No. 1, p. 347—353 (1976).
See also *7*, p. *65*; 22.4.1, p. 379.

L. I. Khandros, G. S. Pekar, M. K. Sheinkman, and E. L. Shtrum
*High-Temperature Investigations of CdS Single Crystals (II): Electron Mobility and Scattering Mechanism
(a) Vol. **34**, No. 1, p. 397—403 (1976).
See also 10.2, p. 90; 22.4.1, p. 379.

Z. Zboiński
*The Tensor of the Drift Mobility of Electrons and Holes in the Anthracene-1,3,5-Trinitrobenzene Charge Transfer Complex
(b) Vol. **74**, No. 2, p. 561–566 (1976).
See also 22.9, p. 435.

H. Köhler
Non-Parabolicity of the Highest Valence Band of Bi_2Te_3 from Shubnikov-de Haas Effect
(b) Vol. **74**, No. 2, p. 591–600 (1976).
See also *13.1*, p. *123*; *13.2*, p. *130*; 22.7, p. 410.

P. Petelenz
Effective Electron–Hole Interaction in Molecular Crystals (II): Phonon Mechanism
(b) Vol. **74**, No. 2, p. 655–658 (1976).
See also *6*, p. *55*; *13*, p. *115*; 13.5.1, p. 152; 22, p. 335.

J. Krempaský and D. Barančok
*The Use of Modified Relaxation Time in the Transport Theory of Non-Crystalline Semiconductors
(b) Vol. **74**, No. 2, p. 741–751 (1976).
See also *2*, p. *37*; 13.4, p. 137; 15, p. 209; 22, p. 335.

E. Litwin-Staszewska and W. Szymańska
*Scattering on Short-Range Potentials in Semiconductors with a Narrow Energy Gap
(b) Vol. **74**, No. 2, p. K89–K92 (1976).
See also 15, p. 209; 22.4.2, p. 383.

H. Böttger and V. V. Bryksin
*On the Equivalence of Different Theoretical Descriptions of the Hopping Conductivity in Disordered Systems
(b) Vol. **74**, No. 2, p. K151–K156 (1976).
See also *2*, p. *37*; 6, p. 55; 13, p. 115.

E. W. Kreutz
*X-Radiation Effects in Bulk Si at Subthreshold Energies
(a) Vol. **34**, No. 2, p. 489–495 (1976).
See also 11, p. 101; 22.1.2, p. 353.

W. Moldenhauer and W. Brückner
*Physical Properties of Nonstoichiometric Iron Sulfide $Fe_{1-x}S$ near the α-Phase Transition
(a) Vol. **34**, No. 2, p. 565–571 (1976).
See also *1.2*, p. *15*; 15, p. 209; 18.4, p. 243; 22, p. 336.

U. Rehberg
Measurement of the Complex Permittivity of the Anion Radical Complex Salt of Tetracyanochinodimethane and Methyltriphenylarsonium
(a) Vol. **34**, No. 2, p. 573–582 (1976).
See also *14.4*, p. *202*; 22.9, p. 435.

E. Gombia, N. Romeo, G. Sberveglieri, and C. Paorici
*Transport Properties of Semiconducting $ZnIn_2S_4$
(a) Vol. **34**, No. 2, p. 651–655 (1976).
See also 15, p. 209; 22.8, p. 414.

B. Jachym, H. Sodolski, T. Słupkowski, and R. Zieliński
*Electric Conduction in the Polyester Resin–Acetylene Carbon Black System
(a) Vol. **34**, No. 2, p. 657–664 (1976).
See also 22.9, p. 435.

J. VAITKUS
*The Nonequilibrium Hall Effect and Related Transport Phenomena in Semiconductors under Inhomogeneous Excitation by a Laser Pulse
(a) Vol. **34**, No. 2, p. 769–775 (1976).
See also *16*, p. *214*; 20.3, p. 282; 22.1.1, p. 346.

P. I. BARANSKII, V. V. KOLOMOETS, and A. V. FEDOSOV
*The Angular Dependences of the Longitudinal Piezoresistance of n-Ge and n-Si at 78 K
(a) Vol. **34**, No. 2, p. K139–K141 (1976).
See also 22.1.1, p. 346; 22.1.2, p. 353.

P. G. ZHUMATII
*Intraband Conductivity and Thermopower of Semiconductors with Slowly Varying Gaussian Random Field
(b) Vol. **75**, No. 1, p. 61–72 (1976).
See also *15*, p. *209*; 22, p. 336.

H. KÖHLER
Anisotropic *g*-Factor of the Conduction Electrons in Bi_2Te_3
(b) Vol. **75**, No. 1, p. 127–136 (1976).
See also *13.1*, p. *123*; *13.2*, p. *131*; 22.7, p. 410.

G. E. PIKE and C. H. SEAGER
*Transport in Microscopically Inhomogeneous Materials
(b) Vol. **75**, No. 1, p. 289–297 (1976).
See also *9*, p. *72*; 22.1.1, p. 347; 22.1.2, p. 353.

V. M. BABICH, P. I. BARANSKII, L. S. DEMIDENKO, V. A. ILCHISHIN, L. S. KOROLYUK, and A. G. SAMOILOVICH
*Impurity Scattering in n-Ge in Quantizing Magnetic Fields
(b) Vol. **75**, No. 1, p. 299–302 (1976).
See also 22.1.1, p. 347.

N. J. DORAN and E. GERLACH
* A New Theory for the Transverse Magneto-Resistance in the Quantum Limit which Includes Screening
(b) Vol. **75**, No. 1, p. K15–K18 (1976).
See also 22.2.3, p. 367.

V. K. ARORA
*Extreme Quantum Limit Magnetotransport in Transverse Configuration
(b) Vol. **75**, No. 1, p. K65–K67 (1976).

YU. A. VODAKOV, G. A. LOMAKINA, E. N. MOKHOV, E. I. RADOVANOVA, V. I. SOKOLOV, M. M. USMANOVA, G. F. YULDASHEV, and B. S. MACHMUDOV
*Silicon Carbide Doped with Gallium
(a) Vol. **35**, No. 1, p. 37–42 (1976).
See also 3, p. 42; *13.4*, p. *138*; 20.1, p. 268; 20.3, p. 283; 22.3, p. 372.

W. SIEGEL, A. HEINRICH, and E. ZIEGLER
*Electron and Hole Mobility in $ZnSiP_2$
(a) Vol. **35**, No. 1, p. 269–279 (1976).
See also 6, p. 56; 13.4, p. 139; 22.8, p. 415.

R. M. HILL
*On the Observation of Variable Range Hopping
(a) Vol. **35**, No. 1, p. K29–K34 (1976).
See also *2*, p. *37*.

S. I. Radautsan, E. K. Arushanov, and A. N. Nateprov
The Conduction Band of Cadmium Phosphide
(a) Vol. **35**, No. 1, p. K53–K56 (1976).
See also *13.1*, p. *124*; 22.3, p. 372.

V. N. Holyavko, V. P. Dragunov, B. V. Morozov, E. M. Skok, and N. B. Velchev
*Study of the Size Quantization in p-Type Silicon Inversion Layers by Means of Magnetoresistance
(b) Vol. **75**, No. 2, p. 423–432 (1976).
See also 13.1, p. 124; 22.1.2, p. 354.

H. Köhler
The *g*-Factor of the Holes in the Highest Valence Band of Bi_2Te_3
(b) Vol. **75**, No. 2, p. 441–449 (1976).
See also *13.1*, p. *124*; *13.2*, p. *131*; 22.7, p. 411.

R. J. Iwanowski and T. Dietl
*Low Temperature Electron Mobility in $Cd_vHg_{1-v}Se$ Mixed Crystals
(b) Vol. **75**, No. 2, p. K83–K87 (1976).
See also 22.4.4, p. 387.

I. I. Farbshtein, L. S. Dubinskaya, and V. G. Krigel
*Resonance Scattering and Piezotransport of Pure Tellurium at Low Temperature
(b) Vol. **75**, No. 2, p. K133–K136 (1976).
See also 22.1.3, p. 359.

V. N. Brudnyi, D. L. Budnitskii, M. A. Krivov, and V. G. Melev
*p–n-Type Conversion and Optical Properties of 2.0 MeV Electron-Irradiated $ZnSnAs_2$
(a) Vol. **35**, No. 2, p. 425–430 (1976).
See also 11, p. 102; 13.4, p. 139; *20.1*, p. *269*; 22.8, p. 415.

S. Mantovani, U. del Pennino, and E. Mazzega
Capacitance Measurements as a New Tool to Investigate the Electronic States of Dislocations in Semiconductors
(a) Vol. **35**, No. 2, p. 451–457 (1976).
See also *13.4*, p. *139*; 22.1.1, p. 347.

A. B. Almazov, A. Pincevičius, and J. Viščakas
*Steady-State Bipolar Phenomena in One-Level Semiconductors
(a) Vol. **35**, No. 2, p. 563–569 (1976).

I. S. Gorban, G. A. Grishchenko, A. P. Sakalas, and I. I. Tychina
*Conducting and Semi-Insulating n- and p-Type $ZnSiP_2$ Single Crystals
(a) Vol. **35**, No. 2, p. 571–574 (1976).
See also *13.4*, p. *139*; 22.8, p. 415.

N. Dupont-Pavlovsky, F. Caralp, P. Delhaes and J. Amiell
Chemical Characterization and Study of the Electronic Properties of Non-Stoichiometric Zinc Oxide. Nature and Repartition of Defects
(a) Vol. **35**, No. 2, p. 615–625 (1976).
See also *10.2*, p. *93*; *13.4*, p. *139*; 18, p. 223; 22.4, p. 376.

H. Menniger, H. Raidt, and G. Voigt
On the Influence of Microdefects on Charge Carriers in Silicon
(a) Vol. **35**, No. 2, p. 639–643 (1976).
See also *1.4*, p. *27*; 22.1.2, p. 354.

K. Somogyi
*Relation between the Activation Energies (ε_1, E_1) and the Temperature of Carrier Mobility Maximum in Semiconductors. A New Method for Determination of E_1
(a) Vol. **35**, No. 2, p. 659—666 (1976).
See also **13.4**, p. 140; 22.1.2, p. 354; 22.2.2, p. 365.

K. L. Chopra, S. K. Barthwal, and D. K. Pandya
*Structural Sensitivity of Transport and Optical Properties of Amorphous GeTe
(a) Vol. **35**, No. 2, p. 761—768 (1976).
See also 2, p. 38; *15*, p. *210*; 20.1, p. 270; 22.7, p. 411.

V. V. Mitin and E. I. Tolpygo
*Piezoresistance of Uniaxially Deformed n-Si
(b) Vol. **76**, No. 1, p. 407—411 (1976).
See also 13.1, p. 125; 22.1.2, p. 355.

J. M. Leger, G. Weill, and I. A. Smirnov
*Resistivity of Sm_3S_4 under High Pressure
(b) Vol. **76**, No. 1, p. K27—K30 (1976).
See also 22, p. 338.

N. J. Doran and E. Gerlach
*Frequency Dependent Conductivity Due to Deformation Potential and Piezoelectric Scattering by Acoustic Phonons in Semiconductors Including Screening
(b) Vol. **76**, No. 1, p. K43—K46 (1976).
See also 6, p. 58; 22.2.1, p. 363.

C. Butti, A. Raymond et F. Bombré
*Effect photo Hall du cinabre
(Photo Hall Effect in Cinnabar)
(a) Vol. **36**, No. 1, p. 133—137 (1976).
See also 22.4.1, p. 381.

V. S. Vavilov, A. E. Kiv, and O. R. Niyazova
Erratum to: The Subthreshold Radiation Effects in Semiconductors
(a) Vol. **36**, No. 1, p. 427 (1976).
See also 9, p. 73; 10.2, p. 94; *11*, p. *102*; 22.1, p. 343; 22.2, p. 361; 22.3, p. 373; 22.4, p. 376.

H. Süssmann und H. Rau
*Dotierungseigenschaften von Pb in Sb_2Te_3
(Doping Properties of Pb in Sb_2Te_3)
(a) Vol. **36**, No. 1, p. K55—K58 (1976).
See also 22.7, p. 411.

Y. Kamiura and F. Hashimoto
Annealing Stage above 400 °C in Quenched Gallium-Doped Germanium
(a) Vol. **36**, No. 1, p. K81—K84 (1976).
See also 9, p. 73; *10.2*, p. *94*; 22.1.1, p. 348.

G. Müller
Optical and Electrical Spectroscopy of Zinc Oxide Crystals Simultaneously Doped with Copper and Donors
(b) Vol. **76**, No. 2, p. 525—532 (1976).
See also *13.4*, p. *141*; 20.1, p. 271; 22.4, p. 377.

D. Barjon, A. Raymond, A. Joullie, and J. L. Robert
*Electron Mobility of $Ga_xIn_{1-x}Sb$ at Room Temperature
(b) Vol. **76**, No. 2, p. 623—627 (1976).
See also 22.2.4, p. 370.

A. Jędrzejczak and T. Dietl
Thermomagnetic Properties of n-Type and p-Type HgTe
(b) Vol. **76**, No. 2, p. 737–751 (1976).
See also *13.1*, p. *126*; *15*, p. *210*; 22.4.3, p. 386.

G. M. Shmelev, I. A. Chaikovskii, and Chan Min Shon
*The Conductivity of Semiconductors with Superlattice
(b) Vol. **76**, No. 2, p. 811–816 (1976).

F. Sprenger
*Magnetoresistance of Frozen-in Photoconductivity in Semiconducting CdF_2:In
(a) Vol. **36**, No. 2, p. 483–490 (1976).
See also 13.4, p. 141; *16*, p. *215*; 22.5, p. 389.

E. Ziegler, W. Siegel, and A. Heinrich
*First-Order NPO Scattering in p-$ZnSiAs_2$ and p-$CdSiAs_2$
(a) Vol. **36**, No. 2, p. 491–494 (1976).
See also *6*, p. *59*; 22.8, p. 417.

A. S. Filipchenko and L. P. Bolshakov
*Mobility of Holes in p-InSb Crystals
(b) Vol. **77**, No. 1, p. 53–58 (1976).
See also *13.1*, p. *127*; 22.2.3, p. 368.

D. O. Welch and K. G. Lynn
*Systematic Variation of the Mean Positron Lifetime and Gaussian Fraction in Annealed Metals and Semiconductors
(b) Vol. **77**, No. 1, p. 277–286 (1976).
See also *13*, p. *118*; 22, p. 339.

I. N. Magden, L. I. Katurina, B. L. Meiler, and P. V. Musienko
*[100] Silicon Doped with Arsenic (I): Electrical Parameters and Their Thermal Stability
(a) Vol. **37**, No. 1, p. 29–38 (1976).
See also 10.2, p. 95; 22.1.2, p. 356.

O. A. Golikova, I. M. Rudnik, V. M. Sergeeva, M. M. Kazanin, and E. N. Tkalenko
*On the Conductivity of Rare-Earth Metal Chalcogenides $Ln_{3-x}X_4$ ($0 \leq x \leq 0.33$)
(a) Vol. **37**, No. 1, p. 199–203 (1976).
See also 22.7, p. 412.

Burma Bansaragtschin, G. Lehmann, and R. Link
Application of the Travelling Solvent Method to Growing of $Pb_{1-x}Sn_xTe$ Single Crystals
(a) Vol. **37**, No. 1, p. K13–K15 (1976).
See also *3*, p. *43*; 22, p. 339.

M. M. Kaila, L. B. Harris, and H. J. Goldsmid
*Thermoelectric and Thermomagnetic Properties of Plastically Bent InSb Samples
(a) Vol. **37**, No. 1, p. K43–K46 (1976).
See also *15*, p. *210*; 22.2.3, p. 368.

K. Ziegler, H.-D. Junker, and U. Birkholz
*Electrical Conductivity and Seebeck Coefficient of Si_2Te_3 Single Crystals
(a) Vol. **37**, No. 1, p. K97–K99 (1976).
See also *15*, p. *211*; 22, p. 340.

J. L. Farvacque and E. Gerlach
*Scattering of Free Carriers Due to the Strain Field of Screw Dislocations in Semiconductors
(b) Vol. **77**, No. 2, p. 651–656 (1976).
See also *10.2*, p. *95*; 22.1.3, p. 360.

D. GENZOW and E. NORMANTAS
*Theory of the Photon-Drag Effect in Semiconductors with Elliptical Isoenergetic Surfaces and Its Application to Tellurium Crystals
(b) Vol. 77, No. 2, p. 667–676 (1976).
See also 16, p. 216; 22.1.3, p. 360.

M. OBIDITSCH and H. KAHLERT
*Electric Field Induced Changes of the Fundamental Absorption and Low-Temperature Energy Relaxation in n-Type GaSb
(b) Vol. 77, No. 2, p. 677–684 (1976).
See also 6, p. 61; *20.1*, p. *274*; 22.2.3, p. 368.

J. BERREHAR, P. DELANOY, and M. SCHOTT
*Drift Mobility of Holes in Crystalline Tetracene
(b) Vol. 77, No. 2, p. K119–K122 (1976).
See also 22.9, p. 438.

B. LORENZ and D. IHLE
*A Model of the Semiconductor–Semiconductor Transition in Ti_4O_7
(b) Vol. 77, No. 2, p. K177–K181 (1976).
See also 22.6, p. 407.

H. SÜSSMANN and K. LOOF
*Kupferdotierung und Fehlordnung im System Bi_2Te_3–Sb_2Te_3
(Copper Doping and Disorder in the System Bi_2Te_3–Sb_2Te_3)
(a) Vol. 37, No. 2, p. 467–471 (1976).
See also 22.7, p. 412.

E. I. TERUKOV, F. A. CHUDNOVSKII, W. REICHELT, H. OPPERMANN, W. BRÜCKNER, H.-P. BRÜCKNER, and W. MOLDENHAUER
*Investigation of the Physical Properties of V_3O_5 at Phase Transition with Consideration of Its Range of Homogeneity
(a) Vol. 37, No. 2, p. 541–546 (1976).
See also 15, p. 211; 18.1, p. 227; 22.6, p. 407.

K. KUMAZAKI, E. MATSUSHIMA, and A. ODAJIMA
*Point Defects and Non-Stoichiometry in HgSe
(a) Vol. 37, No. 2, p. 579–589 (1976).
See also *10.2*, p. *96*; 22.4.2, p. 383.

K. SOMOGYI
*Temperature Dependence of the Hole Mobility in Doped p-GaP
(a) Vol. 37, No. 2, p. 653–661 (1976).
See also 22.2.2, p. 365.

G. HOPPMANN and E. SALJE
*Optical and Electrical Properties of H^+ Doped WO_3 Single Crystals
(a) Vol. 37, No. 2, p. K187–K191 (1976).
See also 13.1, p. 128; *20.1*, p. *275*; 22.6, p. 408.

C. HAMANN, CH. TANTZSCHER, and G. LEHMANN
*A Model of the Electrical Transport Phenomena in Imperfect Crystals of Copper Phthalocyanine (IV): Concentration of Charge Carriers and Current–Voltage Characteristics
(b) Vol. 78, No. 1, p. 79–84 (1976).
See also 13.4, p. 144; *14.3.1*, p. *185*; 22.9, p. 438.

I. N. MAGDEN and V. A. SAFONOV
*[100] Silicon Doped with Arsenic (II): The Connection of Electrical and Structural Parameters with Microplasmas
(a) Vol. 38, No. 1, p. 45–50 (1976).
See also 10.2, p. 97; 22.1.2, p. 357.

W. Brückner, U. Gerlach, W. Moldenhauer, H. P. Brückner, N. Mattern, H. Oppermann, and E. Wolf
Phase Transitions and Semiconductor–Metal Transition in $V_{1-x}Ga_xO_2$ Single Crystals
(a) Vol. 38, No. 1, p. 93–102 (1976).
See also *1.2*, p. *19*; *3*, p. *44*; 15, p. 211; 18.1, p. 227; 22.6.1, p. 409.

F. V. Wald
Self-Compensation in CdTe and ZnTe Crystals Grown from Indium Solvents
(a) Vol. 38, No. 1, p. 253–259 (1976).
See also *3*, p. *44*; 10.2, p. 97; 20.1, p. 277; 22.4.3, p. 386.

T. Boţilă and H. K. Henisch
Study of Trapping in Amorphous Materials
(a) Vol. 38, No. 1, p. 331–337 (1976).
See also *2*, p. *40*; 13.4, p. 145; 16, p. 217; 22.8, p. 418.

E. I. Terukov, F. A. Chudnovskii, W. Brückner, W. Reichelt, H. P.-Brückner, W. Moldenhauer, and H. Oppermann
*Influence of Mo, Ti, and Ga Doping on Conductivity Transition of V_3O_5
(a) Vol. 38, No. 1, p. K23–K26 (1976).
See also *15*, p. *211*; 22.6, p. 408; 22.6.1, p. 409.

P. Byszewski, J. Kowalski, and R. R. Gałązka
*Shubnikov-de Haas Oscillations in HgSe at 77 K in Pulsed Magnetic Fields
(b) Vol. 78, No. 2, p. 477–481 (1976).
See also *13.1*, p. *129*; 22.4.2, p. 384.

K. M. van Vliet and A. H. Marshak
*Conduction Current and Generalized Einstein Relations for Degenerate Semiconductors and Metals
(b) Vol. 78, No. 2, p. 501–517 (1976).
See also 13.1, p. 129; *14.1*, p. *167*; 15, p. 211; 21, p. 300; 22, p. 341.

J. Kossut
*Electron Transport Phenomena in Narrow- and Zero-Gap Semiconductors Containing Magnetic Impurities
(b) Vol. 78, No. 2, p. 537–542 (1976).
See also *13.4*, p. *145*; 15, p. 211; 22.4.3, p. 386.

H. Köhler, W. Haigis, and A. von Middendorff
Shubnikov-de Haas Investigations on n-Type $Bi_2(Se_xTe_{1-x})_3$
(b) Vol. 78, No. 2, p. 637–642 (1976).
See also *13.1*, p. *129*; *13.2*, p. *131*; 22.7, p. 412.

E. J. Fantner, F. Kuchar, and G. Bauer
*Systematic Investigation of the Sign Change of the Hall Effect in n-InSb in the Quantum Limit
(b) Vol. 78, No. 2, p. 643–651 (1976).
See also 22.2.3, p. 369.

H. Biernat and M. Kriechbaum
*Anomalous Hall Effect of n-InSb at High Magnetic Fields
(b) Vol. 78, No. 2, p. 653–657 (1976).
See also 22.2.3, p. 369.

K. M. van Vliet
*Entropy Production and Choice of Heat Flux for Degenerate Semiconductors and Metals
(b) Vol. 78, No. 2, p. 667–676 (1976).
See also *15*, p. *211*; 21, p. 300; 22, p. 342.

V. M. Babich, P. I. Baranskii, L. S. Demidenko, V. A. Ilchishin, and S. L. Korolyuk
*Anisotropy of Magnetoresistance of n-Ge in Transverse Quantizing Magnetic Fields under Conditions of Mixed Scattering
(b) Vol. 78, No. 2, p. K111–K114 (1976).
See also 22.1.1, p. 350.

I. Vilfan
*On the Electron-Intramolecular Vibration Coupling in Aromatic Molecular Crystals
(b) Vol. 78, No. 2, p. K131–K135 (1976).
See also *6*, p. *64*; 13.5.3, p. 159; 22.9, p. 439.

V. R. K. Murthy and J. Sobhanadri
*Electrical Conductivity of Some Nickel-Zinc Ferrites
(a) Vol. 38, No. 2, p. 647–651 (1976).
See also 15, p. 212; 18.3, p. 241; 22.8.2, p. 431.

C. Manfredotti, R. Murri, A. Quirini, and L. Vasanelli
Photoelectronic Properties of n-GaSe
(a) Vol. 38, No. 2, p. 685–693 (1976).
See also 13.4, p. 145; *16*, p. *217*; 22.3, p. 374.

14.3.1 Films

Original Papers and Short Notes

N. Desbarax, Y. Ségui et Bui Ai
*Etude des mécanismes de conduction dans des films minces de polysiloxane
(Study of the Conduction Mechanism in Thin Films of Polyoxisilicone)
(a) Vol. **33**, No. 1, p. 247–254 (1976).
See also **14.3.2**, p. 187; 22.9, p. 433.

K. L. Chopra and P. Nath
*Amorphous Ge Alloy Films (II): Transport Properties
(a) Vol. **33**, No. 1, p. 333–344 (1976).
See also *2*, p. *36*; 15, p. 209; 22.1.1, p. 345.

Yu. A. Vidadi, K. Sh. Kocharli, B. Sh. Barkhalov, and S. A. Sadreddinov
*Direct Current Conductivity of Copper Phthalocyanine Films in the Presence of Blocking Contacts
(a) Vol. **33**, No. 1, p. K67–K71 (1976).
See also **14.3.2**, p. 187; 22.9, p. 433.

P. Müller
*Direct Proof of the Conduction Type in Semi-Insulating Thin Films
(a) Vol. **33**, No. 2, p. 543–553 (1976).
See also 22, p. 334.

V. V. Voskoboynikov, V. A. Gritsenko, V. M. Efimov, V. E. Lesnikovskaya, and F. L. Edelman
*Structure and Electrophysical Properties of Boron Nitride Thin Films
(a) Vol. **34**, No. 1, p. 85–94 (1976).
See also *2*, p. *36*; **14.3.3**, p. 190; 22, p. 335.

Yu. A. Vidadi, K. Sh. Kocharli, B. Sh. Barkhalov, and S. A. Sadreddinov
*Alternating Current Investigation of Copper Phthalocyanine Films in the Presence of Blocking Contacts
(a) Vol. **34**, No. 1, p. K77–K81 (1976).
See also 22.9, p. 435.

K.-D. Ufert
Stress Induced Switching in VO_2 Thin Films
(a) Vol. **34**, No. 1, p. K83–K86 (1976).
See also *8*, p. *68*; *12.2*, p. *110*; 22.6, p. 404.

R. M. Hill
*Variable-Range Hopping
(a) Vol. **34**, No. 2, p. 601–613 (1976).
See also *13.4*, p. *137*; 22.1, p. 343; 22.1.1, p. 346; 22.2.1, p. 362.

C. Albers and H. Gertig
*Magnetoresistance in Thin Tellurium Films
(a) Vol. **34**, No. 2, p. K125–K126 (1976).
See also 22.1.3, p. 359.

M. Janda and A. Kubový
Growth of Vacuum-Deposited Tellurium Films on Glass Substrates and Some of Their Transport Properties
(a) Vol. **35**, No. 1, p. 391–402 (1976).
See also *1.5*, p. *31*; *3*, p. *42*; 22.1.3, p. 359.

S. Balevičius, A. Česnys, and A. Deksnys
*Nanosecond Switching in Amorphous In_2Te_3 Films
(a) Vol. **35**, No. 1, p. K41–K43 (1976).
See also *2*, p. *37*; 22.3, p. 372.

V. V. Mitin, Yu. A. Tkhorik, and Yu. M. Shvarts
*Influence of Stress on the Magnetoresistance of p-Germanium Films
(a) Vol. **35**, No. 1, p. K65–K66 (1976).
See also 22.1.1, p. 347.

A. Kubový and M. Janda
*Transport Properties of Tellurium Thin Films and Their Dependence on Thickness
(a) Vol. **35**, No. 2, p. 471–476 (1976).
See also 22.1.3, p. 359.

A. F. Kravchenko, B. V. Morozov, and E. M. Skok
*Features of Indium Antimonide Films
(a) Vol. **35**, No. 2, p. 755–759 (1976).
See also 22.2.3, p. 368.

J. Hamerský
*Elektrische Eigenschaften von CdSe-Schichten unter dem Einfluß von Sauerstoff
(Electrical Properties of CdSe Films under the Influence of Oxygen)
(a) Vol. **35**, No. 2, p. K97–K99 (1976).
See also 9, p. 73; 22.4.2, p. 383.

P. Thomas
*Addendum to: On Anisotropic Hopping Conduction in Thin Amorphous Germanium Films
(b) Vol. **76**, No. 1, p. 439 (1976).
See also 1.5, p. 31; 22.1.1, p. 347.

A. Ya. Vul, Sh. I. Nabiev, S. G. Petrosyan, and A. Ya. Shik
*Simulation of the Long-Period Conductivity Relaxation (LCR) Phenomenon in Double-Layer Semiconductor Structures
(a) Vol. **36**, No. 1, p. 53–60 (1976).
See also *16*, p. *215*; 22.2.1, p. 363.

K. L. Chopra and D. K. Pandya
*Obliquely Deposited Amorphous Ge Films (II): Electrical Properties
(a) Vol. **36**, No. 1, p. 89–100 (1976).
See also *2*, p. *39*; 15, p. 210; 22.1.1, p. 348.

A. Kubový and M. Janda
*Temperature Dependence of Hole Mobilities in Te Thin Films
(a) Vol. **36**, No. 1, p. 101–105 (1976).
See also 22.1.3, p. 359.

S. Petersson, E. Mgbenu, and P. A. Tove
*Silicide Formation Correlated with Surface Resistivity Measurements
(a) Vol. **36**, No. 1, p. 217–225 (1976).
See also 1.5, p. 32; 22, p. 338.

S. K. Barthwal and K. L. Chopra
*Transport Properties of Amorphous InSb and GaAs
(a) Vol. **36**, No. 1, p. 345–355 (1976).
See also *2*, p. *39*; 15, p. 210; 22.2.1, p. 363; 22.2.3, p. 368.

S. K. Barthwal and K. L. Chopra
*Thermoelectric Power of Amorphous Germanium Films
(a) Vol. **36**, No. 2, p. 533–549 (1976).
See also 2, p. 39; *15*, p. *210*; 22.1.1, p. 348.

B. P. Rai, K. Singh, and R. S. Srivastava
*Current Transport Phenomena in SiO_2 Films
(a) Vol. **36**, No. 2, p. 591–595 (1976).
See also 14.3.4, p. 198; 22.6, p. 406.

D. D. Nedeoglo, R. K. Radu, and A. G. Cheban
*Electrical Properties of n-Type Epitaxial GaP Layers
(a) Vol. **36**, No. 2, p. 659–668 (1976).
See also 22.2.2, p. 365.

E. L. Frankevich, M. M. Tribel, and I. A. Sokolik
*Photoconductivity of Sublimated Rubrene Films during Oxidation. The Magnetic Field Effect
(b) Vol. 77, No. 1, p. 265–276 (1976).
See also 13.5.1, p. 155; *16*, p. *215*; 20.3, p. 286; 22.9, p. 437.

A. Deptuch, M. Grus, A. J. Horodecki, and M. Łepek
*On the Conductivity of Polycrystalline CdS Films
(a) Vol. **37** No. 1, p. K101–K104 (1976).
See also 22.4.1, p. 381.

V. N. Andreev, T. V. Smirnova, and F. A. Chudnovskii
*Electrical Conductivity of Vanadium Dioxide Films in the Vicinity of the Metal–Semiconductor Phase Transition
(b) Vol. 77, No. 2, p. K97–K101 (1976).
See also *14.1*, p. *166*; 22.6, p. 407.

A. Kubový and M. Janda
*Effects of Annealing on Some Transport Properties of Tellurium Thin Films
(a) Vol. **37**, No. 2, p. K127–K129 (1976).
See also 22.1.3, p. 360.

B. Lalevic and N. K. Annamalai
Schottky Barriers Effects on the Dielectric Behavior of the $Nb–Nb_2O_5–Nb$ System
(a) Vol. **37**, No. 2, p. K177–K181 (1976).
See also *14.3.3*, p. *193*; 22.6, p. 407.

R. I. Bolkhovityanova and Yu. B. Bolkhovityanov
*The Influence of Indium on the p–n Junction Formation in Si-Doped LPE Layers of Gallium Arsenide
(a) Vol. **37**, No. 2, p. K193–K195 (1976).
See also *1.5*, p. *32*; 3, p. 44; 22.2.4, p. 371.

C. Hamann, Ch. Tantzscher, and G. Lehmann
*A Model of the Electrical Transport Phenomena in Imperfect Crystals of Copper Phthalocyanine (IV): Concentration of Charge Carriers and Current–Voltage Characteristics
(b) Vol. 78, No. 1, p. 79–84 (1976).
See also 13.4, p. 144; *14.3*, p. *180*; 22.9, p. 438.

J. Świątek
*Electrical Conduction Mechanism in Thin Coronene Films
(a) Vol. **38**, No. 1, p. 285–291 (1976).
See also *13.4*, p. *144*; 22.9, p. 439.

P. C. Mehendru, N. L. Pathak, Satbir Singh, and P. Mehendru
*Electrical Conduction in Polypropylene Thin Films
(a) Vol. **38**, No. 1, p. 355–360 (1976).
See also *14.4*, p. *204*; 22.9, p. 439.

M. K. Rao and S. R. Jawalekar
*Resistivity Measurements of CdSe Films
(a) Vol. **38**, No. 1, p. K93—K95 (1976).
See also 22.4.2, p. 384.

J. Škácha
*Electrical Structure and Transport in Amorphous GeS Layers
(a) Vol. **38**, No. 2, p. K143—K146 (1976).
See also *2*, p. *41*; 22, p. 342.

14.3.2 Surfaces and Interfaces

Original Papers and Short Notes

N. Desbarax, Y. Ségui et Bui Ai
Study of the Conduction Mechanism in Thin Films of Polyoxisilicone
(a) Vol. **33**, No. 1, p. 247–254 (1976).
See also *14.3.1*, p. *183*; 22.9, p. 433.

Yu. A. Vidadi, K. Sh. Kocharli, B. Sh. Barkhalov, and S. A. Sadreddinov
Direct Current Conductivity of Copper Phthalocyanine Films in the Presence of Blocking Contacts
(a) Vol. **33**, No. 1, p. K67–K71 (1976).
See also *14.3.1*, p. *183*; 22.9, p. 433.

B. Korsch, F. Willig, H. J. Gaehrs, and B. Tesche
*Exchange of Holes between Organic Crystals and Metal Films Evaporated in High Vacuum
(a) Vol. **33**, No. 2, p. 461–471 (1976).
See also *14.3.4*, p. *196*; 16, p. 213; 22.9, p. 433.

B. Lepley and S. Ravelet
*Correlation between Open-Circuit Voltage and Interface Parameters at the Au–CdS Illuminated Contact
(a) Vol. **33**, No. 2, p. 517–523 (1976).
See also *16*, p. *213*; 22.4.1, p. 379.

F. Weber
*On the Dispersion Relation for Surface Density Waves in Semiconductors
(a) Vol. **33**, No. 2, p. 651–660 (1976).
See also *13.3*, p. *132*; 14.3.4, p. 196; 22.1.1, p. 345.

R. Sinon, R. Hatert, and F. van de Wiele
*MOS Structures with a P^+P Profile
(a) Vol. **33**, No. 2, p. 661–671 (1976).
See also 14.3.3, p. 190; 22, p. 334.

V. F. Synorov, A. N. Likholet, and B. I. Sysoev
*Analysis of Charge Distribution in MIS-Type Multilayer Systems
(a) Vol. **33**, No. 2, p. K161–K164 (1976).
See also *14.3.3*, p. *190*.

F. Weber
The Influence of Carrier Injection on Helical Density Waves
(a) Vol. **34**, No. 1, p. 107–116 (1976).
See also *14.3.4*, p. *196*; 22.1.1, p. 346.

G. Blasquez
*Etude des phénomènes de conduction et de bruit de fond basse fréquence associés aux régions superficielles des jonctions p–n. Application au transistor à jonctions
(Study of Conduction and Low-Frequency Noise Phenomena Associated with Surface Regions in p–n Junctions. Application to Junction Transistors)
(a) Vol. **34**, No. 1, p. 207–218 (1976).
See also 22.1.2, p. 352.

S. G. Chamberlain, B. Y. Woo, and D. B. Scott
*Modelling of Electrical Charge Injection into Charge-Coupled Devices
(a) Vol. **34**, No. 1, p. 263–275 (1976).
See also *14.3.3*, p. *190*; 14.3.4, p. 196; 22.1.2, p. 352.

A. D. Boardman, B. V. Paranjape, and Y. O. Nakamura
Surface Plasmon-Polaritons in a Spatially Dispersive Inhomogeneous Medium
(b) Vol. 75, No. 1, p. 347–359 (1976).
See also *13.3*, p. *132*; *13.5.2*, p. *157*.

G. Juška and S. Vengris
*Hole Photogeneration in Amorphous Selenium at Low Electric Fields
(a) Vol. 35, No. 1, p. 339–345 (1976).
See also *16*, p. *214*; 22.1.3, p. 359.

S. F. Timashev and M. A. Krykin
*Optical and Electrooptical Properties of the Interfaces of Media
(b) Vol. 76, No. 1, p. 67–76 (1976).
See also 13.3, p. 132; *20.1*, p. *270*.

G. D. Bagratishvili, R. B. Dzhanelidze, N. I. Kurdiani, and O. V. Saksaganskii
*MIS Structure GaAs–Ge_3N_4–Al
(a) Vol. 36, No. 1, p. 73–79 (1976).
See also 13.3, p. 132; 20.3, p. 285; 22.2.1, p. 363.

R. Hatert, R. Sinon, and F. van de Wiele
*Experimental and Theoretical Study of Buried Channel MOS Structures
(a) Vol. 36, No. 1, p. 235–246 (1976).
See also 14.3.3, p. 192; 22.1.2, p. 355.

T. N. Sitenko
*Non-Equilibrium Low-Temperature Depletion
(a) Vol. 36, No. 2, p. K157–K159 (1976).
See also 13.3, p. 133; 22.2.1, p. 363.

J. Buxo, D. Esteve, and G. Sarrabayrouse
A Theoretical Description of Tunneling MIS Device Transport
(a) Vol. 37, No. 1, p. K105–K110 (1976).
See also *14.3.3*, p. *193*.

A. Jucys and V. Kunigelis
*Influence of the Surface Potential upon the Acoustoelectric Interaction in CdS
(b) Vol. 77, No. 2, p. K113–K117 (1976).
See also *7*, p. *66*; 22.4.1, p. 381.

H. Flietner and Ngo Duong Sinh
*Character of Si–SiO_2 Interface States from Analysis of the CV Term Spectra
(a) Vol. 37, No. 2, p. 533–539 (1976).
See also *13.3*, p. *133*; 22.1.2, p. 357.

U. A. Shakirov and M. S. Yunusov
*Influence of X-Irradiation on Silicon Schottky Diodes
(a) Vol. 37, No. 2, p. 681–686 (1976).
See also 9, p. 74; *11*, p. *104*; 22.1.2, p. 357.

R. Hatert, P. Couvreur, R. Sinon, and F. van de Wiele
*Avalanche Multiplication Factor and Reverse Current of Si p–n Junctions
(a) Vol. 38, No. 1, p. 123–129 (1976).
See also 14.3.3, p. 194; 22.1.2, p. 357.

A. E. Gershinskii, L. V. Mironova, and E. I. Cherepov
*SiO_2–Si (n-Type) Interface Investigation with the Electrochemical Method
(a) Vol. 38, No. 1, p. 369–373 (1976).
See also *1.5*, p. *33*; 22.1.2, p. 358; 22.6, p. 408.

D. R. Choudhuri, P. K. Basu, and A. N. Chakravarti
*On the Modification of the Einstein Relation for Semiconductor Inversion Layers
(a) Vol. **38**, No. 1, p. K85–K88 (1976).
See also 22.1.2, p. 358.

A. I. Koifman, A. N. Narkulov, B. L. Oksengendler, and M. S. Yunusov
The Low-Temperature Doping of Silicon by Radiation-Induced Diffusion
(a) Vol. **38**, No. 2, p. 439–446 (1976).
See also *9*, p. *74*; *11*, p. *105*; 22.1.2, p. 358.

14.3.3 Devices. Junctions (Contact Problems See 14.3.4)

Review Article

J. KADLEC and K. H. GUNDLACH
*Results and Problems of Internal Photoemission in Sandwich Structures
(a) Vol. **37**, No. 1, p. 11–28 (1976).
See also **13.1**, p. 120; *16*, p. *213*; 22.6, p. 402.

Original Papers and Short Notes

H.-J. WÜNSCHE and K. HENNEBERGER
*Interband-Current Theory for Esaki Diodes
(b) Vol. **73**, No. 1, p. 245–254 (1976).
See also **13.1**, p. 120; **14.3.4**, p. 195; **22.2.1**, p. 362.

M. LEPPIHALME and T. TUOMI
*Photovoltage Spectra of Metal–Silicon (p-Type) Diodes near the Indirect Absorption Edge of Silicon
(a) Vol. **33**, No. 1, p. 125–133 (1976).
See also **6**, p. 53; **13.5.1**, p. 149; *16*, p. *213*; 20.1, p. 261; 22.1.2, p. 351.

R. SINON, R. HATERT, and F. VAN DE WIELE
MOS Structures with a P^+P Profile
(a) Vol. **33**, No. 2, p. 661–671 (1976).
See also *14.3.2*, p. *187*; 22, p. 334.

V. F. SYNOROV, A. N. LIKHOLET, and B. I. SYSOEV
*Analysis of Charge Distribution in MIS-Type Multilayer Systems
(a) Vol. **33**, No. 2, p. K161–K164 (1976).
See also *14.3.2*, p. *187*.

V. V. VOSKOBOYNIKOV, V. A. GRITSENKO, V. M. EFIMOV, V. E. LESNIKOVSKAYA, and F. L. EDELMAN
Structure and Electrophysical Properties of Boron Nitride Thin Films
(a) Vol. **34**, No. 1, p. 85–94 (1976).
See also *2*, p. *36*; *14.3.1*, p. *183*; 22, p. 335.

M. MANZEL, W. ECKE, and H. BRUCHLOS
Negative Differential Resistance in Gold-Doped High-Resistivity Silicon
(a) Vol. **34**, No. 1, p. 219–226 (1976).
See also **13.4**, p. 137; *14.3.4*, p. *196*; 22.1.2, p. 352.

S. G. CHAMBERLAIN, B. Y. WOO, and D. B. SCOTT
*Modelling of Electrical Charge Injection into Charge-Coupled Devices
(a) Vol. **34**, No. 1, p. 263–275 (1976).
See also *14.3.2*, p. *187*; **14.3.4**, p. 196; 22.1.2, p. 352.

S. G. ATARBEKOV, A. S. FILIPCHENKO, and M. A. RZAEV
*Deep Level in Fused Tunnel Diodes Made from InSb
(a) Vol. **34**, No. 1, p. K7–K8 (1976).
See also *13.4*, p. *137*; 22.2.3, p. 367.

S. MARTINUZZI
*Photopiles solaires a hétérojonction n-CdS–p-CdTe
(n-CdS–p-CdTe Heterojunction Solar Cells)
(a) Vol. **34**, No. 1, p. K21–K26 (1976).
See also *16*, p. *214*; 22.4.1, p. 379; 22.4.3, p. 385.

V. A. GIRII, A. V. KONDRACHUK, S. I. KORNYUSHIN, V. G. LITOVCHENKO, R. O. LITVINOV, and V. I. SHAKHOVTSOV
*Effect of Impurities on the Radiation Sensitivity of the Si–SiO_2 Interface in MIS Structures
(a) Vol. **34**, No. 1, p. K35–K38 (1976).
See also 11, p. 101; *13.3*, p. *132*; 22.1.2, p. 353.

J. E. PARROTT and M. TIENG
*Space Charge Capacitance of an Abrupt p–n Junction
(a) Vol. **34**, No. 1, p. K43–K46 (1976).
See also 14.3.4, p. 196; 22, p. 335.

B. I. SYSOEV, A. N. LIKHOLET, and V. F. SYNOROV
*Study of the Effect of Electric Field Screening on Trap Level Filling by the Capacitor Thermostimulated Discharge Method
(a) Vol. **34**, No. 1, p. K71–K75 (1976).
See also *13.4*, p. *137*.

Y. K. SHARMA
*Surface Flicker Noise in p–i Diodes
(a) Vol. **34**, No. 1, p. K87–K89 (1976).

J. NIEWODNICZAŃSKA-ZAWADZKA and J. RAULUSZKIEWICZ
*The Observation of LO Phonons in HgSe and HgTe by Electron Tunneling
(b) Vol. **74**, No. 2, p. K93–K95 (1976).
See also *6*, p. *55*; 22.4.2, p. 383; 22.4.3, p. 385.

J. JERHOT and V. ŠNEJDAR
*The Influence of Interface States on an Isotype Heterojunction between Two Wide-Gap Semiconductors
(a) Vol. **34**, No. 2, p. 505–511 (1976).
See also *13.3*, p. *132*; 22, p. 335.

I. S. KAULACH and E. A. SILINSH
The Influence of the Hyperfine Range Magnetic Field on Charge Transfer States at the Metal–Organic Semiconductor Interface
(b) Vol. **75**, No. 1, p. 247–254 (1976).
See also *13.3*, p. *132*; 16, p. 214; 22.9, p. 436.

O. I. SHKLYAREVSKII, A. A. LYSYKH, and I. K. YANSON
*Studies on Al–Insulator–Pb Tunnel Junctions with Barriers Prepared in Glow Discharge of CO_2 and CO
(a) Vol. **35**, No. 1, p. 49–53 (1976).
See also 22.6, p. 404.

C. CAPILUPPI, P. G. MERLI, G. POZZI, and I. VECCHI
*Out-of-Focus Observations of p–n Junctions by High-Voltage Electron Microscopy
(a) Vol. **35**, No. 1, p. 165–172 (1976).
See also 22.1.2, p. 353.

A. V. FERRIS-PRABHU
*Tunnelling Theories of Non-Volatile Semiconductor Memories
(a) Vol. **35**, No. 1, p. 243–250 (1976).
See also 13.4, p. 138; 22.1.2, p. 353.

M. PUGNET, J. BARBOLLA, J. C. BRABANT, F. SAINT-YVES, and M. BROUSSEAU
*Energy Levels for Platinum and Palladium in Silicon Mesaured by the Dark Transient Capacitance Technique
(a) Vol. **35**, No. 2, p. 533–543 (1976).
See also *13.4*, p. *139*; 22.1.2, p. 354.

A. J. HARRIS, H. THOMAS, and M. P. CALLAGHAN
*Investigation of the Characteristics of Electroluminescent SiC Diodes (I): Theoretical Model for Transient and Steady-State Conditions
(a) Vol. **35**, No. 2, p. 697–707 (1976).
See also *20.3*, p. *284*; 22.3, p. 373.

Y. KANDA and M. MIGITAKA
*Effect of Mechanical Stress on the Offset Voltages of Hall Devices in Si IC
(a) Vol. **35**, No. 2, p. K115–K118 (1976).
See also 22.1.2, p. 354.

R. K. NAHAR and D. NAGCHOUDHURI
*Depletion Layer Characteristics of Heterojunctions with Built-in Homojunction
(a) Vol. **36**, No. 1, p. 139–145 (1976).
See also 13.3, p. 132; 22, p. 338.

S. MARTINUZZI, O. MALLEM, and T. CABOT
*Dark Currents in Cu_2S–CdS Single Crystal Heterojunctions. Consequences for Photocells
(a) Vol. **36**, No. 1, p. 227–234 (1976).
See also 13.3, p. 133; *16*, p. *215*; 22, p. 338; 22.4.1, p. 381.

R. HATERT, R. SINON, and F. VAN DE WIELE
*Experimental and Theoretical Study of Buried Channel MOS Structures
(a) Vol. **36**, No. 1, p. 235–246 (1976).
See also *14.3.2*, p. *188*; 22.1.2, p. 355.

P. C. MATHUR and VIJENDER SHARMA
*Large-Signal Analysis of IMPATT Diodes
(a) Vol. **36**, No. 1, p. 269–274 (1976).
See also *14.3.4*, p. *198*; 22.2.1, p. 363.

A. J. HARRIS and H. THOMAS
*Investigation of the Characteristics of Electroluminescent SiC Diodes (II): Temperature- and Growth-Dependent Effects
(a) Vol. **36**, No. 1, p. 365–373 (1976).
See also *20.3*, p. *285*; 22.3, p. 373.

B. PETZEL
*Influence of Temperature on Impact Ionization in Gunn Domains
(a) Vol. **36**, No. 1, p. K39–K41 (1976).
See also *14.3.4*, p. *198*; 22.2.1, p. 363.

H.-J. ANKLAM and K.-D. UFERT
*Electrothermal Behaviour of VO_2 Thin Film Switching Devices
(a) Vol. **36**, No. 1, p. K71–K75 (1976).
See also *8*, p. *69*; 14.3.4, p. 198; 22.6, p. 405.

J. BARBOLIA, M. PUGNET, J. C. BRABANT, and M. BROUSSEAU
Thermal Capture Cross-Section of Free Electrons at Neutral Gold Centres in n-Type Silicon
(a) Vol. **36**, No. 2, p. 495–498 (1976).
See also *13.4*, p. *141*; 22.1.2, p. 355.

K. SAMINADAYAR and J. C. PFISTER
*Surface-State Generation on MOS Capacitors Irradiated with UV Light and Electrons
(a) Vol. **36**, No. 2, p. 679–686 (1976).
See also *13.3*, p. *133*; 22.1.2, p. 356.

H. KRAUSE
*Electron Injection and Trap Filling in Insulating Layers
(a) Vol. **36**, No. 2, p. 705–715 (1976).
See also *13.4*, p. *141*; 22.6, p. 406.

J. TOUŠKOVÁ and R. KUŽEL
*Fundamental Properties of Au–CdTe Metal–Semiconductor Contact
(a) Vol. **36**, No. 2, p. 747–755 (1976).
See also *14.3.4*, p. *199*; 22.4.3, p. 386.

W. D. SCHMIDT and D. GEORGE
Switching Behaviour of Nb–Nb_xO_y–Pb Josephson Tunnel Junctions
(a) Vol. **36**, No. 2, p. K153–K155 (1976).
See also *14.2*, p. *170*; 22.6, p. 406.

N. SALEH and A. SOLIMAN
Analysis and Design Considerations of Burstein Narrow-Band Photon Detectors
(a) Vol. **37**, No. 1, p. 39–43 (1976).
See also *16*, p. *216*; 22.2.3, p. 368.

CH. DIPPMANN
*Temperaturmessungen an monostabilen Chalcogenidglas–Schwellwertschaltern
(Temperature Measurements on Mono-Stable Chalcogenide Glass Threshold Switches)
(a) Vol. **37**, No. 1, p. K81–K83 (1976).
See also 22.8, p. 417.

J. BUXO, D. ESTEVE, and G. SARRABAYROUSE
*A Theoretical Description of Tunneling MIS Device Transport
(a) Vol. **37**, No. 1, p. K105–K110 (1976).
See also 14.3.2, p. 188.

J. KADLEC and K. H. GUNDLACH
*Internal Photoemission and Quantum-Mechanical Transmission Coefficient
(a) Vol. **37**, No. 2, p. 385–398 (1976).
See also *16*, p. *216*; 22, p. 340.

PHAM MINH TAN
Bestimmung von Ionenimplantationsprofilen aus der Messung der differentiellen Kapazität am Halbleiter–Elektrolyt-Kontakt
(Determination of Ion Implantation Profiles from Measurements of the Differential Capacity at the Semiconductor–Electrolyte Contact)
(a) Vol. **37**, No. 2, p. 439–444 (1976).
See also 11, p. 104; *13.4*, p. *143*; 22.1.2, p. 356.

P. BASU and H. SAHA
*Possible Performance of Back-Wall Schottky Barrier Solar Cells
(a) Vol. **37**, No. 2, p. 625–631 (1976).
See also 13.3, p. 133; *16*, p. *216*; 22.1.2, p. 357; 22.2.1, p. 364; 22.4.3, p. 386.

W. MOLDENHAUER and W. BRÜCKNER
*Steady-State Switching Characteristics of V_3O_5 Single Crystals
(a) Vol. **37**, No. 2, p. K143–K146 (1976).
See also 14.3.4, p. 200; 22.6, p. 407.

A. SCHLACHETZKI
*Domain Dissolution in Planar Gunn Devices
(a) Vol. **37**, No. 2, p. K151–K155 (1976).
See also 22.2.1, p. 364.

B. LALEVIC and N. K. ANNAMALAI
*Schottky Barriers Effects on the Dielectric Behavior of the Nb–Nb_2O_5–Nb System
(a) Vol. **37**, No. 2, p. K177–K181 (1976).
See also *14.3.1*, p. *185*; 22.6, p. 407.

K.-H. Müller, M. Wolf, and F. Kammler
*Thermal Switching and Current–Voltage Characteristics for Semiconducting Sandwich Samples
(a) Vol. 37, No. 2, p. K201–K203 (1976).
See also 8, p. 71; 14.3.4, p. 200; 22, p. 340.

R. Hatert, P. Couvreur, R. Sinon, and F. van de Wiele
Avalanche Multiplication Factor and Reverse Current of Si p-n Junctions
(a) Vol. 38, No. 1, p. 123–129 (1976).
See also *14.3.2*, p. *188*; 22.1.2, p. 357.

V. S. Lysenko, A. V. Sachenko, A. P. Dubchak, A. N. Nazarov, V. T. Atamanenko, M. P. Lokshin, and Yu. Ya. Polishuk
*Properties of MOS Structures after B^+ Ion Bombardment
(a) Vol. 38, No. 1, p. 131–138 (1976).
See also 11, p. 104; 22.1.2, p. 357.

A. Mansingh, J. K. Vaid, and R. P. Tandon
*Switching and Memory Behaviour of Vanadate Glasses
(a) Vol. 38, No. 1, p. K1–K3 (1976).
See also *2*, p. *41*; 22.8.1, p. 424.

E. P. Trifonova and A. S. Popov
*n–n Heterojunctions $CdSnP_2$–InP
(a) Vol. 38, No. 1, p. K37–K40 (1976).
See also 22.2.2, p. 366; 22.8, p. 418.

Y. Kanda and M. Migitaka
*Design Consideration for Hall Devices in Si IC
(a) Vol. 38, No. 1, p. K41–K44 (1976).
See also 22.1.2, p. 358.

V. A. Gritsenko, W. Leupert, and K. P. Mogilnikov
*On the Influence of Illumination on Charges Trapped in MNOS Structures
(a) Vol. 38, No. 1, p. K57–K59 (1976).
See also 22, p. 341.

B. Joseph
*The Electron Injection from Si into SiO_2 Due to a Mobile Ion Induced Internal Field
(a) Vol. 38, No. 2, p. K127–K130 (1976).
See also *14.3.4*, p. *201*; 22.1.2, p. 358; 22.6, p. 408.

14.3.4 High Field Phenomena, Space-Charge Effects, Inhomogeneities, Injected Carriers (Electroluminescence See 20.3; Junctions See 14.3.3)

Review Article

Z. S. Kachlishvili
*Galvanomagnetic and Recombination Effects in Semiconductors in a Strong Electric Field
(a) Vol. 33, No. 1, p. 15–51 (1976).
See also 13.4, p. 134; 22, p. 333.

Original Papers and Short Notes

H.-J. Wünsche and K. Henneberger
Interband-Current Theory for Esaki Diodes
(b) Vol. 73, No. 1, p. 245–254 (1976).
See also 13.1, p. 120; *14.3.3*, p. *190*; 22.2.1, p. 362.

A. A. Ignatov and Yu. A. Romanov
*Nonlinear Electromagnetic Properties of Semiconductors with a Superlattice
(b) Vol. 73, No. 1, p. 327–333 (1976).
See also 13.1, p. 121; 22, p. 333.

B. Esser and P. Kleinert
*Erratum to: On the Theory of Interband $\varepsilon(\omega)$ and External Electric Field Modulated $\Delta\varepsilon(\omega)$ in Disordered Semiconductors
(b) Vol. 73, No. 1, p. 379 (1976).
See also 2, p. 35; *20.1*, p. *260*; 22, p. 333.

R. B. Hammond, T. C. McGill, and J. W. Mayer
Properties of the Electron–Hole Liquid Luminescence in Si Double Injection Diodes
(a) Vol. 33, No. 1, p. 59–66 (1976).
See also 13.5.1, p. 149; *20.3*, p. *280*; 22.1.2, p. 351.

I. M. Belova
*On the Theory of Tunnelling in Disordered Semiconductors
(b) Vol. 73, No. 2, p. 409–414 (1976).
See also 2, p. 36; 22, p. 334.

J. Röseler, K. Henneberger, and R. Strehlow
*On the Theory of Electroabsorption in Polar Crystals
(b) Vol. 73, No. 2, p. 541–552 (1976).
See also 13.5.3, p. 159; *20.1*, p. *262*; 22, p. 334.

J. von Borzeszkowski
*On the Electric Field Dependence of the Conductivity of Holes in Si at Low Temperatures (I): The Drift Velocity under Consideration of the Non-Equipartition of Acoustic Phonon Energy
(b) Vol. 73, No. 2, p. 607–614 (1976).
See also 6, p. 54; 13.1, p. 121; 22.1.2, p. 351.

W. Fuhs, M. Milleville, and J. Niklas
*Acoustoelectric Current Saturation and Sound Velocity in Tellurium Crystals under Hydrostatic Pressure
(b) Vol. 73, No. 2, p. K127–K130 (1976).
See also 7, p. *65*; 22.1.3, p. 359.

B. KORSCH, F. WILLIG, H. J. GAEHRS, and B. TESCHE
*Exchange of Holes between Organic Crystals and Metal Films Evaporated in High Vacuum
(a) Vol. **33**, No. 2, p. 461–471 (1976).
See also *14.3.2*, p. *187*; 16, p. 213; 22.9, p. 433.

F. WEBER
On the Dispersion Relation for Surface Density Waves in Semiconductors
(a) Vol. **33**, No. 2, p. 651–660 (1976).
See also *13.3*, p. *132*; *14.3.2*, p. *187*; 22.1.1, p. 345.

T. P. LEPÈTRE, J.-M. DUSSEAU, and J.-L. ROBERT
*Expression of Acoustoelectric Net Gain in p-Type GaSb
(a) Vol. **33**, No. 2, p. 715–722 (1976).
See also *7*, p. *65*; 22.2.3, p. 367.

J. K. PRIBRAM and C. M. PENCHINA
Phonon-Assisted Electroabsorption by Contact Excitons
(b) Vol. **74**, No. 1, p. 375–381 (1976).
See also 6, p. 55; *13.5.1*, p. *151*; 20.1, p. 264.

F. WEBER
*The Influence of Carrier Injection on Helical Density Waves
(a) Vol. **34**, No. 1, p. 107–116 (1976).
See also 14.3.2, p. 187; 22.1.1, p. 346.

M. MANZEL, W. ECKE, and H. BRUCHLOS
*Negative Differential Resistance in Gold-Doped High-Resistivity Silicon
(a) Vol. **34**, No. 1, p. 219–226 (1976).
See also 13.4, p. 137; 14.3.3, p. 190; 22.1.2, p. 352.

S. G. CHAMBERLAIN, B. Y. WOO, and D. B. SCOTT
Modelling of Electrical Charge Injection into Charge-Coupled Devices
(a) Vol. **34**, No. 1, p. 263–275 (1976).
See also *14.3.2*, p. *187*; *14.3.3*, p. *190*; 22.1.2, p. 352.

A. KIVERIS and P. PIPINYS
*The Exhaustion of Traps in ZnS Phosphors by Electric Field and the Gudden-Pohl Effect
(a) Vol. **34**, No. 1, p. 377–382 (1976).
See also *20.3*, p. *281*; 22.4.1, p. 379.

P. M. KARAGEORGY-ALKALAEV, I. Z. KARIMOVA P. I. KNIGIN, and A. YU. LEIDERMAN
*Double-Injection Currents in Zn-Compensated Silicon p–i–n Structures
(a) Vol. **34**, No. 1, p. 391–395 (1976).
See also 13.4, p. 137; 22.1.2, p. 352.

J. E. PARROTT and M. TIENG
Space Charge Capacitance of an Abrupt p–n Junction
(a) Vol. **34**, No. 1, p. K43–K46 (1976).
See also *14.3.3*, p. *191*; 22, p. 335.

K. HENNEBERGER
*On the Theory of Photon-Assisted Zener Tunneling
(b) Vol. **74**, No. 2, p. 567–577 (1976).
See also 13.1, p. 123; 22, p. 335.

V. ŠMÍD
*A Contribution to the Interpretation of the High-Field Electrical Conductivity of Amorphous Semiconductors
(b) Vol. **74**, No. 2, p. K169–K171 (1976).
See also *2*, p. *37*.

S. Nešpurek and E. A. Silinsh
*Space-Charge-Limited Current Theory for Molecular Crystals with Gaussian Distribution of Local Trapping States
(a) Vol. **34**, No. 2, p. 747–759 (1976).
See also 13.4, p. 138; 22.9, p. 435.

K.-H. Müller, M. Wolf, H. Hemschik, and K. Hausmann
*Stability of a VO_2 Thermal Switching Sample with Filament Formation
(a) Vol. **34**, No. 2, p. K199–K202 (1976).
See also 22.6, p. 404.

A. Matulionis, J. Požela, and A. Reklaitis
*Monte Carlo Calculations of Hot-Electron Transient Behaviour in CdTe and GaAs
(a) Vol. **35**, No. 1, p. 43–48 (1976).
See also *13*, p. *116*; 22.2.1, p. 363; 22.4.3, p. 385.

S. V. Gantsevich, V. L. Gurevich, V. D. Kagan, and R. Katilius
*Transformation of Incoherent Phonon Flux into a Coherent Acoustical Signal as the Result of the Nonlinear Amplification of Acoustical Phonons in Semiconductors
(b) Vol. **75**, No. 2, p. 407–422 (1976).
See also 6, p. 56; *7*, p. *65*; 22, p. 336.

G. Bauer, M. Overhamm, P. Grosse, W. Müller, E. Gornick, and H. W. Pötzl
*Landau Emission in n-InSb
(b) Vol. **75**, No. 2, p. 543–551 (1976).
See also *13.1*, p. *124*; 20.1, p. 268; 20.3, p. 283; 22.2.3, p. 368.

L. Schenk, H. Bruhns, and K. Hübner
*Numerical Investigation of the *z*-Pinch in Intrinsic Indium Antimonide
(a) Vol. **35**, No. 2, p. 511–520 (1976).
See also 22.2.3, p. 368.

W. Moldenhauer and W. Brückner
*Thermal Switching in Nearly Stoichiometric Iron Sulfide Single Crystals
(a) Vol. **35**, No. 2, p. K109–K112 (1976).
See also 22, p. 337.

V. S. Komissarov
*Calculation of Hot Electron Mobility and Electron Temperature in n-Type InSb under the Assumption of a Maxwellian Distribution
(b) Vol. **76**, No. 1, p. 105–112 (1976).
See also *13*, p. *116*; 22.2.3, p. 368.

D. Barjon, A. Raymond, and J. L. Robert
*Low-Energy Gap Semiconductor Mobilities in Presence of Spatial Fluctuations of Carrier Density
(b) Vol. **76**, No. 1, p. 223–229 (1976).
See also *13*, p. *117*; 22.2.3, p. 368.

K. I. Papunashvilli and E. K. Frolova
*Longitudinal Anisotropy of Conductivity of n-Si in Heating Electric Fields at Low Temperatures
(b) Vol. **76**, No. 1, p. 351–355 (1976).
See also *13*, p. *117*; 22.1.2, p. 354.

I. M. Dykman and P. M. Tomchuk
*Transport Phenomena and Eigenoscillations of Hot Electrons in Semiconductors with Periodic Structure
(b) Vol. **76**, No. 1, p. 385–393 (1976).
Erratum
(b) Vol. **78**, No. 2, p. 843 (1976).
See also *13*, p. *117*; 22, p. 338.

P. C. MATHUR and VIJENDER SHARMA
*Large-Signal Analysis of IMPATT Diodes
(a) Vol. **36**, No. 1, p. 269—274 (1976).
See also *14.3.3*, p. *192*; 22.2.1, p. 363.

G. AERS, A. D. BOARDMAN, and E. D. ISAAC
*A Monte Carlo Study of the Gunn Effect Behaviour of a Ferromagnetic Semiconductor
(a) Vol. **36**, No. 1, p. 357—364 (1976).
See also *13.5.4*, p. *160*; 22.8.2, p. 429.

E. M. AVAKIAN, K. G. BELABAEV, A. A. KAMINSKII, and V. KH. SARKISOV
*Observation of the Spontaneous Electric Breakdown in Pyroelectric $LiNbO_3$ and $LiTaO_3$ Crystals
(a) Vol. **36**, No. 1, p. K25—K27 (1976).
Erratum
(a) Vol. **37**, No. 1, p. K111 (1976).
See also 22.8.1, p. 422.

B. PETZEL
*Influence of Temperature on Impact Ionization in Gunn Domains
(a) Vol. **36**, No. 1, p. K39—K41 (1976).
See also *14.3.3*, p. *192*; 22.2.1, p. 363.

D. ROY CHAUDHURI, P. K. BASU, and A. N. CHAKRAVARTI
Effect of the Surface Electric Field on the Specific Heat of Semiconductor Inversion Layers
(a) Vol. **36**, No. 1, p. K65—K69 (1976).
See also *8*, p. *69*; 22.1.2, p. 355.

H.-J. ANKLAM and K.-D. UFERT
Electrothermal Behaviour of VO_2 Thin Film Switching Devices
(a) Vol. **36**, No. 1, p. K71—K75 (1976).
See also *8*, p. *69*; *14.3.3*, p. *192*; 22.6, p. 405.

S. GUHA and M. SATHE
*Nonlinear Collision-Induced Convective Instability in Solid-State Plasmas
(a) Vol. **36**, No. 2, p. 453—459 (1976).
See also *13*, p. *118*; 22, p. 339.

Ľ. HRIVNÁK
*The Switching Effect in Crystalline Semiconductors Due to the Screening of Donors by Conduction Electrons
(a) Vol. **36**, No. 2, p. 551—556 (1976).
See also 13.4, p. 141; 22, p. 339.

C. MANFREDOTTI, C. DE BLASI, S. GALASSINI, G. MICOCCI, L. RUGGIERO, and A. TEPORE
*Analysis of SCLC Curves by a New Direct Method
(a) Vol. **36**, No. 2, p. 569—577 (1976).
See also 22.3, p. 373.

B. P. RAI, K. SINGH, and R. S. SRIVASTAVA
Current Transport Phenomena in SiO_2 Films
(a) Vol. **36**, No. 2, p. 591—595 (1976).
See also *14.3.1*, p. *185*; 22.6, p. 406.

A. SAMOĆ, M. SAMOĆ, and J. SWORAKOWSKI
Interpretation of One-Carrier Thermally Stimulated Currents and Isothermal Decay Currents (I): Basic Concepts
(a) Vol. **36**, No. 2, p. 735—745 (1976).
See also *13.4*, p. *142*; 14.4, p. 203.

J. Toušková and R. Kužel
*Fundamental Properties of Au–CdTe Metal–Semiconductor Contact
(a) Vol. **36**, No. 2, p. 747—755 (1976).
See also *14.3.3*, p. *193*; 22.4.3, p. 386.

H.-G. Holzhütter and M. Mocker
*Spectral Dependence of the Photocurrent in Semiconductors
(b) Vol. 77, No. 1, p. 203—214 (1976).
See also *16*, p. *215*; 22, p. 339.

H. P. Braun and M. E. Michel-Beyerle
*Current and Fluorescence Oscillations in Anthracene Crystals
(a) Vol. **37**, No. 1, p. 73—82 (1976).
See also 16, p. 216; 20.3, p. 286; 22.9, p. 437.

A. Samoć, M. Samoć, J. Sworakowski, J. M. Thomas, and J. O. Williams
Interpretation of One-Carrier Thermally Stimulated Currents and Isothermal Decay Currents (II): Application to the Evaluation of Trap Parameters in Anthracene
(a) Vol. **37**, No. 1, p. 271—278 (1976).
See also *13.4*, p. *142*; 14.4, p. 204; 22.9, p. 438.

A. Kiveris, Š. Kudžmauskas, and P. Pipinys
Release of Electrons from Traps by an Electric Field with Phonon Participation
(a) Vol. **37**, No. 1, p. 321—327 (1976).
See also 6, p. 61; *13.4*, p. *142*; *20.3*, p. *286*; 22.4.1, p. 381.

Ľ. Hrivnák
*Injection of Electrons into Semi-Insulating GaAs
(a) Vol. **37**, No. 1, p. K1—K4 (1976).
Erratum
(a) Vol. **38**, No. 2, p. K179 (1976).
See also 22.2.1, p. 364.

J. Humlíček and F. Lukeš
*Direct Identification of the $L_3' \rightarrow L_1$ Transition in Ge
(b) Vol. 77, No. 2, p. 731—738 (1976).
See also 13.1, p. 127; *20.1*, p. *274*; 22.1.1, p. 348.

V. A. Brodovoi, N. Z. Derikot, B. R. Kiyak, G. P. Peka, and A. I. Proskura
*Trigger Effect in Electroluminescent Ceramics ZnSe–Cu, I
(a) Vol. **37**, No. 2, p. 363—369 (1976).
See also *13.4*, p. *143*; 20.3, p. 287; 22.4.2, p. 383.

M. Manzel and W. Ecke
*The Influence of Trapped Space Charge on the Double-Injection Switching
(a) Vol. **37**, No. 2, p. 479—484 (1976).
See also 13.4, p. 143; 22.1.2, p. 356.

H. P. Braun
*Current Oscillations in Photo-Oxidized Anthracene Crystals
(a) Vol. **37**, No. 2, p. 553—559 (1976).
See also 13.4, p. 143; 16, p. 216; 22.9, p. 438.

R. Meaudre and M. Meaudre
*Capacitive Effects in Insulators or Semiconductors. Influence of the Density of States in the Gap
(a) Vol. **37**, No. 2, p. 633—643 (1976).
See also *13.4*, p. *143*; 22.4, p. 377.

B. Pohoryles
*Damping of Current Oscillations in Compensated Germanium by a Weak Magnetic Field
(a) Vol. 37, No. 2, p. K117–K119 (1976).
See also 22.1.1, p. 349.

W. Moldenhauer and W. Brückner
Steady-State Switching Characteristics of V_3O_5 Single Crystals
(a) Vol. 37, No. 2, p. K143–K146 (1976).
See also *14.3.3*, p. *193*; 22.6, p. 407.

J. M. Pawlikowski
Negative Resistance of Metal–p-$Cd_xHg_{1-x}Te$ Contacts
(a) Vol. 37, No. 2, p. K183–K186 (1976).
See also 22.4.4, p. 387.

K.-H. Müller, M. Wolf, and F. Kammler
Thermal Switching and Current–Voltage Characteristics for Semiconducting Sandwich Samples
(a) Vol. 37, No. 2, p. K201–K203 (1976).
See also 8, p. 71; *14.3.3*, p. *194*; 22, p. 340.

J. T. Devreese and R. Evrard
*On the Momentum Distribution of Electrons in Polar Semiconductors for High Electric Field
(b) Vol. 78, No. 1, p. 85–92 (1976).
See also 6, p. 62; 22.2.3, p. 369.

I. Neizvestny, N. Ovsyuk, and M. Sinyukov
*Electro-Reflectance Spectra of E_1, $E_2 + \Delta_1$ Transitions in Germanium
(b) Vol. 78, No. 1, p. K19–K22 (1976).
See also 13.1, p. 128; *20.1*, p. *276*; 22.1.1, p. 349.

U. del Pennino and S. Mantovani
Width of the Dislocation Energy Band in Germanium
(a) Vol. 38, No. 1, p. 109–113 (1976).
See also *10.2*, p. *97*; *13.1*, p. *128*; 22.1.1, p. 349.

K. S. Bobev, N. I. Diakonova, and N. V. Mileshkina
*Field Emission through Semiconductor and Dielectric Layers
(a) Vol. 38, No. 1, p. 393–399 (1976).
See also 22.1.1, p. 349; 22.6, p. 408.

F. Bechstedt, R. Enderlein, and K. Peuker
*Resonance Raman Scattering in a Strong Electric Field
(b) Vol. 78, No. 2, p. 711–719 (1976).
See also 6, p. 64; *20.1*, p. *278*; 22, p. 342.

V. Dienys and Ž. Kancleris
*Influence of Uniaxial Stress and Electron–Electron Scattering on the Phenomenological Conductivity Relaxation Time in n-Type Germanium
(b) Vol. 78, No. 2, p. 757–765 (1976).
See also 12.2, p. 112; 22.1.1, p. 349.

V. M. Fomin and E. P. Pokatilov
*Non-Linear Absorption of Electromagnetic Waves by Band Charge Carriers in a Laser Field
(b) Vol. 78, No. 2, p. 831–842 (1976).
See also *20.1*, p. *278*.

A. Conan, M. Zoaeter, and G. Goureaux
*Mechanism of Thermal Switching in Bulk $MoTe_{2-x}$ Compounds
(a) Vol. **38**, No. 2, p. 505—512 (1976).
See also 22, p. 342.

B. Joseph
*The Electron Injection from Si into SiO_2 Due to a Mobile Ion Induced Internal Field
(a) Vol. **38**, No. 2, p. K127—K130 (1976).
See also *14.3.3*, p. *194*; 22.1.2, p. 358; 22.6, p. 408.

Yu. V. Zabara, A. Yu. Kudzin, and K. A. Kolesnichenko
* Space-Charge-Limited Currents in Barium Titanate Single Crystals
(a) Vol. **38**, No. 2, p. K131—K134 (1976).
See also 22.8.1, p. 425.

A. Goswami and S. Radhakrishnan
*Space-Charge-Limited Conduction in Indanthrone Films
(a) Vol. **38**, No. 2, p. K151—K153 (1976).
See also 22.9, p. 439.

14.4 Dielectrics

Original Papers and Short Notes

P. Petelenz
Effect of Polarization on Charge Carrier Transfer in Molecular Crystals
(b) Vol. **73**, No. 1, p. 293–299 (1976).
See also *13*, p. *113*; 22.9, p. 433.

M. Natarajan and E. A. Secco
Electrical Conductivity and Phase Transformation Studies on the Mixed Metal Halides $RbCdX_3$ (X = Cl, Br, I) and Cs_2CuBr_4
(a) Vol. **33**, No. 1, p. 427–433 (1976).
See also *1.2*, p. *14*; 8, p. 67; 22.5.4, p. 401.

A. Tóth
*On the Mechanism of Charge Transport Caused by Moving Dislocations in NaCl Crystals Doped with Ca
(a) Vol. **33**, No. 1, p. K47–K50 (1976).
See also *10.2*, p. *88*; 22.5.2, p. 392.

K. Gesi
*Dielectric Study on the Phase Transitions in Triammonoium Hydrogen Disulfate $(NH_4)_3H(SO_4)_2$
(a) Vol. **33**, No. 2, p. 479–482 (1976).
See also *1.2*, p. *14*; 22.8.1, p. 419.

K. Mazur
*Determination of Real Homocharge in Carnauba Wax Electrets
(a) Vol. **33**, No. 2, p. K155–K159 (1976).
See also 22.9, p. 434.

E. Hartmann and Á. Péter
*The Ionic Conductivity of Plastically Twisted NaCl Whiskers
(a) Vol. **33**, No. 2, p. K169–K171 (1976).
See also *10.2*, p. *89*; 22.5.2, p. 393.

F. Brosens, L. F. Lemmens, and J. T. Devreese
*Frequency-Dependent Exchange Correction to the Dielectric Function of the Electron Gas
(b) Vol. **74**, No. 1, p. 45–55 (1976).
See also *13*, p. *115*.

V. K. Agarwal
*Correlation of Dielectric Breakdown Strength with Basic Parameters in Ionic Crystals
(a) Vol. **34**, No. 1, p. K67–K69 (1976).
See also 22.5.2, p. 394.

U. Rehberg
*Messung der komplexen Dielektrizitätszahl des Anionradikal-Komplexsalzes von Tetracyanochinodimethan mit Methyltriphenylarsonium
(Measurement of the Complex Permittivity of the Anion Radical Complex Salt of Tetracyanochinodimethane and Methyltriphenylarsonium)
(a) Vol. **34**, No. 2, p. 573–582 (1976).
See also 14.3, p. 175; 22.9, p. 435.

C. Schaffrin
*Oxygen Diffusion in $BaTiO_3$ Ceramic
(a) Vol. **35**, No. 1, p. 79–88 (1976).
See also *9*, p. *72*; 22.8.1, p. 421.

J. Bouwma, K. J. de Vries, and A. J. Burggraaf
*Non-Stoichiometry, Defect Structure, and Dielectric Relaxation in Lanthana-Substituted $SrTiO_3$
(a) Vol. **35**, No. 1, p. 281—290 (1976).
See also *10.2*, p. *92*; 22.8, p. 415.

J. Pinguet and S. S. Minn
*Optical Stimulation of Depolarization Currents in Silicon Monoxide Doped with Metallic Impurities
(a) Vol. **35**, No. 2, p. 431—435 (1976).
See also *16*, p. *214*; 22.6, p. 405.

D. I. Sheka and A. N. Korol
*Dielectric Permittivity of Semiconductors with Narrow Forbidden Gap in the Long-Wave Limit
(b) Vol. **76**, No. 1, p. 413—418 (1976).
See also 13.1, p. 125; 22.2, p. 361.

H. Beige, C. Birkholz, E. Ciesla, and G. Schmidt
*Elastic and Electromechanical Properties of Some Crystals near Structural Phase Transitions
(b) Vol. **76**, No. 1, p. K47—K50 (1976).
See also 1.2, p. 17; *12.2*, p. *111*; 22.8.1, p. 422.

A. L. Efros and B. I. Shklovskii
*Critical Behaviour of Conductivity and Dielectric Constant near the Metal–Non-Metal Transition Threshold
(b) Vol. **76**, No. 2, p. 475—485 (1976).
See also *14.1*, p. *165*.

S. Nagel, K. Maschke, and A. Baldereschi
*Pseudopotential Calculation of the Electronic Valence Charge Density and the Optical Dielectric Constant of NaCl and MgO
(b) Vol. **76**, No. 2, p. 629—632 (1976).
See also *13*, p. *117*; 22.5.2, p. 396; 22.6, p. 406.

H. M. Kizilyalli and P. R. Mason
*DC and AC Electrical Conduction in Single Crystal Alumina
(a) Vol. **36**, No. 2, p. 499—508 (1976).
See also 22.6, p. 406.

A. Samoć, M. Samoć, and J. Sworakowski
Interpretation of One-Carrier Thermally Stimulated Currents and Isothermal Decay Currents (I): Basic Concepts
(a) Vol. **36**, No. 2, p. 735—745 (1976).
See also *13.4*, p. *142*; 14.3.4, p. 198.

J. Sárközi and R. Voszka
The Variation of the Surface Charge of NaCl:Pb Monocrystals Due to Doping and Annealing
(a) Vol. **36**, No. 2, p. K105—K107 (1976).
See also *10.2*, p. *94*; 22.5.2, p. 396.

V. R. K. Murthy and J. Sobhanadri
*Dielectric Properties of Some Nickel–Zinc Ferrites at Radio Frequency
(a) Vol. **36**, No. 2, p. K133—K135 (1976).
See also 18.3, p. 241; 22.8.2, p. 430.

A. Samoć, M. Samoć, J. Sworakowski, J. M. Thomas, and J. O. Williams
Interpretation of One-Carrier Thermally Stimulated Currents and Isothermal Decay Currents (II): Application to the Evaluation of Trap Parameters in Anthracene
(a) Vol. **37**, No. 1, p. 271–278 (1976).
See also *13.4*, p. *142*; 14.3.4, p. 199; 22.9, p. 438.

J. Grigas, J. Meshkauskas, and A. Orliukas
*Dielectric Properties of Sb_2S_3 at Microwave Frequencies
(a) Vol. **37**, No. 1, p. K39–K41 (1976).
See also 22, p. 340.

H. G. Wessjohann
Relativistic Excitations of a Thin Graphite Film by Inelastic Electron Scattering in Theory and Experiment
(b) Vol. **77**, No. 2, p. 535–542 (1976).
See also *13.5*, p. *147*; 22.1, p. 344.

M. Taut and W. Hanke
Dielectric Function of Hexagonal Metals (II): Beryllium
(b) Vol. **77**, No. 2, p. 543–550 (1976).
See also *13.1*, p. *127*; 13.5.2, p. 157; *14.1*, p. *166*; 20.1, p. 274; 21.3, p. 324.

M. E. van Hulle and W. Maenhout-van der Vorst
*Ionic Conductivity and Dielectric Loss of AgCl Doped with Pb^{2+}
(a) Vol. **37**, No. 2, p. 473–477 (1976).
See also *10.2*, p. *96*; 22.5.1, p. 391.

M. Taut and P. Ziesche
Dielectric Function of Hexagonal Metals (III): Magnesium
(b) Vol. **78**, No. 1, p. 169–175 (1976).
See also *13.1*, p. *128*; 13.5.2, p. 158; *14.1*, p. *167*; 20.1, p. 275; 21.3, p. 324.

G. Berg and J. Poźniak
Theoretical Analysis of the Dipole Polarization and Depolarization in Me^{++}-Doped NaCl-Type Alkali Halide Crystals (I): Nearest-Neighbour Approach
(a) Vol. **38**, No. 1, p. 115–122 (1976).
See also *10.2*, p. *97*; 22.5.2, p. 398.

N. Starbov, A. Buroff, and J. Malinowski
*Surface Ionic Conductivity, Lattice Disorder, and Space Charge Region in Thin Silver Bromide Layers
(a) Vol. **38**, No. 1, p. 161–170 (1976).
See also 10.2, p. 97; 22.5.1, p. 391.

P. C. Mehendru, N. L. Pathak, Satbir Singh, and P. Mehendru
*Electrical Conduction in Polypropylene Thin Films
(a) Vol. **38**, No. 1, p. 355–360 (1976).
See also *14.3.1*, p. *185*; 22.9, p. 439.

P. K. C. Pillai, Rabinder Nath, and P. K. Nair
*Non-Isothermal Relaxation in ZnO Photoelectrets
(a) Vol. **38**, No. 1, p. K73–K76 (1976).
See also *16*, p. *217*; 22.4, p. 377.

Usha Kapur and V. K. Srivastava
*Thickness Dependence of the Static Dielectric Constant of Some Built-up Langmuir Films
(a) Vol. **38**, No. 1, p. K77–K80 (1976).
See also 22.9, p. 439.

V. I. Zinenko
*On the Theory of the Structural Phase Transition in Squaric Acid
(b) Vol. **78**, No. 2, p. 721–727 (1976).
See also *1.2*, p. *20*; 22.9, p. 439.

T. Gotô, Y. Nigara, and K. Kamiyoshi
*Dielectric Constant of Ammonium and Cesium Halides
(a) Vol. **38**, No. 2, p. K135–K138 (1976).
See also 22.5.2, p. 398; 22.5.4, p. 401.

14.4.1 Ferroelectrics

Original Papers and Short Notes

G. Borchardt, G. Schmidt, and A. Rost
*Non-Linear Dielectric Coefficients of Free and Clamped Barium Titanate Crystals
(a) Vol. **33**, No. 1, p. 143–147 (1976).
See also 22.8.1, p. 419.

A. E. Kruminš and M. Vaněček
*Photoconductivity of Transparent Lanthanum-Modified Lead Zirconate Titanate Ferroelectric Ceramic
(a) Vol. **33**, No. 1, p. K31–K34 (1976).
See also *16*, p. *213*; 22.8.1, p. 419.

S. Suzuki and M. Takagi
Free Rotation of NO_2 Radicals in $NaNO_2$ Crystal
(a) Vol. **33**, No. 2, p. 789–792 (1976).
See also *20.1*, p. *263*; 22.8, p. 413.

V. M. Fridkin, K. D. Kochev, Yu. S. Kusminov, K. A. Verkhovskaya, and T. R. Volk
*The Shift of the Optical Absorption Edge at "Optical Damage" in $LiNbO_3$
(a) Vol. **33**, No. 2, p. K137–K139 (1976).
See also *20.1*, p. *263*; 22.8.1, p. 419.

A. Yoshihara, T. Fujimura, and K. I. Kamiyoshi
*Phase Transition of Ammonium Sulfate
(a) Vol. **34**, No. 1, p. 369–376 (1976).
See also 22.8.1, p. 420.

R. Blinc, H. Arend, and A. Kandušer
Vibrational Spectra, Hydrogen Bonding, and Ferroelectricity in the $PbHPO_4$ Family
(b) Vol. **74**, No. 2, p. 425–435 (1976).
See also 4, p. 45; *6*, p. *55*; *20.1*, p. *265*; 22.8.1, p. 420.

D. V. Kapor
*The Low-Temperature Expansion of the Polarization of the KDP-Like Crystals
(b) Vol. **74**, No. 2, p. 497–505 (1976).

K. N. Pak
*On the Dynamics of Hydrogen-Bonded Ferroelectrics
(b) Vol. **74**, No. 2, p. 673–682 (1976).
See also 22.8.1, p. 420.

E. V. Bursian, Ya. G. Girshberg, and A. V. Ruzhnikov
*The Correlation between Optical Absorption Spectra, Carrier Mobility, and Phase Transition Temperature in Some Ferroelectrics
(b) Vol. **74**, No. 2, p. 689–693 (1976).
See also 13.1, p. 123; *20.1*, p. *266*; 22.8.1, p. 420.

Y. Uesu and J. Kobayashi
*Crystal Structure and Ferroelectricity of Cesium Dihydrogen Phosphate CsH_2PO_4
(a) Vol. **34**, No. 2, p. 475–481 (1976).
See also *1.1*, p. *11*; 22.8.1, p. 421.

J. Klimowski, W. Wanarski, and D. Ożgo
Effect of Hydrostatic Pressure on the Phase Transition Temperatures and Spontaneous Polarization of Thiourea Monocrystals
(a) Vol. **34**, No. 2, p. 697–704 (1976).
See also *1.2*, p. *15*; 22.9, p. 435.

A. S. Janitzki, P. Gerthsen, and B. Hoffmann
*Effective Resistivity of Semiconducting Ceramic $BaTiO_3$
(a) Vol. **34**, No. 2, p. K149—K151 (1976).
See also 22.8.1, p. 421.

K.-D. Toepfer and H. W. Helberg
*Dielectric Dispersion of Thiourea Single Crystals at Microwave Frequencies between 8.25 and 75.0 GHz
(a) Vol. **35**, No. 1, p. 131—136 (1976).
See also 1.2, p. 16; 22.9, p. 436.

R. Beliackas and J. Grigas
*Microwave Dielectric Nonlinearity of SbSI
(a) Vol. **35**, No. 1, p. K61—K64 (1976).
See also 22.8, p. 415.

A. M. Ozorio de Almeida and S. Caticha Ellis
X-Ray Scattering from Crystals with Intracellular Disorder
(b) Vol. **75**, No. 2, p. 657—668 (1976).
See also *1.2*, p. *16*; 22.8.1, p. 421.

A. P. Levanyuk and V. V. Osipov
*Optical Distortion in Crystals
(a) Vol. **35**, No. 2, p. 605—614 (1976).
See also 13.4, p. 139; *20.1*, p. *269*; 22.8.1, p. 422.

K. H. Germann, W. Müller-Lierheim, H. H. Otto, and T. Suski
High Temperature Phase Transition in $Pb_5Ge_3O_{11}$
(a) Vol. **35**, No. 2, p. K165—K167 (1976).
See also *1.2*, p. *17*; 20.1, p. 270; 22.8.1, p. 422.

C. N. W. Darlington
Phase Transitions in Perovskite Crystals
(b) Vol. **76**, No. 1, p. 231—239 (1976).
See also *1.2*, p. *17*; 22.8, p. 416.

J. Albers, R. W. Sailer, and H. E. Müser
*Dielectric, Elastic, and Piezoelectric Properties of Mg–Cl Boracite
(a) Vol. **36**, No. 1, p. 189—195 (1976).
See also *12.2*, p. *111*; 22.8, p. 416.

P. Konsin
*Ferroelectric Phase Transitions in Systems with Broad Electronic Bands Caused by Interband Electron–Phonon Interaction
(b) Vol. **76**, No. 2, p. 487—496 (1976).
See also 6, p. 58; 13.1, p. 125; 22, p. 338.

D. V. Kapor
*On the Interaction of KDP-Like Crystals with an External Electromagnetic Field
(b) Vol. **76**, No. 2, p. K93—K96 (1976).

L. Merten and G. Borstel
*Phenomenological Theory of Coupling of Debye Relaxations with IR-Active Phonons
(b) Vol. **77**, No. 1, p. 221—227 (1976).
See also *6*, p. *60*.

J. C. Peuzin et M. Tasson
*Génération et détection d'hypersons. Transducteur feuilleté
(Generation and Detection of Hypersound. Domain Transducer)
(a) Vol. **37**, No. 1, p. 119—125 (1976).
See also *7*, p. *66*; 22.8.1, p. 423.

H. Veeser, U. Bogner, and W. Eisenmenger
*Analysis of the Acoustic Transients in the Pulse Response of the Linear Electro-Optic Effect
(a) Vol. 37, No. 1, p. 161–170 (1976).
See also *7*, p. *66*; 12.2, p. 111; 20.1, p. 273; 22.8.1, p. 423.

D. Kucharczyk, A. Pietraszko, and K. Łukaszewicz
Temperature Dependence of Lattice Parameters of $NaNO_2$ Single Crystals
(a) Vol. 37, No. 1, p. 287–294 (1976).
See also *1.2*, p. *18*; *8*, p. *70*; 22.8, p. 417.

S. Franck, A. Kühnel, T. Nattermann, and S. Wendt
*Noncritical Fluctuations and the Temperature Dependence of the Order Parameter
(b) Vol. 77, No. 2, p. 631–637 (1976).
See also *1.2*, p. *19*.

K. E. Gauss and H. Happ
*Millimeter Wave Investigation of the Complex Dielectric Constant in DKDP
(b) Vol. 78, No. 1, p. 133–138 (1976).
See also 6, p. 62; 20.1, p. 275; 22.8.1, p. 424.

D. Müller and J. Petersson
*Spin–Lattice Relaxation of Several Nuclei near the Ferroelectric Transitions of TGS and TGFBe
(b) Vol. 78, No. 1, p. 191–198 (1976).
See also 19, p. 255; 22.9, p. 438.

R. S. Momtaz and L. N. Rashkovich
The Distribution of Deuterium between Solid and Liquid Phases during the Crystallization of Crystals of the KDP Group from Aqueous Solutions
(a) Vol. 38, No. 1, p. 401–408 (1976).
See also *3*, p. *44*; 22.8.1, p. 424.

K. H. Germann
Optical Anisotropy Measurements in Multidomain Samples: Application to Phase Transitions in NiO and $Pb_5Ge_3O_{11}$
(a) Vol. 38, No. 1, p. K81–K84 (1976).
See also 1.2, p. 20; 18.4, p. 246; *20.1*, p. *277*; 22.6, p. 408; 22.8.1, p. 424.

I. B. Barkan, S. I. Marennikov, and M. V. Entin
*Holographic Storage in $LiNbO_3$ Crystal at High Temperatures
(a) Vol. 38, No. 2, p. K139–K142 (1976).
See also *16*, p. *217*; 22.8.1, p. 425.

15. Thermoelectric and Thermomagnetic Properties

Original Papers and Short Notes

C. Uher
*Erratum to: Low-Field Thermomagnetic Tenset for the Bi-Like Band Structure
(b) Vol. **73**, No. 1, p. 379 (1976).
See also 6, p. 53; 13.1, p. 121; 21.7, p. 332.

P. P. Bodiul, D. V. Gitsu, and D. F. Migley
*Anisotropy of Thermomagnetic Effects in $Pb_{1-x}Sn_xTe$
(b) Vol. **73**, No. 1, p. K5—K8 (1976).
See also 22.7, p. 410.

Nguyen van Dong, Tran Huu Danh, B. Auguin, and A. Defresne
Transport Properties of Glassy $CdGe_{1-x}Sb_xAs_2$
(a) Vol. **33**, No. 1, p. 295—299 (1976).
See also *2*, p. *35*; *14.3*, p. *173*; 22.8, p. 413.

K. L. Chopra and P. Nath
Amorphous Ge Alloy Films (II): Transport Properties
(a) Vol. **33**, No. 1, p. 333—344 (1976).
See also *2*, p. *36*; *14.3.1*, p. *183*; 22.1.1, p. 345.

F. I. Bekkermann and D. V. Gitsu
*The Nernst-Ettingshausen Effect in Bismuth under the Influence of Directed Deformation
(b) Vol. **74**, No. 1, p. K1—K4 (1976).
See also *12.1*, p. *107*; 21.7, p. 332.

J. Krempaský and D. Barančok
The Use of Modified Relaxation Time in the Transport Theory of Non-Crystalline Semiconductors
(b) Vol. **74**, No. 2, p. 741—751 (1976).
See also *2*, p. *37*; 13.4, p. 137; *14.3*, p. *175*; 22, p. 335.

E. Litwin-Staszewska and W. Szymańska
Scattering on Short-Range Potentials in Semiconductors with a Narrow Energy Gap
(b) Vol. **74**, No. 2, p. K89—K92 (1976).
See also *14.3*, p. *175*; 22.4.2, p. 383.

W. Moldenhauer and W. Brückner
Physical Properties of Nonstoichiometric Iron Sulfide $Fe_{1-x}S$ near the α-Phase Transition
(a) Vol. **34**, No. 2, p. 565—571 (1976).
See also *1.2*, p. *15*; *14.3*, p. *175*; 18.4, p. 243; 22, p. 336.

E. Gombia, N. Romeo, G. Sberveglieri, and C. Paorici
Transport Properties of Semiconducting $ZnIn_2S_4$
(a) Vol. **34**, No. 2, p. 651—655 (1976).
See also *14.3*, p. *175*; 22.8, p. 414.

P. J. P. de Maayer and J. D. Mackenzie
*Effect of Defect Structure on the Seebeck Coefficient of Thin Films of TiN_x
(a) Vol. **34**, No. 2, p. K143—K146 (1976).
See also *10.2*, p. *91*; 22, p. 336.

P. G. Zhumatii
*Intraband Conductivity and Thermopower of Semiconductors with Slowly Varying Gaussian Random Field
(b) Vol. **75**, No. 1, p. 61—72 (1976).
See also *14.3*, p. *176*; 22, p. 336.

W. Brückner, W. Moldenhauer, B. Thuss, and G. Försterling
Crystallographic Changes at the Conductivity Transition of V_3O_5
(a) Vol. **35**, No. 1, p. K49—K51 (1976).
See also *1.2*, p. *16*; 22.6, p. 404.

L. M. Valiev, I. G. Kerimov, N. G. Aliev, and A. A. Abdurragimov
Thermal Expansion and Electrical Properties of $Cu_4Mn_2X_4$(X:S, Se, and Te)
(a) Vol. **35**, No. 1, p. K85—K88 (1976).
See also 8, p. 68; 14, p. 162; *18.1*, p. *226*; *18.4*, p. *244*; 22.8, p. 415.

G. Cubiotti, E. Donato, E. S. Giuliano, R. Ruggeri, and A. Stancanelli
*Transport Properties of Liquid Cadmium
(b) Vol. **75**, No. 2, p. K93—K96 (1976).
See also 2, p. 38; *14.1*, p. *165*; 21, p. 295.

C. Uher and H. J. Goldsmid
*Phonon-Drag Generated Umkehreffekt in Bismuth
(b) Vol. **75**, No. 2, p. K137—K141 (1976).
See also 6, p. 57; 21.7, p. 332.

K. L. Chopra, S. K. Barthwal, and D. K. Pandya
*Structural Sensitivity of Transport and Optical Properties of Amorphous GeTe
(a) Vol. **35**, No. 2, p. 761—768 (1976).
See also 2, p. 38; *14.3*, p. *178*; 20.1, p. 270; 22.7, p. 411.

K. L. Chopra and D. K. Pandya
Obliquely Deposited Amorphous Ge Films (II): Electrical Properties
(a) Vol. **36**, No. 1, p. 89—100 (1976).
See also *2*, p. *39*; *14.3.1*, p. *184*; 22.1.1, p. 348.

S. K. Barthwal and K. L. Chopra
·Transport Properties of Amorphous InSb and GaAs
(a) Vol. **36**, No. 1, p. 345—355 (1976).
See also *2*, p. *39*; *14.3.1*, p. *184*; 22.2.1, p. 363; 22.2.3, p. 368.

A. Jędrzejczak and T. Dietl
*Thermomagnetic Properties of n-Type and p-Type HgTe
(b) Vol. **76**, No. 2, p. 737—751 (1976).
See also *13.1*, p. *126*; 14.3, p. 179; 22.4.3, p. 386.

S. K. Barthwal and K. L. Chopra
*Thermoelectric Power of Amorphous Germanium Films
(a) Vol. **36**, No. 2, p. 533—549 (1976).
See also 2, p. 39; *14.3.1*, p. *184*; 22.1.1, p. 348.

P. Ganguly, Om Parkash, and C. N. R. Rao
Electron Transport and Magnetic Properties of Rare-Earth Ortho-Titanites and -Vanadites, $LnTiO_3$ and $LnVO_3$
(a) Vol. **36**, No. 2, p. 669—678 (1976).
See also *14*, p. *162*; *18*, p. *224*; 22.8.1, p. 423.

P. G. Vassilev
*Annealing Effects on the Thermoelectric Power and Visible Reflectance of Thin Copper Films
(a) Vol. **37**, No. 1, p. K25—K27 (1976).
See also 14.1, p. 166; *20.1*, p. *273*; 21.1, p. 311.

M. M. Kaila, L. B. Harris, and H. J. Goldsmid
*Thermoelectric and Thermomagnetic Properties of Plastically Bent InSb Samples
(a) Vol. **37**, No. 1, p. K43—K46 (1976).
See also *14.3*, p. *179*; 22.2.3, p. 368.

K. ZIEGLER, H.-D. JUNKER, and U. BIRKHOLZ
*Electrical Conductivity and Seebeck Coefficient of Si_2Te_3 Single Crystals
(a) Vol. 37, No. 1, p. K97—K99 (1976).
See also *14.3*, p. *179*; 22, p. 340.

E. I. TERUKOV, F. A. CHUDNOVSKII, W. REICHELT, H. OPPERMANN, W. BRÜCKNER, H.-P. BRÜCKNER, and W. MOLDENHAUER
Investigation of the Physical Properties of V_3O_5 at Phase Transition with Consideration of Its Range of Homogeneity
(a) Vol. 37, No. 2, p. 541—546 (1976).
See also *14.3*, p. *180*; 18.1, p. 227; 22.6, p. 407.

E. MROSAN and G. LEHMANN
Localized Non-Magnetic Impurities in Metals (II): Application to 3d-Transition Metal Impurities in Aluminium
(b) Vol. 78, No. 1, p. 159—167 (1976).
See also 8, p. 71; *13.4*, p. *144*; 14.1, p. 167; 14.2, p. 172; 21, p. 299.

W. BRÜCKNER, U. GERLACH, W. MOLDENHAUER, H. P. BRÜCKNER, N. MATTERN, H. OPPERMANN, and E. WOLF
Phase Transitions and Semiconductor–Metal Transition in $V_{1-x}Ga_xO_2$ Single Crystals
(a) Vol. 38, No. 1, p. 93—102 (1976).
See also *1.2*, p. *19*; *3*, p. *44*; 14.3, p. 181; 18.1, p. 227; 22.6.1, p. 409.

W. BRÜCKNER and W. MOLDENHAUER
*Anisotropy of Thermoelectric Power in $V_{1-x}Me_xO_2$(Me = Al, Ga, Fe)
(a) Vol. 38, No. 1, p. K17—K21 (1976).
See also 1.2, p. 20; 22.6.1, p. 409.

E. I. TERUKOV, F. A. CHUDNOVSKII, W. BRÜCKNER, W. REICHELT, H.-P. BRÜCKNER, W. MOLDENHAUER, and H. OPPERMANN
*Influence of Mo, Ti, and Ga Doping on Conductivity Transition of V_3O_5
(a) Vol. 38, No. 1, p. K23—K26 (1976).
See also *14.3*, p. *181*; 22.6, p. 408; 22.6.1, p. 409.

N. P. KEKELIDZE, Z. V. KVINIKADZE, Z. F. DAVITAYA, G. N. ERITSYAN, V. A. SAAKYAN, and F. K. KARAPETYAN
*Studies of Thermoelectric Characteristics of InP_xAs_{1-x} Alloys
(a) Vol. 38, No. 1, p. K49—K52 (1976).
See also *8*, p. *71*; 22.2.4, p. 371.

K. M. VAN VLIET and A. H. MARSHAK
Conduction Current and Generalized Einstein Relations for Degenerate Semiconductors and Metals
(b) Vol. 78, No. 2, p. 501—517 (1976).
See also 13.1, p. 129; *14.1*, p. *167*; *14.3*, p. *181*; 21, p. 300; 22, p. 341.

J. KOSSUT
Electron Transport Phenomena in Narrow- and Zero-Gap Semiconductors Containing Magnetic Impurities
(b) Vol. 78, No. 2, p. 537—542 (1976).
See also *13.4*, p. *145*; *14.3*, p. *181*; 22.4.3, p. 386.

K. M. VAN VLIET
*Entropy Production and Choice of Heat Flux for Degenerate Semiconductors and Metals
(b) Vol. 78, No. 2, p. 667—676 (1976).
See also *14.3*, p. *181*; 21, p. 300; 22, p. 342.

M. Stordeur and W. Heiliger
*Anisotropy of the Thermopower of p-Sb_2Te_3
(b) Vol. 78, No. 2, p. K103—K106 (1976).
See also 22.7, p. 412.

V. R. K. Murthy and J. Sobhanadri
Electrical Conductivity of Some Nickel-Zinc Ferrites
(a) Vol. 38, No. 2, p. 647—651 (1976).
See also *14.3*, p. *182*; 18.3, p. 241; 22.8.2, p. 431.

16. Photoconductivity. Photovoltaic Effects

Review Article

J. KADLEC and K. H. GUNDLACH
*Results and Problems of Internal Photoemission in Sandwich Structures
(a) Vol. **37**, No. 1, p. 11—28 (1976).
See also 13.1, p. 120; *14.3.3*, p. *190*; 22.6, p. 402.

Original Papers and Short Notes

M. LEPPIHALME and T. TUOMI
*Photovoltage Spectra of Metal–Silicon (p-Type) Diodes near the Indirect Absorption Edge of Silicon
(a) Vol. **33**, No. 1, p. 125—133 (1976).
See also 6, p. 53; 13.5.1, p. 149; *14.3.3*, p. *190*; 20.1, p. 261; 22.1.2, p. 351.

M. TAPIERO, J. P. ZIELINGER, and C. NOGUET
*Photomemory Effect in Cu_2O Single Crystals. Phenomenology and Interpretation
(a) Vol. **33**, No. 1, p. 155—166 (1976).
See also 13.4, p. 135; 22.6, p. 402.

K. MARINOVA, A. OPANOWICZ, and W. RUPPEL
*Photoconductivity of Highly Excited CdSe and $CdS_{0.48}Se_{0.25}$Crystals
(a) Vol. **33**, No. 1, p. 195—204 (1976).
See also 13.5.1, p. 149; 22.4.2, p. 383; 22.4.4, p. 387.

A. E. KRUMINŠ and M. VANĚČEK
*Photoconductivity of Transparent Lanthanum-Modified Lead Zirconate Titanate Ferroelectric Ceramic
(a) Vol. **33**, No. 1, p. K31—K34 (1976).
See also *14.4.1*, p. *206*; 22.8.1, p. 419.

B. KORSCH, F. WILLIG, H. J. GAEHRS, and B. TESCHE
Exchange of Holes between Organic Crystals and Metal Films Evaporated in High Vacuum
(a) Vol. **33**, No. 2, p. 461—471 (1976).
See also *14.3.2*, p. *187*; *14.3.4*, p. *196*; 22.9, p. 433.

B. LEPLEY and S. RAVELET
*Correlation between Open-Circuit Voltage and Interface Parameters at the Au–CdS Illuminated Contact
(a) Vol. **33**, No. 2, p. 517—523 (1976).
See also *14.3.2*, p. *187*; 22.4.1, p. 379.

K. D. GLINCHUK, N. M. LITOVCHENKO, and R. MERKER
The Role of Impurities in the Formation of Quenched-in Recombination Centres in Thermally Treated Silicon
(a) Vol. **33**, No. 2, p. K87—K90 (1976).
See also *13.4*, p. *136*; 22.1.2, p. 352.

I. BIRCHAK, V. V. SERDYUK, and G. G. CHEMERESYUK
*Optical and Temperature Quenching of Photocurrent in Zinc Selenide Single Crystals
(a) Vol. **33**, No. 2, p. K145—K149 (1976).
See also 13.4, p. 136; 22.4.2, p. 383.

P. K. C. PILLAI and R. NATH
*Photo-Induced AC Impedance Measurements on HgI_2 Binder Layers
(a) Vol. **33**, No. 2, p. K151—K154 (1976).
See also 22.5, p. 388.

B. V. NOVIKOV, N. S. SOKOLOV, and S. V. GASTEV
Free-Carrier Generation via Exciton–Phonon and Exciton–Impurity Interaction in Ge Crystals
(b) Vol. 74, No. 1, p. 81–89 (1976).
See also 6, p. 54; *13.4*, p. *136*; *13.5.1*, p. *150*; 22.1.1, p. 345.

K. HAUFFE and A. MEYER-LAACK
*On a Tribo-Photoelectric Effect of Li_2O-Doped Zinc Oxide Single Crystals
(a) Vol. 34, No. 1, p. 145–152 (1976).
See also 22.4, p. 376.

S. MARTINUZZI
*Photopiles solaires à hétérojonction n-CdS–p-CdTe
(n-CdS–p-CdTe Hetrojunction Solar Cells)
(a) Vol. 34, No. 1, p. K21–K26 (1976).
See also *14.3.3*, p. *190*; 22.4.1, p. 379; 22.4.3, p. 385.

J. VAITKUS
*The Nonequilibrium Hall Effect and Related Transport Phenomena in Semiconductors under Inhomogeneous Excitation by a Laser Pulse
(a) Vol. 34, No. 2, p. 769–775 (1976).
See also *14.3*, p. *176*; 20.3, p. 282; 22.1.1, p. 346.

I. S. KAULACH and E. A. SILINSH
The Influence of the Hyperfine Range Magnetic Field on Charge Transfer States at the Metal–Organic Semiconductor Interface
(b) Vol. 75, No. 1, p. 247–254 (1976).
See also *13.3*, p. *132*; 14.3.3, p. 191; 22.9, p. 436.

P. HLÍDEK, I. BARVÍK, V. PROSSER, M. VANĚČEK, and M. ZVÁRA
*Polarization Dependence of Transmission and Photoconductivity in $CdCr_2Se_4$
(b) Vol. 75, No. 1, p. K45–K49 (1976).
See also *20.1*, p. *267*; 22.8.2, p. 428.

A. N. GEORGOBIANI, YU. V. OZEROV, E. E. STRUMBAN, and V. V. ANISIMOV
Investigation of Deep Centres of Gallium-Doped Zinc Sulfide Crystals
(a) Vol. 35, No. 1, p. 251–257 (1976).
See also *13.4*, p. *138*; 20.1, p. 268; 20.3, p. 283; 22.4.1, p. 380.

G. JUŠKA and S. VENGRIS
*Hole Photogeneration in Amorphous Selenium at Low Electric Fields
(a) Vol. 35, No. 1, p. 339–345 (1976).
See also *14.3.2*, p. *188*; 22.1.3, p. 359.

L. BOSI
On the Luminescence and Photoconductivity of the F Centres in NaBr
(b) Vol. 75, No. 2, p. K163–K165 (1976).
See also *10.2*, p. *93*; 20.3, p. 284; 22.5.2, p. 395.

J. PINGUET and S. S. MINN
*Optical Stimulation of Depolarization Currents in Silicon Monoxide Doped with Metallic Impurities
(a) Vol. 35, No. 2, p. 431–435 (1976).
See also *14.4*, p. *203*; 22.6, p. 405.

K. D. GLINCHUK, N. M. LITOVCHENKO, and R. MERKER
Effect of Quenching Rate and Annealing on the Concentration of Quenched-in Recombination Centres in Heat Treated Silicon
(a) Vol. 35, No. 2, p. K157–K160 (1976).
See also *13.4*, p. *140*; 22.1.2, p. 354.

G. DOKTER and H. J. HOFFMANN
Transient Bleaching of the F-Band in Alkali Halides
(b) Vol. 76, No. 1, p. 241–249 (1976).
See also *10.2*, p. *93*; *13.4*, p. *140*; 22.5.2, p. 395.

L. HERNÁNDEZ, O. VIGIL, and F. GÓNZALEZ
Optical Transitions and Distribution of Localized Levels in $ZnIn_2S_4$
(a) Vol. 36, No. 1, p. 33–37 (1976).
See also *13.4*, p. *141*; *20.1*, p. *271*; 22.8, p. 416.

A. YA. VUL, SH. NABIEV, S. G. PETROSYAN, and A. YA. SHIK
*Simulation of the Long-Period Conductivity Relaxation (LCR) Phenomenon in Double-Layer Semiconductor Structures
(a) Vol. 36, No. 1, p. 53–60 (1976).
See also *14.3.1*, p. *184*; 22.2.1, p. 363.

S. MARTINUZZI, O. MALLEM, and T. CABOT
*Dark Currents in Cu_2S–CdS Single Crystal Heterojunctions. Consequences for Photocells
(a) Vol. 36, No. 1, p. 227–234 (1976).
See also 13.3, p. 133; *14.3.3*, p.*192*; 22, p. 338; 22.4.1, p. 381.

V. S. GRIGOREVA, A. A. LEBEDEV, V. D. PROCHUKHAN, YU. V. RUD, and A. A. YAKOVENKO
*Photoelectrical Properties of p-Type $ZnGeP_2$ Single Crystals
(a) Vol. 36, No. 1, p. K51–K54 (1976).
See also 22.8, p. 416.

K. LOCHNER, B. REIMER, and H. BÄSSLER
*Photocarrier Generation in a One-Dimensional Solid: Polydiacetylene-Bis(Toluenesulfonate)
(b) Vol. 76, No. 2, p. 533–540 (1976).
See also 22.9, p. 437.

W. HANUS and M. OSZWAŁDOWSKI
*The Dependence of Photoconductive and Photoelectromagnetic Effects on Slab Thickness in InSb at Room Temperature
(a) Vol. 36, No. 2, p. 445–452 (1976).
See also 13.3, p. 133; 22.2.3, p. 368.

B. SPEIT and G. LEHMANN
Hole Centers in the Feldspar Sanidine
(a) Vol. 36, No. 2, p. 471–481 (1976).
See also *10.2*, p. *94*; 20.1, p. 272; 20.3, p. 285; 22.8.1, p. 423.

F. SPRENGER
*Magnetoresistance of Frozen-in Photoconductivity in Semiconducting CdF_2:In
(a) Vol. 36, No. 2, p. 483–490 (1976).
See also 13.4, p. 141; *14.3*, p. *179*; 22.5, p. 389.

H.-G. HOLZHÜTTER and M. MOCKER
*Spectral Dependence of the Photocurrent in Semiconductors
(b) Vol. 77, No. 1, p. 203–214 (1976).
See also *14.3.4*, p. *199*; 22, p. 339.

E. I. FRANKEVICH, M. M. TRIBEL, and I. A. SOKOLIK
*Photoconductivity of Sublimated Rubrene Films during Oxidation. The Magnetic Field Effect
(b) Vol. 77, No. 1, p. 265–276 (1976).
See also 13.5.1, p. 155; *14.3.1*, p. *185*; 20.3, p. 286; 22.9, p. 437.

N. Saleh and A. Soliman
*Analysis and Design Considerations of Burstein Narrow-Band Photon Detectors
(a) Vol. 37, No. 1, p. 39–43 (1976).
See also 14.3.3, p. 193; 22.2.3, p. 368.

V. A. Vassilev, B. T. Kolomiets, T. N. Mamontova, and E. M. Raspopova
*Photoluminescence and Photoconductivity in Densified Vitreous As_2Se_3
(a) Vol. 37, No. 1, p. 45–49 (1976).
See also 2, p. 40; *20.3*, p. *286*; 22, p. 339.

H. P. Braun and M. E. Michel-Beyerle
Current and Fluorescence Oscillations in Anthracene Crystals
(a) Vol. 37, No. 1, p. 73–82 (1976).
See also *14.3.4*, p. *199*; 20.3, p. 286; 22.9, p. 437.

I. G. Gavrikova, V. P. Migal, and A. L. Rvachev
*Optical Sensitization of the Photoconductivity Bands
(a) Vol. 37, No. 1, p. K21–K24 (1976).
See also 22.4.2, p. 383.

T. Wosiński, T. Figielski, and A. Mąkosa
*Spin-Dependent Photoconductivity Spectrum of Dislocated Silicon
(a) Vol. 37, No. 1, p. K57–K59 (1976).
See also *10.2*, p. *95*; 19, p. 254; 22.1.2, p. 356.

D. Genzow and E. Normantas
Theory of the Photon-Drag Effect in Semiconductors with Elliptical Isoenergetic Surfaces and Its Application to Tellurium Crystals
(b) Vol. 77, No. 2, p. 667–676 (1976).
See also *14.3*, p. *180*; 22.1.3, p. 360.

J. Kadlec and K. H. Gundlach
*Internal Photoemission and Quantum-Mechanical Transmission Coefficient
(a) Vol. 37, No. 2, p. 385–398 (1976).
See also *14.3.3*, p. *193*; 22, p. 340.

P. K. C. Pillai and Rabinder Nath
*Photodielectric Effect in Photoconductors
(a) Vol. 37, No. 2, p. 491–498 (1976).
See also 22.4, p. 377.

H. P. Braun
Current Oscillations in Photo-Oxidized Anthracene Crystals
(a) Vol. 37, No. 2, p. 553–559 (1976).
See also 13.4, p. 143; *14.3.4*, p. *199*; 22.9, p. 438.

P. Basu and H. Saha
*Possible Performance of Back-Wall Schottky Barrier Solar Cells
(a) Vol. 37, No. 2, p. 625–631 (1976).
See also 13.3, p. 133; *14.3.3*, p. *193*; 22.1.2, p. 357; 22.2.1, p. 364; 22.4.3, p. 386.

M. Oszwałdowski
*The Dependence of Spectral Distributions of the Photoconductive and Photo-Electromagnetic Effect on Sample Thickness in InSb
(a) Vol. 37, No. 2, p. 675–679 (1976).
See also 22.2.3, p. 369.

K. Ziegler and U. Birkholz
*Photoconductivity of Si_2Te_3 Single Crystals
(a) Vol. 37, No. 2, p. K147–K149 (1976).
See also 13.4, p. 143; 22, p. 340.

A. N. Georgobiani, R. G. Maev, Yu. V. Ozerov, and E. E. Strumban
Investigation of Deep Centres of Chlorine-Doped Zinc Sulfide Crystals
(a) Vol. **38**, No. 1, p. 77—83 (1976).
See also *13.4*, p. *144*; 20.1, p. 276; 22.4.1, p. 382.

T. Boţilă and H. K. Henisch
Study of Trapping in Amorphous Materials
(a) Vol. **38**, No. 1, p. 331—337 (1976).
See also *2*, p. *40*; 13.4, p. 145; 14.3, p. 181; 22.8, p. 418.

A. M. Mishchenko, I. G. Neizvestnyi, and M. P. Sinyukov
*Spectrum of the Photocapacitance Current near the Fundamental Absorption Edge of Ge
(a) Vol. **38**, No. 1, p. 349—354 (1976).
See also *20.1*, p. *277*; 22.1.1, p. 349.

P. K. C. Pillai, Rabinder Nath, and P. K. Nair
*Non-Isothermal Relaxation in ZnO Photoelectrets
(a) Vol. **38**, No. 1, p. K73—K76 (1976).
See also *14.4*, p. *204*; 22.4, p. 377.

H. Gollnast, G. Richter, and H. Herrmann
*Nitrogen Concentration in GaP Diodes Determined by Photovoltage Measurements
(a) Vol. **38**, No. 1, p. K89—K91 (1976).
See also 13.4, p. 145; 22.2.2, p. 366.

P. Berndt, D. Genzow, and K. H. Herrmann
*Recombination Analysis in 10 μm $Pb_{1-x}Sn_xTe$
(a) Vol. **38**, No. 2, p. 497—503 (1976).
See also 22.8, p. 418.

C. Manfredotti, R. Murri, A. Quirini, and L. Vasanelli
*Photoelectronic Properties of n-GaSe
(a) Vol. **38**, No. 2, p. 685—693 (1976).
See also 13.4, p. 145; 14.3, p. 182; 22.3, p. 374.

I. B. Barkan, S. I. Marennikov, and M. V. Entin
*Holographic Storage in $LiNbO_3$ Crystal at High Temperatures
(a) Vol. **38**, No. 2, p. K139—K142 (1976).
See also *14.4.1*, p. *208*; 22.8.1, p. 425.

17. Emission of Electrons and Ions

Original Papers and Short Notes

R. H. Williams, I. T. McGovern, R. B. Murray, and M. Howells
*An Investigation of the Electronic Structure of GaSe and GaTe by Photoelectron Spectroscopy, Using a Synchrotron Source, and Electron Energy Loss Spectroscopy
(b) Vol. **73**, No. 1, p. 307–316 (1976).
See also *13.1*, p. *120*; 20.1, p. 260; 22.3, p. 372.

M. G. Kaplunov, Yu. M. Shulga, K. I. Pokhodnya, and Yu. G. Borodko
X-Ray Photoelectron Spectra of Quasi-One-Dimensional Conductors
(b) Vol. **73**, No. 1, p. 335–339 (1976).
See also 13, p. 113; *20*, p. *257*; 22.9, p. 433.

W. Schlenk
*Investigations of the Thermally Stimulated Exoelectron Emission (TSEE) from Aluminium Surfaces
(a) Vol. **33**, No. 1, p. 217–228 (1976).
See also 21, p. 290.

Y. Hayakawa, S. Leelaprute, T. Nakazawa, S. Hayakawa, and N. Oda
*Thermally Stimulated Exoelectron Emission from $BaTiO_3$ Powder Coated with LiF Film
(a) Vol. **33**, No. 1, p. K35–K37 (1976).
See also 22.8.1, p. 419.

J. M. Honig, A. P. B. Sinha, W. E. Wahnsiedler, and H. Kuwamoto
Studies of the Band Structure of NbO by X-Ray Photoelectron Spectroscopy
(b) Vol. **73**, No. 2, p. 651–654 (1976).
See also *13.1*, p. *121*; 22.6, p. 403.

N. Wagner and O. Brümmer
XPS Investigations for the Purpose of the Research of Positron Annihilation Effects in Bi_xSb_{1-x} Alloys Concerning Band Structure Modifications
(b) Vol. **74**, No. 1, p. 249–254 (1976).
See also *13.1*, p. *122*; 22.7, p. 410.

U. Berg, G. Dräger, and O. Brümmer
Combined Investigations of the Valence Band Structure of Graphite by K X-Ray Emission Spectroscopy and X-Ray Photoemission Spectroscopy
(b) Vol. **74**, No. 1, p. 341–348 (1976).
See also *13.1*, p. *122*; 20, p. 257; 22.1, p. 343.

M. Euler and A. Scharmann
*Thermally Stimulated Exoelectron Emission (TSEE) of BeO Layers. Auger Electron Spectroscopy Investigations
(a) Vol. **34**, No. 1, p. 297–306 (1976).
See also 1.6, p. 34; 13.3, p. 132; 22.6, p. 403.

T. Sasaki
*ESR Study on Thermally Stimulated Exoelectron Emission from Irradiated K_2SO_4
(a) Vol. **34**, No. 1, p. 339–346 (1976).
See also *19*, p. *249*; 22.8.1, p. 420.

V. Siegel, W. Rasp, and H.-H. Kirchner
*The Influence of Alkyl Groups on Exoelectron Emission from BeO
(a) Vol. **34**, No. 2, p. K203–K204 (1976).
See also 1.6, p. 34; 22.6, p. 404.

R. Nilsson, A. Berndtsson, N. Mårtensson, R. Nyholm, and J. Hedman
The Valence Band of Os Studied by Electron Spectroscopy
(b) Vol. **75**, No. 1, p. 197–203 (1976).
See also *13.1*, p. *123*; 21, p. 293.

H. Kirihata and J. Akutsu
*Exoelectron Emission by Slow Electron Impacts
(a) Vol. **35**, No. 1, p. K73–K75 (1976).
See also 13.3, p. 132; 22.5.2, p. 394; 22.6, p. 405.

U. Berg, G. Dräger, K. Mosebach, and O. Brümmer
Combined XS and XPS Investigations of the FeS_2 Valence Band
(b) Vol. **75**, No. 2, p. K89–K92 (1976).
See also *13.1*, p. *124*; 20, p. 258; 22, p. 337.

S. Rogaschewski and H. Düsterhöft
*z_1-Dependence of Kinetic Ion-Electron Emission from a Gas-Covered Stainless-Steel Surface
(b) Vol. **75**, No. 2, p. K173–K175 (1976).
See also 21.1.1, p. 319.

I. V. Krylova and V. I. Svitov
*Thermostimulated Exoemission and Work Function Measurement of Aluminium
(a) Vol. **35**, No. 2, p. K119–K121 (1976).
See also 21, p. 295.

A. B. Gerasimov, G. M. Dolidze, L. A. Mizandari, and A. A. Tsertsvadze
*On the Physical Mechanism of Exoelectron Emission
(a) Vol. **35**, No. 2, p. K131–K134 (1976).
See also 22.1.1, p. 347.

W. Braun and M. Cardona
Electronic Density of States of Amorphous $CuGaS_2$ and $ZnGeAs_2$
(b) Vol. **76**, No. 1, p. 251–257 (1976).
See also 2, p. 38; *13.1*, p. *125*; 22.8, p. 416.

H. Höchst, A. Goldmann, S. Hüfner, and H. Malter
X-Ray Photoelectron Valence Band Studies on Phthalocyanine Compounds
(b) Vol. **76**, No. 2, p. 559–568 (1976).
See also *13.1*, p. *126*; 22.9, p. 437.

A. Balzarotti and A. Bianconi
Electronic Structure of Aluminium Oxide as Determined by X-Ray Photoemission
(b) Vol. **76**, No. 2, p. 689–694 (1976).
See also *13.1*, p. *126*; 22.6, p. 406.

H. G. Maguire
Investigation of the Band Structure of Diamond Using Low-Energy Electrons
(b) Vol. **76**, No. 2, p. 715–726 (1976).
See also *13.1*, p. *126*; 22.1, p. 344.

H. Biederman
*Emission Patterns at Various Conditions of Imaging the Electron Emission of Al–LiF–Au Structures
(a) Vol. **36**, No. 2, p. 783–789 (1976).
See also 22.5.2, p. 396.

G. Holzapfel and M. Krystek
Thermally Stimulated Luminescence and Exoelectron Emission from Barium and Strontium Sulphate Doped with Europium
(a) Vol. **37**, No. 1, p. 303–312 (1976).
See also *13.4*, p. *142*; 20.3, p. 286; 22.8.1, p. 424.

I. V. Krylova and V. I. Svitov
*On the Role of Adsorption Layers in Thermostimulated Exoemission from Beryllium Oxide
(a) Vol. **37**, No. 1, p. K9–K12 (1976).
See also 22.6, p. 407.

R. Gauthier and P. Pinard
*Pulsed Laser Light Stimulated Desorption
(a) Vol. **38**, No. 1, p. 85–92 (1976).
See also 21.1, p. 313; 22.1.1, p. 349; 22.5.2, p. 397.

J. Marien
*Absolute Method for Work Function Determination of Metals and Semiconductors by Field Emission
(a) Vol. **38**, No. 1, p. 339–347 (1976).
See also 21, p. 300; 22.4, p. 377.

A. Kosakov, K. Unger, and G. Leonhardt
Calculation of Energy-Dependent Photoionization Probabilities for Valence Electrons in Diamond and Silicon Crystals
(b) Vol. **78**, No. 2, p. 681–688 (1976).
See also *13.1*, p. *129*; 20, p. 259; 22.1, p. 344; 22.1.1, p. 349; 22.1.2, p. 358.

R. Gauthier and C. Guittard
*Mechanism Investigations of a Pulsed Laser Light Induced Desorption
(a) Vol. **38**, No. 2, p. 477–486 (1976).
See also 21.1, p. 314; 22.1.1, p. 350; 22.5.2, p. 398.

J. Marien
*Field Emission Study of the Specificity of Zinc Oxide Polar Surfaces (0001) and $(000\bar{1})$ Work Function and Hydrogen Adsorption
(a) Vol. **38**, No. 2, p. 513–522 (1976).
See also 22.4, p. 378.

17.1 Field Emission Microscope Investigation

Original Paper

W. Polanschütz und E. Krautz
*Zur feldionenmikroskopischen Untersuchung von Korngrenzen bei Metallen
(Field Ion Microscopic Study of Grain Boundaries in Metals)
(a) Vol. **38**, No. 1, p. 237–242 (1976).
See also *1.4*, p. *29*; 10.1, p. 85; 21, p. 300.

18. Magnetic Properties

Review Article

H. W. de Wijn, A. M. van Diepen, and K. H. J. Buschow
*Samarium in Crystal Fields, a Case of Strong J-Mixing
(b) Vol. **76**, No. 1, p. 11–30 (1976).
See also 5, p. 48; *13*, p. *113*; 19, p. 247; 21, p. 289; 21.1, p. 302; 21.1.1, p. 316; 21.4, p. 326.

Original Papers and Short Notes

I. V. Chepeleva, E. A. Zhilinskaya, V. N. Lazukin, and A. P. Chernov
Electron Paramagnetic Resonance of Cr_3^+ in Chalcogenide Glasses
(b) Vol. **73**, No. 1, p. 65–72 (1976).
See also 2, p. 35; *19*, p. *247*; 22.8, p. 413.

W. Löser
*A Self-Consistent Cluster Approximation for the Disordered Hubbard Model
(b) Vol. **73**, No. 1, p. 179–185 (1976).
See also *2*, p. *35*; 13, p. 113; 14, p. 162.

I. Moore, M. King, M. Jones, R. J. Deck, and A. M. Hermann
*Microwave Faraday Rotation in Molybdenum Disulfide
(b) Vol. **73**, No. 1, p. K31–K34 (1976).
See also 13, p. 114; *20*, p. *257*; 22, p. 393.

I. N. Douglas
*Magnetic Circular Dichroism of Co^{2+}-Doped Cubic Crystals
(b) Vol. **73**, No. 2, p. 641–650 (1976).
See also 13.4, p. 135; *20.1*, p. *262*; 22.5.2, p. 393; 22.6, p. 403.

J. R. L. de Almeida and S. M. Rezende
Scattering of Light by Magnetoelastic Excitations in Ferromagnets
(b) Vol. **73**, No. 2, p. 661–672 (1976).
See also 13, p. 114; *20*, p. *257*; 22.8.2, p. 427.

C. Manohar and V. K. Kelkar
*Effects of Three-Body Interactions on Hyperfine Fields
(b) Vol. **73**, No. 2, p. 735–740 (1976).
See also *13*, p. *114*; 21.1, p. 303.

R. S. Kim and S. Narita
*Far-Infrared Interband Magnetoabsorption and Band Structure of $Hg_{1-x}Cd_xTe$ Alloys
(b) Vol. **73**, No. 2, p. 741–752 (1976).
See also *13.1*, p. *121*; 20.1, p. 262; 22.4.4, p. 387.

H. E. Flotow, R. Kuentzler, and D. W. Osborne
The Specific Heat of the Nearly Ferromagnetic Compound $(Mn_{0.865}Fe_{0.135})Be_2$ from 0.1 to 23 K
(a) Vol. **34**, No. 1, p. 291–295 (1976).
See also *8*, p. *68*; 22.8, p. 414.

R. Jagannathan and N. R. Ranganathan
*A New Form of Magnetic Bloch Functions and Applications to Magnetic Band Structure Calculations
(b) Vol. **74**, No. 2, p. 667–672 (1976).
See also *13.1*, p. *123*.

G. I. Georgiev and S. A. Kuzmanova
*Effect of Anisotropic Biquadratic Exchange Interaction on the Rigorous Solutions of the One-Dimensional X–Y Spin Model
(b) Vol. 74, No. 2, p. K173–K176 (1976).

G. Debnath and P. Rudra
*Magnetization and Susceptibility in the Truncated Anderson Model
(b) Vol. 75, No. 1, p. 145–153 (1976).
See also *13*, p. *115*.

G. Ya. Selyutin
Mössbauer Spectrum near the Critical Point
(b) Vol. 75, No. 1, p. 233–238 (1976).
See also *5*, p. *49*.

B. Rheinländer
*Volume Plasma Faraday Effect in n-Type GaAs
(b) Vol. 75, No. 1, p. 261–270 (1976).
See also 13.5.2, p. 157; *20.1*, p. *267*; 22.2.1, p. 363.

K. Takanaka and R. Yamamoto
*The Self-Damping of RKKY Interaction in Dilute Magnetic Alloys
(b) Vol. 75, No. 1, p. 279–288 (1976).
See also *13*, p. *116*.

B. Bahadur, S. Chandra, and N. K. Sanyal
*Phase Transition Studies of the Liquid Crystal n-(p-Octyloxybenzylidene)-p-Toluidine
(a) Vol. 35, No. 1, p. 387–390 (1976).
See also *2*, p. *37*; 22.9, p. 436.

F. R. Kessler, J. Metzdorf, and R. H. Müller
*Interband Faraday Rotation in Germanium. Screening of Excitons by Free Carriers
(b) Vol. 75, No. 2, p. 567–572 (1976).
See also 13.5.1, p. 153; *20*, p. *258*; 22.1.1, p. 347.

P. M. Bisch and A. A. Gomes
*Formal Results on the Susceptibility of Dilute Alloys (I): Green's Function Approach
(b) Vol. 75, No. 2, p. 705–715 (1976).
See also *13*, p. *116*; 21.1, p. 308.

N. Dupont-Pavlovsky, F. Caralp, P. Delhaes and J. Amiell
Chemical Characterization and Study of the Electronic Properties of Non-Stoichiometric Zinc Oxide. Nature and Repartition of Defects
(a) Vol. 35, No. 2, p. 615–625 (1976).
See also *10.2*, p. *93*; *13.4*, p. *139*; 14.3, p. 177; 22.4, p. 376.

Z. Biegański and M. Drulis
*Low Temperature Phase Transition in Neodymium Trihydride. Specific Heat Measurements
(a) Vol. 35, No. 2, p. K127–K130 (1976).
See also *8*, p. *69*; 21.4, p. 326.

V. P. Kalantarian and V. I. Chechernikov
Nuclear Magnetic Resonance and Susceptibility Measurements of Transition Metals and Alloys: V–Ta, V–Nb, and Nb–Ta Systems
(b) Vol. 76, No. 1, p. 113–118 (1976).
See also 13, p. 116; *19*, p. *252*; 21, p. 296; 21.1, p. 309.

H. Burkhard, G. Bauer, P. Grosse, and A. Lopez-Otero
*Free Carrier Magneto-Optical Effects in the Reststrahlen Region of n-PbTe
(b) Vol. 76, No. 1, p. 259–266 (1976).
See also 13.1, p. 125; *20.1*, p. *270*; 22, p. 337.

P. M. Bisch and A. A. Gomes
*Formal Results on the Susceptibility of Dilute Alloys (II): Mean-Field Approach and Applications
(b) Vol. **76**, No. 1, p. 395–406 (1976).
See also *13*, p. *117*; 19, p. 252; 21.1, p. 309.

V. Bezák and J. Banský
*A Generalization of the Sondheimer-Wilson Formula for the Averaged Density Matrix and Magnetic Susceptibility of Conduction Electrons in a Model of Disordered Semiconductors
(b) Vol. **76**, No. 2, p. 569–580 (1976).
See also 2, p. 39; *13*, p. *117*; 22, p. 338.

S. Robaszkiewicz
*Magnetic, Charge and Orbital Orderings in a Two-Band Model of Transition Metal Compounds
(b) Vol. **76**, No. 2, p. K127–K130 (1976).
See also *13*, p. *118*.

P. Ganguly, Om Parkash, and C. N. R. Rao
*Electron Transport and Magnetic Properties of Rare-Earth Ortho-Titanites and -Vanadites, $LnTiO_3$ and $LnVO_3$
(a) Vol. **36**, No. 2, p. 669–678 (1976).
See also *14*, p. *162*; 15, p. 210; 22.8.1, p. 423.

V. I. Litvinov
*The Kondo Effect in a Magnetic Field
(b) Vol. **77**, No. 1, p. 71–76 (1976).
See also *14*, p. *162*; 21, p. 297.

J. Donecker and J. Kluge
*Magnetic Circular Dichroism of Indirect Excitonic Transitions in GaP and $In_{1-x}Ga_xP$
(b) Vol. **77**, No. 1, p. 243–248 (1976).
See also 13.5.1, p. 155; *20.1*, p. *273*; 22.2.2, p. 365; 22.2.4, p. 371.

V. P. Zenchenko and E. P. Sinyavskii
*The Johnson-Larsen Effect in the Resonance Laser Field
(b) Vol. **77**, No. 1, p. K39–K42 (1976).
See also *20.1*, p. *273*.

J. Mertsching
*Static Approximation for the Hubbard Model: Magnetic Phase Transitions
(b) Vol. **77**, No. 2, p. 717–724 (1976).
See also *13.1*, p. *127*.

G. I. Georgiev and S. G. Gocheva
*Numerical Computations on a Rigorously Solvable Model
(b) Vol. **78**, No. 1, p. K11–K13 (1976).
See also *8*, p. *71*.

W. Grevendonk, G. Ruymbeek, P. van den Keybus, and A. de Bock
Faraday Rotation in Non-Crystalline As_2S_3
(b) Vol. **78**, No. 2, p. K89–K91 (1976).
See also *2*, p. *41*; 20, p. 259; 22, p. 342.

18.1 Paramagnetic Properties

Review Article

E. Vogt
The Origin of Ferromagnetism and Antiferromagnetism in Alloys of Transition Metals (II): Alloys of Transition Metals with Other Transition Metals and with Polyvalent Non-Transition Elements
(a) Vol. **34**, No. 1, p. 11—67 (1976).
See also 1.3, p. 21; *18.2*, p. *228*; *18.4*, p. *242*; 21, p. 289; 21.1, p. 302; 21.6, p. 329.

Original Papers and Short Notes

D. Scheerlinck and S. Hautecler
*Magnetic Interactions in Co_3O_4
(b) Vol. **73**, No. 1, p. 223—228 (1976).
See also *18.4*, p. *242*; 22.8.2, p. 426.

R. Wirths, P. Runow, and H. W. Schöpgens
*Magnetic Study of the Ordered and Quenched Equiatomic FePt Alloy
(a) Vol. **33**, No. 1, p. 135—142 (1976).
See also 1.3, p. 21; *18.2*, p. *229*; 21.1.1, p. 316.

Y. Aoki and M. Yamamoto
*X-Ray and Magnetic Investigations of the High-Temperature Phase in the Co-Rich Co–V Alloy System
(a) Vol. **33**, No. 2, p. 625—632 (1976).
See also 1.3, p. 21; *18.2*, p. *229*; 21.1, p. 304.

P. J. Becker
*Magnetostriktion und lineare magnetische Doppelbrechung im Jahn-Teller-gekoppelten System $TmAsO_4$ mit effektivem Spin $S = 1$
(Magnetostriction and Linear Magnetic Birefringence in the Jahn-Teller Coupled System $TmAsO_4$ with Effective Spin $S = 1$)
(b) Vol. **74**, No. 1, p. 285—295 (1976).
See also *20.1*, p. *264*; 22.8.1, p. 419.

A. M. Grekhov and A. B. Roitsin
Theory of Low-Symmetry Effects in Paramagnetic Resonance
(b) Vol. **74**, No. 1, p. 323—334 (1976).
See also *19*, p. *249*.

A. Popielewicz and E. Szczepaniak
*On the Dynamics of the XY Model in a Magnetic Field
(b) Vol. **74**, No. 1, p. K81—K84 (1976).

J. Mizia
CPA Treatment of the Hubbard Model with Intra-Atomic Exchange Interaction
(b) Vol. **74**, No. 2, p. 461—470 (1976).
See also *13*, p. *115*; 18.2, p. 230; 21.1, p. 305.

J. Ray and Girish Chandra
*Concentration Dependence of Curie Points of Some Magnetic Binary Alloys
(a) Vol. **34**, No. 2, p. K169—K172 (1976).
See also *18.2*, p. *231*; 21.1, p. 307; 21.1.1, p. 318.

K. Hattori
*Magnetic Properties of a System Consisting of Two Kinds of Magnetic Atoms Randomly Distributed
(b) Vol. **75**, No. 1, p. K39—K43 (1976).
See also *18.2*, p. *232*; 21.1, p. 307.

L. M. Valiev, I. G. Kerimov, N. G. Aliev, and A. A. Abdurragimov
*The Thermal Expansion and Electrical Properties of $Cu_4Mn_2X_4$(X:S, Se, and Te)
(a) Vol. 35, No. 1, p. K85—K88 (1976).
See also 8, p. 68; 14, p. 162; 15, p. 210; *18.4*, p. *244*; 22.8, p. 415.

C. Scherer and I. Aveline
*Reaction Field Approximation for Heisenberg Paramagnets and the Magnetic Specific Heat
(b) Vol. 75, No. 2, p. 465—470 (1976).
See also *8*, p. *68*.

B. J. Fechner
Influence of a Magnetic Field on the Susceptibility of a Heisenberg Ferromagnet near the Critical Point
(b) Vol. 75, No. 2, p. 687—695 (1976).
See also *18.2*, p. *232*; 21.1, p. 308.

G. Kamieniarz
*On First Order Green Function Theory of $S = 1/2$ Transverse Ising Paramagnets
(b) Vol. 75, No. 2, p. K97—K99 (1976).
See also *13*, p. *116*.

F. A. Kassan-Ogly and M. V. Medvedev
*The Theory of a Heisenberg Alloy
(b) Vol. 76, No. 1, p. 45—53 (1976).
See also 18.2, p. 233; 18.4, p. 244; 21.1, p. 308; 21.1.1, p. 319.

B. E. Vugmeister
*Spatial and Spectral Spin Diffusion in Dilute Spin Systems
(b) Vol. 76, No. 1, p. 161—170 (1976).
See also *19*, p. *252*.

N. E. Alekseevskii, E. P. Krasnoperov, and T. Mydlarz
*Susceptibility and Superconductivity of Compounds with A-15 Structure
(b) Vol. 76, No. 1, p. 357—362 (1976).
See also 13.1, p. 125; *14.2*, p. *170*; 22.8, p. 416.

M. Matyáš and J. Horák
*Magnetic Susceptibility of Some Semiconducting V–VI–VII Crystals
(a) Vol. 36, No. 2, p. K137—K139 (1976).
See also *18.2*, p. *234*; 22.8, p. 417.

E. Heiner
*Roth's Method and Moment Conservation of the Spectral Weight Function for the Neutral Hubbard Model
(b) Vol. 77, No. 1, p. 93—103 (1976).
See also 13, p. 118; *18.4*, p. *245*.

K. de Strooper, A. Govaert, C. Dauwe, and G. Robbrecht
Magnetic Properties of the Spinel Series $Co_2Ge_xTi_{1-x}O_4$ ($0 \leqq x \leqq 1$)
(a) Vol. 37, No. 1, p. 127—132 (1976).
See also *18.3*, p. *241*; 18.4, p. 245; 22.8.2, p. 430.

L. K. Aminov, R. Kirmse, and B. V. Solovev
Electron Spin–Lattice Relaxation of Cu^{2+} Ions in Planar Coordination Geometries. Calculation and Experimental Data of T_1 for Cu^{2+} in Ni(II)bis(diethyl-dithiocarbamate) Crystals
(b) Vol. 77, No. 2, p. 505—515 (1976).
See also *4*, p. *46*; 6, p. 61; 19, p. 254; 22.9, p. 438.

E. KÖNIG and R. SCHNAKIG
Zero-Field Splitting of $^6S(d^5)$ Ions in Tetragonal and Rhombic Symmetry (I): Complete d-Electron Calculation of the Crystal-Field Dependence of D and E
(b) Vol. 77, No. 2, p. 657—666 (1976).
See also *13.4*, p. *143*; 19, p. 254.

E. I. TERUKOV, F. A. CHUDNOVSKII, W. REICHELT, H. OPPERMANN, W. BRÜCKNER, H.-P. BRÜCKNER, and W. MOLDENHAUER
Investigation of the Physical Properties of V_3O_4 at Phase Transition with Consideration of Its Range of Homogeneity
(a) Vol. 37, No. 2, p. 541—546 (1976).
See also *14.3*, p. *180*; 15, p. 211; 22.6, p. 407.

L. I. VINOKUROVA, A. V. VLASOV, and M. PARDAVI-HORVÁTH
Pressure Effects on Magnetic Phase Transitions in FeRh and FeRhIr Alloys
(b) Vol. 78, No. 1, p. 353—357 (1976).
See also 12.1, p. 109; *18.2*, p. *236*; *18.4*, p. *245*; 21.1.1, p. 321.

W. BRÜCKNER, U. GERLACH, W. MOLDENHAUER, H. P. BRÜCKNER, N. MATTERN, H. OPPERMANN, and E. WOLF
Phase Transitions and Semiconductor–Metal Transition in $V_{1-x}Ga_xO_2$ Single Crystals
(a) Vol. 38, No. 1, p. 93—102 (1976).
See also *1.2*, p. *19*; *3*, p. *44*; 14.3, p. 181; 15, p. 211; 22.6.1, p. 409.

W. BRÜCKNER, H.-P. BRÜCKNER, U. GERLACH, B. THUSS, and G. FÖRSTERLING
*The Phase Transition M1 → T in $V_{1-x}Ga_xO_2$
(a) Vol. 38, No. 1, p. K13—K16 (1976).
See also *1.2*, p. *19*; 22.6.1, p. 409.

E. KÖNIG and R. SCHNAKIG
Zero-Field Splitting of $^6S(d^5)$ Ions in Tetragonal and Rhombic Symmetry (II): Application of Complete d-Electron Calculations to the Zero-Field Splitting in Mn:MnF_2 and Mn:ZnF_2
(b) Vol. 78, No. 2, p. 543—551 (1976).
See also *13.4*, p. *145*; 19, p. 255; 20.1, p. 277; 22.5, p. 390.

M. V. CHETKIN, A. N. SHALYGIN, and A. V. KIRYUSHIN
*Magnetic Permeability of Ferro- and Paramagnetic Garnets at Optical Frequencies
(b) Vol. 78, No. 2, p. K85—K88 (1976).
See also *18.2*, p. *236*; 22.8.2, p. 431.

A. POPIELEWICZ
*Simple Theory of the Magnetic Properties of $(La_{1-x}Pr_x)_3Tl$
(b) Vol. 78, No. 2, p. K115—K119 (1976).
See also 21.4, p. 327.

S. C. BHARGAVA and P. K. IYENGAR
Study of Hyperfine Fields at Iron Nuclei in $(Co,Fe)_5Ge_3$ Using the Mössbauer Effect
(a) Vol. 38, No. 2, p. 601—609 (1976).
See also *5*, p. *51*; 18.2, p. 236; 21.1.1, p. 322.

18.2 Ferromagnetic Properties

Review Articles

W. Haubenreisser, Th. Klupsch, and F. Voigt
*Quantum Statistical Theory of Parametric Spin-Wave Excitations in Magnetoelastic Ferromagnets under Parallel Pumping (II): Calculation of Macroscopic Observables
(b) Vol. **73**, No. 1, p. 11–64 (1976).
See also 6, p. 52; *13*, p. *113*; 13.5.4, p. 160.

E. Vogt
*The Origin of Ferromagnetism and Antiferromagnetism in Alloys of Transition Metals (II): Alloys of Transition Metals with Other Transition Metals and with Polyvalent Non-Transition Elements
(a) Vol. **34**, No. 1, p. 11–67 (1976).
See also 1.3, p. 21; 18.1, p. 225; *18.4*, p. *242*; 21, p. 289; 21.1, p. 302; 21.6, p. 329.

S. V. Vonsovskii and M. S. Svirskii
Multiplicity and Pseudomultiplicity in Electron–Ion Systems (I)
(b) Vol. 77, No. 1, p. 9–52 (1976).
See also *13*, p. *113*; 14.1, p. 164; 14.2, p. 169; 18.4, p. 242; 21, p. 289.

S. V. Vonsovskii and M. S. Svirskii
Multiplicity and Pseudomultiplicity of Electron–Ion Systems (II)
(b) Vol. 77, No. 2, p. 403–432 (1976).
See also *13*, p. *113*; 14.1, p. 164; 14.2, p. 169; 18.4, p. 242; 21, p. 289.

Original Papers and Short Notes

S. Robaszkiewicz and R. Micnas
*Absence of Magnetic Ordering in One and Two Dimensions in a Model with Localized and Itinerant Electrons
(b) Vol. **73**, No. 1, p. K35–K39 (1976).
See also *13*, p. *114*.

A. Popielewicz
*On the Magnetic Ordering in Induced Magnets
(b) Vol. **73**, No. 1, p. K47–K50 (1976).

F. M. Galperin
*Hyperfine Field on the Nucleus of 3d Transition Metal Atoms
(b) Vol. **73**, No. 1, p. K73–K77 (1976).
See also *13*, p. *114*; 21.1, p. 302; 21.1.1, p. 316.

W. S. Glaunsinger
*Erratum to: Indirect Exchange Interactions in EuB_6
(b) Vol. **73**, No. 1, p. K83 (1976).
See also *19*, p. *247*; 22, p. 334.

S. K. Arif and D. S. P. Bunbury
Mössbauer and X-Ray Studies of $Dy(Fe_{1-x}Ni_x)_3$ and $Y(Fe_{1-x}Ni_x)_3$ Intermetallic Compounds
(a) Vol. **33**, No. 1, p. 91–95 (1976).
See also *5*, p. *48*; 18.3, p. 239; 21.1.1, p. 316.

U. Hofmann
*Physical Properties of Nickel–Rhenium Alloys
(a) Vol. **33**, No. 1, p. 119–123 (1976).
See also 12.1, p. 107; 14.1, p. 164; 21.1, p. 302.

R. Wirths, P. Runow, and H. W. Schöpgens
*Magnetic Study of the Ordered and Quenched Equiatomic FePt Alloy
(a) Vol. **33**, No. 1, p. 135–142 (1976).
See also 1.3, p. 21; *18.1*, p. *225*; 21.1.1, p. 316.

G. Haneczok, J. Ilczuk, R. Kuśka, and J. W. Moroń
*Migrational Relaxations in α-Fe–V–C in the Region of the Snoek Effect
(a) Vol. **33**, No. 1, p. 313–324 (1976).
See also *10.1*, p. *78*; 21.1.1, p. 316.

V. A. Tsurin, A. E. Ermakov, Yu. G. Lebedev, and B. N. Filippov
A Mössbauer Study of the Structural Characteristics of Equiatomic FePd and FePt Alloys
(a) Vol. **33**, No. 1, p. 325–332 (1976).
See also *5*, p. *48*; 21.1.1, p. 316.

P. Novotný and J. Šternberk
*On the Magnetic Properties of U_3As_4 at Low Temperatures
(a) Vol. **33**, No. 1, p. K15–K17 (1976).
See also 22, p. 334.

E. M. Gray and J. H. Smith
*The Effect of Heat Treatment on Micromagnetic Mn<u>Cu</u> Alloys
(a) Vol. **33**, No. 1, p. K43–K46 (1976).
See also 14.1, p. 164; 21.1, p. 303.

A. A. Samokhvalov, Yu. N. Morozov, B. V. Karpenko, and M. I. Simonova
*Temperature Dependence of the Magnetization of the Ferromagnetic Spinels $CdCr_2Se_4$ and $CuCr_2S_4$ and Its Calculation in the Spin-Wave Approximation
(b) Vol. **73**, No. 2, p. 455–459 (1976).
See also 22.8.2, p. 426.

A. Andreeff, H.-J. Hunger, and S. Unterricker
The Magnetic Hyperfine Interaction at Cd^{111} in Iron–Aluminium Alloys
(b) Vol. **73**, No. 2, p. K89–K92 (1976).
See also *5*, p. *48*; 21, p. 291; 21.1.1, p. 317.

H. Rechenberg
*Remarks on the Invar Problem
(b) Vol. **73**, No. 2, p. K119–K122 (1976).
See also 21.1, p. 303; 21.1.1, p. 317.

E. W. Lee and F. Pourarian
Magnetoelastic Properties of (Rare-Earth)–Co_2 Compounds (I): Exchange-Striction
(a) Vol. **33**, No. 2, p. 483–489 (1976).
See also 21.1, p. 303; 21.4, p. 326.

R. Lermer and A. Steyerl
*Investigation of Ferromagnetic Domains and Bloch Walls by Very-Low-Energy Neutron Transmission
(a) Vol. **33**, No. 2, p. 531–541 (1976).
See also 21.1, p. 304; 21.1.1, p. 317.

C. C. Becerra, A. Paduan Filho, and N. F. Oliveira, Jr.
*Differential Magnetization of MnP
(a) Vol. **33**, No. 2, p. 573–577 (1976).
See also *18.4*, p. *242*; 22, p. 334.

Y. Aoki and M. Yamamoto
*X-Ray and Magnetic Investigations of the High-Temperature Phase in the Co-Rich Co–V Alloy System
(a) Vol. **33**, No. 2, p. 625–632 (1976).
See also 1.3, p. 21; *18.1*, p. *225*; 21.1, p. 304.

S. Takahashi and Y. Ishikawa
*Compression-Induced Magnetic Anisotropy Enhanced by Annealing
(a) Vol. 33, No. 2, p. K141–K143 (1976).
See also 21.1, p. 304; 21.1.1, p. 317.

M. A. Ivanov, V. Ya. Mitrovanov, L. D. Falkovskaya, and A. Ya. Fishman
*The Properties of a Threefold Degenerate Orbital Magnetic Impurity in a Ferromagnetic Crystal
(b) Vol. 74, No. 1, p. 57–67 (1976).
See also 13.4, p. 136; 22.1, p. 343.

A. Popielewicz
*On the Magnetic Ordering in Induced Magnets
(b) Vol. 74, No. 1, p. 383–391 (1976).

D. D. Richardson
*Effects of an Impurity on Magnon Sidebands in Insulators
(b) Vol. 74, No. 1, p. K61–K65 (1976).
See also 10.2, p. 90; *20.1*, p. *264*; 22, p. 335.

W.-H. Steeb
*High Temperature Properties of a Doubly-Degenerate Hubbard Model
(b) Vol. 74, No. 1, p. K71–K74 (1976).
See also 13, p. 115; *18.4*, p. *243*.

J. B. A. A. Elemans and K. H. J. Buschow
*Comment on the Crystal and Magnetic Structures of $CaCu_5$-Type Compounds of Th with Fe, Co, and Ni
(a) Vol. 34, No. 1, p. 355–359 (1976).
See also 1.1, p. 11; *18.4*, p. *243*; 21.1, p. 305; 21.1.1, p. 317; 21.5, p. 328.

E. W. Lee and F. Pourarian
*Magnetoelastic Properties of (Rare-Earth)–Co_2 Compounds (II): The Anisotropic Magnetostriction
(a) Vol. 34, No. 1, p. 383–390 (1976).
See also 21.1, p. 305; 21.4, p. 326.

J. Mizia
CPA Treatment of the Hubbard Model with Intra-Atomic Exchange Interaction
(b) Vol. 74, No. 2, p. 461–470 (1976).
See also *13*, p. *115*; 18.1, p. 225; 21.1, p. 305.

H. Pfeiffer and W. Haubenreisser
*Generalized Locator-Coherent-Potential Approach to Ferromagnetic Systems with Random Local Magnetic Fields and Moments
(b) Vol. 74, No. 2, p. 617–626 (1976).
See also 19, p. 250.

J. Zieliński
*On the Existence of Spin Waves in a Model Transition Metal with Finite Anderson Mixing Potential
(b) Vol. 74, No. 2, p. 779–783 (1976).
See also *13*, p. *115*; 21.1, p. 305.

J. D. Patterson and R. C. Weger
*The Specific Heat of Small Magnetic Clusters with Simulated Amorphousness
(b) Vol. 74, No. 2, p. K115–K118 (1976).
See also 2, p. 37; *8*, p. *68*.

S. Krzemiński
*Note on the Mermin-Wagner Theorem
(b) Vol. 74, No. 2, p. K119–K123 (1976).

J. RICHTER, J. SCHREIBER, and K. HANDRICH
*Influence of Structure Fluctuations on Ferromagnetism in the Band Model
(b) Vol. **74**, No. 2, p. K125–K128 (1976).
See also *2*, p. *37*.

A. V. MALAKHOVSKII and I. S. EDELMAN
*Optical and Magneto-Optical Spectra of $FeBO_3$ in the Region of Strong Absorption
(b) Vol. **74**, No. 2, p. K145–K149 (1976).
See also *20.1*, p. *266*; 22.8.1, p. 421.

D. SIEMERS and E. NEMBACH
*The Dependence of the Initial Susceptibility on the Demagnetization Method
(a) Vol. **34**, No. 2, p. 547–551 (1976).
See also 21, p. 293.

G. SCHROEDER and H. KRONMÜLLER
Investigation of the Annealing of the Dislocations in Ni Crystals by Means of the Magneto-Optical Kerr Effect
(b) Vol. **34**, No. 2, p. 623–638 (1976).
See also *10.1*, p. *81*; 20, p. 258; 21.1, p. 306.

G. KAPPEL, G. FISCHER, and A. JAÉGLÉ
*Magnetic Investigation of the System Mn_5Ge_3–Mn_5Si_3
(a) Vol. **34**, No. 2, p. 691–696 (1976).
See also *18.4*, p. *243*; 21, p. 293.

H. OESTERREICHER
*Magnetic Studies on Compounds RCuSi, $R_6Cu_8Si_8$ and RCu_2Si_2 (R = Pr, Gd, Tb)
(a) Vol. **34**, No. 2, p. 723–728 (1976).
See also *18.4*, p. *243*; 21.1, p. 306; 21.4, p. 326.

P. C. M. GUBBENS and K. H. J. BUSCHOW
Mössbauer Effect Study of $Tm_2Fe_{17-x}Co_x$ and $Tm_2Fe_{17-x}Ni_x$ Compounds
(a) Vol. **34**, No. 2, p. 729–735 (1976).
See also *5*, p. *49*; 21.1.1, p. 318; 21.4, p. 326.

A. K. ANDREEV, H. E. KHODENKOV, V. S. POTAPOV, and V. R. RAEV
*Quasistatic Stability of Magnetic Bubble Domains in Potential Wells of Propagating Field
(a) Vol. **34**, No. 2, p. K163–K167 (1976).

J. RAY and GIRISH CHANDRA
*Concentration Dependence of Curie Points of Some Magnetic Binary Alloys
(a) Vol. **34**, No. 2, p. K169–K172 (1976).
See also *18.1*, p. *225*; 21.1, p. 307; 21.1.1, p. 318.

H. E. SCHAEFER and W. DANDER
*The Initial Susceptibility of Cobalt after Electron Irradiation at 4.2 K
(b) Vol. **75**, No. 1, p. 173–178 (1976).
See also 10.1, p. 81; *11*, p. *101*; 14.1, p. 165; 21.1, p. 307.

C. B. DE ARAUJO
On the Coherence Properties of Parametric Magnon States
(b) Vol. **75**, No. 1, p. 327–332 (1976).
See also *13.5.4*, p. *160*.

L. BIEGAŁA
*Thermodynamical Properties of Isotropic Ferromagnets with Biquadratic Interactions
(b) Vol. **75**, No. 1, p. 333–339 (1976).

K. Hattori
*Magnetic Properties of a System Consisting of Two Kinds of Magnetic Atoms Randomly Distributed
(b) Vol. **75**, No. 1, p. K39—K43 (1976).
See also *18.1*, p. *225*; 21.1, p. 307.

L. L. Buishvili, E. B. Volzhan, N. P. Giorgadze, and A. D. Pataraya
*On Weakly Nonlinear Oscillations of Nuclear Spins in Ferromagnets
(b) Vol. 75, No. 1, p. K69—K74 (1976).

A. V. Malakhovskii
*The Form of High Frequency Tensors $\hat{\varepsilon}$ and $\hat{\mu}$ and the Symmetry of Magnetic Crystals
(b) Vol. 75, No. 1, p. K75—K78 (1976).

D. Bastian and E. Biller
*Damping of Ferromagnetic Resonance in Ni–Fe Alloys
(a) Vol. **35**, No. 1, p. 113—120 (1976).
See also *19*, p. *251*; 21.1, p. 307; 21.1.1, p. 318.

A. M. van der Kraan, J. N. J. van der Velden, J. H. F. van Apeldoorn, P. C. M. Gubbens, and K. H. J. Buschow
Mössbauer Effect Study and Magnetization Measurements of RFe_3 Intermetallic Compounds with R = Sm, Gd, Tb, Dy, Ho, and Th
(a) Vol. **35**, No. 1, p. 137—151 (1976).
See also *5*, p. *49*; 21.1.1, p. 318; 21.4, p. 326.

J. Kaczér and Z. Klem
*The Magnetostatic Energy of Coaxial Cylinders and Coils
(a) Vol. **35**, No. 1, p. 235—242 (1976).

L. G. M. M. Lammers and N. J. Wiegman
*Barkhausen Noise and Correlation Measurements of Mumetal Strips
(a) Vol. **35**, No. 1, p. K45—K47 (1976).
See also 21.1.1, p. 318.

H. Bińczycka, A. Szytuła, J. Todorović, T. Zaleski, and A. Zięba
*Metamagnetism of CoMnSi
(a) Vol. **35**, No. 1, p. K69—K72 (1976).
See also *18.4*, p. *244*; 22.8, p. 415.

W. Dander, H. Kronmüller, and H. E. Schaefer
*Magnetic Aftereffect Phenomena in Cobalt between 4.2 and 20 K after Low-Temperature Electron Irradiation
(b) Vol. **75**, No. 2, p. 647—655 (1976).
See also *11*, p. *102*; 21.1, p. 308.

B. J. Fechner
*Influence of a Magnetic Field on the Susceptibility of a Heisenberg Ferromagnet near the Critical Point
(b) Vol. 75, No. 2, p. 687—695 (1976).
See also 18.1, p. 226; 21.1, p. 308.

K. Tsuno and T. Yamamoto
*Observed Depths of Magnetic Domains in High-Voltage Scanning Electron Microscopy
(a) Vol. **35**, No. 2, p. 437—449 (1976).
See also 21.1, p. 308; 21.1.1, p. 319.

D. Bastian and E. Biller
*Anisotropy Constants and *g*-Factors of Ni–Fe Alloys Derived from Ferromagnetic Resonance
(a) Vol. **35**, No. 2, p. 465—470 (1976).
See also *19*, p. *252*; 21.1, p. 308; 21.1.1, p. 319.

F. A. Kassan-Ogly and M. V. Medvedev
The Theory of a Heisenberg Alloy
(b) Vol. **76**, No. 1, p. 45–53 (1976).
See also *18.1*, p. *226*; 18.4, p. 244; 21.1, p. 308; 21.1.1, p. 319.

J. Sivardière
Magnetism in $S = \frac{3}{2}$ Models: Magnetization Jumps and Reorientations, Heat Magnetization, and Tricritical Points
(b) Vol. **76**, No. 1, p. 77–84 (1976).
See also *13*, p. *116*.

V. Kumar and K. C. Sharma
*Green's Function Approach for High-Temperature Ferromagnetic Systems in the Presence of Uniaxial Anisotropy
(b) Vol. **76**, No. 1, p. 337–342 (1976).

F. M. Galperin
*Hyperfine Fields of the Impurities in Ferromagnetic Iron
(b) Vol. **76**, No. 1, p. K9–K13 (1976).
See also *13*, p. *117*; 21.1, p. 309; 21.1.1, p. 319.

V. V. Zolotarev
Magnon Damping Caused by Spin–Spin Exchange Interaction
(b) Vol. **76**, No. 1, p. K31–K33 (1976).
See also *13.5.4*, p. *160*.

A. Wachniewski and L. Biegala
*Erratum to: Magnetostatic Surface Anisotropy in B.C.C. and H.C.P. Ferromagnets and S.C. Antiferromagnets
(b) Vol. **76**, No. 1, p. K73 (1976).
See also *18.4*, p. *244*.

H. Figiel, A. Oppelt, E. Dormann, and K. H. J. Buschow
Transferred Hyperfine Fields at the Y Sites in Y–Co Compounds
(a) Vol. **36**, No. 1, p. 275–283 (1976).
See also *19*, p. *253*; 21.1, p. 309; 21.4, p. 327.

E. I. Ilyashenko, I. G. Avaeva, V. B. Kravchenko, and S. N. Matveev
*A Method for Observation and Measurement of the Velocity of Bubble Propagation in Thin Ferrogarnet Films
(a) Vol. **36**, No. 1, p. K1–K3 (1976).
See also 22.8.2, p. 429.

D. J. Fathers and J. P. Jakubovics
*Magnetic Domain Wall Contrast in the Scanning Electron Microscope
(a) Vol. **36**, No. 1, p. K13–K16 (1976).
See also 21.1.1, p. 320.

J. Schneider, H. Wiesner, and R. Gemperle
*Annealing Effects on the Magnetic Properties of Rapidly Quenched Transition Metal Alloys
(a) Vol. **36**, No. 1, p. K59–K64 (1976).
See also *1.3*, p. *23*; 21.1, p. 309; 21.1.1, p. 320.

J. Schneider and J. Konetzky
*Magnetization Studies of Rapidly Quenched Fe–Ni Alloys
(a) Vol. **36**, No. 1, p. K77–K80 (1976).
See also *1.3*, p. *23*; 21.1, p. 310; 21.1.1, p. 320.

G. Dublon, M. Kroupp, M. P. Dariel, and U. Atzmony
*Crystal Field Induced Spin Rotation Observed by Magnetization Study of Polycrystalline $NdCo_2$
(b) Vol. **76**, No. 2, p. 669–675 (1976).
See also 21.1, p. 310; 21.4, p. 327.

M. N. Botvinko and M. A. Ivanov
*Spectrum of Magnetic Excitations of a Ferromagnet with Weakly Bound Antiferromagnetic Interstitial Impurities
(b) Vol. 76, No. 2, p. 797–809 (1976).
See also *13.4*, p. *141*; 13.5.4, p. 160.

B. S. Tošić and M. M. Marinković
*On the Existence of Local Levels in a Heisenberg Ferromagnet
(b) Vol. 76, No. 2, p. K85–K88 (1976).
See also *13.5.4*, p. *161*.

S. K. Sidorov
*On the Phenomenological Theory of Magnetization of Alloys with Mixed Exchange Interaction
(b) Vol. 76, No. 2, p. K113–K117 (1976).
See also 18.4, p. 244; 21.1, p. 310; 21.1.1, p. 320.

A. Pawlikowski and M. Drzazga
*Magnetization Directions in the Band Theory of Ferromagnetism
(b) Vol. 76, No. 2, p. K123–K126 (1976).

J. F. Dufresne, A. Seeger, P. Groh, and P. Moser
*Hydrogen Relaxation in α-Iron
(a) Vol. 36, No. 2, p. 579–589 (1976).
See also *10.1*, p. *83*; 21.1.1, p. 320.

M. Matyáš and J. Horák
*Magnetic Susceptibility of Some Semiconducting V–VI–VII Crystals
(a) Vol. 36, No. 2, p. K137–K139 (1976).
See also *18.1*, p. *226*; 22.8, p. 417.

Y. Aoki, Y. Gotoh, and Y. Obi
On the Phase Diagram of the Co-Rich Co–V Alloy System
(a) Vol. 36, No. 2, p. K149–K152 (1976).
See also *1.3*, p. *23*; 14.1, p. 166; 21.1, p. 310.

A. Slebarski and A. Chełkowski
Thermal Expansion and the Heat Capacity of $GdAl_2$
(a) Vol. 36, No. 2, p. K171–K174 (1976).
See also *8*, p. *70*; 21.4, p. 327.

F. Sacchetti, P. de Gasperis, and F. Menzinger
*Magnetic Moment Distribution in Cu–Ni Alloys
(b) Vol. 77, No. 1, p. 309–318 (1976).
See also 21.1, p. 311.

S. Zawisz
*Cluster Approximation for Non-Homogeneous Disordered Systems
(b) Vol. 77, No. 1, p. 367–372 (1976).
See also *13.5.2*, p. *157*.

J. Richter
*Influence of Structure Fluctuations on Ferromagnetism of Strongly Correlated Electrons
(b) Vol. 77, No. 1, p. K43–K47 (1976).
See also *2*, p. *39*.

L. Wojtczak and B. Mrygoń
*Dynamical Scaling Hypothesis at the Phase Transition Point
(b) Vol. 77, No. 1, p. K57–K61 (1976).

F. P. Livi, J. D. Rogers, and P. J. Viccaro
Experimental Studies of Hyperfine Interactions and Magnetization in $HfFe_2$
(a) Vol. 37, No. 1, p. 133–138 (1976).
See also *5*, p. *50*; 21.1.1, p. 320.

J. D. Sievert and C. Voigt
*Influence of Domains on the Magnetic Torque in Cubic Crystals
(a) Vol. 37, No. 1, p. 205–209 (1976).
See also 21.1.1, p. 320.

M. Shiga, R. Minakata, T. Fujimoto, and Y. Nakamura
*Forced Volume Magnetostriction of Fe–Al Alloys
(a) Vol. 37, No. 1, p. K33–K36 (1976).
See also 21.1, p. 311; 21.1.1, p. 321.

M. Shiga and Y. Nakamura
Mössbauer Study of B.C.C. Fe–Cr Alloys near the Critical Concentration
(a) Vol. 37, No. 1, p. K89–K92 (1976).
See also *5*, p. *50*; 21.1, p. 311; 21.1.1, p. 321.

E. Beloriżky, R. Casalegno, and P. Fries
*Magnetic Stability at 0 K of a Simple Cubic Ferromagnetic Array of Pseudo-Spin $S = \frac{1}{2}$ with Anisotropic Exchange between Nearest Neighbours
(b) Vol. 77, No. 2, p. 495–504 (1976).

T. Yamamoto
*On the Image Formation of Magnetic Domains and Their Walls in Scanning Electron Microscopy
(a) Vol. 37, No. 2, p. 607–615 (1976).
See also 21.1, p. 312.

H. J. Blythe and L. H. Dunleavy
*Low Temperature Magnetic Disaccommodation in Neutron Irradiated Iron
(a) Vol. 37, No. 2, p. K197–K200 (1976).
See also *10.1*, p. *84*; 21.1.1, p. 321.

A. Oppelt, A. Merkel, and K. H. J. Buschow
*Temperature Dependence of the Y Hyperfine Fields in YFe_3
(a) Vol. 37, No. 2, p. K205–K208 (1976).
See also *19*, p. *255*; 21.1.1, p. 321.

S. K. Godovikov, M. G. Kozin, V. V. Turotsev, and V. S. Shpinel
Hyperfine Fields Acting on Diamagnetic Impurities and the Anisotropic Exchange via the s- and d-Electrons in the Rare Earth Metals
(b) Vol. 78, No. 1, p. 103–111 (1976).
See also *5*, p. *50*; 18.4, p. 245; 21.4, p. 327.

H. E. Schaefer and W. Dander
*The Magnetic After-Effect Spectrum in Cobalt between 20 and 350 K after Electron Irradiation at 4.2 K
(b) Vol. 78, No. 1, p. 139–147 (1976).
See also *11*, p. *104*; 21.1, p. 313.

G. S. Poo and D. M. Edwards
*Electron Interactions and the Spin-Wave Stiffness Constant in Ferromagnetic Two-Component Alloys
(b) Vol. 78, No. 1, p. 287–299 (1976).
See also *13.5.4*, p. *161*; 21.1, p. 313.

J. ZIELIŃSKI
*A Double d-Band Model of Transition Metals
(b) Vol. 78, No. 1, p. 319–324 (1976).
See also *13.1*, p. *128*; 21.1, p. 313.

S. KRAŚNICKI
Neutron Investigation of Magnons in the Cu_2MnAl Heusler Alloy
(b) Vol. 78, No. 1, p. 335–340 (1976).
See also *13.5.4*, p. *161*; 21.1, p. 313.

L. I. VINOKUROVA, A. V. VLASOV, and M. PARDAVI-HORVÁTH
*Pressure Effects on Magnetic Phase Transitions in FeRh and FeRhIr Alloys
(b) Vol. 78, No. 1, p. 353–357 (1976).
See also 12.1, p. 109; 18.1, p. 227; *18.4*, p. *245*; 21.1.1, p. 321.

W.-H. STEEB and E. MARSCH
*Thermodynamics of a Two-Point Doubly Degenerate Hubbard Model in the Half-Filled Case
(b) Vol. 78, No. 1, p. K39–K44 (1976).
See also 13, p. 119; *18.4*, p. *245*.

H. MUGHRABI, R. KÜTTERER, K. LUBITZ, and H. KRONMÜLLER
Magnetic Studies on Cyclically Deformed α-Iron Single Crystals
(a) Vol. 38, No. 1, p. 261–270 (1976).
See also *10.1*, p. *85*; 21.1.1, p. 321.

T. YAMAMOTO
*Measurement of the Emission Area of Backscattered Electrons
(a) Vol. 38, No. 1, p. 361–368 (1976).
See also 21.1.1, p. 321.

K. N. SHRIVASTAVA
*Thermal Effects in Magnetic Exchange Interaction
(b) Vol. 78, No. 2, p. 749–755 (1976).
See also *6*, p. *64*; 22.5.4, p. 401.

R. B. PANDEY and C. M. BHANDARI
*A Green-Function Theory of an Anisotropic Ferromagnet
(b) Vol. 78, No. 2, p. 821–830 (1976).

M. V. CHETKIN, A. N. SHALYGIN, and A. V. KIRYUSHIN
*Magnetic Permeability of Ferro- and Paramagnetic Garnets at Optical Frequencies
(b) Vol. 78, No. 2, p. K85–K88 (1976).
See also *18.1*, p. *227*; 22.8.2, p. 431.

H.-R. HILZINGER
*Computer Simulation of Magnetic Domain Wall Pinning
(a) Vol. 38, No. 2, p. 487–496 (1976).

S. C. BHARGAVA and P. K. IYENGAR
Study of Hyperfine Fields at Iron Nuclei in $(Co, Fe)_5Ge_3$ Using the Mössbauer Effect
(a) Vol. 38, No. 2, p. 601–609 (1976).
See also *5*, p. *51*; 18.1, p. 227; 21.1.1, p. 322.

18.2.1 Ferromagnetic Films

Original Papers and Short Notes

O. G. Ramer and C. H. Wilts
*The Effects of Surface Layers on Spin-Wave Spectra in YIG Films
(b) Vol. **73**, No. 2, p. 443–453 (1976).
See also **19**, p. 247; 22.8.2, p. 426.

A. Duda and W. Maciejewski
*Thickness Dependence of Localized States in a Thin Film with Internal Layer Defect
(b) Vol. **74**, No. 1, p. 235–244 (1976).

O. Bostanjoglo and W. Giese
*Magnetization and Crystallization of Stress-Stabilized Amorphous Iron Films
(a) Vol. **34**, No. 1, p. K1–K4 (1976).
See also *2*, p. *36*; 21.1.1, p. 317.

A. Duda
*Crystal-Field-Induced Surface Anisotropy in Thin Magnetic Films
(b) Vol. **74**, No. 2, p. 701–709 (1976).

V. V. Myakhar, V. G. Pynko, and A. A. Potseluiko
*Stripe Domain Structure of (001) Magnetic Films with Positive Constant of Crystalline Anisotropy
(a) Vol. **34**, No. 2, p. 437–443 (1976).
See also 21.1, p. 306; 21.1.1, p. 317.

I. Khaiyer
*Domain Wall Observation in Permalloy Overlay Bars of Magnetic Bubble Domain Devices
(a) Vol. **34**, No. 2, p. K105–K106 (1976).
See also 21.1, p. 306; 21.1.1, p. 318.

S. Feierabend, U. Kischko, and R. Niedermayer
*Curie Temperature Determination of EuS Films with Structural Disorder
(a) Vol. **34**, No. 2, p. K135–K137 (1976).
See also 22, p. 336.

H. Göbel, E. Wolfgang, and H. Harms
*Properties of MnBi Compounds Partially Substituted with Cu, Zn, Ti, Sb, and Te (II): Stability and Magnetooptic Properties of Thin Films
(a) Vol. **35**, No. 1, p. 89–95 (1976).
See also *1.2*, p. *16*; 1.3, p. 22; 22.7, p. 411.

E. Münster
*Domain Wall Motion in Magnetic Film Strips
(a) Vol. **35**, No. 2, p. 521–531 (1976).
See also 21.1, p. 308.

A. Duda and W. Maciejewski
*On the Thickness Dependence of the Spin-Wave Resonance Spectrum in Thin Films with an Internal Layer Defect
(b) Vol. **76**, No. 1, p. K19–K22 (1976).

P. Schwellinger
*The Analysis of Magnetic Domain Wall Structures in the Transition Region of Néel and Bloch Walls by Lorentz Microscopy
(a) Vol. **36**, No. 1, p. 335–344 (1976).
See also 21.1.1, p. 320.

G. A. Jones
*On the Quality of Type 1 Magnetic Contrast Obtained in the Scanning Electron Microscope
(a) Vol. **36**, No. 2, p. 647–657 (1976).

O. D. Hennemann and E. Siegel
Spin-Wave Measurements of Exchange Constant *A* in Ni–Fe Alloy Thin Films
(b) Vol. 77, No. 1, p. 229–232 (1976).
See also *19*, p. *254*; 21, p. 310; 21.1.1, p. 320.

A. Wachniewski
*Microscopic Theory of Spin-Wave Resonance in Ferromagnetic Films with Surface-Region Imhomogeneities
(b) Vol. 77, No. 1, p. 319–328 (1976).

J. J. Żebrowski
*On Half-Bubble Stability in Platelets with Domain Wall Energy Density Gradients
(a) Vol. **37**, No. 2, p. 407–415 (1976).

S. R. Herd
*Lorentz Microscopy on Amorphous Films with Perpendicular Anisotropy as a Function of Temperature
(a) Vol. **38**, No. 1, p. 305–314 (1976).
See also *2*, p. *40*; 21.1, p. 314; 21.4, p. 327.

Š. Višňovský, B. Knappe, V. Prosser, and H.-R. Müller
*Magnetooptical Kerr Effect Spectra of Amorphous Gadolinium–Cobalt Films
(a) Vol. **38**, No. 1, p. K53–K56 (1976).
See also *20.1*, p. *277*; 21.1, p. 314; 21.4, p. 327.

18.3 Ferrimagnetic Properties

Original Papers and Short Notes

P. Gluck
*Variation Calculation for an Anisotropic Néel Ferrimagnet
(b) Vol. 73, No. 1, p. K79—K82 (1976).

S. K. Arif and D. S. P. Bunbury
Mössbauer and X-Ray Studies of $Dy(Fe_{1-x}Ni_x)_3$ and $Y(Fe_{1-x}Ni_x)_3$ Intermetallic Compounds
(a) Vol. 33, No. 1, p. 91—95 (1976).
See also *5*, p. *48*; 18.2, p. 228; 21.1.1, p. 316.

R. Parker and C. J. Tinsley
Electrical Conduction in Magnetite
(a) Vol. 33, No. 1, p. 189—194 (1976).
See also *14.3*, p. *173*; 22.6, p. 402.

M. Wurlitzer and F. Plümer
*X-Ray Induced Magnetic After-Effect in Si-Doped YIG
(a) Vol. 33, No. 1, p. K55—K57 (1976).
See also 22.8.2, p. 426.

E. De Grave, C. Dauwe, A. Govaert, and J. De Sitter
*Cation Distribution and Magnetic Exchange Interactions in $MgFe_2O_4$ Studied by the Mössbauer Effect
(b) Vol. 73, No. 2, p. 527—532 (1976).
See also *5*, p. *48*; 22.8.2, p. 426.

P. Novák and I. Veltruský
*Overlap and Covalency Contribution to the Zero-Field Splitting of S-State Ions
(b) Vol. 73, No. 2, p. 575—586 (1976).
See also *13*, p. *114*; 22.8.2, p. 426.

J. Baszyński and Z. Frait
*Spin-Wave Resonance in Thin Monocrystalline Ferrite Films
(b) Vol. 73, No. 2, p. K85—K88 (1976).
See also 19, p. 248; 22.8.2, p. 427.

V. G. Vologin, V. D. Parkhomenko, S. F. Dubinin, Yu. G. Chukalkan, B. N. Goshchitskii, S. K. Sidorov, and V. V. Petrov
Exchange Fe^{3+}–Fe^{3+} Interactions in Ferrites with Spinel Structure
(a) Vol. 33, No. 2, p. K83—K86 (1976).
See also 22.8.2, p. 427.

V. G. Kleparskii, N. P. Dymchenko, and S. K. Kukharskaya
*Temperature Dependence of Domain Walls on Pulse Displacement Dynamics in Magnetic Uniaxial Bubble Materials
(a) Vol. 33, No. 2, p. K117—K120 (1976).
See also 22.8:2, p. 427.

M. Maryško and J. Šimšová
*Angular Dependence of FMR Linewidth of Single Crystal Ferrite Films
(a) Vol. 33, No. 2, p. K133—K136 (1976).
See also *19*, p. *248*; 22.8.2, p. 427.

M. B. Kozintseva and V. E. Zilbervarg
The Electron and Magnon Spectra of Ferrimagnetic Semiconductors
(b) Vol. 74, No. 1, p. 199—208 (1976).
See also *13.1*, p. *122*; *13.5.4*, p. *160*; 22.8.2, p. 427.

J. FINK-FINOWICKI
*Magnetooptical Properties of Samarium Orthoferrite in the Spin-Reorientation Region
(b) Vol. 74, No. 1, p. K27—K29 (1976).
See also *20.1*, p. *264*; 22.8.2, p. 427.

K. I. ARAI, K. OHMORI, N. TSUYA, and S. IIDA
*Magnetostriction of Magnetite in the Vicinity of the Low-Temperature Transition
(a) Vol. 34, No. 1, p. 325—330 (1976).
See also 1.2, p. 15; 22.8.2, p. 427.

R. E. VANDENBERGHE, G. G. ROBBRECHT, and V. A. M. BRABERS
*Structure and Ionic Configuration of Oxidic Copper–Manganese Spinels ($Cu_xMn_{3-x}O_4$)
(a) Vol. 34, No. 2, p. 583—592 (1976).
See also *1*, p. *9*; 1.2, p. 15; 22.8.2, p. 428.

Z. ŠIMŠA, J. ŠIMŠOVÁ, K. SUK, E. KRATOCHVÍLOVÁ, and M. MARYŠKO
*Magnetic Bubbles in Spinel Ferrite Films
(a) Vol. 34, No. 2, p. 639—644 (1976).
See also 22.8.2, p. 428.

S. SAITO, S. TAKEMOTO, and Y. WATANABE
*Skin Effect Contributions in FMR of Conductive Ferrite Spheres
(a) Vol. 34, No. 2, p. 671—677 (1976).
See also *19*, p. *250*; 22.8.2, p. 428.

K. HISATAKE
*Photo-Induced Permeability Effect in Lithium Ferrite with Small Amounts of Cobalt
(a) Vol. 34, No. 2, p. K131—K134 (1976).
See also 22.8.2, p. 428.

S. LIGENZA
A Study of the $^6S_5/_2$-Term Splitting of an Fe^{3+} Ion in Zinc Ferrite by Neutron Spectroscopy
(b) Vol. 75, No. 1, p. 315—326 (1976).
See also 5, p. 49; *13.4*, p. *138*; 22.8.2, p. 428.

F. BASILE and P. POIX
Mössbauer Effect and Magnetic Measurement on the Solid Solution $Fe_{2-2x}Mg_{1-x}Sn_xO_4$
(a) Vol. 35, No. 1, p. 153—164 (1976).
See also *5*, p. *49*; 22.8.2, p. 428.

T. MERCERON and P. BERNSTEIN
*Photomagnetic Effect in Ni–Zn Ferrite with Small Amount of Cobalt
(a) Vol. 35, No. 2, p. 681—686 (1976).
See also 22.8.2, p. 429.

J. LOPITAUX et M. LENGLET
*Détermination de la distribution d'un spinelle à quatre cations par diffraction des rayons X et des neutrons et par analyse magnétique
(Determination of the Distribution in a Four Cation Spinel by X-Ray and Neutron Diffraction and by Magnetic Analysis)
(a) Vol. 35, No. 2, p. K141—K143 (1976).
See also 22.8.2, p. 429.

G. BORDIN, A. CECCHETTI, G. CECCHI, M. POPPI, and F. RONCONI
*Magnetostrictive Hysteresis Loop of Ferrites in Alternating Fields
(a) Vol. 35, No. 2, p. K161—K164 (1976).
See also 22.8.2, p. 429.

G. WENDE
*Radiation Damping in FMR Measurements in a Nonresonant Rectangular Waveguide
(a) Vol. 36, No. 2, p. 557—567 (1976).
See also *19*, p. *253*; 22.8.2, p. 430.

H.-R. Müller, B. Knappe, R. Kosicik, P. Rosemann, D. Schmidt, and A. Funke
*Magnetization Reversal in Thin Magnetic Films with In-Plane Compensation Walls
(a) Vol. **36**, No. 2, p. 617–626 (1976).
See also 22.8.2, p. 430.

V. R. K. Murthy and J. Sobhanadri
Dielectric Properties of Some Nickel–Zinc Ferrites at Radio Frequency
(a) Vol. **36**, No. 2, p. K133–K135 (1976).
See also *14.4*, p. *203*; 22.8.2, p. 430.

M. Guillot and H. Le Gall
*Low and Very High Magnetic Field Dependence of the Magnetization and the Faraday Rotation Induced by the Magnetic Dipole Transitions in YIG
(b) Vol. 77, No. 1, p. 121–130 (1976).
See also *20*, p. *258*; 22.8.2, p. 430.

N. D. J. Miller
Magnetostatic Volume Wave Propagation in a Dielectric Layered Structure
(a) Vol. **37**, No. 1, p. 83–91 (1976).
See also 22.8.2, p. 430.

K. De Strooper, A. Govaert, C. Dauwe, and G. Robbrecht
*Magnetic Properties of the Spinel Series $Co_2Ge_xTi_{1-x}O_4$ ($0 \leqq x \leqq 1$)
(a) Vol. **37**, No. 1, p. 127–132 (1976).
See also 18.1, p. 226; 18.4, p. 245; 22.8.2, p. 430.

F. Walz and J. Rivas
*Magnetic Aftereffects in Manganese Ferrites
(a) Vol. **37**, No. 1, p. 151–160 (1976).
See also 22.8.2, p. 430.

V. D. Parkhomenko, S. F. Dubinin, B. N. Goshchitskii, Yu. G. Chukalkin, S. K. Sidorov, V. G. Vologin, and V. V. Petrov
*Peculiarities of Radiation Damage in Ferrites with Spinel Structure
(a) Vol. **38**, No. 1, p. 57–66 (1976).
See also **11**, p. 104; 22.8.2, p. 431.

R. Morineau
*Structural Defects and Oxidation–Reduction Equilibrium in Mn, Zn Ferrites
(a) Vol. **38**, No. 2, p. 559–568 (1976).
See also *10.2*, p. *98*; 22.8.2, p. 431.

V. R. K. Murthy and J. Sobhanadri
Electrical Conductivity of Some Nickel–Zinc Ferrites
(a) Vol. **38**, No. 2, p. 647–651 (1976).
See also *14.3*, p. *182*; 15, p. 212; 22.8.2, p. 431.

A. I. Belyaeva, M. M. Kotlyarskii, and Yu. N. Stelmakhov
*Visual Investigation of Spin Reorientation in $TmFeO_3$
(a) Vol. **38**, No. 2, p. K103–K105 (1976).
See also 22.8.2, p. 431.

V. G. Vologin, S. F. Dubinin, V. D. Parkhomenko, Yu. G. Chukalkin, B. N. Goshchitskii, S. K. Sidorov, and V. V. Petrov
*Experimental Determination of the Preference Energy in Ferrites–Spinels
(a) Vol. **38**, No. 2, p. K147–K150 (1976).
See also 11, p. 105; 22.8.2, p. 432.

J. Šimšová, M. Maryško, and K. Suk
*Spin Wave Modes in Single Crystal Manganese Ferrite Films
(a) Vol. **38**, No. 2, p. K163–K166 (1976).
See also *19*, p. *256*; 22.8.2, p. 432.

18.4 Antiferromagnetic Properties

Review Articles

E. Vogt
*The Origin of Ferromagnetism and Antiferromagnetism in Alloys of Transition Metals (II): Alloys of Transition Metals with Other Transition Metals and with Polyvalent Non-Transition Elements
(a) Vol. **34**, No. 1, p. 11–67 (1976).
See also 1.3, p. 21; 18.1, p. 225; *18.2*, p. *228*; 21, p. 289; 21.1, p. 302; 21.6, p. 329.

S. V. Vonsovskii and M. S. Svirskii
*Multiplicity and Pseudomultiplicity in Electron–Ion Systems (I)
(b) Vol. 77, No. 1, p. 9–52 (1976).
See also *13*, p. *113*; 14.1, p. 164; 14.2, p. 169; 18.2, p. 228; 21, p. 289.

S. V. Vonsovskii and M. S. Svirskii
Multiplicity and Pseudomultiplicity of Electron–Ion Systems (II)
(b) Vol. 77, No. 2, p. 403–432 (1976).
See also *13*, p. *113*; 14.1, p. 164; 14.2, p. 169; 18.2, p. 228; 21, p. 289.

Original Papers and Short Notes

D. Scheerlinck and S. Hautecler
*Magnetic Interactions in Co_3O_4
(b) Vol. **73**, No. 1, p. 223–228 (1976).
See also *18.1*, p. *225*; 22.8.2, p. 426.

A. V. Zalesskii, A. K. Zvezdin, I. S. Zheludev, A. M. Savvinov, and A. F. Lebedev
*A New Type of Domain Wall NMR Spectrum in Canted Antiferromagnets: $YFeO_3$ and $LuFeO_3$ Crystals
(b) Vol. **73**, No. 1, p. 317–325 (1976).
See also *19*, p. *247*; 22.8.2, p. 426.

S. Kobe and K. Handrich
*Correlation Function and Misfit in a Computer-Simulated Two-Dimensional Amorphous Ising Antiferromagnet
(b) Vol. **73**, No. 1, p. K65–K67 (1976).
See also *2*, p. *35*.

I. D. Mikhailov and I. Yu. Shelestina
*One-Particle Excitation Spectrum for the One-Dimensional Hubbard Model at Low Temperature
(b) Vol. **73**, No. 2, p. K145–K148 (1976).

C. C. Becerra, A. Paduan Filho, and N. F. Oliveira, Jr.
*Differential Magnetization of MnP
(a) Vol. **33**, No. 2, p. 573–577 (1976).
See also *18.2*, p. *229*; 22, p. 334.

H. Arend, H. von Känel, and P. Wachter
*Magneto-Optical Measurements on Two-Dimensional Magnetic Model Systems
(b) Vol. **74**, No. 1, p. 151–157 (1976).
See also *20.1*, p. *263*; 22.9, p. 434.

G. A. Gehring, H. G. Kahle, W. Nägele, A. Simon, and W. Wüchner
*Two-Level Singlet Magnetism in $TbAsO_4$ and $TbVO_4$
(b) Vol. **74**, No. 1, p. 297–309 (1976).
See also 8, p. 67; 20.1, p. 264; 22.8.1, p. 420.

S. Arajs, G. Kote, C. A. Moyer, J. R. Kelly, K. V. Rao, and E. E. Anderson
*Antiferromagnetic Susceptibility of Cr–Re Alloys
(b) Vol. **74**, No. 1, p. K23–K26 (1976).
See also 21.1, p. 304.

W. Löser and W. Tiersch
*Thermodynamics of Finite Non-Half-Filled-Band Hubbard Rings
(b) Vol. **74**, No. 1, p. K43–K47 (1976).
See also *8*, p. *67*.

J. Monecke
*On the Phase Diagram of the Hubbard Model
(b) Vol. **74**, No. 1, p. K57–K60 (1976).
See also *13*, p. *115*.

W.-H. Steeb
*High Temperature Properties of a Doubly-Degenerate Hubbard Model
(b) Vol. **74**, No. 1, p. K71–K74 (1976).
See also 13, p. 115; *18.2*, p. *230*.

J. B. A. A. Elemans and K. H. J. Buschow
*Comment on the Crystal and Magnetic Structures of $CaCu_5$-Type Compounds of Th with Fe, Co, and Ni
(a) Vol. **34**, No. 1, p. 355–359 (1976).
See also 1.1, p. 11; *18.2*, p. *230*; 21.1, p. 305; 21.1.1, p. 317; 21.5, p. 328.

S. Deonarine and S. J. Joshua
Symmetry Properties of the Planar Antiferromagnet K_2NiF_4
(b) Vol. **74**, No. 2, p. 659–665 (1976).
See also 13.5.1, p. 152; *13.5.4*, p. *160*; 22.5.4, p. 401.

K. Hattori
*Properties of Randomly Diluted Heisenberg Antiferromagnets with Spin 1/2
(b) Vol. **74**, No. 2, p. 773–777 (1976).

V. S. Ostrovskii
*On the Spin Configuration of Highly Anisotropic Antiferrodielectrics (AFD) in the Magnetic Field
(b) Vol. **74**, No. 2, p. K157–K159 (1976).

W. Moldenhauer and W. Brückner
Physical Properties of Nonstoichiometric Iron Sulfide $Fe_{1-x}S$ near the α-Phase Transition
(a) Vol. **34**, No. 2, p. 565–571 (1976).
See also *1.2*, p. *15*; *14.3*, p. *175*; 15, p. 209; 22, p. 336.

G. Kappel, G. Fischer, and A. Jaéglé
*Magnetic Investigation of the System Mn_5Ge_3–Mn_5Si_3
(a) Vol. **34**, No. 2, p. 691–696 (1976).
See also *18.2*, p. *231*; 21, p. 293.

H. Oesterreicher
*Magnetic Studies on Compounds RCuSi, $R_6Cu_8Si_8$ and RCu_2Si_2 (R = Pr, Gd, Tb)
(a) Vol. **34**, No. 2, p. 723–728 (1976).
See also *18.2*, p. *231*; 21.1, p. 306; 21.4, p. 326.

A. M. Murasik
*Erratum to: An Uncertainty in Magnetic Moment Determination Imposed by Powder Neutron-Diffraction Method
(b) Vol. **75**, No. 1, p. K79 (1976).
See also 22, p. 336.

H. Bińczycka, A. Szytuła, J. Todorović, T. Zaleski, and A. Zięba
*Metamagnetism of CoMnSi
(a) Vol. 35, No. 1, p. K69–K72 (1976).
See also *18.2*, p. *232*; 22.8, p. 415.

L. M. Valiev, I. G. Kerimov, N. G. Aliev, and A. A. Abdurragimov
*The Thermal Expansion and Electrical Properties of $Cu_4Mn_2X_4$(X: S, Se, and Te)
(a) Vol. 35, No. 1, p. K85–K88 (1976).
See also 8, p. 68; 14, p. 162; 15, p. 210; *18.1*, p. *226*; 22.8, p. 415.

M. A. Boubel, R. P. Mainard, H. G. Fousse, A. J. Pointon, and F. R. Jeannot
*The Specific Heat Anomaly of Solid Solutions of Isostructural Antiferromagnetic Oxides (pFeO–qCoO)
(a) Vol. 35, No. 2, p. 459–464 (1976).
See also 8, p. 68; 22.6, p. 405.

A. Heidemann, K. Kosuge, and S. Kachi
*Investigation of the Hyperfine Splitting in V_3O_5 and V_4O_7 by Inelastic Neutron Scattering
(a) Vol. 35, No. 2, p. 481–485 (1976).
See also 22.6, p. 405.

F. A. Kassan-Ogly and M. V. Medvedev
*The Theory of a Heisenberg Alloy
(b) Vol. 76, No. 1, p. 45–53 (1976).
See also *18.1*, p. *226*; 18.2, p. 233; 21.1, p. 308; 21.1.1, p. 319.

F. Brouers
*Alloy Analogy Approximation of the Degenerate Hubbard Model
(b) Vol. 76, No. 1, p. 145–151 (1976).
See also *13*, p. *117*.

A. Wachniewski and L. Biegala
*Erratum to: Magnetostatic Surface Anisotropy in B.C.C. and H.C.P. Ferromagnets and S.C. Antiferromagnets
(b) Vol. 76, No. 1, p. K73 (1976).
See also *18.2*, p. *233*.

K. P. Belov, A. M. Kadomtseva, N. M. Kovtun, V. N. Derkachenko, V. N. Melov and V. A. Khokhlov
*On the Character of Phase Transitions in $ErFeO_3$
(a) Vol. 36, No. 1, p. 415–425 (1976).
See also 12.2, p. 111; 22.8.2, p. 429.

P. Arnold
*Measurements of Lattice Distortions in Antiferromagnetic $Al_2Mn(SiO_4)_3$
(a) Vol. 36, No. 1, p. K37–K38 (1976).
See also 22.8.2, p. 429.

G. A. Petrakovskii and A. I. Pankrats
*The Magnetoelastic Coupling in $FeBO_3$
(a) Vol. 36, No. 1, p. K43–K46 (1976).
See also *12.2*, p. *111*; 22.8.1, p. 423.

R. Micnas
*Erratum to: Spin-One Heisenberg Antiferromagnet with Biquadratic Exchange
(b) Vol. 76, No. 2, p. 851 (1976).

S. K. Sidorov
*On the Phenomenological Theory of Magnetization of Alloys with Mixed Exchange Interaction
(b) Vol. 76, No. 2, p. K113–K117 (1976).
See also *18.2*, p. *234*; 21.1, p. 310; 21.1.1, p. 320.

N. M. SALANSKII, E. A. GLOZMAN, and V. N. SELEZNEV
NMR and Domain Structure in Thin Single Crystals of $FeBO_3$ and $Fe_{1-x}Ga_xBO_3$
(a) Vol. **36**, No. 2, p. 779–782 (1976).
See also *19*, p. *253*; 22.8.1, p. 423.

E. HEINER
*Roth's Method and Moment Conservation of the Spectral Weight Function for the Neutral Hubbard Model
(b) Vol. 77, No. 1, p. 93–103 (1976).
See also 13, p. 118; *18.1*, p. *226*.

K. DE STROOPER, A. GOVAERT, C. DAUWE, and G. ROBBRECHT
Magnetic Properties of the Spinel Series $Co_2Ge_xTi_{1-x}O_4$ ($0 \leqq x \leqq 1$)
(a) Vol. **37**, No. 1, p. 127–132 (1976).
See also 18.1, p. 226; *18.3*, p. *241*; 22.8.2, p. 430.

T. RUSKOV, T. TOMOV, and S. GEORGIEV
Mössbauer Investigation of the Morin Transition in Hematite
(a) Vol. **37**, No. 1, p. 295–302 (1976).
See also *5*, p. *50*; 22.6, p. 407.

N. P. KOLMAKOVA, D. I. SIROTA, and M. ACQUARONE
*Anharmonicity of Sound Propagation near First-Order Orientational Phase Transitions in Antiferromagnets
(b) Vol. 77, No. 2, p. K143–K146 (1976).
See also *7*, p. *66*.

K. V. REDDY and S. C. CHETTY
Mössbauer Studies on the Fe–Te System
(a) Vol. **37**, No. 2, p. 687–694 (1976).
See also *5*, p. *50*; 21.1.1, p. 321.

S. K. GODOVIKOV, M. G. KOZIN, V. V. TUROVTSEV, and V. S. SHPINEL
Hyperfine Fields Acting on Diamagnetic Impurities and the Anisotropic Exchange via the s- and d-Electrons in the Rare Earth Metals
(b) Vol. 78, No. 1, p. 103–111 (1976).
See also *5*, p. *50*; 18.2, p. 235; 21.4, p. 327.

S. LIGENZA
*A Study of Spin Interaction in a Manganese-Substituted Zinc Ferrite by Neutron Spectroscopy
(b) Vol. 78, No. 1, p. 301–310 (1976).
See also 5, p. 50; 22.8.2, p. 431.

L. I. VINOKUROVA, A. V. VLASOV, and M. PARDAVI-HORVÁTH
*Pressure Effects on Magnetic Phase Transitions in FeRh and FeRhIr Alloys
(b) Vol. 78, No. 1, p. 353–357 (1976).
See also 12.1, p. 109; 18.1, p. 227; *18.2*, p. *236*; 21.1.1, p. 321.

K. SUGAWARA and C. Y. HUANG
*Microwave Resistivities of Some Rare Earth-Group VA Intermetallic Compounds
(b) Vol. 78, No. 1, p. K7–K9 (1976).
See also *14.1*, p. *167*; 21, p. 299.

W.-H. STEEB and E. MARSCH
*Thermodynamics of a Two-Point Doubly Degenerate Hubbard Model in the Half-Filled Case
(b) Vol. 78, No. 1, p. K39–K44 (1976).
See also 13, p. 119; *18.2*, p. *236*.

K. H. GERMANN
Optical Anisotropy Measurements in Multidomain Samples: Application to Phase Transitions in NiO and $Pb_5Ge_3O_{11}$
(a) Vol. **38**, No. 1, p. K81—K84 (1976).
See also **1.2**, p. 20; **14.4.1**, p. 208; *20.1*, p. *277*; 22.6, p. 408; 22.8.1, p. 424.

K. DRÄGER and R. GERLING
*Die Stabilität antiferromagnetischer Cluster im magnetisch verdünnten System $(1-x)\ Cr_2O_3 \cdot xAl_2O_3$
(The Stability of Antiferromagnetic Clusters in the Magnetically Dilute System $(1-x)\ Cr_2O_3 \cdot xAl_2O_3$)
(a) Vol. **38**, No. 2, p. 547—552 (1976).
See also **19**, p. 256; 22.6.1, p. 409.

W. BAŻELA, A. SZYTUŁA, J. TODOROVIĆ, Z. TOMKOWICZ, and A. ZIĘBA
*Crystal and Magnetic Structure of NiMnGe
(a) Vol. **38**, No. 2, p. 721—729 (1976).
See also 22.8, p. 418.

19. Magnetic Resonance

Review Article

H. W. de Wijn, A. M. van Diepen, and K. H. J. Buschow
Samarium in Crystal Fields, a Case of Strong *J*-Mixing
(b) Vol. **76**, No. 1, p. 11–30 (1976).
See also 5, p. 48; *13*, p. *113*; *18*, p. *222*; 21, p. 289; 21.1, p. 302; 21.1.1, p. 316; 21.4, p. 326.

Original Papers and Short Notes

I. V. Chepeleva, E. A. Zhilinskaya, V. N. Lazukin, and A. P. Chernov
*Electron Paramagnetic Resonance of Cr^{3+} in Chalcogenide Glasses
(b) Vol. **73**, No. 1, p. 65–72 (1976).
See also 2, p. 35; 18, p. 222; 22.8, p. 413.

W. Baumhoer, J. Thiemann, and O. Kanert
*Monte Carlo Simulation of NMR Line Shapes in Distorted Non-Metallic Solids
(b) Vol. **73**, No. 1, p. 155–164 (1976).
See also 10.2, p. 87; 22.5.2, p. 392.

A. V. Zalesskii, A. K. Zvezdin, I. S. Zheludev, A. M. Savvinov, and A. F. Lebedev
*A New Type of Domain Wall NMR Spectrum in Canted Antiferromagnets: $YFeO_3$ and $LuFeO_3$ Crystals
(b) Vol. **73**, No. 1, p. 317–325 (1976).
See also *18.4*, p. *242*; 22.8.2, p. 426.

M. Kumeda, N. Kobayashi, E. Maruyama, and T. Shimizu
*Electron Spin Resonance and Optical Absorption in Amorphous Se–As
(b) Vol. **73**, No. 1, p. K19–K22 (1976).
See also 2, p. 35; *20.1*, p. *261*; 22, p. 333.

A. V. Alekseev and U. Kh. Kopvillem
The Collapse of Radiation Stimulated by the Dipole–Dipole Temperature
(b) Vol. **73**, No. 1, p. K41–K46 (1976).
See also *6*, p. *53*; 22.8, p. 413.

J. Dietrich
*Anisotropy of the EPR of Yb^{3+} in Cubic Zinc Sulphide Due to Stacking Faults
(b) Vol. **73**, No. 1, p. K57–K59 (1976).
See also 22.4.1, p. 379.

W. S. Glaunsinger
*Erratum to: Indirect Exchange Interactions in EuB_6
(b) Vol. **73**, No. 1, p. K83 (1976).
See also *18.2*, p. *228*; 22, p. 334.

G. K. Semin and I. M. Alymov
*NQR Study of Molecular Crystals of Solid Solution of Methyl Bromide–Methyl Iodide
(a) Vol. **33**, No. 1, p. K63–K66 (1976).
See also 22.9, p. 433.

O. G. Ramer and C. H. Wilts
The Effects of Surface Layers on Spin-Wave Spectra in YIG Films
(b) Vol. **73**, No. 2, p. 443–453 (1976).
See also *18.2.1*, p. *237*; 22.8.2, p. 426.

B. Andlauer, J. Schneider, and W. Tolksdorf
Optical Absorption, Fluorescence, and Electron Spin Resonance of Ir^{4+} on Octahedral Sites in $Y_3Ga_5O_{12}$
(b) Vol. **73**, No. 2, p. 533–540 (1976).
See also *13.4*, p. *135*; 20.1, p. 261; 20.3, p. 280; 22.8.2, p. 426.

D. Vivien, A. Kahn, A. M. Lejus, and J. Livage
*EPR Investigation of the Very High Zero-Field Splitting of Gd^{3+} in La_2O_3 Single Crystals
(b) Vol. **73**, No. 2, p. 593–596 (1976).
See also 22.6, p. 402.

G. Oelgart, R. Herrmann, H. Krüger, and R. Stegmann
Fermi Surface and Helicons in Semiconducting $Bi_{1-x}Sb_x$ Alloys
(b) Vol. **73**, No. 2, p. 615–624 (1976).
See also *13.2*, p. *130*; 14.3, p. 174; 22.7, p. 410.

J. J. Rousseau, M. Rousseau, and J. C. Fayet
*EPR Investigations of a Structural Phase Change in $RbCaF_3$, $RbCdF_3$, and $TlCdF_3$
(b) Vol. **73**, No. 2, p. 625–631 (1976).
See also *1.2*, p. *14*; 22.8, p. 413.

J. Baszyński and Z. Frait
Spin-Wave Resonance in Thin Monocrystalline Ferrite Films
(b) Vol. **73**, No. 2, p. K85–K88 (1976).
See also *18.3*, p. *239*; 22.8.2, p. 427.

C. M. Scala and D. R. Hutton
Site Assignment of Fe^{3+} in α-Quartz
(b) Vol. **73**, No. 2, p. K115–K117 (1976).
See also *10.2*, p. *89*; 22.6, p. 403.

H. Nakata and H. Morishima
*Dose Dependence of ESR Spectra for On-Channel-Irradiated Silicon
(a) Vol. **33**, No. 2, p. 633–638 (1976).
See also 11, p. 100; 22.1.2, p. 352.

G. K. Semin and I. M. Alymov
*NQR Study of Molecular Crystals of Solid Solutions of Chlorobenzene–Bromobenzene
(a) Vol. **33**, No. 2, p. K111–K115 (1976).
See also 22.9, p. 434.

M. Maryško and J. Šimšová
*Angular Dependence of FMR Linewidth of Single Crystal Ferrite Films
(a) Vol. **33**, No. 2, p. K133–K136 (1976).
See also *18.3*, p. *239*; 22.8.2, p. 427.

P. Reineker
*To the Theory of the ESR Lineshape of Triplet Excitons in Molecular Crystals. Influence of the Quasi-Incoherent Exciton Motion and Comparison with Experimental Data
(b) Vol. **74**, No. 1, p. 121–134 (1976).
See also *13.5.1*, p. *150*; 22.9, p. 434.

D. L. Huber and M. S. Seehra
*Electron Paramagnetic Resonance in Anisotropic Magnets
(b) Vol. **74**, No. 1, p. 145–149 (1976).
See also 22.5, p. 388.

U. Kaufmann, A. Räuber, and J. Schneider
*ESR Analysis of IIB–IV–V_2 Semiconductors with Mn^{2+} as a Paramagnetic Probe
(b) Vol. **74**, No. 1, p. 169–175 (1976).
See also 13.4, p. 136; 22.8, p. 414.

P. BRAUN and S. GRANDE
*Spin–Lattice Relaxation Times and Knight Shift in InSb and InAs
(b) Vol. 74, No. 1, p. 209—215 (1976).
See also 22.2.1, p. 362; 22.2.3, p. 367.

J. ALIZON, G. BERTHET, J. P. BLANC, J. GALLICE, and H. ROBERT
*Exciton Dynamics Study by Magnetic Resonance in $TCNQ^- \cdot NH_4^+$
(b) Vol. 74, No. 1, p. 217—224 (1976).
See also 13.5.1, p. 151; 22.9, p. 434.

C. A. BATES and H. SZYMCZAK
*The Phonon-Induced Temperature Dependences of Spin-Hamiltonian Parameters for S-State Ions
(b) Vol. 74, No. 1, p. 225—233 (1976).
See also *6*, p. *55*; 22.4.1, p. 379; 22.6, p. 403.

A. M. GREKHOV and A. B. ROITSIN
*Theory of Low-Symmetry Effects in Paramagnetic Resonance
(b) Vol. 74, No. 1, p. 323—334 (1976).
See also 18.1, p. 225.

S. I. FĂRCAS, A. DARABANT, and A. NICULA
*Electron Spin Resonance Study of Cr^{3+} and Fe^{3+} in Alums
(b) Vol. 74, No. 1, p. 335—339 (1976).
See also 22.8.1, p. 420.

J. Y. BUZARÉ and J. C. FAYET
*EPR Investigations on the $(FeOF_5)^{4-}$ Cluster in $KZnF_3$
(b) Vol. 74, No. 1, p. 393—399 (1976).
See also 22.8, p. 414.

A. JASIŃSKI and Z. SUŁEK
*^{31}P Chemical Shift Tensor in Solid KH_2PO_4, KD_2PO_4, and $NH_4H_2PO_4$
(b) Vol. 74, No. 1, p. K5—K8 (1976).
See also 22.8.1, p. 420.

A. R. KESSEL and V. A. ZHIKHAREV
*The Formation of Echo Signals from the Spins Diffusing in Inhomogeneous Magnetic Field
(b) Vol. 74, No. 1, p. K35—K38 (1976).

B. D. BHATTACHARYYA
*Tunnel Splitting in a Trigonal O^- Center
(b) Vol. 74, No. 1, p. K53—K56 (1976).
See also *10.2*, p. *90*; 22.5.3, p. 399.

G. OELGART, G. SCHNEIDER, W. KRAAK, and R. HERRMANN
The Semiconductor–Semimetal Transition in $Bi_{1-x}Sb_x$ Alloys
(b) Vol. 74, No. 1, p. K75—K78 (1976).
See also *13.2*, p. *130*; 14, p. 162; 22.7, p. 410.

A. I. ZHERNOVOY
*On Fast Adiabatic Passage in Nuclear Magnetic Resonance
(b) Vol. 74, No. 1, p. K79—K80 (1976).

T. SASAKI
*ESR Study on Thermally Stimulated Exoelectron Emission from Irradiated K_2SO_4
(a) Vol. 34, No. 1, p. 339—346 (1976).
See also *17*, p. *218*; 22.8.1, p. 420.

W. S. GLAUNSINGER
*Variable-Temperature EPR Study of the Semiconducting Ferromagnet EuB_6
(b) Vol. 74, No. 2, p. 443—452 (1976).
See also 22, p. 335.

W. DIETRICH and D. SCHMID
*Electron Spin Resonance of the Triplet State of Pure and Doped $NaNO_2$ Single Crystals
(b) Vol. **74**, No. 2, p. 609–616 (1976).
See also 13.5.1, p. 152; 22.8.1, p. 420.

H. PFEIFFER and W. HAUBENREISSER
*Generalized Locator-Coherent-Potential Approach to Ferromagnetic Systems with Random Local Magnetic Fields and Moments
(b) Vol. **74**, No. 2, p. 617–626 (1976).
See also *18.2*, p. *230*.

J. KURIATA, N. GUSKOS, T. REWAJ, J. SZUMOWSKI, and K. FALKOWSKI
*Spin–Lattice Coefficients of Eu^{2+} in SrF_2
(b) Vol. **74**, No. 2, p. 683–687 (1976).
See also 22.5.3, p. 399.

B. D. BHATTACHARYYA
A New Aspect of Jahn-Teller Effect in the Low-Spin Ferrihaemoglobin Azide
(b) Vol. **74**, No. 2, p. 695–700 (1976).
See also *13.4*, p. *137*; 20.1, p. 266; 22.9, p. 435.

V. A. GASPAROV and M. H. HARUTUNIAN
*Precision Measurements of the Fermi Surface of Copper
(b) Vol. **74**, No. 2, p. K107–K110 (1976).
See also *13.2*, p. *130*; 21.1, p. 305.

Y. V. G. S. MURTI, G. SAMUEL, and S. SIVARAMAN
*Studies on Tin Impurity Centres in KCl Crystals
(a) Vol. **34**, No. 2, p. 615–621 (1976).
See also *10.2*, p. *91*; 13.4, p. 137; 20.1, p. 266; 22.5.2, p. 394.

S. SAITO, S. TAKEMOTO, and Y. WATANABE
*Skin Effect Contributions in FMR of Conductive Ferrite Spheres
(a) Vol. **34**, No. 2, p. 671–677 (1976).
See also *18.3*, p. *240*; 22.8.2, p. 428.

C. J. GADSDEN and M. HEATH
*Ferromagnetic Resonance in Nickel Single Crystals
(a) Vol. **34**, No. 2, p. K159–K162 (1976).
See also 21.1, p. 307.

G. SHARP-DENT, M. HARDIMAN, J. R. SAMBLES, and J. E. COUSINS
*Sample Size Effects on the Spin Resonance of Conduction Electrons in Aluminium
(b) Vol. **75**, No. 1, p. 155–161 (1976).
See also 21, p. 293.

M. VILLA and A. AVOGADRO
*Indirect Interactions and Nuclear Relaxation in $TlNO_3$
(b) Vol. **75**, No. 1, p. 179–188 (1976).
See also 22.8.1, p. 421.

G. OELGART and R. HERRMANN
Cyclotron Masses in Semiconducting $Bi_{1-x}Sb_x$Alloys
(b) Vol. **75**, No. 1, p. 189–196 (1976).
See also *13.1*, p. *123*; 13.2, p. 131; 22.7, p. 411.

M. O. M. DE MATOS, H. S. BRANDI, and B. MAFFEO
*Wave Function for the Interstitial Hydrogen Atom in SrF_2
(b) Vol. **75**, No. 1, p. 271–278 (1976).
See also 13.4, p. 138; 22.5.3, p. 399.

L. Pawlak, S. Pokrzywnicki, K. Falkowski, and K. Chęciński
*EPR of Mn^{2+} in CdY_2S_4 and CdY_2Se_4 Spinels
(b) Vol. 75, No. 1, p. K7—K10 (1976).
See also 22.8.2, p. 428.

S. V. Murthy
*Shifts in PMR Spectral Lines of Hydrates
(b) Vol. 75, No. 1, p. K11—K13 (1976).
See also 22.8.1, p. 421.

G. Thummes and H. H. Mende
*Observation of the Radio-Frequency Size Effect in Copper Whiskers
(b) Vol. 75, No. 1, p. K61—K63 (1976).
See also 21.1, p. 307.

D. Bastian and E. Biller
*Damping of Ferromagnetic Resonance in Ni–Fe Alloys
(a) Vol. 35, No. 1, p. 113—120 (1976).
See also *18.2*, p. *232*; 21.1, p. 307; 21.1.1, p. 318.

I. K. Amanis and J. G. Kliava
*Temperature Dependence of the Hyperfine Coupling Constants of Copper Atoms in α-Quartz
(a) Vol. 35, No. 1, p. K67—K68 (1976).
See also 22.6, p. 405.

B. Voigtsberger and H. Rosenberger
*High Resolution Proton Magnetic Resonance in Monocrystals of α-Glycine
(a) Vol. 35, No. 1, p. K89—K92 (1976).
See also 22.9, p. 436.

S. K. Misra and G. R. Sharp
*Electron Paramagnetic Resonance of Gd^{3+} in $PrCl_3 \cdot 7\,H_2O$
(b) Vol. 75, No. 2. p. 607—612 (1976).
See also 22.5, p. 389.

M. L. Falin, V. P. Meiklyar, V. A. Ulanov, and M. M. Zaripov
*Electron Nuclear Double Resonance of Mn^{2+} in SrF_2
(b) Vol. 75, No. 2, p. 731—734 (1976).
See also 22.5.3, p. 399.

M. L. Falin, V. P. Meiklyar, and V. P. Zentsov
*Endor and Covalent Model of the $[TiF_8]^{6-}$ Complex
(b) Vol. 75, No. 2, p. K111—K114 (1976).
Erratum
(b) Vol. 77, No. 1, p. K77 (1976).
See also 22.5.3, p. 400.

S. Radhakrishna, B. V. R. Chowdari, and A. Kasi Viswanath
*EPR Study of Mo^{5+} Ions in NaCl Single Crystal
(b) Vol. 75, No. 2, p. K119—K121 (1976).
See also 22.5.2, p. 395.

D. R. Hutton and C. M. Scala
*An EPR Study of Fe^{3+} in Scolecite
(b) Vol. 75, No. 2, p. K167—K168 (1976).
See also 22.8, p. 415.

S. V. Murthy
*The Pake-Murthy Effect
(b) Vol. 75, No. 2, p. K169—K171 (1976).
See also 22.8.1, p. 421.

D. BASTIAN and E. BILLER
*Anisotropy Constants and g-Factors of Ni–Fe Alloys Derived from Ferromagnetic Resonance
(a) Vol. **35**, No. 2, p. 465–470 (1976).
See also *18.2*, p. *232*; 21.1, p. 308; 21.1.1, p. 319.

D. BALL
*Paramagnetic Resonance of Er^{3+} and Dy^{3+} in Synthetic Single Crystals of Hafnon
(a) Vol. **35**, No. 2, p. 651–658 (1976).
See also 22.8.1, p. 422.

S. RADHAKRISHNA and T. BHASKAR RAO
Radiation Damage of Chromium Doped Potassium Alum
(a) Vol. **35**, No. 2, p. 715–718 (1976).
See also *11*, p. *102*; 20.1, p. 269; 22.8.1, p. 422.

J.-P. RIVERA, H. BILL, and R. LACROIX
*EPR of Powder Mg-OH:Mn^{2+} Boracite Suggesting a Rhombohedral Phase
(a) Vol. **35**, No. 2, p. K105–K108 (1976).
See also 1.2, p. 17; 22.8, p. 416.

P. MATTHYS, M. VANHAELST, and E. BOESMAN
*A Modified Doping Procedure for High O_2^- Concentrations in Potassium and Rubidium Halides
(a) Vol. **35**, No. 2, p. K137–K139 (1976).
See also *3*, p. *42*; 22.5.2, p. 395.

V. P. KALANTARIAN and V. I. CHECHERNIKOV
*Nuclear Magnetic Resonance and Susceptibility Measurements of Transition Metals and Alloys: V–Ta, V–Nb, and Nb–Ta Systems
(b) Vol. **76**, No. 1, p. 113–118 (1976).
See also 13, p. 116; 18, p. 223; 21, p. 296; 21.1, p. 309.

B. E. VUGMEISTER
*Spatial and Spectral Spin Diffusion in Dilute Spin Systems
(b) Vol. **76**, No. 1, p. 161–170 (1976).
See also *18.1*, p. *226*.

D. L. LYFAR, V. E. GONCHARUK, and S. M. RYABCHENKO
*Temperature Dependence of Nuclear Quadrupole Resonance in Layer-Type Crystals
(b) Vol. **76**, No. 1, p. 183–189 (1976).
See also 22.5.3, p. 400.

H. J. VON BARDELEBEN, A. GOLTZENÉ, and C. SCHWAB
*Electron Paramagnetic Resonance of Two Stoichiometry-Induced Intrinsic Defect Centres in $CuGaS_2$
(b) Vol. **76**, No. 1, p. 363–370 (1976).
See also 22.8, p. 416.

P. M. BISCH and A. A. GOMES
Formal Results on the Susceptibility of Dilute Alloys (II): Mean-Field Approach and Applications
(b) Vol. **76**, No. 1, p. 395–406 (1976).
See also *13*, p. *117*; *18*, p. *224*; 21.1, p. 309.

B. BLANZAT, F. RAYNAL, R. PARROT, C. BARTHOU, and B. CANNY
*Electron Paramagnetic Resonance of Gd^{3+} in Bismuth Germanate
(b) Vol. **76**, No. 1, p. K5–K8 (1976).
See also 22.8.1, p. 422.

H. Figiel, A. Oppelt, E. Dormann, and K. H. J. Buschow
*Transferred Hyperfine Fields at the Y Sites in Y–Co Compounds
(a) Vol. **36**, No. 1, p. 275–283 (1976).
See also 18.2, p. 233; 21.1, p. 309; 21.4, p. 327.

D. Ball and B. M. Wanklyn
*Coloured Synthetic Zircon Crystals. Flux Growth and EPR of V^{4+} Impurity Centres
(a) Vol. **36**, No. 1, p. 307–316 (1976).
See also *3*, p. *43*; 22.8.1, p. 422.

G. O. Berim and A. R. Kessel
*The Model Hamiltonian and Some Properties of Magnetic Resonance of Triplet Excitons (I): Non-Localized Exciton Approximation
(b) Vol. **76**, No. 2, p. 827–839 (1976).
See also *13.5.1*, p. *154*.

L. L. Buishvili, T. G. Vardosanidze, and A. I. Ugulava
*On the Theory of the Transverse Overhauser Effect
(b) Vol. **76**, No. 2, p. K97–K100 (1976).

P. G. Baranov, R. A. Zhitnikov, and V. A. Khramtsov
*EPR of Au° Atoms in NaCl Crystals
(b) Vol. **76**, No. 2, p. K109–K112 (1976).
See also 22.5.2, p. 396.

C. J. F. Graf, B. Maffeo, and H. S. Brandi
*Calculation of Magnetic Hyperfine Parameters for the H_i^0 Center in CaF_2, SrF_2 and BaF_2
(b) Vol. **76**, No. 2, p. K137–K140 (1976).
See also *13.4*, p. *141*; 22.5.3, p. 400.

G. Wende
*Radiation Damping in FMR Measurements in a Nonresonant Rectangular Waveguide
(a) Vol. **36**, No. 2, p. 557–567 (1976).
See also *18.3*, p. *240*; 22.8.2, p. 430.

N. M. Salanskii, E. A. Glozman, and V. N. Seleznev
*NMR and Domain Structure in Thin Single Crystals of $FeBO_3$ and $Fe_{1-x}Ga_xBO_3$
(a) Vol. **36**, No. 2, p. 779–782 (1976).
See also 18.4, p. 245; 22.8.1, p. 423.

V. S. Grunin, G. D. Davtyan, V. A. Ioffe, and I. B. Patrina
*EPR of Cu^{2+} and Radiation Centres in Anatase (TiO_2)
(b) Vol. 77, No. 1, p. 85–92 (1976).
See also *13.4*, p. *142*; 22.6, p. 406.

D. Zwingel
*The Electronic Structure of Trapped Hole Centres in SnO_2
(b) Vol. 77, No. 1, p. 171–180 (1976).
See also 22.6, p. 406.

J. J. Rousseau and J. C. Fayet
Influence of the Crystal Field on Bent Configuration of V_K Centres in AMF_3 Compounds
(b) Vol. 77, No. 1, p. 195–201 (1976).
See also *4*, p. *46*; *13.4*, p. *142*; 22.8, p. 417.

R. Mathai, K. Narahari Rao, A. Kumaraswamy, T. Rajasekharan, and K. V. S. Rama Rao
*ESR Study of Paramagnetic Centres in X-Irradiated $AgClO_3$
(b) Vol. 77, No. 1, p. 215–219 (1976).
See also 22.8.1, p. 423.

O. D. Hennemann and E. Siegel
*Spin-Wave Measurements of Exchange Constant A in Ni–Fe Alloy Thin Films
(b) Vol. 77, No. 1, p. 229–232 (1976).
See also 18.2.1, p. 238; 21.1, p. 310; 21.1.1, p. 320.

P. Matthys, M. Vanhaelst and E. Boesman
*Spin–Lattice Relaxation of the O_2^- Centre in RbI and RbBr
(b) Vol. 77, No. 1, p. 249–257 (1976).
See also 22.5.2, p. 396.

C. Larica and A. P. Guimarães
*EPR of $GdAg_{1-x}In_x$ Intermetallic Compounds
(b) Vol. 77, No. 1, p. K11–K16 (1976).
See also 22.7, p. 412.

N. G. Cherenda and D. M. Yudin
*Additives and Cage Effect in Glasses
(b) Vol. 77, No. 1, p. K27–K30 (1976).
See also *2*, p. *39*; 11, p. 103; 22.8.1, p. 423.

W. Brunner and G. Völkel
*The Temperature Dependence of Spin–Lattice Relaxation in CdS: Mn^{2+}
(b) Vol. 77, No. 1, p. K31–K33 (1976).
See also 22.4.1, p. 381.

S. N. Karyagin, P. K. Kashkarov, V. F. Kiselev, and S. N. Kozlov
*On the Nature of Electron and Hole Traps in Germanium Dioxide Thermally Grown on Germanium
(a) Vol. 37, No. 1, p. K17–K19 (1976).
See also *13.3*, p. *133*; 22.6, p. 407.

N. T. Cherpak, V. V. Mishenko, and T. A. Smirnova
*Inversion Ratio in Andalusite Crystals at 6 mm Wave Band
(a) Vol. 37, No. 1, p. K37–K38 (1976).
See also 13.4, p. 142; 22.8, p. 417.

T. Wosiński, T. Figielski, and A. Mąkosa
Spin-Dependent Photoconductivity Spectrum of Dislocated Silicon
(a) Vol. 37, No. 1, p. K57–K59 (1976).
See also *10.2*, p. *95*; *16*, p. *216*; 22.1.2, p. 356.

L. K. Aminov, R. Kirmse, and B. V. Solovev
Electron Spin–Lattice Relaxation of Cu^{2+} Ions in Planar Coordination Geometries. Calculation and Experimental Data of T_1 for Cu^{2+} in Ni(II)bis(diethyl-dithiocarbamate) Crystals
(b) Vol. 77, No. 2, p. 505–515 (1976).
See also *4*, p. *46*; 6, p. 61; 18.1, p. 226; 22.9, p. 438.

E. König and R. Schnakig
Zero-Field Splitting of $^6S(d^5)$ Ions in Tetragonal and Rhombic Symmetry (I): Complete d-Electron Calculation of the Crystal-Field Dependence of D and E
(b) Vol. 77, No. 2, p. 657–666 (1976).
See also *13.4*, p. *143*; 18.1, p. 227.

V. Soriano and D. Galland
*Photosensitivity of the EPR Spectrum of the F^+ Center in ZnO
(b) Vol. 77, No. 2, p. 739–743 (1976).
See also *13.4*, p. *143*; 22.4, p. 377.

R. Kuhl, W. Kraak, H. Haefner, and R. Herrmann
The Semimetal–Semiconductor Transition in Bismuth–Antimony Alloys
(b) Vol. 77, No. 2, p. K109–K111 (1976).
See also *13.2*, p. *131*; *14.1*, p. *166*; 21.7, p. 332.

A. Ramakanth and Ş. K. Ghatak
*Electron-Mediated Nuclear Spin–Phonon Interaction in Metals
(b) Vol. 77, No. 2, p. K123–K127 (1976).
See also *6*, p. *62*.

B. Nowak, N. Piślewski, and W. Leszczyński
*Proton Spin–Lattice Relaxation Time in the Dihydride Phase of the Ternary Ti–Nb–H Alloy
(a) Vol. 37, No. 2, p. 669–674 (1976).
See also 21.1, p. 312.

L. I. Horváth and T. Szörényi
*EPR Studies on Vanadium Phosphate Glasses
(a) Vol. 37, No. 2, p. K169–K171 (1976).
See also 2, p. 40; 22.8.1, p. 424.

A. Oppelt, A. Merkel, and K. H. J. Buschow
*Temperature Dependence of the Y Hyperfine Fields in YFe_3
(a) Vol. 37, No. 2, p. K205–K208 (1976).
See also *18.2*, p. *235*; 21.1.1, p. 321.

L. Kubler, G. Gewinner, J. J. Koulmann, and A. Jaéglé
*EPR Study on Amorphous Germanium. Oxygen, Deposition Angle, Annealing, and Substrate Temperature Effects
(b) Vol. 78, No. 1, p. 149–158 (1976).
See also 2, p. 40; 22.1.1, p. 349.

D. Müller and J. Petersson
Spin–Lattice Relaxation of Several Nuclei near the Ferroelectric Transitions of TGS and TGFBe
(b) Vol. 78, No. 1, p. 191–198 (1976).
See also *14.4.1*, p. *208*; 22.9, p. 438.

L. Rusiniak
*Electric Paramagnetic Resonance and Combined Resonance of Conductivity Electrons in InSb at 24 GHz
(b) Vol. 78, No. 1, p. 265–270 (1976).
See also 13.1, p. 128; 20.1, p. 276; 22.2.3, p. 369.

S. V. Nistor, V. Ghiordănescu, and M. Vodă
*ESR of Mn^{2+} Doped $CdCl_2$ Single Crystals
(b) Vol. 78, No. 1, p. K31–K33 (1976).
See also 22.5, p. 390.

J. Kuriata and T. Rewaj
*Superhyperfine Structure in the EPR of $SrF_2:Eu^{2+}$ Single Crystals
(b) Vol. 78, No. 1, p. K55–K58 (1976).
See also 22.5, p. 390.

Y. Ochiai and E. Matsuura
*ESR in Heavily Doped n-Type Silicon near a Metal–Nonmetal Transition
(a) Vol. 38, No. 1, p. 243–252 (1976).
See also 13.4, p. 144; 22.1.2, p. 357.

E. König and R. Schnakig
Zero-Field Splitting of $^6S(d^5)$ Ions in Tetragonal and Rhombic Symmetry (II): Application of Complete d-Electron Calculations to the Zero-Field Splitting in $Mn:MnF_2$ and $Mn:ZnF_2$
(b) Vol. 78, No. 2, p. 543–551 (1976).
See also *13.4*, p. *145*; 18.1, p. 227; 20.1, p. 277; 22.5, p. 390.

M. Vanhaelst, P. Matthys, and E. Boesman
*Temperature Dependence of the EPR and Optical Spectra in V^{++}-Doped Alkali Halides
(b) Vol. 78, No. 2, p. 553–562 (1976).
See also *20.1*, p. *277*; 22.5.2, p. 398.

A. M. Ziatdinov, M. M. Zaripov, Yu. V. Yablokov, and R. L. Davidovich
The Depression of the Dynamic Jahn-Teller Effect
(b) Vol. 78, No. 2, p. K69–K71 (1976).
See also 1.2, p. 20; *6*, p. *64*; *13.4*, p. *145*; 22.8, p. 418.

R. S. Saraswat and G. C. Upreti
*Electron Paramagnetic Resonance of Mn^{2+} in $ZnSO_4 \cdot 7\,H_2O$ Single Crystals
(b) Vol. 78, No. 2, p. K73–K75 (1976).
See also 22.8.1, p. 424.

F. J. López and F. Agulló-López
*ESR of Conduction Electrons in Electrolytically Coloured Cation-Doped NaCl
(b) Vol. 78, No. 2, p. K81–K83 (1976).
See also 10.2, p. 98; 22.5.2, p. 398.

H. R. Khan, K. Lüders, Ch. J. Raub, and Z. Szücs
*Knight-Shift and Superconductivity of Nb–Al σ- and A-15 Phases
(b) Vol. 78, No. 2, p. K107–K109 (1976).
See also *14.2*, p. *172*; 21, p. 300.

G. A. Stewart and J. A. Barclay
The Effects of Clustering on the Hyperfine Field of Indium in Iron
(a) Vol. 38, No. 2, p. 533–538 (1976).
See also 21.1.1, p. 321.

K. Dräger and R. Gerling
The Stability of Antiferromagnetic Clusters in the Magnetically Dilute System
$(1-x)\,Cr_2O_3 \cdot xAl_2O_3$
(a) Vol. 38, No. 2, p. 547–552 (1976).
See also *18.4*, p. *246*; 22.6.1, p. 409.

J. Šimšová, M. Maryško, and K. Suk
*Spin Wave Modes in Single Crystal Manganese Ferrite Films
(a) Vol. 38, No. 2, p. K163–K166 (1976).
See also *18.3*, p. *241*; 22.8.2, p. 432.

20. Optical Properties

Original Papers and Short Notes

M. G. KAPLUNOV, YU. M. SHULGA, K. I. POKHODNYA, and YU. G. BORODKO
*X-Ray Photoelectron Spectra of Quasi-One-Dimensional Conductors
(b) Vol. **73**, No. 1, p. 335–339 (1976).
See also 13, p. 113; 17, p. 218; 22.9, p. 433.

I. MOORE, M. KING, M. JONES, R. J. DECK, and A. M. HERMANN
*Microwave Faraday Rotation in Molybdenum Disulfide
(b) Vol. **73**, No. 1, p. K31–K34 (1976).
See also 13, p. 114; *18*, p. *222*; 22, p. 333.

J. R. L. DE ALMEIDA and S. M. REZENDE
*Scattering of Light by Magnetoelastic Excitations in Ferromagnets
(b) Vol. **73**, No. 2, p. 661–672 (1976).
See also 13, p. 114; 18, p. 222; 22.8.2, p. 427.

YA. A. IOSILEVSKII
*Construction of the Reciprocal Fourth-Rank Tensors in Crystals
(b) Vol. **73**, No. 2, p. 673–680 (1976).
See also *14*, p. *162*.

K. BREUER, W. TEWS, and E. STEINBEISS
*Energy Loss Measurements on Thin VO_2 Films
(b) Vol. **73**, No. 2, p. K95–K97 (1976).
See also *13*, p. *114*; 22.6, p. 403.

G. DLUBEK and O. BRÜMMER
*A Comparison of Positron Annihilation and Compton Effect Measurements on Si Single Crystals
(b) Vol. **73**, No. 2, p. K107–K110 (1976).
See also *13*, p. *115*; 22.1.2, p. 352.

H. W. HELBERG
*Indicatrix Orientation in the Quasi-One-Dimensional Organic Semiconductor (MTPA)·$(TCNQ)_2$
(a) Vol. **33**, No. 2, p. 453–459 (1976).
See also 22.9, p. 433.

L. HULDT
Phonon-Assisted Auger Recombination in Germanium
(a) Vol. **33**, No. 2, p. 607–614 (1976).
See also *6*, p. *54*; *13.1*, p. *122*; 22.1.1, p. 345.

N. L. KRAMARENKO, I. V. KURNOSOV, and YU. V. NABOIKIN
*Light Scattering by Liquid Crystals in an Electric Field
(a) Vol. **33**, No. 2, p. 773–781 (1976).
See also *2*, p. *36*; 22.9, p. 434.

U. BERG, G. DRÄGER, and O. BRÜMMER
*Combined Investigations of the Valence Band Structure of Graphite by K X-Ray Emission Spectroscopy and X-Ray Photoemission Spectroscopy
(b) Vol. **74**, No. 1, p. 341–348 (1976).
See also *13.1*, p. *122*; 17, p. 218; 22.1, p. 343.

G. SCHROEDER and H. KRONMÜLLER
Investigation of the Annealing of the Dislocations in Ni Crystals by Means of the Magneto-Optical Kerr Effect
(a) Vol. 34, No. 2, p. 623–638 (1976).
See also *10.1*, p. *81*; 18.2, p. 231; 21.1, p. 306.

V. KONDAWAR and C. MANDE
Chemical Shifts in the K X-Ray Absorption Discontinuity of Arsenic
(b) Vol. 75, No. 1, p. 79–83 (1976).
See also *4*, p. *45*; 22, p. 336.

M. BRAUWERS and F. BROUERS
*Optical Relaxation Time and Effective Mass in Noble Metal Alloys
(b) Vol. 75, No. 2, p. 519–525 (1976).
See also *13.1*, p. *124*; 21.6, p. 330.

KWOK LEUNG YIP, A. B. KUNZ, and W. S. WILLIAMS
Electronic Structure of TiC and VC from the Iterative Extended Hückel Theory
(b) Vol. 75, No. 2, p. 533–541 (1976).
See also *13.1*, p. *124*; 22.3, p. 373.

F. R. KESSLER, J. METZDORF, and R. H. MÜLLER
*Interband Faraday Rotation in Germanium. Screening of Excitons by Free Carriers
(b) Vol. 75, No. 2, p. 567–572 (1976).
See also 13.5.1, p. 153; *18*, p. *223*; 22.1.1, p. 347.

U. BERG, G. DRÄGER, K. MOSEBACH, and O. BRÜMMER
Combined XS and XPS Investigations of the FeS_2 Valence Band
(b) Vol. 75, No. 2, p. K89–K92 (1976).
See also *13.1*, p. *124*; 17, p. 219; 22, p. 337.

P. A. BEZIRGANIAN and M. A. TSERUNIAN
*On the Problem of Phase Jumps at Total Internal Reflection of X-Rays
(a) Vol. 35, No. 2, p. K181–K182 (1976).

A. M. RADWAN and M. TAUT
Band Structure of Al and the Laves Phase $CaAl_2$
(b) Vol. 76, No. 2, p. 605–612 (1976).
See also *13.1*, p. *126*; 21, p. 297; 22.7, p. 411.

A. E. LEE and J. WEBSTER
The Soft X-Ray Appearance Potential Spectra of Diamond and Graphite
(b) Vol. 76, No. 2, p. 661–667 (1976).
See also *13.1*, p. *126*; 13.3, p. 133; 22.1, p. 344.

M. GUILLOT and H. LE GALL
*Low and Very High Magnetic Field Dependence of the Magnetization and the Faraday Rotation Induced by the Magnetic Dipole Transitions in YIG
(b) Vol. 77, No. 1, p. 121–130 (1976).
See also *18.3*, p. *241*; 22.8.2, p. 430.

R. HIRSCH and G. STÜHMER
*A Simple Technique to Measure Optical SHG in Powder Samples by Means of Electronic Integration
(a) Vol. 37, No. 1, p. 105–108 (1976).
See also *1.2*, p. *18*; 22.5.4, p. 401.

J. DAUNAY, J. BAILLOU, M. LEMERCIER, JAC. DAUNAY, P. BUGNET, and G. BATAILLER
*Dispersion of the Parasitic Birefringence Induced by Lattice Plane Defects in ZnSe Single Crystals
(a) Vol. 37, No. 2, p. 521–527 (1976).
See also 10.2, p. 96; 22.4.2, p. 383.

J. S. Ahearn, Jr., C. A. B. Ball, and C. Laird
*Stress-Induced Birefringence of Mismatching III–V Heterojunctions
(a) Vol. 38, No. 1, p. 315–320 (1976).
See also *10.2*, p. *98*; 22.2.1, p. 364; 22.2.2, p. 365.

H. Neddermeyer
*X-Ray Spectroscopic Investigation of the Electronic Structure of Mg–Cu Alloys
(b) Vol. 78, No. 2, p. 609–614 (1976).
See also *13.1*, p. *129*; 21.1, p. 314; 21.3, p. 324.

K. Berndt and O. Brümmer
X-Ray Compton Scattering from LiMg Alloys
(b) Vol. 78, No. 2, p. 659–666 (1976).
See also *13.2*, p. *131*; 21.2, p. 323; 21.3, p. 325.

A. Kosakov, K. Unger, and G. Leonhardt
Calculation of Energy-Dependent Photoionization Probabilities for Valence Electrons in Diamond and Silicon Crystals
(b) Vol. 78, No. 2, p. 681–688 (1976).
See also *13.1*, p. *129*; 17, p. 220; 22.1, p. 344; 22.1.1, p. 349; 22.1.2, p. 358.

G. Dräger and O. Brümmer
X-Ray Spectroscopic Investigation of Beryllium by Polarized K-Emission Valence Bands
(b) Vol. 78, No. 2, p. 729–735 (1976).
See also *13.1*, p. *129*; 21.3, p. 325.

W. Grevendonk, G. Buymbeek, P. van den Keybus, and A. de Bock
Faraday Rotation in Non-Crystalline As_2S_3
(b) Vol. 78, No. 2, p. K89–K91 (1976).
See also *2*, p. *41*; 18, p. 224; 22, p. 342.

20.1 Spectra. Optical Constants (X-Ray Spectra See 20)

Review Article

M. N. Sapozhnikov
Zero-Phonon Transitions in the Optical Spectra of Impurity Molecular Crystals
(b) Vol. **75**, No. 1, p. 11–51 (1976).
See also *6*, p. *52*; 13.4, p. 134; *20.3*, p. *280*; 22.9, p. 433.

Original Papers and Short Notes

H. Meinhold and G. Weiser
*Modulation Spectroscopy on MoS_2 and $MoSe_2$
(b) Vol. **73**, No. 1, p. 105–115 (1976).
See also 13.1, p. 120; *13.5.1*, p. *149*; 22, p. 333.

F. Bechstedt, R. Enderlein, A. Kolpakov, and K. Peuker
*Theory of Resonance Raman Scattering near Critical Points
(b) Vol. **73**, No. 1, p. 141–149 (1976).
See also *6*, p. *52*; 13.1, p. 120; 13.5.1, p. 149.

P. Leyral, D. Bois, and P. Pinard
*Strain-Associated Band Tailing Effect in Semiconductors
(b) Vol. **73**, No. 1, p. 187–192 (1976).
See also *13.1*, p. *120*; 22.2.1, p. 362.

J. Höner zu Siederdissen
*Impurity-Induced Raman Spectra of Potassium Halides
(b) Vol. **73**, No. 1, p. 239–244 (1976).
See also *6*, p. *52*; 13.4, p. 134; 22.5.2, p. 392.

W. Rühle, W. Jakowetz, C. Wölk, R. Linnebach, and M. Pilkuhn
Optical Studies of Free and Bound Excitonic States in GaSb. Evidence for Deep A^+-Complexes
(b) Vol. **73**, No. 1, p. 255–264 (1976).
See also 13.4, p. 134; *13.5.1*, p. *149*; *20.3*, p. *280*; 22.2.3, p. 367.

J. W. J. Timans and H. W. den Hartog
*Local-Mode Absorption of H^- and D^- in $SrCl_2$:RE and SrF_2:RE Single Crystals
(b) Vol. **73**, No. 1, p. 283–292 (1976).
See also *6*, p. *53*; 22.5.3, p. 399.

R. H. Williams, I. T. McGovern, R. B. Murray, and M. Howells
An Investigation of the Electronic Structure of GaSe and GaTe by Photoelectron Spectroscopy, Using a Synchrotron Source, and Electron Energy Loss Spectroscopy
(b) Vol. **73**, No. 1, p. 307–316 (1976).
See also *13.1*, p. *120*; *17*, p. *218*; 22.3, p. 372.

T. Fukumoto, S. Nakashima, K. Tabuchi, and A. Mitsuishi
*Temperature Dependence of Raman Spectra of Cuprous Halides
(b) Vol. **73**, No. 1, p. 341–351 (1976).
See also *6*, p. *53*; 22.5, p. 388.

B. Esser and P. Kleinert
*Erratum to: On the Theory of Interband $\varepsilon(\omega)$ and External Electric Field Modulated $\Delta\varepsilon(\omega)$ in Disordered Semiconductors
(b) Vol. **73**, No. 1, p. 379 (1976).
See also 2, p. 35; *14.3.4*, p. *195*; 22, p. 333.

M. Kumeda, N. Kobayashi, E. Maruyama, and T. Shimizu
*Electron Spin Resonance and Optical Absorption in Amorphous Se–As
(b) Vol. **73**, No. 1, p. K19–K22 (1976).
See also 2, p. 35; *19*, p. *247*; 22, p. **333**.

R. Krishnan
*Optical Absorption of Co^{3+} in YIG Crystals
(b) Vol. **73**, No. 1, p. K51–K56 (1976).
See also 13.4, p. 134; 22.8.2, p. 426.

G. B. Abdullaev, T. G. Kerimova, Sh. S. Mamedov, T. R. Mechtiev, R. Kh. Nani, and E. Y. Salaev
*Study of the Fundamental Absorption Edge of $CdInGaS_4$ by Wavelength Modulation
(b) Vol. **73**, No. 1, p. K69–K72 (1976).
See also 13.5.1, p. 149; 22.8, p. 413.

M. Leppihalme and T. Tuomi
Photovoltage Spectra of Metal–Silicon (p-Type) Diodes near the Indirect Absorption Edge of Silicon
(a) Vol. **33**, No. 1, p. 125–133 (1976).
See also 6, p. 53; 13.5.1, p. 149; *14.3.3*, p. *190*; *16*, p. *213*; 22.1.2, p. 351.

M. Fritsche, M. Pöhler, M. Schubert, and G. Staupendahl
*Investigation of the Absorption of Intense CO_2 Laser Radiation by the p-Band of Tellurium
(a) Vol. **33**, No. 1, p. 149–154 (1976).
See also *13.1*, p. *121*; 22.1.3, p. 359.

Hj. Bernhardt
Comparative Studies of the LNT Coloration of $Y_3Al_5O_{12}$, $YAlO_3$, and $LiNbO_3$ Crystals
(a) Vol. **33**, No. 1, p. 211–214 (1976).
See also *10.2*, p. *87*; 20.3, p. 280; 22.8.1, p. 419; 22.8.2, p. 426.

T. J. Magee, N. M. Johnson, and J. Peng
The Effect of Illumination on the 10.6 μm Absorption Coefficient of Electron-Irradiated Single Crystal KCl
(a) Vol. **33**, No. 1, p. 415–419 (1976).
See also *10.2*, p. *88*; *11*, p. *100*; 22.5.3, p. 399.

Yu. B. Gaididei, V. M. Loktev, A. F. Prikhotko, and L. I. Shanskii
*α-Oxygen Doublets (II): Experiment
(b) Vol. **73**, No. 2, p. 415–425 (1976).
See also *13.5.1*, p. *149*; 23, p. 440.

W. Ekardt and M. I. Sheboul
The Influence of the Electron–Hole Exchange Interaction on the Biexciton Ground State in CuCl and CuBr and Related Optical Transitions
(b) Vol. **73**, No. 2, p. 475–482 (1976).
See also 13.1, p. 121; *13.5.1*, p.*150*; 20.3, p. 280; 22.5, p. 388.

M. Giehler and E. Jahne
Effect of Damping on the Plasmon–Phonon Coupling in CdS and GaP
(b) Vol. **73**, No. 2, p. 503–516 (1976).
See also *6*, p. *54*; *13.5.2*, p. *157*; 22.2.2, p. 365; 22.4.1, p. 379.

B. Andlauer, J. Schneider, and W. Tolksdorf
*Optical Absorption, Fluorescence, and Electron Spin Resonance of Ir^{4+} on Octahedral Sites in $Y_3Ga_5O_{12}$
(b) Vol. **73**, No. 2, p. 533–540 (1976).
See also *13.4*, p. *135*; 19, p. 248; 20.3, p. 280; 22.8.2, p. 426.

J. RÖSELER, K. HENNEBERGER, and R. STREHLOW
*On the Theory of Electroabsorption in Polar Crystals
(b) Vol. **73**, No. 2, p. 541–552 (1976).
See also **13.5.3**, p. 159; *14.3.4*, p. *195*; 22, p. 334.

E. P. POKATILOV and V. M. FOMIN
*Multiphoton Absorption by Free Charge Carriers in Semiconductors
(b) Vol. **73**, No. 2, p. 553–564 (1976).

A. GRISEL and PH. SCHMID
Polytypism and Lattice Vibrations of PbI_2
(b) Vol. **73**, No. 2, p. 587–591 (1976).
See also *6*, p. *54*; 22.5, p. 388.

R. A. NEVILLE and B. L. EVANS
The Band Edge Excitons in 2H-MoS_2
(b) Vol. **73**, No. 2, p. 597–606 (1976).
See also *13.5.1*, p. *150*; 22, p. 334.

H. LANGE, J. DONECKER, and H. FRIEDRICH
*Electroreflectance and Wavelength Modulation Study of the Direct and Indirect Fundamental Transition Region of $In_{1-x}Ga_xP$
(b) Vol. **73**, No. 2, p. 633–639 (1976).
See also **13.1**, p. 121; 22.2.4, p. 370.

I. N. DOUGLAS
*Magnetic Circular Dichroism of Co^{2+}-Doped Cubic Crystals
(b) Vol. **73**, No. 2, p. 641–650 (1976).
See also **13.4**, p. 135; *18*, p. *222*; 22.5.2, p. 393; 22.6, p. 403.

B. REIMER, H. BAESSLER, J. HESSE, and G. WEISER
*Optical Properties of a Polydiacetylene Single Crystal — Evidence for van Hove Singularities
(b) Vol. **73**, No. 2, p. 709–716 (1976).
See also **13.1**, p. 121; 22.9, p. 433.

R. S. KIM and S. NARITA
Far-Infrared Interband Magnetoabsorption and Band Structure of $Hg_{1-x}Cd_xTe$ Alloys
(b) Vol. **73**, No. 2, p. 741–752 (1976).
See also *13.1*, p. *121*; 18, p. 222; 22.4.4, p. 387.

R. K. BAGAI and A. V. R. WARRIER
*Ultraviolet Absorption Spectrum of $CaF_2:Mn^{2+}$
(b) Vol. **73**, No. 2, p. K123–K126 (1976).
See also 22.5.3, p. 399.

N. N. SIROTA, I. V. BODNAR, A. I. LUKOMSKII, G. F. SMIRNOVA, and L. M. FINKELSTEIN
*Indirect Absorption Edge of $(GaP)_x(InAs)_{1-x}$ Mixed Crystals
(b) Vol. **73**, No. 2, p. K135–K136 (1976).
See also *13.1*, p. *121*; 22.2.4, p. 370.

I. A. SMIRNOV, R. SURYANARAYANAN, and S. G. SHULMAN
*Optical Absorption of Metallic SmS Films near the Interband Transitions
(b) Vol. **73**, No. 2, p. K137–K140 (1976).
See also 13.1, p. 122; 21.4, p. 326.

K. LÖSCH and J. U. FISCHBACH
*Measurement of High Near-Edge Absorption Coefficients in Surface Layers of InP
(a) Vol. **33**, No. 2, p. 473–478 (1976).
See also 13.1, p. 122; 13.5.1, p. 150; 22.2.2, p. 365.

P. A. Arsenev, K. Bienert, and A. V. Potemkin
The Spectroscopic Investigation of Neodymium in Mixed Aluminates of Yttrium and Gadolinium
(a) Vol. **33**, No. 2, p. 511–515 (1976).
See also *20.2*, p. *279*; 20.3, p. 281; 22.8.1, p. 419.

E. V. K. Rao
*Optical Properties of Oxygen-Implanted GaAs
(a) Vol. **33**, No. 2, p. 683–690 (1976).
See also 11, p. 100; 22.2.1, p. 362.

A. A. Kaminskii, D. Schultze, B. Hermoneit, S. E. Sarkisov, L. Li, J. Bohm, P. Reiche, R. Ehlert, A. A. Mayer, V. A. Lomonov, and V. A. Balashov
Spectroscopic Properties and Stimulated Emission in the $^4F_{3/2} \rightarrow {}^4I_{11/2}$ and $^4F_{3/2} \rightarrow {}^4I_{13/2}$ Transitions of Nd^{3+} Ions from Cubic $Bi_4Ge_3O_{12}$ Crystals
(a) Vol. **33**, No. 2, p. 737–753 (1976).
See also *20.2*, p. *279*; *20.3*, p. *281*; 22.8, p. 413.

S. Suzuki and M. Takagi
*Free Rotation of NO_2 Radicals in $NaNO_2$ Crystal
(a) Vol. **33**, No. 2, p. 789–792 (1976).
See also *14.4.1*, p. *206*; 22.8, p. 413.

E. M. Rockar and R. A. Forman
*$Fe^{4+}(3d^4)$ Optical Absorption in Oxidized $SrTiO_3$
(a) Vol. **33**, No. 2, p. K91–K95 (1976).
See also 13.4, p. 136; 22.8, p. 413.

V. M. Fridkin, K. D. Kochev, Yu. S. Kusminov, K. A. Verkhovskaya, and T. R. Volk
*The Shift of the Optical Absorption Edge at "Optical Damage" in $LiNbO_3$
(a) Vol. **33**, No. 2, p. K137–K139 (1976).
See also *14.4.1*, p. *206*; 22.8.1, p. 419.

E. Salje
*New Crystals $Pb_3(P_xV_{1-x}O_4)_2$ for Acousto-Optical Device Applications
(a) Vol. **33**, No. 2, p. K165–K167 (1976).
See also *7*, p. *65*; 22.8, p. 413.

T. Iida and R. Monnier
Static Properties of the Self-Trapped Hole in Simple Cubic Cesium Halides
(b) Vol. **74**, No. 1, p. 91–102 (1976).
See also 6, p. 54; *13.4*, p. *136*; 22.5.2, p. 393.

H. Arend, H. von Känel, and P. Wachter
*Magneto-Optical Measurements on Two-Dimensional Magnetic Model Systems
(b) Vol. **74**, No. 1, p. 151–157 (1976).
See also *18.4*, p. *242*; 22.9, p. 434.

J. Winter, K. Rössler, J. Bolz, and J. Pelzl
Mössbauer Effect and Raman Studies of the Structural Phase Transitions in $K_2[SnCl_6]$
(b) Vol. **74**, No. 1, p. 193–198 (1976).
See also *1.2*, p. *15*; 5, p. 48; 22.8, p. 414.

N. Kristianpoller, A. Rehavy, and M. Israeli
Effects of Ultraviolet Irradiation on KBr:Gd
(b) Vol. **74**, No. 1, p. 245–248 (1976).
See also *10.2*, p. *89*; 13.5.1, p. 151; 20.3, p. 281; 22.5.2, p. 393.

E. Jahne
*Long-Wavelength Optical Phonons in Mixed Crystals (I): A System of Two Coupled Modes
(b) Vol. **74**, No. 1, p. 275–283 (1976).
See also *6*, p. *55*; 22.2.4, p. 370; 22.3.1, p. 375.

P. J. BECKER
*Magnetostriktion und lineare magnetische Doppelbrechung im Jahn-Teller-gekoppelten System $TmAsO_4$ mit effektivem Spin $S = 1$
(Magnetostriction and Linear Magnetic Birefringence in the Jahn-Teller Coupled System $TmAsO_4$ with Effective Spin $S = 1$)
(b) Vol. **74**, No. 1, p. 285–295 (1976).
See also *18.1*, p. *225*; 22.8.1, p. 419.

G. A. GEHRING, H. G. KAHLE, W. NÄGELE, A. SIMON, and W. WÜCHNER
Two-Level Singlet Magnetism in $TbAsO_4$ and $TbVO_4$
(b) Vol. **74**, No. 1, p. 297–309 (1976).
See also 8, p. 67; *18.4*, p. *242*; 22.8.1, p. 420.

U. GIORGIANNI, G. MONDIO, G. SAITTA, and G. VERMIGLIO
*Role of the Jahn-Teller Effect in the Excited States of KBr:Tl-Type Crystals
(b) Vol. **74**, No. 1, p. 317–322 (1976).
See also *13.4*, p. *136*; 22.5.2, p. 393.

J. K. PRIBRAM and C. M. PENCHINA
Phonon-Assisted Electroabsorption by Contact Excitons
(b) Vol. **74**, No. 1, p. 375–381 (1976).
See also 6, p. 55; *13.5.1*, p. *151*; 14.3.4, p. 196.

N. UHLE and J. WINDSCHEIF
*Optical Absorption in V_2O_5 Doped with Li and Na
(b) Vol. **74**, No. 1, p. K15–K18 (1976).
See also 13.5.3, p. 159; 22.6, p. 403.

J. FINK-FINOWICKI
*Magnetooptical Properties of Samarium Orthoferrite in the Spin-Reorientation Region
(b) Vol. **74**, No. 1, p. K27–K29 (1976).
See also *18.3*, p. *240*; 22.8.2, p. 427.

H. POLLAK
*Charge Transfer in Cordierite
(b) Vol. **74**, No. 1, p. K31–K34 (1976).
See also *5*, p. *49*; 22.8.1, p. 420.

D. D. RICHARDSON
*Effects of an Impurity on Magnon Sidebands in Insulators
(b) Vol. **74**, No. 1, p. K61–K65 (1976).
See also 10.2, p. 90; *18.2*, p. *230*; 22, p. 335.

G. JUNGK
*Determination of Optical Constants. A Null-Method for Non-Absorbing Surface Films
(a) Vol. **34**, No. 1, p. 69–72 (1976).
See also 1.5, p. 31; 22.6, p. 403.

W. BARTH, K. ELSAESSER, and W. GÜTH
*The Optical Absorption of 60° Dislocations in Germanium
(a) Vol. **34**, No. 1, p. 153–163 (1976).
See also *13.4*, p. *137*; 22.1.1, p. 346.

S. S. LI, D. W. SCHOENFELD, and R. T. OWEN
*Optical Transmission Measurements and Electron Microprobe Analysis of a Bulk $InAs_{1-x}P_x$ Alloy System
(a) Vol. **34**, No. 1, p. 255–261 (1976).
See also *13.1*, p. *122*; 22.2.4, p. 370.

P. NATH, D. K. PANDYA, and K. L. CHOPRA
Amorphous Ge Alloy Films (III): Optical Properties
(a) Vol. **34**, No. 1, p. 405–411 (1976).
See also *2*, p. *36*; 22.1.1, p. 346.

P. C. DIGGLE, K. A. GEHRING, and R. M. MAC FARLANE
*Atomic Absorption and Emission in Laser Produced Plasmas
(a) Vol. **34**, No. 1, p. K13–K16 (1976).
See also *20.3*, p. *281*; 22.8, p. 414; 22.8.1, p. 420.

H. NEUMANN and W. JUNGE
*Temperature Dependence of the Indirect Absorption Edge in $Al_xGa_{1-x}As$
(a) Vol. **34**, No. 1, p. K39–K41 (1976).
See also 22.2.4, p. 370.

W. HOFBERGER
*The Birefringence of Crystalline Tetracene Derived from Diffraction Properties of Polycrystalline Films
(a) Vol. **34**, No. 1, p. K55–K58 (1976).
See also 1.5, p. 31; 22.9, p. 434.

R. BLINC, H. AREND, and A. KANDUŠER
*Vibrational Spectra, Hydrogen Bonding, and Ferroelectricity in the $PbHPO_4$ Family
(b) Vol. **74**, No. 2, p. 425–435 (1976).
See also 4, p. 45; *6*, p. *55*; 14.4.1, p. 206; 22.8.1, p. 420.

P. KOIDL
Jahn-Teller Effect in the $^4T_1(l)$ and $^4T_2(l)$ States of Tetrahedrally Coordinated Mn^{2+}
(b) Vol. **74**, No. 2, p. 477–484 (1976).
See also *13.4*, p. *137*; 20.3, p. 282; 22.2.2, p. 365; 22.4.1, p. 380; 22.4.2, p. 383.

D. PUDEWILL, F.-J. HIMPSEL, V. SAILE, N. SCHWENTNER, M. SKIBOWSKI, and E. E. KOCH
Excitons in Pure and Doped Solid Neon
(b) Vol. **74**, No. 2, p. 485–496 (1976).
See also *13.1*, p. *122*; *13.5.1*, p. *151*; 23, p. 440.

H.-J. HAGEMANN, W. GUDAT, and C. KUNZ
*Photoabsorption Coefficient of Alloys of Al with Transition Metals V, Fe, Ni and with Cu and Pr from 30 to 150 eV Photon Energy
(b) Vol. **74**, No. 2, p. 507–521 (1976).
See also 21, p. 292; 21.1, p. 305; 21.1.1, p. 317; 21.4, p. 326.

M. I. DYKMAN
Theory of Nondegenerate Two-Dimensional Electrons Interacting with Phonons in Quantizing Magnetic Fields
(b) Vol. **74**, No. 2, p. 547–559 (1976).
See also *6*, p. *55*; *13.1*, p. *122*.

B. ULRICI and E. JAHNE
*Phonon Frequencies of InP Derived from Two-Phonon Absorption
(b) Vol. **74**, No. 2, p. 601–607 (1976).
See also *6*, p. *55*; 22.2.2, p. 365.

H. MALTER
*On the Electronic Molecular Structure of the Organic Semiconductor Copper Phthalocyanine
(b) Vol. **74**, No. 2, p. 627–633 (1976).
See also *13.1*, p. *123*; 22.9, p. 435.

R. RUPPIN and H. YATOM
*Size and Shape Effects on the Broadening of the Plasma Resonance Absorption in Metals
(b) Vol. **74**, No. 2, p. 647–654 (1976).
See also *13.5.2*, p. *157*; 21, p. 292.

E. V. Bursian, Ya. G. Girshberg, and A. V. Ruzhnikov
*The Correlation between Optical Absorption Spectra, Carrier Mobility, and Phase Transition Temperature in Some Ferroelectrics
(b) Vol. **74**, No. 2, p. 689–693 (1976).
See also **13.1**, p. 123; *14.4.1*, p. *206*; 22.8.1, p. 420.

B. D. Bhattacharyya
A New Aspect of Jahn-Teller Effect in the Low-Spin Ferrihaemoglobin Azide
(b) Vol. **74**, No. 2, p. 695–700 (1976).
See also *13.4*, p. *137*; 19, p. 250; 22.9, p. 435.

S. K. Bahl and H. D. Drew
*Optical Properties of Mo–Re Alloys
(b) Vol. **74**, No. 2, p. 721–726 (1976).
See also **14.1**, p. 165; 21, p. 292.

S. Auluck
On the Pseudopotential Coefficients for Zinc and Cadmium
(b) Vol. **74**, No. 2, p. K129–K133 (1976).
See also *13.1*, p. *123*; *13.2*, p. *130*; 21, p. 292; 21.1, p. 305.

A. V. Malakhovskii and I. S. Edelman
*Optical and Magneto-Optical Spectra of $FeBO_3$ in the Region of Strong Absorption
(b) Vol. **74**, No. 2, p. K145–K149 (1976).
See also *18.2*, p. *231*; 22.8.1, p. 421.

D. Casasent and F. Caimi
Suppressive Writing in Lithium-Doped NaF
(a) Vol. **34**, No. 2, p. 461–466 (1976).
See also *10.2*, p *91*; 22.5.2, p. 394.

D. Chakraborty and G. Lehmann
On the Structures and Orientations of Hydrogen Defects in Natural and Synthetic Quartz Crystals
(a) Vol. **34**, No. 2, p. 467–474 (1976).
See also *10.2*, p. *91*; 22.6, p. 404; 22.8, p. 414.

N. Karl and H. Feederle
*Absorption Spectrum of the Tetracene Monopositive Ion in the Anthracene Lattice
(a) Vol. **34**, No. 2, p. 497–503 (1976).
See also *13.4*, p. *137*; 22.9, p. 435.

Y. V. G. S. Murti, G. Samuel, and S. Sivaraman
Studies on Tin Impurity Centres in KCl Crystals
(a) Vol. **34**, No. 2, p. 615–621 (1976).
See also *10.2*, p. *91*; 13.4, p. 137; 19, p. 250; 22.5.2, p. 394.

R. Metselaar, J. P. M. Damen, P. K. Larsen, and M. A. H. Huyberts
Investigation of Colour Centres in Gadolinium Gallium Garnet Crystals
(a) Vol. **34**, No. 2, p. 665–670 (1976).
See also *10.2*, p. *91*; 22.8.2, p. 428.

G. B. Abdullaev, K. R. Allakhverdiev, R. Kh. Nani, E. Yu. Salaev, and R. M. Sardarly
*Far-Infrared Reflection Spectra of $TlGaSe_2$
(a) Vol. **34**, No. 2, p. K115–K117 (1976).
See also *6*, p. *56*; 22.8, p. 414.

H. Schaumburg and F. Willmann
*Optical Absorption of Plastically Deformed Germanium
(a) Vol. **34**, No. 2, p. K173–K177 (1976).
See also *10.2*, p. *92*; 13.4, p. 138; 22.1.1, p. 346.

V. P. Bhola
*Thermoreflectance of Cd_3As_2
(a) Vol. **34**, No. 2, p. K179—K180 (1976).
See also *13.1*, p. *123*; 22.3, p. 372.

E. L. Bokhenkov, I. Natkaniec, and E. F. Sheka
Neutron Spectroscopy of Naphthalene Crystal Internal Phonon Modes
(b) Vol. **75**, No. 1, p. 105—116 (1976).
See also *6*, p. *56*; 22.9, p. 435.

D. Billard, F. Gervais, and B. Piriou
*Analysis of Multiphonon Absorption in Corundum
(b) Vol. **75**, No. 1, p. 117—126 (1976).
See also *6*, p. *56*; 22.6, p. 404.

P. Loose, M. Rosenzweig, and M. Wöhlecke
*Zeeman Effect of Bound Exciton Complexes in ZnO
(b) Vol. **75**, No. 1, p. 137—144 (1976).
See also *13.5.1*, p. *152*; 20.3, p. 282; 22.4, p. 376.

J. Voigt, M. Senoner, and I. Rückmann
Quantitative Analysis of Interference Structures in the Transmission Spectrum of Very Thin CdS Platelets
(b) Vol. **75**, No. 1, p. 213—219 (1976).
See also *13.5.1*, p. *152*; 22.4.1, p. 380.

E. Jahne
*Long-Wavelength Optical Phonons in Mixed Crystals (II): The Persistence of Common Gaps
(b) Vol. **75**, No. 1, p. 221—232 (1976).
See also *6*, p. *56*; 22.2.4, p. 370.

L. B. Coleman, C. R. Fincher, Jr., A. F. Garito, and A. J. Heeger
*Far-Infrared Single Crystal Studies of TTF-TCNQ
(b) Vol. **75**, No 1, p. 239—246 (1976).
See also 13.1, p. 123; 22.9, p. 436.

B. Rheinländer
*Volume Plasma Faraday Effect in n-Type GaAs
(b) Vol. **75**, No. 1, p. 261—270 (1976).
See also 13.5.2, p. 157; *18*, p. *223*; 22.2.1, p. 363.

J. Camassel, M. Schlüter, S. Kohn, J. P. Voitchovsky, Y. R. Shen, and M. L. Cohen
*Wavelength Modulation Spectra nad Electronic Band Structure of SnS_2 and $SnSe_2$
(b) Vol. **75**, No. 1, p. 303—314 (1976).
See also *13.1*, p. *124*; 22.7, p. 411.

T. R. Rao, M. L. Bansal, V. C. Sahni, and A. P. Roy
Study of Internal Modes of SO_4^{2-} in $LiKSO_4$
(b) Vol. **75**, No. 1, p. K31—K34 (1976).
See also *6*, p. *56*; 22.8.1, p. 421.

P. Hlídek, I. Barvík, V. Prosser, M. Vaněček, and M. Zvára
*Polarization Dependence of Transmission and Photoconductivity in $CdCr_2Se_4$
(b) Vol. **75**, No. 1, p. K45—K49 (1976).
See also *16*, p. *214*; 22.8.2, p. 428.

M. P. Lisitsa, A. P. Zakharchuk, S. F. Terekhova, G. G. Tsebulya, L. K. Mladov, and S. M. Todorov
*Anisotropy of the Absorption Edge in Germanium Selenide Single Crystals
(b) Vol. **75**, No. 1, p. K51—K54 (1976).
See also 13.5.1, p. 152; 22, p. 336.

Yu. A. Vodakov, G. A. Lomakina, E. N. Mokhov, E. I. Radovanova, V. I. Sokolov, M. M. Usmanova, G. F. Yuldashev, and B. S. Machmudov
Silicon Carbide Doped with Gallium
(a) Vol. **35**, No. 1, p. 37–42 (1976).
See also 3, p. 42; *13.4*, p. *138*; 14.3, p. 176; 20.3, p. 283; 22.3, p. 372.

R. Nink and H.-J. Kos
*On the Role of Z Centers in the Trapping Mechanism of Thermoluminescent Lithium Fluoride
(a) Vol. **35**, No. 1, p. 121–129 (1976).
See also *10.2*, p. *92*; 13.4, p. 138; 22.5.2, p. 394.

A. N. Georgobiani, Yu. V. Ozerov, E. E. Strumban, and V. V. Anisimov
Investigation of Deep Centres of Gallium-Doped Zinc Sulfide Crystals
(a) Vol. **35**, No. 1, p. 251–257 (1976).
See also *13.4*, p. *138*; 16, p. 214; 20.3, p. 283; 22.4.1, p. 380.

W. F. Braerman, K. W. Böer, and G. A. Dussel
*Observation of Deep-Trap Modulation with "Near" Band-Edge Light in Copper-Doped CdS Crystals
(a) Vol. **35**, No. 1, p. 309–313 (1976).
See also *13.4*, p. *139*; 22.4.1, p. 380.

F. Matsuura, S. Teratani, S. Kishida, and I. Tsurumi
Temperature Dependence of the Half-Width of the 2.3 eV Optical Absorption Band in Neutron-Irradiated ZnS
(a) Vol. **35**, No. 1, p. K57–K60 (1976).
See also 10.2, p. 92; *11*, p. *101*; 22.4.1, p. 380.

W. Henrion, M. Závětová, and T. Mazets
*Infrared Optical Properties of Pseudobinary Chalcogenide Glasses
(a) Vol. **35**, No. 1, p. K77–K79 (1976).
See also *2*, p. *38*; 22.8, p. 415.

N. Matsunami, Y. Yokoyama, and N. Itoh
*Displacement of Impurities through Interaction with Radiation-Induced Interstitials in Alkali Halides
(b) Vol. **75**, No. 2, p. 483–492 (1976).
See also 10.2, p. 93; *11*, p. *101*; 22.5.2, p. 395.

H. Ueba and S. Ichimura
*Plasmon–Phonon-Coupled Mode Sideband of Excitons
(b) Vol. **75**, No. 2, p. 501–506 (1976).
See also 6, p. 57; *13.5.1*, p. *153*; 13.5.2, p. 157; 22, p. 336.

G. Bauer, M. Overhamm, P. Grosse, W. Müller, E. Gornick, and H. W. Pötzl
Landau Emission in n-InSb
(b) Vol. **75**, No. 2, p. 543–551 (1976).
See also *13.1*, p. *124*; *14.3.4*, p. *197*; 20.3, p. 283; 22.2.3, p. 368.

E. Gerlach, P. Grosse, M. Rautenberg, and W. Senske
*Dynamical Conductivity and Plasmon Excitation in Bi
(b) Vol. **75**, No. 2, p. 553–558 (1976).
See also 13.5.2, p. 157; *14.1*, p. *165*; 21.7, p. 332.

F. F. Sokolov and V. V. Hizhnyakov
*Fading Memory in the Theory of Electronic Excitation Transfer
(b) Vol. **75**, No. 2, p. 669–676 (1976).
See also 13.4, p. 139; 22, p. 337.

C. M. Goel and B. Dayal
*Second-Order Raman Spectra of RbCl, RbBr, and RbI Crystals
(b) Vol. **75**, No. 2, p. 697–703 (1976).
See also *6*, p. *57*; 22.5.2, p. 395.

E. Wiener-Avnear
Polaritons in an Ensemble of CuI Crystallites
(b) Vol. **75**, No. 2, p. 717–720 (1976).
See also *13.5*, p. *147*; 22.5, p. 389.

R. Vacher and J. Pelous
Measurement of the Intensity of the Brillouin Lines in Amorphous Media at Very Low Temperature
(b) Vol. **75**, No. 2, p. K101–K103 (1976).
See also *2*, p. *38*; *6*, p. *57*; 22.9, p. 436.

A. I. Belyaeva, M. M. Kotlyarskii, E. A. Popov, and B. V. Beznosikov
*Spectroscopic Investigation of the Low Temperature Phase Transition in $NaMnCl_3$
(b) Vol. **75**, No. 2, p. K123–K127 (1976).
See also 13.5.4, p. 160; 22.8, p. 415.

J. Zirke, G. Frahm, A. Tausend, and D. Wobig
*Infrared and Raman Studies of TlSe
(b) Vol. **75**, No. 2, p. K149–K152 (1976).
See also *6*, p. *57*; 22, p. 337.

I. A. Smirnov, R. Suryanarayanan, and S. G. Shulman
*Optical Absorption of Metallic SmS Films
(b) Vol. **75**, No. 2, p. K153–K156 (1976).
See also 21, p. 295.

V. N. Brudnyi, D. L. Budnitskii, M. A. Krivov, and V. G. Melev
*p–n-Type Conversion and Optical Properties of 2.0 MeV Electron-Irradiated $ZnSnAs_2$
(a) Vol. **35**, No. 2, p. 425–430 (1976).
See also 11, p. 102; 13.4, p. 139; *14.3*, p. *177*; 22.8, p. 415.

E. Ruske
*Photoexpansion of Amorphous As_2S_3 Films
(a) Vol. **35**, No. 2, p. 477–480 (1976).
See also *1.5*, p. *31*; 2, p. 38; 22, p. 337.

A. P. Levanyuk and V. V. Osipov
*Optical Distortion in Crystals
(a) Vol. **35**, No. 2, p. 605–614 (1976).
See also 13.4, p. 139; *14.4.1*, p. *207*; 22.8.1, p. 422.

M. Kh. Ashurov, Yu. K. Voronko, V. V. Osiko, A. A. Sobol, B. P. Starikov, M. I. Timoshechkin, and A. Ya. Yablonskii
Inequivalent Luminescence Centres of Er^{3+} in Gallium Garnet Single Crystals
(a) Vol. **35**, No. 2, p. 645–649 (1976).
See also *13.4*, p. *140*; 20.3, p. 284; 22.8.2, p. 429.

E. F. Harris and J. H. Crawford, Jr.
The Influence of OH^- Impurity on the Oxidation of Iron Impurity in MgO above 1000 °C
(a) Vol. **35**, No. 2, p. 667–673 (1976).
See also *10.2*, p. *93*; 22.6, p. 405.

S. Radhakrishna and T. Bhaskar Rao
Radiation Damage of Chromium Doped Potassium Alum
(a) Vol. **35**, No. 2, p. 715–718 (1976).
See also *11*, p. *102*; 19, p. 252; 22.8.1, p. 422.

D. K. Pandya and K. L. Chopra
*Obliquely Deposited Amorphous Ge Films (I): Optical Properties
(a) Vol. **35**, No. 2, p. 725–734 (1976).
See also *2*, p. *38*; 22.1.1, p. 347.

K. L. Chopra, S. K. Barthwal, and D. K. Pandya
Structural Sensitivity of Transport and Optical Properties of Amorphous GeTe
(a) Vol. **35**, No. 2, p. 761–768 (1976).
See also 2, p. 38; *14.3*, p. *178*; *15*, p. *210*; 22.7, p. 411.

E. Ruske
*Reversibility of Photoinduced Edge Shift in Evaporated Amorphous As_2S_3 Films
(a) Vol. **35**, No. 2, p. K113–K114 (1976).
See also 2, p. 38; 22, p. 337.

K. H. Germann, W. Müller-Lierheim, H. H. Otto, and T. Suski
High Temperature Phase Transition in $Pb_5Ge_3O_{11}$
(a) Vol. **35**, No. 2, p. K165–K167 (1976).
See also *1.2*, p. *17*; 14.4.1, p. 207; 22.8.1, p. 422.

V. A. Antonov and P. A. Arsenev
*Spectroscopic Properties of Single Crystals of Rare Earth Titanates
(a) Vol. **35**, No. 2, p. K169–K172 (1976).
See also 20.2, p. 279; 22.8.1, p. 422.

V. V. Samartsev, R. G. Usmanov, I. Kh. Khadiev, E. F. Kustov, and M. N. Baranov
*Photon Echo in $CaWO_4$:Nd^{3+} and $LiAl_5O_8$:Cr^{3+}
(b) Vol. **76**, No. 1, p. 55–66 (1976).
See also 13.4, p. 140; *20.3*, p. *284*; 22.8.1, p. 422.

S. F. Timashev and M. A. Krykin
*Optical and Electrooptical Properties of the Interfaces of Media
(b) Vol. **76**, No. 1, p. 67–76 (1976).
See also 13.3, p. 132; *14.3.2*, p. *188*.

A. J. H. Eijkelenkamp
*Powder Reflection Measurements of Some Divalent Lead Halides
(b) Vol. **76**, No. 1, p. 153–160 (1976).
See also 13.5.1, p. 153; 22.5, p. 389.

A. K. Garg
*Pressed-Pellet Infrared Absorption Spectra of Molybdenum Dichalcogenides
(b) Vol. **76**, No. 1, p. 171–174 (1976).
See also *6*, p. *57*; 22, p. 337.

H. Burkhard, G. Bauer, P. Grosse, and A. Lopez-Otero
*Free Carrier Magneto-Optical Effects in the Reststrahlen Region of n-PbTe
(b) Vol. **76**, No. 1, p. 259–266 (1976).
See also 13.1, p. 125; *18*, p. *223*; 22, p. 337.

W. Kühn
*Stress Dependence of the (E_1, $E_1 + \Delta_1$) Transition in Germanium
(b) Vol. **76**, No. 1, p. 289–299 (1976).
See also 12.2, p. 110; *13.1*, p. *125*; 22.1.1, p. 347.

F. Henneberger and J. Voigt
*Polariton Effects in the Optical Spectra of Excitonic Molecules
(b) Vol. **76**, No. 1, p. 313–323 (1976).
See also 13.5, p. 147; *13.5.1*, p. *153*; 20.3, p. 284; 22, p. 337.

E. Sigmund and M. Wagner
*An Operator Method for Optical Lineshape Calculations in Non-Adiabatic Electron–Phonon Systems
(b) Vol. **76**, No. 1, p. 325–336 (1976).
See also 6, p. 58; 13.4, p. 140.

A. I. BELYAEVA and M. M. KOTLYARSKII
*Spectroscopic Investigations of Magnons in Weak Ferromagnetic $RbMnCl_3$
(b) Vol. 76, No. 1, p. 419–426 (1976).
See also *13.5.4*, p. *160*; 22.5, p. 389.

G. RUSZCZYNSKI and R. BOYN
*Jahn-Teller Coupling Parameters from Stress-Dichroism Spectra for CdS:Ti^{2+} and CdS:Co^{2+}
(b) Vol. 76, No. 1, p. 427–436 (1976).
See also 13.4, p. 140; 22.4.1, p. 380.

J. SZYMAŃSKI
*Erratum to: Theory of Free Carrier Absorption in n-PbSe
(b) Vol. 76, No. 1, p. 439 (1976).
See also 6, p. 58; 13.1, p. 125; 22, p. 338.

H. KÖHLER and C. R. BECKER
*Infrared-Active Lattice Vibrations in $SnSe_2$
(b) Vol. 76, No. 1, p. K15–K17 (1976).
See also *6*, p. *58*; 22, p. 338.

NAWAB SINGH and A. W. JOSHI
*Pressure Dependence of the Peak F-Band Absorption Energy in Alkali Halides
(b) Vol. 76, No. 1, p. K39–K42 (1976).
See also *10.2*, p. *93*; 22.5.2, p. 395.

K. HÜBNER and M. SCHÄFER
*Bond Charge Model for the Chemical Shift of the Photoemission in Semiconductors
(b) Vol. 76, No. 1, p. K63–K67 (1976).
See also *4*, p. *46*; 22.2, p. 361.

L. HERNÁNDEZ, O. VIGIL, and F. GÓNZALEZ
*Optical Transitions and Distribution of Localized Levels in $ZnIn_2S_4$
(a) Vol. 36, No. 1, p. 33–37 (1976).
See also *13.4*, p. *141*; 16, p. 215; 22.8, p. 416.

L. HOLZ, K. KNEIPP, A. LAU, and W. WERNCKE
Stimulated Raman Scattering of Picosecond Laser Pulses by Polaritons in $LiIO_3$ and $LiNbO_3$ Single Crystals
(a) Vol. 36, No. 1, p. K5–K8 (1976).
See also *6*, p. *58*; 13.5, p. 147; 20.2, p. 279; 22.8.1, p. 422.

G. MÜLLER
Optical and Electrical Spectroscopy of Zinc Oxide Crystals Simultaneously Doped with Copper and Donors
(b) Vol. 76, No. 2, p. 525–532 (1976).
See also *13.4*, p. *141*; 14.3, p. 178; 22.4, p. 377.

G. JUNGK and R. GRÜNDLER
*Determination of Optical Constants: Au
(b) Vol. 76, No. 2, p. 541–546 (1976).
See also *13.1*, p. *126*; 21.6, p. 330.

E. RIEFLIN, G. STÜHMER, and E. SCHWAB
Exciton Band Width versus Peak Shift in NH_4Br in Phases II and III as Measured by Thermoreflectance
(b) Vol. 76, No. 2, p. 547–550 (1976).
See also *13.5.1*, p. *154*; 22.5, p. 389.

R. DINGES, D. FRÖHLICH, and CH. UIHLEIN
*Two-Photon Absorption of Anisotropic Excitons in β-AgI
(b) Vol. 76, No. 2, p. 613–622 (1976).
See also *13.5.1*, p. *154*; 22.5.1, p. 391.

H.-J. Wünsche and F. Henneberger
*Transmission and Reflectivity of Highly Excited Semiconductors Due to the Optical Conversion of Excitons into Excitonic Molecules
(b) Vol. 76, No. 2, p. 639–646 (1976).
See also *13.5.1*, p. *154*; 22, p. 339.

P. Clauws and J. Vennik
*Lattice Vibrations of V_2O_5. Determination of TO and LO Frequencies from Infrared Reflection and Transmission
(b) Vol. 76, No. 2, p. 707–713 (1976).
See also *6*, p. *59*; 22.6, p. 406.

S. V. Vlachos, A. P. Lambros, A. Thanailakis, and N. A. Economou
*Anisotropic Indirect Absorption Edge in GeSe
(b) Vol. 76, No. 2, p. 727–735 (1976).
See also 6, p. 59; 22.7, p. 411.

G. Borstel, E. Schuller, and H. J. Falge
*Surface Phonon Polaritons on Absorbing Crystals
(b) Vol. 76, No. 2, p. 759–768 (1976).
See also *13.5*, p. *147*; 22.6, p. 406.

A. J. H. Eijkelenkamp and K. Vos
*Reflectance Measurements on Single Crystals of PbFCl, PbFBr, and $PbBr_2$
(b) Vol. 76, No. 2, p. 769–778 (1976).
See also 13.1, p. 126; 13.5.1, p. 154; 22.5, p. 389.

K. Suzuki, N. Miura, S. Uchida, and S. Tanaka
Far-Infrared Magneto-Reflection in HgTe
(b) Vol. 76, No. 2, p. 787–796 (1976).
See also *13.1*, p. *126*; 22.4.3, p. 386.

H. J. Hesse, W. Fuhs, G. Weiser, and L. von Szentpaly
Directional Dispersion of an Exciton Polariton in a Pentamethinium Cyanine Dye
(b) Vol. 76, No. 2, p. 817–826 (1976).
See also *13.5.1*, p. *154*; 22.9, p. 437.

R. Boyn
Tetrahedral Model for Stress Dichroism Due to Deep Impurities in Wurtzite Type Crystals
(b) Vol. 76, No. 2, p. K119–K122 (1976).
See also 13.4, p. 141; 22.4, p. 377.

B. Speit and G. Lehmann
Hole Centers in the Feldspar Sanidine
(a) Vol. 36, No. 2, p. 471–481 (1976).
See also *10.2*, p. *94*; 16, p. 215; 20.3, p. 285; 22.8.1, p. 423.

D. Lapraz and P. Iacconi
*On Some Luminescent and Optical Properties of Synthetic Calcite Single Crystals
(a) Vol. 36, No. 2, p. 603–616 (1976).
See also 13.4, p. 141; *20.3*, p. *285*; 22.8.1, p. 423.

P. Brüesch, M. Bösch, W. Känzig, M. Ziegler, and W. Bührer
Lattice Dynamics of the Phases I, II, and III of Sodium Hyperoxide NaO_2
(b) Vol. 77, No. 1, p. 153–164 (1976).
See also *6*, p. *60*; 22.6, p. 406.

F. Spiegelberg, E. Gutsche, and J. Voigt
Exciton–Phonon Interaction in CdS
(b) Vol. 77, No. 1, p. 233–242 (1976).
See also *6*, p. *60*; *13.5.1*, p. *154*; 22.4.1, p. 381.

J. Donecker and J. Kluge
*Magnetic Circular Dichroism of Indirect Excitonic Transitions in GaP and $In_{1-x}Ga_xP$
(b) Vol. 77, No. 1, p. 243—248 (1976).
See also 13.5.1, p. 155; *18*, p. *224*; 22.2.2, p. 365; 22.2.4, p. 371.

U. M. Grassano, A. Tanga, and M. Valli
Electric Field Effect on the Thallium Emission Bands in KBr and KI
(b) Vol. 77, No. 1, p. 341—346 (1976).
See also *10.2*, p. *95*; *20.3*, p. *286*; 22.5.2, p. 396.

L. Mărculescu
*Optical Spectra of Aggregate Centres in KCl:Pb Crystals
(b) Vol. 77, No. 1, p. 353—357 (1976).
See also *10.2*, p. *95*; 22.5.2, p. 396.

V. G. Kvachadze and Z. G. Akhvlediani
Influence of γ-Irradiation on the Thermal Conductivity of KCl:KOH Crystals
(b) Vol. 77, No. 1, p. K21—K25 (1976).
See also *6*, p. *60*; *8*, p. *70*; 11, p. 103; 22.5.2, p. 396.

V. P. Zenchenko and E. P. Sinyavskii
*The Johnson-Larsen Effect in the Resonance Laser Field
(b) Vol. 77, No. 1, p. K39—K42 (1976).
See also *18*, p. *224*.

Th. Kokkinakis and G. C. Papavassiliou
Surface Plasmons in Small Cu Particles
(b) Vol. 77, No. 1, p. K49—K51 (1976).
See also *13.5.2*, p. *157*; 21.1, p. 311.

R. Martens and F. Freund
*The Potential Energy Curve of the Proton and the Dissociation Energy of the OH^- Ion in $Mg(OH)_2$
(a) Vol. 37, No. 1, p. 97—104 (1976).
See also *6*, p. *61*; 22.8, p. 417.

S. Govinda
*Coloration and Luminescence in Pure and Chromium-Doped Al_2O_3 Single Crystals Irradiated with X-Rays at Room Temperature
(a) Vol. 37, No. 1, p. 109—117 (1976).
See also *10.2*, p. *95*; 11, p. 103; 20.3, p. 286; 22.6, p. 406.

H. Veeser, U. Bogner, and W. Eisenmenger
Analysis of the Acoustic Transients in the Pulse Response of the Linear Electro-Optic Effect
(a) Vol. 37, No. 1, p. 161—170 (1976).
See also *7*, p. *66*; 12.2, p. 111; *14.4.1*, p. *208*; 22.8.1, p. 423.

P. G. Vassilev
*Annealing Effects on the Thermoelectric Power and Visible Reflectance of Thin Copper Films
(a) Vol. 37, No. 1, p. K25—K27 (1976).
See also 14.1, p. 166; *15*, p. *210*; 21.1, p. 311.

A. T. Futro, T. Łukasiewicz, and J. Zmija
*Circular Birefringence in Bismuth Germanium Oxide ($Bi_{12}GeO_{20}$)
(a) Vol. 37, No. 1, p. K75—K79 (1976).
See also 22.8, p. 417.

V. L. Alperovich, V. M. Zaletin, A. F. Kravchenko, and A. S. Terekhov
*The Influence of Phonons and Impurities on the Broadening of Excitonic Spectra in Gallium Arsenide
(b) Vol. 77, No. 2, p. 465–472 (1976).
See also *13.5.1*, p. *155*; 22.2.1, p. 364.

M. Taut and W. Hanke
Dielectric Function of Hexagonal Metals (II): Beryllium
(b) Vol. 77, No. 2, p. 543–550 (1976).
See also *13.1*, p. *127*; 13.5.2, p. 157; *14.1*, p. *166*; 14.4, p. 204; 21.3, p. 324.

M. Obiditsch and H. Kahlert
*Electric Field Induced Changes of the Fundamental Absorption and Low-Temperature Energy Relaxation in n-Type GaSb
(b) Vol. 77, No. 2, p. 677–684 (1976).
See also 6, p. 61; *14.3*, p. *180*; 22.2.3, p. 368.

J. C. Castro, H. C. Basso, and M. de Souza
*Local Mode of H^- Substitutional Ion in KCN
(b) Vol. 77, No. 2, p. 685–692 (1976).
See also 1.2, p. 19; *6*, p. *61*; 10.2, p. 95; 22, p. 340.

L. Marculescu, M. Popescu, I. Baltog, and C. Ghita
The State of Dispersion of Pb^{2+} in Highly Doped KCl Crystals
(b) Vol. 77, No. 2, p. 701–706 (1976).
See also *10.2*, p. *95*; 20.3, p. 287; 22.5.2, p. 397.

N. Banaii and G. Jacobs
The Parallel and Perpendicular Stark Effect of the F-Center in RbBr and RbI
(b) Vol. 77, No. 2, p. 707–715 (1976).
See also *10.2*, p. *96*; 22.5.2, p. 397.

J. Krizek, K. N. R. Taylor, and H. Krizek
*Room Temperature Optical Properties of Rare Earths
(b) Vol. 77, No. 2, p. 725–730 (1976).
See also 14.1, p. 166; 21.4, p. 327.

J. Humlíček and F. Lukeš
*Direct Identification of the $L_3' \rightarrow L_1$ Transition in Ge
(b) Vol. 77, No. 2, p. 731–738 (1976).
See also 13.1, p. 127; *14.3.4*, p. *199*; 22.1.1, p. 348.

J. T. Gourley and E. R. Vance
Sharp Infrared Absorption Lines in Luminescent CaO
(b) Vol. 77, No. 2, p. K85–K89 (1976).
See also *10.2*, p. *96*; 22.6, p. 407.

P. C. Mehendru and Narendra Kumar
*Optical Absorption Studies of KCl:Ru and NaCl: Ru Crystals
(b) Vol. 77, No. 2, p. K133–K134 (1976).
See also 22.5.2, p. 397.

G. M. Arutyunyan and S. M. Shahinyan
*Multiphoton Absorption of Intensive Electromagnetic Waves in Semiconductors
(b) Vol. 77, No. 2, p. K171–K176 (1976).
See also *13*, p. *119*; 22, p. 340.

D. Bloor and F. H. Preston
*Optical Properties of Bis(p-Toluene Sulphonate) Diacetylene (I): Crystal Optics of Partially and Fully Polymerized Crystals
(a) Vol. 37, No. 2, p. 427–438 (1976).
See also 1.1, p. 13; 22.9, p. 438.

HJ. BERNHARDT
The Manganese-Induced O^- Centres in Yttrium Aluminium Garnet
(a) Vol. 37, No. 2, p. 445–448 (1976).
See also *10.2*, p. *96*; 20.3, p. 287; 22.8.2, p. 430.

H. G. LIPSON, P. LIGOR, and J. J. MARTIN
The Effect of Ionizing Radiation on the 10.6 μm Absorption of KCl and NaCl
(a) Vol. 37, No. 2, p. 547–552 (1976).
See also *10.2*, p. *96*; *11*, p. *104*; 22.5.2, p. 397.

F. M. CAIMI, J. O. ARTMAN, W. W. LIN, and D. P. CASASENT
Interpretation of Direct Writing M Center H-D Curve Characteristics of Photodichroic NaF Crystals
(a) Vol. 37, No. 2, p. 617–624 (1976).
See also *10.2*, p. *96*; 13.4, p. 143; 22.5.2, p. 397.

A. S. FILIPCHENKO and A. NAURIZBAEV
Moss-Burstein Effect in n-Type Crystals of Indium Antimonide
(a) Vol. 37, No. 2, p. K139–K142 (1976).
See also *13.4*, p. *143*; 22.2.3, p. 369.

I. V. BODNAR, A. I. LUKOMSKII, and G. F. SMIRNOVA
Absorption Edge of InP_xAs_{1-x} Mixed Crystals
(a) Vol. 37, No. 2, p. K173–K176 (1976).
See also *13.1*, p. *128*; 22.2.4, p. 371.

G. HOPPMANN and E. SALJE
Optical and Electrical Properties of H^+ Doped WO_3 Single Crystals
(a) Vol. 37, No. 2, p. K187–K191 (1976).
See also 13.1, p. 128; *14.3*, p. *180*; 22.6, p. 408.

A. M. BON, C. BENOIT, and O. BERNARD
*Dynamical Properties of Crystals of $Sr(NO_3)_2$, $Ba(NO_3)_2$, and $Pb(NO_3)_2$ (I): Infrared Spectra and Structure
(b) Vol. 78, No. 1, p. 67–78 (1976).
See also *6*, p. *62*; 22.8.1, p. 424.

V. P. FEDOROV, V. V. KIND, and A. V. KORSHUNOV
*The Multiplet Structure of the Electron Absorption Spectra of Naphthalene–β-Naphthol Mixed Molecular Crystals
(b) Vol. 78, No. 1, p. 93–101 (1976).
See also 13.4, p. 144; 13.5.1, p. 155; 22.9, p. 438.

H. J. FALGE, E. SCHULLER, and G. BORSTEL
Directional Dispersion of Polar Phonons Measured by Infrared Reflection
(b) Vol. 78, No. 1, p. 123–132 (1976).
See also *6*, p. *62*; 22.6, p. 408.

K. E. GAUSS and H. HAPP
Millimeter Wave Investigation of the Complex Dielectric Constant in DKDP
(b) Vol. 78, No. 1, p. 133–138 (1976).
See also 6, p. 62; *14.4.1*, p. *208*; 22.8.1, p. 424.

M. TAUT and P. ZIESCHE
Dielectric Function of Hexagonal Metals (III): Magnesium
(b) Vol. 78, No. 1, p. 169–175 (1976).
See also *13.1*, p. *128*; 13.5.2, p. 158; *14.1*, p. *167*; 14.4, p. 204; 21.3, p. 324.

G. A. GEHRING and J. LAUGSCH
*Theory of Side-Bands in a Cooperative Jahn-Teller Material
(b) Vol. 78, No. 1, p. 211–218 (1976).
See also 6, p. 62; 13.5.1, p. 155; 22.8.1, p. 424.

L. Rusiniak
Electric Paramagnetic Resonance and Combined Resonance of Conductivity Electrons in InSb at 24 GHz
(b) Vol. 78, No. 1, p. 265—270 (1976).
See also 13.1, p. 128; *19*, p. *255*; 22.2.3, p. 369.

H. J. Stolz, H. Wendel, A. Otto, L. Pintschovius, and H. Kahlert
*Acoustic and Optical Phonons in $(SN)_x$
(b) Vol. 78, No. 1, p. 277—286 (1976).
See also 4, p. 47; *6*, p. *63*; 22, p. 341.

V. L. Broude, L. M. Umarov, and E. F. Sheka
*Multiphonon Absorption in Molecular Crystals
(b) Vol. 78, No. 1, p. 325—333 (1976).
See also *6*, p. *63*; 22.9, p. 438.

E. V. Chisler, V. Yu. Davydov, I. N. Goncharuk, and E. A. Ivanova
*On Fermi Resonance Theory
(b) Vol. 78, No. 1, p. 359—370 (1976).
See also *6*, p. *63*.

K. P. Meletov and E. F. Sheka
Asymmetry of Phonon Side Bands of Optical Transitions in Impurity Molecular Crystals Caused by a Change in External Phonon Frequencies
(b) Vol. 78, No. 1, p.K1—K5 (1976).
See also *6*, p. *63*; *13.4*, p. *144*; 20.3, p. 287; 22.9, p. 439.

I. Neizvestny, N. Ovsyuk, and M. Sinyukov
*Electro-Reflectance Spectra of E_1, $E_1 + \Delta_1$ Transitions in Germanium
(b) Vol. 78, No. 1, p. K19—K22 (1976).
See also 13.1, p. 128; *14.3.4*, p. *200*; 22.1.1, p. 349.

B. Brückel, U. Birkholz, and K. Ziegler
*Fundamental Absorption and Franz-Keldysh Effect in Silicon Telluride
(b) Vol. 78, No. 1, p. K23—K25 (1976).
See also 13.1, p. 128; 22, p. 341.

H. Finkenrath, K. Krug, and N. Uhle
Optical Absorption by Free Polarons in ZnO
(b) Vol. 78, No. 1, p. K27—K30 (1976).
See also *13.5.3*, p. *159*; 22.4, p. 377.

G. V. Papayan, L. S. Agroskin, R. M. Vlasova, L. D. Rozenshtein, and S. M. Khursandova
*Optical Constants and Dielectric Permittivity Dispersion of Quasi-One-Dimensional $K_2Pt(CN)_4Br_{0.3} \cdot 2.6\ H_2O$ Crystals in the Plasma Edge Region (from Transmission and Reflection Data)
(b) Vol. 78, No. 1, p. K59—K63 (1976).
See also 22.8, p. 418.

H. Maier
Jahn-Teller Effect, Stacking Faults, and the Formation of Polytypes in ZnS
(b) Vol. 78, No. 1, p. K65—K68 (1976).
See also *6*, p. *63*; *13.4*, p. *144*; 20.3, p. 287; 22.4.1, p. 382.

A. N. Georgobiani, R. G. Maev, Yu. V. Ozerov, and E. E. Strumban
Investigation of Deep Centres of Chlorine-Doped Zinc Sulfide Crystals
(a) Vol. 38, No. 1, p. 77—83 (1976).
See also *13.4*, p. *144*; 16, p. 217; 22.4.1, p. 382.

F. V. Wald
Self-Compensation in CdTe and ZnTe Crystals Grown from Indium Solvents
(a) Vol. 38, No. 1, p. 253–259 (1976).
See also *3*, p. *44*; 10.2, p. 97; 14.3, p. 181; 22.4.3, p. 386.

A. M. Mishchenko, I. G. Neizvestnyi, and M. P. Sinyukov
*Spectrum of the Photocapacitance Current near the Fundamental Absorption Edge of Ge
(a) Vol. 38, No. 1, p. 349–354 (1976).
See also *16*, p. *217*; 22.1.1, p. 349.

Š. Višňovský, B. Knappe, V. Prosser, and H.-R. Müller
*Magnetooptical Kerr Effect Spectra of Amorphous Gadolinium–Cobalt Films
(a) Vol. 38, No. 1, p. K53–K56 (1976).
See also *18.2.1*, p. *238*; 21.1, p. 314; 21.4, p. 327.

K. H. Germann
*Optical Anisotropy Measurements in Multidomain Samples: Application to Phase Transitions in NiO and $Pb_5Ge_3O_{11}$
(a) Vol. 38, No. 1, p. K81–K84 (1976).
See also 1.2, p. 20; 14.4.1, p. 208; 18.4, p. 246; 22.6, p. 408; 22.8.1, p. 424.

K. W. Blazey
*Optical Absorption of $SrTiO_3:Fe^{4+}$
(a) Vol. 38, No. 1, p. K97–K99 (1976).
See also 13.4, p. 145; 22.8.1, p. 424.

A. M. Bon, C. Benoit, and J. Giordano
Dynamical Properties of Crystals of $Sr(NO_3)_2$, $Ba(NO_3)_2$, and $Pb(NO_3)_2$ (II): Temperature Dependence of the Infrared Spectra
(b) Vol. 78, No. 2, p. 453–464 (1976).
See also *6*, p. *63*; 22.8.1, p. 424.

P. Petelenz
Mixing of Frenkel Excitons and Ionic Excited States of a Linear Molecular Crystal with Two Molecules in the Unit Cell (I): General Model
(b) Vol. 78, No. 2, p. 489–500 (1976).
See also *13.5.1*, p. *156*; 22, p. 341.

A. A. Elango, and T. N. Nurakhmetov
Structure and Generation Mechanism of Br_3^- Centers in X-Rayed KBr
(b) Vol. 78, No. 2, p. 529–536 (1976).
See also *10.2*, p. *98*; 22.5.2, p. 398.

E. König and R. Schnakig
Zero-Field Splitting of $^6S(d^5)$ Ions in Tetragonal and Rhombic Symmetry (II): Application of Complete d-Electron Calculations to the Zero-Field Splitting in $Mn:MnF_2$ and $Mn:ZnF_2$
(b) Vol. 78, No. 2, p. 543–551 (1976).
See also *13.4*, p. *145*; 18.1, p. 227; 19, p. 255; 22.5, p. 390.

M. Vanhaelst, P. Matthys, and E. Boesman
*Temperature Dependence of the EPR and Optical Spectra in V^{++}-Doped Alkali Halides
(b) Vol. 78, No. 2, p. 553–562 (1976).
See also *19*, p. *256*; 22.5.2, p. 398.

A. Miller, G. D. Holah, W. D. Dunnett, and G. W. Iseler
Optical Phonons in $AgGaSe_2$
(b) Vol. 78, No. 2, p. 569–576 (1976).
See also *6*, p. *63*; 22.8, p. 418.

R. K. Singh and K. P. Singh
Lattice Dynamics of AgBr at 295 K
(b) Vol. 78, No. 2, p. 677–680 (1976).
See also *6*, p. *64*; 22.5.1, p. 391.

F. Bechstedt, R. Enderlein, and K. Peuker
*Resonance Raman Scattering in a Strong Electric Field
(b) Vol. 78, No. 2, p. 711–719 (1976).
See also 6, p. 64; *14.3.4*, p. *200*; 22, p. 342.

M. Schlüter, M. L. Cohen, S. E. Kohn, and C. Y. Fong
Electronic Structure of BiI_3
(b) Vol. 78, No. 2, p. 737–747 (1976).
See also 4, p. 47; *13.1*, p. *129*; 22.5, p. 390.

M. Durand, B. Ayrault, Y. Marqueton, and E. A. Decamps
*Lattice Dynamics of Paratellurite
(b) Vol. 78, No. 2, p. 767–777 (1976).
See also *6*, p. *64*; 22.6, p. 408.

J. P. Buisson, S. Lefrant, A. Sadoc, L. Taurel, and M. Billardon
Raman Scattering by KI Containing F Centres
(b) Vol. 78, No. 2, p. 779–790 (1976).
See also 6, p. 64; *10.2*, p. *98*; 13.4, p. 145; 22.5.2, p. 398.

V. M. Fomin and E. P. Pokatilov
*Non-Linear Absorption of Electromagnetic Waves by Band Charge Carriers in a Laser Field
(b) Vol. 78, No. 2, p. 831–842 (1976).
See also *14.3.4*, p. *200*.

W. Bauhofer, L. Genzel, and W. König
*Far Infrared Studies of the Impurity Mode $NH_4Br{:}Cl^-$
(b) Vol. 78, No. 2, p. K121–K123 (1976).
See also *6*, p. *64*; 22.5.4, p. 401.

V. M. Orera and R. Alcalá
Formation and Size Evolution of Ca Colloids in Additively Colored CaF_2
(a) Vol. 38, No. 2, p. 621–627 (1976).
See also *10.2*, p. *98*; 22.5.3, p. 400.

T. A. McMath and J. C. Irwin
*Indices of Refraction of GaS and GaSe
(a) Vol. 38, No. 2, p. 731–738 (1976).
See also 22.3, p. 374.

M. Galtier, A. Montaner, C. Benoit, J. Deumie, H. Robert, L. Giral, and J. M. Fabre
*Magnetic Field Effects on Infrared Spectra of TTF-TCNQ Thin Films on KBr
(a) Vol. 38, No. 2, p. K159–K161 (1976).
See also 22.9, p. 439.

G. Nimtz and E. Tyssen
*Far Infrared Recombination Radiation from n-Type $Hg_{1-x}Cd_xTe$
(a) Vol. 38, No. 2, p. K171–K173 (1976).
See also 13.4, p. 146; 22.4.4, p. 387.

20.2 Lasers

Original Papers and Short Notes

P. A. Arsenev, K. Bienert, and A. V. Potemkin
*The Spectroscopic Investigation of Neodymium in Mixed Aluminates of Yttrium and Gadolinium
(a) Vol. **33**, No. 2, p. 511–515 (1976).
See also 20.1, p. 263; 20.3, p. 281; 22.8.1, p. 419.

A. A. Kaminski, D. Schultze, B. Hermoneit, S. E. Sarkisov, L. Li, J. Bohm, P. Reiche, R. Ehlert, A. A. Mayer, V. A. Lomonov, and V. A. Balashov
*Spectroscopic Properties and Stimulated Emission in the $^4F_{3/2} \rightarrow {}^4I_{11/2}$ and $^4F_{3/2} \rightarrow {}^4I_{13/2}$ Transitions of Nd^{3+} Ions from Cubic $Bi_4Ge_3O_{12}$ Crystals
(a) Vol. **33**, No. 2, p. 737–753 (1976).
See also 20.1, p. 263; *20.3*, p. *281*; 22.8, p. 413.

A. A. Kaminskii, Kh. S. Bagdasarov, G. A. Bogomolova, M. M. Gritsenko, A. M. Kevorkov, and S. E. Sarkisov
*Luminescence and Stimulated Emission of Nd^{3+} Ions in $Gd_3Sc_2Ga_3O_{12}$ Crystals
(a) Vol. **34**, No. 2, p. K109–K114 (1976).
See also *20.3*, p. *282*; 22.8.2, p. 428.

V. A. Antonov and P. A. Arsenev
Spectroscopic Properties of Single Crystals of Rare Earth Titanates
(a) Vol. **35**, No. 2, p. K169–172 (1976).
See also *20.1*, p. *270*; 22.8.1, p. 422.

L. Holz, K. Kneipp, A. Lau, and W. Werncke
Stimulated Raman Scattering of Picosecond Laser Pulses by Polaritons in $LiIO_3$ and $LiNbO_3$ Single Crystals
(a) Vol. **36**, No. 1, p. K5–K8 (1976).
See also *6*, p. *58*; 13.5, p. 147; 20.1, p. 271; 22.8.1, p. 422.

A. A. Kaminskii, G. A. Bogomolova, D. N. Vylegzhanin, Kh. S. Bagdasarov, A. M. Kevorkov, and M. M. Gritsenko
*Spectroscopic Properties of Nd^{3+} Ions in Garnet Compounds Forming in the Y_2O_3–Ga_2O_3 System
(a) Vol. **38**, No. 1, p. 409–422 (1976).
See also *20.3*, p. *288*; 22.8.2, p. 431.

H. van Cong
*Linewidth of Spontaneous Emission in Laser Diodes
(a) Vol. **38**, No. 1, p. K69–K72 (1976).
See also 22.2.1, p. 364.

20.3 Luminescence (X-Ray Spectra See 20)

Review Article

M. N. Sapozhnikov
*Zero-Phonon Transitions in the Optical Spectra of Impurity Molecular Crystals
(b) Vol. **75**, No. 1, p. 11–51 (1976).
See also *6*, p. *52*; 13.4, p. 134; 20.1, p. 260; 22.9, p. 433.

Original Papers and Short Notes

W. Rühle, W. Jakowetz, C. Wölk, R. Linnebach, and M. Pilkuhn
*Optical Studies of Free and Bound Excitonic States in GaSb. Evidence for Deep A^+-Complexes
(b) Vol. **73**, No. 1, p. 255–264 (1976).
See also 13.4, p. 134; *13.5.1*, p. *149*; 20.1, p. 260; 22.2.3, p. 367.

R. B. Hammond, T. C. McGill, and J. W. Mayer
*Properties of the Electron–Hole Liquid Luminescence in Si Double Injection Diodes
(a) Vol. **33**, No. 1, p. 59–66 (1976).
See also 13.5.1, p. 149; 14.3.4, p. 195; 22.1.2, p. 351.

S. H. Hagen and A. W. C. van Kemenade
*On the Role of Boron in the Luminescence of Silicon Carbide Doped with Nitrogen and Boron
(a) Vol. **33**, No. 1, p. 97–105 (1976).
See also 13.4, p. 135; 22.3, p. 372.

Hj. Bernhardt
Comparative Studies of the LNT Coloration of $Y_3Al_5O_{12}$, $YAlO_3$, and $LiNbO_3$ Crystals
(a) Vol. **33**, No. 1, p. 211–216 (1976).
See also *10.2*, p. *87*; *20.1*, p. *261*; 22.8.1, p. 419; 22.8.2, p. 426.

R. P. Khare
*Thermoluminescence of CaO(Ce, Gd) Phosphors
(a) Vol. **33**, No. 1, p. K51–K53 (1976).
See also 22.6, p. 402.

L. Nagli
*$In^+v_a^+v_c^-$ Centres in a KCl–In Crystal
(b) Vol. **73**, No. 2, p. 427–430 (1976).
See also *10.2*, p. *88*; 22.5.2, p. 393.

W. Ekardt and M. I. Sheboul
The Influence of the Electron–Hole Exchange Interaction on the Biexciton Ground State in CuCl and CuBr and Related Optical Transitions
(b) Vol. **73**, No. 2, p. 475–482 (1976).
See also 13.1, p. 121; *13.5.1*, p. *150*; 20.1, p. 261; 22.5, p. 388.

B. Andlauer, J. Schneider, and W. Tolksdorf
Optical Absorption, Fluorescence, and Electron Spin Resonance of Ir^{4+} on Octahedral Sites in $Y_3Ga_5O_{12}$
(b) Vol. **73**, No. 2, p. 533–540 (1976).
See also *13.4*, p. *135*; 19, p. 248; 20.1, p. 261; 22.8.2, p. 426.

R. Brodmann, G. Tolkiehn, and G. Zimmerer
*Luminescence Originating from Free and Self-Trapped Excitons in Solid Xenon
(b) Vol. **73**, No. 2, p. K99–K101 (1976).
See also *13.5.1*, p. *150*; 23, p. 440.

P. A. Arsenev, K. Bienert, and A. V. Potemkin
The Spectroscopic Investigation of Neodymium in Mixed Aluminates of Yttrium and Gadolinium
(a) Vol. **33**, No. 2, p. 511–515 (1976).
See also 20.1, p. 263; *20.2*, p. *279*; 22.8.1, p. 419.

A. A. Kaminskii, D. Schultze, B. Hermoneit, S. E. Sarkisov, L. Li, J. Bohm, P. Reiche, R. Ehlert, A. A. Mayer, V. A. Lomonov, and V. A. Balashov
*Spectroscopic Properties and Stimulated Emission in the $^4F_{3/2} \rightarrow {}^4I_{11/2}$ and $^4F_{3/2} \rightarrow {}^4I_{13/2}$ Transitions of Nd^{3+} Ions from Cubic $Bi_4Ge_3O_{12}$ Crystals
(a) Vol. **33**, No. 2, p. 737–753 (1976).
See also 20.1, p. 263; *20.2*, p. *279*; 22.8, p. 413.

N. Takeuchi, K. Inabe, and H. Nanto
*Kinetics Order of Thermoluminescent Red Emission from MgO Single Crystals
(a) Vol. **33**, No. 2, p. K125–K127 (1976).
See also 10.2, p. 89; 22.6, p. 403.

T. Chong and N. Itoh
*Interaction between Excitons and Radiation-Induced Radicals in Naphthalene Single Crystals
(b) Vol. **74**, No. 1, p. 111–120 (1976).
See also 13.5.1, p. 150; 22.9, p. 434.

N. Kristianpoller, A. Rehavy, and M. Israeli
Effects of Ultraviolet Irradiation on KBr:Gd
(b) Vol. **74**, No. 2, p. 245–248 (1976).
See also *10.2*, p. *89*; 13.5.1, p. 151; 20.1, p. 263; 22.5.2, p. 393.

W. B. Whitten and S. Arnold
Pressure Modulation of Exciton Fission in Tetracene
(b) Vol. **74**, No. 1, p. 401–407 (1976).
See also *13.5.1*, p. *151*; 22.9, p. 434.

A. M. White, P. J. Dean, B. Day, and F. Mansfield
Polaron, Zeeman, and Diamagnetic Effects in the Recombination Spectra of Donor Bound Excitons in Cadmium Telluride
(b) Vol. **74**, No. 1, p. K9–K13 (1976).
See also *13.4*, p. *136*; *13.5.1*, p. *151*; 13.5.3, p. 159; 22.4.3, p. 385.

R. A. Elliott
*Erratum to: Comment on "Diffusion Theory of Luminescent Emission from a Doped Organic Solid"
(b) Vol. **74**, No. 1, p. K87–K88 (1976).
See also *13.5.1*, p. *151*; 22.9, p. 434.

A. Kiveris and P. Pipinys
*The Exhaustion of Traps in ZnS Phosphors by Electric Field and the Gudden-Pohl Effect
(a) Vol. **34**, No. 1, p. 377–382 (1976).
See also *14.3.4*, p. *196*; 22.4.1, p. 379.

P. C. Diggle, K. A. Gehring, and R. M. Mac Farlane
*Atomic Absorption and Emission in Laser Produced Plasmas
(a) Vol. **34**, No. 1, p. K13–K16 (1976).
See also *20.1*, p. *265*; 22.8, p. 414; 22.8.1, p. 420.

R. Bindemann, H. Fischer, and K. Kreher
*Observation of Nonexponential Long Time Decay of Donor–Acceptor Pair Luminescence in GaP
(a) Vol. **34**, No. 1, p. K63–K65 (1976).
See also 22.2.2, p. 365.

M. S. Brodin, M. A. Dudinskii, S. V. Marisova, and E. N. Myasnikov
*Polariton Scattering and Formation of Exciton Luminescence Spectrum in Anthracene Crystals
(b) Vol. 74, No. 2, p. 453–460 (1976).
See also *13.5.1*, p. *151*; 22.9, p. 435.

P. Koidl
Jahn-Teller Effect in the $^4T_1(l)$ and $^4T_2(l)$ States of Tetrahedrally Coordinated Mn^{2+}
(b) Vol. 74, No. 2, p. 477–484 (1976).
See also *13.4*, p. *137*; 20.1, p. 265; 22.2.2, p. 365; 22.4.1, p. 380; 22.4.2, p. 383.

W. Ekardt and M. I. Sheboul
On the First Excited State of the Biexciton in CuCl
(b) Vol. 74, No. 2, p. 523–529 (1976).
See also *13.5.1*, p. *151*; 22.5, p. 388.

Ch. Ackermann, R. Brodmann, U. Hahn, A. Suzuki, and G. Zimmerer
*Photoluminescence Excitation Spectra of Solid Krypton
(b) Vol. 74, No. 2, p. 579–590 (1976).
See also *13.5.1*, p. *151*; 23, p. 440.

I. A. Markova-Osorgina and S. Z. Shmurak
*Deformation-Induced Exciton Luminescence of Alkali Halide Crystals
(b) Vol. 74, No. 2, p. 753–760 (1976).
See also 10.2, p. 90; 13.5.1, p. 152; 22.5.2, p. 394.

Hj. Bernhardt
The Structure of Coloration Centres in $Y_3Al_5O_{12}$ Crystals after X-Ray Irradiation at Room Temperature
(a) Vol. 34, No. 2, p. 483–487 (1976).
See also *10.2*, p. *91*; 22.8.2, p. 427.

J. Vaitkus
The Nonequilibrium Hall Effect and Related Transport Phenomena in Semiconductors under Inhomogeneous Excitation by a Laser Pulse
(a) Vol. 34, No. 2, p. 769–775 (1976).
See also *14.3*, p. *176*; *16*, p. *214*; 22.1.1, p. 346.

K. D. Glinchuk, A. V. Prokhorovich, and V. I. Vovnenko
*The Internal Radiative Efficiency and the Mechanism of Temperature Quenching of the 1.03, 1.20, and 1.30 eV Emission Bands in n-GaAs
(a) Vol. 34, No. 2, p. 777–786 (1976).
See also *13.4*, p. *138*; 22.2.1, p. 362.

A. A. Kaminskii, Kh. S. Bagdasarov, G. A. Bogomolova, M. M. Gritsenko, A. M. Kevorkov, and S. E. Sarkisov
*Luminescence and Stimulated Emission of Nd^{3+} Ions in $Gd_3Sc_2Ga_3O_{12}$ Crystals
(a) Vol. 34, No. 2, p. K109–K114 (1976).
See also *20.2*, p. *279*; 22.8.2, p. 428.

Yu. Suleymanov, I. G. Pichugin, S. I. Teslenko, and L. A. Marasina
*Photoluminescence Excitation Spectra of Gallium Nitride Single Crystals
(a) Vol. 34, No. 2, p. K195–K197 (1976).
See also 22.2, p. 361.

P. Loose, M. Rosenzweig, and M. Wöhlecke
*Zeeman Effect of Bound Exciton Complexes in ZnO
(b) Vol. 75, No. 1, p. 137–144 (1976).
See also *13.5.1*, p. *152*; *20.1*, p. *267*; 22.4, p. 376.

W. Wardzyński and K. Pataj
*Bound Exciton–Neutral Acceptor Complex in ZnTe
(b) Vol. **75**, No. 1, p. 341–346 (1976).
See also 12.2, p. 110; *13.5.1*, p. *152*; 22.4.3, p. 385.

L. Mărculescu and C. Ghiţă
*On the Luminescence of KCl:Pb and KBr:Pb Crystals
(b) Vol. **75**, No. 1, p. 375–379 (1976).
See also 22.5.2, p. 394.

H. Becherer and P. Kirsten
*Cathodoluminescence of $ZnSiP_2$ in the Energy Region of the Pseudodirect Band Gap
(b) Vol. **75**, No. 1, p. K19–K21 (1976).
See also 22.8, p. 415.

E. Skaistys and L. Valkūnas
*Screening Effects in the Exciton–Electron Luminescence Spectra
(b) Vol. **75**, No. 1, p. K23–K26 (1976).
See also *13.5.1*, p. *152*; 22.4.1, p. 380.

Yu. A. Vodakov, G. A. Lomakina, E. N. Mokhov, E. I. Radovanova, V. I. Sokolov, M. M. Usmanova, G. F. Yuldashev, and B. S. Machmudov
Silicon Carbide Doped with Gallium
(a) Vol. **35**, No. 1, p. 37–42 (1976).
See also 3, p. 42; *13.4*, p. *138*; 14.3, p. 176; 20.1, p. 268; 22.3, p. 372.

A. N. Georgobiani, Yu. V. Ozerov, E. E. Strumban, and V. V. Anisimov
Investigation of Deep Centres of Gallium-Doped Zinc Sulfide Crystals
(a) Vol. **35**, No. 1, p. 251–257 (1976).
See also *13.4*, p. *138*; 16, p. 214; 20.1, p. 268; 22.4.1, p. 380.

R. Visocekas, T. Ceva, C. Marti, F. Lefaucheux, and M. C. Robert
*Tunneling Processes in Afterglow of Calcite
(a) Vol. **35**, No. 1, p. 315–327 (1976).
See also 10.2, p. 92; 22.8.1, p. 421.

M. Egée
*Spectral Distribution of the Green Emission Band of ZnS:Cu in Dependence on the Localization Diversity of the Radiative Recombination Processes
(a) Vol. **35**, No. 1, p. 361–370 (1976).
See also 13.4, p. 139; 22.4.1, p. 380.

L. F. Komolova, V. M. Efremenkova, and G. V. Saparin
*The Investigation of Cathodoluminescence (CL) in CdS after Ion Bombardment in the Scanning Electron Microscope (SEM)
(a) Vol. **35**, No. 1, p. K5–K8 (1976).
See also 22.4.1, p. 380.

J. L. Staehli
Electron–Hole Droplet Formation Kinetics in Pure Germanium
(b) Vol. **75**, No. 2, p. 451–463 (1976).
See also *13.5.1*, p. *152*; 22.1.1, p. 347.

A. Opanowicz, K. Marinova, H. Liebing, and W. Ruppel
*Luminescence of Highly Excited ZnTe Crystals
(b) Vol. **75**, No. 2, p. 471–481 (1976).
See also *13.5.1*, p. *152*; 22.4.3, p. 385.

G. Bauer, M. Overhamm, P. Grosse, W. Müller, E. Gornick, and H. W. Pötzl
Landau Emission in n-InSb
(b) Vol. **75**, No. 2, p. 543–551 (1976).
See also *13.1*, p. *124*; *14.3.4*, p. *197*; 20.1, p. 268; 22.2.3, p. 368.

H. Bill and W. von der Osten
Raman and Vibronic Spectra of a New Oxygen Molecule Ion in CaF_2
(b) Vol. 75, No. 2, p. 613–619 (1976).
See also 6, p. 57; *10.2*, p. *93*; 22.5.3, p. 399.

W. Arden, M. Kotani, and L. M. Peter
Triplet Exciton Decay Processes in Crystalline Tetracene
(b) Vol. 75, No. 2, p. 621–631 (1976).
See also *13.5.1*, p. *153*; 22.9, p. 436.

R. Zimmermann and M. Rösler
*Theory of Electron–Hole Plasma in CdS
(b) Vol. 75, No. 2, p. 633–645 (1976).
See also *13.5*, p. *147*; 13.5.2, p. 157; 22.4.1, p. 380.

U. Gösele
*Comment on "Diffusion Theory of Luminescent Emission from a Doped Organic Solid"
(b) Vol. 75, No. 2, p. K129–K132 (1976).
See also *13.5.1*, p. *153*; 22.9, p. 436.

L. Bosi
On the Luminescence and Photoconductivity of the F Centres in NaBr
(b) Vol. 75, No. 2, p. K163–K165 (1976).
See also *10.2*, p. *93*; 16, p. 214; 22.5.2, p. 395.

M. Kh. Ashurov, Yu. K. Voronko, V. V. Osiko, A. A. Sobol, B. P. Starikov, M. I. Timoshechkin, and A. Ya. Yablonskii
Inequivalent Luminescence Centres of Er^{3+} in Gallium Garnet Single Crystals
(a) Vol. 35, No. 2, p. 645–649 (1976).
See also *13.4*, p. *140*; 20.1, p. 269; 22.8.2, p. 429.

A. J. Harris, H. Thomas, and M. P. Callaghan
*Investigation of the Characteristics of Electroluminescent SiC Diodes (I): Theoretical Model for Transient and Steady-State Conditions
(a) Vol. 35, No. 2, p. 697–707 (1976).
See also *14.3.3*, p. *192*; 22.3, p. 373.

V. V. Samartsev, R. G. Usmanov, I. Kh. Khadiev, E. F. Kustov, and M. N. Baranov
*Photon Echo in $CaWO_4:Nd^{3+}$ and $LiAl_5O_8:Cr^{3+}$
(b) Vol. 76, No. 1, p. 55–66 (1976).
See also 13.4, p. 140; *20.1*, p. *270*; 22.8.1, p. 422.

F. El Akkad
*Free-to-Bound Transitions in Lithium-Doped Zinc Telluride
(b) Vol. 76, No. 1, p. 85–92 (1976).
See also *13.4*, p. *140*; 22.4.3, p. 385.

O. Goede, M. Blaschke, and K. H. Klohs
*Piezoemission Study of I_1- and I_2-Type Bound Excitons in CdS
(b) Vol. 76, No. 1, p. 267–280 (1976).
See also *13.5.1*, p. *153*; 22.4.1, p. 380.

E. Schuberth, M. Creutzburg, and W. Müller-Lierheim
Evidence for a Partial F.C.C.–H.C.P. Phase Transition in Solid Neon from Raman Scattering
(b) Vol. 76, No. 1, p. 301–306 (1976).
See also *1.2*, p. *17*; 6, p. 58; 23, p. 440.

F. Henneberger and J. Voigt
Polariton Effects in the Optical Spectra of Excitonic Molecules
(b) Vol. 76, No. 1, p. 313–323 (1976).
See also 13.5, p. 147; *13.5.1*, p. *153*; *20.1*, p. *270*; 22, p. 337.

V. V. Hizhyakov
Effect of Hot Transfer on Vibrational Relaxation of Electronic Excitation
(b) Vol. **76**, No. 1, p. K69–K72 (1976).
See also *13.4*, p. *140*; 22, p. 338.

G. D. Bagratishvili, R. B. Dzhanelidze, N. I. Kurdiani, and O. V. Saksaganskii
MIS Structure GaAs–Ge_3N_4–Al
(a) Vol. **36**, No. 1, p. 73–79 (1976).
See also 13.3, p. 132; *14.3.2*, p. *188*; 22.2.1, p. 363.

A. J. Harris and H. Thomas
*Investigation of the Characteristics of Electroluminescent SiC Diodes (II): Temperature- and Growth-Dependent Effects
(a) Vol. **36**, No. 1, p. 365–373 (1976).
See also *14.3.3*, p. *192*; 22.3, p. 373.

C. Paorici, C. Pelosi, N. Romeo, G. Sberveglieri, and L. Tarricone
*Electroluminescence in $CdIn_2S_4$
(a) Vol. **36**, No. 1, p. K33–K36 (1976).
See also 22.8, p. 416.

P. I. Khadzhi, K. G. Petrashuk, and S. A. Moskalenko
*Biexciton Luminescence with Interserial Transition of the Formed Exciton
(b) Vol. **76**, No. 2, p. 497–504 (1976).
See also *13.5.1*, p. *154*; 22.4.1, p. 381.

B. Speit and G. Lehmann
Hole Centers in the Feldspar Sanidine
(a) Vol. **36**, No. 2, p. 471–481 (1976).
See a lso *10.2*, p. *94*; 16, p. 215; 20.1, p. 272; 22.8.1, p. 423.

D. Lapraz and P. Iacconi
*On Some Luminescent and Optical Properties of Synthetic Calcite Single Crystals
(a) Vol. **36**, No. 2, p. 603–616 (1976).
See also 13.4, p. 141; *20.1*, p. *272*; 22.8.1, p. 423.

V. V. Ratnam and H. D. Banerjee
On the Glow Curves and TL Centres in CaF_2:Ce, Mn Crystals Irradiated by Cathode Rays
(a) Vol. **36**, No. 2, p. 693–697 (1976).
See also *10.2*, p. *94*; 22.5.3, p. 400.

C. L. Reynolds, Jr. and F. E. Wawner
*On the Origin of Triboluminescence in SiC Filaments
(a) Vol. **36**, No. 2, p. K109–K111 (1976).
See also 22.3, p. 373.

A. Suzuki, H. Matsunami, and T. Tanaka
*Very Slow Decays of the Luminescence from Boron-Doped 4H-SiC
(a) Vol. **36**, No. 2, p. K129–K131 (1976).
See also 22.3, p. 373.

J. N. Das and B. P. Chandra
*Estimation of Mechanical Relaxation Time of Sugar Çrystals from Their Triboluminescence
(a) Vol. **36**, No. 2, p. K175–K177 (1976).
See also 22.9, p. 437.

A. V. Lelyakov
Bose-Einstein Condensation of Free Excitons in AgBr-Type Crystals at High Excitation
(b) Vol. **77**, No. 1, p. 59–70 (1976).
See also 6, p. 60; 13.1, p. 127; *13.5.1*, p. *154*; 22.5.1, p. 391.

E. L. Frankevich, M. M. Tribel, and I. A. Sokolik
Photoconductivity of Sublimated Rubrene Films during Oxidation. The Magnetic Field Effect
(b) Vol. 77, No. 1, p. 265–276 (1976).
See also 13.5.1, p. 155; *14.3.1*, p. *185*; *16*, p. *215*; 22.9, p. 437.

U. M. Grassano, A. Tanga, and M. Valli
*Electric Field Effect on the Thallium Emission Bands in KBr and KI
(b) Vol. 77, No. 1, p. 341–346 (1976).
See also *10.2*, p. *95*; 20.1, p. 273; 22.5.2, p. 396.

R. Levy, C. Klingshirn, E. Ostertag, Vu Duy Phach, and J. B. Grun
*Luminescence of a "Cold" Gas of Biexcitons in CuCl
(b) Vol. 77, No. 1, p. 381–391 (1976).
See also *13.5.1*, p. *155*; 22.5, p. 390.

A. Opanowicz and K. Marinova
*Luminescence of ZnTe Single Crystals under High Two-Photon Excitation
(b) Vol. 77, No. 1, p. K35–K38 (1976).
See also *13.5.1*, p. *155*; 22.4.3, p. 386.

V. A. Vassilev, B. T. Kolomiets, T. N. Mamontova, and E. M. Raspopova
*Photoluminescence and Photoconductivity in Densified Vitreous As_2Se_3
(a) Vol. 37, No. 1, p. 45–49 (1976).
See also 2, p. 40; *16*, p. *216*; 22, p. 339.

M. Riggin, S. Radhakrishna, and P. W. Whippey
*Thermoluminescence of X-Irradiated Alkali Fluoride Crystals
(a) Vol. 37, No. 1, p. 51–56 (1976).
See also 13.4, p. 142; 22.5.2, p. 396.

H. P. Braun and M. E. Michel-Beyerle
Current and Fluorescence Oscillations in Anthracene Crystals
(a) Vol. 37, No. 1, p. 73–82 (1976).
See also *14.3.4*, p. *199*; 16, p. 216; 22.9, p. 437.

S. Govinda
Coloration and Luminescence in Pure and Chromium-Doped Al_2O_3 Single Crystals Irradiated with X-Rays at Room Temperature
(a) Vol. 37, No. 1, p. 109–117 (1976).
See also *10.2*, p. *95*; 11, p. 103; *20.1*, p. *273*; 22.6, p. 406.

G. Holzapfel and M. Krystek
Thermally Stimulated Luminescence and Exoelectron Emission from Barium and Strontium Sulphate Doped with Europium
(a) Vol. 37, No. 1, p. 303–312 (1976).
See also *13.4*, p. *142*; 17, p. 219; 22.8.1, p. 424.

A. Kiveris, Š. Kudžmauskas, and P. Pipinys
*Release of Electrons from Traps by an Electric Field with Phonon Participation
(a) Vol. 37, No. 1, p. 321–327 (1976).
See also 6, p. 61; *13.4*, p. *142*; 14.3.4, p. 199; 22.4.1, p. 381.

A. S. Popov
*Fundamental Photoluminescence of Undoped InP Crystals Cleaned by the Floating Zone Method
(a) Vol. 37, No. 1, p. K53–K56 (1976).
See also 22.2.2, p. 365.

L. MARCULESCU, M. POPESCU, I. BALTOG, and C. GHITA
The State of Dispersion ob Pb^{2+} in Highly Doped KCl Crystals
(b) Vol. 77, No. 2, p. 701–706 (1976).
See also *10.2*, p. *95*; 20.1, p. 274; 22.5.2, p. 397.

V. A. BRODOVOI, N. Z. DERIKOT, B. R. KIYAK, G. P. PEKA, and A. I. PROSKURA
Trigger Effect in Electroluminescent Ceramics ZnSe–Cu, I
(a) Vol. 37, No. 2, p. 363–369 (1976).
See also *13.4*, p. *143*; *14.3.4*, p. *199*; 22.4.2, p. 383.

E. GRILLI, M. GUZZI, and R. MOLTENI
*Luminescence of $ZnIn_2Se_4$ Crystals
(a) Vol. 37, No. 2, p. 399–406 (1976).
See also 22.8, p. 417.

HJ. BERNHARDT
The Manganese-Induced O^- Centres in Yttrium Aluminium Garnet
(a) Vol. 37, No. 2, p. 445–448 (1976).
See also *10.2*, p. *96*; 20.1, p. 275; 22.8.2, p. 430.

R. K. GARTIA
Thermoluminescence Studies of Z_1 Centers in X-Irradiated Ca-Doped NaCl Crystals
(a) Vol. 37, No. 2, p. 571–578 (1976).
See also *10.2*, p. *96*; 22.5.2, p. 397.

H. KUWABARA, K. YAMANAKA, and S. YAMADA
*Donor–Acceptor Pair Emission from β-SiC Doped with Gallium
(a) Vol. 37, No. 2, p. K157–K161 (1976).
See also 13.4, p. 144; 22.3, p. 374.

W. RÜHLE and E. GÖBEL
New Aspects of the Magnetoluminescence of a Band to Acceptor Transition in GaAs
(b) Vol. 78, No. 1, p. 311–317 (1976).
See also *13.1*, p. *128*; 22.2.1, p. 364.

S. PERMOGOROV and V. TRAVNIKOV
Resonant Scattering of Exciton-Polaritons in CdS Crystals
(b) Vol. 78, No. 1, p. 389–398 (1976).
See also *6*, p. *63*; *13.5*, p. *148*; 22.4.1, p. 381.

K. P. MELETOV and E. F. SHEKA
Asymmetry of Phonon Side Bands of Optical Transitions in Impurity Molecular Crystals Caused by a Change in External Phonon Frequencies
(b) Vol. 78, No. 1, p. K1–K5 (1976).
See also *6*, p. *63*; *13.4*, p. *144*; 20.1, p. 276; 22.9, p. 439.

E. A. MENESES, N. JANNUZZI, J. R. FREITAS, and A. GOUSKOV
*Photoluminescence of Layered $GaSe_{1-x}Te_x$ Crystals
(b) Vol. 78, No. 1, p. K35–K38 (1976).
See also 22.3.1, p. 375.

H. MAIER
Jahn-Teller Effect, Stacking Faults, and the Formation of Polytypes in ZnS
(b) Vol. 78, No. 1, p. K65–K68 (1976).
See also *6*, p. *63*; *13.4*, p. *144*; 20.1, p. 276; 22.4.1, p. 382.

S. S. PALIWAL and O. P. SINHA
*Calculation of Activation Energy and Frequency Factor from Glow Curves of SrS:Hf Phosphors
(a) Vol. 38, No. 1, p. 73–76 (1976).
See also *10.2*, p. *97*; 13.4, p. 144; 22, p. 341.

D. F. Crabtree
*Luminescence and Charge Compensation in $SnO_2:Eu^{3+}$
(a) Vol. 38, No. 1, p. 217–222 (1976).
See also 22.6, p. 408.

A. A. Kaminskii, G. A. Bogomolova, D. N. Vylegzhanin, Kh. Bagdasarov, A. M. Kevorkov, and M. M. Gritsenko
*Spectroscopic Properties of Nd^{3+} Ions in Garnet Compounds Forming in the Y_2O_3–Ga_2O_3 System
(a) Vol. 38, No. 1, p. 409–422 (1976).
See also *20.2*, p. *279*; 22.8.2, p. 431.

S. Bhushan and M. Saleem
*Photo- and Electroluminescence of ZnO:Er Phosphors
(a) Vol. 38, No. 1, p. K27–K29 (1976).
See also 22.4, p. 377.

V. K. Jain
*Effect of Annealing on Thermoluminescence of LiF (TLD-100)
(a) Vol. 38, No. 1, p. K65–K68 (1976).
See also 10.2, p. 98; 22.5.2, p. 398.

V. E. Bozhevolnov, L. N. Ivanov, V. K. Kozlov, Yu. V. Voronov, Yu. P. Timofeev, and V. V. Karelin
*Energy Transfer in CaF_2–Gd, Mn Crystals
(b) Vol. 78, No. 2, p. 483–487 (1976).
See also 13.4, p. 145; 22.5.4, p. 401.

S. Fujiwara, T. Nakayama, and N. Itoh
Exciton Interactions in Naphthalene Single Crystals
(b) Vol. 78, No. 2, p. 519–528 (1976).
See also *13.5.1*, p. *156*; 22.9, p. 439.

B. Hönerlage, C. Klingshirn, and J. B. Grun
*Spontaneous Emission Due to Exciton–Electron Scattering in Semiconductors
(b) Vol. 78, No. 2, p. 599–608 (1976).
See also *13.5.1*, p. *156*; 22.4, p. 378; 22.4.1, p. 382; 22.4.2, p. 384.

J. Weber
*Indirect Near Edge Emission in Pure AgBr
(b) Vol. 78, No. 2, p. 699–709 (1976).
See also 6, p. 64; *13.5.1*, p. *156*; 22.5.1, p. 391.

S. Zazubovich and V. Osminin
*Hole Mechanisms of H and F Centre Production in KCl Crystals
(b) Vol. 78, No. 2, p. 813–819 (1976).
See also *10.2*, p. *98*; 22.5.2, p. 398.

L. A. Marasina, A. N. Pikhtin, I. G. Pichugin, and A. V. Solomonov
*Cathodoluminescence of Undoped and Zn-Doped GaN Epitaxial Layers
(a) Vol. 38, No. 2, p. 753–760 (1976).
See also 22.2, p. 361.

21. Metals. Alloys

Review Articles

E. Vogt

The Origin of Ferromagnetism and Antiferromagnetism in Alloys of Transition Metals (II): Alloys of Transition Metals with other Transition Metals and with Polyvalent Non-Transition Elements
(a) Vol. **34**, No. 1, p. 11–67 (1976).
See also 1.3, p. 21; 18.1, p. 225 ; *18.2*, p. *228*; *18.4*, p. *242*; 21.1, p. 302; 21.6, p. 329.

H. W. de Wijn, A. M. van Diepen, and K. H. J. Buschow

Samarium in Crystal Fields, a Case of Strong J-Mixing
(b) Vol. **76**, No. 1, p. 11–30 (1976).
See also 5, p. 48; *13*, p. *113*; *18*, p. *222*; 19, p. 247; 21.1, p. 302; 21.1.1, p. 316 ; 21.4, p. 326.

S. V. Tsivinskii

Dislocation Multiplication in Single and Polycrystals in the Initial Stage of Deformation
(a) Vol. **36**, No. 1, p. 11–31 (1976).
See also *10.1*, p. *78*; *10.2*, p. *87*; 21.1, p. 302; 21.1.1, p. 316; 21.6, p. 329; 22.1.1, p. 345; 22.2.3, p. 367; 22.5.2, p. 392.

S. V. Vonsovskii and M. S. Svirskii

Multiplicity and Pseudomultiplicity in Electron–Ion Systems (I)
(b) Vol. 77, No. 1, p. 9–52 (1976).
See also *13*, p. *113*; 14.1, p. 164; 14.2, p. 169; 18.2, p. 228; 18.4, p. 242.

S. V. Vonsovskii and M. S. Svirskii

Multiplicity and Pseudomultiplicity of Electron–Ion Systems (II)
(b) Vol. 77, No. 2, p. 403–432 (1976).
See also *13*, p. *113*; 14.1, p. 164; 14.2, p. 169; 18.2, p. 228; 18.4, p. 242.

Original Papers and Short Notes

A. Čížek, M. Šob, and K. Krásenský

The Determination of Ionic Radii of Some Metals by the Positron Annihilation Method
(b) Vol. **73**, No. 1, p. K1–K4 (1976).
See also *13*, p. *114*; 21.1, p. 302; 21.1.1, p. 316; 21.6, p. 329.

A. I. Akimenko, V. S. Solovev, and I. K. Yanson

Harmonics Mixing in Superconducting Point Contacts
(a) Vol. **33**, No. 1, p. 53–58 (1976).
See also *14.2*, p. *169*.

V. A. Goltsov and L. I. Smirnov

Diffusion of Interstitial Atoms in Ordered Pt_2Mo-Type Alloys (II)
(a) Vol. **33**, No. 1, p. 67–73 (1976).
See also *1.3*, p. *21*; *9*, p. *72*; 21.6, p. 329.

M. Jergel, S. Takács, and V. Černuško

Influence of Gas Impurities on the Angular Dependence of the Critical Currents in Nb_3Sn Vapour-Deposited Tapes
(a) Vol. **33**, No. 1, p. 85–90 (1976).
See also *14.2*, p. *169*.

D. H. Sastry, M. J. Luton, and J. J. Jonas

Thermally Activated Flow of H.C.P. Metals at Low Temperatures
(a) Vol. **33**, No. 1, p. 173–187 (1976).
See also *10.1*, p. *78*; 21.1, p. 302; 21.3, p. 324.

W. SCHLENK
Investigations of the Thermally Stimulated Exoelectron Emission (TSEE) from Aluminium Surfaces
(a) Vol. **33**, No. 1, p. 217–228 (1976).
See also *17*, p. *218*.

B. WIELKE
Easy-Glide Hardening of Zn and Cd between 4.2 and 400 K
(a) Vol. **33**, No. 1, p. 241–246 (1976).
See also *10.1*, p. *78*; 21.1, p. 302.

A. MELANDER
Strong- and Weak-Beam Contrast from Cuboidal Inclusions
(a) Vol. **33**, No. 1, p. 255–264 (1976).
See also *1.4*, p. *25*; 10.1, p. 78; 21.1, p. 303.

H. LÖFFLER and H.-G. FABIAN
Application of the Jump Model to the Decomposition of Some Binary and Ternary Aluminium Alloys
(a) Vol. **33**, No. 1, p. 345–353 (1976).
See also *1.3*, p. *21*; *10.1*, p. *78*; 14.1, p. 164; 21.1, p. 303.

F. A. MOHAMED and T. G. LANGDON
The Determination of the Activation Energy for Superplastic Flow
(a) Vol. **33**, No. 1, p. 375–381 (1976).
See also *10.1*, p. *78*; 21.1, p. 303.

A. TAMURA and T. KAWAMURA
Quantum Theory of Rosette-Motion Channeling
(b) Vol. **73**, No. 2, p. 391–400 (1976).
See also *11*, p. *100*; 13, p. 114; 22.5.2, p. 393.

S. I. MASHAROV
Anharmonic Effects in Heat Capacity and Saturation Vapour Pressure of Dilute Substitutional Alloys
(b) Vol. **73**, No. 2, p. 401–408 (1976).
See also *6*, p. *53*; *8*, p. *67*.

J. W. ANDREW and S. S. SHEININ
Approximate Methods for Taking Absorption into Account in the Dynamical Theory of Electron Diffraction
(b) Vol. **73**, No. 2, p. 431–442 (1976).
See also *1.1*, p. *10*; 21.6, p. 329.

M. AUSLOOS
Renormalized Density of States by Conduction Electron–Localized Spin Scattering
(b) Vol. **73**, No. 2, p. 469–474 (1976).
See also *13*, p. *114*; 14.1, p. 164.

S. TANIGAWA and M. DOYAMA
The Formulation of the Formation Energy of a Vacancy as an Alloy Problem
(b) Vol. **73**, No. 2, p. 517–526 (1976).
See also 1.3, p. 21; *10.1*, p. *78*.

H. SATO and K. YONEMITSU
Anisotropic Size Effect in Thin Aluminium Films
(b) Vol. **73**, No. 2, p. 723–733 (1976).
See also 13.2, p. 130; *14.1*, p. *164*.

S. BENCKERT
Erratum to: Third-Order Contribution to the Energy and the Elastic Constants of Simple Metals
(b) Vol. **73**, No. 2, p. 753 (1976).
See also *4*, p. *45*; *12.1*, p. *107*; 21.2, p. 323.

A. ANDREEFF, H.-J. HUNGER, and S. UNTERRICKER
The Magnetic Hyperfine Interaction at Cd^{111} in Iron–Aluminium Alloys
(b) Vol. **73**, No. 2, p. K89—K92 (1976).
See also *5*, p. *48*; 18.2, p. 229; 21.1.1, p. 317.

R. YAMAMOTO
Comments on Faber's Expression for the Vacancy Formation Energy in Metals
(b) Vol. **73**, No. 2, p. K103—K105 (1976).
See also *10.1*, p. *79*.

P. GONDI, R. TOGNATO, and E. EVANGELISTA
High-Temperature Relaxation Peaks in Dispersion-Hardened S.A.P. Alloys
(a) Vol. **33**, No. 2, p. 579—586 (1976).
See also *10.1*, p. *79*; 12.1, p. 107.

T. SCHOBER and H. WENZL
β-Phase Melting and Solidification Phenomena in the Niobium–Hydrogen System
(a) Vol. **33**, No. 2, p. 673—681 (1976).
See also *1.2*, p. *14*; *1.4*, p. *25*; 10.1, p. 79.

S. KISS, I. Z. HARANGOZÓ, and F. J. KEDVES
Solubility Study of Cu in Al Using Zener Relaxation
(a) Vol. **33**, No. 2, p. K107—K109 (1976).
See also *1.3*, p. *21*; 21.1, p. 304.

R. YAMAMOTO
The Binding Energy of an Impurity–Vacancy Pair in Simple Metals
(b) Vol. **74**, No. 1, p. 159—168 (1976).
See also *10.1*, p. *79*.

E. SCHNEIDER and H. KRONMÜLLER
The Elementary Interaction between a Crystal Dislocation and the Flux Line Lattice of a Type II Superconductor
(b) Vol. **74**, No. 1, p. 261—273 (1976).
See also *10.1*, p. *79*; *14.2*, p. *169*.

A. J. PINDOR
On the Coulomb Pseudopotential in Superconducting PdH
(b) Vol. **74**, No. 1, p. K19—K22 (1976).
See also *14.2*, p. *169*.

T. SOMA
Elastic Constant of Hypothetical Face-Centered Cubic In
(b) Vol. **74**, No. 1, p. K67—K70 (1976).
See also 1.1, p. 11; *12.1*, p. *107*.

V. I. BETEKHTIN, A. G. KADOMTSEV, A. I. PETROV, and V. I. VLADIMIROV
Reversibility of the First Stage of Fracture in Metals
(a) Vol. **34**, No. 1, p. 73—78 (1976).
See also *10.1*, p. *79*.

E. BERNARD and P. DUVAL
On the Period of Antiphase Domain Structures in X- and α_2-Phases of Copper–Aluminium Alloys
(a) Vol. **34**, No. 1, p. 135—144 (1976).
See also *1.2*, p. *15*; *1.3*, p. *22*; 1.4, p. 26; 21.1, p. 304.

F. APPEL and U. MESSERSCHMIDT
The Estimation of the Effective Stress in Aluminium Foils Deformed in the High-Voltage Electron Microscope (HVEM)
(a) Vol. **34**, No. 1, p. 175–181 (1976).
See also *10.1*, p. *79*.

D. PREININGER
Work Hardening of F.C.C. Metals Subjected to Internal Stresses. Dispersion and Irradiation Hardening
(a) Vol. **34**, No. 1, p. 227–242 (1976).
See also 1.3, p. 22; *10.1*, p. *80*.

R. H. J. FASTENAU, L. M. CASPERS, and A. VAN VEEN
Small Clusters of Vacancies and Helium in Molybdenum
(a) Vol. **34**, No. 1, p. 277–289 (1976).
See also *10.1*, p. *80*.

B. L. MORDIKE
Anomalous Slip and Discontinuous Yielding
(a) Vol. **34**, No. 1, p. K59–K61 (1976).
See also *10.1*, p. *80*.

B. KOZARZEWSKI, J. NOWAK, and J. ZIELIŃSKI
The Coherent Potential Approximation in the Theory of Paramagnetic Alloys
(b) Vol. **74**, No. 2, p. 437–442 (1976).
See also 13.4, p. 137; *14.1*, p. *165*.

I. YA. DEKHTYAR, V. I. SILANTEV, S. G. SAKHAROVA, R. G. FEDCHENKO, V. I. PATOKA, and V. N. KOLESNIK
Fermi Surface Anisotropy and Properties of Tungsten Crystals
(b) Vol. 74, No. 2, p. 471–476 (1976).
See also *13.2*, p. *130*.

H.-J. HAGEMANN, W. GUDAT, and C. KUNZ
Photoabsorption Coefficient of Alloys of Al with Transition Metals V, Fe, Ni and with Cu and Pr from 30 to 150 eV Photon Energy
(b) Vol. 74, No. 2, p. 507–521 (1976).
See also *20.1*, p. *265*; 21.1, p. 305; 21.1.1, p. 317; 21.4, p. 326.

R. RUPPIN and H. YATOM
Size and Shape Effects on the Broadening of the Plasma Resonance Absorption in Metals
(b) Vol. 74, No. 2, p. 647–654 (1976).
See also *13.5.2*, p. *157*; *20.1*, p. *265*.

B. M. LATTA and P. J. SCANLON
Average Electronic Energy Loss in the Modified Firsov Theory
(b) Vol. **74**, No. 2, p. 711–719 (1976).
See also *11*, p. *101*.

S. K. BAHL and H. D. DREW
Optical Properties of Mo–Re Alloys
(b) Vol. 74, No. 2, p. 721–726 (1976).
See also 14.1, p. 165; *20.1*, p. *266*.

L. M. SCHAFE, P. L. ROSSITER, and W. A. RACHINGER
Field History Effects in Superconducting Pb–Sn Alloys
(b) Vol. **74**, No. 2, p. 785–794 (1976).
See also 10.1, p. 80; *14.2*, p. *169*.

S. AULUCK
On the Pseudopotential Coefficients for Zinc and Cadmium
(b) Vol. 74, No. 2, p. K129–K133 (1976).
See also *13.1*, p. *123*; *13.2*, p. *130*; 20.1, p. 266; 21.1, p. 305.

M. Lannoo and J. P. Muscat
An Application of First-Order Perturbation Theory to Metal Surfaces
(b) Vol. **74**, No. 2, p. K141–K143 (1976).
See also *13.3*, p. *132*.

L. Skála
On the Electron Structure of Dilute Substitutional Alloys
(b) Vol. **74**, No. 2, p. K177–K180 (1976).
See also *13*, p. *115*.

F. Maury and P. Lucasson
On the Configurational Anisotropy of Frenkel Pairs in B.C.C. Molybdenum
(a) Vol. **34**, No. 2, p. 513–518 (1976).
See also *10.1*, p. *80*.

N. E. Alekseevskii, V. M. Sakosarenko, K. Blüthner, and H.-J. Köhler
Superconducting Properties of Vanadium Films
(a) Vol. **34**, No. 2, p. 541–546 (1976).
See also *14.2*, p. *169*.

D. Siemers and E. Nembach
The Dependence of the Initial Susceptibility on the Demagnetization Method
(a) Vol. **34**, No. 2, p. 547–551 (1976).
See also *18.2*, p. *231*.

K. D. Schotte and D. Jacob
Linear Magnetoresistance of a Metal with Holes
(a) Vol. **34**, No. 2, p. 593–600 (1976).
See also *14.1*, p. *165*.

G. Kappel, G. Fischer, and A. Jaéglé
Magnetic Investigation of the System Mn_5Ge_3-Mn_5Si_3
(a) Vol. **34**, No. 2, p. 691–696 (1976).
See also *18.2*, p. *231*; *18.4*, p. *243*.

V. B. Vykhodets, V. B. Demin, and P. V. Geld
Hydrogen Isotope Diffusivity and Solubility in Nickel, Palladium, and Copper
(a) Vol. **34**, No. 2, p. 787–795 (1976).
See also *9*, p. *72*; 10.1, p. 81; 21.1, p. 306.

H. S. Chen
Positron Lifetime Study on the Structure of an Electron Irradiated Metallic Glass
(a) Vol. **34**, No. 2, p. K127–K129 (1976).
See also *1.4*, p. *26*; 2, p. 37; 21.1, p. 306.

L. N. Bystrov and S. N. Shakhmaev
The Effect of Thermal Treatment on Stress Relaxation in Metals
(a) Vol. **34**, No. 2, p. K187–K190 (1976).
See also *10.1*, p. *81*; 21.6, p. 330.

G. Sharp-Dent, M. Hardiman, J. R. Sambles, and J. E. Cousins
Sample Size Effects on the Spin Resonance of Conduction Electrons in Aluminium
(b) Vol. **75**, No. 1, p. 155–161 (1976).
See also *19*, p. *250*.

R. Nilsson, A. Berndtsson, N. Mårtensson, R. Nyholm, and J. Hedman
The Valence Band of Os Studied by Electron Spectroscopy
(b) Vol. **75**, No. 1, p. 197–203 (1976).
See also *13.1*, p. *123*; 17, p. 219.

H.-W. POHL and H. TEICHLER
Impurity Effects in the H_{c2} Anisotropy of Cubic Superconductors
(b) Vol. **75**, No. 1, p. 205–212 (1976).
See also *14.2*, p. *169*.

K. H. WESTMACOTT
The Origin and Nature of Multiple Loop Configurations in Quenched Aluminium and Aluminium Alloys
(a) Vol. **35**, No. 1, p. 185–196 (1976).
See also *10.1*, p. *81*; 21.3, p. 324.

P. MERLE, F. FOUQUET, J. MERLIN and P. F. GOBIN
Relations between the Variations of the Mechanical Properties and the Morphologic Characteristics of Phase Precipitation in an Aluminium-Copper Alloy
(a) Vol. **35**, No. 1, p. 213–222 (1976).
See also *10.1*, p. *82*; 12.1, p. 108; 21.1, p. 307.

B. L. MORDIKE
Athermal Solid Solution Hardening in Niobium Alloy Single Crystals
(a) Vol. **35**, No. 1, p. 303–308 (1976).
See also *10.1*, p. *82*; 12.1, p. 108.

O. A. KAIBYSHEV, R. Z. VALIEV, and V. V. ASTANIN
On the Nature of Superplastic Deformation
(a) Vol. **35**, No. 1, p. 403–413 (1976).
See also *10.1*, p. *82*; 21.1, p. 307.

A. S. RUDNEV, V. I. ROSLYAKOV, E. I. SIROTININ, and A. F. TULINOV
A Study of Relative Energy Losses of Channeled Protons in a Tungsten Single Crystal
(a) Vol. **35**, No. 1, p. K23–K27 (1976).
See also *11*, p. *101*.

I. G. RATISHVILI
On a Peculiarity of Ordering in F.C.C. Binary Alloys with a Distribution Function Containing Two Long-Range Order Parameters
(b) Vol. **75**, No. 2, p. 401–406 (1976).
See also *1.2*, p. *16*; *1.3*, p. *22*.

P. TZANETAKIS, J. HILLAIRET, and G. REVEL
The Formation Energy of Vacancies in Aluminium and Magnesium
(b) Vol. **75**, No. 2, p. 433–439 (1976).
See also *10.1*, p. *82*; 14.1, p. 165; 21.3, p. 324.

J. F. GRACZYK and P. CHAUDHARI
Scattering Properties of Dense Random Packings of Hard Spheres
(b) Vol. **75**, No. 2, p. 593–606 (1976).
See also *1.3*, p. *22*; *2*, p. *38*.

M. KULIĆ and L. DOBROSAVLJEVIĆ
Pinning in Superconducting Films with Modulated Concentration of Impurities
(b) Vol. **75**, No. 2, p. 677–686 (1976).
See also 10.1, p. 82; *14.2*, p. *170*.

E. KOLLEY and W. KOLLEY
Effect of Thermal Fluctuations on the Electronic Densities of States in Disordered Binary Alloys. Numerical Results of CPA Model Calculations
(b) Vol. **75**, No. 2, p. 721–730 (1976).
See also *2*, p. *38*; 6, p. 57; *13*, p. *116*.

L. C. Davis
A Reply to "Observations on Quasi-Classical Quantization Near Self-Intersecting Orbits"
(b) Vol. 75, No. 2, p. K81—K82 (1976).
See also *13*, p. *116*.

G. Cubiotti, E. Donato, E. S. Giuliano, R. Ruggeri, and A. Stancanelli
Transport Properties of Liquid Cadmium
(b) Vol. 75, No. 2, p. K93—K96 (1976).
See also 2, p. 38; *14.1*, p. *165*; *15*, p. *210*.

I. A. Smirnov, R. Suryanarayanan, and S. G. Shulman
Optical Absorption of Metallic SmS Films
(b) Vol. 75, No. 2, p. K153—K156 (1976).
See also *20.1*, p. *269*.

H. P. Aubauer
Elastic Energy of Isotropic Solid Solutions
(a) Vol. 35, No. 2, p. 487—497 (1976).
See also *1.2*, p. *16*; 4, p. 45.

J. P. Muscat, M. Lannoo, and G. Allan
Study of Growth of Guinier-Preston Zones in Al–Ag Compounds
(a) Vol. 35, No. 2, p. 499—509 (1976).
See also *1.3*, p. *22*; 10.1, p. 82; 21.6, p. 330.

D. Fargeot, J. C. Glandus, and P. Boch
Elastic Constants of Copper-Rich Copper–Aluminium Alloys between 77 and 293 K
(a) Vol. 35, No. 2, p. 687—695 (1976).
See also *1.3*, p. *22*; *12.1*, p. *108*; 21.1, p. 308.

Kishore and K. I. Vasu
On the Low-Temperature Deformation Behaviour of a Quenched Aluminium–2% Germanium Alloy
(a) Vol. 35, No. 2, p. 769—775 (1976).
See also *10.1*, p. *82*.

V. N. Bykov, V. A. Troyan, G. G. Zdorovtseva, and V. S. Khaimovich
Erratum to: Phase Transformations at Bombardment of Thin Films with Ions
(a) Vol. 35, No. 2, p. 779 (1976).
See also *1.2*, p. *17*; *1.5*, p. *31*; 11, p. 102; 21.1, p. 308; 21.1.1, p. 319; 22.1.1, p. 347.

I. V. Krylova and V. I. Svitov
Thermostimulated Exoemission and Work Function Measurement of Aluminium
(a) Vol. 35, No. 2, p. K119—K121 (1976).
See also *17*, p. *219*.

W. A. Ortiz and F. P. Missell
Superconducting Properties of Ta–19 at% Zr and Ta–29 at % Zr
(a) Vol. 35, No. 2, p. K135—K136 (1976).
See also *14.2*, p. *170*.

O. A. Troitskii
Isotope Enrichment of Metal in the Process of Zone Melting
(a) Vol. 35, No. 2, p. K151—K155 (1976).
See also *3*, p. *42*; 21.1, p. 308.

E. P. Balsamo, G. Paternò, A. Barone, M. Russo, and R. Vaglio
Temperature and Magnetic Field Dependence of the Critical Current in $Sn-Sn_xO_y-In$ Josephson Junctions
(a) Vol. 35, No. 2, p. K173—K175 (1976).
See also *14.2*, p. *170*; 22.6, p. 405.

V. P. KALANTARIAN and V. I. CHECHERNIKOV
Nuclear Magnetic Resonance and Susceptibility Measurements of Transition Metals and Alloys: V–Ta, V–Nb, and Nb–Ta Systems
(b) Vol. **76**, No. 1, p. 113–118 (1976).
See also 13, p. 116; 18, p. 223; *19*, p. *252*; 21, p. 309.

E. HEGENBARTH and B. SCHMIDT
Heat Conductivity Measurement at V_3Si in the Temperature Range 4.2 to 80 K
(b) Vol. **76**, No. 1, p. 307–311 (1976).
See also *8*, p. *69*; 13.1, p. 125; 14.2, p. 170.

V. SEETHARAMAN, S. BANERJEE, and R. KRISHNAN
The Precipitation of $Ta_{64}C$ in Dilute Tantalum–Carbon Alloys
(a) Vol. **36**, No. 1, p. 39–52 (1976).
See also *1.3*, p. *23*; *1.4*, p. *27*.

H. D. WIESINGER
Phase Diagram and Transition Temperature to Superconductivity of A15 Phases
(a) Vol. **36**, No. 1, p. 61–72 (1976).
See also *1.3*, p. *23*; *14.2*, p. *170*.

H.-G. FABIAN and H. LÖFFLER
On the Interpretation of the Decomposition Kinetics of Some Al–Zn(6 at%)–Mg(0.1 to 3.5 at%) Alloys by Means of the Enlarged Jump Model
(a) Vol. **36**, No. 1, p. 127–132 (1976).
See also *1.3*, p. *23*; *10.1*, p. *83*; 14.1, p. 165; 21.1, p. 309.

A. MAGERL, B. BERRE, amd G. ALEFELD
Changes of the Elastic Constants of V, Nb, and Ta by Hydrogen and Deuterium
(a) Vol. **36**, No. 1, p. 161–171 (1976).
See also *12.1*, p. *108*; 21.1, p. 309.

K. MURATA
Exit Angle Dependence of Penetration Depth of Backscattered Electrons in the Scanning Electron Microscope
(a) Vol. **36**, No. 1, p. 197–208 (1976).
See also *11*, p. *102*; 21.1, p. 309; 21.6, p. 330.

C. BIRAC and D. LESUEUR
Lithium Diffusion in an Amorphous Metallic Pd–Si Alloy
(a) Vol. **36**, No. 1, p. 247–251 (1976).
See also 2, p. 39; *9*, p. *73*.

S. IKENO
Direct Observation of Cell Formation in Niobium by a High-Voltage Electron Microscope
(a) Vol. **36**, No. 1, p. 317–328 (1976).
See also *1.4*, p. *28*; *10.1*, p. *83*.

H. DÜSTERHÖFT, R. MANNS, and D. HILDEBRANDT
Experimental Proof of a Theoretical Relation for the Probability of Positive Ion Excitation by Bombardment of Solid Surfaces with 12 keV Ar^+ Ions
(a) Vol. **36**, No. 1, p. K93–K97 (1976).
See also *11*, p. *102*; 21.1, p. 310; 21.1.1, p. 320; 22.1.1, p. 348; 22.1.2, p. 355; 22.6, p. 405

G. WOLF and C. HOHLFELD
The Molar Heat Capacity of Superconductive $(PdRh_{0.06})H_x$ in the Temperature Range from 2 to 7.5 K
(a) Vol. **36**, No. 1, p. K99–K103 (1976).
See also *8*, p. *69*; 14.2, p. 170.

A. M. RADWAN and M. TAUT
Band Structure of Al and the Laves Phase $CaAl_2$
(b) Vol. **76**, No. 2, p. 605–612 (1976).
See also *13.1*, p. *126*; 20, p. 258; 22.7, p. 411.

G. KUKA, W. BRAUNE, and R. HERRMANN
Cyclotron Resonance with a Slightly Anomalous Skin Effect
(b) Vol. **76**, No. 2, p. 633–638 (1976).
See also *13.2*, p. *131*; 14.1, p. 165.

TH. EIFRIG
Cluster-CPA without Breaking of Translational Symmetry
(b) Vol. **76**, No. 2, p. K101–K104 (1976).
See also *13*, p. *118*.

C. G. CHEN and H. K. BIRNBAUM
Low-Temperature H–O and H–N Relaxations: Confirmation of the Flynn-Stoneham Tunneling Theory
(a) Vol. **36**, No. 2, p. 687–692 (1976).
See also *9*, p. *73*.

K. YONEMITSU and T. MATSUDA
Electrical Resistivity of Spherical Guinier-Preston Zone
(a) Vol. **36**, No. 2, p. 791–798 (1976).
See also 13.2, p. 131; *14.2*, p. *166*; 21.6, p. 331.

G. WOLF, J. JAHNKE, and K. BOHMHAMMEL
The Molar Heat Capacity of $(PdAg_{0.2})H_x$ in the Temperature Range from 2 to 6 K
(a) Vol. **36**, No. 2, p. K125–K128 (1976).
See also *8*, p. *70*; 14.2, p. 170.

V. I. LITVINOV
The Kondo Effect in a Magnetic Field
(b) Vol. **77**, No. 1, p. 71–76 (1976).
See also *14*, p. *162*; *18*, p. *224*.

M. TAUT
Dielectric Function of Hexagonal Metals (I): General Considerations
(b) Vol. **77**, No. 1, p. 165–170 (1976).
See also *13.1*, p. *127*; *14.1*, p. *166*.

M. H. YOO and W. H. BUTLER
Steady-State Diffusion of Point Defects in the Interaction Force Field
(b) Vol. **77**, No. 1, p. 181–193 (1976).
See also *9*, p. *74*; *10.1*, p. *83*; 21.1, p. 310.

D. ECKERT and A. HANDSTEIN
Experimental Investigation of the Flux Density Profile in Hard Superconductors
(a) Vol. **37**, No. 1, p. 171–178 (1976).
See also *14.2*, p. *171*; 21.1, p. 311.

G. WOLF, M. ZIMMERMANN, and K. BOHMHAMMEL
The Molar Heat Capacity of Non-Superconducting PdH_x in the Temperature Range from 2 to 7 K
(a) Vol. **37**, No. 1, p. 179–182 (1976).
See also *8*, p. *70*; 13.1, p. 127.

F. PRINZ, H. P. KARNTHALER, and H. O. K. KIRCHNER
Asymmetric Dislocation Nodes in Cu–Al Alloys
(a) Vol. **37**, No. 1, p. 223–233 (1976).
See also *10.1*, p. *84*; 12.1, p. 108; 21.1, p. 311.

H. Zabel and H. Peisl
X-Ray Study of the Phase Diagram of Hydrogen in Niobium
(a) Vol. 37, No. 1, p. K67—K70 (1976).
See also *1.2*, p. *18*; 1.3, p. 23; 10.1, p. 84.

R. Kroggel, O. Kabisch, and H. Löffler
On the Connection between Microhardness and the Mean Radius of the Guinier-Preston Zones in Al–Zn (12, 15, 22 at%) Alloys
(a) Vol. 37, No. 1, p. K85—K88 (1976).
See also *10.1*, p. *84*; 21.1, p. 311.

L. A. Bertolo and M. M. Shukla
A Calculation of Debye-Waller Factors for F.C.C. Metals
(b) Vol. 77, No. 2, p. 433—439 (1976).
See also *6*, p. *61*; 21.1, p. 312; 21.5, p. 328; 21.6, p. 331.

N. E. Alekseevskii, C. Bazan, A. B. Mitin, T. Mydlarz, E. P. Krasnoperov, and B. Rączka
Influence of Magnetic Impurities on Superconducting Properties of Some Compounds with A-15 Structure
(b) Vol. 77, No. 2, p. 451—455 (1976).
See also *14.2*, p. *171*.

H. Kronmüller and H. Riedel
The Critical Field for the Nucleation of n-Domains in Type I Superconductors
(b) Vol. 77, No. 2, p. 581—590 (1976).
See also *10.1*, p. *84*; *14.2*, p. *171*.

E. Mrosan and G. Lehmann
Localized Non-Magnetic Impurities in Metals (I): General Theory
(b) Vol. 77, No. 2, p. 607—614 (1976).
See also *13.4*, p. *142*.

P. Rennert and A. M. Radwan
Calculation of the Electrostatic Energy of the Laves Phases C14, C15, and C36
(b) Vol. 77, No. 2, p. 615—621 (1976).
See also *1.3*, p. *23*.

C. M. Kachhava and D. Parihar
Residual Resistivity of Dilute Binary Alloys
(b) Vol. 77, No. 2, p. K135—K137 (1976).
See also *14.1*, p. *167*; 21.1, p. 312; 21.2, p. 323.

T. Nakajima
On the Isotopic Volume Effect in Superconducting PdH(D) Systems
(b) Vol. 77, No. 2, p. K147—K150 (1976).
See also *14.2*, p. *171*.

L. M. Caspers, A. van Veen, A. A. van Gorkum, A. van den Beukel, and C. M. van Baal
Helium Desorption from a (110) Mo Crystal. Evidence for the Vacancy Model in Stage III Annealing
(a) Vol. 37, No. 2, p. 371—383 (1976).
See also *10.1*, p. *84*; *11*, p. *103*.

H. Bethge, E. Pippel, and J. Woltersdorf
Problems of Epitaxial Growth in Systems of Weak Interaction Observed in the Case of Al on MgO
(a) Vol. 37, No. 2, p. 457—466 (1976).
See also 1.4, p. 28; *1.5*, p. *32*; *3*, p. *44*; 22.6, p. 407.

G. Wolf and M. Zimmermann
The Molar Heat Capacity of Superconducting PdD_x in the Temperature Range from 2 to 12 K
(a) Vol. **37**, No. 2, p. 485–489 (1976).
See also *8*, p. *70*; 14.2, p. 171.

R. F. Egerton
Measurement of Inelastic/Elastic Scattering Ratio for Fast Electrons and Its Use in the Study of Radiation Damage
(a) Vol. **37**, No. 2, p. 663–668 (1976).
See also *11*, p. *104*; 13.5.2, p. 158; 21.3, p. 324; 21.7, p. 332; 22.9, p. 438.

H. Ö. Pamuk and T. Halicioğlu
Evaluation of Morse Parameters for Metals
(a) Vol. **37**, No. 2, p. 695–699 (1976).
See also *4*, p. *46*; 12.1, p. 108.

G. Splitt
Magnetic Properties of Superconducting Lead Films Investigated with Hypersonic Waves
(a) Vol. **37**, No. 2, p. K131–K134 (1976).
See also *7*, p. *66*; *14.2*, p. *171*.

A. W. Szafrański, T. Skośkiewicz, and B. Baranowski
Superconductivity in the $Pd_{1-x}M_xH_c$ (M = Noble Metal) and Pd_xBH_c Alloy Systems
(a) Vol. **37**, No. 2, p. K163–K164 (1976).
See also *14.2*, p. *172*.

E. M. Savitskii, M. I. Bychkova, V. V. Baron, B. E. Dzevitskii, and N. I. Savateev
Lattice Instability of Compound Nb_3Sn in Strained State
(a) Vol. **37**, No. 2, p. K165–K167 (1976).
See also *1.2*, p. *19*; 5, p. 50.

G. A. Hassan and F. H. Hammad
A Resistivity Decrement in Deformed Aluminium
(a) Vol. **37**, No. 2, p. K209–K211 (1976).
See also *10.1*, p. *85*; *14.1*, p. *167*.

G. A. Hassan, K. H. Georgy, F. H. Hammad, and J. Cadek
Effect of Pre-Torsional Deformation on the Transient Creep of Aluminium
(a) Vol. **37**, No. 2, p. K213–K216 (1976).
See also *10.1*, p. *85*.

E. Mrosan and G. Lehmann
Localized Non-Magnetic Impurities in Metals (II): Application to 3d-Transition Metal Impurities in Aluminium
(b) Vol. **78**, No. 1, p. 159–167 (1976).
See also 8, p. 71; *13.4*, p. *144*; 14.1, p. 167; 14.2, p. 172; 15, p. 211.

K. M. Miller and P. T. Heald
The Lattice Distortion around Interstitials in F.C.C. Metals
(b) Vol. **78**, No. 1, p. 341–351 (1976).
See also *10.1*, p. *85*; 12.1, p. 109; 21.1, p. 313; 21.6, p. 331.

K. Sugawara and C. Y. Huang
Microwave Resistivities of Some Rare Earth-Group VA Intermetallic Compounds
(b) Vol. **78**, No. 1, p. K7–K9 (1976).
See also *14.1*, p. *167*; *18.4*, p. *245*.

Y. Li
The Anisotropic Behavior of Poisson's Ratio, Young's Modulus, and Shear Modulus in Hexagonal Materials
(a) Vol. **38**, No. 1, p. 171–175 (1976).
See also *12.1*, p. *109*; 21.3, p. 324.

P. GUYOT and J. P. SIMON
Symmetrical High-Angle Tilt Boundary Energy Calculation in Aluminium and Lithium
(a) Vol. **38**, No. 1, p. 207–216 (1976).
See also 10.1, p. 85; *13.4*, p. *144*; 21.2, p. 323.

W. POLANSCHÜTZ and E. KRAUTZ
Field Ion Microscopic Study of Grain Boundaries in Metals
(a) Vol. **38**, No. 1, p. 237–242 (1976).
See also *1.4*, p. *29*; 10.1, p. 85; *17.1*, p. *221*.

J. MARIEN
Absolute Method for Work Function Determination of Metals and Semiconductors by Field Emission
(a) Vol. **38**, No. 1, p. 339–347 (1976).
See also *17*, p. *220*; 22.4, p. 377.

S. D. MAMCHICH and V. N. DROBYASIN
On the Analysis of the Creep Dynamics in the Stress Relaxation Regime
(a) Vol. **38**, No. 1, p. K5–K7 (1976).
See also *10.1*, p. *85*; 21.1, p. 314.

S. KISS
*Coherency Loss of θ′ Phase Due to Deformation
(a) Vol. **38**, No. 1, p. K9–K11 (1976).
See also *10.1*, p. *85*; 12.1, p. 109.

F. J. KEDVES and L. GERGELY
Hall Coefficient and Electrical Resistivity of Dilute AlCr, AlMn, and AlCrMn Alloys
(a) Vol. **38**, No. 1, p. K31–K35 (1976).
See also *14.1*, p. *167*; 21.1, p. 314.

K. M. VAN VLIET and A. H. MARSHAK
Conduction Current and Generalized Einstein Relations for Degenerate Semiconductors and Metals
(b) Vol. **78**, No. 2, p. 501–517 (1976).
See also 13.1, p. 129; *14.1*, p. *167*; *14.3*, p. *181*; 15, p. 211; 22, p. 341.

K. M. VAN VLIET
Entropy Production and Choice of Heat Flux for Degenerate Semiconductors and Metals
(b) Vol. **78**, No. 2, p. 667–676 (1976).
See also *14.3*, p. *181*; *15*, p. *211*; 22, p. 342.

J. TH. M. DE HOSSON
Th e Electronic States in Molybdenum
(b) Vol. **78**, No. 2, p. 791–801 (1976).
See also *13*, p. *119*.

H. R. KHAN, K. LÜDERS, CH. J. RAUB, and Z. SZÜCS
Knight-Shift and Superconductivity of Nb-Al σ- and A-15 Phases
(b) Vol. **78**, No. 2, p. K107–K109 (1976).
See also *14.2*, p. *172*; *19*, p. *256*.

A. E. KARKIN, V. E. ARKHIPOV, B. N. GOSHCHITSKII, E. P. ROMANOV, and S. K. SIDOROV
Radiation Effects in the Superconductor Nb_3Sn
(a) Vol. **38**, No. 2, p. 433–438 (1976).
See also *11*, p. *104*; *14.2*, p. *172*.

S. M. OHR
Direction of the Black–White Contrast of Dislocation Loops
(a) Vol. **38**, No. 2, p. 553–557 (1976).
See also *1.4*, p. *29*; *10.1*, p. *86*.

N. Kiesewetter and P. Schiller
The Acoustic Emission from Moving Dislocations in Aluminium
(a) Vol. **38**, No. 2, p. 569–576 (1976).
See also *10.1*, p. *86*.

K. Rosan and H. Wipf
Nitrogen Traps for Hydrogen in Tantalum
(a) Vol. **38**, No. 2, p. 611–620 (1976).
See also *10.1*, p. *86*; 14.1, p. 167.

V. M. Chernov and M. M. Savin
Elastic Interaction of Dislocations with Tetragonal Point Defects in Anisotropic H.C.P. Crystals
(a) Vol. **38**, No. 2, p. 761–768 (1976).
See also *10.1*, p. *86*; 21.1, p. 315; 21.3, p. 325; 21.6, p. 331.

S. M. Ashimov, T. P. Batsankalashvili, N. L. Nedzelyak, and J. S. Tsakadze
Torque Anisotropy Due to Crystal Defect Structure of the Superconducting Niobium Single Crystal
(a) Vol. **38**, No. 2, p. 769–779 (1976).
See also 10.1, p. 86; *14.2*, p. *172*.

F. W. Schapink
Weak-Beam Electron Microscopy of Nearly Transverse Grain Boundaries
(a) Vol. **38**, No. 2, p. K107–K109 (1976).
See also *1.4*, p. *30*.

F. J. Kedves, M. Hordós, and F. Schuszter
Precipitation in Al–Mn and Al–Cr Alloys
(a) Vol. **38**, No. 2, p. K123–K126 (1976).
See also *10.1*, p. *86*; *14.1*, p. *168*; 21.1, p. 315.

B. E. Smith
(101) Order-Twinning Mode in a CuAuI Superlattice
(a) Vol. **38**, No. 2, p. K167–K170 (1976).
See also *1.4*, p. *30*; 12.1, p. 109.

21.1 Transition Metals (Group Sc to Zn) and Their Alloys

Review Articles

E. Vogt
The Origin of Ferromagnetism and Antiferromagnetism in Alloys of Transition Metals (II): Alloys of Transition Metals with Other Transition Metals and with Polyvalent Non-Transition Elements
(a) Vol. **34**, No. 1, p. 11–67 (1976).
See also **1.3**, p. 21; **18.1**, p. 225; *18.2*, p. *228*; *18.4*, p. *242*; **21**, p. 289; **21.6**, p. 329.

H. W. de Wijn, A. M. van Diepen, and K. H. J. Buschow
Samarium in Crystal Fields, a Case of Strong J-Mixing
(b) Vol. **76**, No. 1, p. 11–30 (1976).
See also **5**, p. 48; *13*, p. *113*; *18*, p. *222*; **19**, p. 247; **21**, p. 289; **21.1.1**, p. 316; **21.4**, p. 326.

S. V. Tsivinskii
Dislocation Multiplication in Single and Polycrystals in the Initial Stage of Deformation
(a) Vol. **36**, No. 1, p. 11–31 (1976).
See also *10.1*, p. *78*; *10.2*, p. *87*; **21**, p. 289; **21.1.1**, p. 316; **21.6**, p. 329; **22.1.1**, p. 345; **22.2.3**, p. 367; **22.5.2**, p. 392.

Original Papers and Short Notes

L. M. Falicov and W. Y. Hsu
Phonons and Quasi-Periodic Static Fluctuations
(b) Vol. **73**, No. 1, p. 301–306 (1976).
See also *6*, p. *53*; *13.2*, p. *130*; **14.1**, p. 164.

A. Čížek, M. Šob, and K. Krásenský
The Determination of Ionic Radii of Some Metals by the Positron Annihilation Method
(b) Vol. **73**, No. 1, p. K1–K4 (1976).
See also *13*, p. *114*; **21**, p. 289; **21.1.1**, p. 316; **21.6**, p. 329.

F. M. Galperin
Hyperfine Field on the Nucleus of 3d Transition Metal Atoms
(b) Vol. **73**, No. 1, p. K73–K77 (1976).
See also *13*, p. *114*; *18.2*, p. *228*; **21.1.1**, p. 316.

U. Hofmann
Physical Properties of Nickel–Rhenium Alloys
(a) Vol. **33**, No. 1, p. 119–123 (1976).
See also **12.1**, p. 107; **14.1**, p. 164; *18.2*, p. *228*.

D. H. Sastry, M. J. Luton, and J. J. Jonas
Thermally Activated Flow of H.C.P. Metals at Low Temperatures
(a) Vol. **33**, No. 1, p. 173–187 (1976).
See also *10.1*, p. *78*; **21**, p. 289; **21.3**, p. 324.

T. Seidenkranz and E. Hegenbarth
Single-Crystal Elastic Constants of $MgZn_2$ in the Temperature Range from 4.2 to 300 K
(a) Vol. **33**, No. 1, p. 205–210 (1976).
See also *12.1*, p. *107*; **21.3**, p. 324.

B. Wielke
Easy-Glide Hardening of Zn and Cd between 4.2 and 400 K
(a) Vol. **33**, No. 1, p. 241–246 (1976).
See also *10.1*, p. *78*; **21**, p. 290.

A. MELANDER
Strong- and Weak-Beam Contrast from Cuboidal Inclusions
(a) Vol. **33**, No. 1, p. 255–264 (1976).
See also *1.4*, p. *25*; 10.1, p. 78; 21, p. 290.

M. L. SWANSON, L. M. HOWE, and A. F. QUENNEVILLE
The Combined Effects of Lattice Vibrations and Irradiation-Induced Defects on Dechanneling in Copper
(a) Vol. **33**, No. 1, p. 265–274 (1976).
See also 6, p. 53; *11*, p. *100*.

G. J. DAVIES and C. M. SARGENT
Preferred Orientation and Elastic Anisotropy in Sheets of Copper–Nickel Alloys
(a) Vol. **33**, No. 1, p. 281–288 (1976).
See also *1.3*, p. *21*; *12.1*, p. *107*.

I. V. GEKTINA, F. F. LAVRENTEV, and V. I. STARTSEV
Effect of Structure Imperfections on the Drag of Dislocations in Case of Their Viscous Motion in Zn Crystals
(a) Vol. **33**, No. 1, p. 289–294 (1976).
See also *10.1*, p. *78*.

H. LÖFFLER and H.-G. FABIAN
Application of the Jump Model to the Decomposition of Some Binary and Ternary Aluminium Alloys
(a) Vol. **33**, No. 1, p. 345–353 (1976).
See also *1.3*, p. *21*; *10.1*, p. *78*; 14.1, p. 164; 21, p. 290.

L. M. CLAREBROUGH and C. T. FORWOOD
Diffraction from Single and Overlapping Stacking Faults in F.C.C. Crystals
(a) Vol. **33**, No. 1, p. 355–366 (1976).
See also *1.4*, p. *25*; *10.1*, p. *78*.

F. A. MOHAMED and T. G. LANGDON
*The Determination of the Activation Energy for Superplastic Flow
(a) Vol. **33**, No. 1, p. 375–381 (1976).
See also *10.1*, p. *78*; 21, p. 290.

A. J. MORTON
Inversion Anti-Phase Domains in Cu-Rich γ-Brasses (II): Defects in the Domain Structure
(a) Vol. **33**, No. 1, p. 395–403 (1976).
See also 1.3, p. 21; *1.4*, p. *25*.

E. M. GRAY and J. H. SMITH
The Effect of Heat Treatment on Micromagnetic Mn<u>Cu</u> Alloys
(a) Vol. **33**, No. 1, p. K43–K46 (1976).
See also 14.1, p. 164; *18.2*, p. *229*.

C. MANOHAR and V. K. KELKAR
Effects of Three-Body Interactions on Hyperfine Fields
(b) Vol. **73**, No. 2, p. 735–740 (1976).
See also *13*, p. *114*; *18*, p. *222*.

H. RECHENBERG
Remarks on the Invar Problem
(b) Vol. **73**, No. 2, p. K119–K122 (1976).
See also *18.2*, p. *229*; 21.1.1, p. 317.

E. W. LEE and F. POURARIAN
Magnetoelastic Properties of (Rare-Earth)–Co_2 Compounds (I): Exchange-Striction
(a) Vol. **33**, No. 2, p. 483–489 (1976).
See also *18.2*, p. *229*; 21.4, p. 326.

R. Heller and H. Wipf
Diffusion Coefficient, Heat of Transport, and Solubility Limit of H in V
(a) Vol. **33**, No. 2, p. 525—529 (1976).
See also 8, p. 67; *9*, p. *72*; **14.1**, p. 164.

R. Lermer and A. Steyerl
Investigation of Ferromagnetic Domains and Bloch Walls by Very-Low-Energy Neutron Transmission
(a) Vol. **33**, No. 2, p. 531—541 (1976).
See also *18.2*, p. *229*; 21.1.1, p. 317.

Y. Aoki and M. Yamamoto
X-Ray and Magnetic Investigations of the High-Temperature Phase in the Co-Rich Co–V Alloy System
(a) Vol. **33**, No. 2, p. 625—632 (1976).
See also **1.3**, p. 21; *18.1*, p. *225*; *18.2*, p. *229*.

K. G. Banerjee and B. Mills
The Zener Relaxation and Diffusion in the Copper–Nickel–Zinc System
(a) Vol. **33**, No. 2, p. 707—714 (1976).
See also *9*, p. *72*; *10.1*, p. *79*.

B. Hertel and J. Diehl
On the Formation of Vacancy Clusters in Copper Irradiated with 5 keV Argon Ions
(a) Vol. **33**, No. 2, p. K73—K75 (1976).
See also *10.1*, p. *79*; *11*, p. *100*.

B. K. Ray and G. D. Nigam
Variation of Elastic Constants of Partially Ordered Cu_3Au Alloy with Temperature
(a) Vol. **33**, No. 2, p. K101—K105 (1976).
See also **1.3**, p. 21; *12.1*, p. *107*; 21.6, p. 329.

S. Kiss, I. Z. Harangozó, and F. J. Kedves
Solubility Study of Cu in Al Using Zener Relaxation
(a) Vol. **33**, No. 2, p. K107—K109 (1976).
See also *1.3*, p. *21*; 21, p. 291.

S. Takahashi and Y. Ishikawa
Compression-Induced Magnetic Anisotropy Enhanced by Annealing
(a) Vol. **33**, No. 2, p. K141—K143 (1976).
See also *18.2*, p. *230*; 21.1.1, p. 317.

O. B. Sokolov, V. K. Finashkin, and E. A. Turov
Spin Polarization of X-Ray Photoelectrons in Nickel
(b) Vol. **74**, No. 1, p. 35—43 (1976).
See also *13*, p. *115*.

S. Arajs, G. Kote, C. A. Moyer, J. R. Kelly, K. V. Rao, and E. E. Anderson
Antiferromagnetic Susceptibility of Cr–Re Alloys
(b) Vol. **74**, No. 1, p. K23—K26 (1976).
See also *18.4*, p. *243*.

A. W. Smith and R. D. Rawlings
A Neutron Diffraction Study of Ordering in an Iron–Cobalt–1.8% Vanadium Alloy
(a) Vol. **34**, No. 1, p. 117—123 (1976).
See also *1.3*, p. *22*; 21.1.1, p. 317.

E. Bernard and P. Duval
On the Period of Antiphase Domain Structures in X- and α_2-Phases of Copper–Aluminium Alloys
(a) Vol. **34**, No. 1, p. 135—144 (1976).
See also *1.2*, p. *15*; *1.3*, p. *22*; 1.4, p. 26; 21, p. 291.

A. I. Bondarenko and Yu. V. Konobeev
Void Growth Kinetics in Irradiated Metals
(a) Vol. **34**, No. 1, p. 195–205 (1976).
See also *10.1*, p. *80*; 11, p. 100.

E. Hegenbarth, S. V. Pirogova, E. M. Savitskii, and B. Göldner
Thermal Conductivity of the Intermetallic Compound NiAl in the Temperature Range from 20 to 80 K
(a) Vol. **34**, No. 1, p. 319–323 (1976).
See also *8*, p. *68*; 10.1, p. 80.

J. B. A. A. Elemans and K. H. J. Buschow
Comment on the Crystal and Magnetic Structures of $CaCu_5$-Type Compounds of Th with Fe, Co, and Ni
(a) Vol. **34**, No. 1, p. 355–359 (1976).
See also 1.1, p. 11; *18.2*, p. *230*; *18.4*, p. *243*; 21.1.1, p. 317; 21.5, p. 328.

E. W. Lee and F. Pourarian
Magnetoelastic Properties of (Rare-Earth)-Co_2 Compounds (II): The Anisotropic Magnetostriction
(a) Vol. **34**, No. 1, p. 383–390 (1976).
See also *18.2*, p. *230*; 21.4, p. 326.

R. Gonzalez, J. Piqueras, and L. Bru
Annealing of Fatigued Copper in the Electron Microscope
(a) Vol. **34**, No. 1, p. K9–K12 (1976).
See also *10.1*, p. *80*.

J. Ullrich and H. Müllner
Temperature Dependence of Internal Friction in Copper Due to Oxygen Contamination
(a) Vol. **34**, No. 1, p. K31–K34 (1976).
See also *10.1*, p. *80*; 12.1, p. 107.

J. Mizia
CPA Treatment of the Hubbard Model with Intra-Atomic Exchange Interaction
(b) Vol. **74**, No. 2, p. 461–470 (1976).
See also *13*, p. *115*; 18.1, p. 225; 18.2, p. 230.

H.-J. Hagemann, W. Gudat, and C. Kunz
Photoabsorption Coefficient of Alloys of Al with Transition Metals V, Fe, Ni and with Cu and Pr from 30 to 150 eV Photon Energy
(b) Vol. **74**, No. 2, p. 507–521 (1976).
See also *20.1*, p. *265*; 21, p. 292; 21.1.1, p. 317; 21.4, p. 326.

J. Zieliński
On the Existence of Spin Waves in a Model Transition Metal with Finite Anderson Mixing Potential
(b) Vol. **74**, No. 2, p. 779–783 (1976).
See also *13*, p. *115*; *18.2*, p. *230*.

V. A. Gasparov and M. H. Harutunian
Precision Measurements of the Fermi Surface of Copper
(b) Vol. **74**, No. 2, p. K107–K110 (1976).
See also *13.2*, p. *130*; 19, p. 250.

S. Auluck
On the Pseudopotential Coefficients for Zinc and Cadmium
(b) Vol. **74**, No. 2, p. K129–K133 (1976).
See also *13.1*, p. *123*; *13.2*, p. *130*; 20.1, p. 266; 21, p. 292.

V. V. MYAKHAR, V. G. PYNKO, and A. A. POTSELUIKO
Stripe Domain Structure of (001) Magnetic Films with Positive Constant of Crystalline Anisotropy
(a) Vol. **34**, No. 2, p. 437–443 (1976).
See also *18.2.1*, p. *237*; 21.1.1, p. 317.

P. CHARSLEY and M. H. S. RAHIM
The Effect of Ageing on the Lattice Thermal Conductivity of Copper Alloys Deformed at Liquid Helium Temperatures
(a) Vol. **34**, No. 2, p. 533–539 (1976).
See also 6, p. 56; *8*, p. *68*; *10.1*, p. *80*.

G. SCHROEDER and H. KRONMÜLLER
Investigation of the Annealing of the Dislocations in Ni Crystals by Means of the Magneto-Optical Kerr Effect
(a) Vol. **34**, No. 2, p. 623–638 (1976).
See also *10.1*, p. *81*; 18.2, p. 231; 20, p. 258.

L. J. CHEN and A. J. ARDELL
The Observation of Multiple-Layer Loops in Nickel Base Alloys under Ion Bombardment
(a) Vol. **34**, No. 2, p. 679–690 (1976).
See also *10.1*, p. *81*; *11*, p. *101*.

H. OESTERREICHER
Magnetic Studies on Compounds RCuSi, $R_6Cu_8Si_8$ and RCu_2Si_2 (R = Pr, Gd, Tb)
(a) Vol. **34**, No. 2, p. 723–728 (1976).
See also *18.2*, p. *231*; *18.4*, p. *243*; 21.4, p. 326.

G. DLUBEK, O. BRÜMMER, and E. HENSEL
Positron Annihilation Investigation for an Estimation of the Dislocation Density and Vacancy Concentration of Plastically Deformed Polycrystalline Ni of Different Purity
(a) Vol. **34**, No. 2, p. 737–746 (1976).
See also *10.1*, p. *81*.

V. B. VYKHODETS, V. B. DEMIN, and P. V. GELD
Hydrogen Isotope Diffusivity and Solubility in Nickel, Palladium, and Copper
(a) Vol. **34**, No. 2, p. 787–795 (1976).
See also *9*, p. *72*; 10.1, p. 81; 21, p. 293.

M. L. GROSSBECK, P. WILLIAMS, C. A. EVANS, JR., and H. K. BIRNBAUM
Application of Ion Probe Analysis to Studies of Hydrogen Behavior in Solids
(a) Vol. **34**, No. 2, p. K97–K99 (1976).
See also *10.1*, p. *81*.

I. KHAIYER
Domain Wall Observation in Permalloy Overlay Bars of Magnetic Bubble Domain Devices
(a) Vol. **34**, No. 2, p. K105–K106 (1976).
See also *18.2.1*, p. *237*; 21.1.1, p. 318.

H. S. CHEN
Positron Lifetime Study on the Structure of an Electron Irradiated Metallic Glass
(a) Vol. **34**, No. 2, p. K127–K129 (1976).
See also *1.4*, p. *26*; 2, p. 37; 21, p. 293.

M. ANTONINI and H. D. CARSTANJEN
Location of Interstitial Deuterium in $TaD_{0.067}$ by Channeling
(a) Vol. **34**, No. 2, p. K153–K157 (1976).
See also *10.1*, p. *81*.

C. J. Gadsden and M. Heath
Ferromagnetic Resonance in Nickel Single Crystals
(a) Vol. **34**, No. 2, p. K159–K162 (1976).
See also *19*, p. *250*.

J. Ray and Girish Chandra
Concentration Dependence of Curie Points of Some Magnetic Binary Alloys
(a) Vol. **34**, No. 2, p. K169–K172 (1976).
See also *18.1*, p. *225*; *18.2*, p. *231*; 21.1.1, p. 318.

H. E. Schaefer and W. Dander
The Initial Susceptibility of Cobalt after Electron Irradiation at 4.2 K
(b) Vol. **75**, No. 1, p. 173–178 (1976).
See also 10.1, p. 81; *11*, p. *101*; 14.1, p. 165; *18.2*, p. *231*.

K. Hattori
Magnetic Properties of a System Consisting of Two Kinds of Magnetic Atoms Randomly Distributed
(b) Vol. **75**, No. 1, p. K39–K43 (1976).
See also *18.1*, p. *225*; *18.2*, p. *232*.

G. Thummes and H. H. Mende
Observation of the Radio-Frequency Size Effect in Copper Whiskers
(b) Vol. **75**, No. 1, p. K61–K63 (1976).
See also *19*, p. *251*.

G. Chandra, C. Bansal, and J. Ray
Temperature-Dependent Mössbauer Study of Equiatomic Disordered CoFe and NiFe Alloys
(a) Vol. **35**, No. 1, p. 73–77 (1976).
See also *5*, p. *49*; 21.1.1, p. 318.

D. Bastian and E. Biller
Damping of Ferromagnetic Resonance in Ni–Fe Alloys
(a) Vol. **35**, No. 1, p. 113–120 (1976).
See also *18.2*, p. *232*; *19*, p. *251*; 21.1.1, p. 318.

P. Merle, F. Fouquet, J. Merlin, and P. F. Gobin
Relations between the Variations of the Mechanical Properties and the Morphologic Characteristics of Phase Precipitation in an Aluminium-Copper Alloy
(a) Vol. **35**, No. 1, p. 213–222 (1976).
See also *10.1*, p. *82*; 12.1, p. 108; 21, p. 294.

O. A. Kaibyshev, R. Z. Valiev, and V. V. Astanin
On the Nature of Superplastic Deformation
(a) Vol. **35**, No. 1, p. 403–413 (1976).
See also *10.1*, p. *82*; 21, p. 294.

S. J. Rothman and Nghi Q. Lam
Interstitial Retention into Stage III and Radiation-Enhanced Diffusion in Concentrated Alloys
(a) Vol. **35**, No. 1, p. K13–K16 (1976).
See also *9*, p. *73*; *11*, p. *101*.

F. Itoh, K. Kai, M. Kuroha, and S. Takeuchi
Positron Annihilation in Liquid Copper–Bismuth and Copper–Tin Alloys
(b) Vol. **75**, No. 2, p. 559–565 (1976).
See also *2*, p. *38*; 13, p. 116; 21.7, p. 332.

W. DANDER, H. KRONMÜLLER, and H. E. SCHAEFER
Magnetic Aftereffect Phenomena in Cobalt between 4.2 and 20 K after Low-Temperature Electron Irradiation
(b) Vol. **75**, No. 2, p. 647–655 (1976).
See also *11*, p. *102*; *18.2*, p. *232*.

B. J. FECHNER
Influence of a Magnetic Field on the Susceptibility of a Heisenberg Ferromagnet near the Critical Point
(b) Vol. **75**, No. 2, p. 687–695 (1976).
See also 18.1, p. 226; *18.2*, p. *232*.

P. M. BISCH and A. A. GOMES
Formal Results on the Susceptibility of Dilute Alloys (I): Green's Function Approach
(b) Vol. **75**, No. 2, p. 705–715 (1976).
See also *13*, p. *116*; *18*, p. *223*.

K. TSUNO and T. YAMAMOTO
Observed Depths of Magnetic Domains in High-Voltage Scanning Electron Microscopy
(a) Vol. **35**, No. 2, p. 437–449 (1976).
See also *18.2*, p. *232*; 21.1.1, p. 319.

D. BASTIAN and E. BILLER
Anisotropy Constants and g-Factors of Ni–Fe Alloys Derived from Ferromagnetic Resonance
(a) Vol. **35**, No. 2, p. 465–470 (1976).
See also *18.2*, p. *232*; *19*, p. *252*; 21.1.1, p. 319.

E. MÜNSTER
Domain Wall Motion in Magnetic Film Strips
(a) Vol. **35**, No. 2, p. 521–531 (1976).
See also *18.2.1*, p. *237*.

D. FARGEOT, J. C. GLANDUS and P. BOCH
Elastic Constants of Copper-Rich Copper–Aluminium Alloys between 77 and 293 K
(a) Vol. **35**, No. 2, p. 687–695 (1976).
See also *1.3*, p. *22*; *12.1*, p. *108*; 21, p. 295.

M. M. SAVIN, V. M. CHERNOV, and A. M. STROKOVA
Energy Factor of Dislocations in Hexagonal Crystals
(a) Vol. **35**, No. 2, p. 747–754 (1976).
See also *10.1*, p. *82*; 12.1, p. 108; 21.3, p. 324.

A. J. MORTON
Erratum to: Inversion Anti-Phase Domains in Cu-Rich γ-Brasses (I)
(a) Vol. **35**, No. 2, p. 777 (1976).
See also 1.3, p. 23; *1.4*, p. *27*.

V. N. BYKOV, V. A. TROYAN, G. G. ZDOROVTSEVA, and V. S. KHAIMOVICH
Erratum to: Phase Transformations at Bombartment of Thin Films with Ions
(a) Vol. **35**, No. 2, p. 779 (1976).
See also *1.2*, p. *17*; *1.5*, p. *31*; 11, p. 102; 21, p. 295; 21.1.1, p. 319; 22.1.1, p. 347.

O. A. TROITSKII
Isotope Enrichment of Metal in the Process of Zone Melting
(a) Vol. **35**, No. 2, p. K151–K155 (1976).
See also *3*, p. *42*; 21, p. 295.

F. A. KASSAN-OGLY and M. V. MEDVEDEV
The Theory of a Heisenberg Alloy
(b) Vol. **76**, No. 1, p. 45–53 (1976).
See also *18.1*, p. *226*; 18.2, p. 233; 18.4, p. 244; 21.1.1, p. 319.

V. P. Kalantarian and V. I. Chechernikov
Nuclear Magnetic Resonance and Susceptibility Measurements of Transition Metals and Alloys: V–Ta, V–Nb, and Nb–Ta Systems
(b) Vol. **76**, No. 1, p. 113–118 (1976).
See also **13**, p. 116; 18, p. 223; *19*, p. *252*; 21, p. 296.

P. M. Bisch and A. A. Gomes
Formal Results on the Susceptibility of Dilute Alloys (II): Mean-Field Approach and Applications
(b) Vol. **76**, No. 1, p. 395–406 (1976).
See also *13*, p. *117*; *18*, p. *224*; 19, p. 252.

F. M. Galperin
Hyperfine Fields of the Impurities in Ferromagnetic Iron
(b) Vol. **76**, No. 1, p. K9–K13 (1976).
See also *13*, p. *117*; *18.2*, p. *233*; 21.1.1, p. 319.

H.-G. Fabian and H. Löffler
On the Interpretation of the Decomposition Kinetics of Some Al–Zn(6 at%)–Mg(0.1 to 3.5 at%) Alloys by Means of the Enlarged Jump Model
(a) Vol. **36**, No. 1, p. 127–132 (1976).
See also *1.3*, p. *23*; *10.1*, p. *83*; 14.1, p. 165; 21, p. 296.

A. Magerl, B. Berre, and G. Alefeld
Changes of the Elastic Constants of V, Nb, and Ta by Hydrogen and Deuterium
(a) Vol. **36**, No. 1, p. 161–171 (1976).
See also *12.1*, p. *108*; 21, p. 296.

K. Murata
Exit Angle Dependence of Penetration Depth of Backscattered Electrons in the Scanning Electron Microscope
(a) Vol. **36**, No. 1, p. 197–208 (1976).
See also *11*, p. *102*; 21, p. 296; 21.6, p. 330.

M. Wendt, Th. Krajewski, and R. Bimberg
Detection of Thin Surface Films by Electron Beam Microanalysis. A Comparison between Wavelength-Dispersive and Energy-Dispersive Microprobes
(a) Vol. **36**, No. 1, p. 253–261 (1976).
See also *1.5*, p. *32*; 21.1.1, p. 319; 22.1.1, p. 348.

H. Figiel, A. Oppelt, E. Dormann, and K. H. J. Buschow
Transferred Hyperfine Fields at the Y Sites in Y–Co Compounds
(a) Vol. **36**, No. 1, p. 275–283 (1976).
See also 18.2, p. 233; *19*, p. *253*; 21.4, p. 327.

M. Guymont and D. Gratias
On the Stability of Periodically Antiphased Alloys
(a) Vol. **36**, No. 1, p. 329–334 (1976).
See also 1.2, p. 18; *1.3*, p. *23*; *1.4*, p. *28*; 21.6, p. 330.

A. Prodan, F. W. Boswell, and J. M. Corbett
An Electron Microscopic Investigation of $CoTe_2$ Single Crystals
(a) Vol. **36**, No. 1, p. K21–K23 (1976).
See also *1.4*, p. *28*; 10.1, p. 83.

J. Schneider, H. Wiesner, and R. Gemperle
Annealing Effects on the Magnetic Properties of Rapidly Quenched Transition Metal Alloys
(a) Vol. **36**, No. 1, p. K59–K64 (1976).
See also *1.3*, p. *23*; *18.2*, p. *233*; 21.1.1, p. 320.

J. SCHNEIDER and J. KONETZKY
Magnetization Studies of Rapidly Quenched Fe–Ni Alloys
(a) Vol. 36, No. 1, p. K77–K80 (1976).
See also *1.3*, p. *23*; *18.2*, p. *233*; 21.1.1, p. 320.

B. I. FOMIN, A. E. GERSHINSKII, E. I. CHEREPOV, and F. L. EDELMAN
Investigation of Phase Growth Kinetics in the System of Si Single Crystals and V Thin Films
(a) Vol. 36, No. 1, p. K89–K91 (1976).
See also *3*, p. *43*; 22.1.2, p. 355.

H. DÜSTERHÖFT, R. MANNS, and D. HILDEBRANDT
Experimental Proof of a Theoretical Relation for the Probability of Positive Ion Excitation by Bombardment of Solid Surfaces with 12 keV Ar^+ Ions
(a) Vol. 36, No. 1, p. K93–K97 (1976).
See also *11*, p. *102*; 21, p. 296; 21.1.1, p. 320; 22.1.1, p. 348; 22.1.2, p. 355; 22.6, p. 405.

G. DUBLON, M. KROUPP, M. P. DARIEL, and U. ATZMONY
Crystal Field Induced Spin Rotation Observed by Magnetization Study of Polycrystalline $NdCo_2$
(b) Vol. 76, No. 2, p. 669–675 (1976).
See also *18.2*, p. *233*; 21.4, p. 327.

S. K. SIDOROV
On the Phenomenological Theory of Magnetization of Alloys with Mixed Exchange Interaction
(b) Vol. 76, No. 2, p. K113–K117 (1976).
See also *18.2*, p. *234*; 18.4, p. 244; 21.1.1, p. 320.

K. MURATA
Depth Resolution of the Low- and High-Deflection Backscattered Electron Images in the Scanning Electron Microscope
(a) Vol. 36, No. 2, p. 527–532 (1976).
See also *11*, p. *103*.

K. ISHIDA
Direct Estimation of Stacking Fault Energy by Thermodynamic Analysis
(b) Vol. 36, No. 2, p. 717–728 (1976).
See also *10.1*, p. *83*; 21.1.1, p. 320; 21.6, p. 331.

J. M. ANDERSON and A. D. STEWART
The Internal Friction of Electron Irradiated and Cold-Worked Chromium
(a) Vol. 36, No. 2, p. K119–K123 (1976).
See also *10.1*, p. *83*; 11, p. 103; 12.1, p. 108.

Y. AOKI, Y. GOTOH, and Y. OBI
On the Phase Diagram of the Co-Rich Co–V Alloy System
(a) Vol. 36, No. 2, p. K149–K152 (1976).
See also *1.3*, p. *23*; 14.1, p. 166; 18.2, p. 234.

M. H. YOO and W. H. BUTLER
Steady-State Diffusion of Point Defects in the Interaction Force Field
(b) Vol. 77, No. 1, p. 181–193 (1976).
See also *9*, p. *74*; *10.1*, p. *83*; 21, p. 297.

O. D. HENNEMANN and E. SIEGEL
Spin-Wave Measurements of Exchange Constant A in Ni–Fe Alloy Thin Films
(b) Vol. 77, No. 1, p. 229–232 (1976).
See also 18.2.1, p. 238; *19*, p. *254*; 21.1.1, p. 320.

B. Riccò
Density of States of Layer Compounds: 2H-$NbSe_2$
(b) Vol. 77, No. 1, p. 287—293 (1976).
See also *13*, p. *118*.

F. Sacchetti, P. de Gasperis, and F. Menzinger
Magnetic Moment Distribution in Cu–Ni Alloys
(b) Vol. 77, No. 1, p. 309—318 (1976).
See also *18.2*, p. *234*.

Th. Kokkinakis and G. C. Papavassiliou
Surface Plasmons in Small Cu Particles
(b) Vol. 77, No. 1, p. K49—K51 (1976).
See also *13.5.2*, p. *157*; 20.1, p. 273.

D. Eckert and A. Handstein
Experimental Investigation of the Flux Density Profile in Hard Superconductors
(a) Vol. 37, No. 1, p. 171—178 (1976).
See also *14.2*, p. *171*; 21, p. 297.

W. Wycisk and M. Feller-Kniepmeier
Quenching Experiments on High-Purity Nickel
(a) Vol. 37, No. 1, p. 183—191 (1976).
See also *10.1*, p. *83*; 14.1, p. 166.

F. Prinz, H. P. Karnthaler, and H. O. K. Kirchner
Asymmetric Dislocation Nodes in Cu–Al Alloys
(a) Vol. 37, No. 1, p. 223—233 (1976).
See also *10.1*, p. *84*; 12.1, p. 108; 21, p. 297.

J. Petit and M. Quintard
Internal Friction of Cold-Worked Titanium at Low Temperatures
(a) Vol. 37, No. 1, p. 257—262 (1976).
See also *10.1*, p. *84*; *12.1*, p. *108*.

P. G. Vassilev
Annealing Effects on the Thermoelectric Power and Visible Reflectance of Thin Copper Films
(a) Vol. 37, No. 1, p. K25—K27 (1976).
See also 14.1, p. 166; *15*, p. *210*; *20.1*, p. *273*.

I. M. Chapnik
On Possible Formation of Closed Electron Shells with Coupling Spins outside the Inert-Gas Shells in Some Transition Metals and Their Alloys
(a) Vol. 37, No. 1, p. K29—K32 (1976).
See also *13*, p. *118*; 21.1.1, p. 321; 22.7, p. 412.

M. Shiga, R. Minakata, T. Fujimoto, and Y. Nakamura
Forced Volume Magnetostriction of Fe–Al Alloys
(a) Vol. 37, No. 1, p. K33—K36 (1976).
See also *18.2*, p. *235*; 21.1.1, p. 321.

R. Kroggel, O. Kabisch, and H. Löffler
On the Connection between Microhardness and the Mean Radius of the Guinier-Preston Zones in Al–Zn (12, 15, 22 at%) Alloys
(a) Vol. 37, No. 1, p. K85—K88 (1976).
See also *10.1*, p. *84*; 21, p. 298.

M. Shiga and Y. Nakamura
Mössbauer Study of B.C.C. Fe–Cr Alloys near the Critical Concentration
(a) Vol. 37, No. 1, p. K89—K92 (1976).
See also *5*, p. *50*; 18.2, p. 235; 21.1.1, p. 321.

R. Rodloff and H. Neuhäuser
Application of an "Evolution Strategy" to Calculate Static and Dynamic Dislocation Group Configurations
(a) Vol. **37**, No. 1, p. K93—K96 (1976).
See also *10.1*, p. *84*; 11, p. 103.

L. A. Bertolo and M. M. Shukla
A Calculation of Debye-Waller Factors for F.C.C. Metals
(b) Vol. 77, No. 2, p. 433—439 (1976).
See also *6*, p. *61*; 21, p. 298; 21.5, p. 328; 21.6, p. 331.

B. Bosacchi and P. Franzosi
The Electron-Phonon Enhancement Factor on the Fermi Surface of Copper
(b) Vol. 77, No. 2, p. 457—464 (1976).
See also *13.2*, p. *131*.

C. M. Kachhava and D. Parihar
Residual Resistivity of Dilute Binary Alloys
(b) Vol. 77, No. 2, p. K135—K137 (1976).
See also *14.1*, p. *167*; 21, p. 298; 21.2, p. 323.

K. Elk and V. Christoph
CPA Treatment of the Electrical Resistivity of Disordered Alloys
(b) Vol. 77, No. 2, p. K151—K155 (1976).
See also *14.1*, p. *167*.

E. Mrosan and G. Lehmann
Calculation of Residual Resistivity and Single Temperatures for Cu-Based 3d-Transition Metal Dilute Alloys
(b) Vol. 77, No. 2, p. K161—K164 (1976).
See also *13.4*, p. *143*; *14.1*, p. *167*.

P. Spindler and K. Nachtrieb
Diffusion of Phosphorus in Copper
(a) Vol. **37**, No. 2, p. 449—456 (1976).
See also *9*, p. *74*.

T. Yamamoto
On the Image Formation of Magnetic Domains and Their Walls in Scanning Electron Microscopy
(a) Vol. **37**, No. 2, p. 607—615 (1976).
See also *18.2*, p. *235*.

B. Nowak, N. Piślewski, and W. Leszczyński
Proton Spin-Lattice Relaxation Time in the Dihydride Phase of the Ternary Ti-Nb-H Alloy
(a) Vol. **37**, No. 2, p. 669—674 (1976).
See also *19*, p. *255*.

L. E. Popov, Yu. Sharkeev, N. A. Koneva, N. A. Enshina, and G. A. Perov
Conservative Jog Slip on Superdislocations and Work-Hardening of Ordered Alloys
(a) Vol. **37**, No. 2, p. 701—708 (1976).
See also *1.3*, p. *24*; *10.1*, p. *84*.

A. I. Landau and V. I. Dotsenko
Power-Like Dependence of Effective Dislocation Velocity on Load Resulting from the Stochastic Character of Motion through a Random Array of Point Obstacles
(a) Vol. **37**, No. 2, p. 709—718 (1976).
See also *10.1*, p. *84*.

M. I. Zakharova and H. Fuentes
Structural Phase Transformations in the Alloys Ti–25%Nb–15%V and Ti–40%Nb
(a) Vol. 37, No. 2, p. K121–K125 (1976).
See also *1.2*, p. *19*; *1.3*, p. *24*.

H. E. Schaefer and W. Dander
The Magnetic After-Effect Spectrum in Cobalt between 20 and 350 K after Electron Irradiation at 4.2 K
(b) Vol. 78, No. 1, p. 139–147 (1976).
See also *11*, p. *104*; *18.2*, p. *235*.

H. Müller and Ch. Opitz
Quantum Chemical Investigations on Vacancy Diffusion and Vacancy Lattices in Crystallites
(b) Vol. 78, No. 1, p. 177–182 (1976).
See also *9*, p. *74*; *10.1*, p. *85*; 21.1.1, p. 321.

G. S. Poo and D. M. Edwards
Electron Interactions and the Spin-Wave Stiffness Constant in Ferromagnetic Two-Component Alloys
(b) Vol. 78, No. 1, p. 287–299 (1976).
See also *13.5.4*, p. *161*; *18.2*, p. *235*.

J. Zieliński
A Double d-Band Model of Transition Metals
(b) Vol. 78, No. 1, p. 319–324 (1976).
See also *13.1*, p. *128*; *18.2*, p. *236*.

S. Kraśnicki
Neutron Investigation of Magnons in the Cu_2MnAl Heusler Alloy
(b) Vol. 78, No. 1, p. 335–340 (1976).
See also *13.5.4*, p. *161*; 18.2, p. 236.

K. M. Miller and P. T. Heald
The Lattice Distortion around Interstitials in F.C.C. Metals
(b) Vol. 78, No. 1, p. 341–351 (1976).
See also *10.1*, p. *85*; 12.1, p. 109; 21, p. 299; 21.6, p. 331.

Yu. M. Plishkin and I. E. Podchinenov
Vacancy Migration Energy Calculation in an F.C.C. Copper Lattice by Computer Simulation
(a) Vol. 38, No. 1, p. 51–55 (1976).
See also *10.1*, p. *85*.

R. Gauthier and P. Pinard
Pulsed Laser Light Stimulated Desorption
(a) Vol. 38, No. 1, p. 85–92 (1976).
See also *17*, p. *220*; 22.1.1, p. 349; 22.5.2, p. 397.

J. M. Corbett and S. S. Sheinin
Electron Microscope Images of Twin Boundaries (I): Comparison of Experiment with Theory
(a) Vol. 38, No. 1, p. 151–159 (1976).
See also *1.4*, p. *28*.

H. Kousek, P. Bajons, and H. Müllner
Analysis of Structural Changes in High-Frequency Fatigued Copper by Means of Damping and Modulus Studies
(a) Vol. 38, No. 1, p. 199–205 (1976).
See also *10.1*, p. *85*; *12.1*, p. *109*.

S. R. HERD
Lorentz Microscopy on Amorphous Films with Perpendicular Anisotropy as a Function of Temperature
(a) Vol. 38, No. 1, p. 305–314 (1976).
See also *2*, p. *40*; *18.2.1*, p. *238*; 21.4, p. 327.

S. D. MAMCHICH and V. N. DROBYASIN
On the Analysis of the Creep Dynamics in the Stress Relaxation Regime
(a) Vol. 38, No. 1, p. K5–K7 (1976).
See also *10.1*, p. *85*; 21, p. 300.

F. J. KEDVES and L. GERGELY
Hall Coefficient and Electrical Resistivity of Dilute AlCr, AlMn, and AlCrMn Alloys
(a) Vol. 38, No. 1, p. K31–K35 (1976).
See also *14.1*, p. *167*; 21, p. 300.

Š. VIŠŇOVSKÝ, B. KNAPPE, V. PROSSER, and H.-R. MÜLLER
Magnetooptical Kerr Effect Spectra of Amorphous Gadolinium–Cobalt Films
(a) Vol. 38, No. 1, p. K53–K56 (1976).
See also *18.2.1*, p. *238*; *20.1*, p. *277*; 21.4, p. 327.

H. NEDDERMEYER
X-Ray Spectroscopic Investigation of the Electronic Structure of Mg–Cu Alloys
(b) Vol. 78, No. 2, p. 609–614 (1976).
See also *13.1*, p. *129*; *20*, p. *259*; 21.3, p. 324.

K. MAIER, H. MEHRER, E. LESSMANN, and W. SCHÜLE
Self Diffusion in Nickel at Low Temperatures
(b) Vol. 78, No. 2, p. 689–698 (1976).
See also *9*, p. *74*; 10.1, p. 86.

D. C. GUPTA and N. D. SHARMA
Pressure Dependence of the Mössbauer Fraction for ^{57}Fe in Ti, V, and Cu
(b) Vol. 78, No. 2, p. K125–K129 (1976).
See also *5*, p. *51*; 6, p. 64.

O. V. PRESNYAKOVA and V. I. ZAITSEV
Effect of High Pressure on Nickel Polygonization
(a) Vol. 38, No. 2, p. 447–452 (1976).
See also *1.4*, p. *29*; *10.1*, p. *86*.

R. GAUTHIER and C. GUITTARD
Mechanism Investigations of a Pulsed Laser Light Induced Desorption
(a) Vol. 38, No. 2, p. 477–486 (1976).
See also *17*, p. *220*; 22.1.1, p. 350; 22.5.2, p. 398.

K. AOKI and O. IZUMI
The Relation between Defect Hardening and Substitutional Solid Solution Hardening in an Intermetallic Compound Ni_3Al
(a) Vol. 38, No. 2, p. 587–594 (1976).
See also *10.1*, p. *86*; 12.1, p. 109.

B. A. GREENBERG, M. A. IVANOV, YU. N. GORNOSTIREV, and L. E. KARKINA
Some Aspects of Plastic Deformation Theory with an Account for Thermally Activated Dislocation Transformations
(a) Vol. 38, No. 2, p. 653–662 (1976).
See also *10.1*, p. *86*.

S. S. SHEININ and J. M. CORBETT
Application of the Multi-Beam Dynamical Theory to Crystals Containing Twins
(a) Vol. 38, No. 2, p. 675–684 (1976).
See also *1.4*, p. *29*.

V. M. Chernov and M. M. Savin
Elastic Interaction of Dislocations with Tetragonal Point Defects in Anisotropic H.C.P. Crystals
(a) Vol. **38**, No. 2, p. 761–768 (1976).
See also *10.1*, p. *86*; 21, p. 301; 21.3, p. 325; 21.6, p. 331.

J. G. Antonopoulos
Contrast from Atomic Surface Steps Using TEM
(a) Vol. **38**, No. 2, p. K115–K117 (1976).
See also *1.6*, p. *34*; 21.6, p. 331.

F. J. Kedves, M. Hordós, and F. Schusztek
Precipitation in Al–Mn and Al–Cr Alloys
(a) Vol. **38**, No. 2, p. K123–K126 (1976).
See also *10.1*, p. *86*; *14.1*, p. *168*; 21, p. 301.

21.1.1 Fe and Fe Alloys

Review Articles

H. W. DE WIJN, A. M. VAN DIEPEN, and K. H. J. BUSCHOW
Samarium in Crystal Fields, a Case of Strong *J*-Mixing
(b) Vol. **76**, No. 1, p. 11–30 (1976).
See also 5, p. 48; *13*, p. *113*; *18*, p. *222*; 19, p. 247; 21, p. 289; 21.1, p. 302; 21.4, p. 326.

S. V. TSIVINSKII
Dislocation Multiplication in Single and Polycrystals in the Initial Stage of Deformation
(a) Vol. **36**, No. 1, p. 11–31 (1976).
See also *10.1*, p. *78*; *10.2*, p. *87*; 21, p. 289; 21.1, p. 302; 21.6, p. 329; 22.1.1, p. 345; 22.2.3, p. 367; 22.5.2, p. 392.

Original Papers and Short Notes

A. ČÍŽEK, M. ŠOB, and K. KRÁSENSKÝ
The Determination of Ionic Radii of Some Metals by the Positron Annihilation Method
(b) Vol. **73**, No. 1, p. K1–K4 (1976).
See also *13*, p. *114*; 21, p. 289; 21.1, p. 302; 21.6, p. 329.

R. MEHROTRA and M. M. PANT
Static and Dynamic Displacements near a Crystal Surface
(b) Vol. **73**, No. 1, p. K27–K30 (1976).
See also *1.6*, p. *34*; *6*, p. *53*; 8, p. 67.

F. M. GALPERIN
Hyperfine Field on the Nucleus of 3d Transition Metal Atoms
(b) Vol. **73**, No. 1, p. K73–K77 (1976).
See also *13*, p. *114*; *18.2*, p. *228*; 21.1, p. 302.

S. K. ARIF and D. S. P. BUNBURY
Mössbauer and X-Ray Studies of $Dy(Fe_{1-x}Ni_x)_3$ and $Y(Fe_{1-x}Ni_x)_3$ Intermetallic Compounds
(a) Vol. **33**, No. 1, p. 91–95 (1976).
See also *5*, p. *48*; 18.2, p. 228; 18.3, p. 239.

A. R. JONES, P. R. HOWELL, and B. RALPH
Weak Beam Microscopy of Grain Boundaries
(a) Vol. **33**, No. 1, p. 107–117 (1976).
See also *1.4*, p. *25*.

R. WIRTHS, P. RUNOW, and H. W. SCHÖPGENS
Magnetic Study of the Ordered and Quenched Equiatomic FePt Alloy
(a) Vol. **33**, No. 1, p. 135–142 (1976).
See also 1.3, p. 21; *18.1*, p. *225*; *18.2*, p. *229*.

G. HANECZOK, J. ILCZUK, R. KUŚKA, and J. W. MOROŃ
Migrational Relaxations in α-Fe–V–C in the Region of the Snoek Effect
(a) Vol. **33**, No. 1, p. 313–324 (1976).
See also *10.1*, p. *78*; *18.2*, p. *229*.

V. A. TSURIN, A. E. ERMAKOV, YU. G. LEBEDEV, and B. N. FILIPPOV
A Mössbauer Study of the Structural Characteristics of Equiatomic FePd and FePt Alloys
(a) Vol. **33**, No. 1, p. 325–332 (1976).
See also *5*, p. *48*; 18.2, p. 229.

A. ANDREEFF, H.-J. HUNGER, and S. UNTERRICKER
The Magnetic Hyperfine Interaction at Cd^{111} in Iron–Aluminium Alloys
(b) Vol. **73**, No. 2, p. K89—K92 (1976).
See also *5*, p. *48*; 18.2, p. 229; 21, p. 291.

H. RECHENBERG
Remarks on the Invar Problem
(b) Vol. **73**, No. 2, p. K119—K122 (1976).
See also *18.2*, p. *229*; 21.1, p. 303.

R. LERMER and A. STEYERL
Investigation of Ferromagnetic Domains and Bloch Walls by Very-Low-Energy Neutron Transmission
(a) Vol. **33**, No. 2, p. 531—541 (1976).
See also *18.2*, p. *229*; 21.1, p. 304.

F. KROUPA and V. PAIDAR
Structure of DO_3 Superlattice Dislocations in Fe–Si Alloys
(a) Vol. **33**, No. 2, p. 555—561 (1976).
See also *10.1*, p. *79*.

S. TAKAHASHI and Y. ISHIKAWA
Compression-Induced Magnetic Anisotropy Enhanced by Annealing
(a) Vol. **33**, No. 2, p. K141—K143 (1976).
See also *18.2*, p. *230*; 21.1, p. 304.

A. W. SMITH and R. D. RAWLINGS
A Neutron Diffraction Study of Ordering in an Iron–Cobalt–1.8% Vanadium Alloy
(a) Vol. **34**, No. 1, p. 117—123 (1976).
See also *1.3*, p. *22*; 21.1, p. 304.

J. B. A. A. ELEMANS and K. H. J. BUSCHOW
Comment on the Crystal and Magnetic Structures of $CaCu_5$-Type Compounds of Th with Fe, Co, and Ni
(a) Vol. **34**, No. 1, p. 355—359 (1976).
See also 1.1, p. 11; *18.2*, p. *230*; *18.4*, p. *243*; 21.1, p. 305; 21.5, p. 328.

O. BOSTANJOGLO and W. GIESE
Magnetization and Crystallization of Stress-Stabilized Amorphous Iron Films
(a) Vol. **34**, No. 1, p. K1—K4 (1976).
See also *2*, p. *36*; *18.2.1*, p. *237*.

A. FERNANDEZ, L. TEJEDOR, and L. BRU
Electron Microscopy Study of the Phase Change $DO_{19} \rightarrow L1_2$ in the Fe_3Ge Compound
(a) Vol. **34**, No. 1, p. K17—K19 (1976).
See also *1.2*, p. *15*.

H.-J. HAGEMANN, W. GUDAT, and C. KUNZ
Photoabsorption Coefficient of Alloys of Al with Transition Metals V, Fe, Ni and with Cu and Pr from 30 to 150 eV Photon Energy
(b) Vol. **74**, No. 2, p. 507—521 (1976).
See also *20.1*, p. *265*; 21, p. 292; 21.1, p. 305; 21.4, p. 326.

V. V. MYAKHAR, V. G. PYNKO, and A. A. POTSELUIKO
Stripe Domain Structure of (001) Magnetic Films with Positive Constant of Crystalline Anisotropy
(a) Vol. **34**, No. 2, p. 437—443 (1976).
See also *18.2.1*, p. *237*; 21.1, p. 306.

R. A. Varin, J. A. Kozubowski, M. W. Grabski, and R. Buczkowski
Analysis of Transmission Electron Microscopy Contrast at Extrinsic Grain Boundary Dislocations
(a) Vol. **34**, No. 2, p. 711–722 (1976).
See also *1.4*, p. *26*; *10.1*, p. *81*.

P. C. M. Gubbens and K. H. J. Buschow
Mössbauer Effect Study of $Tm_2Fe_{17-x}Co_x$ and $Tm_2Fe_{17-x}Ni_x$ Compounds
(a) Vol. **34**, No. 2, p. 729–735 (1976).
See also *5*, p. *49*; 18.2, p. 231; 21.4, p. 326.

I. Khaiyer
Domain Wall Observation in Permalloy Overlay Bars of Magnetic Bubble Domain Devices
(a) Vol. **34**, No. 2, p. K105–K106 (1976).
See also *18.2.1*, p. *237*; 21.1, p. 306.

J. Ray and Girish Chandra
Concentration Dependence of Curie Points of Some Magnetic Binary Alloys
(a) Vol. **34**, No. 2, p. K169–K172 (1976).
See also *18.1*, p. *225*; *18.2*, p. *231*; 21.1, p. 307.

G. Chandra, C. Bansal, and J. Ray
Temperature-Dependent Mössbauer Study of Equiatomic Disordered CoFe and NiFe Alloys
(a) Vol. **35**, No. 1, p. 73–77 (1976).
See also *5*, p. *49*; 21.1, p. 307.

D. Bastian and E. Biller
Damping of Ferromagnetic Resonance in Ni–Fe Alloys
(a) Vol. **35**, No. 1, p. 113–120 (1976).
See also *18.2*, p. *232*; *19*, p. *251*; 21.1, p. 307.

A. M. van der Kraan, J. N. J. van der Velden, J. H. F. van Apeldoorn, P. C. M. Gubbens, and K. H. J. Buschow
Mössbauer Effect Study and Magnetization Measurements of RFe_3 Intermetallic Compounds with R = Sm, Gd, Tb, Dy, Ho, and Th
(a) Vol. **35**, No. 1, p. 137–151 (1976).
See also *5*, p. *49*; 18.2, p. 232; 21.4, p. 326.

P. Duhaj, J. Sitek, M. Prejsa, and P. Butvin
Electron Microscopy and Mössbauer Study of the Amorphous Metallic Alloys $Pd_{80-x}Fe_xSi_{20}$
(a) Vol. **35**, No. 1, p. 223–233 (1976).
See also *1.2*, p. *16*; *2*, p. *37*; 5, p. 49.

J. Topolsky, S. Y. Chuang, and S. J. Tao
Positron Lifetimes in Heat Treated and Plastically Deformed Iron
(a) Vol. **35**, No. 1, p. K17–K22 (1976).
See also *10.1*, p. *82*.

L. G. M. M. Lammers and N. J. Wiegman
Barkhausen Noise and Correlation Measurements of Mumetal Strips
(a) Vol. **35**, No. 1, p. K45–K47 (1976).
See also *18.2*, p. *232*.

E. Ya. Mikhlin and V. V. Nelaev
On the Increase of the Frenkel Defect Recombination Zone in α-Iron Caused by Hydrostatic Compression
(a) Vol. **35**, No. 1, p. K81–K84 (1976).
Erratum
(a) Vol. **38**, No. 1, p. K101 (1976).
See also *10.1*, p. *82*.

R. Lindsey
Measurement of Mössbauer Diffusion Broadening in α-Iron Stabilized with Vanadium
(b) Vol. **75**, No. 2, p. 583–592 (1976).
See also *5*, p. *49*; *9*, p. *73*.

S. Rogaschewski and H. Düsterhöft
Z_1-Dependence of Kinetic Ion–Electron Emission from a Gas-Covered Stainless-Steel Surface
(b) Vol. **75**, No. 2, p. K173–K175 (1976).
See also *17*, p. *219*.

K. Tsuno and T. Yamamoto
Observed Depths of Magnetic Domains in High-Voltage Scanning Electron Microscopy
(a) Vol. **35**, No. 2, p. 437–449 (1976).
See also *18.2*, p. *232*; 21.1, p. 308.

D. Bastian and E. Biller
Anisotropy Constants and *g*-Factors of Ni–Fe Alloys Derived from Ferromagnetic Resonance
(a) Vol. **35**, No. 2, p. 465–470 (1976).
See also *18.2*, p. *232*; *19*, p. *252*; 21.1, p. 308.

V. N. Bykov, V. A. Troyan, G. G. Zdorovtseva, and V. S. Khaimovich
Erratum to: Phase Transformations at Bombardment of Thin Films with Ions
(a) Vol. **35**, No. 2, p. 779 (1976).
See also *1.2*, p. *17*; *1.5*, p. *31*; 11, p. 102; 21, p. 295; 21.1, p. 308; 22.1.1, p. 347.

W. Hauffe
Faceting Mechanism in the Sputtering Process
(a) Vol. **35**, No. 2, p. K93–K96 (1976).
See also *1.4*, p. *27*; 11, p. 102; 21.6, p. 330.

F. A. Kassan-Ogly and M. V. Medvedev
The Theory of a Heisenberg Alloy
(b) Vol. **76**, No. 1, p. 45–53 (1976).
See also *18.1*, p. *226*; 18.2, p. 233; 18.4, p. 244; 21.1, p. 308.

F. M. Galperin
Hyperfine Fields of the Impurities in Ferromagnetic Iron
(b) Vol. **76**, No. 1, p. K9–K13 (1976).
See also *13*, p. *117*; *18.2*, p. *233*; 21.1, p. 309.

C. W. Allen and P. R. Okamoto
An Imaging Technique in Transmission Electron Microscopy Utilizing Diffraction Fine Structure
(a) Vol. **36**, No. 1, p. 107–116 (1976).
See also *1.1*, p. *12*; *1.4*, p. *27*.

R. J. Salzbrenner and S. H. Carpenter
Anelastic Relaxations in the Fe–Mn–N System
(a) Vol. **36**, No. 1, p. 117–126 (1976).
See also *10.1*, p. *83*.

M. Wendt, Th. Krajewski, and R. Bimberg
Detection of Thin Surface Films by Electron Beam Microanalysis. A Comparison between Wavelength-Dispersive an Energy-Dispersive Microprobes
(a) Vol. **36**, No. 1, p. 253–261 (1976).
See also *1.5*, p. *32*; 21.1, p. 309; 22.1.1, p. 348.

P. SCHWELLINGER
The Analysis of Magnetic Domain Wall Structures in the Transition Region of Néel and Bloch Walls by Lorentz Microscopy
(a) Vol. **36**, No. 1, p. 335–**344** (1976).
See also *18.2.1*, p. *237*.

D. J. FATHERS and J. P. JAKUBOVICS
Magnetic Domain Wall Contrast in the Scanning Electron Microscope
(a) Vol. **36**, No. 1, p. K13–K16 (1976).
See also *18.2*, p. *233*.

D. LEPSKI and R. MICHAEL
Comparison of the Properties of Electron Microscope Images from Dislocations and Rows of Tetragonal Point Defects
(a) Vol. **36**, No. 1, p. K17–K20 (1976).
See also *10.1*, p. *83*.

J. SCHNEIDER, H. WIESNER, and R. GEMPERLE
Annealing Effects on the Magnetic Properties of Rapidly Quenched Transition Metal Alloys
(a) Vol. **36**, No. 1, p. K59–K64 (1976).
See also *1.3*, p. *23*; *18.2*, p. *233*; 21.1, p. 309.

J. SCHNEIDER and J. KONETZKY
Magnetization Studies of Rapidly Quenched Fe–Ni Alloys
(a) Vol. **36**, No. 1, p. K77–K80 (1976).
See also *1.3*, p. *23*; *18.2*, p. *233*; 21.1, p. 310.

H. DÜSTERHÖFT, R. MANNS, and D. HILDEBRANDT
Experimental Proof of a Theoretical Relation for the Probability of Positive Ion Excitation by Bombardment of Solid Surfaces with 12 keV Ar^+ Ions
(a) Vol. **36**, No. 1, p. K93–K97 (1976).
See also *11*, p. *102*; 21, p. 296; 21.1, p. 310; 22.1.1, p. 348; 22.1.2, p. 355; 22.6, p. 405.

S. K. SIDOROV
On the Phenomenological Theory of Magnetization of Alloys with Mixed Exchange Interaction
(b) Vol. **76**, No. 2, p. K113–K117 (1976).
See also *18.2*, p. *234*; 18.4, p. 244; 21.1, p. 310.

J. F. DUFRESNE, A. SEEGER, P. GROH, and P. MOSER
Hydrogen Relaxation in α-Iron
(a) Vol. **36**, No. 2, p. 579–589 (1976).
See also *10.1*, p. *83*; *18.2*, p. *234*.

K. ISHIDA
Direct Estimation of Stacking Fault Energy by Thermodynamic Analysis
(a) Vol. **36**, No. 2, p. 717–728 (1976).
See also *10.1*, p. *83*; 21.1, p. 310; 21.6, p. 331.

O. D. HENNEMANN and E. SIEGEL
Spin-Wave Measurements of Exchange Constant *A* in Ni–Fe Alloy Thin Films
(b) Vol. **77**, No. 1, p. 229–232 (1976).
See also 18.2.1, p. 238; *19*, p. *254*; 21.1, p. 310.

F. P. LIVI, J. D. ROGERS, and P. J. VICCARO
Experimental Studies of Hyperfine Interactions and Magnetization in $HfFe_2$
(a) Vol. **37**, No. 1, p. 133–138 (1976).
See also *5*, p. *50*; *18.2*, p. *235*.

J. D. SIEVERT and C. VOIGT
Influence of Domains on the Magnetic Torque in Cubic Crystals
(a) Vol. **37**, No. 1, p. 205–209 (1976).
See also *18.2*, p. *235*.

I. M. Chapnik
On Possible Formation of Closed Electron Shells with Coupling Spins outside the Inert-Gas Shells in Some Transition Metals and Their Alloys
(a) Vol. **37**, No. 1, p. K29–K32 (1976).
See also *13*, p. *118*; 21.1, p. 311; 22.7, p. 412.

M. Shiga, R. Minakata, T. Fujimoto, and Y. Nakamura
Forced Volume Magnetostriction of Fe–Al Alloys
(a) Vol. **37**, No. 1, p. K33–K36 (1976).
See also *18.2*, p. *235*; 21.1, p. 311.

M. Shiga and Y. Nakamura
Mössbauer Study of B.C.C. Fe–Cr Alloys near the Critical Concentration
(a) Vol. **37**, No. 1, p. K89–K92 (1976).
See also *5*, p. *50*; 18.2, p. 235; 21.1, p. 311.

K. V. Reddy and S. C. Chetty
Mössbauer Studies on the Fe–Te System
(a) Vol. **37**, No. 2, p. 687–694 (1976).
See also *5*, p. *50*; 18.4, p. 245.

H. J. Blythe and L. H. Dunleavy
Low Temperature Magnetic Disaccommodation in Neutron Irradiated Iron
(a) Vol. **37**, No. 2, p. K197–K200 (1976).
See also *10.1*, p. *84*; *18.2*, p. *235*.

A. Oppelt, A. Merkel, and K. H. J. Buschow
Temperature Dependence of the Y Hyperfine Fields in YFe_3
(a) Vol. **37**, No. 2, p. K205–K208 (1976).
See also *18.2*, p. *235*; *19*, p. *255*.

H. Müller and Ch. Opitz
Quantum Chemical Investigations on Vacancy Diffusion and Vacancy Lattices in Crystallites
(b) Vol. 78, No. 1, p. 177–182 (1976).
See also *9*, p. *74*; *10.1*, p. *85*; 21.1, p. 313.

L. I. Vinokurova, A. V. Vlasov, and M. Pardavi-Horváth
Pressure Effects on Magnetic Phase Transitions in FeRh and FeRhIr Alloys
(b) Vol. 78, No. 1, p. 353–357 (1976).
See also 12.1, p. 109; 18.1, p. 227; *18.2*, p. *236*; *18.4*, p. *245*.

H. Mughrabi, R. Kütterer, K. Lubitz, and H. Kronmüller
Magnetic Studies on Cyclically Deformed α-Iron Single Crystals
(a) Vol. **38**, No. 1, p. 261–270 (1976).
See also *10.1*, p. *85*; 18.2, p. 236.

T. Yamamoto
Measurement of the Emission Area of Backscattered Electrons
(a) Vol. **38**, No. 1, p. 361–368 (1976).
See also *18.2*, p. *236*.

K. W. Ingle and A. G. Crocker
A Computer Simulation Study of the Effect of Applied Stress on Divacancy Binding Energies in Body-Centred Cubic Crystals
(a) Vol. **38**, No. 2, p. 523–531 (1976).
See also *10.1*, p. *86*; 12.1, p. 109.

G. A. Stewart and J. A. Barclay
The Effects of Clustering on the Hyperfine Field of Indium in Iron
(a) Vol. **38**, No. 2, p. 533–538 (1976).
See also *19*, p. *256*.

S. C. Bhargava and P. K. Iyengar
Study of Hyperfine Fields at Iron Nuclei in (Co, Fe)$_5$Ge$_3$ Using the Mössbauer Effect
(a) Vol. **38**, No. 2, p. 601–609 (1976).
See also *5*, p. *51*; 18.1, p. 227; 18.2, p. 236.

J. P. Rivière and J. Grilhé
Neutron Irradiation Damage in a Plastically Deformed Fe–40 at% Al Ordered Alloy
(a) Vol. **38**, No. 2, p. 713–720 (1976).
See also 10.1, p. 86; *11*, p. *105*; 14.1, p. 168.

21.2 Alkali Metals

Original Papers and Short Notes

S. Benckert
Erratum to: Third-Order Contribution to the Energy and the Elastic Constants of Simple Metals
(b) Vol. 73, No. 2, p. 753 (1976).
See also *4*, p. *45*; *12.1*, p. *107*; 21, p. 291.

A. M. Brodskii and M. I. Urbakh
Adsorption-Induced Changes of Electron Levels
(b) Vol. 76, No. 1, p. 93–104 (1976).
See also *13*, p. *116*; *13.1*, p. *124*.

R. V. Sharma and K. C. Sharma
A Possible Closed Form Expression for the Static Structure Factor for Liquid Metals
(b) Vol. 76, No. 2, p. K75–K79 (1976).
See also *2*, p. *39*; *13*, p. *118*.

A. Faust, C. Merkel, and H. Bross
Energy Band Structure and Fermi Surface of Potassium
(b) Vol. 77, No. 2, p. 591–600 (1976).
See also *13.1*, p. *127*; 13.2, p. 131.

C. M. Kachhava and D. Parihar
Residual Resistivity of Dilute Binary Alloys
(b) Vol. 77, No. 2, p. K135–K137 (1976).
See also *14.1*, p. *167*; 21, p. 298; 21.1, p. 312.

V. P. Singh and M. P. Hemkar
Dynamical Study of the Temperature Dependence of Grüneisen Parameters in Alkali Metals
(b) Vol. 78, No. 1, p. 381–387 (1976).
See also *6*, p. *63*; 8, p. 71; 12.1, p. 109.

R. N. Singh and P. L. Srivastava
Liquid Electrical Resistivity of Alkali Metals: Exchange and Correlation Effects
(b) Vol. 78, No. 1, p. K45–K49 (1976).
See also 2, p. 40; *14.1*, p. *167*.

P. Guyot and J. P. Simon
Symmetrical High-Angle Tilt Boundary Energy Calculation in Aluminium and Lithium
(a) Vol. 38, No. 1, p. 207–216 (1976).
See also 10.1, p. 85; *13.4*, p. *144*; 21, p. 300.

K. Berndt and O. Brümmer
X-Ray Compton Scattering from LiMg Alloys
(b) Vol. 78, No. 2, p. 659–666 (1976).
See also *13.2*, p. *131*; 20, p. 259; 21.3, p. 325.

J. P. Day and A. L. Ruoff
Erratum to: The Variation of the Elastic Constants of Lithium with Temperature and Pressure
(a) Vol. 38, No. 2, p. 781 (1976).
See also *12.1*, p. *109*.

21.3 Alkali-Earth Metals

Original Papers and Short Notes

M. Taut and H. Eschrig
Electron and Phonon Spectra of Calcium
(b) Vol. **73**, No. 1, p. 151–154 (1976).
See also *6*, p. *52*; *13.1*, p. *120*; 13.2, p. 130.

D. H. Sastry, M. J. Luton, and J. J. Jonas
Thermally Activated Flow of H.C.P. Metals at Low Temperatures
(a) Vol. **33**, No. 1, p. 173–187 (1976).
See also *10.1*, p. *78*; 21, p. 289; 21.1, p. 302.

T. Seidenkranz and E. Hegenbarth
Single-Crystal Elastic Constants of $MgZn_2$ in the Temperature Range from 4.2 to 300 K
(a) Vol. **33**, No. 1, p. 205–210 (1976).
See also *12.1*, p. *107*; 21.1, p. 302.

K. H. Westmacott
The Origin and Nature of Multiple Loop Configurations in Quenched Aluminium and Aluminium Alloys
(a) Vol. **35**, No. 1, p. 185–196 (1976).
See also *10.1*, p. *81*; 21, p. 294.

P. Tzanetakis, J. Hillairet, and G. Revel
The Formation Energy of Vacancies in Aluminium and Magnesium
(b) Vol. **75**, No. 2, p. 433–439 (1976).
See also *10.1*, p. *82*; 14.1, p. 165; 21, p. 294.

M. M. Savin, V. M. Chernov, and A. M. Strokova
Energy Factor of Dislocations in Hexagonal Crystals
(a) Vol. **35**, No. 2, p. 747–754 (1976).
See also *10.1*, p. *82*; 12.1, p. 108; 21.1, p. 308.

M. Taut and W. Hanke
Dielectric Function of Hexagonal Metals (II): Beryllium
(b) Vol. 77, No. 2, p. 543–550 (1976).
See also *13.1*, p. *127*; 13.5.2, p. 157; *14.1*, p. *166*; 14.4, p. 204; 20.1, p. 274.

R. F. Egerton
Measurement of Inelastic/Elastic Scattering Ratio for Fast Electrons and Its Use in the Study of Radiation Damage
(a) Vol. **37**, No. 2, p. 663–668 (1976).
See also *11*, p. *104*; 13.5.2, p. 158; 21, p. 299; 21.7, p. 332; 22.9, p. 438.

M. Taut and P. Ziesche
Dielectric Function of Hexagonal Metals (III): Magnesium
(b) Vol. 78, No. 1, p. 169–175 (1976).
See also *13.1*, p. *128*; 13.5.2, p. 158; *14.1*, p. *167*; 14.4, p. 204; 20.1, p. 275.

Y. Li
The Anisotropic Behavior of Poisson's Ratio, Young's Modulus, and Shear Modulus in Hexagonal Materials
(a) Vol. **38**, No. 1, p. 171–175 (1976).
See also *12.1*, p. *109*; 21, p. 299.

H. Neddermeyer
X-Ray Spectroscopic Investigation of the Electronic Structure of Mg–Cu Alloys
(b) Vol. 78, No. 2, p. 609–614 (1976).
See also *13.1*, p. *129*; *20*, p. *259*; 21.1, p. 314.

K. Berndt and O. Brümmer
X-Ray Compton Scattering from LiMg Alloys
(b) Vol. 78, No. 2, p. 659–666 (1976).
See also *13.2*, p. *131*; 20, p. 259; 21.2, p. 323.

G. Dräger and O. Brümmer
X-Ray Spectroscopic Investigation of Beryllium by Polarized K-Emission Valence Bands
(b) Vol. 78, No. 2, p. 729–735 (1976).
See also *13.1*, p. *129*; 20, p. 259.

V. M. Chernov and M. M. Savin
Elastic Interaction of Dislocations with Tetragonal Point Defects in Anisotropic H.C.P. Crystals
(a) Vol. 38, No. 2, p. 761–768 (1976).
See also *10.1*, p. *86*; 21, p. 301; 21.1, p. 315; 21.6, p. 331.

21.4 Rare-Earth Metals

Review Article

H. W. DE WIJN, A. M. VAN DIEPEN, and K. H. J. BUSCHOW
Samarium in Crystal Fields, a Case of Strong *J*-Mixing
(b) Vol. **76**, No. 1, p. 11–30 (1976).
See also 5, p. 48; *13*, p. *113*; *18*, p. *222*; 19, p. 247; 21, p. 289; 21.1, p. 302; 21.1.1, p. 316.

Original Papers and Short Notes

M. AUSLOOS
On a Structural Phase Transition in Ytterbium
(b) Vol. **73**, No. 1, p. 211–221 (1976).
See also *1.2*, p. *14*; 13.1, p. 120; *14.1*, p. *164*.

I. A. SMIRNOV, R. SURYANARAYANAN, and S. G. SHULMAN
Optical Absorption of Metallic SmS Films near the Interband Transitions
(b) Vol. **73**, No. 2, p. K137–K140 (1976).
See also 13.1, p. 122; *20.1*, p. *262*.

E. W. LEE and F. POURARIAN
Magnetoelastic Properties of (Rare-Earth)-Co_2 Compounds (I): Exchange-Striction
(a) Vol. **33**, No. 2, p. 483–489 (1976).
See also *18.2*, p. *229*; 21.1, p. 303.

E. W. LEE and F. POURARIAN
Magnetoelastic Properties of (Rare-Earth)-Co_2 Compounds (II): The Anisotropic Magnetostriction
(a) Vol. **34**, No. 1, p. 383–390 (1976).
See also *18.2*, p. *230*; 21.1, p. 305.

H.-J. HAGEMANN, W. GUDAT, and C. KUNZ
Photoabsorption Coefficient of Alloys of Al with Transition Metals V, Fe, Ni and with Cu and Pr from 30 to 150 eV Photon Energy
(b) Vol. **74**, No. 2, p. 507–521 (1976).
See also *20.1*, p. *265*; 21, p. 292; 21.1, p. 305; 21.1.1, p. 317.

H. OESTERREICHER
Magnetic Studies on Compounds RCuSi, $R_6Cu_8Si_8$ and RCu_2Si_2 (R = Pr, Gd, Tb)
(a) Vol. **34**, No. 2, p. 723–728 (1976).
See also *18.2*, p. *231*; *18.4*, p. *243*; 21.1, p. 306.

P. C. M. GUBBENS and K. H. J. BUSCHOW
Mössbauer Effect Study of $Tm_2Fe_{17-x}Co_x$ and $Tm_2Fe_{17-x}Ni_x$ Compounds
(a) Vol. **34**, No. 2, p. 729–735 (1976).
See also *5*, p. *49*; 18.2, p. 231; 21.1.1, p. 318.

A. M. VAN DER KRAAN, J. N. J. VAN DER VELDEN, J. H. F. VAN APELDOORN, P. C. M. GUBBENS, and K. H. J. BUSCHOW
Mössbauer Effect Study and Magnetization Measurements of RFe_3 Intermetallic Compounds with R = Sm, Gd, Tb, Dy, Ho, and Th
(a) Vol. **35**, No. 1, p. 137–151 (1976).
See also *5*, p. *49*; 18.2, p. 232; 21.1.1, p. 318.

Z. BIEGAŃSKI and M. DRULIS
Low Temperature Phase Transition in Neodymium Trihydride. Specific Heat Measurements
(a) Vol. **35**, No. 2, p. K127–K130 (1976).
See also *8*, p. *69*; *18*, p. *223*.

H. Figiel, A. Oppelt, E. Dormann, and K. H. J. Buschow
Transferred Hyperfine Fields at the Y Sites in Y–Co Compounds
(a) Vol. **36**, No. 1, p. 275–283 (1976).
See also 18.2, p. 233; *19*, p. *253*; 21.1, p. 309.

G. Dublon, M. Kroupp, M. P. Dariel, and U. Atzmony
Crystal Field Induced Spin Rotation Observed by Magnetization Study of Polycrystalline $NdCo_2$
(b) Vol. **76**, No. 2, p. 669–675 (1976).
See also *18.2*, p. *233*; 21.1, p. 310.

A. Slebarski and A. Chełkowski
Thermal Expansion and the Heat Capacity of $GdAl_2$
(a) Vol. **36**, No. 2, p. K171–K174 (1976).
See also *8*, p. *70*; 18.2, p. 234.

J. E. Lowther
Mechanisms of Energy Transfer between Rare Earth Ions
(b) Vol. **77**, No. 1, p. 359–366 (1976).
See also *2*, p. *39*; 6, p. 60; 13.4, p. 142.

J. Krizek, K. N. R. Taylor, and H. Krizek
Room Temperature Optical Properties of Rare Earths
(b) Vol. **77**, No. 2, p. 725–730 (1976).
See also 14.1, p. 166; *20.1*, p. *274*.

Š. K. Godovikov, M. G. Kozin, V. V. Turovtsev, and V. S. Shpinel
Hyperfine Fields Acting on Diamagnetic Impurities and the Anisotropic Exchange via the s- and d-Electrons in the Rare Earth Metals
(b) Vol. **78**, No. 1, p. 103–111 (1976).
See also *5*, p. *50*; 18.2, p. 235; 18.4, p. 245.

S. R. Herd
Lorentz Microscopy on Amorphous Films with Perpendicular Anisotropy as a Function of Temperature
(a) Vol. **38**, No. 1, p. 305–314 (1976).
See also *2*, p. *40*; *18.2.1*, p. *238*; 21.1, p. 314.

Š. Višňovský, B. Knappe, V. Prosser, and H.-R. Müller
Magnetooptical Kerr, Effect Spectra of Amorphous Gadolinium–Cobalt Films
(a) Vol. **38**, No. 1, p. K53–K56 (1976).
See also *18.2.1*, p. *238*; *20.1*, p. *277*; 21.1, p. 314.

A. Popielewicz
Simple Theory of the Magnetic Properties of $(La_{1-x}Pr_x)_3Tl$
(b) Vol. **78**, No. 2 p. K115–K119 (1976).
See also *18.1*, p. *227*.

21.5 Actinides

Original Papers and Short Notes

J. B. A. A. Elemans and K. H. J. Buschow
Comment on the Crystal and Magnetic Structures of $CaCu_5$-Type Compounds of Th with Fe, Co, and Ni
(a) Vol. **34**, No. 1, p. 355–359 (1976).
See also 1.1, p. 11; *18.2*, p. *230*; *18.4*, p. *243*; 21.1, p. 305; 21.1.1, p. 317.

L. A. Bertolo and M. M. Shukla
A Calculation of Debye-Waller Factors for F.C.C. Metals
(b) Vol. 77, No. 2, p. 433–439 (1976).
See also *6*, p. *61*; 21, p. 298; 21.1, p. 312; 21.6, p. 331.

21.6 Noble Metals

Review Articles

E. VOGT
The Origin of Ferromagnetism and Antiferromagnetism in Alloys of Transition Metals (II): Alloys of Transition Metals with Other Transition Metals and with Polyvalent Non-Transition Elements
(a) Vol. **34**, No. 1, p. 11–67 (1976).
See also 1.3, p. 21; 18.1, p. 225; *18.2*, p. *228*; *18.4*, p. *242*; 21, p. 289; 21.1, p. 302.

S. V. TSIVINSKII
Dislocation Multiplication in Single and Polycrystals in the Initial Stage of Deformation
(a) Vol. **36**, No. 1, p. 11–31 (1976).
See also *10.1*, p. *78*; *10.2*, p. *87*; 21, p. 289; 21.1, p. 302; 21.1.1, p. 316; 22.1.1, p. 345; 22.2.3, p. 367; 22.5.2, p. 392.

Original Papers and Short Notes

A. ČÍŽEK, M. ŠOB, and K. KRÁSENSKÝ
The Determination of Ionic Radii of Some Metals by the Positron Annihilation Method
(b) Vol. **73**, No. 1, p. K1–K4 (1976).
See also *13*, p. *114*; 21, p. 289; 21.1, p. 302; 21.6, p. 316.

V. A. GOLTSOV and L. I. SMIRNOV
Diffusion of Interstitial Atoms in Ordered Pt_2Mo-Type Alloys (II)
(a) Vol. **33**, No. 1, p. 67–73 (1976).
See also *1.3*, p. *21*; *9*, p. *72*; 21, p. 289.

J. W. ANDREW and S. S. SHEININ
Approximate Methods for Taking Absorption into Account in the Dynamical Theory of Electron Diffraction
(b) Vol. **73**, No. 2, p. 431–442 (1976).
See also *1.1*, p. *10*; 21, p. 290.

M. S. BAPNA and V. R. PARAMESWARAN
The Elastic Interaction of Divacancies and Extended Dislocations in F.C.C. Metals
(a) Vol. **33**, No. 2, p. 639–650 (1976).
See also *10.1*, p. *79*.

B. K. RAY and G. D. NIGAM
Variation of Elastic Constants of Partially Ordered Cu_3Au Alloy with Temperature
(a) Vol. **33**, No. 2, p. K101–K105 (1976).
See also 1.3, p. 21; *12.1*, p. *107*; 21.1, p. 304.

P. APEL and U. MÜLLER-JAHREIS
Critical Angles of Low-Energy Light Ions in Gold
(a) Vol. **33**, No. 2, p. K129–K132 (1976).
See also *11*, p. *100*.

G. ALLAN and M. LANNOO
Influence of Self-Consistency on the Relaxation near a Vacancy or a Surface in Transition Metals
(b) Vol. **74**, No. 1, p. 409–414 (1976).
See also *10.1*, p. *79*; *13.4*, p. *136*.

J. Woltersdorf
Interface Structure in Systems of Small Misfit Studied in the Case of Gold on Silver
(a) Vol. **34**, No. 1, p. 183–194 (1976).
See also *1.4*, p. *26*; *1.5*, p. *31*.

J. M. Galligan
Growth Characteristics of Crystals Containing Disclinations
(a) Vol. **34**, No. 2, p. K101–K103 (1976).
See also *3*, p. *42*; *10.1*, p. *81*.

L. N. Bystrov and S. N. Shakhmaev
The Effect of Thermal Treatment on Stress Relaxation in Metals
(a) Vol. **34**, No. 2, p. K187–K190 (1976).
See also *10.1*, p. *81*; 21, p. 293.

O. P. Bansal
On the Pressure Dependence of Electron Transport in Noble Metals
(b) Vol. **75**, No. 1, p. 85–90 (1976).
See also 6, p. 56; 12.1, p. 107; *13.2*, p. *130*.

M. Brauwers and F. Brouers
Optical Relaxation Time and Effective Mass in Noble Metal Alloys
(b) Vol. **75**, No. 2, p. 519–525 (1976).
See also *13.1*, p. *124*; *20*, p. *258*.

J. P. Muscat, M. Lannoo, and G. Allan
Study of Growth of Guinier-Preston Zones in Al–Ag Compounds
(a) Vol. **35**, No. 2, p. 499–509 (1976).
See also *1.3*, p. *22*; 10.1, p. 82; 21, p. 295.

W. Hauffe
Faceting Mechanism in the Sputtering Process
(a) Vol. **35**, No. 2, p. K93–K96 (1976).
See also *1.4*, p. *27*; 11, p. 102; 21.1.1, p. 319.

R. Cavalheiro and M. M. Shukla
Lattice Dynamics of Gold on an Extended de Launay Model
(b) Vol. **76**, No. 1, p. 371–375 (1976).
See also *6*, p. *58*.

K. Murata
Exit Angle Dependence of Penetration Depth of Backscattered Electrons in the Scanning Electron Microscope
(a) Vol. **36**, No. 1, p. 197–208 (1976).
See also *11*, p. *102*; 21, p. 296; 21.1, p. 309.

M. Guymont and D. Gratias
On the Stability of Periodically Antiphased Alloys
(a) Vol. **36**, No. 1, p. 329–334 (1976).
See also 1.2, p. 18; *1.3*, p. *23*; *1.4*, p. *28*; 21.1, p. 309.

E. Kröger and E. Kretschmann
Surface Plasmon and Polariton Dispersion at Rough Boundaries
(b) Vol. **76**, No. 2, p. 515–523 (1976).
See also *13.3*, p. *133*; *13.5*, p. *147*.

G. Jungk and R. Gründler
Determination of Optical Constants: Au
(b) Vol. **76**, No. 2, p. 541–546 (1976).
See also *13.1*, p. *126*; *20.1*, p. *271*.

K. Ishida
Direct Estimation of Stacking Fault Energy by Thermodynamic Analysis
(a) Vol. **36**, No. 2, p. 717—728 (1976).
See also *10.1*, p. *83*; 21.1, p. 310; 21.1.1, p. 320.

K. Yonemitsu and T. Matsuda
Electrical Resistivity of Spherical Guinier-Preston Zone
(a) Vol. **36**, No. 2, p. 791—798 (1976).
See also 13.2, p. 131; *14.1*, p. *166*; 21, p. 297.

B. L. Mordike
Single Crystal Elastic Constants of Silver Base Alloys
(a) Vol. **36**, No. 2, p. K161—K163 (1976).
See also *12.1*, p. *108*.

P. Vašek
Negative Deviations from Matthiessen's Rule in Quenched Platinum at Low Temperatures
(b) Vol. 77, No. 1, p. K67—K70 (1976).
See also *10.1*, p. *83*; *14.1*, p. *166*.

L. A. Bertolo and M. M. Shukla
A Calculation of Debye-Waller Factors for F.C.C. Metals
(b) Vol. 77, No. 2, p. 433—439 (1976).
See also *6*, p. *61*; 21, p. 298; 21.1, p. 312; 21.5, p. 328.

K. M. Miller and P. T. Heald
The Lattice Distortion around Interstitials in F.C.C. Metals
(b) Vol. 78, No. 1, p. 341—351 (1976).
See also *10.1*, p. *85*; 12.1, p. 109; 21, p. 299; 21.1, p. 313.

V. M. Chernov and M. M. Savin
Elastic Interaction of Dislocations with Tetragonal Point Defects in Anisotropic H.C.P. Crystals
(a) Vol. **38**, No. 2, p. 761—768 (1976).
See also *10.1*, p. *86*; 21, p. 301; 21.1, p. 315; 21.3, p. 325.

J. G. Antonopoulos
Contrast from Atomic Surface Steps Using TEM
(a) Vol. **38**, No. 2, p. K115—K117 (1976).
See also *1.6*, p. *34*; 21.1, p. 315.

21.7 Semi-Metals (Semiconducting Alloys See 22.7)

Original Papers and Short Notes

C. Uher
Erratum to: Low-Field Thermomagnetic Tensor for the Bi-Like Band Structure
(b) Vol. **73**, No. 1, p. 379 (1976).
See also 6, p. 53; 13.1, p. 121; *15*, p. *209*.

F. I. Bekkerman and D. V. Gitsu
The Nernst-Ettingshausen Effect in Bismuth under the Influence of Directed Deformation
(b) Vol. **74**, No. 1, p. K1—K4 (1976).
See also *12.1*, p. *107*; *15*, p. *209*.

E. Gerlach, P. Grosse, M. Rautenberg, and W. Senske
Dynamical Conductivity and Plasmon Excitation in Bi
(b) Vol. **75**, No. 2, p. 553—558 (1976).
See also 13.5.2, p. 157; *14.1*, p. *165*; *20.1*, p. *268*.

F. Itoh, K. Kai, M. Kuroha, and S. Takeuchi
Positron Annihilation in Liquid Copper–Bismuth and Copper–Tin Alloys
(b) Vol. **75**, No. 2, p. 559—565 (1976).
See also *2*, p. *38*; 13, p. 116; 21.1, p. 307.

C. Uher and H. J. Goldsmid
Phonon-Drag Generated Umkehreffekt in Bismuth
(b) Vol. **75**, No. 2, p. K137—K141 (1976).
See also 6, p. 57; *15*, p. *210*.

D. W. Field
An X-Ray Diffraction Study of White Tin
(a) Vol. **36**, No. 2, p. 597—601 (1976).
See also *1.1*, p. *12*; 6, p. 59.

G. Quelard and D. Lesueur
Neutron Radiation Defects in Bismuth and Tellurium-Doped Bismuth
(a) Vol. **36**, No. 2, p. 729—734 (1976).
See also *11*, p. *103*; 14.1, p. 165.

R. Kuhl, W. Kraak, H. Haefner, and R. Herrmann
The Semimetal–Semiconductor Transition in Bismuth–Antimony Alloys
(b) Vol. **77**, No. 2, p. K109—K111 (1976).
See also *13.2*, p. *131*; *14.1*, p. *166*; 19, p. 254.

R. F. Egerton
Measurement of Inelastic/Elastic Scattering Ratio for Fast Electrons and Its Use in the Study of Radiation Damage
(a) Vol. **37**, No. 2, p. 663—668 (1976).
See also *11*, p. *104*; 13.5.2, p. 158; 21, p. 299; 21.3, p. 324; 22.9, p. 438.

P. P. Pal-Val, V. Ya. Platkov, and V. I. Startsev
Temperature Dependence of the Dislocation Drag Constant in Antimony
(a) Vol. **38**, No. 1, p. 383—391 (1976).
See also *10.1*, p. *85*.

22. Semiconductors and Ionic Crystals

Review Articles

Z. S. Kachlishvili
Galvanomagnetic and Recombination Effects in Semiconductors in a Strong Electric Field
(a) Vol. **33**, No. 1, p. 15—51 (1976).
See also 13.4, p. 134; *14.3.4*, p. *195*.

V. L. Bonch-Bruevich
Problems of the Electron Theory of Small-Gap Disordered Semiconductors
(b) Vol. **74**, No. 1, p. 11—22 (1976).
See also 2, p. 35; *13.1*, p. *120*.

H. Böttger and V. V. Bryksin
Hopping Conductivity in Ordered and Disordered Solids (I)
(b) Vol. **78**, No. 1, p. 9—56 (1976).
See also 2, p. 35; 6, p. 52; 13, p. 113; 13.5.3, p. 159; *14.3*, p. *173*.

H. Böttger and V. V. Bryksin
Hopping Conductivity in Ordered and Disordered Solids (II)
(b) Vol. **78**, No. 2, p. 415—451 (1976).
See also 2, p. 35; 6, p. 52; 13, p. 113; 13.5.3, p. 159; *14.3*, p. *173*.

Original Papers and Short Notes

H. Meinhold and G. Weiser
Modulation Spectroscopy on MoS_2 and $MoSe_2$
(b) Vol. **73**, No. 1, p. 105—115 (1976).
See also 13.1, p. 120; *13.5.1*, p. *149*; *20.1*, p. *260*.

A. A. Ignatov and Yu. A. Romanov
Nonlinear Electromagnetic Properties of Semiconductors with a Superlattice
(b) Vol. **73**, No. 1, p. 327—333 (1976).
See also 13.1, p. 121; *14.3.4*, p. *195*.

G. Grosso, G. Pastori Parravicini, and R. Resta
Isotropic Compton Profile of LiH in the Tight-Binding Approximation
(b) Vol. **73**, No. 1, p. 371—378 (1976).
See also *13*, p. *114*.

B. Esser and P. Kleinert
Erratum to: On the Theory of Interband $\varepsilon(\omega)$ and External Electric Field Modulated $\Delta\varepsilon(\omega)$ in Disordered Semiconductors
(b) Vol. **73**, No. 1, p. 379 (1976).
See also 2, p. 35; *14.3.4*, p. *195*; *20.1*, p. *260*.

M. Kumeda, N. Kobayashi, E. Maruyama, and T. Shimizu
Electron Spin Resonance and Optical Absorption in Amorphous Se–As
(b) Vol. **73**, No. 1, p. K19—K22 (1976).
See also 2, p. 35; *19*, p. *247*; *20.1*, p. *261*.

I. Moore, M. King, M. Jones, R. J. Deck, and A. M. Hermann
Microwave Faraday Rotation in Molybdenum Disulfide
(b) Vol. **73**, No. 1, p. K31—K34 (1976).
See also 13, p. 114; *18*, p. *222*; *20*, p. *257*.

W. S. Glaunsinger
Erratum to: Indirect Exchange Interactions in EuB_6
(b) Vol. **73**, No. 1, p. K83 (1976).
See also *18.2*, p. *228*; *19*, p. *247*.

R. de Ridder, G. van Tendeloo, and S. Amelinckx
Electron Microscopic Study of the Chimney Ladder Structures $MnSi_{2-x}$ and $MoGe_{2-x}$
(a) Vol. **33**, No. 1, p. 383–393 (1976).
See also *1.1*, p. *10*; 1.4, p. 25.

P. Novotný and J. Šternberk
On the Magnetic Properties of U_3As_4 at Low Temperatures
(a) Vol. **33**, No. 1, p. K15–K17 (1976).
See also *18.2*, p. *229*.

I. M. Belova
On the Theory of Tunnelling in Disordered Semiconductors
(b) Vol. **73**, No. 2, p. 409–414 (1976).
See also 2, p. 36; *14.3.4*, p. *195*.

B. Esser
X-Ray Scattering from Electrons in Random Semiclassical Potential Wells
(b) Vol. **73**, No. 2, p. 499–502 (1976).
See also 2, p. 36; *13.4*, p. *135*.

J. Röseler, K. Henneberger, and R. Strehlow
On the Theory of Electroabsorption in Polar Crystals
(b) Vol. **73**, No. 2, p. 541–552 (1976).
See also 13.5.3, p. 159; *14.3.4*, p. *195*; *20.1*, p. *262*.

P. Petelenz
Effective Electron–Hole Interaction in Molecular Crystals (I)
(b) Vol. **73**, No. 2, p. 565–573 (1976).
See also 6, p. 54; *13*, p. *114*; *13.5.1*, p. *150*; 14.3, p. 174.

R. A. Neville and B. L. Evans
The Band Edge Excitons in 2H-MoS_2
(b) Vol. **73**, No. 2, p. 597–606 (1976).
See also *13.5.1*, p. *150*; 20.1, p. 262.

J. Vicens and P. Delavignette
A Particular Domain Configuration Observed in a New Phase of the Ta–N System
(a) Vol. **33**, No. 2, p. 497–509 (1976).
See also *1.2*, p. *14*; *1.4*, p. *25*.

P. Müller
Direct Proof of the Conduction Type in Semi-Insulating Thin Films
(a) Vol. **33**, No. 2, p. 543–553 (1976).
See also *14.3.1*, p. *183*.

C. C. Becerra, A. Paduan Filho, and N. F. Oliveira, Jr.
Differential Magnetization of MnP
(a) Vol. **33**, No. 2, p. 573–577 (1976).
See also *18.2*, p. *229*; *18.4*, p. *242*.

R. Sinon, R. Hatert, and F. van de Wiele
MOS Structures with a P^+P Profile
(a) Vol. **33**, No. 2, p. 661–671 (1976).
See also *14.3.2*, p. *187*; 14.3.3, p. 190.

C. Manolikas, P. Delavignette, and S. Amelinckx
Electron Microscopy of Phases in the $Cu_{2-x}S$ System
(a) Vol. **33**, No. 2, p. K77—K79 (1976).
See also *1.4*, p. *26*.

S. Matthies
On the Calculation of the Electrostatic Potential in Crystal Lattices (I): Basic Formulae of the Method
(b) Vol. **74**, No. 1, p. 69—80 (1976).
See also *14*, p. *162*.

J. Kurittu and M. Merisalo
Mean-Square Displacements for NH_4Cl and ND_4Cl Crystals
(b) Vol. **74**, No. 1, p. 177—181 (1976).
See also *6*, p. *54*.

D. D. Richardson
Effects of an Impurity on Magnon Sidebands in Insulators
(b) Vol. **74**, No. 1, p. K61—K65 (1976).
See also 10.2, p. 90; *18.2*, p. *230*; *20.1*, p. *264*.

V. V. Voskoboynikov, V. A. Gritsenko, V. M. Efimov, V. E. Lesnikovskaya, and F. L. Edelman
Structure and Electrophysical Properties of Boron Nitride Thin Films
(a) Vol. **34**, No. 1, p. 85—94 (1976).
See also *2*, p. *36*; *14.3.1*, p. *183*; 14.3.3, p. 190.

J. E. Parrott and M. Tieng
Space Charge Capacitance of an Abrupt p-n Junction
(a) Vol. **34**, No. 1, p. K43—K46 (1976).
See also *14.3.3*, p. *191*; 14.3.4, p. 196.

W. S. Glaunsinger
Variable-Temperature EPR Study of the Semiconducting Ferromagnet EuB_6
(b) Vol. **74**, No. 2, p. 443—452 (1976).
See also *19*, p. *249*.

S. Matthies
On the Calculation of the Electrostatic Potential in Crystal Lattices (II): Cubic Lattice — Calculation Method for the Sum $\sigma_{l,m}^{(n_0)}$
(b) Vol. **74**, No. 2, p. 531—546 (1976).
See also *14*, p. *162*.

K. Henneberger
On the Theory of Photon-Assisted Zener Tunneling
(b) Vol. **74**, No. 2, p. 567—577 (1976).
See also 13.1, p. 123; *14.3.4*, p. *196*.

P. Petelenz
Effective Electron–Hole Interaction in Molecular Crystals (II): Phonon Mechanism
(b) Vol. **74**, No. 2, p. 655—658 (1976).
See also *6*, p. *55*; *13*, p. *115*; 13.5.1, p. 152; 14.3, p. 175.

J. Krempaský and D. Barančok
The Use of Modified Relaxation Time in the Transport Theory of Non-Crystalline Semiconductors
(b) Vol. **74**, No. 2, p. 741—751 (1976).
See also *2*, p. *37*; 13.4, p. 137; *14.3*, p. *175*; 15, p. 209.

J. Jerhot and V. Šnejdar
The Influence of Interface States on an Isotype Heterojunction between Two Wide-Gap Semiconductors
(a) Vol. **34**, No. 2, p. 505—511 (1976).
See also *13.3*, p. *132*; *14.3.3*, p. *191*.

W. MOLDENHAUER and W. BRÜCKNER
Physical Properties of Nonstoichiometric Iron Sulfide $Fe_{1-x}S$ near the α-Phase Transition
(a) Vol. 34, No. 2, p. 565–571 (1976).
See also *1.2*, p. *15*; *14.3*, p. *175*; 15, p. 209; 18.4, p. 243.

S. FEIERABEND, U. KISCHKO, and R. NIEDERMAYER
Curie Temperature Determination of EuS Films with Structural Disorder
(a) Vol. 34, No. 2, p. K135–K137 (1976).
See also *18.2.1*, p. *237*.

P. J. P. DE MAAYER and J. D. MACKENZIE
Effect of Defect Structure on the Seebeck Coefficient of Thin Films of TiN_x
(a) Vol. 34, No. 2, p. K143–K146 (1976).
See also *10.2*, p. *91*; *15*, p. *209*.

B. N. OSHCHERIN
On Surface Energies of A^NB^{8-N} Semiconducting Compounds
(a) Vol. 34, No. 2, p. K181–K186 (1976).
See also *1*, p. *9*; *4*, p. *45*.

P. G. ZHUMATII
Intraband Conductivity and Thermopower of Semiconductors with Slowly Varying Gaussian Random Field
(b) Vol. 75, No. 1, p. 61–72 (1976).
See also *14.3*, p. *176*; *15*, p. *209*.

V. KONDAWAR and C. MANDE
Chemical Shifts in the K X-Ray Absorption Discontinuity of Arsenic
(b) Vol. 75, No. 1, p. 79–83 (1976).
See also *4*, p. *45*; 20, p. 258.

V. S. LUBIMOV, G. V. IONOVA, O. M. PACHEV, L. A. MANAKOVA, E. F. MAKAROV, and S. P. IONOV
Low-Lying Collective Electronic Excitations in Semiconducting Charge-Ordered Crystals
(b) Vol. 75, No. 1, p. 91–99 (1976).
See also *13.5*, p. *147*.

M. P. LISITSA, A. P. ZAKHARCHUK, S. F. TEREKHOVA, G. G. TSEBULYA, L. K. MLADOV, and S. M. TODOROV
Anisotropy of the Absorption Edge in Germanium Selenide Single Crystals
(b) Vol. 75, No. 1, p. K51–K54 (1976).
See also 13.5.1, p. 152; *20.1*, p. *267*.

A. M. MURASIK
Erratum to: An Uncertainty Magnetic Moment Determination Imposed by Powder Neutron-Diffraction Method
(b) Vol. 75, No. 1, p. K79 (1976).
See also *18.4*, p. *243*.

S. V. GANTSEVICH, V. L. GUREVICH, V. D. KAGAN, and R. KATILIUS
Transformation of Incoherent Phonon Flux into a Coherent Acoustical Signal as the Result of the Nonlinear Amplification of Acoustical Phonons in Semiconductors
(b) Vol. 75, No. 2, p. 407–422 (1976).
See also 6, p. 56; *7*, p. *65*; *14.3.4*, p. *197*.

H. UEBA and S. ICHIMURA
Plasmon–Phonon-Coupled Mode Sideband of Excitons
(b) Vol. 75, No. 2, p. 501–506 (1976).
See also 6, p. 57; *13.5.1*, p. *153*; 13.5.2, p. 157; *20.1*, p. *268*.

F. F. Sokolov and V. V. Hizhnyakov
Fading Memory in the Theory of Electronic Excitation Transfer
(b) Vol. **75**, No. 2, p. 669—676 (1976).
See also 13.4, p. 139; *20.1*, p. *268*.

U. Berg, G. Dräger, K. Mosebach, and O. Brümmer
Combined XS and XPS Investigations of the FeS_2 Valence Band
(b) Vol. **75**, No. 2, p. K89—K92 (1976).
See also *13.1*, p. *124*; 17, p. 219; 20, p. 258.

J. Zirke, G. Frahm, A. Tausend, and D. Wobig
Infrared and Raman Studies of TlSe
(b) Vol. **75**, No. 2, p. K149—K152 (1976).
See also *6*, p. *57*; *20.1*, p. *269*.

E. Ruske
Photoexpansion of Amorphous As_2S_3 Films
(a) Vol. **35**, No. 2, p. 477—480 (1976).
See also *1.5*, p. *31*; 2, p. 38; *20.1*, p. *269*.

F. W. Boswell, A. Prodan, and J. M. Corbett
Electron Diffraction Study of the Structures of NbS_2 and Order–Disorder in Cu-Intercalated NbS_2
(a) Vol. **35**, No. 2, p. 591—604 (1976).
See also 1.2, p. 16; *1.4*, p. *27*.

W. Moldenhauer and W. Brückner
Thermal Switching in Nearly Stoichiometric Iron Sulfide Single Crystals
(a) Vol. **35**, No. 2, p. K109—K112 (1976).
See also *14.3.4*, p. *197*.

E. Ruske
Reversibility of Photoinduced Edge Shift in Evaporated Amorphous As_2S_3 Films
(a) Vol. **35**, No. 2, p. K113—K114 (1976).
See also 2, p. 38; *20.1*, p. *270*.

G. I. Peresada, E. G. Ponyatovskii, and Zh. D. Sokolovskaya
Pressure Dependence of the Elastic Constants of PbS
(a) Vol. **35**, No. 2, p. K177—K180 (1976).
See also *12.2*, p. *110*.

A. K. Garg
Pressed-Pellet Infrared Absorption Spectra of Molybdenum Dichalcogenides
(b) Vol. **76**, No. 1, p. 171—174 (1976).
See also *6*, p. *57*; *20.1*, p. *270*.

R. Zimmermann
Excitons and Electron–Hole Plasma: A Ground State Calculation
(b) Vol. **76**, No. 1, p. 191—205 (1976).
See also *13.5.1*, p. *153*; 13.5.2, p. 157.

H. Burkhard, G. Bauer, P. Grosse, and A. Lopez-Otero
Free Carrier Magneto-Optical Effects in the Reststrahlen Region of n-PbTe
(b) Vol. **76**, No. 1, p. 259—266 (1976).
See also 13.1, p. 125; *18*, p. *223*; *20.1*, p. *270*.

F. Henneberger and J. Voigt
Polariton Effects in the Optical Spectra of Excitonic Molecules
(b) Vol. **76**, No. 1, p. 313—323 (1976).
See also 13.5, p. 147; *13.5.1*, p. *153*; *20.1*, p. *270*; 20.3, p. 284.

I. M. Dykman and P. M. Tomchuk
Transport Phenomena and Eigenoscillations of Hot Electrons in Semiconductors with Periodic Structure
(b) Vol. **76**, No. 1, p. 385–393 (1976).
Erratum
(b) Vol. **78**, No. 2, p. 843 (1976).
See also *13*, p. *117*; *14.3.4*, p. *197*.

J. Szymański
Erratum to: Theory of Free Carrier Absorption in n-PbSe
(b) Vol. **76**, No. 1, p. 439 (1976).
See also 6, p. 58; 13.1, p. 125; *20.1*, p. *271*.

H. Köhler and C. R. Becker
Infrared-Active Lattice Vibrations in $SnSe_2$
(b) Vol. **76**, No. 1, p. K15–K17 (1976).
See also *6*, p. *58*; *20.1*, p. *271*.

J. M. Leger, G. Weill, and I. A. Smirnov
Resistivity of Sm_3S_4 under High Pressure
(b) Vol. **76**, No. 1, p. K27–K30 (1976).
See also *14.3*, p. *178*.

A. I. Bobrysheva, S. A. Moskalenko, and V. I. Vybornov
The Shift of the Exciton Level and the Biexciton Formation in Crystals under High Density Excitation
(b) Vol. **76**, No. 1, p. K51–K56 (1976).
See also *13.5.1*, p. *153*.

V. V. Hizhnyakov
Effect of Hot Transfer on Vibrational Relaxation of Electronic Excitation
(b) Vol. **76**, No. 1, p. K69–K72 (1976).
See also *13.4*, p. *140*; 20.3, p. 285.

R. K. Nahar and D. Nagchoudhuri
Depletion Layer Characteristics of Heretojunctions with Built-in Homojunction
(a) Vol. **36**, No. 1, p. 139–145 (1976).
See also 13.3, p. 132; *14.3.3*, p. *192*.

S. Petersson, E. Mgbenu, and P. A. Tove
Silicide Formation Correlated with Surface Resistivity Measurements
(a) Vol. **36**, No. 1, p. 217–225 (1976).
See also 1.5, p. 32; *14.3.1*, p. *184*.

S. Martinuzzi, O. Mallem, and T. Cabot
Dark Currents in Cu_2S–CdS Single Crystal Heterojunctions. Consequences for Photocells
(a) Vol. **36**, No. 1, p. 227–234 (1976).
See also 13.3, p. 133; *14.3.3*, p. *192*; *16*, p. *215*; 22.4.1, p. 381.

P. Konsin
Ferroelectric Phase Transitions in Systems with Broad Electronic Bands Caused by Interband Electron–Phonon Interaction
(b) Vol. **76**, No. 2, p. 487–496 (1976).
See also 6, p. 58; 13.1, p. 125; *14.4.1*, p. *207*.

V. Bezák and J. Banský
A Generalization of the Sondheimer-Wilson Formula for the Averaged Density Matrix and Magnetic Susceptibility of Conduction Electrons in a Model of Disordered Semiconductors
(b) Vol. **76**, No. 2, p. 569–580 (1976).
See also 2, p. 39; *13*, p. *117*; *18*, p. *224*.

H.-J. Wünsche and F. Henneberger
Transmission and Reflectivity of Highly Excited Semiconductors Due to the Optical Conversion of Excitons into Excitonic Molecules
(b) Vol. **76**, No. 2, p. 639–646 (1976).
See also *13.5.1*, p. *154*; *20.1*, p. *272*.

R. Pässler
Calculation of Nonradiative Multiphonon Capture Coefficients and Ionization Rates for Neutral Centres According to the Static Coupling Scheme (II): Alternative Trap Models
(b) Vol. **76**, No. 2, p. 647–659 (1976).
See also *6*, p. *59*; 13.1, p. 126; *13.4*, p. *141*; 22.4.1, p. 381.

S. Guha and M. Sathe
Nonlinear Collision-Induced Convective Instability in Solid-State Plasmas
(a) Vol. **36**, No. 2, p. 453–459 (1976).
See also *13*, p. *118*; *14.3.4*, p. *198*.

Ľ. Hrivnák
The Switching Effect in Crystalline Semiconductors Due to the Screening of Donors by Conduction Electrons
(a) Vol. **36**, No. 2, p. 551–556 (1976).
See also 13.4, p. 141; *14.3.4*, p. *198*.

J. van Landuyt, G. van Tendeloo, and S. Amelinckx
Electron Diffraction Study of Inter- and Intrapolytypic Phase Transitions in Transition Metal Dichalcogenides (III): Complementary Diffraction Studies and Lattice Imaging of the Deformation Waves
(a) Vol. **36**, No. 2, p. 757–777 (1976).
See also *1.1*, p. *12*; *1.2*, p. *18*.

H.-G. Holzhütter and M. Mocker
Spectral Dependence of the Photocurrent in Semiconductors
(b) Vol. 77, No. 1, p. 203–214 (1976).
See also *14.3.4*, p. *199*; *16*, p. *215*.

D. O. Welch and K. G. Lynn
Systematic Variation of the Mean Positron Lifetime and Gaussian Fraction in Annealed Metals and Semiconductors
(b) Vol. 77, No. 1, p. 277–286 (1976).
See also *13*, p. *118*; *14.3*, p. *179*.

J. Szymański
Deformation Potential Theory for Optical and Acoustic Phonons in PbSe-Like Semicon ductors
(b) Vol. 77, No. 1, p. 347–351 (1976).
See also *6*, p. *60*.

V. A. Vassilev, B. T. Kolomiets, T. N. Mamontova, and E. M. Raspopova
Photoluminescence and Photoconductivity in Densified Vitreous As_2Se_3
(a) Vol. 37, No. 1, p. 45–49 (1976).
See also 2, p. 40; *16*, p. *216*; *20.3*, p. *286*.

V. Koos, U. Beck, and H.-G. Neumann
Structure and Intrinsic Stress of Nickel Silicide Films
(a) Vol. 37, No. 1, p. 193–198 (1976).
See also *1.5*, p. *32*; *12.2*, p. *111*.

Burma Bansaragtschin, G. Lehmann, and R. Link
Application of the Travelling Solvent Method to Growing of $Pb_{1-x}Sn_xTe$ Single Crystals
(a) Vol. 37, No. 1, p. K13–K15 (1976).
See also *3*, p. *43*; 14.3, p. 179.

J. GRIGAS, J. MESHKAUSKAS, and A. ORLIUKAS
Dielectric Properties of Sb_2S_3 at Microwave Frequencies
(a) Vol. **37**, No. 1, p. K39–K41 (1976).
See also *14.4*, p. *204*.

K. ZIEGLER, H.-D. JUNKER, and U. BIRKHOLZ
Electrical Conductivity and Seebeck Coefficient of Si_2Te_3 Single Crystals
(a) Vol. **37**, No. 1, p. K97–K99 (1976).
See also *14.3*, p. *179*; *15*, p. *211*.

F. PERROT
Bulk Properties of Lithium Hydride up to 100 Mbar
(b) Vol. **77**, No. 2, p. 517–525 (1976).
See also 12.2, p. 112; *13.1*, p. *127*.

H. BÖTTGER
On the Theory of Lattice Vibrations in Mixed Crystals. The Connection between the REI Model, ATA, and CPA
(b) Vol. **77**, No. 2, p. 561–570 (1976).
See also *6*, p. *61*.

S. C. GOYAL and A. K. GOYAL
Dispersion of Phonons in Metal Carbides (UC)
(b) Vol. **77**, No. 2, p. 639–643 (1976).
See also *6*, p. *61*; 12.2, p. 112.

J. C. CASTRO, H. C. BASSO, and M. DE SOUZA
Local Mode of H^- Substitutional Ion in KCN
(b) Vol. **77**, No. 2, p. 685–692 (1976).
See also 1.2, p. 19; *6*, p. *61*; 10.2, p. 95; *20.1*, p. *274*.

M. KERTÉSZ, J. KOLLER, and A. AŽMAN
Ab Initio Treatment of a Polysulfur Nitride $(SN)_x$ Chain
(b) Vol. **77**, No. 2, p. K157–K160 (1976).
See also *13.1*, p. *128*.

G. M. ARUTYUNYAN and S. M. SHAHINYAN
Multiphoton Absorption of Intensive Electromagnetic Waves in Semiconductors
(b) Vol. **77**, No. 2, p. K171–K176 (1976).
See also *13*, p. *119*; *20.1*, p. *274*.

J. KADLEC and K. H. GUNDLACH
Internal Photoemission and Quantum-Mechanical Transmission Coefficient
(a) Vol. **37**, No. 2, p. 385–398 (1976).
See also *14.3.3*, p. *193*; *16*, p. *216*.

R. DE RIDDER, G. VAN TENDELOO, J. VAN LANDUYT, D. VAN DYCK, and S. AMELINCKX
Microstructural Study of Copper-Intercalated Niobium Disulphide and Tantalum Disulphide Cu_xNbS_2 and Cu_xTaS_2 (I): Phase Transitions in $Cu_{0.5}NbS_2$ and $Cu_{0.5}TaS_2$
(a) Vol. **37**, No. 2, p. 591–606 (1976).
See also *1.2*, p. *19*; 1.4, p. 28.

K. ZIEGLER and U. BIRKHOLZ
Photoconductivity of Si_2Te_3 Single Crystals
(a) Vol. **37**, No. 2, p. K147–K149 (1976).
See also 13.4, p. 143; *16*, p. *216*.

K.-H. MÜLLER, M. WOLF, and F. KAMMLER
Thermal Switching and Current–Voltage Characteristics for Semiconducting Sandwich Samples
(a) Vol. **37**, No. 2, p. K201–K203 (1976).
See also 8, p. 71; *14.3.3*, p. *194*; 14.3.4, p. 200.

W. Ebeling, W. D. Kraeft, D. Kremp, and K. Kilimann
Coexisting Phases in an Electron-Hole Plasma
(b) Vol. 78, No. 1, p. 241–253 (1976).
See also *13.5*, p. *148*; *13.5.1*, p. *155*.

R. H. Prince
Low-Temperature Diffusion of Electrons in Ice. Evidence for Polaron Bands
(b) Vol. 78, No. 1, p. 271–276 (1976).
See also 6, p. 62; *13.5.3*, p. *159*.

H. J. Stolz, H. Wendel, A. Otto, L. Pintschovius, and H. Kahlert
Acoustic and Optical Phonons in $(SN)_x$
(b) Vol. 78, No. 1, p. 277–286 (1976).
See also 4, p. 47; *6*, p. *63*; *20.1*, p. *276*.

B. Brückel, U. Birkholz, and K. Ziegler
Fundamental Absorption and Franz-Keldysh Effect in Silicon Telluride
(b) Vol. 78, No. 1, p. K23–K25 (1976).
See also 13.1, p. 128; *20.1*, p. *276*.

S. S. Paliwal and O. P. Sinha
Calculation of Activation Energy and Frequency Factor from Glow Curves of SrS:Hf Phosphors
(a) Vol. 38, No. 1, p. 73–76 (1976).
See also *10.2*, p. *97*; 13.4, p. 144; *20.3*, p. *287*.

J. Salik and S. Nadiv
Structural Studies of the Photodissolution of Ag in Amorphous As_2S_3 Films
(a) Vol. 38, No. 1, p. 177–184 (1976).
See also *2*, p. *40*; 11, p. 104.

G. van Tendeloo, R. de Ridder, D. van Dyck, J. van Landuyt, and S. Amelinckx
Microstructural Study of Copper-Intercalated Niobium Disulphide and Tantalum Disulphide Cu_xNbS_2 and Cu_xTaS_2 (II): Order–Disorder Transitions and Incommensurate Superstructures in Cu_xNbS_2 $(0.8 < x < 1)$
(a) Vol. 38, No. 1, p. 185–198 (1976).
See also *1.2*, p. *19*; 1.4, p. 29.

V. A. Gritsenko, W. Leupert, and K. P. Mogilnikov
On the Influence of Illumination on Charges Trapped in MNOS Structures
(a) Vol. 38, No. 1, p. K57–K59 (1976).
See also *14.3.3*, p. *194*.

P. Petelenz
Mixing of Frenkel Excitons and Ionic Excited States of a Linear Molecular Crystal with Two Molecules in the Unit Cell (I): General Model
(b) Vol. 78, No. 2, p. 489–500 (1976).
See also *13.5.1*, p. *156*; 20.1, p. 277.

K. M. van Vliet and A. H. Marshak
Conduction Current and Generalized Einstein Relations for Degenerate Semiconductors and Metals
(b) Vol. 78, No. 2, p. 501–517 (1976).
See also 13.1, p. 129; *14.1*, p. *167*; *14.3*, p. *181*; 15, p. 211; 21, p. 300.

W. John and J. Schreiber
Localization Theory for the Random Weaire Model
(b) Vol. 78, No. 2, p. 577–583 (1976).
See also *2*, p. *41*; *13*, p. *119*.

H. P. Niedermann and M. Wagner
On the Quantum Mechanical Foundation of Classical Shell Models in Lattice Dynamics
(b) Vol. 78, No. 2, p. 615–624 (1976).
See also 4, p. 47; *6*, p. *64*.

R. Pässler
Relationships between the Nonradiative Multiphonon Carrier-Capture Properties of Deep Charged and Neutral Centres in Semiconductors
(b) Vol. 78, No. 2, p. 625–635 (1976).
See also 6, p. 64; 13, p. 119; *13.4*, p. *145*; 22.1, p. 344; 22.2, p. 361; 22.4, p. 378.

K. M. van Vliet
Entropy Production and Choice of Heat Flux for Degenerate Semiconductors and Metals
(b) Vol. 78, No. 2, p. 667–676 (1976).
See also *14.3*, p. *181*; *15*, p. *211*; 21, p. 300.

F. Bechstedt, R. Enderlein, and K. Peuker
Resonance Raman Scattering in a Strong Electric Field
(b) Vol. 78, No. 2, p. 711–719 (1976).
See also 6, p. 64; *14.3.4*, p. *200*; *20.1*, p. *278*.

W. Grevendonk, G. Ruymbeek, P. van den Keybus, and A. de Bock
Faraday Rotation in Non-Crystalline As_2S_3
(b) Vol. 78, No. 2, p. K89–K91 (1976).
See also *2*, p. *41*; 18, p. 224; 20, p. 259.

A. F. Lubchenko and V. N. Pavlovich
Laser Stimulation of Light Interstitial Diffusion in Semiconductors and Insulators
(b) Vol. 78, No. 2, p. K97–K101 (1976).
See also *9*, p. *74*.

A. Conan, M. Zoaeter, and G. Goureaux
Mechanism of Thermal Switching in Bulk $MoTe_{2-x}$ Compounds
(a) Vol. 38, No. 2, p. 505–512 (1976).
See also *14.3.4*, p. *201*.

R. de Ridder, G. van Tendeloo, D. van Dyck, and S. Amelinckx
A Cluster Model for the Transition State and Its Study by Means of Electron Diffraction (I): Theoretical Model
(a) Vol. 38, No. 2, p. 663–674 (1976).
See also *1.4*, p. *29*; 10.2, p. 99.

R. Moret, M. Huber, and R. Comès
Diffuse Scattering and Titanium Short-Range Order in $Ti_{1+x}S_2$
(a) Vol. 38, No. 2, p. 695–700 (1976).
See also *1.2*, p. *20*; 1.3, p. 24.

K. Borggreen and A. R. Thölén
Diffraction Contrast from Misfitting Spherical Particles Precipitated Close Together on a Dislocation (I): Individual Spheres Close Together on a Straight Line
(a) Vol. 38, No. 2, p. 739–752 (1976).
See also *1.4*, p. *30*; 10.2, p. 99.

J. Škácha
Electrical Structure and Transport in Amorphous GeS Layers
(a) Vol. 38, No. 2, p. K143–K146 (1976).
See also *2*, p. *41*; *14.3.1*, p. *186*.

22.1 Elements

Original Papers and Short Notes

R. I. Cottam and G. A. Saunders
An Extension of Keyes' Correlation
(a) Vol. **33**, No. 1, p. 367–373 (1976).
See also *12.2*, p. *110*; 22.2, p. 361; 22.4, p. 376.

K. Hübner
Strain Dependence of the Baldereschi Point and of the Bond Charge
(b) Vol. **73**, No. 2, p. K141–K144 (1976).
See also *4*, p. *45*; *13.1*, p. *122*.

M. J. Tricker, R. K. Thorpe, J. H. Freeman, and G. A. Gard
Conversion Electron Mössbauer Spectroscopy Study of ^{57}Fe Ion Implanted into Single Crystals of Graphite
(a) Vol. **33**, No. 2, p. K97–K100 (1976).
See also *5*, p. *48*.

M. A. Ivanov, V. Ya. Mitrofanov, L. D. Falkovskaya, and A. Ya. Fishman
The Properties of a Threefold Degenerate Orbital Magnetic Impurity in a Ferromagnetic Crystal
(b) Vol. **74**, No. 1, p. 57–67 (1976).
See also 13.4, p. 136; *18.2*, p. *230*.

U. Berg, G. Dräger, and O. Brümmer
Combined Investigations of the Valence Band Structure of Graphite by K X-Ray Emission Spectroscopy and X-Ray Photoemission Spectroscopy
(b) Vol. **74**, No. 1, p. 341–348 (1976).
See also *13.1*, p. *122*; 17, p. 218; 20, p. 257.

G. M. de'Munari, L. Gabba, F. Giusiano, and G. Mambriani
Energy of Vacancy Formation in Semiconductors
(a) Vol. **34**, No. 1, p. 95–105 (1976).
See also *10.2*, p. *90*; 22.1.1, p. 346; 22.1.2, p. 352.

G. M. de'Munari, L. Gabba, F. Giusiano, and G. Mambriani
Migration Energy of Vacancies in Semiconductors
(a) Vol. **34**, No. 2, p. 455–460 (1976).
See also *10.2*, p. *91*; 22.1.1, p. 346; 22.1.2, p. 353.

R. M. Hill
Variable-Range Hopping
(a) Vol. **34**, No. 2, p. 601–613 (1976).
See also *13.4*, p. *137*; *14.3.1*, p. *183*; 22.1.1, p. 346; 22.2.1, p. 362.

A. P. Rusakov and M. V. Anfimov
Non-Central Forces and Stability of the A^NB^{8-N} Tetrahedral Compounds (I): Bond Charges and the Valence-Force-Field Model
(b) Vol. **75**, No. 1, p. 73–77 (1976).
See also *4*, p. *45*; 12.2, p. 110; 22.2, p. 361; 22.4, p. 376; 22.5, p. 389.

V. S. Vavilov, A. E. Kiv, and O. R. Niyazova
Erratum to: The Subthreshold Radiation Effects in Semiconductors
(a) Vol. **36**, No. 1, p. 427 (1976).
See also 9, p. 73; 10.2, p. 94; *11*, p. *102*; 14.3, p. 178; 22.2, p. 361; 22.3, p. 373; 22.4, p. 376.

A. E. LEE and J. WEBSTER
The Soft X-Ray Appearance Potential Spectra of Diamond and Graphite
(b) Vol. 76, No. 2, p. 661—667 (1976).
See also *13.1*, p. *126*; 13.3, p. 133; 20, p. 258.

H. G. MAGUIRE
Investigation of the Band Structure of Diamond Using Low-Energy Electrons
(b) Vol. 76, No. 2, p. 715—726 (1976).
See also *13.1*, p. *126*; 17, p. 219.

T. SOMA
Modified Perturbation Method for Diamond-Type Crystals
(b) Vol. 76, No. 2, p. 753—758 (1976).
See also *6*, p. *59*; 12.2, p. 111; 13.1, p. 126; 22.1.1, p. 348; 22.1.2, p. 355.

P. R. W. HUDSON and P. P. PHAKEY
Observations of Planar Faults or Giant Platelets in Natural Gem Diamond
(a) Vol. 36, No. 2, p. 627—636 (1976).
See also *1.4*, p. *28*.

H. G. WESSJOHANN
Relativistic Excitations of a Thin Graphite Film by Inelastic Electron Scattering in Theory and Experiment
(b) Vol. 77, No. 2, p. 535—542 (1976).
See also *13.5*, p. *147*; 14.4, p. 204.

P. R. W. HUDSON
The Significance of the Frequency Dependence of Phonon Scattering by Platelets in Type I_a Diamond
(a) Vol. 37, No. 2, p. 645—651 (1976).
See also *6*, p. *62*; 8, p. 71.

V. S. ZASIMOV, N. N. LOBANOV, J. RÜDIGER und R. N. KUSMIN
Study of Thermal Diffusion Scattering in Graphite Crystal by the Mössbauer Effect
(a) Vol. 38, No. 1, p. K45—K47 (1976).
See also *5*, p. *51*.

W. WEGENER and I. FRITSCHE
A Study of the Band Structure of Graphite Based on the Rigorous Cellular Method
(b) Vol. 78, No. 2, p. 585—597 (1976).
See also *13.1*, p. *129*.

R. PÄSSLER
Relationships between the Nonradiative Multiphonon Carrier-Capture Properties of Deep Charged and Neutral Centres in Semiconductors
(b) Vol. 78, No. 2, p. 625—635 (1976).
See also 6, p. 64; 13, p. 119; *13.4*, p. *145*; 22, p. 342; 22.2, p. 361; 22.4, p. 378.

A. KOSAKOV, K. UNGER, and G. LEONHARDT
Calculation of Energy-Dependent Photoionization Probabilities for Valence Electrons in Diamond and Silicon Crystals
(b) Vol. 78, No. 2, p. 681—688 (1976).
See also *13.1*, p. *129*; 17, p. 220; 20, p. 259; 22.1.1, p. 349; 22.1.2, p. 358.

22.1.1 Germanium

Review Articles

P. V. Pavlov, E. I. Zorin, D. I. Tetelbaum, and A. F. Khokhlov
Nitrogen as Dopant in Silicon and Germanium
(a) Vol. **35**, No. 1, p. 11–36 (1976).
See also 11, p. 100; *13.4*, p. *134*; *14.3*, p. *173*; 22.1.2, p. 351.

S. V. Tsivinskii
Dislocation Multiplication in Single and Polycrystals in the Initial Stage of Deformation
(a) Vol. **36**, No. 1, p. 11–31 (1976).
See also *10.1*, p. *78*; *10.2*, p. *87*; 21, p. 289; 21.1, p. 302; 21.1.1, p. 316; 21.6, p. 329; 22.2.3, p. 367; 22.5.2, p. 392.

Original Papers and Short Notes

D. Fink and R. Braunstein
Determination of the Conduction Band Effective Masses of the Germanium-Silicon Alloy by Infrared Cyclotron Resonance
(b) Vol. **73**, No. 1, p. 361–370 (1976).
See also *13.1*, p. *121*; 22.1.2, p. 351.

S. V. Tsivinskii
Statistical Theory of Thermally Activated Dislocation Motion in Crystals in the Case of Zero Internal Stresses
(a) Vol. **33**, No. 1, p. 301–311 (1976).
See also *10.2*, p. *88*; 22.1.2, p. 351; 22.2.3, p. 367; 22.5.2, p. 392.

K. L. Chopra and P. Nath
Amorphous Ge Alloy Films (II): Transport Properties
(a) Vol. **33**, No. 1, p. 333–344 (1976).
See also *2*, p. *36*; *14.3.1*, p. *183*; 15, p. 209.

E. M. Trukhanov, E. B. Gorokhov, and S. I. Stenin
Specific Features of the Dislocation Structure of Germanium in the System Germanium–Pyrolitic Silica Film
(a) Vol. **33**, No. 1, p. 435–442 (1976).
See also *1.5*, p. *31*; 2, p. 36; 10.2, p. 88; 22.6, p. 402.

H.-J. Möller and P. Haasen
Thermally Activated Cross Slip of Screw Dislocations in Ge
(a) Vol. **33**, No. 1, p. K59–K62 (1976).
See also *10.2*, p. *88*.

L. Huldt
Phonon-Assisted Auger Recombination in Germanium
(a) Vol. **33**, No. 2, p. 607–614 (1976).
See also *6*, p. *54*; *13.1*, p. *122*; 20, p. 257.

F. Weber
On the Dispersion Relation for Surface Density Waves in Semiconductors
(a) Vol. **33**, No. 2, p. 651–660 (1976).
See also *13.3*, p. *132*; *14.3.2*, p. *187*; 14.3.4, p. 196.

B. V. Novikov, N. S. Sokolov, and S. V. Gastev
Free-Carrier Generation via Exciton–Phonon and Exciton–Impurity Interaction in Ge Crystals
(b) Vol. **74**, No. 1, p. 81–89 (1976).
See also 6, p. 54; *13.4*, p. *136*; *13.5.1*, p. *150*; 16, p. 214.

G. M. de'Munari, L. Gabba, F. Giusiano, and G. Mambriani
Energy of Vacancy Formation in Semiconductors
(a) Vol. **34**, No. 1, p. 95–105 (1976).
See also *10.2*, p. *90*; 22.1, p. 343; 22.1.2, p. 352.

F. Weber
The Influence of Carrier Injection on Helical Density Waves
(a) Vol. **34**, No. 1, p. 107–116 (1976).
See also 14.3.2, p. 187; *14.3.4*, p. *196*.

W. Barth, K. Elsaesser, and W. Güth
The Optical Absorption of 60° Dislocations in Germanium
(a) Vol. **34**, No. 1, p. 153–163 (1976).
See also *13.4*, p. *137*; *20.1*, p. *264*.

P. Nath, D. K. Pandya, and K. L. Chopra
Amorphous Ge Alloy Films (III): Optical Properties
(a) Vol. **34**, No. 1, p. 405–411 (1976).
See also *2*, p. *36*; *20.1*, p. *265*.

S. Popović and M. Šljukić
X-Ray Diffraction Measurements of Unit Cell Parameters of Differently Crystallized Samples
(a) Vol. **34**, No. 1, p. K91–K95 (1976).
See also *1.1*, p. *11*; 22.8, p. 414.

R. M. Westervelt
Nucleation Phenomena in Electron-Hole Drop Formation in Ge and Si (I): Nucleation Rates
(b) Vol. **74**, No. 2, p. 727–739 (1976).
See also *13.5.1*, p. *152*; 22.1.2, p. 353.

V. V. Bhokare and R. M. Singru
Valence Electron Momentum Distribution in Ge
(b) Vol. **74**, No. 2, p. K161–K163 (1976).
See also *13.1*, p. *123*.

G. M. de'Munari, L. Gabba, F. Giusiano, and G. Mambriani
Migration Energy of Vacancies in Semiconductors
(a) Vol. **34**, No. 2, p. 455–460 (1976).
See also *10.2*, p. *91*; 22.1, p. 343; 22.1.2, p. 353.

R. M. Hill
Variable-Range Hopping
(a) Vol. **34**, No. 2, p. 601–613 (1976).
See also *13.4*, p. *137*; *14.3.1*, p. *183*; 22.1, p. 343; 22.2.1, p. 362.

J. Vaitkus
The Nonequilibrium Hall Effect and Related Transport Phenomena in Semiconductors under Inhomogeneous Excitation by a Laser Pulse
(a) Vol. **34**, No. 2, p. 769–775 (1976).
See also *14.3*, p. *176*; *16*, p. *214*; 20.3, p. 282.

P. I. Baranskii, V. V. Kolomoets, and A. V. Fedosov
The Angular Dependences of the Longitudinal Piezoresistance of n-Ge and n-Si at 78 K
(a) Vol. **34**, No. 2, p. K139–K141 (1976).
See also *14.3*, p. *176*; 22.1.2, p. 353.

H. Schaumburg and F. Willmann
Optical Absorption of Plastically Deformed Germanium
(a) Vol. **34**, No. 2, p. K173–K177 (1976).
See also *10.2*, p. *92*; 13.4, p. 138; *20.1*, p. *266*.

G. E. PIKE and C. H. SEAGER
Transport in Microscopically Imhomogeneous Materials
(b) Vol. **75**, No. 1, p. 289–297 (1976).
See also *9*, p. *72*; 14.3, p. 176; 22.1.2, p. 353.

V. M. BABICH, P. I. BARANSKII, L. S. DEMIDENKO, V. A. ILCHISHIN, L. S. KOROLYUK, and A. G. SAMOILOVICH
Impurity Scattering in n-Ge in Quantizing Magnetic Fields
(b) Vol. **75**, No. 1, p. 299–302 (1976).
See also *14.3*, p. *176*.

V. V. MITIN, YU. A. TKHORIK, and YU. M. SHVARTS
Influence of Stress on the Magnetoresistance of p-Germanium Films
(a) Vol. **35**, No. 1, p. K65–K66 (1976).
See also *14.3.1*, p. *184*.

J. L. STAEHLI
Electron-Hole Droplet Formation Kinetics in Pure Germanium
(b) Vol. **75**, No. 2, p. 451–463 (1976).
See also *13.5.1*, p. *152*; 20.3, p. 283.

F. R. KESSLER, J. METZDORF, and R. H. MÜLLER
Interband Faraday Rotation in Germanium. Screening of Excitons by Free Carriers
(b) Vol. **75**, No. 2, p. 567–572 (1976).
See also **13.5.1**, p. **153**; *18*, p. *223*; *20*, p. *258*.

S. MANTOVANI, U. DEL PENNINO, and E. MAZZEGA
Capacitance Measurements as a New Tool to Investigate the Electronic States of Dislocations in Semiconductors
(a) Vol. **35**, No. 2, p. 451–457 (1976).
See also *13.4*, p. *139*; 14.3, p. 177.

D. K. PANDYA and K. L. CHOPRA
Obliquely Deposited Amorphous Ge Films (I): Optical Properties
(a) Vol. **35**, No. 2, p. 725–734 (1976).
See also *2*, p. *38*; *20.1*, p. *269*.

V. N. BYKOV, V. A. TROYAN, G. G. ZDOROVTSEVA, and V. S. KHAIMOVICH
Erratum to: Phase Transformations at Bombardment of Thin Films with Ions
(a) Vol. **35**, No. 2, p. 779 (1976).
See also *1.2*, p. *17*; *1.5*, p. *31*; 11, p. 102; 21, p. 295; 21.1, p. 308; 21.1.1, p. 319.

A. B. GERASIMOV, G. M. DOLIDZE, L. A. MIZANDARI, and A. A. TSERTSVADZE
On the Physical Mechanism of Exoelectron Emission
(a) Vol. **35**, No. 2, p. K131–K134 (1976).
See also *17*, p. *219*.

R. M. WESTERVELT
Nucleation Phenomena in Electron-Hole Drop Formation in Ge and Si (II): Application to Observable Phenomena
(b) Vol. **76**, No. 1, p. 31–43 (1976).
See also *13.5.1*, p. *153*; 22.1.2, p. 354.

W. KÜHN
Stress Dependence of the (E_1, $E_1 + \Delta_1$) Transition in Germanium
(b) Vol. **76**, No. 1, p. 289–299 (1976).
See also **12.2**, p. 110; *13.1*, p. *125*; *20.1*, p. *270*.

P. THOMAS
Addendum to: On Anisotropic Hopping Conduction in Thin Amorphous Germanium Films
(b) Vol. **76**, No. 1, p. 439 (1976).
See also 1.5, p. 31; *14.3.1*, p. *184*.

K. L. CHOPRA and D. K. PANDYA
Obliquely Deposited Amorphous Ge Films (II): Electrical Properties
(a) Vol. **36**, No. 1, p. 89—100 (1976).
See also *2*, p. *39*; *14.3.1*, p. *184*; 15, p. 210.

M. WENDT, TH. KRAJEWSKI, and R. BIMBERG
Detection of Thin Surface Films by Electron Beam Microanalysis. A Comparison between Wavelength-Dispersive and Energy-Dispersive Microprobes
(a) Vol. **36**, No. 1, p. 253—261 (1976).
See also *1.5*, p. *32*; 21.1, p. 309; 21.1.1, p. 319.

Y. KAMIURA and F. HASHIMOTO
Annealing Stage above 400 ° C in Quenched Gallium-Doped Germanium
(a) Vol. **36**, No. 1, p. K81—K84 (1976).
See also 9, p. 73; *10.2*, p. *94*; 14.3, p. 178.

H. DÜSTERHÖFT, R. MANNS, and D. HILDEBRANDT
Experimental Proof of a Theoretical Relation for the Probability of Positive Ion Excitation by Bombardment of Solid Surfaces with 12 keV Ar^+ Ions
(a) Vol. **36**, No. 1, p. K93—K97 (1976).
See also *11*, p. *102*; 21, p. 296; 21.1, p. 310; 21.1.1, p. 320; 22.1.2, p. 355; 22.6, p. 405.

T. SOMA
Modified Perturbation Method for Diamond-Type Crystals
(b) Vol. 76, No. 2, p. 753—758 (1976).
See also *6*, p. *59*; 12.2, p. 111; 13.1, p. 126; 22.1, p. 344; 22.1.2, p. 355.

S. K. BARTHWAL and K. L. CHOPRA
Thermoelectric Power of Amorphous Germanium Films
(a) Vol. **36**, No. 2, p. 533—549 (1976).
See also 2, p. 39; *14.3.1*, p. *184*; *15*, p. *210*.

R. H. WADE
Tilted-Beam Electron Microscopy of Amorphous Films
(a) Vol. **37**, No. 1, p. 247—256 (1976).
See also *1.5*, p. *32*; *2*, p. *40*.

H. S. RANDHAWA, L. K. MALHOTRA, H. K. SEHGAL, and K. L. CHOPRA
Structural Transformations in a-Ge Alloy Films
(a) Vol. **37**, No. 1, p. 313—320 (1976).
See also *1.2*, p. *18*; 1.5, p. 32; *2*, p. *40*.

L. N. ALEKSANDROV and R. N. LOVYAGIN
Epitaxial Germanium–Silicon Structures Obtained in Ultrahigh Vaccuum
(a) Vol. **37**, No. 1, p. 341—352 (1976).
See also *1.5*, p. *32*; 3, p. 43; 22.1.2, p. 356.

D. KLÄHN and K. KILIMANN
The Damping of Excitons by Phonons
(b) Vol. 77, No. 2, p. 483—493 (1976).
See also 6, p. 61; *13.5.1*, p. *155*.

R. PICKENHAIN and A. MILCHEV
The Quadratic Response of a Fermi Gas. Electronic Charge Density of Valence Electrons in Zincblende Semiconductors
(b) Vol. 77, No. 2, p. 571—579 (1976).
See also *4*, p. *46*; 13, p. 119.

J. HUMLÍČEK and F. LUKEŠ
Direct Identification of the $L_3' \rightarrow L_1$ Transition in Ge
(b) Vol. 77, No. 2, p. 731—738 (1976).
See also 13.1, p. 127; *14.3.4*, p. *199*; *20.1*, p. *274*.

J. H. YEE and G. MYERS
Pseudopotential Calculation of the Acoustic Deformation Potential for Several Group IV, II–VI, and III–V Compounds
(b) Vol. 77, No. 2, p. K81—K84 (1976).
See also *13.1*, p. *128*; 22.1.2, p. 356; 22.2, p. 361; 22.4, p. 377.

M. UMENO
Six-Beam Anomalous Transmission of Ge Crystals (I): Anomalous Transmission Pattern
(a) Vol. 37, No. 2, p. 561—570 (1976).
See also *1.1*, p. *13*.

B. POHORYLES
Damping of Current Oscillations in Compensated Germanium by a Weak Magnetic Field
(a) Vol. 37, No. 2, p. K117—K119 (1976).
See also *14.3.4*, p. *200*.

L. KUBLER, G. GEWINNER, J. J. KOULMANN, and A. JAÉGLÉ
EPR Study on Amorphous Germanium. Oxygen, Deposition Angle, Annealing, and Substrate Temperature Effects
(b) Vol. 78, No. 1, p. 149—158 (1976).
See also 2, p. 40; *19*, p. *255*.

I. NEIZVESTNY, N. OVSYUK, and M. SINYUKOV
Electro-Reflectance Spectra of E_1, $E_1 + \Delta_1$ Transitions in Germanium
(b) Vol. 78, No. 1, p. K19—K22 (1976).
See also 13.1, p. 128; *14.3.4*, p. *200*; *20.1*, p. *276*.

R. GAUTHIER and P. PINARD
Pulsed Laser Light Stimulated Desorption
(a) Vol. 38, No. 1, p. 85—92 (1976).
See also *17*, p. *220*; 21.1, p. 313; 22.5.2, p. 397.

U. DEL PENNINO and S. MANTOVANI
Width of the Dislocation Energy Band in Germanium
(a) Vol. 38, No. 1, p. 109—113 (1976).
See also *10.2*, p. *97*; *13.1*, p. *128*; 14.3.4, p. 200.

A. M. MISHCHENKO, I. G. NEIZVESTNYI, and M. P. SINYUKOV
Spectrum of the Photocapacitance Current near the Fundamental Absorption Edge of Ge
(a) Vol. 38, No. 1, p. 349—354 (1976).
See also *16*, p. *217*; *20.1*, p. *277*.

K. S. BOBEV, N. I. DIAKONOVA, and N. V. MILESHKINA
Field Emission through Semiconductor and Dielectric Layers
(a) Vol. 38, No. 1, p. 393—399 (1976).
See also *14.3.4*, p. *200*; 22.6, p. 408.

A. KOSAKOV, K. UNGER, and G. LEONHARDT
Calculation of Energy-Dependent Photoionization Probabilities for Valence Electrons in Diamond and Silicon Crystals
(b) Vol. 78, No. 2, p. 681—688 (1976).
See also *13.1*, p. *129*; 17, p. 220; 20, p. 259; 22.1, p. 344; 22.1.2, p. 358.

V. DIENYS and Ž. KANCLERIS
Influence of Uniaxial Stress and Electron–Electron Scattering on the Phenomenological Conductivity Relaxation Time in n-Type Germanium
(b) Vol. 78, No. 2, p. 757—765 (1976).
See also 12.2, p. 112; *14.3.4*, p. *200*.

K. SUZUKI
Piezo-Thermal Conductivity of n-Ge
(b) Vol. 78, No. 2, p. K77—K79 (1976).
See also *8*, p. *71*.

V. M. Babich, P. I. Baranskii, L. S. Demidenko. V. A. Ilchishin, and S. L. Korolyuk
Anisotropy of Magnetoresistance of n-Ge in Transverse Quantizing Magnetic Fields under Conditions of Mixed Scattering
(b) Vol. 78, No. 2, p. K111–K114 (1976).
See also *14.3*, p. *182*.

T. Shishido and M. Tanaka
Determination of Structure Factors of Germanium by the Critical-Voltage and Convergent-Beam Diffraction Methods
(a) Vol. 38, No. 2, p. 453–461 (1976).
See also *1.1*, p. *13*; 1.4, p. 29.

R. Gauthier and C. Guittard
Mechanism Investigations of a Pulsed Laser Light Induced Desorption
(a) Vol. 38, No. 2, p. 477–486 (1976).
See also *17*, p. *220*; 21.1, p. 314; 22.5.2, p. 398.

M. Umeno
Six-Beam Anomalous Transitions of Ge Crystals (II): Interpretations from Eigensolutions
(a) Vol. 38, No. 2, p. 701–711 (1976).
See also *1.1*, p. *13*.

22.1.2 Silicon

Review Article

P. V. Pavlov, E. I. Zorin, D. I. Tetelbaum, and A. F. Khokhlov
Nitrogen as Dopant in Silicon and Germanium
(a) Vol. **35**, No. 1, p. 11–36 (1976).
See also 11, p. 100; *13.4*, p. *134*; *14.3*, p. *173*; 22.1.1, p. 345.

Original Papers and Short Notes

J. Hansen-Schmidt
Investigations of the Anomalous Absorption of Electrons in Si near the Zone Axes
(b) Vol. **73**, No. 1, p. 117–130 (1976).
See also *1.1*, p. *10*.

D. Fink and R. Braunstein
Determination of the Conduction Band Effective Masses of the Germanium–Silicon Alloy by Infrared Cyclotron Resonance
(b) Vol. **73**, No. 1, p. 361–370 (1976).
See also *13.1*, p. *121*; 22.1.1, p. 345.

R. B. Hammond, T. C. McGill, and J. W. Mayer
Properties of the Electron–Hole Liquid Luminescence in Si Double Injection Diodes
(a) Vol. **33**, No. 1, p. 59–66 (1976).
See also 13.5.1, p. 149; 14.3.4, p. 195; *20.3*, p. *280*.

M. Leppihalme and T. Tuomi
Photovoltage Spectra of Metal–Silicon (p-Type) Diodes near the Indirect Absorption Edge of Silicon
(a) Vol. **33**, No. 1, p. 125–133 (1976).
See also 6, p. 53; 13.5.1, p. 149; *14.3.3*, p. *190*; *16*, p. *213*; 20.1, p. 261.

E. N. Kislovskii, L. I. Datsenko, and A. S. Vasilkovskii
Effect of Crystal Net Plane Curvature on X-Ray Intensity in the Case of Laue Diffraction
(a) Vol. **33**, No. 1, p. 275–280 (1976).
See also *1.1*, p. *10*; *10.2*, p. *87*.

S. V. Tsivinskii
Statistical Theory of Thermally Activated Dislocation Motion in Crystals in the Case of Zero Internal Stresses
(a) Vol. **33**, No. 1, p. 301–311 (1976).
See also *10.2*, p. *88*; 22.1.1, p. 345; 22.2.3, p. 367; 22.5.2, p. 392.

E. Nes
On the Nucleation of Precipitate Colonies from Swirl Defects in Silicon
(a) Vol. **33**, No. 1, p. K5–K8 (1976).
See also *10.2*, p. *88*.

J. Hansen-Schmidt
Application of Group-Theoretical Methods to the Solution of the Equations of Electron Diffraction in Single Crystals (Si)
(b) Vol. **73**, No. 2, p. 461–467 (1976).
See also *1.1*, p. *10*.

J. von Borzeszkowski
On the Electric Field Dependence of the Conductivity of Holes in Si at Low Temperatures (I): The Drift Velocity under Consideration of the Non-Equipartition of Acoustic Phonon Energy
(b) Vol. **73**, No. 2, p. 607–614 (1976).
See also 6, p. 54; 13.1, p. 121; *14.3.4*, p. *195*.

G. Dlubek and O. Brümmer
A Comparison of Positron Annihilation and Compton Effect Measurements on Si Single Crystals
(b) Vol. **73**, No. 2, p. K107–K110 (1976).
See also *13*, p. *115*; *20*, p. *257*.

O. Brümmer, H. R. Höche, and J. Nieber
Influence of the Total Reflection on the X-Ray Diffraction in the Strongly Asymmetric Bragg Case
(a) Vol. **33**, No. 2, p. 587–593 (1976).
See also *1.1*, p. *11*.

H. Nakata and H. Morishima
Dose Dependence of ESR Spectra for On-Channel-Irradiated Silicon
(a) Vol. **33**, No. 2, p. 633–638 (1976).
See also 11, p. 100; *19*, p. *248*.

S. Mader and A. E. Michel
Dislocation Reactions in Arsenic-Implanted and Annealed Silicon
(a) Vol. **33**, No. 2, p. 793–805 (1976).
See also *10.2*, p. *89*; 11, p. 100.

G. Filscher
On the Theoretical Calculation of the Intensity Distribution in X-Ray Section Topographs of Crystals with Superposed Films
(a) Vol. **33**, No. 2, p. K81–K82 (1976).
See also *1.6*, p. *34*; 10.2, p. 89.

K. D. Glinchuk, N. M. Litovchenko, and R. Merker
The Role of Impurities in the Formation of Quenched-in Recombination Centres in Thermally Treated Silicon
(a) Vol. **33**, No. 2, p. K87–K90 (1976).
See also *13.4*, p. *136*; 16, p. 213.

G. M. de'Munari, L. Gabba, F. Giusiano, and G. Mambriani
Energy of Vacancy Formation in Semiconductors
(a) Vol. **34**, No. 1, p. 95–105 (1976).
See also *10.2*, p. *91*; 22.1, p. 343; 22.1.1, p. 346.

G. Blasquez
Study of Conduction and Low-Frequency Noise Phenomena Associated with Surface Regions in p–n Junctions. Application to Junction Transistors
(a) Vol. **34**, No. 1, p. 207–218 (1976).
See also *14.3.2*, p. *187*.

M. Manzel, W. Ecke, and H. Bruchlos
Negative Differential Resistance in Gold-Doped High-Resistivity Silicon
(a) Vol. **34**, No. 1, p. 219–226 (1976).
See also 13.4, p. 137; 14.3.3, p. 190; *14.3.4*, p. *196*.

S. G. Chamberlain, B. Y. Woo, and D. B. Scott
Modelling of Electrical Charge Injection into Charge-Coupled Devices
(a) Vol. **34**, No. 1, p. 263–275 (1976).
See also *14.3.2*, p. *187*; *14.3.3*, p. *190*; 14.3.4, p. 196.

P. M. Karageorgy-Alkalaev, I. Z. Karimova, P. I. Knigin, and A. Yu. Leiderman
Double-Injection Currents in Zn-Compensated Silicon p–i–n Structures
(a) Vol. **34**, No. 1, p. 391–395 (1976).
See also 13.4, p. 137; *14.3.4*, p. *196*.

V. A. Girii, A. V. Kondrachuk, S. I. Kornyushin, V. G. Litovchenko, R. O. Litvinov, and V. I. Shakhovtsov
Effect of Impurities on the Radiation Sensitivity of the Si–SiO_2 Interface in MIS Structures
(a) Vol. **34**, No. 1, p. K35–K38 (1976).
See also **11**, p. 101; *13.3*, p. *132*; *14.3.3*, p. *191*.

R. M. Westervelt
Nucleation Phenomena in Electron–Hole Drop Formation in Ge and Si (I): Nucleation Rates
(b) Vol. **74**, No. 2, p. 727–739 (1976).
See also *13.5.1*, p. *152*; 22.1.1, p. 346.

G. M. de'Munari, L. Gabba, F. Giusiano, and G. Mambriani
Migration Energy of Vacancies in Semiconductors
(a) Vol. **34**, No. 2, p. 455–460 (1976).
See also *10.2*, p. *91*; 22.1, p. 343; 22.1.1, p. 346.

E. W. Kreutz
X-Radiation Effects in Bulk Si at Subthreshold Energies
(a) Vol. **34**, No. 2, p. 489–495 (1976).
See also **11**, p. 101; *14.3*, p. *175*.

K. Godwod, A. T. Nagy, and Z. Rek
The Application of the X-Ray Triple-Crystal Spectrometer for Measuring the Radius of Curvature of Bent Single Crystals
(a) Vol. **34**, No. 2, p. 705–710 (1976).
See also *1.1*, p. *11*; 10.2, p. 91.

T. Warminski
On the Detectability of Impurity Atoms in a Thin Surface Layer by an Electron Microprobe Analyser
(a) Vol. **34**, No. 2, p. K119–K123 (1976).
See also *10.2*, p. *91*.

P. I. Baranskii, V. V. Kolomoets, and A. V. Fedosov
The Angular Dependences of the Longitudinal Piezoresistance of n-Ge and n-Si at 78 K
(a) Vol. **34**, No. 2, p. K139–K141 (1976).
See also *14.3*, p. *176*; 22.1.1, p. 346.

G. E. Pike and C. H. Seager
Transport in Microscopically Inhomogeneous Materials
(b) Vol. **75**, No. 1. p 289–297 (1976).
See also *9*, p. *72*; 14.3, p. 176; 22.1.1, p. 347.

R. Bubáková and Z. Šourek
Contrast in the Dislocation Images for the Bragg Case
(a) Vol. **35**, No. 1, p. 55–60 (1976).
See also *10.2*, p. *92*.

A. Prasad, R. Beserman, P. Germain, and J. C. Bourgoin
Kinetics of Recrystallization in Ion-Implanted Silicon Layers
(a) Vol. **35**, No. 1, p. 109–112 (1976).
See also *3*, p. *42*.

C. Capiluppi, P. G. Merli, G. Pozzi, and I. Vecchi
Out-of-Focus Observations of p–n Junctions by High-Voltage Electron Microscopy
(a) Vol. **35**, No. 1, p. 165–172 (1976).
See also *14.3.3*, p. *191*.

A. V. Ferris-Prabhu
Tunnelling Theories of Non-Volatile Semiconductor Memories
(a) Vol. **35**, No. 1, p. 243–250 (1976).
See also **13.4**, p. 138; *14.3.3*, p. *191*.

M. LEFELD-SOSNOWSKA and E. ZIELIŃSKA-ROHOZIŃSKA
Symmetrical and Asymmetrical X-Ray Section Topographs of Ion-Implanted Silicon
(a) Vol. **35**, No. 1, p. K1–K3 (1976).
See also *1.4*, p. *27*; 10.2, p. 92.

T. WARMINSKII, E. MIZERA, M. KLIMKIEWICZ, and M. IIYAMA
On the Energy Dependence of K_αSi X-Ray Production in Silicon by Electron Bombardment
(a) Vol. **35**, No. 1, p. K9–K11 (1976).
See also *10.2*, p. *92*.

V. N. HOLYAVKO, V. P. DRAGUNOV, B. V. MOROZOV, E. M. SKOK, and N. B. VELCHEV
Study of the Size Quantization in p-Type Silicon Inversion Layers by Means of Magnetoresistance
(b) Vol. **75**, No. 2, p. 423–432 (1976).
See also 13.1, p. 124; *14.3*, p. *177*.

M. PUGNET, J. BARBOLLA, J. C. BRABANT, F. SAINT-YVES, and M. BROUSSEAU
Energy Levels for Platinum and Palladium in Silicon Measured by the Dark Transient Capacitance Technique
(a) Vol. **35**, No. 2, p. 533–543 (1976).
See also *13.4*, p. *139*; *14.3.3*, p. *191*.

H. MENNIGER, H. RAIDT, and G. VOIGT
On the Influence of Microdefects on Charge Carriers in Silicon
(a) Vol. **35**, No. 2, p. 639–643 (1976).
See also *1.4*, p. *27*; 14.3, p. 177.

K. SOMOGYI
Relation between the Activation Energies (ε_1, E_1) and the Temperature of Carrier Mobility Maximum in Semiconductors. A New Method for Determination of E_1
(a) Vol. **35**, No. 2, p. 659–666 (1976).
See also 13.4, p. 140; *14.3*, p. *178*; 22.2.2, p. 365.

Y. KANDA and M. MIGITAKA
Effect of Mechanical Stress on the Offset Voltages of Hall Devices in Si IC
(a) Vol. **35**, No. 2, p. K115–K118 (1976).
See also *14.3.3*, p. *192*.

M. S. YUNUSOV, M. A. ZAIKOVSKAYA, B. L. OKSENGENDLER, and K. R. TOKHIROV
Subthreshold Defect Production in Silicon
(a) Vol. **35**, No. 2, p. K145–K149 (1976).
See also *11*, p. *102*.

K. D. GLINCHUK, N. M. LITOVCHENKO, and R. MERKER
Effect of Quenching Rate and Annealing on the Concentration of Quenched-in Recombination Centres in Heat Treated Silicon
(a) Vol. **35**, No. 2, p. K157–K160 (1976).
See also *13.4*, p. *140*; 16, p. 214.

R. M. WESTERVELT
Nucleation Phenomena in Electron–Hole Drop Formation in Ge and Si (II): Application to Observable Phenomena
(b) Vol. **76**, No. 1, p. 31–43 (1976).
See also *13.5.1*, p. *153*; 22.1.1, p. 347.

K. I. PAPUNASHVILI and E. K. FROLOVA
Longitudinal Anisotropy of Conductivity of n-Si in Heating Electric Fields at Low Temperatures
(b) Vol. **76**, No. 1, p. 351–355 (1976).
See also *13*, p. *117*; *14.3.4*, p. *197*.

V. V. Mitin and E. I. Tolpygo
Piezoresistance of Uniaxially Deformed n-Si
(b) Vol. **76**, No. 1, p. 407–411 (1976).
See also **13.1**, p. 125; *14.3*, p. *178*.

P. V. Pavlov, T. A. Kruze, D. I. Tetelbaum, E. I. Zorin, E. V. Shitova, and N. V. Gudkova
Electron Microscopic Studies of Silicon Layers Irradiated with High Doses of Nitrogen Ions
(a) Vol. **36**, No. 1, p. 81–88 (1976).
See also *1.5*, p. *31*; 2, p. 39; *11*, p. *102*.

W. H. Hubrig, J. Auleytner, and M. Maciaszek
Changes of X-Ray Topographic Contrast Due to Annealing of Boron-Implanted Silicon
(a) Vol. **36**, No. 1, p. 209–215 (1976).
See also *1.1*, p. *12*; 10.2, p. 93.

R. Hatert, R. Sinon, and F. van de Wiele
Experimental and Theoretical Study of Buried Channel MOS Structures
(a) Vol. **36**, No. 1, p. 235–246 (1976).
See also *14.3.2*, p. *188*; 14.3.3, p. 192.

V. Koos and H.-G. Neumann
Effect of Heat Treatment on Stress in Amorphous Silicon Films
(a) Vol. **36**, No. 1, p. K47–K49 (1976).
See also *1.5*, p. *32*; *12.2*, p. *111*.

D. Roy Chaudhuri, P. K. Basu, and A. N. Chakravarti
Effect of the Surface Electric Field on the Specific Heat of Semiconductor Inversion Layers
(a) Vol. **36**, No. 1, p. K65–K69 (1976).
See also *8*, p. *69*; 14.3.4, p. 198.

B. I. Fomin, A. E. Gershinskii, E. I. Cherepov, and F. L. Edelman
Investigation of Phase Growth Kinetics in the System of Si Single Crystals and V Thin Films
(a) Vol. **36**, No. 1, p. K89–K91 (1976).
See also *3*, p. *43*; 21.1, p. 310.

H. Düsterhöft, R. Manns, and D. Hildebrandt
Experimental Proof of a Theoretical Relation for the Probability of Positive Ion Excitation by Bombardment of Solid Surfaces with 12 keV Ar^+ Ions
(a) Vol. **36**, No. 1, p. K93–K97 (1976).
See also *11*, p. *102*; 21, p. 296; 21.1, p. 310; 21.1.1, p. 320; 22.1.1, p. 348; 22.6, p. 405.

T. Soma
Modified Perturbation Method for Diamond-Type Crystals
(b) Vol. **76**, No. 2, p. 753–758 (1976).
See also *6*, p. *59*; 12.2, p. 111; 13.1, p. 126; 22.1, p. 344; 22.1.1, p. 348.

V. V. Kaplin, D. E. Popov, and S. A. Vorobev
Bound States of Fast Electrons in Planar Channeling
(b) Vol. **76**, No. 2, p. 779–786 (1976).
See also *11*, p. *103*; 13, p. 118.

J. Barbolla, M. Pugnet, J. C. Brabant, and M. Brousseau
Thermal Capture Cross-Section of Free Electrons at Neutral Gold Centres in n-Type Silicon
(a) Vol. **36**, No. 2, p. 495–498 (1976).
See also *13.4*, p. *141*; 14.3.3, p. 192.

T. Bedyńska, R. Bubáková, and Z. Šourek
Comparison between Experimental and Theoretical Dislocation Image for the Bragg Case
(a) Vol. **36**, No. 2, p. 509–516 (1976).
See also *10.2*, p. *94*.

K. SAMINADAYAR and J. C. PFISTER
Surface-State Generation on MOS Capacitors Irradiated with UV Light and Electrons
(a) Vol. **36**, No. 2, p. 679—686 (1976).
See also *13.3*, p. *133*; *14.3.3*, p. *192*.

S. ETIENNE, C. MAI, G. FANTOZZI and P. F. GOBIN
Anelastic Phenomena in Silicon at Low Temperatures
(a) Vol. **36**, No. 2, p. K113—K118 (1976).
See also *10.2*, p. *95*.

U. VOLAND, P. DEUS, and H. A. SCHNEIDER
Debye Temperature of Polycrystalline Silicon
(a) Vol. **36**, No. 2, p. K165—K169 (1976).
See also *6*, p. *60*; *8*, p. *70*.

J. SAWICKI, B. SAWICKA, J. STANEK, and J. KOWALSKI
Mössbauer Study of Heat Treated ^{57}Fe Implanted Silicon
(b) Vol. **77**, No. 1, p. K1—K4 (1976).
See also *5*, p. *50*.

I. N. MAGDEN, L. I. KATURINA, B. L. MEILER, and P. V. MUSIENKO
[100] Silicon Doped with Arsenic (I): Electrical Parameters and Their Thermal Stability
(a) Vol. **37**, No. 1, p. 29—38 (1976).
See also 10.2, p. 95; *14.3*, p. *179*.

N. P. MOROZOV, D. I. TETELBAUM, P. V. PAVLOV, and E. I. ZORIN
The Calculation of Secondary Defect Formation at Ion Implantation of Silicon
(a) Vol. **37**, No. 1, p. 57—64 (1976).
See also *11*, p. *103*.

L. N. ALEKSANDROV and R. N. LOVYAGIN
Epitaxial Germanium-Silicon Structures Obtained in Ultrahigh Vaccuum
(a) Vol. **37**, No. 1, p. 341—352 (1976).
See also *1.5*, p. *32*; 3, p. 43; 22.1.1, p. 348.

T. WOSIŃSKI, T. FIGIELSKI, and A. MĄKOSA
Spin-Dependent Photoconductivity Spectrum of Dislocated Silicon
(a) Vol. **37**, No. 1, p. K57—K59 (1976).
See also *10.2*, p. *95*; *16*, p. *216*; 19, p. 254.

J. H. YEE and G. MYERS
Pseudopotential Calculation of the Acoustic Deformation Potential for Several Group IV, II–VI, and III–V Compounds
(b) Vol. 77, No. 2, p. K81—K84 (1976).
See also *13.1*, p. *128*; 22.1.1, p. 349; 22.2, p. 361; 22.4, p. 377.

J. HÄRTWIG
X-Ray Diffraction in an Extremely Asymmetric Laue Case
(a) Vol. **37**, No. 2, p. 417—425 (1976).
See also *1.1*, p. *13*.

PHAM MINH TAN
Determination of Ion Implantation Profiles from Measurements of the Differential Capacity at the Semiconductor–Electrolyte Contact
(a) Vol. **37**, No. 2, p. 439—444 (1976).
See also 11, p. 104; *13.4*, p. *143*; *14.3.3*, p. *193*.

M. MANZEL and W. ECKE
The Influence of Trapped Space Charge on the Double-Injection Switching
(a) Vol. **37**, No. 2, p. 479—484 (1976).
See also 13.4, p. 143; *14.3.4*, p. *199*.

O. BRÜMMER, H. R. HÖCHE, and J. NIEBER
Study of Extremely Asymmetric X-Ray Diffraction in the Bragg Case of Grazing Incidence
(a) Vol. **37**, No. 2, p. 529–532 (1976).
See also *1.1*, p. *13*.

H. FLIETNER and NGO DUONG SINH
Character of Si–SiO_2 Interface States from Analysis of the CV Term Spectra
(a) Vol. **37**, No. 2, p. 533–539 (1976).
See also *13.3*, p. *133*; 14.3.2, p. 188.

P. BASU and H. SAHA
Possible Performance of Back-Wall Schottky Barrier Solar Cells
(a) Vol. **37**, No. 2, p. 625–631 (1976).
See also 13.3, p. 133; *14.3.3*, p. *193*; *16*, p. *216*; 22.2.1, p. 364; 22.4.3, p. 386.

U. A. SHARIKOV and M. S. YUNUSOV
Influence of X-Irradiation on Silicon Schottky Diodes
(a) Vol. **37**, No. 2, p. 681–686 (1976).
See also 9, p. 74; *11*, p. *104*; *14.3.2*, p. *188*.

J. KIES, R. KÖHLER, and W. MÖHLING
Double Crystal X-Ray Topographs of Microdefects in Silicon
(a) Vol. **37**, No. 2, p. K113–K115 (1976).
See also *1.4*, p. *28*; 10.2, p. 97.

T. SOMA and T. UMENAI
Madelung Energies of Tetragonal and Orthorhombic Systems
(b) Vol. 78, No. 1, p. 229–240 (1976).
See also *1.1*, p. *13*; 1.2, p. 19.

I. N. MAGDEN and V. A. SAFONOV
[100] Silicon Doped with Arsenic (II): The Connection of Electrical and Structural Parameters with Microplasmas
(a) Vol. **38**, No. 1, p. 45–50 (1976).
See also 10.2, p. 97; *14.3*, p. *180*.

R. HATERT, P. COUVREUR, R. SINON, and F. VAN DE WIELE
Avalanche Multiplication Factor and Reverse Current of Si p–n Junctions
(a) Vol. **38**, No. 1, p. 123–129 (1976).
See also *14.3.2*, p. *188*; 14.3.3, p. 194.

V. S. LYSENKO, A. V. SACHENKO, A. P. DUBCHAK, A. N. NAZAROV, V. T. ATAMANENKO, M. P. LOKSHIN, and YU. YA. POLISHUK
Properties of MOS Structures after B^+ Ion Bombardment
(a) Vol. **38**, No. 1, p. 131–138 (1976).
See also 11, p. 104; *14.3.3*, p. *194*.

A. GOLAŃSKI, A. FIDERKIEWICZ, H. RZEWUSKI, M. LEFELD-SOSNOWSKA, J. GRONKOWSKI, R. GRÖTZSCHEL, U. KREISSIG, and H. BARTSCH
Particularities of Crystalline to Amorphous State Conversion in Silicon Heavily Damaged by 140 keV Si^{++} Ions
(a) Vol. **38**, No. 1, p. 139–149 (1976).
See also *1.2*, p. *19*; *11*, p. *104*.

Y. OCHIAI and E. MATSUURA
ESR in Heavily Doped n-Type Silicon near a Metal–Nonmetal Transition
(a) Vol. **38**, No. 1, p. 243–252 (1976).
See also 13.4, p. 144; *19*, p. *255*.

L. J. Cheng, M. Süli, J. P. Kowall, P. W. Wallace, and J. W. Corbett
Effects of Radiation on Gold Diffusion in Silicon
(a) Vol. **38**, No. 1, p. 321–330 (1976).
See also *9*, p. *74*; 11, p. 104.

A. E. Gershinskii, L. V. Mironova, and E. I. Cherepov
SiO_2–Si (n-Type) Interface Investigation with the Electrochemical Method
(a) Vol. **38**, No. 1, p. 369–373 (1976).
See also *1.5*, p. *33*; *14.3.2*, p. *188*; 22.6, p. 408.

Y. Kanda and M. Migitaka
Design Consideration for Hall Devices in Si IC
(a) Vol. **38**, No. 1, p. K41–K44 (1976).
See also *14.3.3*, p. *194*.

D. R. Choudhuri, P. K. Basu, and A. N. Chakravarti
On the Modification of the Einstein Relation for Semiconductor Inversion Layers
(a) Vol. **38**, No. 1, p. K85–K88 (1976).
See also *14.3.2*, p. *189*.

A. Kosakov, K. Unger, and G. Leonhardt
Calculation of Energy-Dependent Photoionization Probabilities for Valence Electrons in Diamond and Silicon Crystals
(b) Vol. 78, No. 2, p. 681–688 (1976).
See also *13.1*, p. *129*; 17, p. 220; 20, p. 259; 22.1, p. 344; 22.1.1, p. 349.

P. Fahrentholtz and J. Mimkes
Diffusion of Sb along Dislocations in Si
(b) Vol. 78, No. 2, p. K137–K140 (1976).
See also *9*, p. *74*; 10.2, p. 98.

A. I. Koifman, A. N. Narkulov, B. L. Oksengendler, and M. S. Yunusov
The Low-Temperature Doping of Silicon by Radiation-Induced Diffusion
(a) Vol. **38**, No. 2, p. 439–446 (1976).
See also *9*, p. *74*; *11*, p. *105*; 14.3.2, p. 189.

B. Joseph
The Electron Injection from Si into SiO_2 Due to a Mobile Ion Induced Internal Field
(a) Vol. **38**, No. 2, p. K127–K130 (1976).
See also *14.3.3*, p. *194*; *14.3.4*, p. *201*; 22.6, p. 408.

22.1.3 Group VI Elements

Original Papers and Short Notes

M. Fritsche, M. Pöhler, M. Schubert, and G. Staupendahl
Investigation of the Absorption of Intense CO_2 Laser Radiation by the p-Band of Tellurium
(a) Vol. **33**, No. 1, p. 149–154 (1976).
See also *13.1*, p. *121*; *20.1*, p. *261*.

W. Fuhs, M. Milleville, and J. Niklas
Acoustoelectric Current Saturation and Sound Velocity in Tellurium Crystals under Hydrostatic Pressure
(b) Vol. **73**, No. 2, p. K127–K129 (1976).
See also *7*, p. *65*; *14.3.4*, p. *195*.

C. Albers and H. Gertig
Magnetoresistance in Thin Tellurium Films
(a) Vol. **34**, No. 2, p. K125–K126 (1976).
See also *14.3.1*, p. *183*.

G. Juška and S. Vengris
Hole Photogeneration in Amorphous Selenium at Low Electric Fields
(a) Vol. **35**, No. 1, p. 339–345 (1976).
See also *14.3.2*, p. *188*; *16*, p. *214*.

M. Janda and A. Kubový
Growth of Vacuum-Deposited Tellurium Films on Glass Substrates and Some of Their Transport Properties
(a) Vol. **35**, No. 1, p. 391–402 (1976).
See also *1.5*, p. *31*; *3*, p. *42*; 14.3.1, p. 183.

I. I. Farbhstein, L. S. Dubinskaya, and V. G. Krigel
Resonance Scattering and Piezotransport of Pure Tellurium at Low Temperature
(b) Vol. **75**, No. 2, p. K133–K136 (1976).
See also *14.3*, p. *177*.

A. Kubový and M. Janda
Transport Properties of Tellurium Thin Films and Their Dependence on Thickness
(a) Vol. **35**, No. 2, p. 471–476 (1976).
See also *14.3.1*, p. *184*.

A. Kubový and M. Janda
Temperature Dependence of Hole Mobilities in Te Thin Films
(a) Vol. **36**, No. 1, p. 101–105 (1976).
See also *14.3.1*, p. *184*.

H. Rogass, J. Lauckner, and J. Finke
On the Influence of an Electric Field on the Condensation of Selenium Layers
(a) Vol. **36**, No. 1, p. K9–K11 (1976).
See also *3*, p. *43*.

H.-J. Unger
An Analytically Tractable Structure Model for Disordered Systems (II): Complicated Structures and Application to Amorphous Selenium
(b) Vol. **76**, No. 2, p. 841–850 (1976).
See also *2*, p. *39*.

J. L. FARVACQUE, J. C. DOUKHAN, B. ESCAIG, and T. TUOMI
Plastic Deformation of Selenium Single Crystals in **a** Glide and X-Ray Observation of the Induced Substructure
(a) Vol. **36**, No. 2, p. 461–470 (1976).
See also *10.2*, p. *94*.

D. W. BULLETT
Non-Empirical Calculation of Structural Properties and Valence Force Fields. Application to Se and Te
(b) Vol. 77, No. 2, p. 645–650 (1976).
See also 1, p. 9; *4*, p. *46*; 6, p. 61.

J. L. FARVACQUE and E. GERLACH.
Scattering of Free Carriers Due to the Strain Field of Screw Dislocations in Semiconductors
(b) Vol. 77, No. 2, p. 651–656 (1976).
See also *10.2*, p. *95*; *14.3*, p. *179*.

D. GENZOW and E. NORMANTAS
Theory of the Photon-Drag Effect in Semiconductors with Elliptical Isoenergetic Surfaces and Its Application to Tellurium Crystals
(b) Vol. 77, No. 2, p. 667–676 (1976).
See also *14.3*, p. *180*; 16, p. 216.

A. KUBOVÝ and M. JANDA
Effects of Annealing on Some Transport Properties of Tellurium Thin Films
(a) Vol. **37**, No. 2, p. K127–K129 (1976).
See also *14.3.1*, p. *185*.

J. L. FARVACQUE, J. C. DOUKHAN, and B. ESCAIG
The Asymmetry and Orientation Dependence of **c**-Slip in Tellurium
(a) Vol. **38**, No. 1, p. 271–284 (1976).
See also *10.2*, p. *97*.

22.2 III–V Compounds

Original Papers and Short Notes

R. I. Cottam and G. A. Saunders
An Extension of Keyes'Correlation
(a) Vol. **33**, No. 1, p. 367–373 (1976).
See also *12.2*, p. *110*; 22.1, p. 343; 22.4, p. 376.

Yu. Suleymanov, I. G. Pichugin, S. I. Teslenko, and L. A. Marasina
Photoluminescence Excitation Spectra of Gallium Nitride Single Crystals
(a) Vol. **34**, No. 2, p. K195–K197 (1976).
See also *20.3*, p. *282*.

A. P. Rusakov and M. V. Anfimov
Non-Central Forces and Stability of the $A^N B^{8-N}$ Tetrahedral Compounds (I): Bond Charges and the Valence-Force-Field Model
(b) Vol. **75**, No. 1, p. 73–77 (1976).
See also *4*, p. *45*; 12.2, p. 110; 22.1, p. 343; 22.4, p. 376; 22.5, p. 389.

B. N. Oshcherin
Debye Temperature and Melting Criterion of Ternary Chalcopyrite $A^1B^3C^6_2$- and $A^2B^4C^5_2$-Type Semiconductors
(a) Vol. **35**, No. 1, p. K35–K39 (1976).
See also *1*, p. *9*; *8*, p. *68*; 22.4, p. 376; 22.8, p. 415.

D. I. Sheka and A. N. Korol
Dielectric Permittivity of Semiconductors with Narrow Forbidden Gap in the Long-Wave Limit
(b) Vol. **76**, No. 1, p. 413–418 (1976).
See also 13.1, p. 125; *14.4*, p. *203*.

K. Hübner and M. Schäfer
Bond Charge Model for the Chemical Shift of the Photoemission in Semiconductors
(b) Vol. **76**, No. 1, p. K63–K67 (1976).
See also *4*, p. *46*; *20.1*, p. *271*.

V. S. Vavilov, A. E. Kiv, and O. R. Niyazova
Erratum to: The Subthreshold Radiation Effects in Semiconductors
(a) Vol. **36**, No. 1, p. 427 (1976).
See also 9, p. 73; 10.2, p. 94; *11*, p. *102*; 14.3, p. 178; 22.1, p. 343; 22.3, p. 373; 22.4, p. 376.

K. Hübner and V. K. Bashenov
Ionicity and Structure of Solids
(b) Vol. **77**, No. 2, p. 473–482 (1976).
See also 1.1, p. 13; 2, p. 40; *4*, p. *46*; *13.1*, p. *127*; 22.2.4, p. 371.

J. H. Yee and G. Myers
Pseudopotential Calculation of the Acoustic Deformation Potential for Several Group IV, II–VI, and III–V Compounds
(b) Vol. **77**, No. 2, p. K81–K84 (1976).
See also *13.1*, p. *128*; 22.1.1, p. 349; 22.1.2, p. 356; 22.4, p. 377.

R. Pässler
Relationships between the Nonradiative Multiphonon Carrier-Capture Properties of Deep Charged and Neutral Centres in Semiconductors
(b) Vol. **78**, No. 2, p. 625–635 (1976).
See also 6, p. 64; 13, p. 119; *13.4*, p. *145*; 22, p. 342; 22.1, p. 344; 22.4, p. 378.

L. A. Marasina, A. N. Pikhtin, I. G. Pichugin, and A. V. Solomonov
Cathodoluminescence of Undoped and Zn-Doped GaN Epitaxial Layers
(a) Vol. **38**, No. 2, p. 753–760 (1976).
See also *20.3*, p. *288*.

22.2.1 Arsenides

Original Papers and Short Notes

P. Leyral, D. Bois, and P. Pinard
Strain-Associated Band Tailing Effect in Semiconductors
(b) Vol. **73**, No. 1, p. 187–192 (1976).
See also *13.1*, p. *120*; *20.1*, p. *260*.

N. J. Doran and E. Gerlach
A Theory for Electron Transport Due to Phonon Scattering Including Screening
(b) Vol. **73**, No. 1, p. 203–209 (1976).
See also *6*, p. *52*; *14.3*, p. *173*.

H.-J. Wünsche and K. Henneberger
Interband-Current Theory for Esaki Diodes
(b) Vol. **73**, No. 1, p. 245–254 (1976).
See also **13.1**, p. **120**; *14.3.3*, p. *190*; **14.3.4**, p. **195**.

P. Hiesinger
Hall-Effect Levels in Ag- and Au-Doped p-Type GaAs
(a) Vol. **33**, No. 1, p. K39–K41 (1976).
See also *13.4*, p. *135*; *14.3*, p. *174*.

E. V. K. Rao
Optical Properties of Oxygen-Implanted GaAs
(a) Vol. **33**, No. 2, p. 683–690 (1976).
See also **11**, p. **100**; *20.1*, p. *263*.

V. D. Verner, S. K. Maksimov, and D. K. Nichugovskii
The Nature of Defects of the Crystalline Structure in GaAs Heavily Doped with Te
(a) Vol. **33**, No. 2, p. 755–763 (1976).
See also *10.2*, p. *89*.

J. C. Jones and P. R. Wallace
On the Anomalous Magneto-Resistance of Gallium Arsenide in Intense Magnetic Fields
(b) Vol. **74**, No. 1, p. 23–33 (1976).
See also **13.1**, p. **122**; *14.3*, p. *174*.

P. Braun and S. Grande
Spin–Lattice Relaxation Times and Knight Shift in InSb and InAs
(b) Vol. **74**, No. 1, p. 209–215 (1976).
See also *19*, p. *249*; **22.2.3**, p. **367**.

D. Lemoine, C. Pelletier, S. Rolland, and R. Granger
Electronic Conduction in GaAs at Low Temperatures
(a) Vol. **34**, No. 1, p. 307–317 (1976).
See also *14.3*, p. *174*.

R. M. Hill
Variable-Range Hopping
(a) Vol. **34**, No. 2, p. 601–613 (1976).
See also *13.4*, p. *137*; *14.3.1*, p. *183*; **22.1**, p. **343**; **22.1.1**, p. **346**.

K. D. Glinchuk, A. V. Prokhorovich, and V. I. Vovnenko
The Internal Radiative Efficiency and the Mechanism of Temperature Quenching of the 1.03, 1.20, and 1.30 eV Emission Bands in n-GaAs
(a) Vol. **34**, No. 2, p. 777–786 (1976).
See also *13.4*, p. *138*; *20.3*, p. *282*.

B. Rheinländer
Volume Plasma Faraday Effect in n-Type GaAs
(b) Vol. **75**, No. 1, p. 261–270 (1976).
See also 13.5.2, p. 157; *18*, p. *223*; *20.1*, p. *267*.

D. Mukherji and B. R. Nag
Electron Effective Mass in Semiconductor Superlattices
(b) Vol. **75**, No. 1, p. K35–K37 (1976).
See also *13.1*, p. *124*.

A. Matulionis, J. Požela, and A. Reklaitis
Monte Carlo Calculations of Hot-Electron Transient Behaviour in CdTe and GaAs
(a) Vol. **35**, No. 1, p. 43–48 (1976).
See also *13*, p. *116*; *14.3.4*, p. *197*; 22.4.3, p. 385.

K. Hattori
Exciton Binding Energies in Polar Crystals
(b) Vol. **76**, No. 1, p. 281–287 (1976).
See also *13.5.1*, p. *153*; 22.4, p. 376; 22.5, p. 389.

N. J. Doran and E. Gerlach
Frequency Dependent Conductivity Due to Deformation Potential and Piezoelectric Scattering by Acoustic Phonons in Semiconductors Including Screening
(b) Vol. **76**, No. 1, p. K43–K46 (1976).
See also 6, p. 58; *14.3*, p. *178*.

A. Ya. Vul, Sh. I. Nabiev, S. G. Petrosyan, and A. Ya. Shik
Simulation of the Long-Period Conductivity Relaxation (LCR) Phenomenon in Double-Layer Semiconductor Structures
(a) Vol. **36**, No. 1, p. 53–60 (1976).
See also *14.3.1*, p. *184*; *16*, p. *215*.

G. D. Bagratishvili, R. B. Dzhanelidze, N. I. Kurdiani, and O. V. Saksaganskii
MIS Structure GaAs–Ge_3N_4–Al
(a) Vol. **36**, No. 1, p. 73–79 (1976).
See also 13.3, p. 132; *14.3.2*, p. *188*; 20.3, p. 285.

P. C. Mathur and Vijender Sharma
Large-Signal Analysis of IMPATT Diodes
(a) Vol. **36**, No. 1, p. 269–274 (1976).
See also *14.3.3*, p. *192*; *14.3.4*, p. *198*.

S. K. Barthwal and K. L. Chopra
Transport Properties of Amorphous InSb and GaAs
(a) Vol. **36**, No. 1, p. 345–355 (1976).
See also *2*, p. *39*; *14.3.1*, p. *184*; 15, p. 210; 22.2.3, p. 368.

B. Petzel
Influence of Temperature on Impact Ionization in Gunn Domains
(a) Vol. **36**, No. 1, p. K39–K41 (1976).
See also *14.3.3*, p. *192*; *14.3.4*, p. *198*.

T. D. Dzhafarov and Yu. P. Demakov
Diffusion of Group I Impurities in Epitaxial Layers AlAs–GaAs Solid Solutions
(a) Vol. **36**, No. 2, p. 439–443 (1976).
See also *9*, p. *73*; 10.2, p. 94; 22.2.4, p. 370.

T. N. Sitenko
Non-Equilibrium Low-Temperature Depletion
(a) Vol. **36**, No. 2, p. K157–K159 (1976).
See also 13.3, p. 133; *14.3.2*, p. *188*.

P. R. W. HUDSON
Heat Flow Rectification
(a) Vol. **37**, No. 1, p. 93–96 (1976).
See also *8*, p. *70*.

Ľ. HRIVNÁK
Injection of Electrons into Semi-Insulating GaAs
(a) Vol. **37**, No. 1, p. K1–K4 (1976).
Erratum
(a) Vol. **38**, No. 2, p. K179 (1976).
See also *14.3.4*, p. *199*.

Cz. MARUCHA, J. MUCHA, and J. RAFAŁOWICZ
Phenomenological Interpretation of Heat Flux Volume Rectification in Non-Homogeneous Media
(a) Vol. **37**, No. 1, p. K5–K7 (1976).
See also *8*, p. *70*.

V. L. ALPEROVICH, V. M. ŻALETIN, A. F. KRAVCHENKO, and A. S. TEREKHOV
The Influence of Phonons and Impurities on the Broadening of Excitonic Spectra in Gallium Arsenide
(b) Vol. 77, No. 2, p. 465–472 (1976).
See also *13.5.1*, p. *155*; *20.1*, p. *274*.

P. BASU and H. SAHA
Possible Performance of Back-Wall Schottky Barrier Solar Cells
(a) Vol. **37**, No. 2, p. 625–631 (1976).
See also 13.3, p. 133; *14.3.3*, p. *193*; *16*, p. *216*; 22.1.2, p. 357; 22.4.3, p. 386.

A. SCHLACHETZKI
Domain Dissolution in Planar Gunn Devices
(a) Vol. **37**, No. 2, p. K151–K155 (1976).
See also *14.3.3*, p. *193*.

W. RÜHLE and E. GÖBEL
New Aspects of the Magnetoluminescence of a Band to Acceptor Transition in GaAs
(b) Vol. 78, No. 1, p. 311–317 (1976).
See also *13.1*, p. *128*; 20.3, p. 287.

J. S. AHEARN, JR., C. A. B. BALL, and C. LAIRD
Stress-Induced Birefringence of Mismatching III–V Heterojunctions
(a) Vol. **38**, No. 1, p. 315–320 (1976).
See also *10.2*, p. *98*; *20*, p. *259*; 22.2.2, p 365.

H. VAN CONG
Linewidth of Spontaneous Emission in Laser Diodes
(a) Vol. **38**, No. 1, p. K69–K72 (1976).
See also *20.2*, p. *279*.

22.2.2 Phosphides

Original Papers and Short Notes

M. GIEHLER and E. JAHNE
Effect of Damping on the Plasmon–Phonon Coupling in CdS and GaP
(b) Vol. **73**, No. 2, p. 503–516 (1976).
See also *6*, p. *54*; *13.5.2*, p. *157*; 20.1, p. 261; 22.4.1, p. 379.

K. LÖSCH and J. U. FISCHBACH
Measurement of High Near-Edge Absorption Coefficients in Surface Layers of InP
(a) Vol. **33**, No. 2, p. 473–478 (1976).
See also 13.1, p. 122; 13.5.1, p. 150; *20.1*, p. *262*.

R. BINDEMANN, H. FISCHER, and K. KREHER
Observation of Nonexponential Long Time Decay of Donor–Acceptor Pair Luminescence in GaP
(a) Vol. **34**, No. 1, p. K63–K65 (1976).
See also *20.3*, p. *281*.

P. KOIDL
Jahn-Teller Effect in the $^4T_1(1)$ and $^4T_2(1)$ States of Tetrahedrally Coordinated Mn^{2+}
(b) Vol. **74**, No. 2, p. 477–484 (1976).
See also *13.4*, p. *137*; 20.1, p. 265; 20.3, p. 282; 22.4.1, p. 380; 22.4.2, p. 383.

B. ULRICI and E. JAHNE
Phonon Frequencies of InP Derived from Two-Phonon Absorption
(b) Vol. **74**, No. 2, p. 601–607 (1976).
See also *6*, p. *55*; *20.1*, p. *265*.

K. SOMOGYI
Relation between the Activation Energies (ε_1, E_1) and the Temperature of Carrier Mobility Maximum in Semiconductors. A New Method for Determination of E_1
(a) Vol. **35**, No. 2, p. 659–666 (1976).
See also 13.4, p. 140; *14.3*, p. *178*; 22.1.2, p. 354.

D. D. NEDEOGLO, R. K. RADU, and A. G. CHEBAN
Electrical Properties of n-Type Epitaxial GaP Layers
(a) Vol. **36**, No. 2, p. 659–668 (1976).
See also *14.3.1*, p. *185*.

J. DONECKER and J. KLUGE
Magnetic Circular Dichroism of Indirect Excitonic Transitions in GaP and $In_{1-x}Ga_xP$
(b) Vol. **77**, No. 1, p. 243–248 (1976).
See also 13.5.1, p. 155; *18*, p. *224*; *20.1*, p. *273*; 22.2.4, p. 371.

A. S. POPOV
Fundamental Photoluminescence of Undoped InP Crystals Cleaned by the Floating Zone Method
(a) Vol. **37**, No. 1, p. K53–K56 (1976).
See also *20.3*, p. *286*.

K. SOMOGYI
Temperature Dependence of the Hole Mobility in Doped p-GaP
(a) Vol. **37**, No. 2, p. 653–661 (1976).
See also *14.3*, p. *180*.

J. S. AHEARN, JR., C. A. B. BALL, and C. LAIRD
Stress-Induced Birefringence of Mismatching III–V Heterojunctions
(a) Vol. **38**, No. 1, p. 315–320 (1976).
See also *10.2*, p. *98*; *20*, p. *259*; 22.2.1, p. 364.

E. P. Trifonova and A. S. Popov
n–n Heterojunctions $CdSnP_2$–InP
(a) Vol. 38, No. 1, p. K37–K40 (1976).
See also *14.3.3*, p. *194*; 22.8, p. 418.

H. Gollnast, G. Richter, and H. Herrmann
Nitrogen Concentration in GaP Diodes Determined by Photovoltage Measurements
(a) Vol. 38, No. 1, p. K89–K91 (1976).
See also 13.4, p. 145; *16*, p. *217*.

22.2.3 Antimonides

Review Article

S. V. TSIVINSKII
Dislocation Multiplication in Single and Polycrystals in the Initial Stage of Deformation
(a) Vol. **36**, No. 1, p. 11–31 (1976).
See also *10.1*, p. *78*; *10.2*, p. *87*; 21, p. 289; 21.1, p. 302; 21.1.1, p. 316; 21.6, p. 329; 22.1.1, p. 345; 22.5.2, p. 392.

Original Papers and Short Notes

W. RÜHLE, W. JAKOWETZ, C. WÖLK, R. LINNEBACH, and M. PILKUHN
Optical Studies of Free and Bound Excitonic States in GaSb. Evidence for Deep A^+-Complexes
(b) Vol. **73**, No. 1, p. 255–264 (1976).
See also **13.4**, p. **134**; *13.5.1*, p. *149*; 20.1, p. 260; *20.3*, p. *280*.

P. K. BASU and B. R. NAG
Iterative Calculation of the Polar Optical Mobility in Semiconductors in the Extreme Quantum Limit
(b) Vol. **73**, No. 1, p. K15–K17 (1976).
See also 6, p. 53; *14.3*, p. *173*.

S. V. TSIVINSKII
Statistical Theory of Thermally Activated Dislocation Motion in Crystals in the Case of Zero Internal Stresses
(a) Vol. **33**, No. 1, p. 301–311 (1976).
See also *10.2*, p. *88*; 22.1.1, p. 345; 22.1.2, p. 351; 22.5.2, p. 392.

V. K. ARORA
Effect of Inelasticity on Magnetoresistance
(b) Vol. **73**, No. 2, p. K93–K94 (1976).
See also 6, p. 54; *14.3*, p. *174*.

L. DMOWSKI, M. KOŃCZYKOWSKI, R. PIOTRZKOWSKI, and S. POROWSKI
Pressure-Induced Slow Relaxation of the Free Electron Concentration in Undoped n-Type InSb
(b) Vol. **73**, No. 2, p. K131–K134 (1976).
See also 12.2, p. 110; *14.3*, p. *174*.

T. P. LEPÈTRE, J.-M. DUSSEAU, and J.-L. ROBERT
Expression of Acoustoelectric Net Gain in p-Type GaSb
(a) Vol. **33**, No. 2, p. 715–722 (1976).
See also 7, p. *65*; *14.3.4*, p. *196*.

P. BRAUN and S. GRANDE
Spin–Lattice Relaxation Times and Knight Shift in InSb and InAs
(b) Vol. **74**, No. 1, p. 209–215 (1976).
See also *19*, p. *249*; 22.2.1, p. 362.

S. G. ATARBEKOV, A. S. FILIPCHENKO, and M. A. RZAEV
Deep Level in Fused Tunnel Diodes Made from InSb
(a) Vol. **34**, No. 1, p. K7–K8 (1976).
See also *13.4*, p. *137*; *14.3.3*, p. *190*.

N. J. DORAN and E. GERLACH
A New Theory for the Transverse Magneto-Resistance in the Quantum Limit which Includes Screening
(b) Vol. **75**, No. 1, p. K15–K18 (1976).
See also *14.3*, p. *176*.

G. Bauer, M. Overhamm, P. Grosse, W. Müller, E. Gornick, and H. W. Pötzl
Landau Emission in n-InSb
(b) Vol. 75, No. 2, p. 543–551 (1976).
See also *13.1*, p. *124*; *14.3.4*, p. *197*; 20.1, p. 268; 20.3, p. 283.

L. Schenk, H. Bruhns, and K. Hübner
Numerical Investigation of the *z*-Pinch in Intrinsic Indium Antimonide
(a) Vol. 35, No. 2, p. 511–520 (1976).
See also *14.3.4*, p. *197*.

A. F. Kravchenko, B. V. Morozov, and E. M. Skok
Features of Indium Antimonide Films
(a) Vol. 35, No. 2, p. 755–759 (1976).
See also *14.3.1*, p. *184*.

V. S. Komissarov
Calculation of Hot Electron Mobility and Electron Temperature in n-Type InSb under the Assumption of a Maxwellian Distribution
(b) Vol. 76, No. 1, p. 105–112 (1976).
See also *13*, p. *116*; *14.3.4*, p. *197*.

D. Barjon, A. Raymond, and J. L. Robert
Low-Energy Gap Semiconductor Mobilities in Presence of Spatial Fluctuations of Carrier Density
(b) Vol. 76, No. 1, p. 223–229 (1976).
See also *13*, p. *117*; *14.3.4*, p. *197*.

S. K. Barthwal and K. L. Chopra
Transport Properties of Amorphous InSb and GaAs
(a) Vol. 36, No. 1, p. 345–355 (1976).
See also *2*, p. *39*; *14.3.1*, p. *184*; 15, p. 210; 22.2.1, p. 363.

W. Hanus and M. Oszwałdowski
The Dependence of Photoconductive and Photoelectromagnetic Effects on Slab Thickness in InSb at Room Temperature
(a) Vol. 36, No. 2, p. 445–452 (1976).
See also 13.3, p. 133; *16*, p. *215*.

A. S. Filipchenko and L. P. Bolshakov
Mobility of Holes in p-InSb Crystals
(b) Vol. 77, No. 1, p. 53–58 (1976).
See also *13.1*, p. *127*; *14.3*, p. *179*.

N. Saleh and A. Soliman
Analysis and Design Considerations of Burstein Narrow-Band Photon Detectors
(a) Vol. 37, No. 1, p. 39–43 (1976).
See also 14.3.3, p. 193; *16*, p. *216*.

M. M. Kaila, L. B. Harris, and H. J. Goldsmid
Thermoelectric and Thermomagnetic Properties of Plastically Bent InSb Samples
(a) Vol. 37, No. 1, p. K43–K46 (1976).
See also *14.3*, p. *179*; *15*, p. *210*.

M. Obiditsch and H. Kahlert
Electric Field Induced Changes of the Fundamental Absorption and Low-Temperature Energy Relaxation in n-Type GaSb
(b) Vol. 77, No. 2, p. 677–684 (1976).
See also 6, p. 61; *14.3*, p. *180*; *20.1*, p. *274*.

M. Oszwałdowski
The Dependence of Spectral Distributions of the Photoconductive and Photo-Electromagnetic Effect on Sample Thickness in InSb
(a) Vol. **37**, No. 2, p. 675–679 (1976).
See also *16*, p. *216*.

A. S. Filipchenko and A. Naurizbaev
Moss-Burstein Effect in n-Type Crystals of Indium Antimonide
(a) Vol. **37**, No. 2, p. K139–K142 (1976).
See also *13.4*, p. *143*; 20.1, p. 275.

J. T. Devreese and R. Evrard
On the Momentum Distribution of Electrons in Polar Semiconductors for High Electric Field
(b) Vol. 78, No. 1, p. 85–92 (1976).
See also 6, p. 62; *14.3.4*, p. *200*.

L. Rusiniak
Electric Paramagnetic Resonance and Combined Resonance of Conductivity Electrons in InSb at 24 GHz
(b) Vol. 78, No. 1, p. 265–270 (1976).
See also 13.1, p. 128; *19*, p. *255*; 20.1, p. 276.

E. J. Fantner, F. Kuchar, and G. Bauer
Systematic Investigation of the Sign Change of the Hall Effect in n-InSb in the Quantum Limit
(b) Vol. 78, No. 2, p. 643–651 (1976).
See also *14.3*, p. *181*.

H. Biernat and M. Kriechbaum
Anomalous Hall Effect of n-InSb at High Magnetic Fields
(b) Vol. 78, No. 2, p. 653–657 (1976).
See also *14.3*, p. *181*.

22.2.4 Mixed Crystals

Original Papers and Short Notes

H. LANGE, J. DONECKER, and H. FRIEDRICH
Electroreflectance and Wavelength Modulation Study of the Direct and Indirect Fundamental Transition Region of $In_{1-x}Ga_xP$
(b) Vol. **73**, No. 2, p. 633–639 (1976).
See also **13.1**, p. **121**; *20.1*, p. *262*.

N. N. SIROTA, I. V. BODNAR, A. I. LUKOMSKII, G. F. SMIRNOVA, and L. M. FINKELSTEIN
Indirect Absorption Edge of $(GaP)_x(InAs)_{1-x}$ Mixed Crystals
(b) Vol. **73**, No. 2, p. K135–K136 (1976).
See also *13.1*, p. *121*; *20.1*, p. *262*.

E. JAHNE
Long-Wavelength Optical Phonons in Mixed Crystals (I): A System of Two Coupled Modes
(b) Vol. **74** No. 1, p. 275–283 (1976).
See also *6*, p. *55*; *20.1*, p. *263*; 22.3.1, p. 375.

S. S. LI, D. W. SCHOENFELD, and R. T. OWEN
Optical Transmission Measurements and Electron Microprobe Analysis of a Bulk $InAs_{1-x}P_x$ Alloy System
(a) Vol. **34** No. 1 p. 255–261 (1976).
See also *13.1*, p. *122*; *20.1*, p. *264*.

H. NEUMANN and W. JUNGE
Temperature Dependence of the Indirect Absorption Edge in $Al_xGa_{1-x}As$
(a) Vol. **34** No. 1, p. K39–K41 (1976).
See also *20.1*, p. *265*.

H.-G. BRÜHL, L. HILDISCH, H. MORWINSKI, and W. SCHMIDT
X-Ray Study of Order in the Solid Solution $Ga_xIn_{1-x}P$
(a) Vol. **34**, No. 2, p. K147–K148 (1976).
See also *1.1*, p. *11*.

E. JAHNE
Long-Wavelength Optical Phonons in Mixed Crystals (II): The Persistence of Common Gaps
(b) Vol. **75**, No. 1, p. 221–232 (1976).
See also *6*, p. *56*; *20.1*, p. *267*.

K.-R. SCHULZE, H. NEUMANN, and K. UNGER
Band Structure of $Ga_{1-x}In_xAs$
(b) Vol. **75**, No. 2, p. 493–500 (1976).
See also *13.1*, p. *124*.

G. VOIGT, R. RAIDT, H. PEIBST, H. MENNIGER, and L. HILDISCH
Investigations of Growth Defects in Solution-Grown $Ga_xIn_{1-x}P$ with a SEM Operating in the Cathodoluminescence Mode
(a) Vol. **36**, No. 1, p. 173–179 (1976).
See also *1.4*, p. *27*; *3*, p. *43*.

D. BARJON, A. RAYMOND, A. JOULLIE, and J. L. ROBERT
Electron Mobility of $Ga_xIn_{1-x}Sb$ at Room Temperature
(b) Vol. **76**, No. 2, p. 623–627 (1976).
See also *14.3*, p. *178*.

T. D. DZHAFAROV and YU. P. DEMAKOV
Diffusion of Group I Impurities in Epitaxial Layers of AlAs–GaAs Solid Solutions
(a) Vol. **36**, No. 2, p. 439–443 (1976).
See also *9*, p. *73*; 10.2, p. 94; 22.2.1, p. 363.

I. V. Bodnar, E. E. Matyas, and L. A. Makovetskaya
Phase Diagram of $Ga_xIn_{1-x}P$ Solid Solution
(a) Vol. **36**, No. 2, p. K141—K144 (1976).
See also *3*, p. *43*.

J. Donecker and J. Kluge
Magnetic Circular Dichroism of Indirect Excitonic Transitions in GaP and $In_{1-x}Ga_xP$
(b) Vol. 77, No. 1, p. 243—248 (1976).
See also 13.5.1, p. 155; *18*, p. *224*; *20.1*, p. *273*; 22.2.2, p. 365.

L. G. Sakvarelidze, A. A. Mirtskhulava, I. A. Mirtskhulava, and G. G. Gegiadze
Production of Bulk Single Crystals of $Ga_{1-x}In_xAs$ Solid Solutions
(a) Vol. **37**, No. 1, p. K71—K73 (1976).
See also *3*, p. *43*.

K. Hübner and V. K. Bashenov
Ionicity and Structure of Solids
(b) Vol. 77, No. 2, p. 473—482 (1976).
See also 1.1, p. 13; 2, p. 40; *4*, p. *46*; *13.1*, p. *127*; 22.2, p. 361.

I. V. Bodnar, A. I. Lukomskii, and G. F. Smirnova
Absorption Edge of InP_xAs_{1-x}Mixed Crystals
(a) Vol. **37**, No. 2, p. K173—K176 (1976).
See also *13.1*, p. *128*; 20.1, p. 275.

R. I. Bolkhovityanova and Yu. B. Bolkhovityanov
The Influence of Indium on the p–n Junction Formation in Si-Doped LPE Layers of Gallium Arsenide
(a) Vol. **37**, No. 2, p. K193—K196 (1976).
See also *1.5*, p. *32*; **3**, p. 44; *14.3.1*, p. *185*.

N. P. Kekelidze, Z. V. Kvinikadze, Z. F. Davitaya, G. N. Eritsyan, V. A. Saakyan, and F. K. Karapetyan
Studies of Thermoelectric Characteristics of InP_xAs_{1-x} Alloys
(a) Vol. **38**, No. 1, p. K49—K52 (1976).
See also *8*, p. *71*; *15*, p. *211*.

B. Baranov and L. Däweritz
Vapour-Phase Epitaxial Growth of $Ga_{1-x}Al_xN$ on Sapphire
(a) Vol. **38**, No. 2, p. K111—K113 (1976).
See also *1.5*, p. *33*; *3*, p. *44*.

22.3 IV–IV, II–V, and III–VI Compounds

Review Article

R. PRASAD
Present State of Polytypism in Cadmium Iodide Crystals
(a) Vol. **38**, No. 1, p. 11–44 (1976).
See also *1.1*, p. *10*; *1.2*, p. *14*; 10.2, p. 87; 22.4.1, p. 379; 22.5, p. 388; 22.5.3, p. 399; 22.8, p. 413.

Original Papers and Short Notes

R. H. WILLIAMS, I. T. MCGOVERN, R. B. MURRAY, and M. HOWELLS
An Investigation of the Electronic Structure of GaSe and GaTe by Photoelectron Spectroscopy, Using a Synchrotron Source, and Electron Energy Loss Spectroscopy
(b) Vol. **73**, No. 1, p. **307–316** (1976).
See also *13.1*, p. *120*; *17*, p. *218*; 20.1, p. 260.

S. H. HAGEN and A. W. C. VAN KEMENADE
On the Role of Boron in the Luminescence of Silicon Carbide Doped with Nitrogen and Boron
(a) Vol. **33**, No. 1, p. 97–105 (1976).
See also 13.4, p. 135; *20.3*, p. *280*.

S. POPOVIĆ, B. ČELUSTKA, and D. BIDJIN
A Remark on the Paper "Phase Transitions in In_2Se_3 as Studied by Electron Microscopy and Electron Diffraction"
(a) Vol. **33**, No. 1, p. K23–K24 (1976).
See also *1.1*, p. *10*; 1.2, p. 14.

T. SHIMIZU and N. ISHII
Electronic Structure of As_2S_3, As_2Se_3, and As_2Te_3 Glasses by the Extended Hückel Theory
(b) Vol. **74**, No. 1, p. K39–K42 (1976).
See also *2*, p. *36*; *13*, p. *115*.

V. P. BHOLA
Thermoreflectance of Cd_3As_2
(a) Vol. **34**, No. 2, p. K179–K180 (1976).
See also *13.1*, p. *123*; *20.1*, p. *267*.

N. TAJABOR and D. R. LOVETT
Etching and Chemical Polishing of {112} Surfaces of Cd_3As_2 and Zn_3As_2 Crystals
(a) Vol. **34**, No. 2, p. K191–K194 (1976).
See also *1.6*, p. *34*.

YU. A. VODAKOV, G. A. LOMAKINA, E. N. MOKHOV, E. I. RADOVANOVA, V. I. SOKOLOV, M. M. USMANOVA, G. F. YULDASHEV, and B. S. MACHMUDOV
Silicon Carbide Doped with Gallium
(a) Vol. **35**, No. 1, p. 37–42 (1976).
See also **3**, p. 42; *13.4*, p. *138*; 14.3, p. 176; 20.1, p. 268; 20.3, p. 283.

S. BALEVIČIUS, A. ČESNYS, and A. DEKSNYS
Nanosecond Switching in Amorphous In_2Te_3 Films
(a) Vol. **35**, No. 1, p. K41–K43 (1976).
See also *2*, p. *37*; *14.3.1*, p. *184*.

S. I. RADAUTSAN, E. K. ARUSHANOV, and A. N. NATEPROV
The Conduction Band of Cadmium Phosphide
(a) Vol. **35**, No. 1, p. K53–K56 (1976).
See also *13.1*, p. *124*; 14.3, p. 177.

Kwok Leung Yip, A. B. Kunz, and W. S. Williams
Electronic Structure of TiC and VC from the Iterative Extended Hückel Theory
(b) Vol. **75**, No. 2, p. 533–541 (1976).
See also *13.1*, p. *124*; 20, p. 258.

A. J. Harris, H. Thomas, and M. P. Callaghan
Investigation of the Characteristics of Electroluminescent SiC Diodes (I): Theoretical Model for Transient and Steady-State Conditions
(a) Vol. **35**, No. 2, p. 697–707 (1976).
See also *14.3.3*, p. *192*; *20.3*, p. *284*.

L. Opilska and A. Opilski
Relation between Energy Gap and Velocity of Ultrasonic Wave Propagation in $A^{II}B^{V}$ Group Semiconductors
(a) Vol. **35**, No. 2, p. K183–K187 (1976).
See also *7*, p. *65*; 22.3.1, p. 375.

A. J. Harris and H. Thomas
Investigation of the Characteristics of Electroluminescent SiC Diodes (II): Temperature- and Growth-Dependent Effects
(a) Vol. **36**, No. 1, p. 365–373 (1976).
See also *14.3.3*, p. *192*; *20.3*, p. *285*.

V. S. Vavilov, A. E. Kiv, and O. R. Niyazova
Erratum to: The Subthreshold Radiation Effects in Semiconductors
(a) Vol. **36**, No. 1, p. 427 (1976).
See also 9, p. 73; 10.2, p. 94; *11*, p. *102*; 14.3, p. 178; 22.1, p. 343; 22.2, p. 361; 22.4, p. 376.

K. D. Kundra and S. Z. Ali
X-Ray Study of Thermal Expansion and Phase Transformation in β-In_2S_3
(a) Vol. **36**, No. 2, p. 517–525 (1976).
See also *1.2*, p. *18*; 8, p. 70.

C. Manfredotti, C. de Blasi, S. Galassini, G. Micocci, L. Ruggiero, and A. Tepore
Analysis of SCLC Curves by a New Direct Method
(a) Vol. **36**, No. 2, p. 569–577 (1976).
See also *14.3.4*, p. *198*.

R. S. Rai, S. R. Singh, and G. Singh
On the Origin of Large Spiral Steps on SiC Growth Faces
(a) Vol. **36**, No. 2, p. 699–703 (1976).
See also *1.4*, p. *28*; *3*, p. *43*.

C. L. Reynolds, Jr. and F. E. Wawner
On the Origin of Triboluminescence in SiC Filaments
(a) Vol. **36**, No. 2, p. K109–K111 (1976).
See also *20.3*, p. *285*.

A. Suzuki, H. Matsunami, and T. Tanaka
Very Slow Decays of the Luminescence from Boron-Doped 4H-SiC
(a) Vol. **36**, No. 2, p. K129–K131 (1976).
See also *20.3*, p. *285*.

S. R. Singh, R. S. Rai, and G. Singh
K-Surfaces of Cubic Silicon Carbide (β-SiC)
(a) Vol. **37**, No. 1, p. 139–142 (1976).
See also *1.1*, p. *12*; *12.2*, p. *111*.

S. K. Lilov, Yu. M. Tairov, V. F. Tsvetkov, and M. A. Chernov
Structural and Morphological Peculiarities of the Epitaxial Layers and Monocrystals of Silicon Carbide Highly Doped by Nitrogen
(a) Vol. **37**, No. 1, p. 143–150 (1976).
See also *1.5*, p. *32*.

H. Kuwabara, K. Yamanaka, and S. Yamada
Donor–Acceptor Pair Emission from β-SiC Doped with Gallium
(a) Vol. **37**, No. 2, p. K157–K161 (1976).
See also 13.4, p. 144; *20.3*, p. *287*.

L. P. Mokhracheva, V. A. Tskhai, and P. V. Geld
One-Centre Exchange Potentials in a Crystal (TiC)
(b) Vol. **78**, No. 2, p. 465–469 (1976).
See also *13.1*, p. *129*.

C. Manfredotti, R. Murri, A. Quirini, and L. Vasanelli
Photoelectronic Properties of n-GaSe
(a) Vol. **38**, No. 2, p. 685–693 (1976).
See also 13.4, p. 145; 14.3, p. 182; *16*, p. *217*.

T. A. McMath and J. C. Irwin
Indices of Refraction of GaS and GaSe
(a) Vol. **38**, No. 2, p. 731–738 (1976).
See also *20.1*, p. *278*.

22.3.1 Mixed Crystals

Original Papers and Short Notes

G. Arancia, M. Grandolfo, C. Manfredotti, and A. Rizzo
Electron Diffraction Study of Melt- and Vapour-Grown $GaSe_{1-x}S_x$ Single Crystals
(a) Vol. **33**, No. 2, p. 563–571 (1976).
See also *1.1*, p. *11*; *3*, p. *42*.

E. Jahne
Long-Wavelength Optical Phonons in Mixed Crystals (I): A System of Two Coupled Modes
(b) Vol. **74**, No. 1, p. 275–283 (1976).
See also *6*, p. *55*; *20.1*, p. *263*; 22.2.4, p. 370.

L. Opilska and A. Opilski
Relation between Energy Gap and Velocity of Ultrasonic Wave Propagation in $A^{II}B^{V}$ Group Semiconductors
(a) Vol. **35**, No. 2, p. K183–K187 (1976).
See also *7*, p. *65*; 22.3, p. 373.

A. Kuhn, A. Chevy, and R. Chevalier
Interatomic Distances in $GaSe_xS_{1-x}$
(a) Vol. **36**, No. 1, p. 181–187 (1976).
See also *1.1*, p. *12*.

E. A. Meneses, N. Jannuzzi, J. R. Freitas, and A. Gouskov
Photoluminescence of Layered $GaSe_{1-x}Te_x$ Crystals
(b) Vol. 78, No. 1, p. K35–K38 (1976).
See also *20.3*, p. *287*.

22.4 II–VI Compounds

Original Papers and Short Notes

P. Loose and M. Wöhlecke
A Simple Approach to the Energy of Bound Exciton Complexes
(b) Vol. **73**, No. 1, p. 265–270 (1976).
See also *13.4*, p. *134*; *13.5.1*, p. *149*.

R. I. Cottam and G. A. Saunders
An Extension of Keyes' Correlation
(a) Vol. **33**, No. 1, p. 367–373 (1976).
See also *12.2*, p. *110*; 22.1, p. **343**; 22.2, p. 361.

K. Kumazaki
Elastic Properties and Ionicity of Zero-Gap Semiconductors
(a) Vol. **33**, No. 2, p. 615–623 (1976).
See also *4*, p. *45*; 7, p. 65; *12.2*, p. *110*.

K. Hauffe and A. Meyer-Laack
On a Tribo-Photoelectric Effect of Li_2O-Doped Zinc Oxide Single Crystals
(a) Vol. **34**, No. 1, p. 145–152 (1976).
See also *16*, p. *214*.

A. P. Rusakov and M. V. Anfimov
Non-Central Forces and Stability of the A^NB^{8-N} Tetrahedral Compounds (I): Bond Charges and the Valence-Force-Field Model
(b) Vol. **75**, No. 1, p. 73–77 (1976).
See also *4*, p. *45*; 12.2, p. 110; 22.1, p. 343; 22.2, p. 361; 22.5, p. 389.

P. Loose, M. Rosenzweig, and M. Wöhlecke
Zeeman Effect of Bound Exciton Complexes in ZnO
(b) Vol. **75**, No. 1, p. 137–144 (1976).
See also *13.5.1*, p. *152*; *20.1*, p. *267*; 20.3, p. 282.

B. N. Oshcherin
Debye Temperature and Melting Criterion of Ternary Chalcopyrite $A^1B^3C_2^6$- and $A^2B^4C_2^5$-Type Semiconductors
(a) Vol. **35**, No. 1, p. K35–K39 (1976).
See also *1*, p. *9*; *8*, p. *68*; 22.2, p. 361; 22.8, p. 415.

N. Dupont-Pavlovsky, F. Caralp, P. Delhaes, and J. Amiell
Chemical Characterization and Study of the Electronic Properties of Non-Stoichiometric Zinc Oxide. Nature and Repartition of Defects
(a) Vol. **35**, No. 2, p. 615–625 (1976).
See also *10.2*, p. *93*; *13.4*, p. *139*; 14.3, p. 177; 18, p. 223.

K. Hattori
Exciton Bindung Energies in Polar Crystals
(b) Vol. **76**, No. 1, p. 281–287 (1976).
See also *13.5.1*, p. *153*; 22.2.1, p. 363; 22.5, p. 389.

V. S. Vavilov, A. E. Kiv, and O. R. Niyazova
Erratum to: The Subthreshold Radiation Effects in Semiconductors
(a) Vol. **36**, No. 1, p. 427 (1976).
See also 9, p. 73; 10.2, p. 94; *11*, p. *102*; 14.3, p. 178; 22.1, p. 343; 22.2, p. 361; 22.3, p. 373.

G. Müller
Optical and Electrical Spectroscopy of Zinc Oxide Crystals Simultaneously Doped with Copper and Donors
(b) Vol. 76, No. 2, p. 525–532 (1976).
See also *13.4*, p. *141*; 14.3, p. 178; 20.1, p. 271.

R. Boyn
Tetrahedral Model for Stress Dichroism Due to Deep Impurities in Wurtzite Type Crystals
(b) Vol. 76, No. 2, p. K119–K122 (1976).
See also 13.4, p. 141; *20.1*, p. *272*.

W. E. Hagston and H. H. Sutherland
Properties of an Exciton Bound in an Asymmetric Electric Dipole Field
(b) Vol. 77, No. 1, p. K53–K56 (1976).
See also 13.4, p. 142; *13.5.1*, p. *155*.

V. Soriano and D. Galland
Photosensitivity of the EPR Spectrum of the F^+ Center in ZnO
(b) Vol. 77, No. 2, p. 739–743 (1976).
See also *13.4*, p. *143*; *19*, p. *254*.

J. H. Yee and G. Myers
Pseudopotential Calculation of the Acoustic Deformation Potential for Several Group IV, II–VI, and III–V Compounds
(b) Vol. 77, No. 2, p. K81–K84 (1976).
See also *13.1*, p. *128*; 22.1.1, p. 349; 22.1.2, p. 356; 22.2, p. 361.

P. K. C. Pillai and Rabinder Nath
Photodielectric Effect in Photoconductors
(a) Vol. 37, No. 2, p. 491–498 (1976).
See also *16*, p. *216*.

R. Meaudre and M. Meaudre
Capacitive Effects in Insulators or Semiconductors. Influence of the Density of States in the Gap
(a) Vol. 37, No. 2, p. 633–643 (1976).
See also *13.4*, p. *143*; *14.3.4*, p. *199*.

H. Finkenrath, K. Krug, and N. Uhle
Optical Absorption by Free Polarons in ZnO
(b) Vol. 78, No. 1, p. K27–K30 (1976).
See also *13.5.3*, p. *159*; *20.1*, p. *276*.

J. Marien
Absolute Method for Work Function Determination of Metals and Semiconductors by Field Emission
(a) Vol. 38, No. 1, p. 339–347 (1976).
See also *17*, p. *220*; 21, p. 300.

S. Bhushan and M. Saleem
Photo- and Electroluminescence of ZnO:Er Phosphors
(a) Vol. 38, No. 1, p. K27–K29 (1976).
See also *20.3*, p. *288*.

P. K. C. Pillai, Rabinder Nath, and P. K. Nair
Non-Isothermal Relaxation in ZnO Photoelectrets
(a) Vol. 38, No. 1, p. K73–K76 (1976).
See also *14.4*, p. *204*; *16*, p. *217*.

B. Hönerlage, C. Klingshirn, and J. B. Grun
Spontaneous Emission Due to Exciton–Electron Scattering in Semiconductors
(b) Vol. 78, No. 2, p. 599–608 (1976).
See also *13.5.1*, p. *156*; *20.3*, p. *288*; 22.4.1, p. 382; 22.4.2, p. 384.

R. Pässler
Relationships between the Nonradiative Multiphonon Carrier-Capture Properties of Deep Charged and Neutral Centres in Semiconductors
(b) Vol. 78, No. 2, p. 625–635 (1976).
See also 6, p. 64; 13, p. 119; *13.4*, p. *145*; 22, p. 342; 22.1, p. 344; 22.2, p. 361.

J. Marien
Field Emission Study of the Specificity of Zinc Oxide Polar Surfaces (0001) and (000$\bar{1}$). Work Function and Hydrogen Adsorption
(a) Vol. 38, No. 2, p. 513–522 (1976).
See also *17*, p. *220*.

H. Iwanaga, K. Suzuki, and S. Takeuchi
Direct Observation of Dislocation Motion in ZnO
(a) Vol. 38, No. 2, p. K119–K122 (1976).
See also *10.2*, p. *99*.

22.4.1 Sulphides

Review Article

R. Prasad
Present State of Polytypism in Cadmium Iodide Crystals
(a) Vol. **38**, No. 1, p. 11–44 (1976).
See also *1.1*, p. *10*; *1.2*, p. *14*; 10.2, p. 87; 22.3, p. 372; 22.5, p. 388; 22.5.3, p. 399; 22.8, p. 413.

Original Papers and Short Notes

J. Dietrich
Anisotropy of the EPR of Yb^{3+} in Cubic Zinc Sulphide Due to Stacking Faults
(b) Vol. **73**, No. 1, p. K57–K59 (1976).
See also *19*, p. *247*.

M. Giehler and E. Jahne
Effect of Damping on the Plasmon–Phonon Coupling in CdS and GaP
(b) Vol. **73**, No. 2, p. 503–516 (1976).
See also *6*, p. *54*; *13.5.2*, p. *157*; 20.1, p. 261; 22.2.2, p. 365.

B. Lepley and S. Ravelet
Correlation between Open-Circuit Voltage and Interface Parameters at the Au–CdS Illuminated Contact
(a) Vol. **33**, No. 2, p. 517–523 (1976).
See also *14.3.2*, p. *187*; *16*, p. *213*.

L. I. Khandros, G. S. Pekar, M. K. Sheinkman, and E. L. Shtrum
High-Temperature Investigations of CdS Single Crystals (I): Electron Concentration, Cd Solubility, and Charge of Intrinsic Defects
(a) Vol. **33**, No. 2, p. 765–771 (1976).
See also 10.2, p. 89; *14.3*, p. *174*.

C. A. Bates and H. Szymczak
The Phonon-Induced Temperature Dependences of Spin-Hamiltonian Parameters for S-State Ions
(b) Vol. **74**, No. 1, p. 225–233 (1976).
See also *6*, p. *55*; *19*, p. *249*; 22.6, p. 403.

H. Kuzmany
Light Scattering Experiments and Nonlinear Behaviour of the Acoustoelectric Oscillator
(a) Vol. **34**, No. 1, p. 347–353 (1976).
See also *7*, p. *65*; *14.3*, p. *174*.

A. Kiveris and P. Pipinys
The Exhaustion of Traps in ZnS Phosphors by Electric Field and the Gudden-Pohl Effect
(a) Vol. **34**, No. 1, p. 377–382 (1976).
See also *14.3.4*, p. *196*; *20.3*, p. *281*.

L. I. Khandros, G. S. Pekar, M. K. Sheinkman, and E. L. Shtrum
High-Temperature Investigations of CdS Single Crystals (II): Electron Mobility and Scattering Mechanism
(a) Vol. **34**, No. 1, p. 397–403 (1976).
See also 10.2, p. 90; *14.3*, p. *174*.

S. Martinuzzi
n-CdS–p-CdTe Heterojunction Solar Cells
(a) Vol. **34**, No. 1, p. K21–K26 (1976).
See also *14.3.3*, p. *190*; *16*, p. *214*; 22.4.3, p. 385.

P. Koidl
Jahn-Teller Effect in the $^4T_1(l)$ and $^4T_2(l)$ States of Tetrahedrally Coordinated Mn^{2+}
(b) Vol. **74**, No. 2, p. 477–484 (1976).
See also *13.4*, p. *137*; 20.1, p. 265; 20.3, p. 282; 22.2.2, p. 365; 22.4.2, p. 383.

J. Voigt, M. Senoner, and I. Rückmann
Quantitative Analysis of Interference Structures in the Transmission Spectrum of Very Thin CdS Platelets
(b) Vol. **75**, No. 1, p. 213–219 (1976).
See also *13.5.1*, p. *152*; 20.1, p. 267.

E. Skaistys and L. Valkūnas
Screening Effects in the Exciton–Electron Luminescence Spectra
(b) Vol. **75**, No. 1, p. K23–K26 (1976).
See also *13.5.1*, p. *152*; *20.3*, p. *283*.

A. N. Georgobiani, Yu. V. Ozerov, E. E. Strumban, and V. V. Anisimov
Investigation of Deep Centres of Gallium-Doped Zinc Sulfide Crystals
(a) Vol. **35**, No. 1, p. 251–257 (1976).
See also *13.4*, p. *138*; 16, p. 214; 20.1, p. 268; 20.3, p. 283.

W. F. Braerman, K. W. Böer, and G. A. Dussel
Observation of Deep-Trap Modulation with "Near" Band-Edge Light in Copper-Doped CdS Crystals
(a) Vol. **35**, No. 1, p. 309–313 (1976).
See also *13.4*, p. *139*; *20.1*, p. *268*.

M. Egée
Spectral Distribution of the Green Emission Band of ZnS:Cu in Dependence on the Localization Diversity of the Radiative Recombination Processes
(a) Vol. **35**, No. 1, p. 361–370 (1976).
See also 13.4, p. 139; *20.3*, p. *283*.

L. F. Komolova, V. M. Efremenkova, and G. V. Saparin
The Investigation of Cathodoluminescence (CL) in CdS after Ion Bombardment in the Scanning Electron Microscope (SEM)
(a) Vol. **35**, No. 1, p. K5–K8 (1976).
See also *20.3*, p. *283*.

F. Matsuura, S. Teratani, S. Kishida, and I. Tsurumi
Temperature Dependence of the Half-Width of the 2.3 eV Optical Absorption Band in Neutron-Irradiated ZnS
(a) Vol. **35**, No. 1, p. K57–K60 (1976).
See also 10.2, p. 92; *11*, p. *101*; *20.1*, p. *268*.

R. Zimmermann and M. Rösler
Theory of Electron–Hole Plasma in CdS
(b) Vol. **75**, No. 2, p. 633–645 (1976).
See also *13.5*, p. *147*; 13.5.2, p. 157; *20.3*, p. *284*.

O. Goede, M. Blaschke, and K. H. Klohs
Piezoemission Study of I_1- and I_2-Type Bound Excitons in CdS
(b) Vol. **76**, No. 1, p. 267–280 (1976).
See also *13.5.1*, p. *153*; *20.3*, p. *284*.

G. Ruszczynski and R. Boyn
Jahn-Teller Coupling Parameters from Stress-Dichroism Spectra for CdS:Ti^{2+} and CdS:Co^{2+}
(b) Vol. **76**, No. 1, p. 427–436 (1976).
See also 13.4, p. 140; *20.1*, p. *271*.

C. Butti, A. Raymond, and F. Bombré
Photo Hall Effect in Cinnabar
(a) Vol. **36**, No. 1, p. 133–137 (1976).
See also *14.3*, p. *178*.

S. Martinuzzi, O. Mallem, and T. Cabot
Dark Currents in Cu_2S-CdS Single Crystal Heterojunctions. Consequences for Photocells
(a) Vol. **36**, No. 1, p. 227–234 (1976).
See also 13.3, p. 133; *14.3.3*, p. *192*; *16*, p. *215*; 22, p. 338.

A. Anedda and E. Fortin
Temperature Shift of Excitons in CdS
(a) Vol. **36**, No. 1, p. 385–389 (1976).
See also *13.5.1*, p. *153*.

P. I. Khadzhi, K. G. Petrashku, and S. A. Moskalenko
Biexciton Luminescence with Interserial Transition of the Formed Exciton
(b) Vol. **76**, No. 2, p. 497–504 (1976).
See also *13.5.1*, p. *154*; *20.3*, p. *285*.

R. Pässler
Calculation of Nonradiative Multiphonon Capture Coefficients and Ionization Rates for Neutral Centres According to the Static Coupling Scheme (II): Alternative Trap Models
(b) Vol. **76**, No. 2, p. 647–659 (1976).
See also *6*, p. *59*; 13.1, p. 126; *13.4*, p. *141*; 22, p. 339.

W. Ekardt and M. I. Sheboul
Influence of the Electron–Hole Exchange Interaction on the Biexciton Ground State in CdS
(b) Vol. **76**, No. 2, p. K89–K91 (1976).
See also *13.5.1*, p. *154*.

F. Spiegelberg, E. Gutsche, and J. Voigt
Exciton–Phonon Interaction in CdS
(b) Vol. **77**, No. 1, p. 233–242 (1976).
See also *6*, p. *60*; *13.5.1*, p. *154*; 20.1, p. 272.

W. Brunner and G. Völkel
The Temperature Dependence of Spin–Lattice Relaxation in CdS:Mn^{2+}
(b) Vol. **77**, No. 1, p. K31–K33 (1976).
See also *19*, p. *254*.

A. Kiveris, Š. Kudžmauskas, and P. Pipinys
Release of Electrons from Traps by an Electric Field with Phonon Participation
(a) Vol. **37**, No. 1, p. 321–327 (1976).
See also 6, p. 61; *13.4*, p. *142*; 14.3.4, p. 199; *20.3*, p. *286*.

A. Deptuch, M. Grus, A. J. Horodecki, and M. Łepek
On the Conductivity of Polycrystalline CdS Films
(a) Vol. **37**, No. 1, p. K101–K104 (1976).
See also *14.3.1*, p. *185*.

A. Jucys and V. Kunigelis
Influence of the Surface Potential upon the Acoustoelectric Interaction in CdS
(b) Vol. **77**, No. 2, p. K113–K117 (1976).
See also *7*, p. *66*; *14.3.2*, p. *188*.

S. Permogorov and V. Travnikov
Resonant Scattering of Exciton-Polaritons in CdS Crystals
(b) Vol. **78**, No. 1, p. 389–398 (1976).
See also *6*, p. *63*; *13.5*, p. *148*; 20.3, p. 287.

H. Maier
Jahn-Teller Effect, Stacking Faults, and the Formation of Polytypes in ZnS
(b) Vol. 78, No. 1, p. K65–K68 (1976).
See also *6*, p. *63*; *13.4*, p. *144*; 20.1, p. 276; 20.3, p. 287.

A. N. Georgobiani, R. G. Maev, Yu. V. Ozerov, and E. E. Strumban
Investigation of Deep Centres of Chlorine-Doped Zinc Sulfide Crystals
(a) Vol. 38, No. 1, p. 77–83 (1976).
See also *13.4*, p. *144*; 16, p. 217; 20.1, p. 276.

B. Hönerlage, C. Klingshirn, and J. B. Grun
Spontaneous Emission Due to Exciton–Electron Scattering in Semiconductors
(b) Vol. 78, No. 2, p. 599–608 (1976).
See also *13.5.1*, p. *156*; *20.3*, p. *288*; 22.4, p. 378; 22.4.2, p. 384.

22.4.2 Selenides

Original Papers and Short Notes

J. Varvas and T. Nirk
High-Temperatures Conductivity of Undoped and Copper-Doped Cadmium Selenide
(a) Vol. **33**, No. 1, p. 75–84 (1976).
See also *10.2*, p. *87*; 13.4, p. 135; *14.3*, p. *173*.

K. Marinova, A. Opanowicz, and W. Ruppel
Photoconductivity of Highly Excited CdSe and $CdS_{0.48}Se_{0.52}$ Crystals
(a) Vol. **33**, No. 1, p. 195–204 (1976).
See also 13.5.1, p. 149; *16*, p. *213*; 22.4.4, p. 387.

I. Birchak, V. V. Serdyuk, and G. G. Chemeresyuk
Optical and Temperature Quenching of Photocurrent in Zinc Selenide Single Crystals
(a) Vol. **33**, No. 2, p. K145–K149 (1976).
See also 13.4, p. 136; *16*, p. *213*.

P. Koidl
Jahn-Teller Effect in the $^4T_2(l)$ and $^4T_1(l)$ States of Tetrahedrally Coordinated Mn^{2+}
(b) Vol. **74**, No. 2, p. 477–484 (1976).
See also *13.4*, p. *137*; 20.1, p. 265; 20.3, p. 282; 22.2.2, p. 365; 22.4.1, p. 380.

E. Litwin-Staszewska and W. Szymańska
Scattering on Short-Range Potentials in Semiconductors with a Narrow Energy Gap
(b) Vol. **74**, No. 2, p. K89–K92 (1976).
See also *14.3*, p. *175*; 15, p. 209.

J. Niewodniczańska-Zawadzka and J. Raułuszkiewicz
The Observation of LO Phonons in HgSe and HgTe by Electron Tunneling
(b) Vol. **74**, No. 2, p. K93–K95 (1976).
See also *6*, p. *55*; *14.3.3*, p. *191*; 22.4.3, p. 385.

J. Hamerský
Electrical Properties of CdSe Films under the Influence of Oxygen
(a) Vol. **35**, No. 2, p. K97–K99 (1976).
See also 9, p. 73; *14.3.1*, p. *184*.

I. G. Gavrikova, V. P. Migal, and A. L. Rvachev
Optical Sensitization of the Photoconductivity Bands
(a) Vol. **37**, No. 1, p. K21–K24 (1976).
See also *16*, p. *216*.

V. A. Brodovoi, N. Z. Derikot, B. R. Kiyak, G. P. Peka, and A. I. Proskura
Trigger Effect in Electroluminescent Ceramics ZnSe–Cu, I
(a) Vol. **37**, No. 2, p. 363–369 (1976).
See also *13.4*, p. *143*; *14.3.4*, p. *199*; 20.3, p. 287.

J. Daunay, J. Baillou, M. Lemercier, Jac. Daunay, P. Bugnet, and G. Batailler
Dispersion of the Parasitic Birefringence Induced by Lattice Plane Defects in ZnSe Single Crystals
(a) Vol. **37**, No. 2, p. 521–527 (1976).
See also 10.2, p. 96; *20*, p. *258*.

K. Kumazaki, E. Matsushima, and A. Odajima
Point Defects and Non Stoichiometry in HgSe
(a) Vol. **37**, No. 2, p. 579–589 (1976).
See also *10.2*, p. *96*; *14.3*, p. *180*.

M. K. Rao and S. R. Jawalekar
Resistivity Measurements of CdSe Films
(a) Vol. 38, No. 1, p. K93–K95 (1976).
See also *14.3.1*, p. *186*.

P. Byszewski, J. Kowalski, and R. R. Gałązka
Shubnikov-de Haas Oscillations in HgSe at 77 K in Pulsed Magnetic Fields
(b) Vol. 78, No. 2, p. 477–481 (1976).
See also *13.1*, p. *129*; 14.3, p. 181.

B. Hönerlage, C. Klingshirn, and J. B. Grun
Spontaneous Emission Due to Exciton–Electron Scattering in Semiconductors
(b) Vol. 78, No. 2, p. 599–608 (1976).
See also *13.5.1*, p. *156*; *20.3*, p. *288*; 22.4, p. 378; 22.4.1, p. 382.

22.4.3 Tellurides

Original Papers and Short Notes

P. Plumelle and M. Vandevyver
Lattice Dynamics of ZnTe and CdTe
(b) Vol. **73**, No. 1, p. 271–281 (1976).
See also *6*, p. *52*.

M. Baj, L. Dmowski, M. Kończykowski, and S. Porowski
A New Explanation of the Cl Donor Deionization and Its Kinetics Observed in CdTe:Cl under Hydrostatic Pressure
(a) Vol. **33**, No. 1, p. 421–426 (1976).
See also 13.4, p. 135; *14.3*, p. *174*.

A. M. White, P. J. Dean, B. Day, and F. Mansfield
Polaron, Zeeman, and Diamagnetic Effects in the Recombination Spectra of Donor Bound Excitons in Cadmium Telluride
(b) Vol. **74**, No. 1, p. K9–K13 (1976).
See also *13.4*, p. *136*; *13.5.1*, p. *151*; 13.5.3, p. 159; 20.3, p. 281.

Z. Liliental and H. Bartsch
The Effect of Diffusion of Ag on the Real Structure of ZnTe Crystals
(a) Vol. **34**, No. 1, p. K5–K6 (1976).
See also *10.2*, p. *90*.

S. Martinuzzi
n-CdS–p-CdTe Heterojunction Solar Cells
(a) Vol. **34**, No. 1, p. K21–K26 (1976).
See also *14.3.3*, p. *190*; *16*, p. *214*; 22.4.1, p. 379.

J. Niewodniczańska-Zawadzka and J. Raułuszkiewicz
The Observation of LO Phonons in HgSe and HgTe by Electron Tunneling
(b) Vol. **74**, No. 2, p. K93–K95 (1976).
See also *6*, p. *55*; *14.3.3*, p. *191*; 22.4.2, p. 383.

E. Mizera and M. Klimkiewicz
Some Remarks on the Defect Structure in HgTe Crystals Revealed by Transmission Electron Microscopy
(a) Vol. **34**, No. 2, p. K107–K108 (1976).
See also *10.2*, p. *91*.

W. Wardzyński and K. Pataj
Bound Exciton–Neutral Acceptor Complex in ZnTe
(b) Vol. **75**, No. 1, p. 341–346 (1976).
See also 12.2, p. 110; *13.5.1*, p. *152*; *20.3*, p. *283*.

A. Matulionis, J. Požela, and A. Reklaitis
Monte Carlo Calculations of Hot-Electron Transient Behaviour in CdTe and GaAs
(a) Vol. **35**, No. 1, p. 43–48 (1976).
See also *13*, p. *116*; *14.3.4*, p. *197*; 22.2.1, p. 363.

A. Opanowicz, K. Marinova, H. Liebing, and W. Ruppel
Luminescence of Highly Excited ZnTe Crystals
(b) Vol. **75**, No. 2, p. 471–481 (1976).
See also *13.5.1*, p. *152*; *20.3*, p. *283*.

F. El Akkad
Free-to-Bound Transitions in Lithium-Doped Zinc Telluride
(b) Vol. **76**, No. 1, p. 85–92 (1976).
See also *13.4*, p. *140*; *20.3*, p. *284*.

A. Jędrzejczak and T. Dietl
Thermomagnetic Properties of n-Type and p-Type HgTe
(b) Vol. **76**, No. 2, p. 737—751 (1976).
See also *13.1*, p. *126*; 14.3, p. 179; *15*, p. *210*.

K. Suzuki, N. Miura, S. Uchida, and S. Tanaka
Far-Infrared Magneto-Reflection in HgTe
(b) Vol. **76**, No. 2, p. 787—796 (1976).
See also *13.1*, p. *126*; 20.1, p. 272.

J. Toušková and R. Kužel
Fundamental Properties of Au–CdTe Metal–Semiconductor Contact
(a) Vol. **36**, No. 2, p. 747—755 (1976).
See also *14.3.3*, p. *193*; *14.3.4*, p. *199*.

A. Opanowicz and K. Marinova
Luminescence of ZnTe Single Crystals under High Two-Photon Excitation
(b) Vol. **77**, No. 1, p. K35—K38 (1976).
See also *13.5.1*, p. *155*; *20.3*, p. *286*.

P. Basu and H. Saha
Possible Performance of Back-Wall Schottky Barrier Solar Cells
(a) Vol. **37**, No. 2, p. 625—631 (1976).
See also 13.3, p. 133; *14.3.3*, p. *193*; *16*, p. *216*; 22.1.2, p. 357; 22.2.1, p. 364.

F. V. Wald
Self Compensation in CdTe and ZnTe Crystals Grown from Indium Solvents
(a) Vol. **38**, No. 1, p. 253—259 (1976).
See also *3*, p. *44*; 10.2, p. 97; 14.3, p. 181; 20.1, p. 277.

J. Kossut
Electron Transport Phenomena in Narrow- and Zero-Gap Semiconductors Containing Magnetic Impurities
(b) Vol. **78**, No. 2, p. 537—542 (1976).
See also *13.4*, p. *145*; *14.3*, p. *181*; 15, p. 211.

22.4.4 Mixed Crystals

Original Papers and Short Notes

K. Marinova, A. Opanowicz, and W. Ruppel
Photoconductivity of Highly Excited CdSe and $CdS_{0.48}Se_{0.52}$ Crystals
(a) Vol. **33**, No. 1, p. 195–204 (1976).
See also 13.5.1, p. 149; *16*, p. *213*; 22.4.2, p. 383.

R. S. Kim and S. Narita
Far-Infrared Interband Magnetoabsorption and Band Structure of $Hg_{1-x}Cd_xTe$ Alloys
(b) Vol. **73**, No. 2, p. 741–752 (1976).
See also *13.1*, p. *121*; 18, p. 222; 20.1, p. 262.

R. J. Iwanowski and T. Dietl
Low Temperature Electron Mobility in $Cd_vHg_{1-v}Se$ Mixed Crystals
(b) Vol. **75**, No. 2, p. K83–K87 (1976).
See also *14.3*, p. *177*.

K. Sierański, J. Szatkowski, and E. Majchrowska
Effective Mass of Heavy Holes in $Cd_xHg_{1-x}Te$ at 4.2 to 300 K
(b) Vol. **77**, No. 2, p. K91–K95 (1976).
See also *13.1*, p. *128*.

J. M. Pawlikowski
Negative Resistance of Metal-p-$Cd_xHg_{1-x}Te$ Contacts
(a) Vol. **37**, No. 2, p. K183–K186 (1976).
See also *14.3.4*, p. *200*.

G. Nimtz und E. Tyssen
Far Infrared Recombination Radiation from n-Type $Hg_{1-x}Cd_xTe$
(a) Vol. **38**, No. 2, p. K171–K173 (1976).
See also 13.4, p. 146; *20.1*, p. *278*.

22.5 Halides

Review Article

R. PRASAD
Present State of Polytypism in Cadmium Iodide Crystals
(a) Vol. **38**, No. 1, p. 11–44 (1976).
See also *1.1*, p. *10*; *1.2*, p. *14*; 10.2, p. 87; 22.3, p. 372; 22.4.1, p. 379; 22.5.3, p. 399; 22.8, p. 413.

Original Papers and Short Notes

M. I. SHEBOUL and W. EKARDT
Theory of Biexcitons Using the Boson Formalism. The Ground State Binding Energy
(b) Vol. **73**, No. 1, p. 165–178 (1976).
See also *13.5.1*, p. *149*.

T. FUKUMOTO, S. NAKASHIMA, K. TABUCHI, and A. MITSUISHI
Temperature Dependence of Raman Spectra of Cuprous Halides
(b) Vol. **73**, No. 1, p. 341–351 (1976).
See also *6*, p. *53*; *20.1*, p. *260*.

W. EKARDT and M. I. SHEBOUL
The Influence of the Electron–Hole Exchange Interaction on the Biexciton Ground State in CuCl and CuBr and Related Optical Transitions
(b) Vol. **73**, No. 2, p. 475–482 (1976).
See also 13.1, p. 121; *13.5.1*, p. *150*; 20.1, p. 261; *20.3*, p. *280*.

A. GRISEL and PH. SCHMID
Polytypism and Lattice Vibrations of PbI_2
(b) Vol. **73**, No. 2, p. 587–591 (1976).
See also *6*, p. *54*; 20.1, p. 262.

B. DORNER, R. E. GHOSH, and G. HARBEKE
Phonon Dispersion in the Layered Compound PbI_2
(b) Vol. **73**, No. 2, p. 655–659 (1976).
See also *6*, p. *54*; 7, p. 65; 12.2, p. 110.

P. K. C. PILLAI and R. NATH
Photo-Induced AC Impedance Measurements on HgI_2 Binder Layers
(a) Vol. **33** No. 2, p. K151–K154 (1976).
See also *16*, p. *213*.

D. L. HUBER and M. S. SEEHRA
Electron Paramagnetic Resonance in Anisotropic Magnets
(b) Vol. **74**, No. 1, p. 145–149 (1976).
See also *19*, p. *248*.

Č. BÁRTA, J. P. CHAPELLE, G. HAURET, CAO-XUÂN AN, A. FOUSKOVÁ, and Č. KOŇÁK
Brillouin Scattering in Hg_2Cl_2
(a) Vol. **34**, No. 1, p. K51–K54 (1976).
See also *1.2*, p. *15*.

W. EKARDT and M. I. SHEBOUL
On the First Excited State of the Biexciton in CuCl
(b) Vol. **74**, No. 2, p. 523–529 (1976).
See also *13.5.1*, p. *151*; 20.3, p. 282.

A. P. Rusakov and M. V. Anfimov
Non-Central Forces and Stability of the A^NB^{8-N} Tetrahedral Compounds (I): Bond Charges and the Valence-Force-Field Model
(b) Vol. **75**, No. 1, p. 73—77 (1976).
See also *4*, p. *45*; 12.2, p. 110; 22.1, p. **343**; 22.2, p. 361; 22.4, p. 376.

V. G. Vaks and V. E. Schneider
On the Theory of Phase Transitions in the Ammonium Halides
(a) Vol. **35**, No. 1, p. 61—72 (1976).
See also *1.2*, p. *16*; 22.5.4, p. 401.

P. Ramasamy and T. Nagarajan
Positron Angular Correlation Studies of Phase Transitions in Ammonium Halides
(a) Vol. **35**, No. 1, p. 379—386 (1976).
See also *1.2*, p. *16*; 4, p. 45.

S. K. Misra and G. R. Sharp
Electron Paramagnetic Resonance of Gd^{3+} in $PrCl_3 \cdot 7\,H_2O$
(b) Vol. **75**, No. 2, p. 607—612 (1976).
See also *19*, p. *251*.

E. Wiener-Avnear
Polaritons in an Ensemble of CuI Crystallites
(b) Vol. **75**, No. 2, p. 717—720 (1976).
See also *13.5*, p. *147*; 20.1, p. 269.

A. K. Rai R. S. Tiwari, and O. N. Srivastava
Intrapolytypic and Interpolytypic Transformation in Cadmium Iodide Crystals
(a) Vol. **35** No. 2 p. 719—724 (1976).
See also *1*, p. *9*.

A. J. H. Eijkelenkamp
Powder Reflection Measurements of Some Divalent Lead Halides
(b) Vol. **76**, No. 1, p. 153—160 (1976).
See also 13.5.1, p. 153; *20.1*, p. *270*.

K. Hattori
Exciton Binding Energies in Polar Crystals
(b) Vol. **76**, No. 1, p. 281—287 (1976).
See also *13.5.1*, p. *153*; 22.2.1, p. **363**; 22.4, p. 376.

A. I. Belyaeva and M. M. Kotlyarskii
Spectroscopic Investigations of Magnons in Weak Ferromagnetic $RbMnCl_3$
(b) Vol. **76**, No. 1, p. 419—426 (1976).
See also *13.5.4*, p. *160*; *20.1*, p. *271*.

E. Rieflin, G. Stühmer, and E. Schwab
Exciton Band Width versus Peak Shift in NH_4Br in Phases II and III as Measured by Thermoreflectance
(b) Vol. **76**, No. 2, p. 547—550 (1976).
See also *13.5.1*, p. *154*; 20.1, p. 271.

A. J. H. Eijkelenkamp and K. Vos
Reflectance Measurements on Single Crystals of PbFCl, PbFBr, and $PbBr_2$
(b) Vol. **76**, No. 2, p. 769—778 (1976).
See also 13.1, p. 126; 13.5.1, p. 154; *20.1*, p. *272*.

F. Sprenger
Magnetoresistance of Frozen-in Photoconductivity in Semiconducting CdF_2:In
(a) Vol. **36**, No. 2, p. 483—490 (1976).
See also 13.4, p. 141; *14.3*, p. *179*; *16*, p. *215*.

R. LEVY, C. KLINGSHIRN, E. OSTERTAG, VU DUY PHACH, and J. B. GRUN
Luminescence of a "Cold" Gas of Biexcitons in CuCl
(b) Vol. 77, No. 1, p. 381–391 (1976).
See also *13.5.1*, p. *155*; *20.3*, p. *286*.

M. B. HAXTON
Parameter Calculation from Zinc Blende Zone Boundary Conditions of CuI
(b) Vol. 78, No. 1, p. 259–264 (1976).
See also *6*, p. *62*.

S. V. NISTOR, V. GHIORDĂNESCU, and M. VODĂ
ESR of Mn^{2+} Doped $CdCl_2$ Single Crystals
(b) Vol. 78, No. 1, p. K31–K33 (1976).
See also *19*, p. *255*.

J. KURIATA and T. REWAJ
Superhyperfine Structure in the EPR of $SrF_2:Eu^{2+}$ Single Crystals
(b) Vol. 78, No. 1, p. K55–K58 (1976).
See also *19*, p. *255*.

E. KÖNIG and R. SCHNAKIG
Zero-Field Splitting of $^6S(d^5)$ Ions in Tetragonal and Rhombic Symmetry (II): Application of Complete d-Electron Calculations to the Zero-Field Splitting in $Mn:MnF_2$ and $Mn:ZnF_2$
(b) Vol. 78, No. 2, p. 543–551 (1976).
See also *13.4*, p. *145*; 18.1, p. 227; 19, p. 255; 20.1, p. 277.

M. SCHLÜTER, M. L. COHEN, S. E. KOHN, and C. Y. FONG
Electronic Structure of BiI_3
(b) Vol. 78, No. 2, p. 737–747 (1976).
See also 4, p. 47; *13.1*, p. *129*; 20.1, p. 278.

R. V. SHARMA and K. C. SHARMA
The Structure of Molten Cuprous Chloride
(b) Vol. 78, No. 2, p. K93–K95 (1976).
See also *2*, p. *41*.

M. CONSTANTINESCU, C. GHITA, L. GHITA, and I. BALTOG
Localized Levels in the Energy Gap of PbI_2 Crystals Evidenced by the TSC Method
(a) Vol. 38, No. 2, p. K175–K178 (1976).
See also *13.4*, p. *146*.

22.5.1 Silver Halides

Original Papers and Short Notes

D. J. Mickish and A. B. Kunz
Self-Consistent Local Orbitals for Silver Halide Crystals
(b) Vol. **73**, No. 1, p. 193–202 (1976).
See also *13.4*, p. *134*.

K. Hattori
Polarizability of an Electron Trapped by an Interstitial Ion
(b) Vol. **73**, No. 2, p. 717–721 (1976).
See also *13.4*, p. *135*; 22.5.2, p. 393.

M. Roulet, H. Huber, and C. Jaccard
Thermal Disorder of Silver Bromide Studied by 100 keV Proton Channeling
(a) Vol. **35**, No. 1, p. 97–108 (1976).
See also *10.2*, p. *92*; 11, p. 101.

R. Dinges, D. Fröhlich, and Ch. Uihlein
Two-Photon Absorption of Anisotropic Excitons in β-AgI
(b) Vol. **76**, No. 2, p. 613–622 (1976).
See also *13.5.1*, p. *154*; *20.1*, p. *271*.

A. V. Lelyakov
Bose-Einstein Condensation of Free Excitons in AgBr-Type Crystals at High Excitation
(b) Vol. 77, No. 1, p. 59–70 (1976).
See also 6, p. 60; 13.1, p. 127; *13.5.1*, p. *154*; 20.3, p. 285.

J. Shy-Yih Wang, M. Schlüter, and M. L. Cohen
The Electronic Structure of AgCl. Energy Bands, Densities of States, and Charge Densities
(b) Vol. 77, No. 1, p. 295–300 (1976).
See also *13.1*, p. *127*.

M. E. van Hulle and W. Maenhout-van der Vorst
Ionic Conductivity and Dielectric Loss of AgCl Doped with Pb^{2+}
(a) Vol. **37**, No. 2, p. 473–477 (1976).
See also *10.2*, p. *96*; *14.4*, p. *204*.

N. Starbov, A. Buroff, and J. Malinowski
Surface Ionic Conductivity, Lattice Disorder, and Space Charge Region in Thin Silver Bromide Layers
(a) Vol. **38**, No. 1, p. 161–170 (1976).
See also 10.2, p. 97; *14.4*, p. *204*.

R. K. Singh and K. P. Singh
Lattice Dynamics of AgBr at 295 K
(b) Vol. **78**, No. 2, p. 677–680 (1976).
See also *6*, p. *64*; 20.1, p. 277.

J. Weber
Indirect Near Edge Emission in Pure AgBr
(b) Vol. **78**, No. 2, p. 699–709 (1976).
See also 6, p. 64; *13.5.1*, p. *156*; *20.3*, p. *288*.

22.5.2 Alkali Halides

Review Article

S. V. Tsivinskii
Dislocation Multiplication in Single and Polycrystals in the Initial Stage of Deformation
(a) Vol. **36**, No. 1, p. 11–31 (1976).
See also *10.1*, p. *78*; *10.2*, p. *87*; 21, p. 289; 21.1, p. 302; 21.1.1, p. 316; 21.6, p. 329; 22.1.1, p. 345; 22.2.3, p. 367.

Original Papers and Short Notes

W. Baumhoer, J. Thiemann, and O. Kanert
Monte Carlo Simulation of NMR Line Shapes in Distorted Non-Metallic Solids.
(b) Vol. **73**, No. 1, p. 155–164 (1976).
See also 10.2, p. 87; *19*, p. *247*.

J. Höner zu Siederdissen
Impurity-Induced Raman Spectra of Potassium Halides
(b) Vol. **73**, No. 1, p. 239–244 (1976).
See also *6*, p. *52*; 13.4, p. 134; *20.1*, p. *260*.

R. Sati
Calculations of Optical Transitions of the F_3^+ Center
(b) Vol. **73**, No. 1, p. 353–359 (1976).
See also *10.2*, p. *87*; 13.4, p. 134.

W. Bausch
Accuracy of the Boundary Relaxation Time Approximation
(b) Vol. **73**, No. 1, p. K23–K25 (1976).
See also *6*, p. *53*; 8, p. 67.

L. Bosi
A Study of Electric Field Effect on the Radiative Lifetime of the M Center in NaCl and KCl
(b) Vol. **73**, No. 1, p. K61–K64 (1976).
See also *10.2*, p. *87*.

J. E. Strutt and E. Lilley
Structural Aspects of Clustering Reactions in Alkali Halides Doped with Divalent Impurities
(a) Vol. **33**, No. 1, p. 229–239 (1976).
See also *10.2*, p. *87*.

S. V. Tsivinskii
Statistical Theory of Thermally Activated Dislocation Motion in Crystals in the Case of Zero Internal Stresses
(a) Vol. **33**, No. 1, p. 301–311 (1976).
See also *10.2*, p. *88*; 22.1.1, p. 345; 22.1.2, p. 351; 22.2.3, p. 367.

V. Z. Bengus and T. P. Kovalenko
Dependence of the Dislocation Distribution in the Glide Band on the Density of Immobile Dislocations
(a) Vol. **33**, No. 1, p. K1–K3 (1976).
See also *10.2*, p. *88*.

A. Tóth
On the Mechanism of Charge Transport Caused by Moving Dislocations in NaCl Crystals Doped with Ca
(a) Vol. **33**, No. 1, p. K47–K50 (1976).
See also *10.2*, p. *88*; *14.4*, p. *202*.

A. TAMURA and T. KAWAMURA
Quantum Theory of Rosette-Motion Channeling
(b) Vol. **73**, No. 2, p. 391—400 (1976).
See also *11*, p. *100*; 13, p. 114; 21, p. 290.

L. NAGLI
$In^{+}v_a^{+}v_c^{-}$ Centres in a KCl–In Crystal
(b) Vol. **73**, No. 2, p. 427—430 (1976).
See also *10.2*, p. *88*; 20.3, p. 280.

E. M. LEDOVSKAYA and M. I. PETRASHEN
The Determination of the Symmetry Group of Molecular Centres in Crystals
(b) Vol. **73**, No. 2, p. 495—498 (1976).
See also 6, p. 54; *10.2*, p. *89*.

I. N. DOUGLAS
Magnetic Circular Dichroism of Co^{2+}-Doped Cubic Crystals
(b) Vol. **73**, No. 2, p. 641—650 (1976).
See also 13.4, p. 135; *18*, p. *222*; *20.1*, p. *262*; 22.6, p. 403.

K. HATTORI
Polarizability of an Electron Trapped by an Interstitial Ion
(b) Vol. **73**, No. 2, p. 717—721 (1976).
See also *13.4*, p. *135*; 22.5.1, p. 391.

L. B. HARRIS and J. FIASSON
Cation Diffusion along Boundaries in Sodium Chloride
(a) Vol. **33**, No. 2, p. 697—706 (1976).
See also *1.4*, p. *25*; *9*, p. *72*.

E. HARTMANN and Á. PÉTER
The Ionic Conductivity of Plastically Twisted NaCl Whiskers
(a) Vol. **33**, No. 2, p. K169—K171 (1976).
See also *10.2*, p. *89*; *14.4*, p. *202*.

T. IIDA and R. MONNIER
Static Properties of the Self-Trapped Hole in Simple Cubic Cesium Halides
(b) Vol. **74**, No. 1, p. 91—102 (1976).
See also 6, p. 54; *13.4*, p. *136*; 20.1, p. 263.

A. G. MADDOCK, A. F. WILLIAMS, K. E. SIEKIERSKA, and J. FENGER
The Mössbauer Emission Spectra of Impurity ^{57}Cobalt in a Halide Matrix
(b) Vol. **74**, No. 1, p. 183—192 (1976).
See also *5*, p. *48*; 22.5.3, p. 399.

N. KRISTIANPOLLER, A. REHAVY, and M. ISRAELI
Effects of Ultraviolet Irradiation on KBr:Gd
(b) Vol. **74**, No. 1, p. 245—248 (1976).
See also *10.2*, p. *89*; 13.5.1, p. 151; 20.1, p. 263; 20.3, p. 281.

U. GIORGIANNI, G. MONDIO, G. SAITTA, and G. VERMIGLIO
Role of the Jahn-Teller Effect in the Excited States of KBr:Tl-Type Crystals
(b) Vol. **74**, No. 1, p. 317—322 (1976).
See also *13.4*, p. *136*; *20.1*, p. *264*.

F. FRÖHLICH, P. GRAU, and M. SUSZYNSKA
The Simple Power Law of the Critical Flow Stress by Solution Hardening with Respect to the Aggregation of Ca Impurities in NaCl
(a) Vol. **34**, No. 1, p. 165—173 (1976).
See also *10.2*, p. *90*.

V. K. Agarwal
Correlation of Dielectric Breakdown Strength with Basic Parameters in Ionic Crystals
(a) Vol. **34**, No. 1, p. K67–K69 (1976).
See also *14.4*, p. *202*.

I. A. Markova-Osorgina and S. Z. Shmurak
Deformation-Induced Exciton Luminescence of Alkali Halide Crystals
(b) Vol. **74**, No. 2, p. 753–760 (1976).
See also 10.2, p. 90; 13.5.1, p. 152; *20.3*, p. *282*.

R. Balzer and F. R. Engler
Mechanical Anisotropy of H Centers in CsBr
(b) Vol. **74**, No. 2, p. K135–K140 (1976).
See also *10.2*, p. *90*; 11, p. 101.

V. Z. Bengus and T. P. Kovalenko
Mobile and Immobile Dislocations in LiF Crystals under Straining
(a) Vol. **34**, No. 2, p. 445–454 (1976).
See also *10.2*, p. *90*.

D. Casasent and F. Caimi
Suppressive Writing in Lithium-Doped NaF
(a) Vol. **34**, No. 2, p. 461–466 (1976).
See also *10.2*, p. *91*; 20.1, p. 266.

Y. V. G. S. Murti, G. Samuel, and S. Sivaraman
Studies on Tin Impurity Centres in KCl Crystals
(a) Vol. **34**, No. 2, p. 615–621 (1976).
See also *10.2*, p. *91*; 13.4, p. 137; 19, p. 250; 20.1, p. 266.

S. Duraiswamy and T. M. Haridasan
Moments of the Phonon Spectrum in Lithium Chloride
(b) Vol. **75**, No. 1, p. 53–59 (1976).
See also *6*, p. *56*; *8*, p. *68*.

L. Mărculescu and C. Ghiță
On the Luminescence of KCl:Pb and KBr:Pb Crystals
(b) Vol. **75**, No. 1, p. 375–379 (1976).
See also *20.3*, p. *283*.

L. Bosi
Lifetime Studies on Excited M and M^+ Centers in NaF
(b) Vol. **75**, No. 1, p. K27–K30 (1976).
See also *10.2*, p. *92*.

R. Nink and H.-J. Kos
On the Role of Z Centers in the Trapping Mechanism of Thermoluminescent Lithium Fluoride
(a) Vol. **35**, No. 1, p. 121–129 (1976).
See also *10.2*, p. *92*; 13.4, p. 138; *20.1*, p. *268*.

L. B. Harris, D. A. Moncrieff, and V. N. E. Robinson
High-Resolution Examination of Uncoated Insulators by SEM Applied to Grain Boundaries in Sodium Chloride
(a) Vol. **35**, No. 1, p. 371–377 (1976).
See also *1.4*, p. *27*; *10.2*, p. *92*.

H. Kirihata and J. Akutsu
Exoelectron Emission by Slow Electron Impacts
(a) Vol. **35**, No. 1, p. K73–K75 (1976).
See also 13.3, p. 132; *17*, p. *219*; 22.6, p. 405.

N. MATSUNAMI, Y. YOKOYAMA, and N. ITOH
Displacement of Impurities through Interaction with Radiation-Induced Interstitials in Alkali Halides
(b) Vol. **75**, No. 2, p. 483–492 (1976).
See also 10.2, p. 93; *11*, p. *101*; *20.1*, p. *268*.

H. MICHEL and M. WAGNER
Non-Thermal Stimulation of Three-Level Impurities by Phonons
(b) Vol. **75**, No. 2, p. 507–517 (1976).
See also *6*, p. *57*; 8, p. 68; *13.4*, p. *139*.

C. M. GOEL and B. DAYAL
Second-Order Raman Spectra of RbCl, RbBr, and RbI Crystals
(b) Vol. **75**, No. 2, p. 697–703 (1976).
See also *6*, p. *57*; *20.1*, p. *268*.

S. RADHAKRISHNA, B. V. R. CHOWDARI, and A. KASI VISWANATH
EPR Study of Mo^{5+} Ions in NaCl Single Crystal
(b) Vol. **75**, No. 2, p. K119–K121 (1976).
See also *19*, p. *251*.

L. BOSI
On the Luminescence and Photoconductivity of the F Centres in NaBr
(b) Vol. **75**, No. 2, p. K163–K165 (1976).
See also *10.2*, p. *93*; 16, p. 214; 20.3, p. 284.

P. MATTHYS, M. VANHAELST, and E. BOESMAN
A Modified Doping Procedure for High O_2^- Concentrations in Potassium and Rubidium Halides
(a) Vol. **35**, No. 2, p. K137–K139 (1976).
See also *3*, p. *42*; *19*, p. *252*.

G. DOKTER and H. J. HOFFMANN
Transient Bleaching of the F-Band in Alkali Halides
(b) Vol. **76**, No. 1, p. 241–249 (1976).
See also *10.2*, p. *93*; *13.4*, p. *140*; 16, p. 215.

R. A. EVARESTOV, M. I. PETRASHEN, and E. M. LEDOVSKAYA
Comparison of the Large Unit Cell and Small Periodic Cluster Approaches to the Quasi-molecular Calculations of the Band Spectra of Crystals
(b) Vol. **76**, No. 1, p. 377–383 (1976).
See also *13.1*, p. *125*; *13.4*, p. *140*.

NAWAB SINGH and A. W. JOSHI
Pressure Dependence of the Peak F-Band Absorption Energy in Alkali Halides
(b) Vol. **76**, No. 1, p. K39–K42 (1976).
See also *10.2*, p. *93*; *20.1*, p. *271*.

V. P. KISEL
Mobility of Dislocations in NaCl Single Crystals at Low Temperatures (I): 77 to 298 K
(a) Vol. **36**, No. 1, p. 297–306 (1976).
See also *10.2*, p. *93*.

A. A. URUSOVSKAYA, G. G. KNAB, and YU. Z. ESTRIN
Comparison of Dislocation Mobility in LiF Measured by Direct and Indirect Methods
(a) Vol. **36**, No. 1, p. 397–402 (1976).
See also *10.2*, p. *94*.

A. LOIDL, H. JEX, J. DAUBERT, and M. MÜLLNER
Zero and First Sound in NaF and RbI
(b) Vol. **76**, No. 2, p. 581–592 (1976).
See also *6*, p. *59*; 7, p. 66; 8, p. 69; *12.2*, p. *111*.

S. NAGEL, K. MASCHKE, and A. BALDERESCHI
Pseudopotential Calculation of the Electronic Valence Charge Density and the Optical Dielectric Constant of NaCl and MgO
(b) Vol. 76, No. 2, p. 629–632 (1976).
See also *13*, p. *117*; *14.4*, p. *203*; 22.6, p. 406.

A. BATANA and E. R. GONZALEZ
On the Pressure Dependence of the Effective Ionic Charge of Lithium, Sodium, and Rubidium Halides
(b) Vol. 76, No. 2, p. K105–K108 (1976).
See also *4*, p. *46*; *12.2*, p. *111*.

P. G. BARANOV, R. A. ZHITNIKOV, and V. A. KHRAMTSOV
EPR of Au° Atoms in NaCl Crystals
(b) Vol. 76, No. 2, p. K109–K112 (1976).
See also *19*, p. *253*.

H. BIEDERMAN
Emission Patterns at Various Conditions of Imaging the Electron Emission of Al–LiF–Au Structures
(a) Vol. 36, No. 2, p. 783–789 (1976).
See also *17*, p. *219*.

J. SÁRKÖZI and R. VOSZKA
The Variation of the Surface Charge of NaCl:Pb Monocrystals Due to Doping and Annealing
(a) Vol. 36, No. 2, p. K105–K107 (1976).
See also *10.2*, p. *94*; 14.4, p. 203.

P. MATTHYS, M. VANHAELST, and E. BOESMAN
Spin-Lattice Relaxation of the O_2^- Centre in RbI and RbBr
(b) Vol. 77, No. 1, p. 249–257 (1976).
See also *19*, p. *254*.

S. K. CHAKRABARTI, S. K. SARKAR, and S. SENGUPTA
On the Change of Electronic Polarisability of an Ion in a Crystal
(b) Vol. 77, No. 1, p. 329–340 (1976).
See also *4*, p. *46*.

U. M. GRASSANO, A. TANGA, and M. VALLI
Electric Field Effect on the Thallium Emission Bands in KBr and KI
(b) Vol. 77, No. 1, p. 341–346 (1976).
See also *10.2*, p. *95*; 20.1, p. 273; *20.3*, p. *286*.

L. MĂRCULESCU
Optical Spectra of Aggregate Centres in KCl:Pb Crystals
(b) Vol. 77, No. 1, p. 353–357 (1976).
See also *10.2*, p. *95*; *20.1*, p. *273*.

V. G. KVACHADZE and Z. G. AKHVLEDIANI
Influence of γ-Irradiation on the Thermal Conductivity of KCl:KOH Crystals
(b) Vol. 77, No. 1, p. K21–K25 (1976).
See also *6*, p. *60*; *8*, p. *70*; 11, p. 103; 20.1, p. 273.

M. RIGGIN, S. RADHAKRISHNA, and P. W. WHIPPEY
Thermoluminescence of X-Irradiated Alkali Fluoride Crystals
(a) Vol. 37, No. 1, p. 51–56 (1976).
See also 13.4, p. 142; *20.3*, p. *286*.

I. M. Boswarva
The Lattice Energies of Precipitate Phases and the Heats of Solution of Divalent Cations in Alkali Halide Crystals
(a) Vol. **37**, No. 1, p. 65–72 (1976).
See also *10.2*, p. *95*.

L. Marculescu, M. Popescu, I. Baltog, and C. Ghita
The State of Dispersion of Pb^{2+} in Highly Doped KCl Crystals
(b) Vol. 77, No. 2, p. 701–706 (1976).
See also *10.2*, p. *95*; 20.1, p. 274; 20.3, p. 287.

N. Banaii and G. Jacobs
The Parallel and Perpendicular Stark Effect of the F-Center in RbBr and RbI
(b) Vol. 77, No. 2, p. 707–715 (1976).
See also *10.2*, p. *96*; 20.1, p. 274.

P. C. Mehendru and Narendra Kumar
Optical Absorption Studies of KCl:Ru and NaCl:Ru Crystals
(b) Vol. 77, No. 2, p. K133–K134 (1976).
See also *20.1*, p. *274*.

H. G. Lipson, P. Ligor, and J. J. Martin
The Effect of Ionizing Radiation on the 10.6 μm Absorption of KCl and NaCl
(a) Vol. **37**, No. 2, p. 547–552 (1976).
See also *10.2*, p. *96*; *11*, p. *104*; 20.1, p. 275.

R. K. Gartia
Thermoluminescence Studies of Z_1 Centers in X-Irradiated Ca-Doped NaCl Crystals
(a) Vol. **37**, No. 2, p. 571–578 (1976).
See also *10.2*, p. *96*; 20.3, p. 287.

F. M. Caimi, J. O. Artman, W. W. Lin, and D. P. Casasent
Interpretation of Direct Writing M Center H-D Curve Characteristics of Photodichroic NaF Crystals
(a) Vol. **37**, No. 2, p. 617–624 (1976).
See also *10.2*, p. *96*; 13.4, p. 143; 20.1, p. 275.

D. S. Puri, V. K. Garg, and M. P. Verma
Many-Body Effects on the Third-Order Elastic Constants and Pressure Derivatives of the Second-Order Elastic Constants of CsCl Structure Solids
(b) Vol. 78, No. 1, p. 113–122 (1976).
See also *12.2*, p. *112*.

W. Schulz and W. Gründler
Quantum Mechanical ab initio Calculations of the Surface Structure of Crystalline Lithium Hydride
(b) Vol. 78, No. 1, p. 183–189 (1976).
See also *1.6*, p. *34*.

N. Banaii and G. Jacobs
Stark Effect of the F Center in NaBr
(b) Vol. 78, No. 1, p. K51–K53 (1976).
See also *10.2*, p. *97*.

J. Montemayor, R. Gómez-Ramírez, and E. Carrillo
High-Temperature Creep on KBr Single Crystals
(a) Vol. **38**, No. 1, p. 67–72 (1976).
See also *10.2*, p. *97*.

R. Gauthier and P. Pinard
Pulsed Laser Light Stimulated Desorption
(a) Vol. **38**, No. 1, p. 85–92 (1976).
See also *17*, p. *220*; 21.1, p. 313; 22.1.1, p. 349.

G. Berg and J. Poźniak
Theoretical Analysis of the Dipole Polarization and Depolarization in Me^{++}-Doped NaCl Type Alkali Halide Crystals (I): Nearest-Neighbour Approach
(a) Vol. 38, No. 1, p. 115–122 (1976).
See also *10.2*, p. *97*; 14.4, p. 204.

Ah Mee Hor, P. W. M. Jacobs, and K. S. Moodie
Thermal Depolarization in Crystals of Potassium Chloride Containing Strontium Chloride
(a) Vol. 38, No. 1, p. 293–298 (1976).
See also *8*, p. *71*; 10.2, p. 98.

R. W. Whitworth
A Charged-Dislocation Model for the Effect of an Electric Field on the Flow Stress of an Ionic Crystal
(a) Vol. 38, No. 1, p. 299–304 (1976).
See also *10.2*, p. *98*.

V. K. Jain
Effect of Annealing on Thermoluminescence of LiF (TLD-100)
(a) Vol. 38, No. 1, p. K65–K68 (1976).
See also 10.2, p. 98; *20.3*, p. *288*.

A. A. Elango and T. N. Nurakhmetov
Structure and Generation Mechanism of Br_3^- Centers in X-Rayed KBr
(b) Vol. 78, No. 2, p. 529–536 (1976).
See also *10.2*, p. *98*; 20.1, p. 277.

M. Vanhaelst, P. Matthys, and E. Boesman
Temperature Dependence of the EPR and Optical Spectra in V^{++}-Doped Alkali Halides
(b) Vol. 78, No. 2, p. 553–562 (1976).
See also *19*, p. *256*; *20.1*, p. *277*.

J. P. Buisson, S. Lefrant, A. Sadoc, L. Taurel, and M. Billardon
Raman Scattering by KI Containing F Centres
(b) Vol. 78, No. 2, p. 779–790 (1976).
See also 6, p. 64; *10.2*, p. *98*; 13.4, p. 145; 20.1, p. 278.

S. Zazubovich and V. Osminin
Hole Mechanisms of H and F Centre Production in KCl Crystals
(b) Vol. 78, No. 2, p. 813–819 (1976).
See also *10.2*, p. *98*; *20.3*, p. *288*.

F. J. López and F. Agulló-López
ESR of Conduction Electrons in Electrolytically Coloured Cation-Doped NaCl
(b) Vol. 78, No. 2, p. K81–K83 (1976).
See also 10.2, p. 98; *19*, p. *256*.

R. Gauthier and C. Guittard
Mechanism Investigations of a Pulsed Laser Light Induced Desorption
(a) Vol. 38, No. 2, p. 477–486 (1976).
See also *17*, p. *220*; 21.1, p. 314; 22.1.1, p. 350.

T. Gotô, Y. Nigara, and K. Kamiyoshi
Dielectric Constant of Ammonium and Cesium Halides
(a) Vol. 38, No. 2, p. K135–K138 (1976).
See also *14.4*, p. *205*; 22.5.4, p. 401.

22.5.3 Alkali-Earth Halides

Review Article

R. Prasad
Present State of Polytypism in Cadmium Iodide Crystals
(a) Vol. 38, No. 1, p. 11–44 (1976).
See also *1.1*, p. *10*; *1.2*, p. *14*; 10.2, p. 87; 22.3, p. 372; 22.4.1, p. 379; 22.5, p. 388; 22.8, p. 413.

Original Papers and Short Notes

J. W. J. Timans and H. W. den Hartog
Local-Mode Absorption of H^- and D^- in $SrCl_2$:RE and SrF_2:RE Single Crystals
(b) Vol. 73, No. 1, p. 283–292 (1976).
See also *6*, p. *53*; *20.1*, p. *260*.

T. J. Magee, N. M. Johnson, and J. Peng
The Effect of Illumination on the 10.6 μm Absorption Coefficient of Electron-Irradiated Single Crystal KCl
(a) Vol. 33, No. 1, p. 415–419 (1976).
See also *10.2*, p. *88*; *11*, p. *100*; 20.1, p. 261.

R. K. Bagai and A. V. R. Warrier
Ultraviolet Absorption Spectrum of CaF_2:Mn^{2+}
(b) Vol. 73, No. 2, p. K123–K126 (1976).
See also *20.1*, p. *262*.

A. G. Maddock, A. F. Williams, K. E. Siekierska, and J. Fenger
The Mössbauer Emission Spectra of Impurity ^{57}Cobalt in a Halide Matrix
(b) Vol. 74, No. 1, p. 183–192 (1976).
See also *5*, p. *48*; 22.5.2, p. 393.

B. D. Bhattacharyya
Tunnel Splitting in a Trigonal O^- Center
(b) Vol. 74, No. 1, p. K53–K56 (1976).
See also *10.2*, p. *90*; *19*, p. *249*.

J. Kuriata, N. Guskos, T. Rewaj, J. Szumowski, and K. Falkowski
Spin–Lattice Coefficients of Eu^{2+} in SrF_2
(b) Vol. 74, No. 2, p. 683–687 (1976).
See also *19*, p. *250*.

M. O. M. de Matos, H. S. Brandi, and B. Maffeo
Wave Function for the Interstitial Hydrogen Atom in SrF_2
(b) Vol. 75, No. 1, p. 271–278 (1976).
See also 13.4, p. 138; *19*, p. *250*.

H. Bill and W. von der Osten
Raman and Vibronic Spectra of a New Oxygen Molecule Ion in CaF_2
(b) Vol. 75, No. 2, p. 613–619 (1976).
See also 6, p. 57; *10.2*, p. *93*; 20.3, p. 284.

M. L. Falin, V. P. Meiklyar, V. A. Ulanov, and M. M. Zaripov
Electron Nuclear Double Resonance of Mn^{2+} in SrF_2
(b) Vol. 75, No. 2, p. 731–734 (1976).
See also *19*, p. *251*.

M. L. Falin, V. P. Meiklyar, and V. P. Zentsov
ENDOR and Covalent Model of the $[TiF_8]^{6-}$ Complex
(b) Vol. 75, No. 2, p. K111–K114 (1976).
Erratum
(b) Vol. 77, No. 1, p. K77 (1976).
See also *19*, p. *251*.

D. L. Lyfar, V. E. Goncharuk, and S. M. Ryabchenko
Temperature Dependence of Nuclear Quadrupole Resonance in Layer-Type Crystals
(b) Vol. 76, No. 1, p. 183–189 (1976).
See also *19*, p. *252*.

C. J. F. Graf, B. Maffeo, and H. S. Brandi
Calculation of Magnetic Hyperfine Parameters for the H_i^0 Center in CaF_2, SrF_2 and BaF_2
(b) Vol. 76, No. 2, p. K137–K140 (1976).
See also *13.4*, p. *141*; *19*, p. *253*.

V. V. Ratnam and H. D. Banerjee
On the Glow Curves and TL Centres in CaF_2:Ce, Mn Crystals Irradiated by Cathode Rays
(a) Vol. 36, No. 2, p. 693–697 (1976).
See also *10.2*, p. *94*; 20.3, p. 285.

V. M. Orera and R. Alcalá
Formation and Size Evolution of Ca Colloids in Additively Colored CaF_2
(a) Vol. 38, No. 2, p. 621–627 (1976).
See also *10.2*, p. *98*; 20.1, p. 278.

22.5.4 Mixed Crystals

Original Papers and Short Notes

T. ROMANOVSKIS
Concentration Dependence of the Debye Temperature in Ideal Solid Solutions
(b) Vol. **73**, No. 1, p. K13–K14 (1976).
See also *8*, p. *67*.

M. NATARAJAN and E. A. SECCO
Electrical Conductivity and Phase Transformation Studies on the Mixed Metal Halides $RbCdX_3$ (X=Cl, Br, I) and Cs_2CuBr_4
(a) Vol. **33**, No. 1, p. 427–433 (1976).
See also *1.2*, p. *14*; 8, p. 67; 14.4, p. 202.

S. DEONARINE and S. J. JOSHUA
Symmetry Properties of the Planar Antiferromagnet K_2NiF_4
(b) Vol. **74**, No. 2, p. 659–665 (1976).
See also 13.5.1, p. 152; *13.5.4*, p. *160*; 18.4, p. 243.

V. G. VAKS and V. E. SCHNEIDER
On the Theory of Phase Transitions in the Ammonium Halides
(a) Vol. **35**, No. 1, p. 61–72 (1976).
See also *1.2*, p. *16*; 22.5, p. 389.

R. HIRSCH and G. STÜHMER
A Simple Technique to Measure Optical SHG in Powder Samples by Means of Electronic Integration
(a) Vol. **37**, No. 1, p. 105–108 (1976).
See also *1.2*, p. *18*; *20*, p. *258*.

J. ASLAM, S. ROLANDSON, M. M. BEG, N. M. BUTT, and Q. H. KHAN
Phonons in the Mixed Crystal $K_{0.5}Rb_{0.5}I$
(b) Vol. 77, No. 2, p. 693–700 (1976).
See also *6*, p. *61*; 12.2, p. 112.

V. E. BOZHEVOLNOV, L. N. IVANOV, V. K. KOZLOV, YU. V. VORONOV, YU. P. TIMOFEEV, and V. V. KARELIN
Energy Transfer in CaF_2–Gd, Mn Crystals
(b) Vol. 78, No. 2, p. 483–487 (1976).
See also 13.4, p. 145; *20.3*, p. *288*.

K. N. SHRIVASTAVA
Thermal Effects in Magnetic Exchange Interaction
(b) Vol. 78, No. 2, p. 749–755 (1976).
See also *6*, p. *64*; *18.2*, p. *236*.

W. BAUHOFER, L. GENZEL, and W. KÖNIG
Far Infrared Studies of the Impurity Mode $NH_4Br:Cl^-$
(b) Vol. 78, No. 2, p. K121–K123 (1976).
See also *6*, p. *64*; *20.1*, p. *278*.

T. GOTÔ, Y. NIGARA, and K. KAMIYOSHI
Dielectric Constant of Ammonium and Cesium Halides
(a) Vol. **38**, No. 2, p. K135–K138 (1976).
See also *14.4*, p. *205*; 22.5.2, p. 398.

22.6 Simple Oxides (CdO, ZnO See 22.4)

Review Article

J. Kadlec and K. H. Gundlach
Results and Problems of Internal Photoemission in Sandwich Structures
(a) Vol. **37**, No. 1, p. 11—28 (1976).
See also 13.1, p. 120; *14.3.3*, p. *190*; *16*, p. *213*.

Original Papers and Short Notes

A. N. Ermoshkin, R. A. Evarestov, and E. A. Kotomin
Molecular Cluster Approach to Magnesium and Calcium Oxide Crystals (II): F^+ and F Centres
(b) Vol. **73**, No. 1, p. 81—86 (1976).
See also *10.2*, p. *87*; *13.4*, p. *134*.

M. Tapiero, J. P. Zielinger, and C. Noguet
Photomemory Effect in Cu_2O Single Crystals. Phenomenology and Interpretation
(a) Vol. **33**, No. 1, p. 155—166 (1976).
See also 13.4, p. 135; *16*, p. *213*.

R. Parker and C. J. Tinsley
Electrical Conduction in Magnetite
(a) Vol. **33**, No. 1, p. 189—194 (1976).
See also *14.3*, p. *173*; 18.3, p. 239.

E. M. Trukhanov, E. B. Gorokhov, and S. I. Stenin
Specific Features of the Dislocation Structure of Germanium in the System Germanium-Pyrolitic Silica Film
(a) Vol. **33**, No. 1, p. 435—442 (1976).
See also *1.5*, p. *31*; 2, p. 36; 10.2, p. 88; 22.1.1, p. 345.

G. Schiffmacher, P. E. Caro, and C. Boulesteix
Twin Dislocations Observed by Electron Microscopy at High Temperatures
(a) Vol. **33**, No. 1, p. K9—K13 (1976).
See also *10.2*, p. *88*.

C. F. Conde, A. Dominguez-Rodriguez, A. Conde, and R. Marquez
Microhardness Tests in Nickel Oxide Single Crystals
(a) Vol. **33**, No. 1, p. K25—K29 (1976).
See also *10.2*, p. *88*.

R. P. Khare
Thermoluminescence of CaO(Ce, Gd) Phosphors
(a) Vol. **33**, No. 1, p. K51—K53 (1976).
See also *20.3*, p. *280*.

R. A. Evarestov, A. N. Ermoshkin, and E. A. Kotomin
Molecular Cluster Approach to Magnesium and Calcium Oxide Crystals (III): Charge Distribution Analysis of Some Hole Centres
(b) Vol. **73**, No. 2, p. 483—486 (1976).
See also *10.2*, p. *89*; *13.4*, p. *135*.

D. Vivien, A. Kahn, A. M. Lejus, and J. Livage
EPR Investigation of the Very High Zero-Field Splitting of Gd^{3+} in La_2O_3 Single Crystals
(b) Vol. **73**, No. 2, p. 593—596 (1976).
See also *19*, p. *248*.

I. N. Douglas
Magnetic Circular Dichroism of Co^{2+}-Doped Cubic Crystals
(b) Vol. **73**, No. 2, p. 641—650 (1976).
See also **13.4, p. 135**; *18, p. 222*; *20.1, p. 262*; **22.5.2, p. 393.**

J. M. Honig, A. P. B. Sinha, W. E. Wahnsiedler, and H. Kuwamoto
Studies of the Band Structure of NbO by X-Ray Photoelectron Spectroscopy
(b) Vol. **73**, No. 2, p. 651—654 (1976).
See also *13.1, p. 121*; 17, p. 218.

K. Breuer, W. Tews, and E. Steinbeiss
Energy Loss Measurements on Thin VO_2 Films
(b) Vol. **73**, No. 2, p. K95—K97 (1976).
See also *13, p. 114*; *20, p. 257.*

C. M. Scala and D. R. Hutton
Site Assignment of Fe^{3+} in α-Quartz
(b) Vol. **73**, No. 2, p. K115—K117 (1976).
See also *10.2, p. 89*; 19, p. 248.

C. Boulesteix, J. van Landuyt, and S. Amelinckx
Identification of Rotation and Reflection Twins by Diffraction and Contrast Experiments in the Electron Microscope
(a) Vol. **33**, No. 2, p. 595—606 (1976).
See also *1.4, p. 25*; *10.2, p. 89.*

G. van Tendeloo, J. van Landuyt, and S. Amelinckx
The α → β Phase Transition in Quartz and $AlPO_4$ as Studied by Electron Microscopy and Diffraction
(a) Vol. **33**, No. 2, p. 723—735 (1976).
See also *1.2, p. 14*; *1.4, p. 26*; 22.8.1, p. 419.

N. Takeuchi, K. Inabe, and H. Nanto
Kinetics Order of Thermoluminescent Red Emission from MgO Single Crystals
(a) Vol. **33**, No. 2, p. K125—K127 (1976).
See also 10.2, p. 89; *20.3, p. 281.*

C. A. Bates and H. Szymczak
The Phonon-Induced Temperature Dependences of Spin-Hamiltonian Parameters for S-State Ions
(b) Vol. **74**, No. 1, p. 225—233 (1976).
See also *6, p. 55*; *19, p. 249*; 22.4.1, p. 379.

N. Uhle and J. Windscheif
Optical Absorption in V_2O_5 Doped with Li and Na
(b) Vol. **74**, No. 1, p. K15—K18 (1976).
See also 13.5.3, p. 159; *20.1, p. 264.*

G. Jungk
Determination of Optical Constants. A Null-Method for Non-Absorbing Surface Films
(a) Vol. **34**, No. 1, p. 69—72 (1976).
See also 1.5, p. 31; *20.1, p. 264.*

M. Euler and A. Scharmann
Thermally Stimulated Exoelectron Emission (TSEE) of BeO Layers. Auger Electron Spectroscopy Investigations
(a) Vol. **34**, No. 1, p. 297—306 (1976).
See also 1.6, p. 34; 13.3, p. 132; *17, p. 218.*

G. Nihoul-Boutang and C. Boulesteix
Observation of Lattice-Conservating Twins by Electron Microscopy. Study of the Image Contrasts
(a) Vol. **34**, No. 1, p. 331—338 (1976).
See also *1.4, p. 26*; *10.2, p. 90.*

K.-D. Ufert
Stress Induced Switching in VO_2 Thin Films
(a) Vol. **34**, No. 1, p. K83–K86 (1976).
See also *8*, p. *68*; *12.2*, p. *110*; 14.3.1, p. 183.

D. Chakraborty and G. Lehmann
On the Structures and Orientations of Hydrogen Defects in Natural and Synthetic Quartz Crystals
(a) Vol. **34**, No. 2, p. 467–474 (1976).
See also *10.2*, p. *91*; 20.1, p. 266; 22.8, p. 414.

K.-H. Müller, M. Wolf, H. Hemschik, and K. Hausmann
Stability of a VO_2 Thermal Switching Sample with Filament Formation
(a) Vol. **34**, No. 2, p. K199–K202 (1976).
See also *14.3.4*, p. *197*.

V. Siegel, W. Rasp, and H.-H. Kirchner
The Influence of Alkyl Groups on Exoelectron Emission from BeO
(a) Vol. **34**, No. 2, p. K203–K204 (1976).
See also 1.6, p. 34; *17*, p. *218*.

R. P. Goyal, K. K. Sarkar, and S. C. Goyal
Dispersion of Phonons in Transition Metal Oxides (MnO)
(b) Vol. **75**, No. 1, p. 101–104 (1976).
See also *6*, p. *56*.

D. Billard, F. Gervais, and B. Piriou
Analysis of Multiphonon Absorption in Corundum
(b) Vol. **75**, No. 1, p. 117–126 (1976).
See also *6*, p. *56*; *20.1*, p. *267*.

O. I. Shklyarevskii, A. A. Lysykh, and I. K. Yanson
Studies on Al–Insulator–Pb Tunnel Junctions with Barriers Prepared in Glow Discharge of CO_2 and CO
(a) Vol. **35**, No. 1, p. 49–53 (1976).
See also *14.3.3*, p. *191*.

M. Ali, S. J. H. Bukhari, K. A. Shoaib, and F. H. Hashmi
Monoclinic ⇌ Tetragonal Phase Transformation in ZrO_2 Single Crystals
(a) Vol. **35**, No. 1, p. 181–184 (1976).
See also *1.2*, p. *16*.

A. Mazel and F. Ajustron
Influence of Systematic Interactions on the Intensity Profiles of Kikuchi Lines Obtained at Very High Voltage
(a) Vol. **35**, No. 1, p. 197–211 (1976).
See also *1.1*, p. *11*.

R. Dragila, L. Bárdoš, and G. Lončar
Model of SiO_2 Film Growth on Si in Oxygen Microwave Discharge
(a) Vol. **35**, No. 1, p. 291–301 (1976).
See also *1.5*, p. *31*; *3*, p. *42*.

B. Pardo and D. Renard
Electron Microscopic Study on Single Crystals of Non-Uniform Thickness
(a) Vol. **35**, No. 1, p. 329–338 (1976).
See also *1.1*, p. *12*.

W. Brückner, W. Moldenhauer, B. Thuss, and G. Försterling
Crystallographic Changes at the Conductivity Transition of V_3O_5
(a) Vol. **35**, No. 1, p. K49–K51 (1976).
See also *1.2*, p. *16*; 15, p. 210.

I. K. Amanis and J. G. Kliava
Temperature Dependence of the Hyperfine Coupling Constants of Copper Atoms in α-Quartz
(a) Vol. **35**, No. 1, p. K67—K68 (1976).
See also *19*, p. *251*.

H. Kirihata and J. Akutsu
Exoelectron Emission by Slow Electron Impacts
(a) Vol. **35**, No. 1, p. K73—K75 (1976).
See also 13.3, p. 132; *17*, p. *219*; 22.5.2, p. 394.

A. Andreeff and G. Brauer
Positron Annihilation Study of the Metal–Insulator Transition in V_2O_3 and VO_2
(b) Vol. **75**, No. 2, p. K115—K118 (1976).
See also *13*, p. *116*.

J. Pinguet and S. S. Minn
Optical Stimulation of Depolarization Currents in Silicon Monoxide Doped with Metallic Impurities
(a) Vol. **35**, No. 2, p. 431—435 (1976).
See also *14.4*, p. *203*; *16*, p. *214*.

M. A. Boubel, R. P. Mainard, H. G. Fousse, A. J. Pointon, and F. R. Jeannot
The Specific Heat Anomaly of Solid Solutions of Isostructural Antiferromagnetic Oxides (pFeO–qCoO)
(a) Vol. **35**, No. 2, p. 459—464 (1976).
See also 8, p. 68; *18.4*, p. *244*.

A. Heidemann, K. K. Kosuge, and S. Kachi
Investigation of the Hyperfine Splitting in V_3O_5 and V_4O_7 by Inelastic Neutron Scattering
(a) Vol. **35**, No. 2, p. 481—485 (1976).
See also *18.4*, p. *244*.

E. F. Harris and J. H. Crawford, Jr.
The Influence of OH^- Impurity on the Oxidation of Iron Impurity in MgO above 1000 °C
(a) Vol. **35**, No. 2, p. 667—673 (1976).
See also *10.2*, p. *93*; 20.1, p. 269.

E. P. Balsamo, G. Paternò, A. Barone, M. Russo, and R. Vaglio
Temperature and Magnetic Field Dependence of the Critical Current in Sn-Sn_xO_y-In Josephson Junctions
(a) Vol. **35**, No. 2, p. K173—K175 (1976).
See also *14.2*, p. *170*; 21, p. 295.

C. Escribe and E. van der Voort
Lattice Energy Calculations of Complexes Generated by Dissolution of Gd and Fe in Monoclinic Zirconia
(a) Vol. **36**, No. 1, p. 375—383 (1976).
See also *10.2*, p. *94*.

H.-J. Anklam and K.-D. Ufert
Electrothermal Behaviour of VO_2 Thin Film Switching Devices
(a) Vol. **36**, No. 1, p. K71—K75 (1976).
See also *8*, p. *69*; *14.3.3*, p. *192*; 14.3.4, p. 198.

H. Düsterhöft, R. Manns, and D. Hildebrandt
Experimental Proof of a Theoretical Relation for the Probability of Positive Ion Excitation by Bombardment of Solid Surfaces with 12 keV Ar^+ Ions
(a) Vol. **36**, No. 1, p. K93—K97 (1976).
See also *11*, p. *102*; 21, p. 296; 21.1, p. 310; 21.1.1, p. 320; 22.1.1, p. 348; 22.1.2, p. 355.

S. Nagel, K. Maschke, and A. Baldereschi
Pseudopotential Calculation of the Electronic Valence Charge Density and the Optical Dielectric Constant of NaCl and MgO
(b) Vol. **76**, No. 2, p. 629–632 (1976).
See also *13*, p. *117*; *14.4*, p. *203*; 22.5.2, p. 396.

A. Balzarotti and A. Bianconi
Electronic Structure of Aluminium Oxide as Determined by X-Ray Photoemission
(b) Vol. **76**, No. 2, p. 689–694 (1976).
See also *13.1*, p. *126*; 17, p. 219.

P. Clauws and J. Vennik
Lattice Vibrations of V_2O_5. Determination of TO and LO Frequencies from Infrared Reflection and Transmission
(b) Vol. **76**, No. 2, p. 707–713 (1976).
See also *6*, p. *59*; *20.1*, p. *272*.

G. Borstel, E. Schuller, and H. J. Falge
Surface Phonon Polaritons on Absorbing Crystals
(b) Vol. **76**, No. 2, p. 759–768 (1976).
See also *13.5*, p. *147*; *20.1*, p. *272*.

H. M. Kizilyalli and P. R. Mason
DC and AC Electrical Conduction in Single Crystal Alumina
(a) Vol. **36**, No. 2, p. 499–508 (1976).
See also *14.4*, p. *203*.

B. P. Rai, K. Singh, and R. S. Srivastava
Current Transport Phenomena in SiO_2 Films
(a) Vol. **36**, No. 2, p. 591–595 (1976).
See also *14.3.1*, p. *185*; 14.3.4, p. 198.

H. Krause
Electron Injection and Trap Filling in Insulating Layers
(a) Vol. **36**, No. 2, p. 705–715 (1976).
See also *13.4*, p. *141*; *14.3.3*, p. *192*.

W. D. Schmidt and D. George
Switching Behaviour of Nb–Nb_xO_y–Pb Josephson Tunnel Junctions
(a) Vol. **36**, No. 2, p. K153–K155 (1976).
See also *14.2*, p. *170*; 14.3.3, p. 193.

V. S. Grunin, G. D. Davtyan, V. A. Ioffe, and I. B. Patrina
EPR of Cu^{2+} and Radiation Centres in Anatase (TiO_2)
(b) Vol. 77, No. 1, p. 85–92 (1976).
See also *13.4*, p. *142*; *19*, p. *253*.

P. Brüesch, M. Bösch, W. Känzig, M. Ziegler, and W. Bührer
Lattice Dynamics of the Phases I, II, and III of Sodium Hyperoxide NaO_2
(b) Vol. 77, No. 1, p. 153–164 (1976).
See also *6*, p. *60*; 20.1, p. 272.

D. Zwingel
The Electronic Structure of Trapped Hole Centres in SnO_2
(b) Vol. 77, No. 1, p. 171–180 (1976).
See also *19*, p. *253*.

S. Govinda
Coloration and Luminescence in Pure and Chromium-Doped Al_2O_3 Single Crystals Irradiated with X-Rays at Room Temperature
(a) Vol. **37**, No. 1, p. 109–117 (1976).
See also *10.2*, p. *95*; 11, p. 103; *20.1*, p. *273*; 20.3, p. 286.

T. L. ANDERSON, R. E. NEWNHAM, L. E. CROSS, and J. W. LAUGHNER
Laser-Induced Twinning in Quartz
(a) Vol. **37**, No. 1, p. 235–245 (1976).
See also 1.4, p. 28; *10.2*, p. *95*; *12.2*, p. *112*.

T. RUSKOV, T. TOMOV, and S. GEORGIEV
Mössbauer Investigation of the Morin Transition in Hematite
(a) Vol. **37**, No. 1, p. 295–302 (1976).
See also *5*, p. *50*; 18.4, p. 245.

I. V. KRYLOVA and V. I. SVITOV
On the Role of Adsorption Layers in Thermostimulated Exoemission from Beryllium Oxide
(a) Vol. **37**, No. 1, p. K9–K12 (1976).
See also *17*, p. *220*.

S. N. KARYAGIN, P. K. KASHKAROV, V. F. KISELEV, and S. N. KOZLOV
On the Nature of Electron and Hole Traps in Germanium Dioxide Thermally Grown on Germanium
(a) Vol. **37**, No. 1, p. K17–K19 (1976).
See also *13.3*, p. *133*; *19*, p. *254*.

J. T. GOURLEY and E. R. VANCE
Sharp Infrared Absorption Lines in Luminescent CaO
(b) Vol. 77, No. 2, p. K85–K89 (1976).
See also *10.2*, p. *96*; 20.1, p. 274.

V. N. ANDREEFF, T. V. SMIRNOVA, and F. A. CHUDNOVSKII
Electrical Conductivity of Vanadium Dioxide Films in the Vicinity of the Metal–Semiconductor Phase Transition
(b) Vol. 77, No. 2, p. K97–K101 (1976).
See also *14.1*, p. *166*; *14.3.1*, p. *185*.

B. LORENZ and D. IHLE
A Model of the Semiconductor–Semiconductor Transition in Ti_4O_7
(b) Vol. 77, No. 2, p. K177–K181 (1976).
See also *14.3*, p. *180*.

H. BETHGE, E. PIPPEL, and J. WOLTERSDORF
Problems of Epitaxial Growth in Systems of Weak Interaction Observed in the Case of Al on MgO
(a) Vol. **37**, No. 2, p. 457–466 (1976).
See also 1.4, p. 28; *1.5*, p. *32*; *3*, p. *44*; 21, p. 298.

E. I. TERUKOV, F. A. CHUDNOVSKII, W. REICHELT, H. OPPERMANN, W. BRÜCKNER, H.-P. BRÜCKNER, and W. MOLDENHAUER
Investigation of the Physical Properties of V_3O_5 at Phase Transition with Consideration of Its Range of Homogeneity
(a) Vol. **37**, No. 2, p. 541–546 (1976).
See also *14.3*, p. *180*; 15, p. 211; 18.1, p. 227.

W. MOLDENHAUER and W. BRÜCKNER
Steady-State Switching Characteristics of V_3O_5 Single Crystals
(a) Vol. **37**, No. 2, p. K143–K146 (1976).
See also *14.3.3*, p. *193*; 14.3.4, p. 200.

B. LALEVIC and N. K. ANNAMALAI
Schottky Barriers Effects on the Dielectric Behavior of the Nb–Nb_2O_5–Nb System
(a) Vol. **37**, No. 2, p. K177–K181 (1976).
See also *14.3.1*, p. *185*; *14.3.3*, p. *193*.

G. Hoppmann and E. Salje
Optical and Electrical Properties of H^+ Doped WO_3 Single Crystals
(a) Vol. **37**, No. 2, p. K187—K191 (1976).
See also 13.1, p. 128; *14.3*, p. *180*; *20.1*, p. *275*.

H. J. Falge, E. Schuller, and G. Borstel
Directional Dispersion of Polar Phonons Measured by Infrared Reflection
(b) Vol. 78, No. 1, p. 123—132 (1976).
See also *6*, p. *62*; 20.1, p. 275.

A. N. Pandey and K. S. Upadhyaya
Lattice Dynamics of BaO
(b) Vol. 78, No. 1, p. 399—404 (1976).
See also *6*, p. *63*.

F. Appel, H. Bethge, and U. Messerschmidt
Distribution of Point-Obstacle Distances during the Motion of Screw Dislocations in MgO Single Crystals
(a) Vol. **38**, No. 1, p. 103—108 (1976).
See also *10.2*, p. *97*.

D. F. Crabtree
Luminescence and Charge Compensation in $SnO_2:Eu^{3+}$
(a) Vol. **38**, No. 1, p. 217—222 (1976).
See also *20.3*, p. *288*.

A. E. Gershinskii, L. V. Mironova, and E. I. Cherepov
SiO_2–Si (n-Type) Interface Investigation with the Electrochemical Method
(a) Vol. **38**, No. 1, p. 369—373 (1976).
See also *1.5*, p. *33*; *14.3.2*, p. *188*; 22.1.2, p. 358.

K. S. Bobev, N. I. Diakonova, and N. V. Mileshkina
Field Emission through Semiconductor and Dielectric Layers
(a) Vol. **38**, No. 1, p. 393—399 (1976).
See also *14.3.4*, p. *200*; 22.1.1, p. 349.

E. I. Terukov, F. A. Chudnovskii, W. Brückner, W. Reichelt, H.-P. Brückner, W. Moldenhauer, and H. Oppermann
Influence of Mo, Ti, and Ga Doping on Conductivity Transition of V_3O_5
(a) Vol. **38**, No. 1, p. K23—K26 (1976).
See also *14.3*, p. *181*; *15*, p. *211*; 22.6.1, p. 409.

K. H. Germann
Optical Anisotropy Measurements in Multidomain Samples: Application to Phase Transitions in NiO and $Pb_5Ge_3O_{11}$
(a) Vol. **38**, No. 1, p. K81—K84 (1976).
See also 1.2, p. 20; 14.4.1, p. 208; 18.4, p. 246; *20.1*, p. *277*; 22.8.1, p. 424.

M. Durand, B. Ayrault, Y. Marqueton, and E. A. Decamps
Lattice Dynamics of Paratellurite
(b) Vol. 78, No. 2, p. 767—777 (1976).
See also *6*, p. *64*; *20.1*, p. *278*.

B. Joseph
The Electron Injection from Si into SiO_2 Due to a Mobile Ion Induced Internal Field
(a) Vol. **38**, No. 2, p. K127—K130 (1976).
See also *14.3.3*, p. *194*; *14.3.4*, p. *201*; 22.1.2, p. 358.

22.6.1 Mixed Crystals

Original Papers and Short Notes

A. C. McLaren and W. S. MacKenzie
The Spatial Coherence of X-Ray and Electron Beams and Its Influence on the Diffraction Patterns from Materials with Long-Period Superlattices
(a) Vol. **33**, No. 2, p. 491–495 (1976).
See also *1*, p. *9*; *1.4*, p. *25*.

W. Brückner, U. Gerlach, W. Moldenhauer, H. P. Brückner, N. Mattern, H. Oppermann, and E. Wolf
Phase Transitions and Semiconductor–Metal Transition in $V_{1-x}Ga_xO_2$ Single Crystals
(a) Vol. **38**, No. 1, p. 93–102 (1976).
See also *1.2*, p. *19*; *3*, p. *44*; 14.3, p. 181; 15, p. 211; 18.1, p. 227.

W. Brückner, H.-P. Brückner, U. Gerlach, B. Thuss, and G. Försterling
The Phase Transition M1 → T in $V_{1-x}Ga_xO_2$
(a) Vol. **38**, No. 1, p. K13–K16 (1976).
See also *1.2*, p. *19*; *18.1*, p. *227*.

W. Brückner and W. Moldenhauer
Anisotropy of Thermoelectric Power in $V_{1-x}Me_xO_2$ (Me = Al, Ga, Fe)
(a) Vol. **38**, No. 1, p. K17–K21 (1976).
See also 1.2, p. 20; *15*, p. *211*.

E. I. Terukov, F. A. Chudnovskii, W. Brückner, W. Reichelt, H. P. Brückner, W. Moldenhauer, and H. Oppermann
Influence of Mo, Ti, and Ga Doping on Conductivity Transition of V_3O_5
(a) Vol. **38**, No. 1, p. K23–K26 (1976).
See also *14.3*, p. *181*; *15*, p. *211*; 22.6, p. 408.

K. Dräger and R. Gerling
The Stability of Antiferromagnetic Clusters in the Magnetically Dilute System $(1-x)\,Cr_2O_3 \cdot x\,Al_2O_3$
(a) Vol. **38**, No. 2, p. 547–552 (1976).
See also *18.4*, p. *246*; 19, p. 256.

22.7 Semiconducting Intermetallic Compounds

Original Papers and Short Notes

H. Köhler
Non-Parabolic $E(k)$ Relation of the Lowest Conduction Band in Bi_2Te_3
(b) Vol. **73**, No. 1, p. 95–104 (1976).
See also *13.1*, p. *120*; *13.2*, p. *130*; 14.3, p. 173.

P. P. Bodiul, D. V. Gitsu, and D. F. Migley
Anisotropy of Thermomagnetic Effects in $Pb_{1-x}Sn_xTe$
(b) Vol. **73**, No. 1, p. K5–K8 (1976).
See also *15*, p. *209*.

G. Oelgart, R. Herrmann, H. Krüger, and R. Stegmann
Fermi Surface and Helicons in Semiconducting $Bi_{1-x}Sb_x$ Alloys
(b) Vol. **73**, No. 2, p. 615–624 (1976).
See also *13.2*, p. *130*; 14.3, p. 174; 19, p. 248.

Y. Sagara, O. Uemura, Y. Suzuki, and T. Satow
The Amorphous Structure of the Sb–Se System
(a) Vol. **33**, No. 2, p. 691–696 (1976).
See also 1.5, p. 31; *2*, p. *36*.

N. Wagner and O. Brümmer
XPS Investigations for the Purpose of the Research of Positron Annihilation Effects in Bi_xSb_{1-x} Alloys Concerning Band Structure Modifications
(b) Vol. **74**, No. 1, p. 249–254 (1976).
See also *13.1*, p. *122*; 17, p. 218.

G. Oelgart, G. Schneider, W. Kraak, and R. Herrmann
The Semiconductor–Semimetal Transition in $Bi_{1-x}Sb_x$ Alloys
(b) Vol. **74**, No. 1, p. K75–K78 (1976).
See also *13.2*, p. *130*; 14, p. 162; 19, p. 249.

G. L. Bleris, T. Karakostas, J. Stoemenos, and N. A. Economou
On the Short-Range Order Arrangement at the Transition State of β-In_2Te_3
(a) Vol. **34**, No. 1, p. 243–254 (1976).
See also *1.3*, p. *22*; 10.2, p. 90.

H. Köhler
Non-Parabolicity of the Highest Valence Band of Bi_2Te_3 from Shubnikov-de Haas Effect
(b) Vol. **74**, No. 2, p. 591–600 (1976).
See also *13.1*, p. *123*; *13.2*, p. *130*; 14.3, p. 175.

C. Ghezzi
Short-Range Order and Diffuse X-Ray Scattering in Single Crystal $Pb_xSn_{1-x}Te$ Alloys
(a) Vol. **34**, No. 2, p. 525–531 (1976).
See also *1.3*, p. *22*.

H. Göbel, E. Wolfgang, and H. Harms
Properties of MnBi Compounds Partially Substituted with Cu, Zn, Ti, Sb, and Te (I): Formation of Mixed Phases and Crystal Structures
(a) Vol. **34**, No. 2, p. 553–564 (1976).
See also *1.2*, p. *15*; *1.3*, p. *22*.

H. Köhler
Anisotropic g-Factor of the Conduction Electrons in Bi_2Te_3
(b) Vol. **75**, No. 1, p. 127–136 (1976).
See also *13.1*, p. *123*; *13.2*, p. *131*; 14.3, p. 176.

G. Oelgart and R. Herrmann
Cyclotron Masses in Semiconducting $Bi_{1-x}Sb_x$ Alloys
(b) Vol. 75, No. 1, p. 189–196 (1976).
See also *13.1*, p. *123*; 13.2, p. 131; 19, p. 250.

J. Camassel, M. Schlüter, S. Kohn, J. P. Voitchovsky, Y. R. Shen, and M. L. Cohen
Wavelength Modulation Spectra and Electronic Band Structure of SnS_2 and $SnSe_2$
(b) Vol. 75, No. 1, p. 303–314 (1976).
See also *13.1*, p. *124*; *20.1*, p. *267*.

H. Göbel, E. Wolfgang, and H. Harms
Properties of MnBi Compounds Partially Substituted with Cu, Zn, Ti, Sb, and Te (II): Stability and Magnetooptic Properties of Thin Films
(a) Vol. 35, No. 1, p. 89–95 (1976).
See also *1.2*, p. *16*; 1.3, p. 22; *18.2.1*, p. *237*.

M. Popescu
Relaxed Random Network Model of the $GeAs_2Te_7$ Glass
(a) Vol. 35, No. 1, p. 355–360 (1976).
See also *2*, p. *37*; *4*, p. *45*.

H. Köhler
The *g*-Factor of the Holes in the Highest Valence Band of Bi_2Te_3
(b) Vol. 75, No. 2, p. 441–449 (1976).
See also *13.1*, p. *124*; *13.2*, p. *131*; 14.3, p. 177.

N. Wagner and O. Brümmer
Remarks on the Relation between Crystal Structure and Electronic Band Structure in Bismuth-Antimony Alloys
(b) Vol. 75, No. 2, p. K157–K161 (1976).
See also *1.1*, p. *12*; *13.1*, p. *124*.

K. L. Chopra, S. K. Barthwal, and D. K. Pandya
Structural Sensitivity of Transport and Optical Properties of Amorphous GeTe
(a) Vol. 35, No. 2, p. 761–768 (1976).
See also 2, p. 38; *14.3*, p. *178*; *15*, p. *210*; 20.1, p. 270.

B. N. N. Achar and G. R. Barsch
Phonon Dispersion in V_3Si
(b) Vol. 76, No. 1, p. 133–143 (1976).
See also *6*, p. *57*.

H. Süssmann und H. Rau
Doping Properties of Pb in Sb_2Te_3
(a) Vol. 36, No. 1, p. K55–K58 (1976).
See also *14.3*, p. *178*.

A. M. Radwan and M. Taut
Band Structure of Al and the Laves Phase $CaAl_2$
(b) Vol. 76, No. 2, p. 605–612 (1976).
See also *13.1*, p. *126*; 20, p. 258; 21, p. 297.

B. N. N. Achar and G. R. Barsch
Group Theoretical Analysis and Phonon Modes of V_3Si
(b) Vol. 76, No. 2, p. 677–688 (1976).
See also *6*, p. *59*.

S. V. Vlachos, A. P. Lambros, A. Thanailakis, and N. A. Economou
Anisotropic Indirect Absorption Edge in GeSe
(b) Vol. 76, No. 2, p. 727–735 (1976).
See also 6, p. 59; *20.1*, p. *272*.

C. LARICA and A. P. GUIMARÃES
EPR of $GdAg_{1-x}In_x$ Intermetallic Compounds
(b) Vol. 77, No. 1, p. K11–K16 (1976).
See also *19*, p. *254*.

O. A. GOLIKOVA, I. M. RUDNIK, V. M. SERGEEVA, M. M. KAZANIN, and E. N. TKALENKO
On the Conductivity of Rare-Earth Metal Chalcogenides $Ln_{3-x}X_4$ ($0 \leqq x \leqq 0.33$)
(a) Vol. **37**, No. 1, p. 199–203 (1976).
See also *14.3*, p. *179*.

I. M. CHAPNIK
On Possible Formation of Closed Electron Shells with Coupling Spins outside the Inert-Gas Shells in Some Transition Metals and Their Alloys
(a) Vol. **37**, No. 1, p. K29–K32 (1976).
See also *13*, p. *118*; 21.1, p. 311; 21.1.1, p. 321.

H. SÜSSMANN and K. LOOF
Copper Doping and Disorder in the System Bi_2Te_3–Sb_2Te_3
(a) Vol. **37**, No. 2, p. 467–471 (1976).
See also *14.3*, p. *180*.

H. KÖHLER, W. HAIGIS, and A. VON MIDDENDORFF
Shubnikov-de Haas Investigations on n-Type $Bi_2(Se_xTe_{1-x})_3$
(b) Vol. 78, No. 2, p. 637–642 (1976).
See also *13.1*, p. *129*; *13.2*, p. *131*; 14.3, p. 181.

M. STORDEUR and W. HEILIGER
Anisotropy of the Thermopower of p-Sb_2Te_3
(b) Vol. 78, No. 2, p. K103–K106 (1976).
See also *15*, p. *212*.

22.8 Tertiary and Higher Compounds

Review Article

R. Prasad
Present State of Polytypism in Cadmium Iodide Crystals
(a) Vol. **38**, No. 1, p. 11–44 (1976).
See also *1.1*, p. *10*; *1.2*, p. *14*; 10.2, p. 87; 22.3, p. 372; 22.4.1, p. 379; 22.5, p. 388; 22.5.3, p. 399.

Original Papers and Short Notes

I. V. Chepeleva, E. A. Zhilinskaya, V. N. Lazukin, and A. P. Chernov
Electron Paramagnetic Resonance of Cr^{3+} in Chalcogenide Glasses
(b) Vol. **73**, No. 1, p. 65–72 (1976).
See also 2, p. 35; 18, p. 222; *19*, p. *247*.

A. V. Alekseev and U. Kh. Kopvillem
The Collapse of Radiation Stimulated by the Dipole–Dipole Temperature
(b) Vol. **73**, No. 1, p. K41–K46 (1976).
See also *6*, p. *53*; 19, p. 247.

G. B. Abdullaev, T. G. Kerimova, Sh. S. Mamedov, T. R. Mechtiev, R. Kh. Nani, and E. Y. Salaev
Study of the Fundamental Absorption Edge of $CdInGaS_4$ by Wavelength Modulation
(b) Vol. **73**, No. 1, p. K69–K72 (1976).
See also 13.5.1, p. 149; *20.1*, p. *261*.

Nguyen van Dong, Tran Huu Danh, B. Auguin, and A. Defresne
Transport Properties of Glassy $CdGe_{1-x}Sb_xAs_2$
(a) Vol. **33**, No. 1, p. 295–299 (1976).
See also *2*, p. *35*; *14.3*, p. *173*; 15, p. 209.

J. J. Rousseau, M. Rousseau, and J. C. Fayet
EPR Investigations of a Structural Phase Change in $RbCaF_3$, $RbCdF_3$, and $TlCdF_3$
(b) Vol. **73**, No. 2, p. 625–631 (1976).
See also *1.2*, p. *14*; *19*, p. *248*.

A. A. Kaminskii, D. Schultze, B. Hermoneit, S. E. Sarkisov, L. Li, J. Bohm, P. Reiche, R. Ehlert, A. A. Mayer, V. A. Lomonov, and V. A. Balashov
Spectroscopic Properties and Stimulated Emission in the ${}^4F_{3/2} \rightarrow {}^4I_{11/2}$ and ${}^4F_{3/2} \rightarrow {}^4I_{13/2}$ Transitions of Nd^{3+} Ions from Cubic $Bi_4Ge_3O_{12}$ Crystals
(a) Vol. **33**, No. 2, p. 737–753 (1976).
See also 20.1, p. 263; *20.2*, p. *279*; *20.3*, p. *281*.

S. Suzuki and M. Takagi
Free Rotation of NO_2 Radicals in $NaNO_2$ Crystal
(a) Vol. **33**, No. 2, p. 789–792 (1976).
See also 14.4.1, p. 206; *20.1*, p. *263*.

E. M. Rockar and R. A. Forman
$Fe^{4+}(3d^4)$ Optical Absorption in Oxidized $SrTiO_3$
(a) Vol. **33**, No. 2, p. K91–K95 (1976).
See also 13.4, p. 136; *20.1*, p. *263*.

E. Salje
New Crystals $Pb_3(P_xV_{1-x}O_4)_2$ for Acousto-Optical Device Applications
(a) Vol. **33**, No. 2, p. K165–K167 (1976).
See also *7*, p. *65*; *20.1*, p. *263*.

U. KAUFMANN, A. RÄUBER, and J. SCHNEIDER
ESR Analysis of IIB-IV-V_2 Semiconductors with Mn^{2+} as a Paramagnetic Probe
(b) Vol. **74**, No. 1, p. 169–175 (1976).
See also 13.4, p. 136; *19*, p. *248*.

J. WINTER, K. RÖSSLER, J. BOLZ, and J. PELZL
Mössbauer Effect and Raman Studies of the Structural Phase Transitions in $K_2[SnCl_6]$
(b) Vol. **74**, No. 1, p. 193–198 (1976).
See also *1.2*, p. *15*; 5, p. 48; 20.1, p. 263.

J. Y. BUZARÉ and J. C. FAYET
EPR Investigations on the $(FeOF_5)^{4-}$ Cluster in $KZnF_3$
(b) Vol. **74**, No, No. 1, p. 393–399 (1976).
See also *19*, p. *249*.

H. E. FLOTOW, R. KUENTZLER, and D. W. OSBORNE
The Specific Heat of the Nearly Ferromagnetic Compound $(Mn_{0.865}Fe_{0.135})Be_2$ from 0.1 to 23 K
(a) Vol. **34**, No. 1, p. 291–295 (1976).
See also *8*, p. *68*; 18, p. 222.

P. C. DIGGLE, K. A. GEHRING, and R. M. MAC FARLANE
Atomic Absorption and Emission in Laser Produced Plasmas
(a) Vol. **34**, No. 1, p. K13–K16 (1976).
See also *20.1*, p. *265*; *20.3*, p. *281*; 22.8.1, p. 420.

S. POPOVIĆ and M. ŠLJUKIĆ
X-Ray Diffraction Measurements of Unit Cell Parameters of Differently Crystallized Samples
(a) Vol. **34**, No. 1, p. K91–K95 (1976).
See also *1.1*, p. *11*; 22.1.1, p. 346.

D. CHAKRABORTY and G. LEHMANN
On the Structures and Orientations of Hydrogen Defects in Natural and Synthetic Quartz Crystals
(a) Vol. **34**, No. 2, p. 467–474 (1976).
See also *10.2*, p. *91*; 20.1, p. 266; 22.6, p. 404.

E. GOMBIA, N. ROMEO, G. SBERVEGLIERI, and C. PAORICI
Transport Properties of Semiconducting $ZnIn_2S_4$
(a) Vol. **34**, No. 2, p. 651–655 (1976).
See also *14.3*, p. *175*; 15, p. 209.

G. B. ABDULLAEV, K. R. ALLAKHVERDIEV, R. KH. NANI, E. YU. SALAEV, and R. M. SARDARLY
Far-Infrared Reflection Spectra of $TlGaSe_2$
(a) Vol. **34**, No. 2, p. K115–K117 (1976).
See also *6*, p. *56*; *20.1*, p. *266*.

J. EMERY and J. C. FAYET
Computation of the Two-Centre Molecular Integrals Involved in the Exchange Interaction between Ligand and Metal Electrons for the $[VF_6]^{4-}$ Cluster in $KMgF_3$
(b) Vol. **75**, No. 1, p. 255–259 (1976).
See also *13.4*,.p. *138*.

V. KUMAR and K. CHANDRA
A Study of the Nature of the Crystalline Field of the Yb^{3+} Ion in C_{3h} and D_{2d} Site Symmetries
(b) Vol. **75**, No. 1, p. K1–K5 (1976).
See also *13.4*, p. *138*; 22.8.1, p. 421.

H. Becherer and P. Kirsten
Cathodoluminescence of $ZnSiP_2$ in the Energy Region of the Pseudodirect Band Gap
(b) Vol. 75, No. 1, p. K19—K21 (1976).
See also *20.3*, p. *283*.

W. Siegel, A. Heinrich, and E. Ziegler
Electron and Hole Mobility in $ZnSiP_2$
(a) Vol. **35**, No. 1, p. 269—279 (1976).
See also 6, p. 56; 13.4, p. 139; *14.3*, p. *176*.

J. Bouwma, K. J. de Vries, and A. J. Burggraaf
Non-Stoichiometry, Defect Structure, and Dielectric Relaxation in Lanthana-Substituted $SrTiO_3$
(a) Vol. **35**, 1, p. 281—290 (1976).
See also *10.2*, p. *92*; *14.4*, p. *203*.

B. N. Oshcherin
Debye Temperature and Melting Criterion of Ternary Chalcopyrite $A^1B^3C^6_2$- and $A^2B^4C^5_2$-Type Semiconductors
(a) Vol. **35**, No. 1, p. K35—K39 (1976).
See also *1*, p. *9*; *8*, p. *68*; 22.2, p. 361; 22.4, p. 376.

R. Beliackas and J. Grigas
Microwave Dielectric Nonlinearity of SbSI
(a) Vol. **35**, No. 1, p. K61—K64 (1976).
See also *14.4.1*, p. *207*.

H. Bińczycka, A. Szytuła, J. Todorović, J. Zaleski, and A. Zięba
Metamagnetism of CoMnSi
(a) Vol. **35**, No. 1, p. K69—K72 (1976).
See also *18.2*, p. *232*; *18.4*, p. *244*.

W. Henrion, M. Závětová, and T. Mazets
Infrared Optical Properties of Pseudobinary Chalcogenide Glasses
(a) Vol. **35**, No. 1, p. K77—K79 (1976).
See also *2*, p. *38*; *20.1*, p. *268*.

L. M. Valiev, I. G. Kerimov, N. G. Aliev, and A. A. Abdurragimov
The Thermal Expansion and Electrical Properties of $Cu_4Mn_2X_4$(X: S, Se, and Te)
(a) Vol. **35**, No. 1, p. K85—K88 (1976).
See also 8, p. 68; 14, p. 162; 15, p. 210; *18.1*, p. *226*; *18.4*, p. *244*.

A. I. Belyaeva, M. M. Kotlyarskii, E. A. Popov, and B. V. Beznosikov
Spectroscopic Investigation of the Low Temperature Phase Transition in $NaMnCl_3$
(b) Vol. 75, No. 2, p. K123—K127 (1976).
See also 13.5.4, p. 160; *20.1*, p. *269*.

D. R. Hutton and C. M. Scala
An EPR Study of Fe^{3+} in Scolecite
(b) Vol. 75, No. 2, p. K167—K168 (1976).
See also *19*, p. *251*.

V. N. Brudnyi, D. L. Budnitskii, M. A. Krivov, and V. G. Melev
p–n-Type Conversion and Optical Properties of 2.0 MeV Electron-Irradiated $ZnSnAs_2$
(a) Vol. **35**, No. 2, p. 425—430 (1976).
See also 11, p. 102; 13.4, p. 139; *14.3*, p. *177*; *20.1*, p. *269*.

I. S. Gorban, G. A. Grishchenko, A. P. Sakalas, and I. I. Tychina
Conducting and Semi-Insulating n- and p-Type $ZnSiP_2$ Single Crystals
(a) Vol. **35**, No. 2, p. 571—574 (1976).
See also *13.4*, p. *139*; 14.3, p. 177.

M. GOETTEL, J. M. JANIK, J. A. JANIK, and M. RACHWALSKA
Phase Transitions and NH_3 Rotations in $[Co(NH_3)_6]I_3$ Studied by Calorimetry and IR Profile Methods
(a) Vol. **35**, No. 2, p. 675–680 (1976).
See also *1.2*, p. *17*; 8, p. 69.

J.-P. RIVERA, H. BILL, and R. LACROIX
EPR of Powder Mg-OH:Mn^{2+} Boracite Suggesting a Rhombohedral Phase
(a) Vol. **35**, No. 2, p. K105–K108 (1976).
See also 1.2, p. 17; *19*, p. *252*.

C. N. W. DARLINGTON
Phase Transitions in Perovskite Crystals
(b) Vol. **76**, No. 1, p. 231–239 (1976).
See also *1.2*, p. *17*; 14.4.1, p. 207.

W. BRAUN and M. CARDONA
Electronic Density of States of Amorphous $CuGaS_2$ and $ZnGeAs_2$
(b) Vol. **76**, No. 1, p. 251–257 (1976).
See also 2, p. 38; *13.1*, p. *125*; 17, p. 219.

N. E. ALEKSEEVSKII, E. P. KRASNOPEROV, and T. MYDŁARZ
Susceptibility and Superconductivity of Compounds with A-15 Structure
(b) Vol. **76**, No. 1, p. 357–362 (1976).
See also 13.1, p. 125; *14.2*, p. *170*; *18.1*, p. *226*.

H. J. VON BARDELEBEN, A. GOLTZENÉ, and C. SCHWAB
Electron Paramagnetic Resonance of Two Stiochiometry-Induced Intrinsic Defect Centres in $CuGaS_2$
(b) Vol. **76**, No. 1, p. 363–370 (1976).
See also *19*, p. *252*.

L. HERNÁNDEZ, O. VIGIL, and F. GÓNZALEZ
Optical Transitions and Distribution of Localized Levels in $ZnIn_2S_4$
(a) Vol. **36**, No. 1, p. 33–37 (1976).
See also *13.4*, p. *141*; 16, p. 215; *20.1*, p. *271*.

J. ALBERS, R. W. SAILER, and H. E. MÜSER
Dielectric, Elastic, and Piezoelectric Properties of Mg–Cl Boracite
(a) Vol. **36**, No. 1, p. 189–195 (1976).
See also *12.2*, p. *111*; *14.4.1*, p. *207*.

Y. KASHIWADA and Y. FURUHATA
Domains in Magnesium Pyroborate
(a) Vol. **36**, No. 1, p. K29–K31 (1976).
See also *1.2*, p. *18*; *3*, p. *43*.

C. PAORICI, C. PELOSI, N. ROMEO, G. SBERVEGLIERI, and L. TARRICONE
Electroluminescence in $CdIn_2S_4$
(a) Vol. **36**, No. 1, p. K33–K36 (1976).
See also *20.3*, p. *285*.

V. S. GRIGOREVA, A. A. LEBEDEV, V. D. PROCHUKHAN, YU. V. RUD, and A. A. YAKOVENKO
Photoelectrical Properties of p-Type $ZnGeP_2$ Single Crystals
(a) Vol. **36**, No. 1, p. K51–K54 (1976).
See also *16*, p. *215*.

T. V. QUYNH, G. SCHMIDT, N. R. IVANOV, and L. A. SHUVALOV
Determination of the Temperature Dependence of the Stiffness Coefficient c_{55} in $KH_3(SeO_3)_2$ by Ultrasonic Velocity Measurements
(a) Vol. **36**, No. 1, p. K85–K88 (1976).
See also *1.2*, p. *18*; *7*, p. *65*.

E. ZIEGLER, W. SIEGEL, and A. HEINRICH
First-Order NPO Scattering in p-$ZnSiAs_2$ and p-$CdSiAs_2$
(a) Vol. **36**, No. 2, p. 491–494 (1976).
See also *6*, p. *59*; *14.3*, p. *179*.

M. MATYÁŠ and J. HORÁK
Magnetic Susceptibility of Some Semiconducting V–VI–VII Crystals
(a) Vol. **36**, No. 2, p. K137–K139 (1976).
See also *18.1*, p. *226*; *18.2*, p. *234*.

J. J. ROUSSEAU and J. C. FAYET
Influence of the Crystal Field on Bent Configuration of V_K Centres in AMF_3 Compounds
(b) Vol. 77, No. 1, p. 195–201 (1976).
See also *4*, p. *46*; *13.4*, p. *142*; 19, p. 253.

R. MARTENS and F. FREUND
The Potential Energy Curve of the Proton and the Dissociation Energy of the OH^- Ion in $Mg(OH)_2$
(a) Vol. **37**, No. 1, p. 97–104 (1976).
See also *6*, p. *61*; *20.1*, p. *273*.

D. KUCHARCZYK, A. PIETRASZKO, and K. ŁUKASZEWICZ
Temperature Dependence of Lattice Parameters of $NaNO_2$ Single Crystals
(a) Vol. **37**, No. 1, p. 287–294 (1976).
See also *1.2*, p. *18*; *8*, p. *70*; 14.4.1, p. 208.

N. T. CHERPAK, V. V. MISHENKO, and T. A. SMIRNOVA
Inversion Ratio in Andalusite Crystals at 6 mm Wave Band
(a) Vol. **37**, No. 1, p. K37–K38 (1976).
See also 13.4, p. 142; *19*, p. *254*.

A. T. FUTRO, T. ŁUKASIEWICZ, and J. ZMIJA
Circular Birefringence in Bismuth Germanium Oxide ($Bi_{12}GeO_{20}$)
(a) Vol. **37**, No. 1, p. K75–K79 (1976).
See also *20.1*, p. *273*.

CH. DIPPMANN
Temperature Measurements on Mono-Stable Chalcogenide Glass Threshold Switches
(a) Vol. **37**, No. 1, p. K81–K83 (1976).
See also *14.3.3*, p. *193*.

L. PASEMANN, W. CORDTS, A. HEINRICH, and J. MONECKE
Band Structure Calculation of $ZnSiP_2$
(b) Vol. 77, No. 2, p. 527–533 (1976).
See also *13.1*, p. *127*.

N. V. AGEEV, N. E. ALEKSEEVSKII, and V. F. SHAMRAY
Superconductivity of Nb_3Al Base Ternary Solid Solutions
(b) Vol. 77, No. 2, p. K129–K131 (1976).
See also *14.2*, p. *171*.

REVA GARG and V. K. GARG
On the Quadrupole Splitting–Isomer Shift Linear Correlationship and the Sign of Electric Field Gradient
(b) Vol. 77, No. 2, p. K165–K170 (1976).
See also *5*, p. *50*.

E. GRILLI, M. GUZZI, and R. MOLTENI
Luminescence of $ZnIn_2Se_4$ Crystals
(a) Vol. **37**, No. 2, p. 399–406 (1976).
See also *20.3*, p. *287*.

G. V. Papayan, L. S. Agroskin, R. M. Vlasova, L. D. Rozenshtein, and S. M. Khursandova
Optical Constants and Dielectric Permittivity Dispersion of Quasi-One-Dimensional $K_2Pt(CN)_4Br_{0.3} \cdot 2.6\ H_2O$ Crystals in the Plasma Edge Region (from Transmission and Reflection Data)
(b) Vol. 78, No. 1, p. K59—K63 (1976).
See also *20.1*, p. *276*.

T. Boţilă and H. K. Henisch
Study of Trapping in Amorphous Materials
(a) Vol. 38, No. 1, p. 331—337 (1976).
See also *2*, p. *40*; 13.4, p. 145; 14.3, p. 181; 16, p. 217.

E. P. Trifonova and A. S. Popov
n–n Heterojunctions $CdSnP_2$–InP
(a) Vol. 38, No. 1, p. K37—K40 (1976).
See also *14.3.3*, p. *194*; 22.2.2, p. 366.

A. Miller, G. D. Holah, W. D. Dunnett, and G. W. Iseler
Optical Phonons in $AgGaSe_2$
(b) Vol. 78, No. 2, p. 569—576 (1976).
See also *6*, p. *63*; 20.1, p. 277.

A. M. Ziatdinov, M. M. Zaripov, Yu. V. Yablokov, and R. L. Davidovich
The Depression of the Dynamic Jahn-Teller Effect
(b) Vol. 78, No. 2, p. K69—K71 (1976).
See also 1.2, p. 20; *6*, p. *64*; *13.4*, p. *145*; 19, p. 256.

P. Berndt, D. Genzow, and K. H. Herrmann
Recombination Analysis in 10 μm $Pb_{1-x}Sn_xTe$
(a) Vol. 38, No. 2, p. 497—503 (1976).
See also *16*, p. *217*.

W. Bażela, A. Szytuła, J. Todorović, Z. Tomkowicz, and A. Zięba
Crystal and Magnetic Structure of NiMnGe
(a) Vol. 38, No. 2, p. 721—729 (1976).
See also *18.4*, p. *246*.

22.8.1 Salts of Oxy-Acids (Carbonates, Phosphates, Silicates Including Mica etc.)

Original Papers and Short Notes

G. Borchardt, G. Schmidt, and A. Rost
Non-Linear Dielectric Coefficients of Free and Clamped Barium Titanate Crystals
(a) Vol. **33**, No. 1, p. 143—147 (1976).
See also *14.4.1*, p. *206*.

Hj. Bernhardt
Comparative Studies of the LNT Coloration of $Y_3Al_5O_{12}$, $YAlO_3$, and $LiNbO_3$ Crystals
(a) Vol. **33**, No. 1, p. 211—216 (1976).
See also *10.2*, p. *87*; *20.1*, p. *261*; 20.3, p. 280; 22.8.2, p. 426.

P. Braillon and J. Serughetti
Mechanical Twinning in Calcite. Observation of Twinning Dislocations and Planar Defects by Transmission Electron Microscopy
(a) Vol. **33**, No. 1, p. 405—413 (1976).
See also *10.2*, p. *88*.

A. E. Kruminš and M. Vaněček
Photoconductivity of Transparent Lanthanum-Modified Lead Zirconate Titanate Ferroelectric Ceramic
(a) Vol. **33**, No. 1, p. K31—K34 (1976).
See also *14.4.1*, p. *206*; *16*, p. *213*.

Y. Hayakawa, S. Leelaprute, T. Nakazawa, S. Hayakawa, and N. Oda
Thermally Stimulateed Exoelectron Emission from $BaTiO_3$ Powder Coated with LiF Film
(a) Vol. **33**, No. 1, p. K35—K37 (1976).
See also *17*, p. *218*.

K. Gesi
Dielectric Study on the Phase Transitions in Triammonium Hydrogen Disulfate $(NH_4)_3H(SO_4)_2$
(a) Vol. **33**, No. 2, p. 479—482 (1976).
See also *1.2*, p. *14*; *14.4*, p. *202*.

P. A. Arsenev, K. Bienert, and A. V. Potemkin
The Spectroscopic Investigation of Neodymium in Mixed Aluminates of Yttrium and Gadolinium
(a) Vol. **33**, No. 2, p. 511—515 (1976).
See also 20.1, p. 263; *20.2*, p. *279*; 20.3, p. 281.

G. van Tendeloo, J. van Landuyt, and S. Amelinckx
The $\alpha \rightarrow \beta$ Phase Transition in Quartz and $AlPO_4$ as Studied by Electron Microscopy and Diffraction
(a) Vol. **33**, No. 2, p. 723—735 (1976).
See also *1.2*, p. *14*; *1.4*, p. *26*; 22.6, p. 403.

V. M. Fridkin, K. D. Kochev, Yu. S. Kusminov, K. A. Verkhovskaya, and T. R. Volk
The Shift of the Optical Absorption Edge at "Optical Damage" in $LiNbO_3$
(a) Vol. **33**, No. 2, p. K137—K139 (1976).
See also *14.4.1*, p. *206*; *20.1*, p. *263*.

P. J. Becker
Magnetostriction and Linear Magnetic Birefringence in the Jahn-Teller Coupled System $TmAsO_4$ with Effective Spin $S = 1$
(b) Vol. **74**, No. 1, p. 285—295 (1976).
See also *18.1*, p. *225*; *20.1*, p. *264*.

G. A. Gehring, H. G. Kahle, W. Nägele, A. Simon, and W. Wüchner
Two-Level Singlet Magnetism in $TbAsO_4$ and $TbVO_4$
(b) Vol. 74, No. 1, p. 297–309 (1976).
See also 8, p. 67; *18.4*, p. *242*; 20.1, p. 264.

S. I. Fărcas, A. Darabant, and A. Nicula
Electron Spin Resonance Study of Cr^{3+} and Fe^{3+} in Alums
(b) Vol. 74, No. 1, p. 335–339 (1976).
See also *19*, p. *249*.

A. Jasiński and Z. Sułek
^{31}P Chemical Shift Tensor in Solid KH_2PO_4, KD_2PO_4, and $NH_4H_2PO_4$
(b) Vol. 74, No. 1, p. K5–K8 (1976).
See also *19*, p. *249*.

H. Pollak
Charge Transfer in Cordierite
(b) Vol. 74, No. 1, p. K31–K34 (1976).
See also *5*, p. *49*; *20.1*, p. *264*.

T. Sasaki
ESR Study on Thermally Stimulated Exoelectron Emission from Irradiated K_2SO_4
(a) Vol. 34, No. 1, p. 339–346 (1976).
See also *17*, p. *218*; *19*, p. *249*.

J. N. Boland
An Electron Microscope Study of Dislocation Image Contrast in Olivine $(Mg, Fe)_2SiO_4$
(a) Vol. 34, No. 1, p. 361–367 (1976).
See also *10.2*, p. *90*.

A. Yoshihara, T. Fujimura, and K. I. Kamiyoshi
Phase Transition of Ammonium Sulfate
(a) Vol. 34, No. 1, p. 369–376 (1976).
See also *14.4.1*, p. *206*.

P. C. Diggle, K. A. Gehring, and R. M. Mac Farlane
Atomic Absorption and Emission in Laser Produced Plasmas
(a) Vol. 34, No. 1, p. K13–K16 (1976).
See also *20.1*, p. *265*; *20.3*, p. *281*; 22.8, p. 414.

R. Blinc, H. Arend, and A. Kandušer
Vibrational Spectra, Hydrogen Bonding, and Ferroelectricity in the $PbHPO_4$ Family
(b) Vol. 74, No. 2, p. 425–435 (1976).
See also 4, p. 45; *6*, p. *55*; 14.4.1, p. 206; *20.1*, p. *265*.

W. Dietrich and D. Schmid
Electron Spin Resonance of the Triplet State of Pure and Doped $NaNO_2$ Single Crystals
(b) Vol. 74, No. 2, p. 609–616 (1976).
See also 13.5.1, p. 152; *19*, p. *250*.

K. N. Pak
On the Dynamics of Hydrogen-Bonded Ferroelectrics
(b) Vol. 74, No. 2, p. 673–682 (1976).
See also *14.4.1*, p. *206*.

E. V. Bursian, Ya. G. Girshberg, and A. V. Ruzhnikov
The Correlation between Optical Absorption Spectra, Carrier Mobility, and Phase Transition Temperature in Some Ferroelectrics
(b) Vol. 74, No. 2, p. 689–693 (1976).
See also 13.1, p. 123; *14.4.1*, p. *206*; *20.1*, p. *266*.

A. V. Malakhovskii and I. S. Edelman
Optical and Magneto-Optical Spectra of $FeBO_3$ in the Region of Strong Absorption
(b) Vol. **74**, No. 2, p. K145–K149 (1976).
See also *18.2*, p. *231*; *20.1*, p. *266*.

Y. Uesu and J. Kobayashi
Crystal Structure and Ferroelectricity of Cesium Dihydrogen Phosphate CsH_2PO_4
(a) Vol. **34**, No. 2, p. 475–481 (1976).
See also *1.1*, p. *11*; *14.4.1*, p. *206*.

A. S. Janitzki, P. Gerthsen, and B. Hoffmann
Effective Resistivity of Semiconducting Ceramic $BaTiO_3$
(a) Vol. **34**, No. 2, p. K149–K151 (1976).
See also *14.4.1*, p. *207*.

M. Villa and A. Avogadro
Indirect Interactions and Nuclear Relaxation in $TlNO_3$
(b) Vol. **75**, No. 1, p. 179–188 (1976).
See also *19*, p. *250*.

V. Kumar and K. Chandra
A Study of the Nature of the Crystalline Field of the Yb^{3+} Ion in C_{3h} and D_{2d} Site Symmetries
(b) Vol. **75**, No. 1, p. K1–K5 (1976).
See also *13.4*, p. *138*; 22.8, p. 414.

S. V. Murthy
Shifts in PMR Spectral Lines of Hydrates
(b) Vol. **75**, No. 1, p. K11–K13 (1976).
See also *19*, p. *251*.

T. R. Rao, M. L. Bansal, V. C. Sahni, and A. P. Roy
Study of Internal Modes of SO_4^{2-} in $LiKSO_4$
(b) Vol. **75**, No. 1, p. K31–K34 (1976).
See also *6*, p. *56*; 20.1, p. 267.

C. Schaffrin
Oxygen Diffusion in $BaTiO_3$ Ceramic
(a) Vol. **35**, No. 1, p. 79–88 (1976).
See also *9*, p. *72*; 14.4, p. 202.

R. Visocekas, T. Ceva, C. Marti, F. Lefaucheux, and M. C. Robert
Tunneling Processes in Afterglow of Calcite
(a) Vol. **35**, No. 1, p. 315–327 (1976).
See also 10.2, p. 92; *20.3*, p. *283*.

A. M. Ozorio de Almeida and S. Caticha Ellis
X-Ray Scattering from Crystals with Intracellular Disorder
(b) Vol. **75**, No. 2, p. 657–668 (1976).
See also *1.2*, p. *16*; 14.4.1, p. 207.

S. V. Murthy
The Pake-Murthy Effect
(b) Vol. **75**, No. 2, p. K169–K171 (1976).
See also *19*, p. *251*.

F. Gilletta, M. Chabin, and Cao-Xuan An
Thermal Properties of Lead Phosphate
(a) Vol. **35**, No. 2, p. 545–549 (1976).
See also *8*, p. *69*.

A. P. Levanyuk and V. V. Osipov
Optical Distortion in Crystals
(a) Vol. **35**, No. 2, p. 605–614 (1976).
See also 13.4, p. 139; *14.4.1*, p. *207*; *20.1*, p. *269*.

D. Ball
Paramagnetic Resonance of Er^{3+} and Dy^{3+} in Synthetic Single Crystals of Hafnon
(a) Vol. **35**, No. 2, p. 651–658 (1976).
See also *19*, p. *252*.

S. Radhakrishna and T. Bhaskar Rao
Radiation Damage of Chromium Doped Potassium Alum
(a) Vol. **35**, No. 2, p. 715–718 (1976).
See also *11*, p. *102*; 19, p. 252; 20.1, p. 269.

K. H. Germann, W. Müller-Lierheim, H. H. Otto, and T. Suski
High Temperature Phase Transition in $Pb_5Ge_3O_{11}$
(a) Vol. **35**, No. 2, p. K165–K167 (1976).
See also *1.2*, p. *17*; 14.4.1, p. 207; 20.1, p. 270.

V. A. Antonov and P. A. Arsenev
Spectroscopic Properties of Single Crystals of Rare Earth Titanates
(a) Vol. **35**, No. 2, p. K169–K172 (1976).
See also *20.1*, p. *270*; 20.2, p. 279.

V. V. Samartsev, R. G. Usmanov, I. Kh. Khadiev, E. F. Kustov, and M. N. Baranov
Photon Echo in $CaWO_4:Nd^{3+}$ and $LiAl_5O_8:Cr^{3+}$
(b) Vol. **76**, No. 1, p. 55–66 (1976).
See also 13.4, p. 140; *20.1*, p. *270*; *20.3*, p. *284*.

B. Blanzat, F. Raynal, R. Parrot, C. Barthou, and B. Canny
Electron Paramagnetic Resonance of Gd^{3+} in Bismuth Germanate
(b) Vol. **76**, No. 1, p. K5–K8 (1976).
See also *19*, p. *252*.

H. Beige, C. Birkholz, E. Ciesla, and G. Schmidt
Elastic and Electromechanical Properties of Some Crystals near Structural Phase Transitions
(b) Vol. **76**, No. 1, p. K47–K50 (1976).
See also 1.2, p. 17; *12.2*, p. *111*; *14.4*, p. *203*.

D. Ball and B. M. Wanklyn
Coloured Synthetic Zircon Crystals, Flux Growth and EPR of V^{4+} Impurity Centres
(a) Vol. **36**, No. 1, p. 307–316 (1976).
See also *3*, p. *43*; *19*, p. *253*.

L. Holz, K. Kneipp, A. Lau, and W. Werncke
Stimulated Raman Scattering of Picosecond Laser Pulses by Polaritons in $LiIO_3$ and $LiNbO_3$ Single Crystals
(a) Vol. **36**, No. 1, p. K5–K8 (1976).
See also *6*, p. *58*; 13.5, p. 147; 20.1, p. 271; 20.2, p. 279.

E. M. Avakian, K. G. Belabaev, A. A. Kaminskii, and V. Kh. Sarkisov
Observation of the Spontaneous Electric Breakdown in Pyroelectric $LiNbO_3$ and $LiTaO_3$ Crystals
(a) Vol. **36**, No. 1, p. K25–K27 (1976).
Erratum
(a) Vol. **37**, No. 1, p. K111 (1976).
See also *14.3.4*, p. *198*.

G. A. Petrakovskii and A. I. Pankrats
The Magnetoelastic Coupling in $FeBO_3$
(a) Vol. **36**, No. 1, p. K43—K46 (1976).
See also *12.2*, p. *111*; *18.4*, p. *244*.

Y. Luspin and G. Hauret
Study of the Velocity and Damping of Acoustic Waves Obtained by Brillouin Scattering in the Paraelastic Phase in GMO
(b) Vol. **76**, No. 2, p. 551—558 (1976).
See also *1.2*, p. *18*; 7, p. 65.

B. Speit and G. Lehmann
Hole Centers in the Feldspar Sanidine
(a) Vol. **36**, No. 2, p. 471—481 (1976).
See also *10.2*, p. *94*; 16, p. 215; 20.1, p. 272; 20.3, p. 285.

D. Lapraz and P. Iacconi
On Some Luminescent and Optical Properties of Synthetic Calcite Single Crystals
(a) Vol. **36**, No. 2, p. 603—616 (1976).
See also 13.4, p. 141; *20.1*, p. *272*; *20.3*, p. *285*.

P. Braillon and J. Serughetti
Plastic Deformation of Calcite Single Crystals by 001 Compression
(a) Vol. **36**, No. 2, p. 637—646 (1976).
See also *10.2*, p. *94*.

P. Ganguly, Om Parkash, and C. N. R. Rao
Electron Transport and Magnetic Properties of Rare-Earth Ortho-Titanites and -Vanadites, $LnTiO_3$ and $LnVO_3$
(a) Vol. **36**, No. 2, p. 669—678 (1976).
See also *14*, p. *162*; 15, p. 210; *18*, p. *224*.

N. M. Salanskii, E. A. Glozman, and V. N. Seleznev
NMR and Domain Structure in Thin Single Crystals of $FeBO_3$ and $Fe_{1-x}Ga_xBO_3$
(a) Vol. **36**, No. 2, p. 779—782 (1976).
See also 18.4, p. 245; *19*, p. *253*.

R. Mathai, K. Narahari Rao, A. Kumaraswamy, T. Rajasekharan, and K. V. S. Rama Rao
ESR Study of Paramagnetic Centres in X-Irradiated $AgClO_3$
(b) Vol. 77, No. 1, p. 215—219 (1976).
See also *19*, p. *253*.

N. G. Cherenda and D. M. Yudin
Additives and Cage Effect in Glasses
(b) Vol. 77, No. 1, p. K27—K30 (1976).
See also *2*, p. *39*; 11, p. 103; *19*, p. *254*.

J. C. Peuzin and M. Tasson
Generation and Detection of Hypersound. Domain Transducer
(a) Vol. **37**, No. 1, p. 119—125 (1976).
See also *7*, p. *66*; *14.4.1*, p. *207*.

H. Veeser, U. Bogner, and W. Eisenmenger
Analysis of the Acoustic Transients in the Pulse Response of the Linear Electro-Optic Effect
(a) Vol. **37**, No. 1, p. 161—170 (1976).
See also *7*, p. *66*; 12.2, p. 111; *14.4.1*, p. *208*; 20.1, p. 273.

Y. Motohashi, P. Braillon, and J. Serughetti
Elastic Energy, Stress Field of Dislocation, and Dislocation Parameters in Calcite Crystals
(a) Vol. **37**, No. 1, p. 263—270 (1976).
See also *10.2*, p. *95*; 12.2, p. 112.

G. Holzapfel and M. Krystek
Thermally Stimulated Luminescence and Exoelectron Emission from Barium and Strontium Sulphate Doped with Europium
(a) Vol. 37, No. 1, p. 303—312 (1976).
See also *13.4*, p. *142*; 17, p. 219; 20.3, p. 286.

L. I. Horváth and T. Szörényi
EPR Studies on Vanadium Phosphate Glasses
(a) Vol. 37, No. 2, p. K169—K171 (1976).
See also 2, p. 40; *19*, p. *255*.

A. M. Bon, C. Benoit, and O. Bernard
Dynamical Properties of Crystals of $Sr(NO_3)_2$, $Ba(NO_3)_2$, and $Pb(NO_3)_2$ (I): Infrared Spectra and Structure
(b) Vol. 78, No. 1, p. 67—78 (1976).
See also *6*, p. 62; *20.1*, p. *275*.

K. E. Gauss and H. Happ
Millimeter Wave Investigation of the Complex Dielectric Constant in DKDP
(b) Vol. 78, No. 1, p. 133—138 (1976).
See also 6, p. *62*; *14.4.1*, p. *208*; 20.1, p. 275.

G. A. Gehring and J. Laugsch
Theory of Side-Bands in a Cooperative Jahn-Teller Material
(b) Vol. 78, No. 1, p. 211—218 (1976).
See also 6, p. 62; 13.5.1, p. 155; *20.1*, p. *275*.

R. S. Momtaz and L. N. Rashkovich
The Distribution of Deuterium between Solid and Liquid Phases during the Crystallization of Crystals of the KDP Group from Aqueous Solutions
(a) Vol. 38, No. 1, p. 401—408 (1976).
See also *3*, p. *44*; 14.4.1, p. 208.

A. Mansingh, J. K. Vaid, and R. P. Tandon
Switching and Memory Behaviour of Vanadate Glasses
(a) Vol. 38, No. 1, p. K1—K3 (1976).
See also *2*, p. *41*; *14.3.3*, p. *194*.

K. H. Germann
Optical Anisotropy Measurements in Multidomain Samples: Application to Phase Transitions in NiO and $Pb_5Ge_3O_{11}$
(a) Vol. 38, No. 1, p. K81—K84 (1976).
See also 1.2, p. 20; 14.4.1, p. 208; 18.4, p. 246; *20.1*, p. *277*; 22.6, p. 408.

K. W. Blazey
Optical Absorption of $SrTiO_3:Fe^{4+}$
(a) Vol. 38, No. 1, p. K97—K99 (1976).
See also 13.4, p. 145; *20.1*, p. *277*.

A. M. Bon, C. Benoit, and J. Giordano
Dynamical Properties of Crystals of $Sr(NO_3)_2$, $Ba(NO_3)_2$, and $Pb(NO_3)_2$ (II): Temperature Dependence of the Infrared Spectra
(b) Vol. 78, No. 2, p. 453—464 (1976).
See also *6*, p. *63*; 20.1, p. 277.

R. S. Saraswat and G. C. Upreti
Electron Paramagnetic Resonance of Mn^{2+} in $ZnSO_4 \cdot 7\ H_2O$ Single Crystals
(b) Vol. 78, No. 2, p. K73—K75 (1976).
See also *19*, p. *256*.

Yu. V. Zabara, A. Yu. Kudzin, and K. A. Kolesnichenko
Space-Charge-Limited Currents in Barium Titanate Single Crystals
(a) Vol. **38**, No. 2, p. K131–K134 (1976).
See also *14.3.4*, p. *201*.

I. B. Barkan, S. I. Marennikov, and M. V. Entin
Holographic Storage in $LiNbO_3$ Crystal at High Temperatures
(a) Vol. **38**, No. 2, p. K139–K142 (1976).
See also *14.4.1*, p. *208*; *16*, p. *217*.

22.8.2 Spinels, Garnets, Ferrites

Original Papers and and Short Notes

D. SCHEERLINCK and S. HAUTECLER
Magnetic Interactions in Co_3O_4
(b) Vol. **73**, No. 1, p. 223–228 (1976).
See also *18.1*, p. *225*; *18.4*, p. *242*.

A. V. ZALESSKII, A. K. ZVEZDIN, I. S. ZHELUDEV, A. M. SAVVINOV, and A. F. LEBEDEV
A New Type of Domain Wall NMR Spectrum in Canted Antiferromagnets: $YFeO_3$ and $LuFeO_3$ Crystals
(b) Vol. **73**, No. 1, p. 317–325 (1976).
See also *18.4*, p. *242*; *19*, p. *247*.

R. KRISHNAN
Optical Absorption of Co^{3+} in YIG Crystals
(b) Vol. **73**, No. 1, p. K51–K56 (1976).
See also 13.4, p. 134; *20.1*, p. *261*.

HJ. BERNHARDT
Comparative Studies of the LNT Coloration of $Y_3Al_5O_{12}$, $YAlO_3$, and $LiNbO_3$ Crystals
(a) Vol. **33**, No. 1, p. 211–216 (1976).
See also *10.2*, p. *87*; *20.1*, p. *261*; 20.3, p. 280; 22.8.1, p. 419.

M. WURLITZER and F. PLÜMER
X-Ray Induced Magnetic After-Effect in Si-Doped YIG
(a) Vol. **33**, No. 1, p. K55–K57 (1976).
See also *18.3*, p. *239*.

O. G. RAMER and C. H. WILTS
The Effects of Surface Layers on Spin-Wave Spectra in YIG Films
(b) Vol. **73**, No. 2, p. 443–453 (1976).
See also *18.2.1*, p. *237*; 19, p. 247.

A. A. SAMOKHVALOV, YU. N. MOROZOV, B. V. KARPENKO, and M. I. SIMONOVA
Temperature Dependence of the Magnetization of the Ferromagnetic Spinels $CdCr_2Se_4$ and $CuCr_2S_4$ and Its Calculation in the Spin-Wave Approximation
(b) Vol. **73**, No. 2, p. 455–459 (1976).
See also *18.2*, p. *229*.

E. DE GRAVE, C. DAUWE, A. GOVAERT, and J. DE SITTER
Cation Distribution and Magnetic Exchange Interactions in $MgFe_2O_4$ Studied by the Mössbauer Effect
(b) Vol. **73**, No. 2, p. 527–532 (1976).
See also *5*, p. *48*; *18.3*, p. *239*.

B. ANDLAUER, J. SCHNEIDER, and W. TOLKSDORF
Optical Absorption, Fluorescence, and Electron Spin Resonance of Ir^{4+} on Octahedral Sites in $Y_3Ga_5O_{12}$
(b) Vol. **73**, No. 2, p. 533–540 (1976).
See also *13.4*, p. *135*; 19, p. 248; 20.1, p. 261; 20.3, p. 280.

P. NOVÁK and I. VELTRUSKÝ
Overlap and Covalency Contribution to the Zero-Field Splitting of S-State Ions
(b) Vol. **73**, No. 2, p. 575–586 (1976).
See also *13*, p. *114*; *18.3*, p. *239*.

J. R. L. de Almeida and S. M. Rezende
Scattering of Light by Magnetoelastic Excitations in Ferromagnets
(b) Vol. **73**, No. 2, p. 661—672 (1976).
See also **13**, p. **114**; 18, p. 222; *20*, p. *257*.

J. Baszyński and Z. Frait
Spin-Wave Resonance in Thin Monocrystalline Ferrite Films
(b) Vol. **73**, No. 2, p. K85—K88 (1976).
See also *18.3*, p. *239*; 19, p. 248.

V. A. Povitskii, E. F. Makarov, N. V. Murashko, and A. N. Salugin
Mössbauer Study of Superparamagnetic δ-FeOOH and Its Transformation into α-Fe_2O_3
(a) Vol. **33**, No. 2, p. 783—787 (1976).
See also 1.2, p. 15; *5*, p. *48*.

V. G. Vologin, V. D. Parkhomenko, S. F. Dubinin, Yu. G. Chukalkan, B. N. Goshchitskii, S. K. Sidorov, and V. V. Petrov
Exchange Fe^{3+}–Fe^{3+} Interactions in Ferrites with Spinel Structure
(a) Vol. **33**, No. 2, p. K83—K86 (1976).
See also *18.3*, p. *239*.

V. G. Kleparskii, N. P. Dymchenko, and S. K. Kukharskaya
Temperature Dependence of Domain Walls on Pulse Displacement Dynamics in Magnetic Uniaxial Bubble Materials
(a) Vol. **33**, No. 2, p. K117—K120 (1976).
See also *18.3*, p. *239*.

M. Maryško and J. Šimšová
Angular Dependence of FMR Linewidth of Single Crystal Ferrite Films
(a) Vol. **33**, No. 2, p. K133—K136 (1976).
See also *18.3*, p. *239*; *19*, p. *248*.

M. B. Kozintseva and V. E. Zilbervarg
The Electron and Magnon Spectra of Ferrimagnetic Semiconductors
(b) Vol. **74**, No. 1, p. 199—208 (1976).
See also *13.1*, p. *122*; *13.5.4*, p. *160*; 18.3, p. 239.

J. Fink-Finowicki
Magnetooptical Properties of Samarium Orthoferrite in the Spin-Reorientation Region
(b) Vol. **74**, No. 1, p. K27—K29 (1976).
See also *18.3*, p. *240*; *20.1*, p. *264*.

P. Bassoul, A. Dubon, A. Lefebvre and J. C. Gilles
Periodic Antiphase Structures in the Non-Stoichiometric Mg Phase of the System $MgGa_2O_4$–Ga_2O_3
(a) Vol. **34**, No. 1, p. 125—134 (1976).
See also *1.2*, p. *15*; *1.4*, p. *26*.

K. I. Arai, K. Ohmori, N. Tsuya, and S. Iida
Magnetostriction of Magnetite in the Vicinity of the Low-Temperature Transition
(a) Vol. **34**, No. 1, p. 325—330 (1976).
See also 1.2, p. 15; *18.3*, p. *240*.

A. M. Kapitonov and E. M. Smokotin
Elastic Constants of Single-Crystal Lithium Ferrite
(a) Vol. **34**, No. 1, p. K47—K49 (1976).
See also *12.2*, p. *110*.

Hj. Bernhardt
The Structure of Coloration Centres in $Y_3Al_5O_{12}$ Crystals after X-Ray Irradiation at Room Temperature
(a) Vol. **34**, No. 2, p. 483—487 (1976).
See also *10.2*, p. *91*; 20.3, p. 282.

C. BECKER, E. ZSOLDOS, and A. WEBER
Dislocation Configurations at Inclusions in GGG ($Gd_3Ga_5O_{12}$)
(a) Vol. **34**, No. 2, p. 519–523 (1976).
See also *10.2*, p. *91*.

R. E. VANDENBERGHE, G. G. ROBBRECHT, and V. A. BRABERS
Structure and Ionic Configuration of Oxidic Copper–Manganese Spinels ($Cu_xMn_{3-x}O_4$)
(a) Vol. **34**, No. 2, p. 583–592 (1976).
See also *1*, p. *9*; 1.2, p. 15; *18.3*, p. *240*.

Z. ŠIMŠA, J. ŠIMŠOVÁ, K. SUK, E. KRATOCHVÍLOVÁ, and M. MARYŠKO
Magnetic Bubbles in Spinel Ferrite Films
(a) Vol. **34**, No. 2, p. 639–644 (1976).
See also *18.3*, p. *240*.

R. METSELAAR, J. P. M. DAMEN, P. K. LARSEN, and M. A. H. HUYBERTS
Investigation of Colour Centres in Gadolinium Gallium Garnet Crystals
(a) Vol. **34**, No. 2, p. 665–670 (1976).
See also *10.2*, p. *91*; 20.1, p. 266.

S. SAITO, S. TAKEMOTO, and Y. WATANABE
Skin Effect Contributions in FMR of Conductive Ferrite Spheres
(a) Vol. **34**, No. 2, p. 671–677 (1976).
See also *18.3*, p. *240*; *19*, p. *250*.

A. A. KAMINSKII, KH. S. BAGDASAROV, G. A. BOGOMOLOVA, M. M. GRITSENKO, A. M. KEVORKOV, and S. E. SARKISOV
Luminescence and Stimulated Emission of Nd^{3+} Ions in $Gd_3Sc_2Ga_3O_{12}$ Crystals
(a) Vol. **34**, No. 2, p. K109–K114 (1976).
See also *20.2*, p. *279*; *20.3*, p. *282*.

K. HISATAKE
Photo-Induced Permeability Effect in Lithium Ferrite with Small Amounts of Cobalt
(a) Vol. **34**, No. 2, p. K131–K134 (1976).
See also *18.3*, p. *240*.

S. LIGENZA
A Study of the $^6S_{5/2}$-Term Splitting of an Fe^{3+} Ion in Zinc Ferrite by Neutron Spectroscopy
(b) Vol. **75**, No. 1, p. 315–326 (1976).
See also 5, p. 49; *13.4*, p. *138*; 18.3, p. 240.

L. PAWLAK, S. POKRZYWNICKI, K. FALKOWSKI, and K. CHĘCIŃSKI
EPR of Mn^{2+} in CdY_2S_4 and CdY_2Se_4 Spinels
(b) Vol. **75**, No. 1, p. K7–K10 (1976).
See also *19*, p. *251*.

P. HLÍDEK, I. BARVÍK, V. PROSSER, M. VANĚČEK, and M. ZVÁRA
Polarization Dependence of Transmission and Photoconductivity in $CdCr_2Se_4$
(b) Vol. **75**, No. 1, p. K45–K49 (1976).
See also *16*, p. *214*; *20.1*, p. *267*.

F. BASILE and P. POIX
Mössbauer Effect and Magnetic Measurement on the Solid Solution $Fe_{2-2x}Mg_{1+x}Sn_xO_4$
(a) Vol. **35**, No. 1, p. 153–164 (1976).
See also *5*, p. *49*; 18.3, p. 240.

J. RABIER, P. VEYSSIÈRE, and J. GRILHÉ
Possibility of Stacking Faults and Dissociation of Dislocations in the Garnet Structure
(a) Vol. **35**, No. 1, p. 259–268 (1976).
See also *1.4*, p. *27*; *10.2*, p. *92*.

R. Hergt, M. Wendt, P. Görnert, and S. Bornmann
Investigation of the Growth Mechanism in High-Temperature Solutions by the Induced Striation Method (I): Dissolution Striations
(a) Vol. **35**, No. 1, p. 347–353 (1976).
See also *3*, p. *42*.

P. Görnert, S. Bornmann, and R. Hergt
Investigation of the Growth Mechanism in High-Temperature Solutions by the Induced Striation Method (II): Driving Forces in Slow-Cooling Experiments
(a) Vol. **35**, No. 2, p. 583–590 (1976).
See also *3*, p. *42*.

M. Kh. Ashurov, Yu. K. Voronko, V. V. Osiko, A. A. Sobol, B. P. Starikov, M. I. Timoshechkin, and A. Ya. Yablonskii
Inequivalent Luminescence Centres of Er^{3+} in Gallium Garnet Single Crystals
(a) Vol. **35**, No. 2, p. 645–649 (1976).
See also *13.4*, p. *140*; 20.1, p. 269; 20.3, p. 284.

T. Merceron and P. Bernstein
Photomagnetic Effect in Ni–Zn Ferrite with Small Amount of Cobalt
(a) Vol. **35**, No. 2, p. 681–686 (1976).
See also *18.3*, p. *240*.

J. Lopitaux and M. Lenglet
Determination of the Distribution in a Four Cation Spinel by X-Ray and Neutron Diffraction and by Magnetic Analysis
(a) Vol. **35**, No. 2, p. K141–K143 (1976).
See also *18.3*, p. *240*.

G. Bordin, A. Cecchetti, G. Cecchi, M. Poppi, and F. Ronconi
Magnetostrictive Hysteresis Loop of Ferrites in Alternating Fields
(a) Vol. **35**, No. 2, p. K161–K164 (1976).
See also *18.3*, p. *240*.

F. M. Gashimzade and K. A. Rustamov
Erratum to: Selection Rules for the Spinel Structure
(b) Vol. **76**, No. 1, p. 437 (1976).
See also *6*, p. *58*; *13.4*, p. *140*.

G. Aers, A. D. Boardman, and E. D. Isaac
A Monte Carlo Study of the Gunn Effect Behaviour of a Ferromagnetic Semiconductor
(a) Vol. **36**, No. 1, p. 357–364 (1976).
See also *13.5.4*, p. *160*; *14.3.4*, p. *198*.

K. P. Belov, A. M. Kadomtseva, N. M. Kovtun, V. N. Derkachenko, V. N. Melov, and V. A. Khokhlov
On the Character of Phase Transitions in $ErFeO_3$
(a) Vol. **36**, No. 1, p. 415–425 (1976).
See also 12.2, p. 111; *18.4*, p. *244*.

E. I. Ilyashenko, I. G. Avaeva, V. B. Kravchenko, and S. N. Matveev
A Method for Observation and Measurement of the Velocity of Bubble Propagation in Thin Ferrogarnet Films
(a) Vol. **36**, No. 1, p. K1–K3 (1976).
See also *18.2*, p. *233*.

P. Arnold
Measurements of Lattice Distortions in Antiferromagnetic $Al_2Mn(SiO_4)_3$
(a) Vol. **36**, No. 1, p. K37–K38 (1976).
See also *18.4*, p. *244*.

G. Wende
Radiation Damping in FMR Measurements in a Nonresonant Rectangular Waveguide
(a) Vol. **36**, No. 2, p. 557–567 (1976).
See also *18.3*, p. *240*; *19*, p. *253*.

H.-R. Müller, B. Knappe, R. Kosicik, P. Rosemann, D. Schmidt, and A. Funke
Magnetization Reversal in Thin Magnetic Films with In-Plane Compensation
(a) Vol. **36**, No. 2, p. 617–626 (1976).
See also *18.3*, p. *241*.

V. R. K. Murthy and J. Sobhanadri
Dielectric Properties of Some Nickel–Zinc Ferrites at Radio Frequency
(a) Vol. **36**, No. 2, p. K133–K135 (1976).
See also *14.4*, p. *203*; 18.3, p. 241.

M. Guillot and H. Le Gall
Low and Very High Magnetic Field Dependence of the Magnetization and the Faraday Rotation Induced by the Magnetic Dipole Transitions in YIG
(b) Vol. **77**, No. 1, p. 121–130 (1976).
See also *18.3*, p. *241*; *20*, p. *258*.

N. D. J. Miller
Magnetostatic Volume Wave Propagation in a Dielectric Layered Structure
(a) Vol. **37**, No. 1, p. 83–91 (1976).
See also *18.3*, p. *241*.

K. de Strooper, A. Govaert, C. Dauwe, and G. Robbrecht
Magnetic Properties of the Spinel Series $Co_2Ge_xTi_{1-x}O_4$ ($0 \leqq x \leqq 1$)
(a) Vol. **37**, No. 1, p. 127–132 (1976).
See also 18.1, p. 226; *18.3*, p. *241*; 18.4, p. 245.

F. Walz and J. Rivas
Magnetic Aftereffects in Manganese Ferrites
(a) Vol. **37**, No. 1, p. 151–160 (1976).
See also *18.3*, p. *241*.

Z. Jirák, S. Vratislav, and J. Zajíček
Oxygen Parameters and Debye-Waller Factors in $Mn_xCr_{3-x}O_4$ Spinels
(a) Vol. **37**, No. 1, p. K47–K51 (1976).
See also *1.1*, p. *12*.

Hj. Bernhardt
The Manganese-Induced O^- Centres in Yttrium Aluminium Garnet
(a) Vol. **37**, No. 2, p. 445–448 (1976).
See also *10.2*, p. *96*; 20.1, p. 275; 20.3, p. 287.

J. W. Matthews and T. S. Plaskett
Optical Birefringence Images of Edge Dislocations
(a) Vol. **37**, No. 2, p. 499–503 (1976).
See also *10.2*, p. *96*.

P. Görnert and G. Wende
Study of the Mass Transport during Crystal Growth from High-Temperature Solutions by the "High-Resolution Induced Striation Method"
(a) Vol. **37**, No. 2, p. 505–513 (1976).
See also *3*, p. *44*.

W. K. Chen and H. Downing
Cation Self-Diffusion in $CoFe_2O_4$
(a) Vol. **37**, No. 2, p. 515–520 (1976).
See also *9*, p. *74*; 10.2, p. 96.

N. K. Radhakrishnan and A. B. Biswas
A Neutron Diffraction Study of the Cation Migration in $MgMn_2O_4$
(a) Vol. 37, No. 2, p. 719–722 (1976).
See also *1*, p. *9*; 10.2, p. 96.

K. Knoch and H. Dannheim
Temperature Dependence of the Cation Distribution in Magnesium Ferrite
(a) Vol. 37, No. 2, p. K135–K137 (1976).
See also *5*, p. *50*.

S. Ligenza
A Study of Spin Interaction in a Manganese-Substituted Zinc Ferrite by Neutron Spectroscopy
(b) Vol. 78, No. 1, p. 301–310 (1976).
See also 5, p. 50; *18.4*, p. *245*.

V. D. Parkhomenko, S. F. Dubinin, B. N. Goshchitskii, Yu. G. Chukalkin, S. K. Sidorov, V. G. Vologin, and V. V. Petrov
Peculiarities of Radiation Damage in Ferrites with Spinel Structure
(a) Vol. 38, No. 1, p. 57–66 (1976).
See also 11, p. 104; *18.3*, p. *241*.

A. A. Kaminskii, G. A. Bogomolova, D. N. Vylegzhanin, Kh. S. Bagdasarov, A. M. Kevorkov, and M. M. Gritsenko
Spectroscopic Properties of Nd^{3+} Ions in Garnet Compounds Forming in the Y_2O_3–Ga_2O_3 System
(a) Vol. 38, No. 1, p. 409–422 (1976).
See also *20.2*, p. *279*; *20.3*, p. *288*.

M. V. Chetkin, A. N. Shalygin, and A. V. Kiryushin
Magnetic Permeability of Ferro- and Paramagnetic Garnets at Optical Frequencies
(b) Vol. 78, No. 2, p. K85–K88 (1976).
See also *18.1*, p. *227*; *18.2*, p. *236*.

R. Morineau
Structural Defects and Oxidation–Reduction Equilibrium in Mn, Zn Ferrites
(a) Vol. 38, No. 2, p. 559–568 (1976).
See also *10.2*, p. *98*; *18.3*, p. *241*.

J. W. Matthews and T. S. Plaskett
Formation of Large Climb Loops at Decorated Dislocations in Gadolinium Gallium Garnet
(a) Vol. 38, No. 2, p. 577–585 (1976).
See also *10.2*, p. *98*.

D. Gratias, J. P. Boilot, Y. Le Cars, and J. Thery
Syntaxy between β and β'' Structures in Alumina/Sodium Oxide and Gallia/Sodium Oxide Systems
(a) Vol. 38, No. 2, p. 595–600 (1976).
See also *1.4*, p. *29*.

V. R. K. Murthy and J. Sobhanadri
Electrical Conductivity of Some Nickel–Zinc Ferrites
(a) Vol. 38, No. 2, p. 647–651 (1976).
See also *14.3*, p. *182*; 15, p. 212; 18.3, p. 241.

A. I. Belyaeva, M. M. Kotlyarskii, and Yu. N. Stelmakhov
Visual Investigation of Spin Reorientation in $TmFeO_3$
(a) Vol. 38, No. 2, p. K103–K105 (1976).
See also *18.3*, p. *241*.

V. G. Vologin, S. F. Dubinin, V. D. Parkhomenko, Yu. G. Chukalkin, B. N. Goshchitskii, S. K. Sidorov, and V. V. Petrov
Experimental Determination of the Preference Energy in Ferrites-Spinels
(a) Vol. **38**, No. 2, p. K147—K150 (1976).
See also 11, p. 105; *18.3*, p. *241*.

J. Šimšová, M. Maryško, and K. Suk
Spin Wave Modes in Single Crystal Manganese Ferrite Films
(a) Vol. **38**, No. 2, p. K163—K166 (1976).
See also *18.3*, p. *241*; *19*, p. *256*.

22.9 Organic Semiconductors

Review Article

M. N. Sapozhnikov
Zero-Phonon Transitions in the Optical Spectra of Impurity Molecular Crystals
(b) Vol. **75**, No. 1, p. 11–51 (1976).
See also *6*, p. *52*; 13.4, p. 134; 20.1, p. 260; *20.3*, p. *280*.

Original Papers and Short Notes

P. Petelenz
Effect of Polarization on Charge Carrier Transfer in Molecular Crystals
(b) Vol. **73**, No. 1, p. 293–299 (1976).
See also *13*, p. *113*; 14.4, p. 202.

M. G. Kaplunov, Yu. M. Shulga, K. I. Pokhodnya, and Yu. G. Borodko
X-Ray Photoelectron Spectra of Quasi-One-Dimensional Conductors
(b) Vol. **73**, No. 1, p. 335–339 (1976).
See also 13, p. 113; 17, p. 218; *20*, p. *257*.

N. Desbarax, Y. Ségui, and Bui Ai
Study of the Conduction Mechanism in Thin Films of Polyoxisilicone
(a) Vol. **33**, No. 1, p. 247–254 (1976).
See also *14.3.1*, p. *183*; 14.3.2, p. 187.

G. K. Semin and I. M. Alymov
NQR Study of Molecular Crystals of Solid Solution of Methyl Bromide–Methyl Iodide
(a) Vol. **33**, No. 1, p. K63–K66 (1976).
See also *19*, p. *247*.

Yu. A. Vidadi, K. Sh. Kocharli, B. Sh. Barkhalov, and S. A. Sadreddinov
Direct Current Conductivity of Copper Phthalocyanine Films in the Presence of Blocking Contacts
(a) Vol. **33**, No. 1, p. K67–K71 (1976).
See also *14.3.1*, p. *183*; 14.3.2, p. 187.

T. Markvart
Crystal–Impurity Energy Transport by Molecular Excitons in the Wave-Like Limit (I): Capture Mechanism
(b) Vol. **73**, No. 2, p. 689–698 (1976).
See also *13.4*, p. *135*; *13.5.1*, p. *150*.

B. Reimer, H. Baessler, J. Hesse, and G. Weiser
Optical Properties of a Polydiacetylene Single Crystal — Evidence for van Hove Singularities
(b) Vol. **73**, No. 2, p. 709–716 (1976).
See also 13.1, p. 121; *20.1*, p. *262*.

H. W. Helberg
Indicatrix Orientation in the Quasi-One-Dimensional Organic Semiconductor (MTPA) · $(TCNQ)_2$
(a) Vol. **33**, No. 2, p. 453–459 (1976).
See also *20*, p. *257*.

B. Korsch, F. Willig, H. J. Gaehrs, and B. Tesche
Exchange of Holes between Organic Crystals and Metal Films Evaporated in High Vacuum
(a) Vol. **33**, No. 2, p. 461–471 (1976).
See also *14.3.2*, p. *187*; *14.3.4*, p. *196*; 16, p. 213.

N. L. Kramarenko, I. V. Kurnosov, and Yu. V. Naboikin
Light Scattering by Liquid Crystals in an Electric Field
(a) Vol. **33**, No. 2, p. 773–781 (1976).
See also *2*, p. *36*; *20*, p. *257*.

G. K. Semin and I. M. Alymov
NQR Study of Molecular Crystals of Solid Solutions of Chlorobenzene–Bromobenzene
(a) Vol. **33**, No. 2, p. K111–K115 (1976).
See also *19*, p. *248*.

K. Mazur
Determination of Real Homocharge in Carnauba Wax Electrets
(a) Vol. **33**, No. 2, p. K155–K159 (1976).
See also *14.4*, p. *202*.

T. Chong and N. Itoh
Interaction between Excitons and Radiation-Induced Radicals in Naphthalene Single Crystals
(b) Vol. **74**, No. 1, p. 111–120 (1976).
See also 13.5.1, p. 150; *20.3*, p. *281*.

P. Reineker
To the Theory of the ESR Lineshape of Triplet Excitons in Molecular Crystals. Influence of the Quasi-Incoherent Exciton Motion and Comparison with Experimental Data
(b) Vol. **74**, No. 1, p. 121–134 (1976).
See also *13.5.1*, p. *150*; *19*, p. *248*.

T. Markvart
Crystal–Impurity Energy Transport by Molecular Excitons in the Wave-Like Limit (II): Diffusion in Crystals
(b) Vol. **74**, No. 1, p. 135–143 (1976).
Erratum
(b) Vol. **76**, No. 2, p. 853 (1976).
See also *13.4*, p. *136*; *13.5.1*, p. *151*.

H. Arend, H. von Känel, and P. Wachter
Magneto-Optical Measurements on Two-Dimensional Magnetic Model Systems
(b) Vol. **74**, No. 1, p. 151–157 (1976).
See also *18.4*, p. *242*; *20.1*, p. *263*.

J. Alizon, G. Berthet, J. P. Blanc, J. Gallice, and H. Robert
Exciton Dynamics Study by Magnetic Resonance in $TCNQ^- \cdot NH_4^+$
(b) Vol. **74**, No. 1, p. 217–224 (1976).
See also 13.5.1, p. 151; *19*, p. *249*.

W. B. Whitten and S. Arnold
Pressure Modulation of Exciton Fission in Tetracene
(b) Vol. **74**, No. 1, p. 401–407 (1976).
See also *13.5.1*, p. *151*; 20.3, p. 281.

R. A. Elliott
Erratum to: Comment on "Diffusion Theory of Luminescent Emission from a Doped Organic Solid"
(b) Vol. **74**, No. 1, p. K87 –K88(1976).
See also *13.5.1*, p. *151*; *20.3*, p. *281*.

W. Hofberger
The Birefringence of Crystalline Tetracene Derived from Diffraction Properties of Polycrystalline Films
(a) Vol. **34**, No. 1, p. K55–K58 (1976).
See also 1.5, p. 31; *20.1*, p. *265*.

YU. A. VIDADI, K. SH. KOCHARLI, B. SH. BARKHALOV, and S. A. SADREDDINOV
Alternating Current Investigation of Copper Phthalocyanine Films in the Presence of Blocking Contacts
(a) Vol. **34**, No. 1, p. K77–K81 (1976).
See also *14.3.1*, p. *183*.

M. S. BRODIN, M. A. DUDINSKII, S. V. MARISOVA, and E. N. MYASNIKOV
Polariton Scattering and Formation of Exciton Luminescence Spectrum in Anthracene Crystals
(b) Vol. **74**, No. 2, p. 453–460 (1976).
See also *13.5.1*, p. *151*; *20.3*, p. *282*.

Z. ZBOIŃSKI
The Tensor of the Drift Mobility of Electrons and Holes in the Anthracene-1,3,5-Trinitrobenzene Charge Transfer Complex
(b) Vol. **74**, No. 2, p. 561–566 (1976).
See also *14.3*, p. *175*.

H. MALTER
On the Electronic Molecular Structure of the Organic Semiconductor Copper Phthalocyanine
(b) Vol. **74**, No. 2, p. 627–633 (1976).
See also *13.1*, p. *123*; *20.1*, p. *265*.

B. D. BHATTACHARYYA
A New Aspect of Jahn-Teller Effect in the Low-Spin Ferrihaemoglobin Azide
(b) Vol. **74**, No. 2, p. 695–700 (1976).
See also *13.4*, p. *137*; 19, p. 250; 20.1, p. 266.

N. KARL and H. FEEDERLE
Absorption Spectrum of the Tetracene Monopositive Ion in the Anthracene Lattice
(a) Vol. **34**, No. 2, p. 497–503 (1976).
See also *13.4*, p. *137*; *20.1*, p. *266*.

U. REHBERG
Measurement of the Complex Permittivity of the Anion Radical Complex Salt of Tetracyanochinodimethane and Methyltriphenylarsonium
(a) Vol. **34**, No. 2, p. 573–582 (1976).
See also 14.3, p. 175; *14.4*, p. *202*.

B. JACHYM, H. SODOLSKI, T. SŁUPKOWSKI, and R. ZIELIŃSKI
Electric Conduction in the Polyester Resin–Acetylene Carbon Black System
(a) Vol. **34**, No. 2, p. 657–664 (1976).
See also *14.3*, p. *175*.

J. KLIMOWSKI, W. WANARSKI, and D. OŻGO
Effect of Hydrostatic Pressure on the Phase Transition Temperatures and Spontaneous Polarization of Thiourea Monocrystals
(a) Vol. **34**, No. 2, p. 697–704 (1976).
See also *1.2*, p. *15*; 14.4.1, p. 206.

S. NEŠPUREK and E. SILINSH
Space-Charge-Limited Current Theory for Molecular Crystals with Gaussian Distribution of Local Trapping States
(a) Vol. **34**, No. 2, p. 747–759 (1976).
See also 13.4, p. 138; *14.3.4*, p. *197*.

E. L. BOKHENKOV, I. NATKANIEC, and E. F. SHEKA
Neutron Spectroscopy of Naphthalene Crystal Internal Phonon Modes
(b) Vol. **75**, No. 1, p. 105–116 (1976).
See also *6*, p. *56*; 20.1, p. 267.

J. P. MATHIEU and J. CHAPPERT
Electronic Properties and Relaxation Processes of Fe^{2+} in Iron Bisacetylacetonate Dihydrate from Mössbauer Spectroscopy
(b) Vol. 75, No. 1, p. 163—171 (1976).
See also *5*, p. *49*.

L. B. COLEMAN, C. R. FINCHER, JR., A. F. GARITO, and A. J. HEEGER
Far-Infrared Single Crystal Studies of TTF-TCNQ
(b) Vol. 75, No. 1, p. 239—246 (1976).
See also 13.1, p. 123; *20.1*, p. *267*.

I. S. KAULACH and E. A. SILINSH
The Influence of the Hyperfine Range Magnetic Field on Charge Transfer States at the Metal–Organic Semiconductor Interface
(b) Vol. 75, No. 1, p. 247—254 (1976).
See also *13.3*, p. *132*; 14.3.3, p. 191; 16, p. 214.

K.-D. TOEPFER and H. W. HELBERG
Dielectric Dispersion of Thiourea Single Crystals at Microwave Frequencies between 8.25 and 75.0 GHz
(a) Vol. 35, No. 1, p. 131—136 (1976).
See also 1.2, p. 16; *14.4.1*, p. *207*.

B. BAHADUR, S. CHANDRA, and N. K. SANYAL
Phase Transition Studies of the Liquid Crystal n-(p-Octyloxybenzylidene)-p-Toluidine
(a) Vol. 35, No. 1, p. 387—390 (1976).
See also *2*, p. *37*; *18*, p. *223*.

B. VOIGTSBERGER and H. ROSENBERGER
High Resolution Proton Magnetic Resonance in Monocrystals of α-Glycine
(a) Vol. 35, No. 1, p. K89—K92 (1976).
See also *19*, p. *251*.

W. ARDEN, M. KOTANI, and L. M. PETER
Triplet Exciton Decay Processes in Crystalline Tetracene
(b) Vol. 75, No. 2, p. 621—631 (1976).
See also *13.5.1*, p. *153*; 20.3, p. 284.

R. VACHER and J. PELOUS
Measurement of the Intensity of the Brillouin Lines in Amorphous Media at Very Low Temperature
(b) Vol. 75, No. 2, p. K101—K103 (1976).
See also *2*, p. *38*; *6*, p. *57*; 20.1, p. 269.

U. GÖSELE
Comment on "Diffusion Theory of Luminescent Emission from a Doped Organic Solid"
(b) Vol. 75, No. 2, p. K129—K132 (1976).
See also *13.5.1*, p. *153*; *20.3*, p. *284*

G. HEGER, D. MULLEN, and K. KNORR
On the Importance of Hydrogen Bonding for the Structural Phase Transitions in $(CH_3NH_3)_2MnCl_4$. A Single-Crystal Neutron Diffraction Study
(a) Vol. 35, No. 2, p. 627—637 (1976).
See also *1.2*, p. *17*; 4, p. 46.

A. Z. TRIFONOV, M. D. MIKHAILOV, I. A. TSENOV, and B. K. SHISHEDIEV
A Spectroscopic Method for Evaluating the Diffusion of a Liquid into Another One
(a) Vol. 35, No. 2, p. 709—714 (1976).
See also *9*, p. *73*.

T. Wasiutyński
Self-Consistent Phonon Calculation for Hexamethylenetetramine Molecular Crystals
(b) Vol. 76, No. 1, p. 175–181 (1976).
See also 4, p. 46; *6*, p. *57*; 12.2, p. 110.

T. Goworek and C. Rybka
Impurity Induced Positronium in Molecular Crystals
(b) Vol. 76, No. 1, p. K35–K37 (1976).
See also *10.2*, p. *93*.

G. Chapuis, R. Kind, and H. Arend
X-Ray Study of Structural Phase Transitions in the Perovskite-Type Layer Compound $(CH_3NH_3)_2CdCl_4$
(a) Vol. 36, No. 1, p. 285–295 (1976).
See also *1.2*, p. *17*.

M. D. Mikhailov, A. Z. Trifonov, I. A. Tsenov, and M. A. Aladzhem
A Spectroscopic Method for Determination of the Cross-Term Diffusion Coefficients in a Three-Component Liquid System
(a) Vol. 36, No. 1, p. 391–396 (1976).
See also *9*, p. *73*.

K. Lochner, B. Reimer, and H. Bässler
Photocarrier Generation in a One-Dimensional Solid: Polydiacetylene-Bis(Toluenesulfonate)
(b) Vol. 76, No. 2, p. 533–540 (1976).
See also *16*, p. *215*.

H. Höchst, A. Goldmann, S. Hüfner, and H. Malter
X-Ray Photoelectron Valence Band Studies on Phthalocyanine Compounds
(b) Vol. 76, No. 2, p. 559–568 (1976).
See also *13.1*, p. *126*; 17, p. 219.

H. J. Hesse, W. Fuhs, G. Weiser, and L. von Szentpaly
Directional Dispersion of an Exciton Polariton in a Pentamethinium Cyanine Dye
(b) Vol. 76, No. 2, p. 817–826 (1976).
See also *13.5.1*, p. *154*; 20.1, p. 272.

J. N. Das and B. P. Chandra
Estimation of Mechanical Relaxation Time of Sugar Crystals from Their Triboluminescence
(a) Vol. 36, No. 2, p. K175–K177 (1976).
See also *20.3*, p. *285*.

J. C. Phillips
Structure and Conductivity of Bilinear Organic Compounds
(b) Vol. 77, No. 1, p. 259–264 (1976).
See also *1*, p. *9*; *14*, p. *163*.

E. L. Frankevich, M. M. Tribel, and I. A. Sokolik
Photoconductivity of Sublimated Rubrene Films during Oxidation. The Magnetic Field Effect
(b) Vol. 77, No. 1, p. 265–276 (1976).
See also 13.5.1, p. 155; *14.3.1*, p. *185*; *16*, p. *215*; 20.3, p. 286.

H. P. Braun and M. E. Michel-Beyerle
Current and Fluorescence Oscillations in Anthracene Crystals
(a) Vol. 37, No. 1, p. 73–82 (1976).
See also *14.3.4*, p. *199*; 16, p. 216; 20.3, p. 286.

A. SAMOĆ, M. SAMOĆ, J. SWORAKOWSKI, J. M. THOMAS, and J. O. WILLIAMS
Interpretation of One-Carrier Thermally Stimulated Currents and Isothermal Decay Currents (II): Application to the Evaluation of Trap Parameters in Anthracene
(a) Vol. 37, No. 1, p. 271—278 (1976).
See also *13.4*, p. *142*; 14.3.4, p. 199; 14.4, p. 204.

L. K. AMINOV, R. KIRMSE, and B. V. SOLOVEV
Electron Spin-Lattice Relaxation of Cu^{2+} Ions in Planar Coordination Geometries. Calculation and Experimental Data of T_1 for Cu^{2+} in Ni(II)bis(diethyl-dithiocarbamate) Crystals
(b) Vol. 77, No. 2, p. 505—515 (1976).
See also *4*, p. *46*; 6, p. 61; 18.1, p. 226; 19, p. 254.

J. BERREHAR, P. DEDLANOY, and M. SCHOTT
Drift Mobility of Holes in Crystalline Tetracene
(b) Vol. 77, No. 2, p. K119—K122 (1976).
See also *14.3*, p. *180*.

D. BLOOR and F. H. PRESTON
Optical Properties of Bis(p-Toluene Sulphonate)Diacetylene (I): Crystal Optics of Partially and Fully Polymerized Crystals
(a) Vol. 37, No. 2, p. 427—438 (1976).
See also 1.1, p. 13; *20.1*, p. *274*.

H. P. BRAUN
Current Oscillations in Photo-Oxidized Anthracene Crystals
(a) Vol. 37, No. 2, p. 553—559 (1976).
See also 13.4, p. 143; *14.3.4*, p. *199*; 16, p. 216.

R. F. EGERTON
Measurement of Inelastic/Elastic Scattering Ratio for Fast Electrons and Its Use in the Study of Radiation Damage
(a) Vol. 37, No. 2, p. 663—668 (1976).
See also *11*, p. *104*; 13.5.2, p. 158; 21, p. 299; 21.3, p. 324; 21.7, p. 332.

C. HAMANN, CH. TANTZSCHER, and G. LEHMANN
A Model of the Electrical Transport Phenomena in Imperfect Crystals of Copper Phthalocyanine (IV): Concentration of Charge Carriers and Current-Voltage Characteristics
(b) Vol. 78, No. 1, p. 79—84 (1976).
See also 13.4, p. 144; *14.3*, p. *180*; *14.3.1*, p. *185*.

V. P. FEDOROV, V. V. KIND, and A. V. KORSHUNOV
The Multiplet Structure of the Electron Absorption Spectra of Naphthalene-β-Naphthol Mixed Molecular Crystals
(b) Vol. 78, No. 1, p. 93—101 (1976).
See also 13.4, p. 144; 13.5.1, p. 155; *20.1*, p. *275*.

D. MÜLLER and J. PETERSSON
Spin-Lattice Relaxation of Several Nuclei near the Ferroelectric Transitions of TGS and TGFBe
(b) Vol. 78, No. 1, p. 191—198 (1976).
See also *14.4.1*, p. *208*; 19, p. 255.

V. L. BROUDE, L. M. UMAROV, and E. F. SHEKA
Multiphonon Absorption in Molecular Crystals
(b) Vol. 78, No. 1, p. 325—333 (1976).
See also *6*, p. *63*; *20.1*, p. *276*.

J. C. PHILLIPS
Phase Transitions in the Bilinear Organic Compounds (TTF, HMTSF)-TCNQ
(b) Vol. 78, No. 1, p. 371—379 (1976).
See also *1*, p. *9*; *14*, p. *163*.

K. P. Meletov and E. F. Sheka
Asymmetry of Phonon Side Bands of Optical Transitions in Impurity Molecular Crystals Caused by a Change in External Phonon Frequencies
(b) Vol. 78, No. 1, p. K1–K5 (1976).
See also *6*, p. *63*; *13.4*, p. *144*; 20.1, p. 276; 20.3, p. 287.

J. Świątek
Electrical Conduction Mechanism in Thin Coronene Films
(a) Vol. 38, No. 1, p. 285–291 (1976).
See also *13.4*, p. *144*; *14.3.1*, p. *185*.

P. C. Mehendru, N. L. Pathak, Satbir Singh, and P. Mehendru
Electrical Conduction in Polypropylene Thin Films
(a) Vol. 38, No. 1, p. 355–360 (1976).
See also *14.3.1*, p. *185*; *14.4*, p. *204*.

Usha Kapur and V. K. Srivastava
Thickness Dependence of the Static Dielectric Constant of Some Built-up Langmuir Films
(a) Vol. 38, No. 1, p. K77–K80 (1976).
See also *14.4*, p. *204*.

J. C. Phillips
High-Temperature Conductivity of Organic Bilinear Addition Compounds
(b) Vol. 78, No. 2, p. 471–476 (1976).
See also 6, p. 63; *14*, p. *163*.

S. Fujiwara, T. Nakayama, and N. Itoh
Exciton Interactions in Naphthalene Single Crystals
(b) Vol. 78, No. 2, p. 519–528 (1976).
See also *13.5.1*, p. *156*; 20.3, p. 288.

V. I. Zinenko
On the Theory of the Structural Phase Transition in Squaric Acid
(b) Vol. 78, No. 2, p. 721–727 (1976).
See also *1.2*, p. *20*; *14.4*, p. *205*.

I. Vilfan
On the Electron–Intramolecular Vibration Coupling in Aromatic Molecular Crystals
(b) Vol. 78, No. 2, p. K131–K135 (1976).
See also *6*, p. *64*; 13.5.3, p. 159; *14.3*, p. *182*.

A. Goswami and S. Radhakrishnan
Space-Charge-Limited Conduction in Indanthrone Films
(a) Vol. 38, No. 2, p. K151–K153 (1976).
See also *14.3.4*, p. *201*.

M. Galtier, A. Montaner, C. Benoit, J. Deumie, H. Robert, L. Giral, and J. M. Fabre
Magnetic Field Effects on Infrared Spectra of TTF-TCNQ Thin Films on KBr
(a) Vol. 38, No. 2, p. K159–K161 (1976).
See also *20.1*, p. *278*.

23. Solidified Gases

Original Papers and Short Notes

YU. B. GAIDIDEI, V. M. LOKTEV, A. F. PRIKHOTKO, and L. I. SHANSKII
α-Oxygen Doublets (II): Experiment
(b) Vol. 73, No. 2, p. 415–425 (1976).
See also *13.5.1*, p. *149*; *20.1*, p. *261*.

R. BRODMANN, G. TOLKIEHN, and G. ZIMMERER
Luminescence Originating from Free and Self-Trapped Excitons in Solid Xenon
(b) Vol. 73, No. 2, p. K99–K101 (1976).
See also *13.5.1*, p. *150*; *20.3*, p. *280*.

A. N. ALEKSANDROVSKII, V. I. KUCHNEV, V. G. MANZHELII, and A. M. TOLKACHEV
Studies on the Low Temperature Anomaly in the Thermal Expansion of Solid Methane
(b) Vol. 73, No. 2, p. K111–K114 (1976).
See also *8*, p. *67*.

D. PUDEWILL, F.-J. HIMPSEL, V. SAILE, N. SCHWENTNER, M. SKIBOWSKI, and E. E. KOCH
Excitons in Pure and Doped Solid Neon
(b) Vol. 74, No. 2, p. 485–496 (1976).
See also *13.1*, p. *122*; *13.5.1*, p. *151*; 20.1, p. 265.

CH. ACKERMANN, R. BRODMANN, U. HAHN, A. SUZUKI, and G. ZIMMERER
Photoluminescence Excitation Spectra of Solid Krypton
(b) Vol. 74, No. 2, p. 579–590 (1976).
See also *13.5.1*, p. *151*; *20.3*, p. *282*.

E. SCHUBERTH, M. CREUZBURG, and W. MÜLLER-LIERHEIM
Evidence for a Partial F.C.C.–H.C.P. Phase Transition in Solid Neon from Raman Scattering
(b) Vol. 76, No. 1, p. 301–306 (1976).
See also *1.2*, p. *17*; 6, p. 58; 20.3, p. 284.

V. YAKHOT
Theory of Multiphonon Vibrational Relaxation of a Guest Molecule in a Host Crystal
(b) Vol. 76, No. 2, p. 451–456 (1976).
See also *6*, p. *58*.

A. K. BOHRA and R. K. GUPTA
Lattice Dynamics of Rare-Gas Solids by Kihara's Potential (III): Green Functions
(b) Vol. 78, No. 1, p. 255–258 (1976).
See also *6*, p. *62*.

Author Index

G. B. Abdullaev	(b) **73**, K69 (1976)	13.5.1, p. 149; *20.1*, p. *261*; 22.8, p. 413
	(a) **34**, K115 (1976)	*6*, p. *56*; *20.1*, p. *266*; 22.8, p. 414
A. A. Abdurragimov	(a) **35**, K85 (1976)	8, p. 68; 14, p. 162; 15, p. 210; *18.1*, p. *226*; *18.4*, p. *244*; 22.8, p. 415
B. N. N. Achar	(b) **76**, 133 (1976)	*6*, p. *57*; 22.7, p. 411
	(b) **76**, 677 (1976)	*6*, p. *59*; 22.7, p. 411
Ch. Ackermann	(b) **74**, 579 (1976)	*13.5.1*, p. *151*; *20.3*, p. *282*; 23, p. 440
M. Acquarone	(b) **77**, K143 (1976)	*7*, p. *66*; *18.4*, p. *245*
G. Aers	(a) **36**, 357 (1976)	*13.5.4*, p. *160*; *14.3.4*, p. *198*; 22.8.2, p. 429
A. M. Afanasev	(b) **73**, 73 (1976)	*5*, p. *48*
	(b) **76**, 465 (1976)	*5*, p. *50*
V. K. Agarwal	(a) **34**, K67 (1976)	*14.4*, p. *202*; 22.5.2, p. 394
N. V. Ageev	(b) **77**, K129 (1976)	*14.2*, p. *171*; 22.8, p. 417
L. S. Agroskin	(b) **78**, K59 (1976)	*20.1*, p. *276*; 22.8, p. 418
F. Agulló-López	(b) **78**, K81 (1976)	10.2, p. 98; *19*, p. *256*; 22.5.2, p. 398
Ah Mee Hor	(a) **38**, 293 (1976)	*8*, p. *71*; 10.2, p. 98; 22.5.2, p. 398
J. S. Ahearn, Jr.	(a) **38**, 315 (1976)	*10.2*, p. *98*; *20*, p. *259*; 22.2.1, p. 364; 22.2.2 p. 365
Bui Ai	(a) **33**, 247 (1976)	*14.3.1*, p. *183*; 14.3.2, p. 187; 22.9, p. 433
F. Ajustron	(a) **35**, 197 (1976)	*1.1*, p. *11*; 22.6, p. 404
Z. G. Akhvlediani	(b) **77**, K21 (1976)	*6*, p. *60*; *8*, p. *70*; 11, p. 103; 20.1, p. 273; 22.5.2, p. 396
A. I. Akimenko	(a) **33**, 53 (1976)	*14.2*, p. *169*; 21, p. 289
F. El Akkad	(b) **76**, 85 (1976)	*13.4*, p. *140*; *20.3*, p. *284*; 22.4.3, p. 385
J. Akutsu	(a) **35**, K73 (1976)	13.3, p. 132; *17*, p. *219*; 22.5.2, p. 394; 22.6, p. 405
M. A. Aladzhem	(a) **36**, 391 (1976)	*9*, p. *73*; 22.9, p. 437
C. Albers	(a) **34**, K125 (1976)	*14.3.1*, p. *183*; 22.1.3, p. 359
J. Albers	(a) **36**, 189 (1976)	*12.2*, p. *111*; *14.4.1*, p. *207*; 22.8, p. 416
R. Alcalá	(a) **38**, 621 (1976)	*10.2*, p. *98*; 20.1, p. 278; 22.5.3, p. 400
G. Alefeld	(a) **36**, 161 (1976)	*12.1*, p. *108*; 21, p. 296; 21.1, p. 309
L. N. Aleksandrov	(a) **36**, 403 (1976)	1.2, p. 18; *3*, p. *43*
	(a) **37**, 341 (1976)	*1.5*, p. *32*; 3, p. 43; 22.1.1, p. 348; 22.1.2, p. 356
A. N. Aleksandrovskii	(b) **73**, K111 (1976)	*8*, p. *67*; 23, p. 440
A. V. Alekseev	(b) **73**, K41 (1976)	*6*, p. *53*; 19, p. 247; 22.8, p. 413.
N. E. Alekseevskii	(a) **34**, 541 (1976)	*14.2*, p. *169*; 21, p. 293
	(b) **76**, 357 (1976)	131, p. 125; *14.2*, p. *170*; *18.1*, p. *226*; 22.8, p. 416
	(b) **77**, 451 (1976)	*14.2*, p. *171*; 21, p. 298
	(b) **77**, K129 (1976)	*14.2*, p. *171*; 22.8, p. 417
M. Ali	(a) **35**, 181 (1976)	*1.2*, p. *16*; 22.6, p. 404
S. Z. Ali	(a) **36**, 517 (1976)	*1.2*, p. *18*; 8, p. 70; 22.3, p. 373
N. G. Aliev	(a) **35**, K85 (1976)	8, p. 68. 14, p. 162; 15, p. 210; *18.1*, p. *226*; *18.4*, p. *244*; 22.8, p. 415
J. Alizon	(b) **74**, 217 (1976)	13.5.1, p. 151; *19*, p. *249*; 22.9, p. 434
K. R. Allakhverdiev	(a) **34**, K115 (1976)	*6*, p. *56*; *20.1*, p. *266*; 22.8, p. 414

G. ALLAN	(b) **74**, 409 (1976)	*10.1*, p. *79*; *13.4*, p. *136*; 21.6, p. 329
	(a) **35**, 499 (1976)	*1.3*, p. *22*; 10.1, p. 82; 21, p. 295; 21.6, p. 330
G. P. ALLDREDGE	(b) **76**, 457 (1976)	6, p. 58; *8*, p. *69*
	(b) **78**, 219 (1976)	6, p. 62; *8*, p. *71*
C. W. ALLEN	(a) **36**, 107 (1976)	*1.1*, p. *12*; *1.4*, p. *27*; 21.1.1, p. 319
A. B. ALMAZOV	(a) **35**, 563 (1976)	*14.3*, p. *177*
A. M. OZORIO DE ALMEIDA see under letter "O"		
J. R. L. DE ALMEIDA	(b) **73**, 661 (1976)	13, p. 114; 18, p. 222; *20*, p. *257*; 22.8.2, p. 427
V. L. ALPEROVICH	(b) **77**, 465 (1976)	*13.5.1*, p. *155*; *20.1*, p. *274*; 22.2.1, p. 364
V. I. ALSHITS	(b) **74**, K85 (1976)	6, p. 55; *10*, p. *75*
I. M. ALYMOV	(a) **33**, K63 (1976)	*19*, p. *247*; 22.9, p. 433
	(a) **33**, K111 (1976)	*19*, p. *248*; 22.9, p. 434
I. K. AMANIS	(a) **35**, K67 (1976)	*19*, p. *251*; 22.6, p. 405
I. I. AMELIN	(b) **77**, 441 (1976)	*13*, p. *119*
S. AMELINCKX	(a) **33**, 383 (1976)	*1.1*, p. *10*; 1.4, p. 25; 22, p. 334
	(a) **33**, 595 (1976)	*1.4*, p. *25*; *10.2*, p. *89*; 22.6, p. 403
	(a) **33**, 723 (1976)	*1.2*, p. *14*; *1.4*, p. *26*; 22.6, p. 403; 22.8.1, p. 419
	(a) **33**, K77 (1976)	*1.4*, p. *26*; 22, p. 335
	(a) **36**, 757 (1976)	*1.1*, p. *12*; *1.2*, p. *18*; 22, p. 339
	(a) **37**, 591 (1976)	*1.2*, p. *19*; 1.4, p. 28; 22, p. 340
	(a) **38**, 185 (1976)	*1.2*, p. *19*; 1.4, p. 29; 22, p. 341
	(a) **38**, 663 (1976)	*1.4*, p. *29*; 10.2, p. 99; 22, p. 342
J. AMIELL	(a) **35**, 615 (1976)	*10.2*, p. *93*; *13.4*, p. *139*; 14.3, p. 177; 18, p. 223; 22.4, p. 376
L. K. AMINOV	(b) **77**, 505 (1976)	*4*, p. *46*; 6, p. 61; 18.1, p. 226; 19, p. 254; 22.9, p. 438
CAO-XUÂN AN see under letter "C"		
E. E. ANDERSON	(b) **74**, K23 (1976)	*18.4*, p. *243*; 21.1, p. 304
J. M. ANDERSON	(a) **36**, K119 (1976)	*10.1*, p. *83*; 11, p. 103; 12.1, p. 108; 21.1, p. 310
T. L. ANDERSON	(a) **37**, 235 (1976)	1.4, p. 28; *10.2*, p. *95*; *12.2*, p. *112*; 22.6, p. 407
B. ANDLAUER	(b) **73**, 533 (1976)	*13.4*, p. *135*; 19, p. 248; 20.1, p. 261; 20.3, p. 280; 22.8.2, p. 426
A. ANDREEFF	(b) **73**, K89 (1976)	*5*, p. *48*; 18.2, p. 229; 21, p. 291; 21.1.1, p. 317
	(b) **75**, K115 (1976)	*13*, p. *116*; 22.6, p. 405
A. K. ANDREEV	(a) **34**, K163 (1976)	*18.2*, p. *231*
V. N. ANDREEV	(b) **77**, K97 (1976)	*14.1*, p. *166*; *14.3.1*, p. *185*; 22.6, p. 407
J. W. ANDREW	(b) **73**, 431 (1976)	*1.1*, p. *10*; 21, p. 290; 21.6, 329
A. ANEDDA	(a) **36**, 385 (1976)	*13.5.1*, p. *153*; 22.4.1, p. 381
M. V. ANFIMOV	(b) **75**, 73 (1976)	*4*, p. *45*; 12.2, p. 110; 22.1, p. 343; 22.2, p. 361; 22.4, p. 376; 22.5, p. 389
V. V. ANISIMOV	(a) **35**, 251 (1976)	*13.4*, p. *138*; 16, p. 214; 20.1, p. 268; 20.3, p. 283; 22.4.1, p. 380
H.-J. ANKLAM	(a) **36**, K71 (1976)	*8*, p. *69*; *14.3.3*, p. *192*; 14.3.4, p. 198; 22.6, p. 405
N. K. ANNAMALAI	(a) **37**, K177 (1976)	*14.3.1*, p. *185*; *14.3.3*, p. *193*; 22.6, p. 407
M. ANTONINI	(a) **34**, K153 (1976)	*10.1*, p. *81*; 21.1, p. 306
J. G. ANTONOPOULOS	(a) **38**, K115 (1976)	*1.6*, p. *34*; 21.1, p. 315; 21.6, p. 331
V. A. ANTONOV	(a) **35**, K169 (1976)	*20.1*, p. *270*; 20.2, p. 279; 22.8.1, p. 422
K. AOKI	(a) **38**, 587 (1976)	*10.1*, p. *86*; 12.1, p. 109; 21.1, p. 314
Y. AOKI	(a) **33**, 625 (1976)	1.3, p. 21; *18.1*, p. *225*; *18.2*, p. *229*; 21.1, p. 304
	(a) **36**, K149 (1976)	*1.3*, p. *23*; 14.1, p. 166; 28.2, p. 234; 21.1, p. 310
P. APEL	(a) **33**, K129 (1976)	*11*, p. *100*; 21.6, p. 329
J. H. F. VAN APELDOORN	(a) **35**, 137 (1976)	*5*, p. *49*; 18.2, p. 232; 21.1.1, p. 318; 21.4, p. 326

F. Appel	(a) **34**, 175 (1976)	*10.1*, p. *79*; 21, p. 292
	(a) **38**, 103 (1976)	*10.2*, p. *97*; 22.6, p. 408
J. Appel	(b) 77, 373 (1976)	*13*, p. *118*
K. I. Arai	(a) **34**, 325 (1976)	1.2, p. 15; *18.3*, p. *240*; 22.8.2, p. 427
S. Arajs	(b) 74, K23 (1976)	*18.4*, p. *243*; 21.1, p. 304
G. Arancia	(a) **33**, 563 (1976)	*1.1*, p. *11*; *3*, p. *42*; 22.3.1, p. 375
C. B. de Araujo	(b) 75, 327 (1976)	*13.5.4*, p. *160*; 18.2, p. 231
A. J. Ardell	(a) **34**, 679 (1976)	*10.1*, p. *81*; *11*, p. *101*; 21.1, p. 306
W. Arden	(b) 75, 621 (1976)	*13.5.1*, p. *153*; 20.3, p. 284; 22.9, p. 436
H. Arend	(b) 74, 151 (1976)	*18.4*, p. *242*; *20.1*, p. *263*; 22.9, p. 434
	(b) 74, 425 (1976)	4, p. 45; *6*, p. *55*; 14.4.1, p. 206; *20.1*, p. *265*; 22.8.1, p. 420
	(a) **36** 285 (1976)	*1.2*, p. *17*; 22.9, p. 437
S. K. Arif	(a) **33**, 91 (1976)	*5*, p. *48*; 18.2, p. 228; 18.3, p. 239; 21.1.1, p. 316
V. E. Arkhipov	(a) **38**, 433 (1976)	*11*, p. *104*; *14.2*, p. *172*; 21, p. 300
P. Arnold	(a) **36**, K37 (1976)	*18.4*, p. *244*; 22.8.2, p. 429
S. Arnold	(b) 74, 401 (1976)	*13.5.1*, p. *151*; 20.3, p. 281; 22.9, p. 434
V. K. Arora	(b) 73, K93 (1976)	6, p. 54; *14.3*, p. *174*; 22.2.3, p. 367
	(b) 74, K111 (1976)	*14*, p. *162*
	(b) 75, K65 (1976)	*14.3*, p. *176*
	(b) 77, 77 (1976)	*6*, p. *60*; *14*, p. *163*
P. A. Arsenev	(a) **33**, 511 (1976)	20.1, p. 263; *20.2*, p. *279*; 20.3, p. 281; 22.8.1, p. 419
	(a) **35**, K169 (1976)	*20.1*, p. *270*; 20.2, p. 279; 22.8.1, p. 422
J. O. Artman	(a) **37**, 617 (1976)	*10.2*, p. *96*; 13.4, p. 143; 20.1, p. 275; 22.5.2, p. 397
E. K. Arushanov	(a) **35**, K53 (1976)	*13.1*, p. *124*; 14.3, p. 177; 22.3, p. 372
G. M. Arutyunyan	(b) 77, K171 (1976)	*13*, p. *119*; *20.1*, p. *274*; 22, p. 340
S. M. Ashinov	(a) **38**, 769 (1976)	10.1, p. 86; *14.2*, p. *172*; 21, p. 301
M. Kh. Ashurov	(a) **35**, 654 (1976)	*13.4*, p. *140*; 20.1, p. 269; 20.3, p. 284; 22.8.2, p. 429
J. Aslam	(b) 77, 693 (1976)	*6*, p. *61*; 12.2, p. 112; 22.5.4, p. 401
V. V. Astanin	(a) **35**, 403 (1976)	*10.1*, p. *82*; 21, p. 294; 21.1, p. 307
V. T. Atamanenko	(a) **38**, 131 (1976)	11, p. 104; *14.3.3*, p. *194*; 22.1.2, p. 357
S. G. Atarbekov	(a) **34**, K7 (1976)	*13.4*, p. *137*; *14.3.3*, p. *190*; 22.2.3, p. 367
U. Atzmony	(b) 76, 669 (1976)	*18.2*, p. *233*; 21.1, p. 310; 21.4, p. 327
H. P. Aubauer	(a) **35**, 487 (1976)	*1.2*, p. *16*; 4, p. 45; 21, p. 295
B. Auguin	(a) **33**, 295 (1976)	*2*, p. *35*; *14.3*, p. *173*; 15, p. 209; 22.8, p. 413
J. Auleytner	(a) **36**, 209 (1976)	*1.1*, p. *12*; 10.2, p. 93; 22.1.2, p. 355
S. Auluck	(b) 74, K129 (1976)	*13.1*, p. *123*; *13.2*, p. *130*; 20.1, p. 266; 21, p. 292; 21.1, p. 305
M. Ausloos	(b) 73, 211 (1976)	*1.2*, p. *14*; 13.1, p. 120; *14.1*, p. *164*; 21.4, p. 326
	(b) 73, 469 (1976)	*13*, p. *114*; 14.1, p. 164; 21, p. 290
I. G. Avaeva	(a) **36**, K1 (1976)	*18.2*, p. *233*; 22.8.2, p. 429
E. M. Avakian	(a) **36**, K25 (1976)	*14.3.4*, p. *198*; 22.8.1, p. 422
	(a) **37**, K111 (1976)	*14.3.4*, p. *198*; 22.8.1, p. 422
I. Aveline	(b) 75, 465 (1976)	*8*, p. *68*; *18.1*, p. *226*
A. Avogadro	(b) 75, 179 (1976)	19, p. 250; 22.8.1, p. 421
B. Ayrault	(b) 78, 767 (1976)	*6*, p. *64*; *20.1*, p. *278*; 22.6, p. 408
A. Ažman	(b) 77, K157 (1976)	*13.1*, p. *128*; 22, p. 340
C. M. van Baal	(a) **37**, 371 (1976)	*10.1*, p. *84*; *11*, p. *103*; 21, p. 298
V. M. Babich	(b) 75, 299 (1976)	*14.3*, p. *176*; 22.1.1, p. 346
	(b) 78, K111 (1976)	*14.3*, p. *182*; 22.1.1, p. 350

H. Baessler	(b) **73**, 709 (1976)	13.1, p. 121; *20.1*, p. *262*; 22.9, p. 433
R. K. Bagai	(b) **73**, K123 (1976)	*20.1*, p. *262*; 22.5.3, p. 399
Kh. S. Bagdasarov	(a) **34**, K109 (1976)	*20.2*, p. *279*; *20.3*, p. *282*; 22.8.2, p. 428
	(a) **38**, 409 (1976)	*20.2*, p. *279*; *20.3*, p. *288*; 22.8.2, p. 431
G. D. Bagratishvili	(a) **36**, 73 (1976)	13.3, p. 132; *14.3.2*, p. *188*; 20.3, p. 285; 22.2.1, p. 363
B. Bahadur	(a) **35**, 387 (1976)	*2*, p. *37*; *18*, p. *223*; 22.9, p. 436
S. K. Bahl	(b) **74**, 721 (1976)	14.1, p. 165; *20.1*, p. *266*; 21, p. 292
J. Baillou	(a) **37**, 521 (1976)	10.2, p. 96; *20*, p. *258*; 22.4.2, p. 383
M. Baj	(a) **33**, 421 (1976)	13.4, p. 135; *14.3*, p. *174*; 22.4.3, p. 385
P. Bajons	(a) **38**, 199 (1976)	*10.1*, p. *85*; *12.1*, p. *109*; 21.1, p. 313
V. A. Balashov	(a) **33**, 737 (1976)	20.1, p. 263; *20.2*, p. *279*; *20.3*, p. *281*; 22.8, p. 413
A. Baldereschi	(b) **76**, 629 (1976)	*13*, p. *117*; *14.4*, p. *203*; 22.5.2, p. 396; 22.6, p. 406
S. Balevičius	(a) **35**, K41 (1976)	*2*, p. *37*; *14.3.1*, p. *184*; 22.3, p. 372
C. A. B. Ball	(a) **38**, 315 (1976)	*10.2*, p. *98*; *20*, p. *259*; 22.2.1, p. 364; 22.2.2, p. 365
D. Ball	(a) **35**, 651 (1976)	*19*, p. *252*; 22.8.1, p. 422
	(a) **36**, 307 (1976)	*3*, p. *43*; *19*, p. *253*; 22.8.1, p. 422
E. P. Balsamo	(a) **35**, K173 (1976)	*14.2*, p. *170*; 21, p. 295; 22.6, p. 405
I. Baltog	(b) 77, 701 (1976)	*10.2*, p. *95*; 20.1, p. 274; 20.3, p. 287; 22.5.2, p. 397
	(a) 38 K175 (1976)	*13.4*, p. *146*; 22.5, p. 390
A. Balzarotti	(b) **76**, 689 (1976)	*13.1*, p. *126*; 17, p. 219; 22.6, p. 406
R. Balzer	(b) **74**, K135 (1976)	*10.2*, p. *90*; 11, p. 101; 22.5.2, p. 394
N. Banaii	(b) **77**, 707 (1976)	*10.2*, p. *96*; 20.1, p. 274; 22.5.2, p. 397
	(b) **78**, K51 (1976)	*10.2*, p. *97*; 22.5.2, p. 397
H. D. Banerjee	(a) **36**, 39 (1976)	*1.3*, p. *23*; *1.4*, p. *27*; 21, p. 296
	(a) **36**, 693 (1976)	*10.2*, p. *94*; 20.3, p. 285; 22.5.3, p. 400
K. G. Banerjee	(a) **33**, 707 (1976)	*9*, p. *72*; *10.1*, p. *79*; 21.1, p. 304
S. Banerjee	(a) **36**, 39 (1976)	*1.3*, p. *23*; *1.4*, p. *27*; 21, p. 296
C. Bansal	(a) **35**, 73 (1976)	*5*, p. *49*; 21.1, p. 307; 21.1.1, p. 318
M. L. Bansal	(b) **75**, K31 (1976)	*6*, p. *56*; 20.1, p. 267; 22.8.1, p. 421
O. P. Bansal	(b) **75**, 85 (1976)	6, p. 56; 12.1, p. 107; *13.2*, p. *130*; 21.6, p. 330
Burma Bansaragtschin	(a) **37**, K13 (1976)	*3*, p. *43*; 14.3, p. 179; 22, p. 339
J. Banský	(b) **76**, 569 (1976)	2, p. 39; *13*, p. *117*; *18*, p. *224*; 22, p. 338
M. S. Bapna	(a) **33**, 639 (1976)	*10.1*, p. *79*; 21.6, p. 329
D. Barančok	(b) **74**, 741 (1976)	*2*, p. *37*; 13.4, p. 137; *14.3*, p. *175*; 15, p. 209; 22, p. 335
B. Baranov	(a) **38**, K111 (1976)	*1.5*, p. *33*; *3*, p. *44*; 22.2.4, p. 371
M. N. Baranov	(b) **76**, 55 (1976)	13.4, p. 140; *20.1*, p. *270*; *20.3*, p. *284*; 22.8.1, p. 422.
P. G. Baranov	(b) **76**, K109 (1976)	*19*, p. *253*; 22.5.2, p. 396
B. Baranowski	(a) **37**, K163 (1976)	*14.2*, p. *172*; 21, p. 299
P. I. Baranskii	(a) **34**, K139 (1976)	*14.3*, p. *176*; 22.1.1, p. 346; 22.1.2, p. 353
	(b) **75**, 299 (1976)	*14.3*, p. *176*; 22.1.1, p. 346
	(b) **78**, K111 (1976)	*14.3*, p. *182*; 22.1.1, p. 350
J. Barbolla	(a) **35**, 533 (1976)	*13.4*, p. *139*; *14.3.3*, p. *191*; 22.1.2, p. 354
	(a) **36**, 495 (1976)	*13.4*, p. *141*; 14.3.3, p. 192; 22.1.2, p. 355
J. A. Barclay	(a) **38**, 533 (1976)	*19*, p. *256*; 21.1.1, p. 321
H. J. von Bardeleben	(b) **76**, 363 (1976)	*19*, p. *252*; 22.8, p. 416
L. Bárdoš	(a) **35**, 291 (1976)	*1.5*, p. *31*; *3*, p. *42*; 22.6, p. 404
D. Barjon	(b) **76**, 223 (1976)	*13*, p. *117*; *14.3.4*, p. *197*; 22.2.3, p. 368
	(b) **76**, 623 (1976)	*14.3*, p. *178*; 22.2.4, p. 370
I. B. Barkan	(a) **38**, K139 (1976)	*14.4.1*, p. *208*; *16*, p. *217*; 22.8.1, p. 425
B. Sh. Barkhalow	(a) **33**, K67 (1976)	*14.3.1*, p. *183*; 14.3.2, p. 187; 22.9, p. 433

	(a) **34**, K77 (1976)	*14.3.1*, p. *183*; 22.9, p. 435
D. M. Barnett	(b) **75**, 573 (1976)	*10*, p. *76*
	(a) **38**, 637 (1976)	*10*, p. *77*
V. V. Baron	(a) **37**, K165 (1976)	*1.2*, p. *19*; 5, p. 50; 21, p. 299
A. Barone	(a) **35**, K173 (1976)	*14.2*, p. *170*; 21, p. 295; 22.6, p. 405
G. R. Barsch	(b) **76**, 133 (1976)	*6*, p. *57*; 22.7, p. 411
	(b) **76**, 677 (1976)	*6*, p. *59*; 22.7, p. 411
Č. Bárta	(a) **34**, K51 (1976)	*1.2*, p. *15*; 22.5, p. 388
W. Barth	(a) **34**, 153 (1976)	*13.4*, p. *137*; *20.1*, p. *264*; 22.1.1, p. 346
C. Barthou	(b) **76**, K5 (1976)	*19*, p. *252*; 22.8.1, p. 422
S. K. Barthwal	(a) **35**, 761 (1976)	2, p. 38; *14.3*, p. *178*; *15*, p. *210*; 20.1, p. 270; 22.7, p. 411
	(a) **36**, 345 (1976)	*2*, p. *39*; *14.3.1*, p. *184*; 15, p. 210; 22.2.1, p. 363; 22.2.3, p. 368
	(a) **36**, 533 (1976)	2, p. 39; *14.3.1*, p. *184*; *15*, p. *210*; 22.1.1, p. 348
H. Bartsch	(a) **34**, K5 (1976)	*10.2*, p. *90*; 22.4.3, p. 385
	(a) **38**, 139 (1976)	*1.2*, p. *19*; *11*, p. *104*; 22.1.2, p. 357
I. Barvík	(b) **75**, K45 (1976)	*16*, p. *214*; *20.1*, p. *267*; 22.8.2, p. 428
V. K. Bashenov	(b) **77**, 473 (1976)	1.1, p. 13; 2, p. 40; *4*, p. *46*; *13.1*, p. *127*; 22.2, p. 361; 22.2.4, p. 371
F. Basile	(a) **35**, 153 (1976)	*5*, p. *49*; 18.3, p. 240; 22.8.2, p. 428
H, Bässler	(b) **76**, 533 (1976)	*16*, p. *215*; 22.9, p. 437
H. C. Basso	(b) **77**, 685 (1976)	1.2, p. 19; *6*, p. *61*; 10.2, p. 95; *20.1*, p. *274*; 22, p. 340
P. Bassoul	(a) **34**, 125 (1976)	*1.2*, p. *15*; *1.4*, p. *26*; 22.8.2, p. 427
D. Bastian	(a) **35**, 113 (1976)	*18.2*, p. *232*; *19*, p. *251*; 21.1, p. 307; 21.1.1, 318
	(a) **35**, 465 (1976)	*18.2*, p. *232*; *19*, p. *252*; 21.1, p. 308; 21.1.1, p. 319
P. Basu	(a) **37**, 625 (1976)	13.3, p. 133; *14.3.3*, p. *193*; *16*, p. *216*; 22.1.2, p. 357; 22.2.1, p. 364; 22.4.3, p. 386
P. K. Basu	(b) **73**, K15 (1976)	6, p. 53; *14.3*, p. *173*; 22.2.3, p. 367
	(a) **36**, K65 (1976)	*8*, p. *69*; 14.3.4, p. 198; 22.1.2, p. 355
	(a) **38**, K85 (1976)	*14.3.2*, p. *189*; 22.1.2, p. 358
J. Bascyński	(b) **73**, K85 (1976)	*18.3*, p. *239*; 19, p. 248; 22.8.2, p. 427
G. Batailler	(a) **37**, 521 (1976)	10.2, p. 96; *20*, p. *258*; 22.4.2, p. 383
A. Batana	(b) **76**, K105 (1976)	*4*, p. *46*; *12.2*, p. *111*; 22.5.2, p. 396
C. A. Bates	(b) **74**, 225 (1976)	*6*, p. *55*; *19*, p. *249*; 22.4.1, p. 379; 22.6, p. 403
T. P. Batsankalashvili	(a) **38**, 769 (1976)	10.1, p. 86; *14.2*, p. *172*; 21, p. 301
G. Bauer	(b) **75**, 543 (1976)	*13.1*, p. *124*; *14.3.4*, p. *197*; 20.1, p. 268; 20.3, p. 283; 22.2.3, p. 368
	(b) **76**, 259 (1976)	13.1, p. 125; *18*, p. *223*; *20.1*, p. *270*; 22, p. 337
	(b) **78**, 643 (1976)	*14.3*, p. *181*; 22.2.3, p. 369
W. Bauhofer	(b) **78**, K121 (1976)	*6*, p. *64*; *20.1*, p. *278*; 22.5.4, p. 401
W. Baumhoer	(b) **73**, 155 (1976)	10.2, p. 87; *19*, p. *247*; 22.5.2, p. 392
W. Bausch	(b) **73**, K23 (1976)	*6*, p. *53*; 8, p. 67; 22.5.2, p. 392
C. Bazán	(b) **77**, 451 (1976)	*14.2*, p. *171*; 21, p. 298
W. Bażela	(a) **38**, 721 (1976)	*18.4*, p. *246*; 22.8, p. 418
C. C. Becerra	(a) **33**, 573 (1976)	*18.2*, p. *229*; *18.4*, p. *242*; 22, p. 334
H. Becherer	(b) **75**, K19 (1976)	*20.3*, p. *283*; 22.8, p. 415
F. Bechstedt	(b) **73**, 141 (1976)	*6*, p. *52*; 13.1, p. 120; 13.5.1, p. 149; *20.1*, p. *260*
	(b) **78**, 711 (1976)	6, p. 64; *14.3.4*, p. *200*; *20.1*, p. *278*; 22, p. 342
U. Beck	(a) **37**, 193 (1976)	*1.5*, p. *32*; *12.2*, p. *111*; 22, p. 339
C. Becker	(a) **34**, 519 (1976)	*10.2*, p. *91*; 22.8.2, p. 428
C. R. Becker	(b) **76**, K15 (1976)	*6*. p. *58*; *20.1*, p. *271*; 22, p. 338

P. J. Becker	(b) **74**, 285 (1976)	*18.1*, p. *225*; *20.1*, p. *264*; 22.8.1, p. 419
T. Bedyńska	(a) **36**, 509 (1976)	*10.2*, p. *94*; 22.1.2, p. 355
M. M. Beg	(b) **77**, 693 (1976)	*6*, p. *61*; 12.2, p. 112; 22.5.4, p. 401
U. Behn	(b) **76**, K23 (1976)	*13.1*, p. *125*
	(b) **77**, K79 (1976)	*13.1*, p. *125*
H. Beige	(b) **76**, K47 (1976)	1.2, p. 17; *12.2*, p. *111*; *14.4*, p. *203*; 22.8.1, p. 422
F. I. Bekkerman	(b) **74**, K1 (1976)	*12.1*, p. *107*; *15*, p. *209*; 21.7, p. 332
K. G. Belabaev	(a) **36**, K25 (1976)	*14.3.4*, p. *198*; 22.8.1, p. 422
	(a) **37**, K111 (1976)	*14.3.4*, p. *198*; 22.8.1, p. 422
R. Beliackas	(a) **35**, K61 (1976)	*14.4.1*, p. *207*; 22.8, p. 415
E. Belorizky	(b) **77**, 495 (1976)	*18.2*, p. *235*
K. P. Belov	(a) **36**, 415 (1976)	12.2, p. 111; *18.4*, p. *244*; 22.8.2, p. 429
I. M. Belova	(b) **73**, 409 (1976)	2, p. 36; *14.3.4*, p. *195*; 22, p. 334
A. I. Belyaeva	(b) **75**, K123 (1976)	13.5.4, p. 160; *20.1*, p. *269*; 22.8, p. 415
	(b) **76**, 419 (1976)	*13.5.4*, p. *160*; *20.1*, p. *271*; 22.5, p. 389
	(a) **38**, K103 (1976)	*18.3*, p. *241*; 22.8.2, p. 431
S. Benckert	(b) **73**, 753 (1976)	*4*, p. *45*; *12.1*, p. *107*; 21, p. 291; 21.2, p. 323
V. Z. Bengus	(a) **33**, K1 (1976)	*10.2*, p. *88*; 22.5.2, p. 392
	(a) **34**, 445 (1976)	*10.2*, p. *90*; 22.5.2, p. 392
C. Benoit	(b) **78**, 67 (1976)	*6*, p. *62*; *20.1*, p. *275*; 22.8.1, p. 424
	(b) **78**, 453 (1976)	*6*, p. *63*; 20.1, p. 277; 22.8.1, p. 424
	(a) **38**, K159 (1976)	*20.1*, p. *278*; 22.9, p. 439
G. Berg	(a) **38**, 115 (1976)	*10.2*, p. *97*; 14.4, p. 204; 22.5.2, p. 398
U. Berg	(b) **74**, 341 (1976)	*13.1*, p. *122*; 17, p. 218; 20, p. 257; 22.1, p. 343
	(b) **75**, K89 (1976)	*13.1*, p. *124*; 17, p. 219; 20, p. 258; 22, p. 337
G. O. Berim	(b) **76**, 827 (1976)	*13.5.1*, p. *154*; *19*, p. *253*
R. Bernard	(a) **34**, 135 (1976)	*1.2*, p. *15*; *1.3*, p. *22*; 1.4, p. 26; 21, p. 291; 21.1, p. 304
O. Bernard	(b) **78**, 67 (1976)	*6*, p. *62*; *20.1*, p. *275*; 22.8.1, p. 424
K. Berndt	(b) **78**, 659 (1976)	*13.2*, p. *131*; 20, p. 259; 21.2, p. 323; 21.3, p. 325
P. Berndt	(a) **38**, 497 (1976)	*16*, p. *217*; 22.8, p. 418
A. Berndtsson	(b) **75**, 197 (1976)	*13.1*, p. *123*; 17, p. 219; 21, p. 293
Hj. Bernhardt	(a) **33**, 211 (1976)	*10.2*, p. *87*; *20.1*, p. *261*; 20.3, p. 280; 22.8.1, p. 419; 22.8.2, p. 426
	(a) **34**, 483 (1976)	*10.2*, p. *91*; 20.3, p. 282; 22.8.2, p. 427
	(a) **37**, 445 (1976)	*10.2*, p. *96*; 20.1, p. 275; 20.3, p. 287; 22.8.2, p. 430
P. Bernstein	(a) **35**, 681 (1976)	*18.3*, p. *240*; 22.8.2, p. 429
B. Berre	(a) **37**, 161 (1976)	*12.1*, p. *108*; 21, p. 296; 21.1, p. 309
J. Berrehar	(b) **77**, K119 (1976)	*14.3*, p. *180*; 22.9, p. 438
G. Berthet	(b) **74**, 217 (1976)	13.5.1, p. 151; *19*, p. *249*; 22.9, p. 434
L. A. Bertolo	(b) **77**, 433 (1976)	*6*, p. *61*; 21, p. 298; 21.1, p. 312; 21.5, p. 328; 21.6, p. 331
R. Beserman	(a) **35**, 109 (1976)	*3*, p. *42*; 22.1.2, p. 353
V. I. Betekhtin	(a) **34** 73 (1976)	*10.1*, p. *79*; 21, p. 291
H. Bethge	(a) **37**, 457 (1976)	1.4, p. 28; *1.5*, p. *32*; *3*, p. *44*; 21, p. 298; 22.6, p. 407
	(a) **38**, 103 (1976)	*10.2*, p. *97*; 22.6, p. 408
A. van den Beukel	(a) **37**, 371 (1976)	*10.1*, p. *84*; *11*, p. *103*; 21, p. 298
V. Bezák	(b) **76**, 569 (1976)	2, p. 39; *13*, p. *117*; *18*, p. *224*; 22, p. 338
P. A. Bezirganian	(a) **35**, K181 (1976)	*20*, p. *258*
B. V. Beznosikov	(b) **75**, K123 (1976)	13.5.4, p. 160; *20.1*, p. *269*; 22.8, p. 415
C. M. Bhandari	(b) **78**, 821 (1976)	*18.2*, p. *236*
S. C. Bhargava	(a) **38**, 601 (1976)	*5*, p. *51*; 18.1, p. 227; 18.2, p. 236; 21.1.1, p. 322

T. BHASKAR RAO see under letter "R"		
DEBADIDEB BHATTACHARYA	(b) **76**, K81 (1976)	*5*, p. *50*; *10.2*, p. *94*
B. D. BHATTACHARYYA	(b) **74**, K53 (1976)	*10.2*, p. *90*; *19*, p. *249*; 22.5.3, p. 399
	(b) **74**, 695 (1976)	*13.4*, p. *137*; 19, p. 250; 20.1, p. 266; 22.9, p. 435
V. V. BHOKARE	(b) **74**, K161 (1976)	*13.1*, p. *123*; 22.1.1, p. 346
V. P. BHOLA	(a) **34**, K179 (1976)	*13.1*, p. *123*; *20.1*, p. *267*; 22.3, p. 372
S. BHUSHAN	(a) **38**, K27 (1976)	*20.3*, p. *288*; 22.4, p. 377
A. BIANCONI	(b) **76**, 689 (1976)	*13.1*, p. *126*; 17, p. 219; 22.6, p. 406
D. BIDJIN	(a) **33**, K23 (1976)	*1.1*, p. *10*; 1.2, p. 14; 22.3, p. 372
H. BIEDERMAN	(a) **36**, 783 (1976)	*17*, p. *219*; 22.5.2, p. 396
L. BIEGAŁA	(b) **75**, 333 (1976)	*18.2*, p. *231*
	(b) **76**, K73 (1976)	*18.2*, p. *233*; *18.4*, p. *244*
Z. BIEGAŃSKI	(a) **35**, K127 (1976)	*8*, p. *69*; *18*, p. *223*; 21.4, p. 326
K. BIENERT	(a) **33**, 511 (1976)	20.1, p. 263; *20.2*, p. *279*; 20.3, p. 281; 22.8.1, p. 419
H. BIERNAT	(b) **78**, 653 (1976)	*14.3*, p. *181*; 22.2.3, p. 369
H. BILL	(b) **75**, 613 (1976)	6, p. 57; *10.2*, p. *93*; 20.3, p. 284; 22.5.3, p. 399
	(a) **35**, K105 (1976)	1.2, p. 17; *19*, p. *252*; 22.8, p. 416
D. BILLARD	(b) **75**, 117 (1976)	*6*, p. *56*; *20.1*, p. *267*; 22.6, p. 404
M. BILLARDON	(b) **78**, 779 (1976)	6, p. 64; *10.2*, p. *98*; 13.4, p. 145; 20.1, p. 278; 22.5.2, p. 398
E. BILLER	(a) **35**, 113 (1976)	*18.2*, p. *232*; *19*, p. *251*; 21.1, p. 307; 21.1.1, p. 318
	(a) **35**, 465 (1976)	*18.2*, p. *232*; *19*, p. *252*; 21.1, p. 308; 21.1.1, p. 319
R. BIMBERG	(a) **36**, 253 (1976)	*1.5*, p. *32*; 21.1, p. 309; 21.1.1, p. 319; 22.1.1, p. 348
H. BIŃCZYCKA	(a) **35**, K69 (1976)	*18.2*, p. *232*; *18.4*, p. *244*; 22.8, p. 415
R. BINDEMANN	(a) **34**, K63 (1976)	*20.3*, p. *281*; 22.2.2, p. 365
C. BIRAC	(a) **36**, 247 (1976)	2, p. 39; *9*, p. *73*; 21, p. 296
I. BIRCHAK	(a) **33**, K145 (1976)	13.4, p. 136; *16*, p. *213*; 22.4.2, p. 383
C. BIRKHOLZ	(b) **76**, K47 (1976)	1.2, p. 17; *12.2*, p. *111*; *14.4*, p. *203*; 22.8.1, p. 422
U. BIRKHOLZ	(a) **37**, K97 (1976)	*14.3*, p. *179*; *15*, p. *211*; 22, p. 340
	(a) **37**, K147 (1976)	13.4, p. 143; *16*, p. *216*; 22, p. 340
	(b) **78**, K23 (1976)	13.1, p. 128; *20.1*, p. *276*; 22, p. 341
A. R. BIRKS	(b) **76**, 599 (1976)	*13.1*, p. *126*; 14.2, p. 170
H. K. BIRNBAUM	(a) **34**, K97 (1976)	*10.1*, p. *81*; 21.1, p. 306
	(a) **36**, 687 (1976)	*9*, p. *73*; 21, p. 297
P. M. BISCH	(b) **75**, 705 (1976)	*13*, p. *116*; *18*, p. *223*; 21.1, p. 308
	(b) **76**, 395 (1976)	*13*, p. *117*; *18*, p. *224*; 19, p. 252; 21.1, p. 309
A. B. BISWAS	(a) **37**, 719 (1976)	*1*, p. *9*; 10.2, p. 96; 22.8.2, p. 431
J. P. BLANC	(b) **74**, 217 (1976)	13.5.1, p. 151; *19*, p. *249*; 22.9, p. 434
B. BLANZAT	(b) **76**, K5 (1976)	*19*, p. *252*; 22.8.1, p. 422
M. BLASCHKE	(b) **76**, 267 (1976)	*13.5.1*, p. *153*; *20.3*, p. *284*; 22.4.1, p. 380
C. DE BLASI	(a) **36**, 569 (1976)	*14.3.4*, p. *198*; 22.3, p. 373
G. BLASQUEZ	(a) **34**, 207 (1976)	*14.3.2*, p. *187*; 22.1.2, p. 352
P. BLAUBECK	(b) **77**, 601 (1976)	*2*, p. *40*; 11, p. 103
K. W. BLAZEY	(a) **38**, K97 (1976)	13.4, p. 145; *20.1*, p. *277*; 22.8.1, p. 424
G. L. BLERIS	(a) **34**, 243 (1976)	*1.3*, p. *22*; 10.2, p. 90; 22.7, p. 410
R. BLINC	(b) **74**, 425 (1976)	4, p. 45; *6*, p. *55*; 14.4.1, p. 206; *20.1*, p. *265*; 22.8.1, p. 420
D. BLOOR	(a) **37**, 427 (1976)	1.1, p. 13; *20.1*, p. *274*; 22.9, p. 438
K. H. BLUEMEL	(b) **75**, 527 (1976)	*10*, p. *76*; 12, p. 106
K. BLÜTHNER	(a) **34**, 541 (1976)	*14.2*, p. *169*; 21, p. 293

H. J. Blythe	(a) **37**, K197 (1976)	*10.1*, p. *84*; *18.2*, p. *235*; 21.1.1, p. 321
A. D. Boardman	(b) **75**, 347 (1976)	*13.3*, p. *132*; *13.5.2*, p. *157*; 14.3.2, p. 188
	(a) **37**, 357 (1976)	*13.5.4*. p. *160*; *14.3.4*, p. *198*; 22.8.2, p. 429
K. S. Bobev	(a) **38**, 393 (1976)	*14.3.4*, p. *200*; 22.1.1, p. 349; 22.6, p. 408
A. I. Bobrysheva	(b) **76**, K51 (1976)	*13.5.1*, p. *153*; 22, p. 338
P. Boch	(a) **35**, 687 (1976)	*1.3*, p. *22*; *12.1*, p. *108*; 21, p. 295; 21.1, p. 308
A. de Bock	(b) **78**, K89 (1976)	*2*, p. *41*; 18, p. 224; 20, p. 259; 22, p. 342
P. P. Bodiul	(b) **73**, K5 (1976)	*15*, p. *209*; 22.7, p. 410
I. V. Bodnar	(b) **73**, K135 (1976)	*13.1*, p. *121*; *20.1*, p. *262*; 22.2.4, p. 370
	(a) **36**, K141 (1976)	*3*, p. *43*; 22.2.4, p. 371
	(a) **37**, K173 (1976)	*13.1*, p. *128*; 20.1, p. 275; 22.2.4, p. 371
K. W. Böer	(a) **35**, 309 (1976)	*13.4*, p. *139*; *20.1*, p. *268*; 22.4.1, p. 380
E. Boesman	(a) **35**, K137 (1976)	*3*, p. *42*; *19*, p. *252*; 22.5.2, p. 395
	(b) **77**, 249 (1976)	*19*, p. *254*; 22.5.2, p. 396
	(b) **78**, 553 (1976)	*19*, p. *256*; *20.1*, p. *277*; 22.5.2, p. 398
U. Bogner	(a) **37**, 161 (1976	*7*, p. *66*; 12.2, p. 111; *14.4.1*, p. *208*; 20.1, p. 273; 22.8.1, p. 423
G. A. Bogomolova	(a) **34**, K109 (1976)	*20.2*, p. *279*; *20.3*, p. *282*; 22.8.2, p. 428
	(a) **38**, 409 (1976)	*20.2*, p. *279*; *20.3*, p. *288*; 22.8.2, p. 431
J. Bohm	(a) **33**, 737 (1976)	20.1, p. 263; *20.2*, p. *279*; *20.3*, p. *281*; 22.8, p. 413
K. Bohmhammel	(a) **36**, K125 (1976)	*8*, p. *70*; 14.2, p. 170; 21, p. 297
	(a) **37**, 179 (1976)	*8*, p. *70*; 13.1, p. 127; 21, p. 297
A. K. Bohra	(b) **78**, 255 (1976)	*6*, p. *62*; 23, p. 440
J. P. Boilot	(a) **38**, 595 (1976)	*1.4*, p. *29*; 22.8.2, p. 431
D. Bois	(b) **73**, 187 (1976)	*13.1*, p. *120*; *20.1*, p. *260*; 22.2.1, p. 362
E. L. Bokhenkov	(b) **75**, 105 (1976)	*6*, p. *56*; 20.1, p. 267; 22.9, p. 435
J. N. Boland	(a) **34**, 361 (1976)	*10.2*, p. *90*; 22.8.1, p. 420
Yu. B. Bolkhovityanov	(a) **37**, K193 (1976)	*1.5*, p. *32*; 3, p. 44; *14.3.1*, p. *185*; 22.2.4, p. 371
R. I. Bolkhovityanova	(a) **37**, K193 (1976)	*1.5*, p. *32*; 3, p. 44; *14.3.1*, p. *185*; 22.2.4, p. 371
L. P. Bolshakov	(b) **77**, 53 (1976)	*13.1*, p. *127*; *143*. p. *179*; 22.2.3, p. 368
J. Bolz	(b) **74**, 193 (1976)	*1.2*, p. *15*; 5, p. 48; 20.1, p. 263; 22.8, p. 414
F. Bombré	(a) **36**, 133 (1976)	*14.3*, p. *178*; 22.4.1, p. 381
A. M. Bon	(b) **78**, 67(1976)	*6*, p. *62*; *20.1*, p. *275*; 22.8.1, p. 424
	(b) **78**, 453 (1976)	*6*, p. *63*; 20.1, p. 277; 22.8.1, p. 424
V. L. Bonch-Bruevich	(b) **74**, 11 (1976)	2, p. 35; 13.1, p. *120*; 22, p. 333
A. I. Bondarenko	(a) **34**, 195 (1976)	*10.1*, p. *80*; 11. p. 100; 21.1, p. 305
G. Borchardt	(a) **33**, 143 (1976)	*14.4.1*, p. *206*; 22.8.1, p. 419
G. Bordin	(a) **35**, K161 (1976)	*18.3*, p. *240*; 22.8.2, p. 429
K. Borggreen	(a) **38**, 739 (1976)	*1.4*, p. *30*; 10.2, p. 99; 22, p. 342
S. Bornmann	(a) **35**, 347 (1976)	*3*, p. *42*; 22.8.2, p. 429
	(a) **35**, 583 (1976)	*3*, p. *42*; 22.8.2, p. 429
Yu. G. Borodko	(b) **73**, 335 (1976)	13, p. 113; 17, p. 218; *20*, p. *257*; 22.9, p. 433
G. Borstel	(b) **76**, 759 (1976)	*13.5*, p. *147*; *20.1*, p. *272*; 22.6, p. 406
	(b) **77**, 221 (1976)	*6*, p. *60*; *14.4.1*, p. *207*
	(b) **78**, 123 (1976)	*6*, p. *62*; 20.1, p. 275; 22.6, p. 408
J. von Borzeskowski	(b) **73**, 607 (1976)	6, p. 54; 13.1, p. 121; *14.3.4*, p. *195*; 22.1.2, p. 351
B. Bosacchi	(b) **77**, 457 (1976)	*13.2*, p. *131*; 21.1, p. 312
M. Bösch	(b) **77**, 153 (1976)	*6*, p. *60*, 20.1, p. 272; 22.6, p. 406
L. Bosi	(b) **73**, K61 (1976)	*10.2*, p. *87*; 22.5.2, p. 392
	(b) **75**, K27 (1976)	*10.2*, p. *92*; 22.5.2, p. 394
	(b) **75**, K163 (1976)	*10.2*, p. *93*; 16, p. 214; 20.3, p. 284; 22.5.2, p. 395
O. Bostanjoglo	(a) **34**, K1 (1976)	*2*, p. *36*; *18.2.1*, p. *237*; 21.1.1, p. 317
I. M. Boswarva	(a) **37**, 65 (1976)	*10.2*, p. *95*; 22.5.2, p. 397

F. W. Boswell	(a) **35,** 591 (1976)	1.2, p. 16; *1.4*, p. *27*; 22, p. 337
	(a) **36,** K21 (1976)	*1.4*, p. *28*; 10.1, p. 83; 21.1, p. 309
T. Boţilă	(a) **38,** 331 (1976)	*2*, p. *40*; 13.4, p. 145; 14.3, p. 181; 16, p. 217; 22.8, p. 418
H. Böttger	(b) **73,** 229 (1976)	*2*, p. *35*; 6, p. 52; 13, p. 113; *14.3*, p. *173*
	(b) **74,** K151 (1976)	*2*, p. *37*; 6, p. 55; 13, p. 115; *14.3*, p. *175*
	(b) **77,** 561 (1976)	*6*, p. *61*; 22, p. 340
	(b) **78,** 9 (1976)	2, p. 35; 6, p. 52; 13, p. 113; 13.5.3, p. 159; *14.3*, p. *173*; 22, p. 333
	(b) **78,** 415 (1976)	2, p. 35; 6, p. 52; 13, p. 113; 13.5.3, p. 159; *14.3*, p. *173*; 22, p. 333
M. N. Botvinko	(b) **76,** 797 (1976)	*13.4*, p. *141*; 13.5.4, p. 160; *18.2*, p. *234*
M. A. Boubel	(a) **35,** 459 (1976)	8, p. 68; *18.4*, p. *244*; 22.6, p. 405
C. Boulesteix	(a) **33,** K9 (1976)	*10.2*, p. *88*; 22.6, p. 402
	(a) **33,** 595 (1976)	*1.4*, p. *25*; *10.2*, p. *89*; 22.6, p. 403
	(a) **34,** 331 (1976)	*1.4*, p. *26*; *10.2*, p. *90*; 22.6, p. 403
J. C. Bourgoin	(a) **35,** 109 (1976)	*3*, p. *42*; 22.1.2, p. 353
J. Bouwma	(a) **35,** 281 (1976)	*10.2*, p. *92*; *14.4*, p. *203*; 22.8, p. 415
R. Boyn	(b) **76,** 427 (1976)	13.4, p. 140; *20.1*, p. *271*; 22.4.1, p. 380
	(b) **76,** K119 (1976)	13.4, p. 141; *20.1*, p. *272*; 22.4, p. 377
V. E. Bozhevolnov	(b) **78,** 483 (1976)	13.4, p. 145; *20.3*, p. *288*; 22.5.4, p. 401
J. C. Brabant	(a) **35,** 533 (1976)	*13.4*, p. *139*; *14.3.3*, p. *191*; 22.1.2, p. 354
	(a) **36,** 495 (1976)	*13.4*, p. *141*; 14.3.3, p. 192; 22.1.2, p. 355
V. A. M. Brabers	(a) **34,** 583 (1976)	*1*, p. *9*; 1.2, p. 15; *18.3*, p. *240*; 22.8.2, p. 428
W. F. Braerman	(a) **35,** 309 (1976)	*13.4*, p. *139*; *20.1*, p. *268*; 22.4.1, p. 380
P. Braillon	(a) **33,** 405 (1976)	*10.2*, p. *88*; 22.8.1, p. 419
	(a) **36,** 637 (1976)	*10.2*, p. *94*; 22.8.1, p. 423
	(a) **37,** 263 (1976)	*10.2*, p. *95*; 12.2, p. 112; 22.8.1, p. 423
H. S. Brandi	(b) **75,** 271 (1976)	13.4, p. 138; *19*, p. *250*; 22.5.3, p. 399
	(b) **76,** K137 (1976)	*13.4*, p. *141*; *19*, p. *253*; 22.5.3, p. 400
E. H. Brandt	(b) **77,** 105 (1976)	*14.2*, p. *170*
	(b) **77,** 551 (1976)	*14.2*, p. *171*
H. Brauer	(b) **75,** K115 (1976)	*13*, p. *116*; 22.6, p. 405
H. P. Braun	(a) **37,** 73 (1976)	*14.3.4*, p. *199*; 16, p. 216; 20.3, p. 286; 22.9, 437
	(a) **37,** 553 (1976)	13.4, p. 143; *14.3.4*, p. *199*; 16, p. 216; 22.9, p. 438
P. Braun	(b) **74,** 209 (1976)	*19*, p. *249*; 22.2.1, p. 362; 22.2.3, p. 367
W. Braun	(b) **76,** 251 (1976)	2, p. 38; *13.1*, p. *125*; 17, p. 219; 22.8, p. 416
W. Braune	(b) **76,** 633 (1976)	*13.2*, p. *131*; 14.1, p. 165; 21, p. 297
R. Braunstein	(b) **73,** 361 (1976)	*13.1*, p. *121*; 22.1.1, p. 345; 22.1.2, p. 351
M. Brauwers	(b) **75,** 519 (1976)	*13.1*, p. *124*; *20*, p. *258*; 21.6, p. 330
K. Breuer	(b) **73,** K95 (1976)	13, p. 114; *20*, p. *257*; 22.6, p. 403
M. S. Brodin	(b) **74,** 453 (1976)	*13.5.1*, p. *151*; *20.3*, p. *282*; 22.9, p. 435
R. Brodmann	(b) **73,** K99 (1976)	*13.5.1*, p. *150*; *20.3*, p. *280*; 23, p. 440
	(b) **74,** 579 (1976)	*13.5.1*, p. *151*; *20.3*, p. *282*; 23, p. 440
V. A. Brodovoi	(a) **37,** 363 (1976)	*13.4*, p. *143*; *14.3.4*, p. *199*; 20.3, p. 287; 22.4.2, p. 383
A. M. Brodskii	(b) **76,** 93 (1976)	*13*, p. *116*; *13.1*, p. *124*; 21.2, p. 323
F. Brosens	(b) **74,** 45 (1976)	*13*, p. *115*; *14.4*, p. *202*
H. Bross	(b) **77,** 591 (1976)	*13.1*, p. *127*; 13.2, p. 131; 21.2, p. 323
V. L. Broude	(b) **78,** 325 (1976)	*6*, p. *63*; *20.1*, p. *276*; 22.9, p. 438
F. Brouers	(b) **75,** 519 (1976)	*13.1*, p. *124*; *20*, p. *258*; 21.6, p. 330
	(b) **76,** 145 (1976)	*13*, p. *117*; *18.4*, p. *244*
M. Brousseau	(a) **35,** 533 (1976)	*13.4*, p. *139*; *14.3.3*, p. *191*; 22.1.2, p. 354
	(a) **36,** 495 (1976)	*13.4*, p. *141*; 14.3.3, p. 192; 22.1.2, p. 355
L. Bru	(a) **34,** K9 (1976)	*10.1*, p. *80*; 21.1, p. 305

	(a) **34**, K17 (1976)	*1.2*, p. *15*; 21.1.1, p. 317
H. BRUCHLOS	(a) **34**, 219 (1976)	13.4, p. 137; 14.3.3, p. 190; *14.3.4*, p. *196*; 22.1.2, p. 352
B. BRÜCKEL	(b) **78**, K23 (1976)	13.1, p. 128; *20.1*, p. *276*; 22, p. 341
H.-P. BRÜCKNER	(a) **37**, 541 (1976)	*14.3*, p. *180*; 15, p. 211; 18.1, p. 227; 22.6, p. 407
	(a) **38**, 93 (1976)	*1.2*, p. *19*; *3*, p. *44*; 14.3, p. 181; 15, p. 211; 18.1, p. 227; 22.6.1, p. 409
	(a) **38**, K13 (1976)	*1.2*, p. *19*; *18.1*, p. *227*; 22.6.1, p. 409
	(a) **38**, K23 (1976)	*14.3*, p. *181*; *15*, p. *211*; 22.6, p. 408; 22.6.1, p. 409
W. BRÜCKNER	(a) **34**, 565 (1976)	*1.2*, p. *15*; *14.3*, p. *175*; 15, p. 209; 18.4, p. 243; 22, p. 336
	(a) **35**, K49 (1976)	*1.2*, p. *16*; 15, p. 210; 22.6, p. 404
	(a) **35**, K109 (1976)	*14.3.4*, p. *197*; 22, p. 337
	(a) **37**, 541 (1976)	*14.3*, p. *180*; 15, p. 211; 18.1, p. 227; 22.6, p. 407
	(a) **37**, K143 (1976)	*14.3.3*, p. *193*; 14.3.4, p. 200; 22.6, p. 407
	(a) **38**, 93 (1976)	*1.2*, p. *19*; 3, p. *44*; 14.3, p. 181; 15, p. 211; 18.1, p. 227; 22.6.1, p. 409
	(a) **38**, K13 (1976)	*1.2*, p. *19*; *18.1*, p. *227*; 22.6.1, p. 409
	(a) **38**, K17 (1976)	1.2, p. 20; *15*, p. *211*; 22.6.1, p. 409
	(a) **38**, K23 (1976)	*14.3*, p. *181*; *15*, p. *211*; 22.6, p. 408; 22.6.1, p. 409
V. N. BRUDNYI	(a) **35**, 425 (1976)	11, p. 102; 13.4, p. 139; *14.3*, p. *177*; *20.1*, p. *269*; 22.8, p. 415
P. BRÜESCH	(b) **77**, 153 (1976)	*6*, p. *60*; 20.1, p. 272; 22.6, p. 406
H.-G. BRÜHL	(a) **34**, K147 (1976)	*1.1*, p. *11*; 22.2.4, p. 370
H. BRUHNS	(a) **35**, 511 (1976)	*14.3.4*, p. *197*; 22.2.3, p. 368
O. BRÜMMER	(b) **73**, K107 (1976)	*13*, p. *115*; *20*, p. *257*; 22.1.2, p. 352
	(a) **33**, 587 (1976)	*1.1*, p. *11*; 22.1.2, p. 352
	(b) **74**, 249 (1976)	*13.1*, p. *122*; 17, p. 218; 22.7, p. 410
	(b) **74**, 341 (1976)	*13.1*, p. *122*; 17, p. 218; 20, p. 257; 22.1, p. 343
	(a) **34**, 737 (1976)	*10.1*, p. *81*; 21.1, p. 306
	(b) **75**, K89 (1976)	*13.1*, p. *124*; 17, p. 219; 20, p. 258; 22, p. 337
	(b) **75**, K157 (1976)	*1.1*, p. *12*; *13.1*, p. *124*; 22.7, p. 411
	(a) **37**, 529 (1976)	*1.1*, p. *13*; 22.1.2, p. 357
	(b) **78**, 659 (1976)	*13.2*, p. *131*; 20, p. 259; 21.2, p. 323; 21.3, p. 325
	(b) **78**, 729 (1976)	*13.1*, p. *129*; 20, p. 259; 21.3, p. 325
W. BRUNNER	(b) **77**, K31 (1976)	*19*, p. *254*; 22.4.1, p. 381
V. V. BRYKSIN	(b) **73**, 229 (1976)	*2*, p. *35*; 6, p. 52; 13, p. 113; *14.3*, p. *173*
	(b) **74**, K151 (1976)	*2*, p. *37*; 6, p. 55; 13, p. 115; *14.3*, p. *175*
	(b) **78**, 9 (1976)	2, p. 35; 6, p. 52; 13, p. 113; 13.5.3, p. 159; *14.3*, p. *173*; 22, p. 333
	(b) **78**, 415 (1976)	2, p. 35; 6, p. 52; 13, p. 113; 13.5.3, p. 159; *14.3*, p. *173*; 22, p. 333
R. BUBÁKOVÁ	(a) **35**, 55 (1976)	*10.2*, p. *92*; 22.1.2, p. 353
	(a) **36**, 509 (1976)	*10.2*, p. *94*; 22.1.2, p. 355
R. BUCZKOWSKI	(a) **34**, 711 (1976)	*1.4*, p. *26*; *10.1*, p. *81*; 21.1.1, p. 318
D. L. BUDNITSKII	(a) **35**, 425 (1976)	11, p. 102; 13.4, p. 139; *14.3*, p. *177*; *20.1*, p. *269*; 22.8, p. 415
P. BUGNET	(a) **37**, 521 (1976)	10.2, p. 96; *20*, p. *258*; 22.4.2, p. 383
W. BÜHRER	(b) **77**, 153 (1976)	*6*, p. *60*; 20.1, p. 272; 22.6, p. 406
BUI AI see under letter "A"		
L. L. BUISHVILI	(b) **75**, K69 (1976)	*18.2*, p. *232*
	(b) **76**, K97 (1976)	*19*, p. *253*

J. P. BUISSON	(b) **78**, 779 (1976)	6, p. 64; *10.2*, p. *98*; 13.4, p. 145; 20.1, p. 278; 22.5.2, p. 398
S. J. H. BUKHARI	(a) **35**, 181 (1976)	*1.2*, p. *16*; 22.6, p. 404
D. W. BULLETT	(b) **77**, 645 (1976)	1, p. 9; *4*, p. *46*; 6, p. 61; 22.1.3, p. 360
D. S. P. BUNBURY	(a) **33**, 91 (1976)	*5*, p. *48*; 18.2, p. 228; 18.3, p. 239; 21.1.1, p. 316
A. J. BURGGRAAF	(a) **35**, 281 (1976)	*10.2*, p. *92*; *14.4*, p. *203*; 22.8, p. 415
H. BURKHARD	(b) **76**, 259 (1976)	13.1, p. 125; *18*, p. *223*; *20.1*, p. *270*; 22, p. 337
A. BUROFF	(a) **38**, 161 (1976)	10.2, p. 97; *14.4*, p. *204*; 22.5.1, p. 391
E. V. BURSIAN	(b) **74**, 689 (1976)	13.1, p. 123; *14.4.1*, p. *206*; *20.1*, p. *266*; 22.8.1, p. 420
K. H. J. BUSCHOW	(a) **34**, 355 (1976)	1.1, p. 11; *18.2*, p. *230*; *18.4*, p. *243*; 21.1, p. 305; 21.1.1, p. 317; 21.5, p. 328
	(a) **34**, 729 (1976)	*5*, p. *49*; 18.2, p. 231; 21.1.1, p. 318; 21.4, p. 326
	(a) **35**, 137 (1976)	*5*, p. *49*; 18.2, p. 232; 21.1.1, p. 318; 21.4, p. 326
	(b) **76**, 11 (1976)	5, p. 48; *13*, p. *113*; *18*, p. *222*; 19, p. 247; 21, p. 289; 21.1, p. 302; 21.1.1, p. 316; 21.4, p. 326
	(a) **36**, 275 (1976)	18.2, p. 233; *19*, p. *253*; 21.1, p. 309; 21.4, p. 327
	(a) **37**, K205 (1976)	*18.2*, p. *235*; *19*, p. *255*; 21.1.1, p. 321
W. H. BUTLER	(b) **77**, 181 (1976)	*9*, p. *74*; *10.1*, p. *83*; 21, p. 297; 21.1, p. 310
N. M. BUTT	(b) **77**, 693 (1976)	*6*, p. *61*; 12.2, p. 112; 22.5.4, p. 401
C. BUTTI	(a) **36**, 133 (1976)	*14.3*, p. *178*; 22.4.1, p. 381
P. BUTVIN	(a) **35**, 223 (1976)	*1.2*, p. *16*; *2*, p. *37*; 5, p. 49; 21.1.1, p. 318
J. BUXO	(a) **37**, K105 (1976)	14.3.2, p. 188; *14.3.3*, p. *193*
J. Y. BUZÁRE	(b) **74**, 393 (1976)	*19*, p. *249*; 22.8, p. 414
M. I. BYCHKOVA	(a) **37**, K165 (1976)	*1.2*, p. *19*; 5, p. 50; 21, p. 299
V. N. BYKOV	(a) **35**, 779 (1976)	*1.2*, p. *17*; *1.5*, p. *31*; 11, p. 102; 21, p. 295; 21.1, p. 308; 21.1.1, p. 319; 22.1.1, p. 347
L. N. BYSTROV	(a) **34**, K187 (1976)	*10.1*, p. *81*; 21, p. 293; 21.6, p. 330
P. BYSZEWSKI	(b) **78**, 477 (1976)	*13.1*, p. *129*; 14.3, p. 181; 22.4.2, p. 384
T. CABOT	(a) **36**, 227 (1976)	13.3, p. 133; *14.3.3*, p. *192*; *16*, p. *215*; 22, p. 338; 22.4.1, p. 381
E. CABRERA	(a) **33**, K19 (1976)	*1.1*, p. *10*
J. ČADEK	(a) **35**, K123 (1976)	*10.1*, p. *83*; 12.1, p. 108
	(a) **37**, K123 (1976)	*10.1*, p. *85*; 21, p. 299
F. M. CAIMI	(a) **34**, 461 (1976)	*10.2*, p. *91*; 20.1, p. 266; 22.5.2, p. 394
	(a) **37**, 617 (1976)	*10.2*, p. *96*; 13.4, p. 143; 20.1, p. 275; 22.5.2, p. 397
M. P. CALLAGHAN	(a) **35**, 697 (1976)	*14.3.3*, p. *192*; *20.3*, p. *284*; 22.3, p. 373
J. CAMASSEL	(b) **75**, 303 (1976)	*13.1*, p. *124*; *20.1*, p. *267*; 22.7, p. 411
B. CANNY	(b) **76**, K5 (1976)	*19*, p. *252*; 22.8.1, p. 422
CAO-XUÂN AN	(a) **34**, K51 (1976)	*1.2*, p. *15*; 22.5, p. 388
	(a) **35**, 545 (1976)	*8*, p. *69*; 22.8.1, p. 421
C. CAPPILUPPI	(a) **35**, 165 (1976)	*14.3.3*, p. *191*; 22.1.2, p. 353
F. CARALP	(a) **35**, 615 (1976)	*10.2*, p. *93*; *13.4*, p. *139*; 14.3, p. 177; 18, p. 223; 22.4, p. 376
M. CARDONA	(b) **76**, 251 (1976)	2, p. 38; *13.1*, p. *125*; 17, p. 219; 22.8, p. 416
P. E. CARO	(a) **33**, K9 (1976)	*10.2*, p. *88*; 22.6, p. 402
G. J. C. CARPENTER	(a) **37**, K61 (1976)	*10*, p. *76*
S. H. CARPENTER	(a) **36**, 117 (1976)	*10.1*, p. *83*; 21.1.1, p. 319

E. Carrillo	(a) **38**, 67 (1976)	*10.2*, p. *97*; 22.5.2, p. 397
Y. Le Cars	(a) **38**, 595 (1976)	*1.4*, p. *29*; 22.8.2, p. 431
H. D. Carstanjen	(a) **34**, K153 (1976)	*10.1*, p. *81*; 21.1, p. 306
R. Casalegno	(b) **77**, 495 (1976)	*18.2*, p. *235*
D. P. Casasent	(a) **34**, 461 (1976)	*10.2*, p. *91*; 20.1, p. 266; 22.5.2, p. 394
	(a) **37**, 617 (1976)	*10.2*, p. *96*; 13.4, p. 143; 20.1, p. 275; 22.5.2, p. 397
L. M. Caspers	(a) **34**, 277 (1976)	*10.1*, p. *80*; 21, p. 292
	(a) **37**, 371 (1976)	*10.1*, p. *84*; *11*, p. *103*; 21, p. 298
J. C. Castro	(b) **77**, 685 (1976)	1.2, p. 19; *6*, p. *61*; 10.2, p. 95; *20.1*, p. *274*; 22, p. 340
S. Caticha Ellis	(b) **75**, 657 (1976)	*1.2*, p. *16*; 14.4.1, p. 207; 22.8.1, p. 421
R. Cavalheiro	(b) **76**, 371 (1976)	*6*, p. *58*; 21.6, p. 330
A. Ceccetti	(a) **35**, K161 (1976)	*18.3*, p. *240*; 22.8.2, p. 429
G. Cecchi	(a) **35**, K161 (1976)	*18.3*, p. *240*; 22.8.2, p. 429
B. Čelustka	(a) **33**, K23 (1976)	*1.1*, p. *10*; 1.2, p. 14; 22.3, p. 372
V. Černuško	(a) **33**, 85 (1976)	*14.2*, p. *169*; 21, p. 289
A. Česnys	(a) **35**, K41 (1976)	*2*, p. *37*; *14.3.1*, p. *184*; 22.3, p. 372
T. Ceva	(a) **35**, 315 (1976)	10.2, p. 92; *20.3*, p. *283*; 22.8.1, p. 421
M. Chabin	(a) **35**, 545 (1976)	*8*, p. *69*; 22.8.1, p. 421
I. A. Chaikovskii	(b) **76**, 811 (1976)	*14.3*, p. *179*
S. K. Chakrabarti	(b) **77**, 329 (1976)	*4*, p. *46*; 22.5.2, p. 396
D. Chakraborty	(a) **34**, 467 (1976)	*10.2*, p. *91*; 20.1, p. 266; 22.6, p. 404; 22.8, p. 414
A. N. Chakravarti	(a) **36**, K65 (1976)	*8*, p. *69*; 14.3.4, p. 198; 22.1.2, p. 355
	(a) **38**, K85 (1976)	*14.3.2*, p. *189*; 22.1.2, p. 358
S. G. Chamberlain	(a) **34**, 263 (1976)	*14.3.2*, p. *187*; *14.3.3*, p. *190*; 14.3.4, p. 196; 22.1.2, p. 352
Chan Min Shon	(b) **76**, 811 (1976)	*14.3*, p. *179*
B. P. Chandra	(a) **36**, K175 (1976)	*20.3*, p. *285*; 22.9, p. 437
G. Chandra	(a) **35**, 73 (1976)	*5*, p. *49*; 21.1, p. 307; 21.1.1, p. 318
Girish Chandra	(a) **34**, K169 (1976)	*18.1*, p. *225*; *18.2*, p. *231*; 21.1, p. 307; 21.1.1, p. 318
K. Chandra	(b) **75**, K1 (1976)	*13.4*, p. *138*; 22.8, p. 414; 22.8.1, p. 421
S. Chandra	(a) **35**, 387 (1976)	*2*, p. *37*; *18*, p. *223*; 22.9, p. 436
J. P. Chapelle	(a) **34**, K51 (1976)	*1.2*, p. *15*; 22.5, p. 388
I. M. Chapnik	(a) **37**, K29 (1976)	*13*, p. *118*; 21.1, p. 311; 21.1.1, p. 321; 22.7, p. 412
J. Chappert	(b) **75**, 163 (1976)	*5*, p. *49*; 22.9, p. 436
G. Chapuis	(a) **36**, 285 (1976)	*1.2*, p. *17*; 22.9, p. 437
P. Charsley	(a) **34**, 533 (1976)	6, p. 56; *8*, p. *68*; *10.1*, p. *80*; 21.1, p. 306
P. Chaudhari	(b) **75**, 593 (1976)	*1.3*, p. *22*; *2*, p. *38*; 21, p. 294
D. Roy Chaudhuri	(a) **36**, K65 (1976)	*8*, p. *69*; 14.3.4, p. 198; 22.1.2, p. 355
A. G. Cheran	(a) **36**, 659 (1976)	*14.3.1*, p. *185*; 22.2.2, p. 365
V. I. Chechernikov	(b) **76**, 113 (1976)	13, p. 116; 18, p. 223; *19*, p. *252*; 21, p. 296; 21.1, p. 309
K. Chęciński	(b) **75**, K7 (1976)	*19*, p. *251*; 22.8.2, p. 428
A. Chełkowski	(a) **36**, K171 (1976)	*8*, p. *70*; 18.2, p. 234; 21.4, p. 327
G. G. Chemeresyuk	(a) **33**, K145 (1976)	13.4, p. 136; *16*, p. *213*; 22.4.2, p. 383
H. S. Chen	(a) **34**, K127 (1976)	*1.4*, p. *26*; 2, p. 37; 21, p. 293; 21.1, p. 306
C. G. Chen	(a) **36**, 687 (1976)	*9*, p. *73*; 21, p. 297
L. J. Chen	(a) **34**, 679 (1976)	*10.1*, p. *81*; *11*, p. *101*; 21.1, p. 306
W. K. Chen	(a) **37**, 515 (1976)	*9*, p. *74*; 10.2, p. 96; 22.8.2, p. 430
L. J. Cheng	(a) **38**, 321 (1976)	*9*, p. *74*; 11, p. 104; 22.1.2, p. 358
I. V. Chepeleva	(b) **73**, 65 (1976)	2, p. 35; 18, p. 222; *19*, p. *247*; 22.8, p. 413
N. G. Cherenda	(b) **77**, K27 (1976)	*2*, p. *39*; 11, p. 103; *19*, p. *254*; 22.8.1, p. 423
A. E. Cherepov	(a) **36**, K89 (1976)	*3*, p. *43*; 21.1, p. 310; 22.1.2, p. 355

E. I. CHEREPOV	(a) **38**, 369 (1976)	*1.5*, p. *33*; *14.3.2*, p. *188*; 22.1.2, p. 358; 22.6, p. 408
A. P. CHERNOV	(b) **73**, 65 (1976)	2, p. 35; 18, p. 222; *19*, p. *247*; 22.8, p. 413
M. A. CHERNOV	(a) **37**, 143 (1976)	*1.5*, p. *32*; 22.3, p. 374
V. M. CHERNOV	(a) **35**, 747 (1976)	*10.1*, p. *82*; 12.1, p. 108; 21.1, p. 308; 21.3, p. 324
	(a) **38**, 761 (1976)	*10.1*, p. *86*; 21, p. 301; 21.1, p. 315; 21.3, p. 325; 21.6, p. 331
N. T. CHERPAK	(a) **37**, K37 (1976)	13.4, p. 142; *19*, p. *254*; 22.8, p. 417
M. V. CHETKIN	(b) **78**, K85 (1976)	*18.1*, p. *227*; *18.2*, p. *236*; 22.8.2, p. 431
S. C. CHETTY	(a) **37**, 687 (1976)	*5*, p. *50*; 18.4, p. 245; 21.1.1, p. 321
R. CHEVALIER	(a) **36**, 181 (1976)	*1.1*, p. *12*; 22.3.1, p. 375
A. CHEVY	(a) **36**, 181 (1976)	*1.1*, p. *12*; 22.3.1, p. 375
E. V. CHISLER	(b) **78**, 359 (1976)	*6*, p. *63*; *20.1*, p. *276*
T. CHONG	(b) **74**, 111 (1976)	13.5.1, p. 150; *20.3*, p. *281*; 22.9, p. 434
K. L. CHOPRA	(a) **33**, 333 (1976)	*2*, p. *36*; *14.3.1*, p. *183*; 15, p. 209; 22.1.1, p. 345
	(a) **34**, 405 (1976)	*2*, p. *36*; *20.1*, p. *265*; 22.1.1, p. 346
	(a) **35**, 725 (1976)	*2*, p. *38*; *20.1*, p. *269*; 22.1.1, p. 347
	(a) **35**, 761 (1976)	2, p. 38; *14.3*, p. *178*; *15*, p. *210*; 20.1, p. 270; 22.7, p. 411
	(a) **36**, 89 (1976)	*2*, p. *39*; *14.3.1*, p. *184*; 15, p. 210; 22.1.1, p. 348
	(a) **36**, 345 (1976)	*2*, p. *39*; *14.3.1*, p. *183*; 15, p. 210; 22.2.1, p. 363; 22.2.3, p. 368
	(a) **36**, 533 (1976)	2, p. 39; *14.3.1*, p. *184*; *15*, p. *210*; 22.1.1, p. 348
	(a) **37**, 313 (1976)	*1.2*, p. *18*; 1.5, p. 32; *2*, p. *40*; 22.1.1, p. 348
Y. T. CHOU	(a) **34**, 645 (1976)	*10*, p. *75*
D. R. CHOUDHURY	(a) **38**, K85 (1976)	*14.3.2*, p. *189*; 22.1.2, p. 358
B. V. R. CHOWDARI	(b) **75**, K119 (1976)	*19*, p. *251*; 22.5.2, p. 395
V. CHRISTOPH	(b) **76**, 695 (1976)	6, p. 59; 8, p. 69; *14.1*, p. *165*
	(b) **77**, K151 (1976)	*14.1*, p. *167*; 21.1, p. 312
S. Y. CHUANG	(a) **35**, K17 (1976)	*10.1*, p. *82*; 21.1.1, p. 318
F. A. CHUDNOVSKII	(b) **77**, K97 (1976)	*14.1*, p. *166*; *14.3.1*, p. *185*; 22.6, p. 407
	(a) **37**, 541 (1976)	*14.3*, p. *180*; 15, p. 211; 18.1, p. 227; 22.6, p. 407
	(a) **38**, K23 (1976)	*14.3*, p. *181*; *15*, p. *211*; 22.6, p. 408; 22.6.1, p. 409
YU. G. CHUKALKIN	(a) **33**, K83 (1976)	*18.3*, p. *239*; 22.8.2, p. 427
	(a) **38**, 57 (1976)	11, p. 104; *18.3*, p. *241*; 22.8.2, p. 431
	(a) **38**, K147 (1976)	11, p. 105; *18.3*, p. *241*; 22.8.2, p. 432
E. CIESLA	(b) **76**, K47 (1976)	1.2, p. 17; *12.2*, p. *111*; *14.4*, p. *203*; 22.8.1, p. 422
A. ČÍŽEK	(b) **73**, K1 (1976)	*13*, p. *114*; 21, p. 289; 21.1, p. 302; 21.1.1, p. 316; 21.6, p. 329
L. M. CLAREBROUGH	(a) **33**, 355 (1976)	*1.4*, p. *25*; *10.1*, p. *78*; 21.1, p. 303
P. CLAUWS	(b) **76**, 707 (1976)	*6*, p. *59*; *20.1*, p. *272*; 22.6, p. 406
J. A. COCKS	(b) **77**, K5 (1976)	*6*, p. *60*; 10, p. 76
M. L. COHEN	(b) **75**, 303 (1976)	*13.1*, p. *124*; *20.1*, p. *267*; 22.7, p. 411
	(b) **77**, 295 (1976)	*13.1*, p. *127*; 22.5.1, p. 391
	(b) **78**, 737 (1976)	4, p. 47; *13.1*, p. *129*; 20.1, p. 278; 22.5, p. 390
L. B. COLEMAN	(b) **75**, 239 (1976)	13.1, p. 123; *20.1*, p. *267*; 22.9, p. 436
R. COMÈS	(a) **38**, 695 (1976)	*1.2*, p. *20*; 1.3, p. 24; 22, p. 342
A. CONAN	(a) **38**, 505 (1976)	*14.3.4*, p. *201*; 22, p. 342
A. CONDE	(a) **33**, K25 (1976)	*10.2*, p. *88*; 22.6, p. 402

C. F. CONDE	(a) **33**, K25 (1976)	*10.2*, p. *88*; 22.6, p. 402
H. VAN CONG	(a) **38**, K69 (1976)	*20.2*, p. *279*; 22.2.1, p. 364
M. CONSTANTINESCU	(a) **38**, K175 (1976)	*13.4*, p. *146*; 22.5, p. 390
J. M. CORBETT	(a) **35**, 591 (1976)	1.2, p. 16; *1.4*, p. *27*; 22, p. 337
	(a) **36**, K21 (1976)	*1.4*, p. *28*; 10.1, p. 83; 21.1, p. 309
	(a) **38**, 151 (1976)	*1.4*, p. *28*; 21.1, p. 313
	(a) **38**, 675 (1976)	*1.4*, p. *29*; 21.1, p. 314
J. W. CORBETT	(a) **38**, 321 (1976)	*9*, p. *74*; 11, p. 104; 22.1.2, p. 358
W. CORDTS	(b) **77**, 527 (1976)	*13.1*, p. *127*; 22.8, p. 417
R. I. COTTAM	(a) **33**, 367 (1976)	*12.2*, p. *110*; 22.1, p. 343; 22.2, p. 361; 22.4, p. 376
J. E. COUSINS	(b) **75**, 155 (1976)	*19*, p. *250*; 21, p. 293
P. COUVREUR	(a) **38**, 123 (1976)	*14.3.2*, p. *188*; 14.3.3, p. 194; 22.1.2, p. 357
D. F. CRABTREE	(a) **38**, 217 (1976)	*20.3*, p. *288*; 22.6, p. 408
J. H. CRAWFORD, JR.	(a) **35**, 667 (1976)	*10.2*, p. *93*; 20.1, p. 269; 22.6, p. 405
M. CREUZBURG	(b) **76**, 301 (1976)	*1.2*, p. *17*; 6, p. 58; 20.3, p. 284; 23, p. 440
A. G. CROCKER	(a) **38**, 523 (1976)	*10.1*, p. *86*; 12.1, p. 109; 21.1.1, p. 321
L. E. CROSS	(a) **37**, 235 (1976)	1.4, p. 28; *10.2*, p. *95*; *12.2*, p. *112*; 22.6, p. 407
G. CUBIOTTI	(b) **75**, K93 (1976)	2, p. 38; *14.1*, p. *165*; *15*, p. *210*; 21, p. 295
J. P. M. DAMEN	(a) **34**, 665 (1976)	*10.2*, p. *91*; 20.1, p. 266; 22.8.2, p. 428
W. DANDER	(b) **75**, 173 (1976)	10.1, p. 81; *11*, p. *101*; 14.1, p. 165; *18.2*, p. *231*; 21.1, p. 307
	(b) **75**, 647 (1976)	*11*, p. *102*; *18.2*, p. *232*; 21.1, p. 308
	(b) **78**, 139 (1976)	*11*, p. *104*; *18.2*, p. *235*; 21.1, p. 313
TRAN HUU DANH see under letter "T"		
H. DANNHEIM	(a) **37**, K135 (1976)	*5*, p. *50*; 22.8.2, p. 431
A. DARABANT	(b) **74**, 335 (1976)	*19*, p. *249*; 22.8.1, p. 420
M. P. DARIEL	(b) **76**, 669 (1976)	*18.2*, p. *233*; 21.1, p. 310; 21.4, p. 327
C. N. W. DARLINGTON	(b) **76**, 231 (1976)	*1.2*, p. *17*; 14.4.1, p. 207; 22.8, p. 416
J. N. DAS	(a) **36**, K175 (1976)	*20.3*, p. *285*; 22.9, p. 437
L. I. DATSENKO	(a) **33**, 275 (1976)	*1.1*, p. *10*, *10.2*, p. *87*; 22.1.2, p. 351
J. DAUBERT	(b) **76**, 581 (1976)	*6*, p. *59*, 7, p. 66; 8, p. 69; *12.2*, p. *111*; 22.5.2, p. 395
J. DAUNAY	(a) **37**, 521 (1976)	10.2, p. 96; *20*, p. *258*; 22.4.2, p. 383
JAC. DAUNAY	(a) **37**, 521 (1976)	10.2, p. 96; *20*, p. *258*; 22.4.2, p. 383
C. DAUWE	(b) **73**, 527 (1976)	*5*, p. *48*; *18.3*, p. *239*; 22.8.2, p. 426
	(a) **37**, 127 (1976)	18.1, p. 226; *18.3*, p. *241*; 18.4, p. 245; 22.8.2, p. 430
M. DAVID	(b) **74**, 359 (1976)	*11*, p. *11*
R. L. DAVIDOVICH	(b) **78**, K69 (1976)	1.2, p. 20; *6*, p. *64*; *13.4*, p. *145*; 19, p. 256; 22.8, p. 418
G. J. DAVIES	(a) **33**, 281 (1976)	*1.3*, p. *21*; *12.1*, p. *107*; 21.1, p. 303
L. C. DAVIS	(b) **75**, K81 (1976)	*13*, p. *116*; 21, p. 295
Z. F. DAVITAYA	(a) **38**, K49 (1976)	*8*, p. *71*; *15*, p. *211*; 22.2.4, p. 371
G. D. DAVTYAN	(b) **77**, 85 (1976)	13.4, p. 142; *19*, p. *253*; 22.6, p. 406
A. S. DAVYDOV	(b) **75**, 735 (1976)	*13.5*, p. *147*
V. YU. DAVYDOW	(b) **78**, 359 (1976)	*6*, p. *63*; *20.1*, p. *276*
L. DÄWERITZ	(a) **38**, K111 (1976)	*1.5*, p. *33*; *3*, p. *44*; 22.2.4, p. 371
B. DAY	(b) **74**, K9 (1976)	*13.4*, p. *136*; *13.5.1*, p. *151*; 13.5.3, p. 159; 20.3, p. 281; 22.4.3, p. 385
J. P. DAY	(a) **38**, 781 (1976)	*12.1*, p. *109*; 21.2, p. 323
B. DAYAL	(b) **75**, 697 (1976)	*6*, p. *57*; *20.1*, p. *268*; 22.5.2, p. 395
UDAYAN DE	(b) **75**, K55 (1976)	*9*, p. *72*

P. J. Dean	(b) **74**, K9 (1976)	*13.4*, p. *136*; *13.5.1*, p. *151*; 13.5.3, p. 159; 20.3, p. 281; 22.4.3, p. 385
G. Debnath	(b) **75**, 145 (1976)	*13*, p. *115*; *18*, p. *223*
E. A. Decamps	(b) **78**, 767 (1976)	*6*, p. *64*; *20.1*, p. *278*; 22.6, p. 408
R. J. Deck	(b) **73**, K31 (1976)	13, p. 114; *18*, p. *222*; *20*, p. *257*; 22, p. 333
A. Defresne	(a) **33**, 295 (1976)	*2*, p. *35*; *14.3*, p. *173*; 15, p. 209; 22.8, p. 413
P. Deimel	(b) **73**, 87 (1976)	*1.1*, p. *10*
I. Ya. Dekhtyar	(b) **74**, 471 (1976)	*13.2*, p. *130*; 21, p. 292
A. Deksnys	(a) **35**, K41 (1976)	*2*, p. *37*; *14.3.1*, p. *184*; 22.3, p. 372
P. Delannoy	(b) **77**, K119 (1976)	*14.3*, p. *180*; 22.9, p. 438
P. Delavignette	(a) **33**, 497 (1976)	*1.2*, p. *14*; *1.4*, p. *25*; 22, p. 334
	(a) **33**, K77 (1976)	*1.4*, p. *26*; 22, p. 335
P. Delhaes	(a) **35**, 615 (1976)	*10.2*, p. *93*; *13.4*, p. *139*; 14.3, p. 177; 18, p. 223; 22.4, p. 376
Yu. P. Demakov	(a) **36**, 439 (1976)	*9*, p. *73*; 10.2, p. 94; 22.2.1, p. 363; 22.2.4, p. 370
L. S. Demidenko	(b) **75**, 299 (1976)	*14.3*, p. *176*; 22.1.1, p. 346
	(b) **78**, K111 (1976)	*14.3*, p. *182*; 22.1.1, p. 350
V. B. Demin	(a) **34**, 787 (1976)	*9*, p. *72*; 10.1, p. 81; 21, p. 293; 21.1, p. 306
S. Deonarine	(b) **74**, 659 (1976)	13.5.1, p. 152; *13.5.4*, p. *160*; 18.4, p. 243; 22.5.4, p. 401
A. Deptuch	(a) **37**, K101 (1076)	*14.3.1*, p. *185*; 22.4.1, p. 381
N. Z. Derikot	(a) **37**, 363 (1976)	*13.4*, p. *143*; *14.3.4*, p. *199*; 20.3, p. 287; 22.4.2, p. 383
V. N. Derkachenko	(a) **36**, 415 (1976)	12.2, p. 111; *18.4*, p. *244*; 22.8.2, p. 429
N. Desbarax	(a) **33**, 247 (1976)	*14.3.1*, p. *183*; 14.3.2, p. 187; 22.9, p. 433
J. Deumie	(a) **38**, K159 (1976)	*20.1*, p. *278*; 22.9, p. 439
P. Deus	(a) **36**, K165 (1976)	*6*, p. *60*; *8*, p. *70*; 22.1.2, p. 356
J. T. Devreese	(b) **74**, 45 (1976)	*13*, p. *115*; *14.4*, p. *202*
	(b) **78**, 85 (1976)	6, p. 62; *14.3.4*, p. *200*; 22.2.3, p. 369
N. I. Diakonova	(a) **38**, 393 (1976)	*14.3.4*, p. *200*; 22.1.1, p. 349; 22.6, p. 408
J. Diehl	(a) **33**, K73 (1976)	*10.1*, p. *79*; *11*, p. *100*; 21.1, p. 304
V. Dienys	(b) **78**, 757 (1976)	12.2, p. 112, *14.3.4*, p. *200*; 22.1.1, p. 349
A. M. van Diepen	(b) **76**, 11 (1976)	5, p. 48; *13*, p. *113*; *18*, p. *222*; 19, p. 247; 21, p. 289; 21.1, p. 302; 21.1.1, p. 316; 21.4, p. 326
T. Dietl	(b) **75**, K83 (1976)	*14.3*, p. *177*; 22.4.4, p. 387
	(b) **76**, 737 (1976)	*13.1*, p. *126*; 14.3, p. 179; *15*, p. *210*; 22.4.3, p. 386
J. Dietrich	(b) **73**, K57 (1976)	*19*, p. *247*; 22.4.1, p. 379
W. Dietrich	(b) **74**, 609 (1976)	13.5.1, p. 152; *19*, p. *250*; 22.8.1, p. 420
P. C. Diggle	(a) **34**, K13 (1976)	*20.1*, p. *265*; *20.3*, p. *281*; 22.8, p. 414; 22.8.1, p. 420
R. I. DiMelfi	(b) **75**, 573 (1976)	*10*, p. *76*
R. Dinges	(b) **76**, 613 (1976)	*13.5.1*, p. *154*; *20.1*, p. *271*; 22.5.1, p. 391
Ch. Dippmann	(a) **37**, K81 (1976)	*14.3.3*, p. *193*; 22.8, p. 417
G. Dlubek	(b) **73**, K107 (1976)	*13*, p. *115*; *20*, p. *257*; 22.1.2, p. 352
	(a) **34**, 737 (1976)	*10.1*, p. *81*; 21.1, p. 306
L. Dmowski	(b) **73**, K131 (1976)	12.2, p. 110; *14.3*, p. *174*; 22.2.3, p. 367
	(a) **33**, 421 (1976)	13.4, p. 135; *14.3*, p. *174*; 22.4.3, p. 385
F. Dobeš	(a) **34**, K27 (1976)	*10.1*, p. *80*; 12.1, p. 107
	(a) **35**, K123 (1976)	*10.1*, p. *83*; 12.1, p. 108
L. Dobrosavljević	(b) **75**, 677 (1976)	10.1, p. 82; *14.2*, p. *170*; 21, p. 294
G. Dokter	(b) **76**, 241 (1976)	*10.2*, p. *93*; *13.4*, p. *140*; 16, p. 215; 22.5.2, p. 395
G. M. Dolidze	(a) **35**, K131 (1976)	*17*, p. *219*; 22.1.1, p. 347
J. Dominguez-Rodriguez	(a) **33**, K25 (1976)	*10.2*, p. *88*; 22.6, p. 402
E. Donato	(b) **75**, K93 (1976)	2, p. 38; *14.1*, p. *165*; *15*, p. *210*; 21, p. 295

J. Donecker	(b) **73**, 633 (1976)	13.1, p. 121; *20.1*, p. *262*; 22.2.4, p. 370
	(b) **77**, 243 (1976)	13.5.1, p. 155; *18*, p. *224*; *20.1*, p. *273*; 22.2.2, p. 365; 22.2.4, p. 371
Nguyen van Dong	(a) **33**, 295 (1976)	*2*, p. *35*; *14.3*, p. *173*; 15, p. 209; 22.8, p. 413
E. Donth	(b) **74**, K49 (1976)	*2*, p. *36*; *8*, p. *67*
N. J. Doran	(b) **73**, 203 (1976)	*6*, p. *52*; *14.3*, p. *173*; 22.2.1, p. 362
	(b) **75**, K15 (1976)	*14.3*, p. *176*; 22.2.3, p. 367
	(b) **76**, K43 (1976)	6, p. 58; *14.3*, p. *178*; 22.2.1, p. 363
E. Dormann	(a) **36**, 275 (1976)	18.2, p. 233; *19*, p. *253*; 21.1, p. 309; 21.4, p. 327
B. Dorner	(b) **73**, 655 (1976)	*6*, p. *54*; 7, p. 65; 12.2, p. 110; 22.5, p. 388
V. I. Dotsenko	(a) **37**, 709 (1976)	*10.1*, p. *84*; 21.1, p. 312
I. N. Douglas	(b) **73**, 641 (1976)	13.4, p. 135; *18*, p. *222*; *20.1*, p. *262*; 22.5.2, p. 393; 22.6, p. 403
J. C. Doukhan	(a) **36**, 461 (1976)	*10.2*, p. *94*; 22.1.3, p. 360
	(a) **38**, 271 (1976)	*10.2*, p. *97*; 22.1.3, p. 360
H. Downing	(a) **37**, 515 (1976)	*9*, p. *74*; 10.2, p. 96; 22.8.2, p. 430
M. Doyama	(b) **73** 517 (1976)	1.3, p. 21; *10.1*, p. *78*; 21, p. 290
D. Dräger	(b) **74**, 341 (1976)	*13.1*, p. *122*; 17, p. 218; 20, p. 257; 22.1, p. 343
	(b) **75**, K89 (1976)	*13.1*, p. *124*; 17, p. 219; 20, p. 258; 22, p. 337
	(b) **78**, 729 (1976)	*13.1*, p. *129*; 20, p. 259; 21.3, p. 325
K. Dräger	(a) **38**, 547 (1976)	*18.4*, p. *246*; 19, p. 256; 22.6.1, p. 409
R. Dragila	(a) **35**, 291 (1976)	*1.5*, p. *31*; *3*, p. *42*; 22.6, p. 404
V. P. Dragunov	(b) **75**, 423 (1976)	13.1, p. 124; *14.3*, p. *177*; 22.1.2, p. 354
H. D. Drew	(b) **74**, 721 (1976)	14.1, p. 165; *20.1*, p. *266*; 21, p. 292
V. N. Drobyasin	(a) **38**, K5 (1976)	*10.1*, p. *85*; 21, p. 300; 21.1, p. 314
M. Drulis	(a) **35**, K127 (1976)	*8*, p. *69*; *18*, p. *223*; 21.4, p. 326
M. Drzazga	(b) **76**, K123 (1976)	*18.2*, p. *234*
A. P. Dubchack	(a) **38**, 131 (1976)	11, p. 104; *14.3.3*, p. *194*; 22.1.2, p. 357
S. F. Dubinin	(a) **33**, K83 (1976)	*18.3*, p. *239*; 22.8.2, p. 427
	(a) **38**, 57 (1976)	11, p. 104; *18.3*, p. *241*; 22.8.2, p. 431
	(a) **38**, K147 (1976)	11, p. 105; *18.3*, p. *241*; 22.8.2, p. 432
L. S. Dubinskaya	(b) **75**, K133 (1976)	*14.3*, p. *177*; 22.1.3, p. 359
G. Dublon	(b) **76**, 669 (1976)	*18.2*, p. *233*; 21.1, p. 310; 21.4, p. 327
A. Dubon	(a) **34**, 125 (1976)	*1.2*, p. *15*; *1.4*, p. *26*; 22.8.2, p. 427
A. Duda	(b) **74**, 235 (1976)	*18.2.1*, p. *237*
	(b) **74**, 701 (1976)	*18.2.1*, p. *237*
	(b) **76**, K19 (1976)	*18.2.1*, p. *237*
M. A. Dudinskii	(b) **74**, 453 (1976)	*13.5.1*, p. *151*; *20.3*, p. *282*; 22.9, p. 435
J. F. Dufresne	(a) **36**, 579 (1976)	*10.1*, p. *83*; *18.2*, p. *234*; 21.1.1, p. 320
P. Duhaj	(a) **35**, 223 (1976)	*1.2*, p. *16*; *2*, p. *37*; 5, p. 49; 21.1.1, p. 318
L. H. Dunleavy	(a) **37**, K197 (1976)	*10.1*, p. *84*; *18.2*, p. *235*; 21.1.1, p. 321
W. D. Dunnett	(b) **78**, 569 (1976)	*6*, p. *63*; 20.1, p. 277; 22.8, p. 418
Ngo Duong Sinh see under letter "S"		
N. Dupont-Pavlovsky	(a) **35**, 615 (1976)	*10.2*, p. *93*; *13.4*, p. *139*; 14.3, p. 177; 18, p. 223; 22.4, p. 376
S. Duraiswamy	(b) **75**, 53 (1976)	*6*, p. *56*; *8*, p. *68*; 22.5.2, p. 394
M. Durand	(b) **78**, 767 (1976)	*6*, p. *64*; *20.1*, p. *278*; 22.6, p. 408
J.-M. Dusseau	(a) **33**, 715 (1976)	*7*, p. *65*; *14.3.4*, p. *196*; 22.2.3, p. 367
G. A. Dussel	(a) **35**, 309 (1976)	*13.4*, p. *139*; *20.1*, p. *268*; 22.4.1, p. 380
H. Düsterhöft	(b) **75**, K173 (1976)	*17*, p. *219*; 21.1.1, p. 319
	(a) **36**, K93 (1976)	*11*, p. *102*; 21, p. 296; 21.1, p. 310; 21.1.1, p. 320; 22.1.1, p. 348; 22.1.2, p. 355; 22.6, p. 405
P. Duval	(a) **34**, 135 (1976)	*1.2*, p. *15*; *1.3*, p. *22*; 1.4, p. 26; 21, p. 291; 21.1, p. 304

VU DUY PHACH see under letter "V"		
D. VAN DYCK	(b) 77, 301 (1976)	*1.1*, p. *12*; *13*, p. *118*
	(a) **37**, 591 (1976)	*1.2*, p. *19*; 1.4, p. 28; 22, p. 340
	(a) **38**, 185 (1976)	*1.2*, p. *19*; 1.4, p. 29; 22, p. 341
	(a) **38**, 663 (1976)	*1.4*, p. *29*; 10.2, p. 99; 22, p. 342
I. M. DKYMAN	(b) 76, 385 (1976)	*13*, p. *117*; *14.3.4*, p. *197*; 22, p. 338
M. I. DYKMAN	(b) 74, 547 (1976)	*6*, p. *55*; *13.1*, p. *122*; 20.1, p. 265
P. M. DYKMAN	(b) 78, 843 (1976)	*13*, p. *117*; *14.3.4*, p. *197*; 22, p. 338
N. P. DYMCHENKO	(a) **33**, K117 (1976)	*18.3*, p. *239*; 22.8.2, p. 427
B. E. DZEVITSKII	(a) **37**, K165 (1976)	*1.2*, p. *19*; 5, p. 50; 21, p. 299
T. D. DZHAFAROV	(a) **36**, 439 (1976)	*9*, p. *73*; 10.2, p. 94; 22.2.1, p. 363; 22.2.4, p. 370
R. B. DZHANELIDZE	(a) **36**, 73 (1976)	13.3, p. 132; *14.3.2*, p. *188*; 20.3, p. 285; 22.2.1, p. 363
W. EBELING	(b) 78, 241 (1976)	*13.5*, p. *148*; *13.5.1*, p. *155*; 22, p. 341
W. ECKE	(a) **34**, 219 (1976)	13.4, p. 137; 14.3.3, p. 190; *14.3.4*, p. *196*; 22.1.2, p. 352
	(a) **37**, 479 (1976)	13.4, p. 143; *14.3.4*, p. *199*; 22.1.2, p. 356
D. ECKERT	(a) **37**, 171 (1976)	*14.2*, p. *171*; 21, p. 297; 21.1, p. 311
N. A. ECONOMOU	(a) **34**, 243 (1976)	*1.3*, p. *22*; 10.2, p. 90; 22.7, p. 410
	(b) 76, 727 (1976)	6, p. 59; *20.1*, p. *272*; 22.7, p. 411
F. L. EDELMAN	(a) **34**, 85 (1976)	*2*, p. *36*; *14.3.1*, p. *183*; 14.3.3, p. 190; 22, p. 335
	(a) **36**, K89 (1976)	*3*, p. *43*; 21.1, p. 310; 22.1.2, p. 355
I. S. EDELMAN	(b) 74, K145 (1976)	*18.2*, p. *231*; *20.1*, p. *266*; 22.8.1, p. 421
D. M. EDWARDS	(b) 78, 287 (1976)	*13.5.4*, p. *161*; *18.2*, p. *235*; 21.1, p. 313
V. M. EFIMOV	(a) **34**, 85 (1976)	*2*, p. *36*; *14.3.1*, p. *183*; 14.3.3, p. 190; 22, p. 335
V. M. EFREMENKOVA	(a) **35**, K5 (1976)	*20.3*, p. *283*; 22.4.1, p. 380
A. L. EFROS	(b) 76, 475 (1976)	*14.1*, p. *165*; *14.4*, p. *203*
M. EGÉE	(a) **35**, 361 (1976)	13.4, p. 139; *20.3*, p. *283*; 22.4.1, p. 380
R. F. EGERTON	(a) **37**, 663 (1976)	*11*, p. *104*; 13.5.2, p. 158; 21, p. 299; 21.3, p. 324; 21.7, p. 332; 22.9, p. 438
R. EHLERT	(a) **33**, 737 (1976)	20.1, p. 263; *20.2*, p. *279*; *20.3*, p. *281*; 22.8, p. 413
TH. EIFRIG	(b) 76, K23 (1976)	*13.1*, p. *125*
	(b) 76, K101 (1976)	*13*, p. *118*; 21, p. 297
	(b) 77, K79 (1976)	*13.1*, p. *125*
A. J. H. EIJKELENKAMP	(b) 76, 153 (1976)	13.5.1, p. 153; *20.1*, p. *270*; 22.5, p. 389
	(b) 76, 769 (1976)	13.1, p. 126; 13.5.1, p. 154; *20.1*, p. *272*; 22.5 p. 389
W. EISENMENGER	(a) **37**, 161 (1976)	*7*, p. *66*; 12.2, p. 111; *14.4.1*, p. *208*; 20.1, p. 273; 22.8.1, p. 423
W. EKARDT	(b) 73, 165 (1976)	*13.5.1*, p. *149*; 22.5, p. 388
	(b) 73, 475 (1976)	13.1, p. 121; *13.5.1*, p. *150*; 20.1, p. 261; 20.3, p. 280; 22.5, p. 388
	(b) 74, 523 (1976)	*13.5.1*, p. *151*; 20.3, p. 282; 22.5, p. 388
	(b) 76, K89 (1976)	*13.5.1*, p. *154*; 22.4.1, p. 381
A. A. ELANGO	(b) 78, 529 (1976)	*10.2*, p. *98*; 20.1, p. 277; 22.5.2, p. 398
J. B. A. A. ELEMANS	(a) **34**, 355 (1976)	1.1, p. 11; *18.2*, p. *230*; *18.4*, p. *243*; 21.1, p. 305; 21.1.1, p. 317; 21.5, p. 328
K. ELK	(b) 77, K151 (1976)	*14.1*, p. *167*; 21.1, p. 312
R. A. ELLIOT	(b) 74, K87 (1976)	*13.5.1*, p. *151*; *20.3*, p. *281*; 22.9, p. 434

S. Caticha Ellis see under letter "C"		
K. Elsaesser	(a) **34**, 153 (1976)	*13.4*, p. *137*; *20.1*, p. *264*; 22.1.1, p. 346
J. Emery	(b) 75, 255 (1976)	*13.4*, p. *138*; 22.8, p. 414
R. Enderlein	(b) 73, 141 (1976)	*6*, p. *52*; 13.1, p. 120; 13.5.1, p. 149; *20.1*, p. *260*
	(b) 78, 711 (1976)	6, p. 64; *14.3.4*, p. *200*; *20.1*, p. *278*; 22, p. 342
F. R. Engler	(b) 74, K135 (1976)	*10.2*, p. *90*; 11, p. 101; 22.5.2, p. 394
N. A. Enshina	(a) **37**, 701 (1976)	*1.3*, p. *24*; *10.1*, p. *84*; 21.1, p. 312
M. V. Entin	(a) **38**, 139 (1976)	*14.4.1*, p. *208*; *16*, p. *217*; 22.8.1, p. 425
G. N. Eritsyan	(a) **38**, K49 (1976)	*8*, p. *71*; *15*, p. *211*; 22.2.4, p. 371
A. E. Ermakov	(a) **33**, 325 (1976)	*5*, p. *48*; 18.2, p. 229; 21.1.1, p. 316
A. N. Ermoshkin	(b) 73, 81 (1976)	*10.2*, p. *87*; *13.4*, p. *134*; 22.6, p. 402
	(b) 73, 483 (1976)	*10.2*, p. *89*; *13.4*, p. *135*; 22.6, p. 402
B. Escaig	(a) **36**, 461 (1976)	*10.2*, p. *94*; 22.1.3, p. 360
	(a) **38**, 271 (1976)	*10.2*, p. *97*; 22.1.3, p. 360
H. Eschrig	(b) 73, 151 (1976)	*6*, p. *52*; *13.1*, p. *120*; 13.2, p. 130; 21.3, p. 324
C. Escribe	(a) **36**, 375 (1976)	*10.2*, p. *94*; 22.6, p. 405
B. Esser	(b) 73, 379 (1976)	2, p. 35; *14.3.4*, p. *195*; *20.1*, p. *260*; 22, p. 333
	(b) 73, 499 (1976)	2, p. 36; *13.4*, p. *135*; 22, p. 334
D. Esteve	(a) **37**, K105 (1976)	14.3.2, p. 188; *14.3.3*, p. *193*
Yu. Z. Estrin	(a) **36**, 397 (1976)	*10.2*, p. *94*; 22.5.2, p. 395
S. Etienne	(a) **36**, K113 (1976)	*10.2*, p. *95*; 22.1.2, p. 356
M. Euler	(a) **34**, 297 (1976)	1.6, p. 34; 13.3, p. 132; *17*, p. *218*; 22.6, p. 403
E. Evangelista	(a) **33**, 579 (1976)	*10.1*, p. *79*; 12.1, p. 107; 21, p. 291
C. A. Evans, Jr.	(a) **34**, K97 (1976)	*10.1*, p. *81*; 21.1, p. 306
B. L. Evans	(b) 73, 597 (1976)	*13.5.1*, p. *150*; 20.1, p. 262; 22, p. 334
R. A. Evarestov	(b) 73, 81 (1976)	*10.2*, p. *87*; *13.4*, p. *134*; 22.6, p. 402
	(b) 73, 483 (1976)	*10.2*, p. *89*; *13.4*, p. *135*; 22.6, p. 402
	(b) 76, 377 (1976)	*13.1*, p. *125*; *13.4*, p. *140*; 22.5.2, p. 395
R. Evrard	(b) 78, 85 (1976)	6, p. 62; *14.3.4*, p. *200*; 22.2.3, p. 369
H.-G. Fabian	(a) **33**, 345 (1976)	*1.3*, p. *21*; *10.1*, p. *78*; 14.1, p. 164; 21, p. 290; 21.1, p. 303
	(a) **36**, 127 (1976)	*1.3*, p. *23*; *10.1*, p. *83*; 14.1, p. 165; 21, p. 296; 21.1, p. 309
J. M. Fabre	(a) **38**, K159 (1976)	*20.1*, p. *278*; 22.9, p. 439
P. Fahrentholtz	(b) 78, K137 (1976)	*9*, p. *74*; 10.2, p. 98; 22.1.2, p. 358
H. J. Falge	(b) 76, 759 (1976)	*13.5*, p. *147*; *20.1*, p. *272*; 22.6, p. 406
	(b) 78, 123 (1976)	*6*, p. *62*; 20.1, p. 275; 22.6, p. 408
L. M. Falicov	(b) 73, 301 (1976)	*6*, p. *53*; *13.2*, p. *130*; 14.1, p. 164; 21.1, p. 302
M. L. Falin	(b) 75, 731 (1976)	*19*, p. *251*; 22.5.3, p. 399
	(b) 75, K111 (1976)	*19*, p. *251*; 22.5.3, p. 400
	(b) 77, K77 (1976)	*19*, p. *251*; 22.5.3, p. 400
G. L. Falko	(b) 76, 593 (1976)	6, p. 59; *8*, p. *69*
L. D. Falkovskaya	(b) 74, 57 (1976)	13.4, p. 136; *18.2*, p. *230*; 22.1, p. 343
K. Falkowski	(b) 74, 683 (1976)	*19*, p. *250*; 22.5.3, p. 399
	(b) 75, K7 (1976)	*19*, p. *251*; 22.8.2, p. 428
E. J. Fantner	(b) 78, 643 (1976)	*14.3*, p. *181*; 22.2.3, p. 369
G. Fantozzi	(a) **36**, K113 (1976)	*10.2*, p. *95*; 22.1.2, p. 356
I. I. Farbshtein	(b) 75, K133 (1976)	*14.3*, p. *177*; 22.1.3, p. 359
S. I. Fărcas	(b) 74, 335 (1976)	*19*, p. *249*; 22.8.1, p. 420
D. Fargeot	(a) **35**, 687 (1976)	*1.3*, p. *22*; *12.1*, p. *108*; 21, p. 295; 21.1, p. 308
R. M. Mac Farlane see under letter "M"		
J. L. Farvacque	(a) **36**, 461 (1976)	*10.2*, p. *94*; 22.1.3, p. 360
	(b) 77, 651 (1976)	*10.2*, p. *95*; *14.3*, p. *179*; 22.1.3, p. 360

	(a) **38**, 271 (1976)	*10.2*, p. *97*; 22.1.3, p. 360
R. H. J. Fastenau	(a) **34**, 277 (1976)	*10.1*, p. *80*; 21, p. 292
D. J. Fathers	(a) **36**, K13 (1976)	*18.2*, p. *233*; 21.1.1, p. 320
A. Faust	(b) **77**, 591 (1976)	*13.1*, p. *127*; 13.2, p. 131; 21.2, p. 323
M. Fayard	(a) **38**, 629 (1976)	1.2, p. 20; *1.4*, p. *29*
J. C. Fayet	(b) **73**, 625 (1976)	*1.2*, p. *14*; *19*, p. *248*; 22.8, p. 413
	(b) **74**, 393 (1976)	*19*, p. *249*; 22.8, p. 414
	(b) **75**, 255 (1976)	*13.4*, p. *138*; 22.8, p. 414
	(b) **77**, 195 (1976)	*4*, p. *46*; *13.4*, p. *142*; 19, p. 253; 22.8, p. 417
B. J. Fechner	(b) **75**, 687 (1976)	18.1, p. 226; *18.2*, p. *232*; 21.1, p. 308
R. G. Fedchenko	(b) **74**, 471 (1976)	*13.2*, p. *130*; 21, p. 292
A. M. Fedorchenko	(b) **76**, 593 (1976)	6, p. 59; *8*, p. *69*
V. P. Fedorov	(b) **78**, 93 (1976)	13.4, p. 144; 13.5.1, p. 155; *20.1*, p. *275*; 22.9, p. 438
A. V. Fedosov	(a) **34**, K139 (1976)	*14.3*, p. *176*; 22.1.1, p. 346; 22.1.2, p. 353
H. Feederle	(a) **34**, 497 (1976)	*13.4*, p. *137*; *20.1*, p. *266*; 22.9, p. 435
S. Feierabend	(a) **34**, K135 (1976)	*18.2.1*, p. *237*; 22, p. 336
M. Feller-Kniepmeyer	(a) **37**, 183 (1976)	*10.1*, p. *83*; 14.1, p. 166; 21.1, p. 311
J. Fenger	(b) **74**, 183 (1976)	*5*, p. *48*; 22.5.2, p. 393; 22.5.3, p. 399
W. Fennel	(b) **73**, 487 (1976)	*13.5*, p. *147*
A. Fernandez	(a) **34**, K17 (1976)	*1.2*, p. *15*; 21.1.1, p. 317
A. V. Ferris-Prabhu	(a) **35**, 243 (1976)	13.4, p. 138; *14.3.3*, p. *191*; 22.1.2, p. 353
J. Fiasson	(a) **33**, 697 (1976)	*1.4*, p. *25*; *9*, p. *72*; 22.5.2, p. 393
A. Fiderkiewicz	(a) **38**, 139 (1976)	*1.2*, p. *19*; *11*, p. *104*; 22.1.2, p. 357
D. W. Field	(a) **36**, 597 (1976)	*1.1*, p. *12*; 6, p. 59; 21.7, p. 332
H. Figiel	(a) **36**, 275 (1976)	18.2, p. 233; *19*, p. *253*; 21.1, p. 309; 21.4, p. 327
T. Figielski	(a) **37**, K57 (1976)	*10.2*, p. *95*; *16*, p. *216*; 19, p. 254; 22.1.2, p. 356
A. Paduan Filho see under letter "P"		
A. S. Filipchenko	(a) **34**, K7 (1976)	*13.4*, p. *137*; *14.3.3*, p. *190*; 22.2.3, p. 367
	(b) **77**, 53 (1976)	*13.1*, p. *127*; *14.3*, p. *179*; 22.2.3, p. 368
	(a) **37**, K139 (1976)	*13.4*, p. *143*; 20.1, p. 275; 22.2.3, p. 369
B. N. Filippov	(a) **33**, 325 (1976)	*5*, p. *48*; 18.2, p. 229; 21.1.1, p. 316
G. Filscher	(a) **33**, K81 (1976)	*1.6*, p. *34*; 10.2, p. 89; 22.1.2, p. 352
V. K. Finashkin	(b) **74**, 35 (1976)	*13*, p. *115*; 21.1, p. 304
C. R. Fincher, Jr.	(b) **75**, 239 (1976)	13.1, p. 123; *20.1*, p. *267*; 22.9, p. 436
D. Fink	(b) **73**, 361 (1976)	*13.1*, p. *121*; 22.1.1, p. 345; 22.1.2, p. 351
J. Fink-Finowicki	(b) **74**, K27 (1976)	*18.3*, p. *240*; *20.1*, p. *264*; 22.8.2, p. 427
J. Finke	(a) **36**, K9 (1976)	*3*, p. *43*; 22.1.3, p. 359
L. M. Finkelstein	(b) **73**, K135 (1976)	*13.1*, p. *121*; *20.1*, p. *262*; 22.2.4, p. 370
H. Finkenrath	(b) **78**, K27 (1976)	*13.5.3*, p. *159*; *20.1*, p. *276*; 22.4, p. 377
J. U. Fischbach	(a) **33**, 473 (1976)	13.1, p. 122; 13.5.1, p. 150; *20.1*, p. *262*; 22.2.2, p. 365
G. Fischer	(a) **34**, 691 (1976)	*18.2*, p. *231*; *18.4*, p. *243*; 21, p. 293
H. Fischer	(a) **34**, K63 (1976)	*20.3*, p. *281*; 22.2.2, p. 365
A. Ya. Fishman	(b) **74**, 57 (1976)	13.4, p. 136; *18.2*, p. *230*; 22.1, p. 343
H. Flietner	(a) **37**, 533 (1976)	*13.3*, p. *133*; 14.3.2, p. 188; 22.1.2, p. 357
H. E. Flotow	(a) **34**, 291 (1976)	*8*, p. *68*; 18, p. 222; 22.8, p. 414
R. Folk	(a) **36**, 147 (1976)	*1.1*, p. *12*
B. I. Fomin	(a) **36**, K89 (1976)	*3*, p. *43*; 21.1, p. 310; 22.1.2, p. 355
V. M. Fomin	(b) **73**, 553 (1976)	*20.1*, p. *262*
	(b) **78**, 831 (1976)	*14.3.4*, p. *200*; *20.1*, p. *278*
C. Y. Fong	(b) **78**, 737 (1976)	4, p. 47; *13.1*, p. *129*; 20.1, p. 278; 22.5, p. 390
R. A. Forman	(a) **33**, K91 (1976)	*13.4*, p. *136*; 16, p. 213; 22.1.2, p. 352
G. Försterling	(a) **35**, K49 (1976)	*1.2*, p. *16*; 15, p. 210; 22.6, p. 404
	(a) **38**, K13 (1976)	*1.2*, p. *19*; *18.1*, p. *227*; 22.6.1, p. 409

E. Fortin	(a) **36**, 385 (1976)	*13.5.1*, p. *153*; 22.4.1, p. 381
C. T. Forwood	(a) **33**, 355 (1976)	*1.4*, p. *25*; *10.1*, p. *78*; 21.1, p. 303
F. Fouquet	(a) **35**, 213 (1976)	*10.1*, p. *82*; 12.1, p. 108; 21, p. 294; 21.1, p. 307
A. Fousková	(a) **36**, K51 (1976)	*1.2*, p. *15*; 22.5, p. 388
H. G. Fousse	(a) **35**, 459 (1976)	8, p. 68; *18.4*, p. *244*; 22.6, p. 405
G. Frahm	(b) **75**, K149 (1976)	*6*, p. *57*; *20.1*, p. *269*; 22, p. 337
Z. Frait	(b) **73**, K85 (1976)	*18.3*, p. *239*; 19, p. 248; 22.8.2, p. 427
S. Franck	(b) **77**, 631 (1976)	*1.2*, p. *19*; *14.4.1*, p. *208*
E. L. Frankevich	(b) **77**, 265 (1976)	13.5.1, p. 155; *14.3.1*, p. *185*; *16*, p. *215*; 20.3, p. 286; 22.9, p. 437
P. Franzosi	(b) **77**, 457 (1976)	*13.2*, p. *131*; 21.1, p. 312
T. Frauenheim	(b) **74**, K101 (1976)	*14*, p. *162*
J. H. Freeman	(a) **33**, K97 (1976)	*5*, p. *48*; 22.1, p. 343
J. R. Freitas	(b) **78**, K35 (1976)	*20.3*, p. *287*; 22.3.1, p. 375
F. Freund	(a) **37**, 97 (1976)	*6*, p. *61*; *20.1*, p. *273*; 22.8, p. 417
V. M. Fridkin	(a) **33**, K137 (1976)	*14.4.1*, p. *206*; *20.1*, p. *263*; 22.8.1, p. 419
H. Friedrich	(b) **73**, 633 (1976)	13.1, p. 121; *20.1*, p. *262*; 22.2.4, p. 370
P. Fries	(b) **77**, 495 (1976)	*18.2*, p. *235*
L. Fritsche	(b) **78**, 585 (1976)	*13.1*, p. *129*; 22.1, p. 344
M. Fritsche	(a) **33**, 149 (1976)	*13.1*, p. *121*; *20.1*, p. *261*; 22.1.3, p. 359
D. Fröhlich	(b) **76**, 613 (1976)	*13.5.1*, p. *154*; *20.1*, p. *271*; 22.5.1, p. 391
F. Fröhlich	(a) **34**, 165 (1976)	*10.2*, p. *90*; 22.5.2, p. 393
E. K. Frolova	(b) **76**, 351 (1976)	*13*, p. *117*; *14.3.4*, p. *197*; 22.1.2, p. 354
H. Fuentes	(a) **37**, K121 (1976)	*1.2*, p. *19*; *1.3*, p. *24*; 21.1, p. 313
W. Fuhs	(b) **73**, K127 (1976)	*7*, p. *65*; *14.3.4*, p. *195*; 22.1.3, p. 359
	(b) **76**, 817 (1976)	*13.5.1*, p. *154*; 20.1, p. 272; 22.9, p. 437
T. Fujimoto	(a) **37**, K33 (1976)	*18.2*, p. *235*; 21.1, p. 311; 21.1.1, p. 321
T. Fujimura	(a) **34**, 369 (1976)	*14.4.1*, p. *206*; 22.8.1, p. 420
S. Fujiwara	(b) **78**, 519 (1976)	*13.5.1*, p. *156*; 20.3, p. 288; 22.9, p. 439
T. Fukumoto	(b) **73**, 341 (1976)	*6*, p. *53*; *20.1*, p. *260*; 22.5, p. 388
A. Funke	(a) **36**, 617 (1976)	*18.3*, p. *241*; 22.8.2, p. 430
Y. Furuhata	(a) **36**, K29 (1976)	*1.2*, p. *18*,*3*, p. *43*; 22.8, p. 416
A. T. Futro	(a) **37**, K75 (1976)	*20.1*, p. *273*; 22.8, p. 417

L. Gabba	(a) **34**, 95 (1976)	*10.2*, p. *90*; 22.1, p. 343; 22.1.1, p. 346; 22.1.2, p. 352
	(a) **34**, 455 (1976)	*10.2*, p. *91*; 22.1, p. 343; 22.1.1, p. 346; 22.1.2, p. 353
R. G. Gabrielyan	(b) **73**, 681 (1976)	*5*, p. *48*
C. J. Gadsden	(a) **34**, K159 (1976)	*19*, p. *250*; 21.1, p. 307
H. J. Gaehrs	(a) **33**, 461 (1976)	*14.3.2*, p. *187*; *14.3.4*, p. *196*; 16, p. 213; 22.9, p. 433
Yu. B. Gaididei	(b) **73**, 415 (1976)	*13.5.1*, p. *149*; *20.1*, p. *261*; 23, p. 440
S. Galassini	(a) **36**, 569 (1976)	*14.3.4*, p. *198*; 22.3, p. 373
R. R. Gałązka	(b) **78**, 477 (1976)	*13.1*, p. *129*; 14.3, p. 181; 22.4.2, p. 384
H. le Gall	(b) **77**, 121 (1976)	*18.3*, p. *241*; *20*, p. *258*; 22.8.2, p. 430
D. Galland	(b) **77**, 739 (1976)	*13.4*, p. *143*; *19*, p. *254*; 22.4, p. 377
J. Gallice	(b) **74**, 217 (1976)	13.5.1, p. 151; *19*, p. *249*; 22.9, p. 434
J. M. Galligan	(a) **34**, K101 (1976)	*3*, p. *42*; *10.1*, p. *81*; 21.6, p. 330
F. M. Galprin	(b) **73**, K73 (1976)	*13*, p. *114*; *18.2*, p. *228*; 21.1, p. 302; 21.1.1, p. 316
	(b) **76**, K9 (1976)	*13*, p. *117*; *18.2*, p. *233*; 21.1, p. 309; 21.1.1, p. 319
M. Galtier	(a) **38**, K159 (1976)	*20.1*, p. *278*; 22.9, p. 439
P. Ganguly	(a) **36**, 669 (1976)	*14*, p. *162*; 15, p. 210; *18*, p. *224*; 22.8.1, p. 423

S. V. Gantsevich	(b) **75**, 407 (1976)	6, p. 56; *7*, p. *65*; *14.3.4*, p. *197*; 22, p. 336
G. A. Gard	(a) **33**, K97 (1976)	*5*, p. *48*; 22.1, p. 343
A. K. Garg	(b) **76**, 171 (1976)	*6*, p. *57*; *20.1*, p. *270*; 22, p. 337
Reva Garg	(b) **77**, K165 (1976)	*5*, p. *50*; 22.8, p. 417
V. K. Garg	(b) **77**, K165 (1976)	*5*, p. *50*; 22.8, p. 417
	(b) **78**, 113 (1976)	*12.2*, p. *112*; 22.5.2, p. 397
A. F. Garito	(b) **75**, 239 (1976)	13.1, p. 123; *20.1*, p. *267*; 22.9, p. 436
R. K. Gartia	(a) **37**, 571 (1976)	*10.2*, p. *96*; 20.3, p. 287; 22.5.2, p. 397
F. M. Gashimzade	(b) **76**, 437 (1976)	*6*, p. *58*; *13.4*, p. *140*; 22.8.2, p. 429
V. A. Gasparov	(b) **74**, K107 (1976)	*13.2*, p. *130*; 19, p. 250; 21.1, p. 305
P. de Gasperis	(b) **77**, 309 (1976)	*18.2*, p. *234*; 21.1, p. 311
W. Gasser	(b) **77**, K103 (1976)	*14*, p. *163*
S. V. Gastev	(b) **74**, 81 (1976)	6, p. 54; *13.4*, p. *136*; *13.5.1*, p. *150*; 16, p. 214; 22.1.1, p. 345
K. E. Gauss	(b) **78**, 133 (1976)	6, p. 62; *14.4.1*, p. *208*; 20.1, p. 275; 22.8.1, p. 424
R. Gauthier	(a) **38**, 85 (1976)	*17*, p. *220*; 21.1, p. 313; 22.1.1, p. 349; 22.5.2, p. 397
	(a) **38**, 477 (1976)	*17*, p. *220*; 21.1, p. 314; 22.1.1, p. 350; 22.5.2, p. 398
I. G. Gavrikova	(a) **37**, K21 (1976)	*16*, p. *216*; 22.4.2, p. 383
G. G. Gegiadze	(a) **37**, K71 (1976)	*3*, p. *43*; 22.2.4, p. 371
G. A. Gehring	(b) **74**, 297 (1976)	8, p. 67; *18.4*, p. *242*; 20.1, p. 264; 22.8.1, p. 420
	(b) **78**, 211 (1976)	6, p. 62; 13.5.1, p. 155; *20.1* p. *275*; 22.8.1, p. 424
K. A. Gehring	(a) **34**, K13 (1976)	*20.1*, p. *265*; *20.3*, p. *281*; 22.8, p. 414; 22.8.1, p. 420
I. V. Gektina	(a) **33**, 289 (1976)	*10.1*, p. *78*; 21.1, p. 303
P. V. Geld	(a) **34**, 787 (1976)	*9*, p. *72*; 10.1, p. 81; 21, p. 293; 21.1, p. 306
	(b) **78**, 465 (1976)	*13.1*, p. *129*; 22.3, p. 374
R. Gemperle	(a) **36**, K59 (1976)	*1.3*, p. *23*; *18.2*, p. *233*; 21.1, p. 309; 21.1.1, p. 320
L. Genzel	(b) **78**, K121 (1976)	*6*, p. *64*; *20.1*, p. *278*; 22.5.4, p. 401
D. Genzow	(b) **77**, 667 (1976)	*14.3*, p. *180*; 16, p. 216; 22.1.3, p. 360
	(a) **38**, 497 (1976)	*16*, p. *217*; 22.8, p. 418
D. George	(a) **36**, K153 (1976)	*14.2*, p. *170*; 14.3.3, p. 193; 22.6, p. 406
G. I. Georgiev	(b) **74**, K173 (1976)	*18*, p. *223*
	(b) **78**, K11 (1976)	*8*, p. *71*; *18*, p. *224*
S. Georgiev	(a) **37**, 295 (1976)	*5*, p. *50*; 18.4, p. 245; 22.6, p. 407
A. N. Georgobiani	(a) **35**, 251 (1976)	*13.4*, p. *138*; 16, p. 214; 20.1, p. 268; 20.3, p. 283; 22.4.1, p. 380
	(a) **38**, 77 (1976)	*13.4*, p. *144*; 16, p. 217; 20.1, p. 276; 22.4.1, p. 382
K. H. Georgy	(a) **37**, K213 (1976)	*10.1*, p. *85*; 21, p. 299
A. B. Gerasimov	(a) **35**, K131 (1976)	*17*, p. *219*; 22.1.1, p. 347
L. Gergely	(a) **38**, K31 (1976)	*14.1*, p. *167*; 21, p. 300; 21.1, p. 314
E. Gerlach	(b) **73**, 203 (1976)	*6*, p. *52*; *14.3*, p. *173*; 22.2.1, p. 362
	(b) **75**, 553 (1976)	13.5.2, p. 157; *14.1*, p. *165*; *20.1*, p. *268*; 21.7, p. 332
	(b) **75**, K15 (1976)	*14.3*, p. *176*; 22.2.3, p. 367
	(b) **76**, K43 (1976)	6, p. 58; *14.3*, p. *178*; 22.2.1, p. 363
	(b) **77**, 651 (1976)	*10.2*, p. *95*; *14.3*, p. *179*; 22.1.3, p. 360
U. Gerlach	(a) **38**, 93 (1976)	*1.2*, p. *19*; *3*, p. *44*; 14.3, p. 181; 15, p. 211; 18.1, p. 227; 22.6.1, p. 409
	(a) **38**, K13 (1976)	*1.2*, p. *19*; *18.1*, p. *227*; 22.6.1, p. 409
R. Gerling	(a) **38**, 547 (1976)	*18.4*, p. *246*; 19, p. 256; 22.6.1, p. 409
P. Germain	(a) **35**, 109 (1976)	*3*, p. *42*; 22.1.2, p. 353

K. H. Germann	(a) **35**, K165 (1976)	*1.2*, p. *17*; 14.4.1, p. 207; 20.1, p. 270; 22.8.1, p. 422
	(a) **38**, K81 (1976)	1.2, p. 20; 14.4.1, p. 208; 18.4, p. 246; *20.1*, p. *277*; 22.6, p. 408; 22.8.1, p. 424
A. E. Gershinskii	(a) **36**, K89 (1976)	*3*, p. *43*; 21.1, p. 310; 22.1.2, p. 355
	(a) **38**, 369 (1976)	*1.5*, p. *33*; *14.3.2*, p. *188*; 22.1.2, p. 358; 22.6, p. 408
P. Gerthsen	(a) **34**, K149 (1976)	*14.4.1*, p. *207*; 22.8.1, p. 421
H. Gertig	(a) **34**, K125 (1976)	*14.3.1*, p. *183*; 22.1.3, p. 359
F. Gervais	(b) **75**, 117 (1976)	*6*, p. *56*; *20.1*, p. *267*; 22.6, p. 404
K. Gesi	(a) **33**, 479 (1976)	*1.2*, p. *14*; *14.4*, p. *202*; 22.8.1, p. 419
R. Gevers	(b) **74**, 359 (1976)	*1.1*, p. *11*
G. Gewinner	(b) **78**, 149 (1976)	2, p. 40; *19*, p. *255*; 22.1.1, p. 349
S. K. Ghatak	(b) **77**, K123 (1976)	*6*, p. *62*; *19*, p. *255*
C. Ghezzi	(a) **34**, 525 (1976)	*1.3*, p. *22*; 22.7, p. 410
V. Ghiordănescu	(b) **78**, K31 (1976)	*19*, p. *255*; 22.5, p. 390
C. Ghiţă	(b) **75**, 375 (1976)	*20.3*, p. *283*; 22.5.2, p. 394
	(b) **77**, 701 (1976)	*10.2*, p. *95*; 20.1, p. 274; 20.3, p. 287; 22.5.2, p. 397
	(a) **38**, K175 (1976)	*13.4*, p. *146*; 22.5, p. 390
L. Ghita	(a) **38**, K175 (1976)	*13.4*, p. *146*; 22.5, p. 390
R. E. Ghosh	(b) **73**, 655 (1976)	*6*, p. *54*; 7, p. 65; 12.2, p. 110; 22.5, p. 388
M. Giehler	(b) **73**, 503 (1976)	*6*, p. *54*; *13.5.2*, p. *157*; 20.1, p. 261; 22.2.2, p. 365; 22.4.1, p. 379
W. Giese	(a) **34**, K1 (1976)	*2*, p. *36*; *18.2.1*, p. *237*; 21.1.1, p. 317
T. C. McGill see under letter "M"		
J. C. Gilles	(a) **34**, 125 (1976)	*1.2*, p. *15*; *1.4*, p. *26*; 22.8.2, p. 427
F. Gilletta	(a) **35**, 545 (1976)	*8*, p. *69*; 22.8.1, p. 421
N. K. Gilra	(a) **34**, 79 (1976)	*10*, p. *75*
J. Giordano	(b) **78**, 453 (1976)	*6*, p. *63*; 20.1, p. 277; 22.8.1, p. 424
N. P. Giorgadze	(b) **75**, K69 (1976)	*18.2*, p. *232*
U. Giorgianni	(b) **74**, 317 (1926)	*13.4*, p. *136*; *20.1*, p. *264*; 22.5.2, p. 393
L. Giral	(a) **38**, K159 (1976)	*20.1*, p. *278*; 22.9, p. 439
V. A. Girii	(a) **34**, K35 (1976)	11, p. 101; *13.3*, p. *132*; *14.3.3*, p. *191*; 22.1.2, p. 353
Ya. G. Girshberg	(b) **74**, 689 (1976)	13.1, p. 123; *14.4.1*, p. *206*; *20.1*, p. *266*; 22.8.1, p. 420
D. V. Gitsu	(b) **73**, K5 (1976)	*15*, p. *209*; 22.7, p. 410
	(b) **74**, K1 (1976)	*12.1*, p. *107*; *15*, p. *209*; 21.7, p. 332
E. S. Giuliano	(b) **75**, K93 (1976)	2, p. 38; *14.1*, p. *165*; *15*, p. *210*; 21, p. 295
F. Giusiano	(a) **34**, 95 (1976)	*10.2*, p. *90*; 22.1, p. 343; 22.1.1, p. 346; 22.1.2, p. 352
	(a) **34**, 455 (1976)	*10.2*, p. *91*; 22.1, p. 343; 22.1.1, p. 346; 22.1.2, p. 353
J. C. Glandus	(a) **35**, 687 (1976)	*1.3*, p. *22*; *12.1*, p. *108*; 21, p. 295; 21.1, p. 308
W. S. Glaunsinger	(b) **73**, K83 (1976)	*18.2*, p. *228*; *19*, p. *247*; 22, p. 334
	(b) **74**, 443 (1976)	*19*, p. *249*; 22, p. 335
K. D. Glinchuk	(a) **33**, K87 (1976)	*13.4*, p. *136*; 16, p. 213; 22.1.2, p. 352
	(a) **34**, 777 (1976)	*13.4*, p. *138*; *20.3*, p. *282*; 22.2.1, p. 362
	(a) **35**, K157 (1976)	*13.4*, p. *140*; 16, p. 214; 22.1.2, p. 354
E. A. Glozman	(a) **36**, 779 (1976)	18.4, p. 245; *19*, p. *253*; 22.8.1, p. 423
P. Gluck	(b) **73**, K79 (1976)	*18.3*, p. *239*
E. Göbel	(b) **78**, 311 (1976)	*13.1*, p. *128*; 20.3, p. 287; 22.2.1, p. 364
H. Göbel	(a) **34**, 553 (1976)	*1.2*, p. *15*; *1.3*, p. *22*; 22.7, p. 410
	(a) **35**, 89 (1976)	*1.2*, p. *16*; 1.3, p. 22; *18.2.1*, p. *237*; 22.7, p. 411

P. F. Gobin	(a) **35**, 213 (1976)	*10.1*, p. *82*; 12.1, p. 108; 21, p. 294; 21.1, p. 307
	(a) **36**, K113 (1976)	*10.2*, p. *95*; 22.1.2, p. 356
I. G. Gochev	(b) **74**, 311 (1976)	*13.5.4*, p. *160*
S. G. Gocheva	(b) **78**, K11 (1976)	*8*, p. *71*; *18*, p. *224*
S. K. Godovikov	(b) **78**, 103 (1976)	*5*, p. *50*; 18.2, p. 235; 18.4, p. 245; 21.4, p. 327
K. Godwod	(a) **34**, 705 (1976)	*1.1*, p. *11*; 10.2, p. 91; 22.1.2, p. 353
O. Goede	(b) **76**, 267 (1976)	*13.5.1*, p. *153*; *20.3*, p. *284*; 22.4.1, p. 380
C. M. Goel	(b) **75**, 697 (1976)	*6*, p. *57*; *20.1*, p. *268*; 22.5.2, p. 395
M. Goettel	(a) **35**, 675 (1976)	*1.2*, p. *17*; 8, p. 69; 22.8, p. 416
A. Golański	(a) **38**, 139 (1976)	*1.2*, p. *19*; *11*, p. *104*; 22.1.2, p. 357
A. Goldmann	(b) **76**, 559 (1976)	*13.1*, p. *126*; 17, p. 219; 22.9, p. 437
B. Göldner	(a) **34**, 319 (1976)	*8*, p. *68*; 10.1, p. 80; 21.1, p. 305
H. J. Goldsmid	(b) **75**, K137 (1976)	6, p. 57; *15*, p. *210*; 21.7, p. 332
	(a) **37**, K43 (1976)	*14.3*, p. *179*; *15*, p. *210*; 22.2.3, p. 368
O. A. Golikova	(a) **37**, 199 (1976)	*14.3*, p. *179*; 22.7, p. 412
H. Gollnast	(a) **38**, K89 (1976)	13.4, p. 145; *16*, p. *217*; 22.2.2, p. 366
V. A. Goltsov	(a) **33**, 67 (1976)	*1.3*, p. *21*; *9*, p. *72*; 21, p. 289; 21.6, p. 329
A. Goltzené	(b) **76**, 363 (1976)	*19*, p. *252*; 22.8, p. 416
K. L. Gomber	(b) **74**, K111 (1976)	*14*, p. *162*
E. Gombia	(a) **34**, 651 (1976)	*14.3*, p. *175*; 15, p. 209; 22.8, p. 414
A. A. Gomes	(b) **75**, 705 (1976)	*13*, p. *116*; *18*, p. *223*; 21.1, p. 308
	(b) **76**, 395 (1976)	*13*, p. *117*; *18*, p. *224*; 19, p. 252; 21.1, p. 309
R. Gómez-Ramírez	(a) **38**, 67 (1976)	*10.2*, p. *97*; 22.5.2, p. 397
I. N. Goncharuk	(b) **78**, 359 (1976)	*6*, p. *63*; *20.1*, p. *276*
V. E. Goncharuk	(b) **76**, 183 (1976)	*19*, p. *252*; 22.5.3, p. 400
P. Gondi	(a) **33**, 579 (1976)	*10.1*, p. *79*; 12.1, p. 107; 21, p. 291
E. R. Gonzalez	(b) **76**, K105 (1976)	*4*, p. *46*; *12.2*, p. *111*; 22.5.2, p. 396
F. Gónzalez	(a) **36**, 33 (1976)	*13.4*, p. *141*; 16, p. 215; *20.1*, p. *271*; 22.8, p. 416
R. Gonzalez	(a) **34**, K9 (1976)	*10.1*, p. *80*; 21.1, p. 305
I. S. Gorban	(a) **35**, 571 (1976)	*13.4*, p. *139*; 14.3, p. 177; 22.8, p. 415
T. Gordillo	(a) **33**, K19 (1976)	*1.1*, p. *10*
A. A. van Gorkum	(a) **37**, 371 (1976)	*10.1*, p. *84*; *11*, p. *103*; 21, p. 298
P. Görnert	(a) **35**, 347 (1976)	*3*, p. *42*; 22.8.2, p. 429
	(a) **35**, 583 (1976)	*3*, p. *42*; 22.8.2, p. 429
	(a) **37**, 505 (1976)	*3*, p. *44*; 22.8.2, p. 430
E. Gornik	(b) **75**, 543 (1976)	*13.1*, p. *124*; *14.3.4*, p. *197*; 20.1, p. 268; 20.3, p. 283; 22.2.3, p. 368
Yu. N. Gornostirev	(a) **38**, 653 (1976)	*10.1*, p. *86*; 21.1, p. 314
V. D. Gorobchenko	(b) **73**, 73 (1976)	*5*, p. *48*
	(b) **76**, 465 (1976)	*5*, p. *50*
E. B. Gorokhov	(a) **33**, 435 (1976)	*1.5*, p. *31*; 2, p. 36; 10.2, p. 88; 22.1.1, p. 345; 22.6, p. 402
U. Gösele	(b) **75**, K129 (1976)	*13.5.1*, p. *153*; *20.3*, p. *284*; 22.9, p. 436
B. N. Goshchitskii	(a) **33**, K83 (1976)	*18.3*, p. *239*; 22.8.2, p. 427
	(a) **38**, 57 (1976)	11, p. 104; *18.3*, p. *241*; 22.8.2, p. 431
	(a) **38**, 433 (1976)	*11*, p. *104*; *14.2*, p. *172*; 21, p. 300
	(a) **38**, K147 (1976)	11, p. 105; *18.3*, p. *241*; 22.8.2, p. 432
A. Goswami	(a) **38**, K151 (1976)	*14.3.4*, p. *201*; 22.9, p. 439
T. Gotô	(a) **38**, K135 (1976)	*14.4*, p. *205*; 22.5.2, p. 398; 22.5.4, p. 401
Y. Gotoh	(a) **36**, K149 (1976)	*1.3*, p. *23*; 14.1, p. 166; 18.2, p. 234; 21.1, p. 310
G. Goureaux	(a) **38**, 505 (1976)	*14.3.4*, p. *201*; 22, p. 342
J. T. Gourley	(b) **77**, K85 (1976)	*10.2*, p. *96*; 20.1, p. 274; 22.6, p. 407
A. Gouskov	(b) **78**, K35 (1976)	*20.3*, p. *287*; 22.3.1, p. 375
A. Govaert	(b) **73**, 527 (1976)	*5*, p. *48*; *18.3*, p. *239*; 22.8.2, p. 426
	(a) **37**, 127 (1976)	18.1, p. 226; *18.3*, p. *241*; 18.4, p. 245; 22.8.2, p. 430

I. T. McGovern see under letter "M"		
S. Govinda	(a) **37**, 109 (1976)	*10.2*, p. *95*; 11, p. 103; *20.1*, p. *273*; 20.3, p. 286; 22.6, p. 406
T. Goworek	(b) **76**, K35 (1976)	*10.2*, p. *93*; 22.9, p. 437
A. K. Goyal	(b) **77**, 639 (1976)	*6*, p. *61*; 12.2, p. 112; 22, p. 340
R. P. Goyal	(b) **75**, 101 (1976)	*6*, p. *56*; 22.6, p. 404
S. C. Goyal	(b) **75**, 101 (1976)	*6*, p. *56*; 22.6, p. 404
	(b) **77**, 639 (1976)	*6*, p. *61*; 12.2, p. 112; 22, p. 340
M. W. Grabski	(a) **34**, 711 (1976)	*1.4*, p. *26*; *10.1*, p. *81*; 21.1.1, p. 318
J. F. Graczyk	(b) **75**, 593 (1976)	*1.3*, p. *22*; *2*, p. *38*; 21, p. 294
C. J. F. Graf	(b) **76**, K137 (1976)	*13.4*, p. *141*; *19*, p. *253*; 22.5.3, p. 400
S. Grande	(b) **74**, 209 (1976)	*19*, p. *249*; 22.2.1, p. 362; 22.2.3, p. 367
M. Grandolfo	(a) **33**, 563 (1976)	*1.1*, p. *11*; *3*, p. *42*; 22.3.1, p. 375
R. Granger	(a) **34**, 307 (1976)	*14.3*, p. *174*; 22.2.1, p. 362
U. M. Grassãno	(b) **77**, 341 (1976)	*10.2*, p. *95*; 20.1, p. 273; *20.3*, p. *286*; 22.5.2, p. 396
D. Gratias	(a) **36**, 329 (1976)	1.2, p. 18; *1.3*, p. *23*; *1.4*, p. *28*; 21.1, p. 309; 21.6, p. 330
	(a) **38**, 595 (1976)	*1.4*, p. *29*; 22.8.2, p. 431
	(a) **38**, 629 (1976)	1.2, p. 20; *1.4*, p. *29*
P. Grau	(a) **34**, 165 (1976)	*10.2*, p. *90*; 22.5.2, p. 393
E. De Grave	(b) **73**, 527 (1976)	*5*, p. *48*; *18.3*, p. *239*; 22.8.2, p. 426
E. M. Gray	(a) **33**, K43 (1976)	14.1, p. 164; *18.2*, p. *229*; 21.1, p. 303
L. G. Grecho	(b) **76**, 593 (1976)	6, p. 59; *8*, p. *69*
B. A. Greenberg	(a) **38**, 653 (1976)	*10.1*, p. *86*; 21.2, p. 314
A. M. Grekhov	(b) **74**, 323 (1976)	18.1, p. 225; *19*, p. *249*
W. Grevendonk	(b) **78**, K89 (1976)	*2*, p. *41*; 18, p. 224; 20, p. 259; 22, p. 342
J. Grigas	(a) **35**, K61 (1976)	*14.4.1*, p. *207*; 22.8, p. 415
	(a) **37**, K39 (1976)	*14.4*, p. *204*; 22, p. 340
V. S. Grigoreva	(a) **36**, K51 (1976)	*16*, p. *215*; 22.8, p. 416
J. Grilhé	(a) **35**, 259 (1976)	*1.4*, p. *27*; *10.1*, p. *92*; 22.8.2, p. 428
	(a) **38**, 713 (1976)	10.1, p. 86; *11*, p. *105*; 14.1, p. 168; 21.1.1, p. 322
E. Grilli	(a) **37**, 399 (1976)	*20.3*, p. *287*; 22.8, p. 417
A. Grisel	(b) **73**, 587 (1976)	*6*, p. *54*; 20.1, p. 262; 22.5, p. 388
G. A. Grishchenko	(a) **35**, 571 (1976)	*13.4*, p. *139*; 14.3, p. 177; 22.8, p. 415
M. M. Gritsenko	(a) **34**, K109 (1976)	*20.2*, p. *279*; *20.3*, p. *282*; 22.8.2, p. 428
	(a) **38**, 409 (1976)	*20.2*, p. *279*; *20.3*, p. *288*; 22.8.2, p. 431
V. A. Gritsenko	(a) **34**, 85 (1976)	*2*, p. *36*; *14.3.1*, p. *183*; 14.3.3, p. 190; 22, p. 335
	(a) **38**, K57 (1976)	*14.3.3*, p. *194*; 22, p. 341
P. Groh	(a) **36**, 579 (1976)	*10.1*, p. *83*; *18.2*, p. *234*; 21.1.1, p. 320
J. Gronkowski	(a) **38**, 139 (1976)	*1.2*, p. *19*; *11*, p. *104*; 22.1.2, p. 357
M. L. Grossbeck	(a) **34**, K97 (1976)	*10.1*, p. *81*; 21.1, p. 306
P. Grosse	(b) **75**, 543 (1976)	*13.1*, p. *124*; *14.3.4*, p. *197*; 20.1, p. 268; 20.3, p. 283; 22.2.3, p. 368
	(b) **75**, 553 (1976)	13.5.2, p. 157; *14.1*, p. *165*; *20.1*, p. *268*; 21.7, p. 332
	(b) **76**, 259 (1976)	13.1, p. 125; *18*, p. *223*; *20.1*, p. *270*; 22, p. 337
G. Grosso	(b) **73**, 371 (1976)	*13*, p. *114*; 22, p. 333
R. Grötzschel	(a) **38**, 139 (1976)	*1.2*, p. *19*; *11*, p. *104*; 22.1.2, p. 357
J. B. Grun	(b) **77**, 381 (1976)	*13.5.1*, p. *155*; *20.3*, p. *286*; 22.5, p. 390
	(b) **78**, 599 (1976)	*13.5.1*, p. *156*; *20.3*, p. *288*; 22.4, p. 378; 22.4.1, p. 382; 22.4.2, p. 384
R. Gründler	(b) **76**, 541 (1976)	*13.1*, p. *126*; *20.1*, p. *271*; 21.6, p. 330
W. Gründler	(b) **78**, 183 (1976)	*1.6*, p. *34*; 22.5.2, p. 397
V. S. Grunin	(b) **77**, 85 (1976)	*13.4*, p. *142*; *19*, p. *253*; 22.6, p. 406

M. GRUS	(a) **37**, K101 (1976)	*14.3.1*, p. *185*; 22.4.1, p. 381
P. C. M. GUBBENS	(a) **34**, 729 (1976)	*5*, p. *49*; 18.2, p. 231; 21.1.1, p. 318; 21.4, p. 326
	(a) **35**, 137 (1976)	*5*, p. *49*; 18.2, p. 232; 21.1.1, p. 318; 21.4, p. 326
W. GUDAT	(b) **74**, 507 (1976)	*20.1*, p. *265*; 21, p. 292; 21.1, p. 305; 21.1.1, p. 317; 21.4, p. 326
N. V. GUDKOVA	(a) **36**, 81 (1976)	*1.5*, p. *31*; 2, p. 39; *11*, p. *102*; 22.1.2, p. 355
S. GUHA	(a) **36**, 453 (1976)	*13*, p. *118*; *14.3.4*, p. *198*; 22, p. 339
M. GUILLOT	(b) **77**, 121 (1976)	*18.3*, p. *241*; *20*, p. *258*; 22.8.2, p. 430
A. P. GUIMARÃES	(b) **77**, K11 (1976)	*19*, p. *254*; 22.7, p. 412
C. GUITTARD	(a) **38**, 477 (1976)	*17*, p. *220*; 21.1, p. 314; 22.1.1, p. 350; 22.5.2, p. 398
K. H. GUNDLACH	(a) **37**, 11 (1976)	13.1, p. 120; *14.3.3*, p. *190*; *16*, p. *213*; 22.6, p. 402
	(a) **37**, 385 (1976)	*14.3.3*, p. *193*; *16*, p. *216*; 22, p. 340
D. C. GUPTA	(b) **78**, K125 (1976)	*5*, p. *51*; 6, p. 64; 21.1, p. 314
R. K. GUPTA	(b) **78**, 255 (1976)	*6*, p. *62*; 23, p. 440
V. L. GUREVICH	(b) **75**, 407 (1976)	6, p. 56; 7, p. 65; *14.3.4*, p. *197*; 22, p. 336
N. GUSKOS	(b) **74** 683 (1976)	*19*, p. *250*; 22.5.3, p. 399
W. GÜTH	(a) **34**, 153 (1976)	*13.4*, p. *137*; *20.1*, p. *264*; 22.1.1, p. 346
E. GUTSCHE	(b) **77**, 233 (1976)	*6*, p. *60*; *13.5.1*, p. *154*; 20.1, p. 272; 22.4.1, p. 381
M. GUYMONT	(a) **36**, 329 (1976)	1.2, p. 18; *1.3*, p. *23*; *1.4*, p. *28*; 21.1, p. 309; 21.6, p. 330
	(a) **38**, 629 (1976)	1.2, p. 20; *1.4*, p. *29*
P. GUYOT	(a) **38**, 207 (1976)	10.1, p. 85; *13.4*, p. *144*; 21, p. 300; 21.2, p. 323
M. GUZZI	(a) **37**, 399 (1976)	*20.3*, p. *287*; 22.8, p. 417

P. HAASEN	(a) **33**, K59 (1976)	*10.2*, p. *88*; 22.1.1, p. 345
H. HAEFNER	(b) **77**, K109 (1976)	*13.2*, p. *131*; *14.1*, p. *166*; 19, p. 254; 21.7, p. 332
H.-J. HAGEMANN	(b) **74**, 507 (1976)	*20.1*, p. *265*; 21, p. 292; 21.1, p. 305; 21.1.1, p. 317; 21.4, p. 326
S. H. HAGEN	(a) **33**, 97 (1976)	13.4, p. 135; *20.3*, p. *280*; 22.3, p. 372
W. E. HAGSTON	(b) **77**, K53 (1976)	13.4, p. 142; *13.5.1*, p. *155*; 22.4, p. 377
U. HAHN	(b) **74**, 579 (1976)	*13.5.1*, p. *151*; *20.3*, p. *282*; 23, p. 440
W. HAIGIS	(b) **78**, 637 (1976)	*13.1*, p. *129*; *13.2*, p. *131*; 14.3, p. 181; 22.7, p. 412
T. HALICIOĞLU	(a) **37**, 695 (1976)	*4*, p. *46*; 12.1, p. 108; 21, p. 299
C. HAMANN	(b) **78**, 79 (1976)	13.4, p. 144; *14.3*, p. *180*; *14.3.1*, p. *185*; 22.9, p. 438
J. HAMERSKÝ	(a) **35**, K97 (1976)	9, p. 73; *14.3.1*, p. *184*; 22.4.2, p. 383
F. H. HAMMAD	(a) **37**, K209 (1976)	*10.1*, p. *85*; *14.1*, p. *167*; 21, p. 299
	(a) **37**, K213 (1976)	*10.1*, p. *85*; 21, p. 299
R. B. HAMMOND	(a) **33**, 59 (1976)	13.5.1, p. 149; 14.3.4, p. 195; *20.3*, p. *280*; 22.1.2, p. 351
K. HANDRICH	(b) **73**, K65 (1976)	*2*, p. *35*; *18.4*, p. *242*
	(b) **74**, K125 (1976)	*2*, p. *37*; *18.2*, p. *231*
A. HANDSTEIN	(a) **37**, 171 (1976)	*14.2*, p. *171*; 21, p. 297; 21.1, p. 311
G. HANECZOK	(a) **33**, 313 (1976)	*10.1*, p. *78*; *18.2*, p. *229*; 21.1.1, p. 316
W. HANKE	(b) **77**, 543 (1976)	*13.1*, p. *127*; 13.5.2, p. 157; *14.1*, p. *166*; 14.4, p. 204; 20.1, p. 274; 21.3, p. 324
J. HANSEN-SCHMIDT	(b) **73**, 117 (1976)	*1.1*, p. *10*; 22.1.2, p. 351
	(b) **73**, 461 (1976)	*1.1*, p. *10*; 22.1.2, p. 351
W. HANUS	(a) **36**, 445 (1976)	13.3, p. 133; *16*, p. *215*; 22.2.3, p. 368

H. Happ	(b) 78, 133 (1976)	6, p. 62; *14.4.1*, p. *208*; 20.1, p. 275; 22.8.1, p. 424
I. Z. Harangozó	(a) 33, K107 (1976)	*1.3*, p. *21*; 21, p. 291; 21.1, p. 304
G. Harbeke	(b) 73, 655 (1976)	*6*, p. *54*; 7, p. 65; 12.2, p. 110; 22.5, p. 388
M. Hardiman	(b) 75, 155 (1976)	*19*, p. *250*; 21, p. 293
T. M. Haridasan	(b) 75, 53 (1976)	*6*, p. *56*; *8*, p. *68*; 22.5.2, p. 394
H. Harms	(a) 34, 553 (1976)	*1.2*, p. *15*; *1.3*, p. *22*; 22.7, p. 410
	(a) 35, 89 (1976)	*1.2*, p. *16*; 1.3, p. 22; *18.2.1*, p. *237*; 22.7, p. 411
A. J. Harris	(a) 35, 697 (1976)	*14.3.3*, p. *192*; *20.3*, p. *284*; 22.3, p. 373
	(a) 36, 365 (1976)	*14.3.3*, p. *192*; *20.3*, p. *285*; 22.3, p. 373
E. F. Harris	(a) 35, 667 (1976)	*10.2*, p. *93*; 20.1, p. 269; 22.6, p. 405
L. B. Harris	(a) 33, 697 (1976)	*1.4*, p. *25*; *9*, p. *72*; 22.5.2, p. 393
	(a) 35, 371 (1976)	*1.4*, p. *27*; *10.2*, p. *92*; 22.5.2, p. 394
	(a) 37, K43 (1976)	*14.3*, p. *179*; *15*, p. *210*; 22.2.3, p. 368
E. Hartmann	(a) 33, K169 (1976)	*10.2*, p. *89*; *14.4*, p. *202*; 22.5.2, p. 393
H. W. den Hartog	(b) 73, 283 (1976)	*6*, p. *53*; *20.1*, p. *260*; 22.5.3, p. 399
J. Härtwig	(a) 37, 417 (1976)	*1.1*, p. *13*; 22.1.2, p. 356
M. H. Harutunian	(b) 74, K107 (1976)	*13.2*, p. *130*; 19, p. 250; 21.1, p. 305
F. Hashimoto	(a) 36, K81 (1976)	9, p. 73; *10.2*, p. *94*; 14.3, p. 178; 22.1.1, p. 348
F. H. Hashmi	(a) 35, 181 (1976)	*1.2*, p. *16*; 22.6, p. 404
G. A. Hassan	(a) 37, K209 (1976)	*10.1*, p. *85*; *14.1*, p. *167*; 21, p. 299
	(a) 37, K213 (1976)	*10.1*, p. *85*; 21, p. 299
R. Hatert	(a) 36, 235 (1976)	*14.3.2*, p. *188*; 14.3.3, p. 192; 22.1.2, p. 355
	(a) 33, 661 (1976)	*14.3.2*, p. *187*; 14.3.3, p. 190; 22, p. 334
	(a) 38, 123 (1976)	*14.3.2*, p. *188*; 14.3.3, p. 194; 22.1.2, p. 357
K. Hattori	(b) 73, 717 (1976)	*13.4*, p. *135*; 22.5.1, p. 391; 22.5.2, p. 393
	(b) 74, 773 (1976)	*18.4*, p. *243*
	(b) 75, K39 (1976)	*18.1*, p. *225*; *18.2*, p. *232*; 21.1, p. 307
	(b) 76, 281 (1976)	*13.5.1*, p. *153*; 22.2.1, p. 363; 22.4, p. 376; 22.5, p. 389
W. Haubenreisser	(b) 73, 11 (1976)	6, p. 52; *13*, p. *113*; 13.5.4, p. 160; *18.2*, p. *228*
	(b) 74, 617 (1976)	*18.2*, p. *230*; 19, p. 250
K. Hauffe	(a) 34, 145 (1976)	*16*, p. *214*; 22.4, p. 376
W. Hauffe	(a) 35, K93 (1976)	*1.4*, p. *27*; 11, p. 102; 21.1.1, p. 319; 21.6, p. 330
G. Hauret	(a) 34, K51 (1976)	*1.2*, p. *15*; 22.5, p. 388
	(b) 76, 551 (1976)	*1.2*, p. *18*; 7, p. 65; 22.8.1, p. 423
K. Hausmann	(a) 34, K199 (1976)	*14.3.4*, p. *197*; 22.6, p. 404
S. Hautecler	(b) 73, 223 (1976)	*18.1*, p. *225*; *18.4*, p. *242*; 22.8.2, p. 426
M. B. Haxton	(b) 78, 259 (1976)	*6*, p. *62*; 22.5, p. 390
S. Hayakawa	(a) 33, K35 (1976)	*17*, p. *218*; 22.8.1, p. 419
Y. Hayakawa	(a) 33, K35 (1976)	*17*, p. *218*; 22.8.1, p. 419
P. T. Heald	(b) 78, 341 (1976)	*10.1*, p. *85*; 12.1, p. 109; 21, p. 299; 21.1, p. 313; 21.6, p. 331
M. Heath	(a) 34, K159 (1976)	*19*, p. *250*; 21.1, p. 307
J. Hedman	(b) 75, 197 (1976)	*13.1*, p. *123*; 17, p. 219; 21, p. 293
A. J. Heeger	(b) 75, 239 (1976)	13.1, p. 123; *20.1*, p. *267*; 22.9, p. 436
E. Hegenbarth	(a) 33, 205 (1976)	*12.1*, p. *107*; 21.1, p. 302; 21.3, p. 324
	(a) 34, 319 (1976)	*8*, p. *68*; 10.1, p. 80; 21.1, p. 305
	(b) 76, 307 (1976)	*8*, p. *69*; 13.1, p. 125; 14.2, p. 170; 21, p. 296
G. Heger	(a) 35, 627 (1976)	*1.2*, p. *17*; 4, p. 46; 22.9, p. 436
A. Heidemann	(a) 35, 481 (1976)	*18.4*, p. *244*; 22.6, p. 405
W. Heiliger	(b) 78 K103 (1976)	*15*, p. *212*; 22.7, p. 412
E. Heiner	(b) 77, 93 (1976)	13, p. 118; *18.1*, p. *226*; *18.4*, p. *245*
A. Heinrich	(a) 35, 269 (1976)	6, p. 56; 13.4, p. 139; *14.3*, p. *176*; 22.8, p. 415
	(a) 36, 491 (1976)	*6*, p. *59*; *14.3*, p. *179*; 22.8, p. 417
	(b) 77, 527 (1976)	*13.1*, p. *127*; 22.8, p. 417

H. W. Helberg	(a) **33**, 453 (1976)	*20*, p. *257*; 22.9, p. 433
	(a) **35**, 131 (1976)	1.2, p. 16; *14.4.1*, p. *207*; 22.9, p. 436
R. Heller	(a) **33**, 525 (1976)	8, p. 67; *9*, p. *72*; 14.1, p. 164; 21.1, p. 304
M. P. Hemkar	(b) **78**, 381 (1976)	*6*, p. *63*; 8, p. 71; 12.1, p. 109; 21.2, p. 323
H. Hemschik	(a) **34**, K199 (1976)	*14.3.4*, p. *197*; 22.6, p. 404
H. K. Henisch	(a) **38**, 331 (1976)	*2*, p. *40*; 13.4, p. 145; 14.3, p. 181; 16, p. 217; 22.8, p. 418
F. Henneberger	(b) **76**, 313 (1976)	13.5, p. 147; *13.5.1*, p. *153*; *20.1*, p. *270*; 20.3, p. 284; 22, p. 337
	(b) **76**, 639 (1976)	*13.5.1*, p. *154*; *20.1*, p. *272*; 22, p. 339
K. Henneberger	(b) **73**, 245 (1976)	13.1, p. 120; *14.3.3*, p. *190*; 14.3.4, p. 195; 22.2.1, p. 362
	(b) **73**, 541 (1976)	13.5.3, p. 159; *14.3.4*, p. *195*; *20.1*, p. *262*; 22, p. 334
	(b) **74**, 567 (1976)	13.1, p. 123; *14.3.4*, p. *196*; 22, p. 335
O. D. Hennemann	(b) **77**, 229 (1976)	18.2.1, p. 238; *19*, p. *254*; 21.1, p. 310; 21.1.1, p. 320
W. Henrion	(a) **35**, K77 (1976)	*2*, p. *38*; *20.1*, p. *268*; 22.8, p. 415
E. Hensel	(a) **34**, 737 (1976)	*10.1*, p. *81*; 21.1, p. 306
S. R. Herd	(a) **38**, 305 (1976)	*2*, p. *40*; *18.2.1*, p. *238*; 21.1, p. 314; 21.4, p. 327
R. Hergt	(a) **35**, 347 (1976)	*3*, p. *42*; 22.8.2, p. 429
	(a) **35**, 583 (1976)	*3*, p. *42*; 22.8.2, p. 429
A. M. Hermann	(b) **73**, K31 (1976)	13, p. 114; *18*, p. *222*; *20*, p. *257*; 22, p. 333
B. Hermoneit	(a) **33**, 737 (1976)	20.1, p. 263; *20.2*, p. *279*; *20.3*, p. *281*; 22.8, p. 413
R. Hernández	(a) **36**, 33 (1976)	*13.4*, p. *141*; 16, p. 215; *20.1*, p. *271*; 22.8, p. 416
H. Herrmann	(a) **38**, K89 (1976)	13.4, p. 145; *16*, p. *217*; 22.2.2, p. 366
K. H. Herrmann	(a) **38**, 497 (1976)	*16*, p. *217*; 22.8, p. 418
R. Herrmann	(b) **73**, 615 (1976)	*13.2*, p. *130*; 14.3, p. 174; 19, p. 248; 22.7, p. 410
	(b) **74**, K75 (1976)	*13.2*, p. *130*; 14, p. 162; 19, p. 249; 22.7, p. 410
	(b) **75**, 189 (1976)	*13.1*, p. *123*; 13.2, p. 131; 19, p. 250; 22.7, p. 411
	(b) **76**, 633 (1976)	*13.2*, p. *131*; 14.1, p. 165; 21, p. 297
	(b) **77**, K109 (1976)	*13.2*, p. *131*; *14.1*, p. *166*; 19, p. 254; 21.7, p. 332
B. Hertel	(a) **33**, K73 (1976)	*10.1*, p. *79*; *11*, p. *100*; 21.1, p. 304
H. J. Hesse	(b) **76**, 817 (1976)	*13.5.1*, p. *154*; 20.1, p. 272; 22.9, p. 437
J. Hesse	(b) **73**, 709 (1976)	13.1, p. 121; *20.1*, p. *262*; 22.9, p. 433
P. Hiesinger	(a) **33**, K39 (1976)	*13.4*, p. *135*; *14.3*, p. *174*; 22.2.1, p. 362
D. Hildebrandt	(a) **36**, K93 (1976)	*11*, p. *102*; 21, p. 296; 21.1, p. 310; 21.1.1, p. 320; 22.1.1, p. 348; 22.1.2, p. 355; 22.6, p. 405
	(a) **38**, K155 (1976)	*11*, p. *105*
L. Hildisch	(a) **34**, K147 (1976)	*1.1*, p. *11*; 22.2.4, p. 370
	(a) **36**, 173 (1976)	*1.4*, p. *27*; *3*, p. *43*; 22.2.4, p. 370
R. M. Hill	(a) **34**, 601 (1976)	*13.4*, p. *137*; *14.3.1*, p. *183*; 22.1, p. 343; 22.1.1, p. 346; 22.2.1, p. 362
	(a) **35**, K29 (1976)	*2*, p. *37*; *14.3*, p. *176*
J. Hillairet	(b) **75**, 433 (1976)	*10.1*, p. *82*; 14.1, p. 165; 21, p. 294; 21.3, p. 324
H.-R. Hilzinger	(a) **38**, 487 (1976)	*18.2*, p. *236*
F.-J. Himpsel	(b) **74**, 485 (1976)	*13.1*, p. *122*; *13.5.1*, p. *151*; 20.1, p. 265; 23, p. 440
S. P. Hind	(b) **76**, 599 (1976)	*13.1*, p. *126*; 14.2, p. 170
R. Hirsch	(a) **37**, 105 (1976)	*1.2*, p. *18*; *20*, p. *258*; 22.5.4, p. 401
K. Hisatake	(a) **34**, K131 (1976)	*18.3*, p. *240*; 22.8.2, p. 428

V. V. Hizhnyakov	(b) **75**, 669 (1976)	13.4, p. 139; *20.1*, p. *268*; 22, p. 337
	(b) **76**, K69 (1976)	*13.4*, p. *140*; 20.3, p. 285; 22, p. 338
P. Hlídek	(b) **75**, K45 (1976)	*16*, p. *214*; *20.1*, p. *267*; 22.8.2, p. 428
H. R. Höche	(a) **33**, 587 (1976)	*1.1*, p. *11*; 22.1.2, p. 352
	(a) **37**, 529 (1976)	*1.1*, p. *13*; 22.1.2, p. 357
H. Höchst	(b) **76**, 559 (1976)	*13.1*, p. *126*; 17, p. 219; 22.9, p. 437
W. Hofberger	(a) **34**, K55 (1976)	1.5, p. 31; *20.1*, p. *265*; 22.9, p. 434
B. Hoffmann	(a) **34**, K149 (1976)	*14.4.1*, p. *207*; 22.8.1, p. 421
H. J. Hoffmann	(b) **76**, 241 (1976)	*10.2*, p. *93*; *13.4*, p. *140*; 16, p. 215; 22.5.2, p. 395
U. Hofmann	(a) **33**, 119 (1976)	12.1, p. 107; 14.1, p. 164; *18.2*, p. *228*; 21.1, p. 302
C. Hohlfeld	(a) **36**, K99 (1976)	*8*, p. *69*; 14.2, p. 170; 21, p. 296
G. D. Holah	(b) **78**, 569 (1976)	*6*, p. *63*; 20.1, p. 277; 22.8, p. 418
J. H. Holbrook	(b) **75**, 573 (1976)	*10*, p. *76*
V. N. Holyavko	(b) **75**, 423 (1976)	13.1, p. 124; *14.3*, p. *177*; 22.1.2, p. 354
L. Holz	(a) **36**, K5 (1976)	*6*, p. *58*; 13.5, p. 147; 20.1, p. 271; 20.2, p. 279; 22.8.1, p. 422
G. Holzapfel	(a) **37**, 303 (1976)	*13.4*, p. *142*; 17, p. 219; 20.3, p. 286; 22.8.1, p. 424
H.-G. Holzhütter	(b) **77**, 203 (1976)	*14.3.4*, p. *199*; *16*, p. *215*; 22, p. 339
J. Höner zu Siederdissen	(b) **73**, 239 (1976)	*6*, p. *52*; 13.4, p. 134; *20.1*, p. *260*; 22.5.2, p. 392
B. Hönerlage	(b) **78**, 599 (1976)	*13.5.1*, p. *156*; *20.3*, p. *288*; 22.4, p. 378; 22.4.1, p. 382; 22.4.2, p. 384
J. M. Honig	(b) **73**, 651 (1976)	*13.1*, p. *121*; 17, p. 218; 22.6, p. 403
G. Hoppmann	(a) **37**, K187 (1976)	13.1, p. 128; *14.3*, p. *180*; *20.1*, p. *275*; 22.6, p. 408
Ah Mee Hor see under letter "A"		
J. Horák	(a) **36**, K137 (1976)	*18.1*, p. *226*; *18.2*, p. *234*; 22.8, p. 417
M. Hordós	(a) **38**, K123 (1976)	*10.1*, p. *86*; *14.1*, p. *168*; 21, p. 301; 21.1, p. 315
A. J. Horodecki	(a) **37**, K101 (1976)	*14.3.1*, p. *185*; 22.4.1, p. 381
L. I. Horváth	(a) **37**, K169 (1976)	2, p. 40; *19*, p. *255*; 22.8.1, p. 424
J. Th. M. de Hosson	(b) **78**, 791 (1976)	*13*, p. *119*; 21, p. 300
L. M. Howe	(a) **33**, 265 (1976)	6, p. 53; *11*, p. *100*; 21.1, p. 303
P. R. Howell	(a) **33**, 107 (1976)	*1.4*, p. *25*; 21.1.1, p. 316
M. Howells	(b) **73**, 307 (1976)	*13.1*, p. *120*; *17*, p. *218*; 20.1, p. 260; 22.3, p. 372
Ľ. Hrivnák	(a) **36**, 551 (1976)	13.4, p. 141; *14.3.4*, p. *198*; 22, p. 339
	(a) **37**, K1 (1976)	*14.3.4*, p. *199*; 22.2.1, p. 364
	(a) **38**, K179 (1976)	*14.3.4*, p. *199*; 22.2.1, p. 364
W. Y. Hsu	(b) **73**, 301 (1976)	*6*, p. *53*; *13.2*, p. *130*; 14.1, p. 164; 21.1, p. 302
C. Y. Huang	(b) **78**, K7 (1976)	*14.1*, p. *167*; *18.4*, p. *245*; 21, p. 299
D. L. Huber	(b) **74**, 145 (1976)	*19*, p. *248*; 22.5, p. 388
H. Huber	(a) **35**, 97 (1976)	*10.2*, p. *92*; 11, p. 101; 22.5.1, p. 391
M. Huber	(a) **38**, 695 (1976)	*1.2*, p. *20*; 1.3, p. 24; 22, p. 342
K. Hübner	(b) **73**, K141 (1976)	*4*, p. *45*; *13.1*, p. *122*; 22.1, p. 343
	(a) **35**, 511 (1976)	*14.3.4*, p. *197*; 22.2.3, p. 368
	(b) **76**, K63 (1976)	*4*, p. *46*; *20.1*, p. *271*; 22.2, p. 361
	(b) **77**, 473 (1976)	1.1, p. 13; 2, p. 40; *4*, p. *46*; *13.1*, p. *127*; 22.2 p. 361; 22.2.4, p. 371
W. H. Hubrig	(a) **36**, 109 (1976)	*1.1*, p. *12*; 10.2, p. 93; 22.1.2, p. 355
P. R. W. Hudson	(a) **36**, 627 (1976)	*1.4*, p. *28*; 22.1, p. 344
	(a) **37**, 93 (1976)	*8*, p. *70*; 22.2.1, p. 364
	(a) **37**, 645 (1976)	*6*, p. *62*; 8, p. 71; 22.1, p. 344
S. Hüfner	(b) **76**, 559 (1976)	*13.1*, p. *126*; 17, p. 219; 22.9, p. 437
L. Huldt	(a) **33**, 607 (1976)	*6*, p. *54*; *13.1*, p. *122*; 20, p. 257; 22.1.1, p. 345

M. E. van Hulle	(a) **37**, 473 (1976)	*10.2*, p. *96*; *14.4*, p. *204*; 22.5.1, p. 391
J. Humlíček	(b) 77, 731 (1976)	13.1, p. 127; *14.3.4*, p. *199*; *20.1*, p. *274*; 22.1.1, p. 348
H.-J. Hunger	(b) **73**, K89 (1976)	*5*, p. *48*; 18.2, p. 229; 21, p. 291; 21.1.1, p. 317
D. R. Hutton	(b) **73**, K115 (1976)	*10.2*, p. *89*; 19, p. 248; 22.6, p. 403
	(b) **75**, K167 (1976)	*19*, p. *251*; 22.8, p. 415
Tran Huu Danh see under letter "T"		
M. A. H. Huyberts	(a) **34**, 665 (1976)	*10.2*, p. *91*; 20.1, p. 266; 22.8.2, p. 428
P. Iacconi	(a) **36**, 603 (1976)	13.4, p. 141; *20.1*, p. *272*; *20.3*, p. *285*; 22.8.1, p. 423
S. Ichimura	(b) **75**, 501 (1976)	6, p. 57; *13.5.1*, p. *153*; 13.5.2, p. 157; *20.1*, p. *268*; 22, p. 336
A. A. Ignatov	(b) **73**, 327 (1976)	13.1, p. 121; *14.3.4*, p. *195*; 22, p. 333
D. Ihle	(b) 77, K177 (1976)	*14.3*, p. *180*; 22.6, p. 407
S. Iida	(a) **34**, 325 (1976)	1.2, p. 15; *18.3*, p. *240*; 22.8.2, p. 427
T. Iida	(b) **74**, 91 (1976)	6, p. 54; *13.4*, p. *136*; 20.1, p. 263; 22.5.2, p. 393
M. Iiyama	(a) **35**, K9 (1976)	*10.2*, p. *92*; 22.1.2, p. 354
S. Ikenko	(a) **36**, 317 (1976)	*1.4*, p. *28*; *10.1*, p. *83*; 21, p. 296
V. A. Ilchishin	(b) **75**, 299 (1976)	*14.3*, p. *176*; 22.1.1, p. 346
	(b) **78**, K111 (1976)	*14.3*, p. *182*; 22.1.1, p. 350
J. Ilczuk	(a) **33**, 313 (1976)	*10.1*, p. *78*; *18.2*, p. *229*; 21.1.1, p. 316
E. I. Ilyashenko	(a) **36**, K1 (1976)	*18.2*, p. *233*; 22.8.2, p. 429
K. Inabe	(a) **33**, K125 (1976)	10.2, p. 89; *20.3*, p. *281*; 22.6, p. 403
V. L. Indenbom	(b) **73**, K9 (1976)	*1.1*, p. *10*
K. W. Ingle	(a) **38**, 523 (1976)	*10.1*, p. *86*; 12.1, p. 109; 21.1.1, p. 321
V. A. Ioffe	(b) 77, 85 (1976)	*13.4*, p. *142*; *19*, p. *253*; 22.6, p. 406
S. P. Ionov	(b) **75**, 91 (1976)	*13.5*, p. *147*; 22, p. 336
	(b) **76**, 505 (1976)	*8*, p. *69*; *13*, p. *117*
	(b) 77, 441 (1976)	*13*, p. *119*
G. V. Ionova	(b) **75**, 91 (1976)	*13.5*, p. *147*; 22, p. 336
	(b) 77, 441 (1976)	*13*, p. *119*
Ya. A. Iosilevskii	(b) **73**, 673 (1976)	*14*, p. *162*; *20*, p. *257*
J. C. Irwin	(a) **38**, 731 (1976)	*20.1*, p. *278*; 22.3, p. 374
E. D. Isaac	(a) **36**, 357 (1976)	*13.5.4*, p. *160*; *14.3.4*, p. *198*; 22.8.2, p. 429
G. W. Iseler	(b) **78**, 569 (1976)	*6*, p. *63*; 20.1, p. 277; 22.8, p. 418
K. Ishida	(a) **36**, 717 (1976)	*10.1*, p. *83*; 21.1, p. 310; 21.1.1, p. 320; 21.6, p. 331
N. Ishii	(b) **74**, K39 (1976)	*2*, p. *36*; *13*, p. *115*; 22.3, p. 372
Y. Ishikawa	(a) **33**, K141 (1976)	*18.2*, p. *230*; 21.1, p. 304; 21.1.1, p. 317
M. Israeli	(b) **74**, 245 (1976)	*10.2*, p. *89*; 13.5.1, p. 151; 20.1, p. 263; 20.3, p. 281; 22.5.2, p. 393
F. Itoh	(b) **75**, 559 (1976)	*2*, p. *38*; 13, p. 116; 21.1, p. 307; 21.7, p. 332
N. Itoh	(b) **74**, 111 (1976)	13.5.1, p. 150; *20.3*, p. *281*; 22.9, p. 434
	(b) **75**, 483 (1976)	10.2, p. 93; *11*, p. *101*; *20.1*, p. *268*; 22.5.2, p. 395
	(b) **78**, 519 (1976)	*13.5.1*, p. *156*; 20.3, p. 288; 22.9, p. 439
L. N. Ivanov	(b) **78**, 483 (1976)	13.4, p. 145; *20.3*, p. *288*; 22.5.4, p. 401
M. A. Ivanov	(b) **74**, 57 (1976)	13.4, p. 136; *18.2*, p. *230*; 22.1, p. 343
	(b) **76**, 797 (1976)	13.4, p. *141*; 13.5.4, p. 160; *18.2*, p. *234*
	(a) **38**, 653 (1976)	*10.1*, p. *86*; 21.1, p. 314
N. R. Ivanov	(a) **36**, K85 (1976)	*1.2*, p. *18*; *7*, p. *65*; 22.8, p. 416
E. A. Ivanova	(b) **78**, 359 (1976)	*6*, p. *63*; *20.1*, p. *276*
H. Iwanaga	(a) **38**, K119 (1976)	*10.2*, p. *99*; 22.4, p. 378
R. J. Iwanowski	(b) **75**, K83 (1976)	*14.3*, p. *177*; 22.4.4, p. 387

P. K. Iyengar	(a) **38**, 601 (1976)	*5*, p. *51*; 18.1, p. 227; 18.2, p. 236; 21.1.1, p. 322
O. Izumi	(a) **38**, 587 (1976)	*10.1*, p. *86*; 12.1, p. 109; 21.1, p. 314
C. Jaccard	(a) **35**, 97 (1976)	*10.2*, p. *92*; 11, p. 101; 22.5.1, p. 391
B. Jachym	(a) **34**, 657 (1976)	*14.3*, p. *175*; 22.9, p. 435
D. Jacob	(a) **34**, 593 (1976)	*14.1*, p. *165*; 21, p. 293
G. Jacobs	(b) **77**, 707 (1976)	*10.2*, p. *96*; 20.1, p. 274; 22.5.2, p. 397
	(b) **78**, K51 (1976)	*10.2*, p. *97*; 22.5.2, p. 397
P. W. M. Jacobs	(a) **38**, 293 (1976)	*8*, p. *71*; 10.2, p. 98; 22.5.2, p. 398
A. Jaéglé	(a) **34**, 691 (1976)	*18.2*, p. *231*; *18.4*, p. *243*; 21, p. 293
	(b) **78**, 149 (1976)	2, p. 40; *19*, p. *255*; 22.1.1, p. 349
R. Jagannathan	(b) **74**, 667 (1976)	*13.1*, p. *123*; *18*, p. *222*
I. Jäger	(a) **33**, 167 (1976)	*10*, p. *75*
E. Jahne	(b) **73**, 503 (1976)	*6*, p. *54*; *13.5.2*, p. *157*; 20.1, p. 261; 22.2.2, p. 365; 22.4.1, p. 379
	(b) **74**, 275 (1976)	*6*, p. *55*; *20.1*, p. *263*; 22.2.4, p. 370; 22.3.1, p. 375
	(b) **74**, 601 (1976)	*6*, p. *55*; *20.1*, p. *265*; 22.2.2, p. 365
	(b) **75**, 221 (1976)	*6*, p. *56*; *20.1*, p. *267*; 22.2.4, p. 370
J. Jahnke	(a) **36**, K125 (1976)	*8*, p. *70*; 14.2, p. 170; 21, p. 297
V. K. Jain	(a) **38**, K65 (1976)	10.2, p. 98; *20.3*, p. *288*; 22.5.2, p. 398
W. Jakowetz	(b) **73**, 255 (1976)	13.4, p. 134; *13.5.1*, p. *149*; 20.1, p. 260; *20.3*, p. *280*; 22.2.3, p. 367
J. P. Jakubovics	(a) **36**, K13 (1976)	*18.2*, p. *233*; 21.1.1, p. 320
M. Janda	(a) **35**, 391 (1976)	*1.5*, p. *31*; *3*, p. *42*; 14.3.1, p. 183; 22.1.3, p. 359
	(a) **35**, 471 (1976)	*14.3.1*, p. *184*; 22.1.3, p. 359
	(a) **36**, 101 (1976)	*14.3.1*, p. *184*; 22.1.3, p. 359
	(a) **37**, K127 (1976)	*14.3.1*, p. *185*; 22.1.3, p. 360
J. A. Janik	(a) **35**, 675 (1976)	*1.2*, p. *17*; 8, p. 69; 22.8, p. 416
J. M. Janik	(a) **35**, 675 (1976)	*1.2*, p. *17*; 8, p. 69; 22.8, p. 416
A. S. Janitzki	(a) **34**, K149 (1976)	*14.4.1*, p. *207*; 22.8.1, p. 421
N. Jannuzzi	(b) **78**, K35 (1976)	*20.3*, p. *287*; 22.3.1, p. 375
A. Jasiński	(b) **74**, K5 (1976)	*19*, p. *249*; 22.8.1, p. 420
S. R. Jawalekar	(a) **38**, K93 (1976)	*14.3.1*, p. *186*; 22.4.2, p. 384
F. R. Jeannot	(a) **35**, 459 (1976)	8, p. 68; *18.4*, p. *244*; 22.6, p. 405
A. Jędrzejczak	(b) **76**, 737 (1976)	*13.1*, p. *126*; 14.3, p. 179; *15*, p. *210*; 22.4.3, p. 386
M. Jergel	(a) **33**, 85 (1976)	*14.2*, p. *169*; 21, p. 289
J. Jerhot	(a) **34**, 505 (1976)	*13.3*, p. *132*; *14.3.3*, p. *191*; 22, p. 335
H. Jex	(b) **76**, 581 (1976)	*6*, p. *59*; 7, p. 66; 8, p. 69; *12.2*, p. *111*; 22.5.2, p. 395
Z. Jirák	(a) **37**, K47 (1976)	*1.1*, p. *12*; 22.8.2, p. 430
W. John	(b) **78**, 199 (1976)	*2*, p. *40*; *13*, p. *119*
	(b) **78**, 577 (1976)	*2*, p. *41*; *13*, p. *119*; 22, p. 341
N. M. Johnson	(a) **33**, 415 (1976)	*10.2*, p. *88*; *11*, p. *100*; 20.1, p. 261; 22.5.3, p. 399
G. Johri	(a) **38**, K61 (1976)	*7*, p. *66*
J. J. Jonas	(a) **33**, 173 (1976)	*10.1*, p. *78*; 21, p. 289; 21.1, p. 302; 21.3, p. 324
A. R. Jones	(a) **33**, 107 (1976)	*1.4*, p. *25*; 21.1.1, p. 316
G. A. Jones	(a) **36**, 647 (1976)	*18.2.1*, p. *238*
J. C. Jones	(b) **74**, 23 (1976)	13.1, p. 122; *14.3*, p. *174*; 22.2.1, p. 362
M. Jones	(b) **73**, K31 (1976)	13, p. 114; *18*, p. *222*; *20*, p. *257*; 22, p. 333
B. Joseph	(a) **38**, K127 (1976)	*14.3.3*, p. *194*; *14.3.4*, p. *201*; 22.1.2, p. 358; 22.6, p. 408
A. W. Joshi	(b) **76**, K39 (1976)	*10.2*, p. *93*; *20.1*, p. *271*; 22.5.2, p. 395

S. J. JOSHUA	(b) **74**, 659 (1976)	13.5.1, p. 152; *13.5.4*, p. *160*; 18.4, p. 243; 22.5.4, p. 401
A. JOULLIE	(b) **76**, 623 (1976)	*14.3*, p. *178*; 22.2.4, p. 370
A. JUCYS	(b) **77**, K113 (1976)	*7*, p. *66*; *14.3.2*, p. *188*; 22.4.1, p. 381
W. JUNGE	(a) **34**, K39 (1976)	*20.1*, p. *265*; 22.2.4, p. 370
G. JUNGK	(a) **34**, 69 (1976)	*1.5*, p. *31*; *20.1*, p. *264*; 22.6, p. 403
	(b) **76**, 541 (1976)	*13.1*, p. *126*; *20.1*, p. *271*; 21.6, p. 330
H.-D. JUNKER	(a) **37**, K97 (1976)	*14.3*, p. *179*; 15, p. 211; 22, p. 340
G. JUŠKA	(a) **35**, 339 (1976)	*14.3.2*, p. *188*; *16*, p. *214*; 22.1.3, p. 359
O. KABISCH	(a) **37**, K85 (1976)	*10.1*, p. *84*; 21, p. 298; 21.1, p. 311
C. M. KACHHAVA	(b) **77**, K135 (1976)	*14.1*, p. *167*; 21, p. 298; 21.1, p. 312; 21.2, p. 323
S. KACHI	(a) **35**, 481 (1976)	*18.4*, p. *244*; 22.6, p. 405
Z. S. KACHLISHVILI	(a) **33**, 15 (1976)	13.4, p. 134; *14.3.4*, p. *195*; 22, p. 333
J. KACZÉR	(a) **35**, 235 (1976)	*18.2*, p. *232*
J. KADLEC	(a) **37**, 11 (1976)	13.1, p. 120; *14.3.3*, p. *190*; *16*, p. *213*; 22.6, p. 402
	(a) **37**, 385 (1976)	*14.3.3*, p. *193*; *16*, p. *216*; 22, p. 340
A. G. KADOMTSEV	(a) **34**, 73 (1976)	*10.1*, p. *79*; 21, p. 291
A. M. KADOMTSEVA	(a) **36**, 415 (1976)	12.2, p. 111; *18.4*, p. *244*; 22.8.2, p. 429
V. D. KAGAN	(b) **75**, 407 (1976)	6, p. 56; *7*, p. *65*; *14.3.4*, p. *197*; 22, p. 336
H. G. KAHLE	(b) **74**, 297 (1976)	8, p. 67; *18.4*, p. *242*; 20.1, p. 264; 22.8.1, p. 420
H. KAHLERT	(b) **77**, 677 (1976)	6, p. 61; *14.3*, p. *180*; *20.1*, p. *274*; 22.2.3, p. 368
	(b) **78**, 277 (1976)	4, p. 47; *6*, p. *63*; *20.1*, p. *276*; 22, p. 341
A. KAHN	(b) **73**, 593 (1976)	*19*, p. *248*; 22.6, p. 402
K. KAI	(b) **75**, 559 (1976)	*2*, p. *38*; 13, p. 116; 21.1, p. 307; 21.7, p. 332
O. A. KAIBYSHEV	(a) **35**, 403 (1976)	*10.1*, p. *82*; 21, p. 294; 21.1, p. 307
M. M. KAILA	(a) **37**, K43 (1976)	*14.3*, p. *179*; *15*, p. *210*; 22.2.3, p. 368
V. P. KALANTARIAN	(b) **76**, 113 (1976)	13, p. 116; 18, p. 223; *19*, p. *252*; 21, p. 296; 21.1, p. 309
G. KAMIENIARZ	(b) **75**, K97 (1976)	*13*, p. *116*; *18.1*, p. *226*
A. A. KAMINSKII	(a) **33**, 737 (1976)	20.1, p. 263; *20.2*, p. *279*; *20.3*, p. *281*; 22.8, p. 413
	(a) **34**, K109 (1976)	*20.2*, p. *279*; *20.3*, p. *282*; 22.8.2, p. 428
	(a) **36**, K25 (1976)	*14.3.4*, p. *198*; 22.8.1, p. 422
	(a) **37**, K111 (1976)	*14.3.4*, p. *198*; 22.8.1, p. 422
	(a) **38**, 409 (1976)	*20.2*, p. *279*; *20.3*, p. *288*; 22.8.2, p. 431
K. KAMIYOSHI	(a) **38**, K135 (1976)	*14.4*, p. *205*; 22.5.2, p. 398; 22.5.4, p. 401
K. I. KAMIYOSHI	(a) **34**, 369 (1976)	*14.4.1*, p. *206*; 22.8.1, p. 420
F. KAMMLER	(a) **37**, K201 (1976)	8, p. 71; *14.3.3*, p. *194*; 14.3.4, p. 200; 22, p. 340
Y. KAMIURA	(a) **36**, K81 (1976)	9, p. 73; *10.2*, p. *94*; 14.3, p. 178; 22.1.1, p. 348
Ž. KANCLERIS	(b) **78**, 757 (1976)	12.2, p. 112; *14.3.4*, p. *200*; 22.1.1, p. 349
Y. KANDA	(a) **35**, K115 (1976)	*14.3.3*, p. *192*; 22.1.2, p. 354
	(a) **38**, K41 (1976)	*14.3.3*, p. *194*; 22.1.2, p. 358
A. KANDUŠER	(b) **74**, 425 (1976)	4, p. 45; *6*, p. *55*; 14.4.1, p. 206; *20.1*, p. *265*; 22.8.1, p. 420
H. VON KÄNEL	(b) **74**, 151 (1976)	*18.4*, p. *242*; *20.1*, p. *263*; 22.9, p. 434
O. KANERT	(b) **73**, 155 (1976)	10.2, p. 87; *19*, p. *247*; 22.5.2, p. 392
W. KÄNZIG	(b) **77**, 153 (1976)	*6*, p. *60*; 20.1, p. 272; 22.6, p. 406
A. M. KAPITONOV	(a) **34**, K47 (1976)	*12.2*, p. *110*; 22.8.2, p. 427
V. V. KAPLIN	(b) **76**, 779 (1976)	*11*, p. *103*; 13, p. 118; 22.1.2, p. 355
M. G. KAPLUNOV	(b) **73**, 355 (1976)	13, p. 113; 17, p. 218; *20*, p. *257*; 22.9, p. 433
D. V. KAPOR	(b) **74**, 103 (1976)	*6*, p. *54*; *13.5.1*, p. *150*

	(b) **74**, 497 (1976)	*14.4.1*, p. *206*
	(b) **76**, K93 (1976)	*14.4.1*, p. *207*
G. Kappel	(a) **34**, 691 (1976)	*18.2*, p. *231*; *18.4*, p. *243*; 21, p. 293
Usha Kapur	(a) **38**, K77 (1976)	*14.4*, p. *204*; 22.9, p. 439
P. M. Karageorgy-Alkalaev	(a) **34**, 391 (1976)	13.4, p. 137; *14.3.4*, p. *196*; 22.1.2, p. 352
T. Karakostas	(a) **34**, 243 (1976)	*1.3*, p. *22*; 10.2, p. 90; 22.7, p. 410
F. K. Karapetyan	(a) **38**, K49 (1976)	*8*, p. *71*; *15*, p. *211*; 22.2.4, p. 371
V. V. Karelin	(b) **78**, 483 (1976)	13.4, p. 145; *20.3*, p. *288*; 22.5.4, p. 401
I. Z. Karimova	(a) **34**, 391 (1976)	13.4, p. 137; *14.3.4*, p. *196*; 22.1.2, p. 352
A. E. Karkin	(a) **38**, 433 (1976)	*11*, p. *104*; *14.2*, p. *172*; 21, p. 300
L. E. Karkina	(a) **38**, 653 (1976)	*10.1*, p. *86*; 21.1, p. 314
N. Karl	(a) **34**, 497 (1976)	*13.4*, p. *137*; *20.1*, p. *266*; 22.9, p. 435
H. P. Karnthaler	(a) **37**, 223 (1976)	*10.1*, p. *84*; 12.1, p. 108; 21, p. 297; 21.1, p. 311
B. V. Karpenko	(b) **73**, 455 (1976)	*18.2*, p. *229*; 22.8.2, p. 426
S. N. Karyagin	(a) **37**, K17 (1976)	*13.3*, p. *133*; *19*, p. *254*; 22.6, p. 407
Y. Kashiwada	(a) **36**, K29 (1976)	*1.2*, p. *18*; *3*, p. *43*; 22.8, p. 416
P. K. Kashkarov	(a) **37**, K17 (1976)	*13.3*, p. *133*; *19*, p. *254*; 22.6, p. 407
A. Kasi Viswanath	see under letter "V"	
F. A. Kassan-Ogly	(b) **76**, 45 (1976)	*18.1*, p. *226*; 18.2, p. 233; 18.4, p. 244; 21.1, p. 308; 21.1.1, p. 319
K.-H. Katerbau	(a) **38**, 463 (1976)	*10*, p. *76*
R. Katilius	(b) **75**, 407 (1976)	6, p. 56; *7*, p. *65*; *14.3.4*, p. *197*; 22, p. 336
L. I. Katurina	(a) **37**, 29 (1976)	10.2, p. 95; *14.3*, p. *179*; 22.1.2, p. 356
U. Kaufmann	(b) **74**, 169 (1976)	13.4, p. 136; *19*, p. *248*; 22.8, p. 414
I. S. Kaulach	(b) **75**, 247 (1976)	*13.3*, p. *132*; 14.3.3, p. 191; 16, p. 214; 22.9, p. 436
T. Kawamura	(b) **73**, 391 (1976)	*11*, p. *100*; 13, p. 114; 21, p. 290; 22.5.2, p. 393
M. M. Kazanin	(a) **37**, 199 (1976)	*14.3*, p. *179*; 22.7, p. 412
F. J. Kedves	(a) **33**, K107 (1976)	*1.3*, p. *21*; 21, p. 291; 21.1, p. 304
	(a) **38**, K31 (1976)	*14.1*, p. *167*; 21, p. 300; 21.1, p. 314
	(a) **38**, K123 (1976)	*10.1*, p. *86*; *14.1*, p. *168*; 21, p. 301; 21.1, p. 315
N. P. Kekelidze	(a) **38**, K49 (1976)	*8*, p. *71*; *15*, p. *211*; 22.2.4, p. 371
V. K. Kelkar	(b) **73**, 735 (1976)	*13*, p. *114*; *18*, p. *222*; 21.1, p. 303
J. R. Kelly	(b) **74**, K23 (1976)	*18.4*, p. *243*; 21.1, p. 304
A. W. C. van Kemenade	(a) **33**, 97 (1976)	13.4, p. 135; *20.3*, p. *280*; 22.3, p. 372
W. S. MacKenzie see under letter "M"		
I. G. Kerimov	(a) **35**, K85 (1976)	8, p. 68; 14, p. 162; 15, p. 210; *18.1*, p. *226*; *18.4*, p. *244*; 22.8, p. 415
T. G. Kerimova	(b) **73**, K69 (1976)	13.5.1, p. 149; *20.1*, p. *261*; 22.8, p. 413
M. Kertész	(b) **77**, K157 (1976)	*13.1*, p. *128*; 22, p. 340
A. R. Kessel	(b) **74**, K35 (1976)	*19*, p. *249*
	(b) **76**, 827 (1976)	*13.5.1*, p. *154*; *19*, p. *253*
F. R. Kessler	(b) **75**, 567 (1976)	13.5.1, p. 153; *18*, p. *223*; *20*, p. *258*; 22.1.1, p. 347
A. M. Kevorkov	(a) **34**, K109 (1976)	*20.2*, p. *279*; *20.3*, p. *282*; 22.8.2, p. 428
	(a) **38**, 409 (1976)	*20.2*, p. *279*; *20.3*, p. *288*; 22.8.2, p. 431
P. van den Keybus	(b) **78**, K89 (1976)	*2*, p. *41*; 18, p. 224; 20, p. 259; 22, p. 342
I. Kh. Khadiev	(b) **76**, 55 (1976)	13.4, p. 140; *20.1*, p. *270*; *20.3*, p. *284*; 22.8.1, p. 422
P. I. Khadzhi	(b) **76**, 497 (1976)	*13.5.1*, p. *154*; *20.3*, p. *285*; 22.4.1, p. 381
V. S. Khaimovich	(a) **35**, 779 (1976)	*1.2*, p. *17*; *1.5*, p. *31*; 11, p. 102; 21, p. 295; 21.1, p. 308; 21.1.1, p. 319; 22.1.1, p. 347
I. Khaiyer	(a) **34**, K105 (1976)	*18.2.1*, p. *237*; 21.1, p. 306; 21.1.1, p. 318
H. R. Khan	(b) **78**, K107 (1976)	*14.2*, p. *172*; *19*, p. *256*; 21, p. 300
Q. H. Khan	(b) **77**, 693 (1976)	*6*, p. *61*; 12.2, p. 112; 22.5.4, p. 401

L. I. KHANDROS	(a) **33**, 765 (1976)	10.2, p. 89; *14.3*, p. *174*; 22.4.1, p. 379
	(a) **34**, 397 (1976)	10.2, p. 90; *14.3*, p. *174*; 22.4.1, p. 379
R. P. KHARE	(a) **33**, K51 (1976)	*20.3*, p. *280*; 22.6, p. 402
H. E. KHODENKOV	(a) **34**, K163 (1976)	*18.2*, p. *231*
A. F. KHOKHLOV	(a) **35**, 11 (1976)	11, p. 100; *13.4*, p. *134*; *14.3*, p. *173*; 22.1.1, p. 345; 22.1.2, p. 351
V. A. KHOKHLOV	(a) **36**, 415 (1976)	12.2, p. 111; *18.4*, p. *244*; 22.8.2, p. 429
V. A. KHRAMTSOV	(b) **76**, K109 (1976)	*19*, p. *253*; 22.5.2, p. 396
S. M. KHURSANDOVA	(b) **78**, K59 (1976)	*20.1*, p. *276*; 22.8, p. 418
B. I. KIDYAROV	(a) **36**, 403 (1976)	1.2, p. 18; *3*, p. *43*
J. KIES	(a) **37**, K113 (1976)	*1.4*, p. *28*; 10.2, p. 97; 22.1.2, p. 357
N. KIESEWETTER	(a) **38**, 569 (1976)	*10.1*, p. *86*; 21, p. 301
K. KILIMANN	(b) **77**, 483 (1976)	6, p. 61; *13.5.1*, p. *155*; 22.1.1, p. 348
	(b) **78**, 241 (1976)	*13.5*, p. *148*; *13.5.1*, p. *155*; 22, p. 341
R. S. KIM	(b) **73**, 741 (1976)	*13.1*, p. *121*; 18, p. 222; 20.1, p. 262; 22.4.4, p. 387
R. KIND	(a) **36**, 285 (1976)	*1.2*, p. *17*; 22.9, p. 437
V. V. KIND	(b) **78**, 93 (1976)	13.4, p. 144; 13.5.1, p. 155; *20.1*, p. *275*; 22.9, p. 438
M. KING	(b) **73**, K31 (1976)	13, p. 114; *18*, p. *222*; *20*, p. *257*; 22, p. 333
H.-H. KIRCHNER	(a) **34**, K203 (1976)	1.6, p. 34; *17*, p. *218*; 22.6, p. 404
H. O. K. KIRCHNER	(b) **74**, 255 (1976)	*10*, p. *75*
	(b) **75**, 527 (1976)	*10*. p. *76*; 12, p. 106
	(a) **37**, 223 (1976)	*10.1*, p. *84*; 12.1, p. 108; 21, p. 297; 21.1, p. 311
H. KIRIHATA	(a) **35**, K73 (1976)	13.3, p. 132; *17*, p. *219*; 22.5.2, p. 394; 22.6, p. 405
R. KIRMSE	(b) **77**, 505 (1976)	*4*, p. *46*; 6, p. 61; 18.1, p. 226; 19, p. 254; 22.9, p. 438
H. O. K. KIRSCHNER	(b) **74**, 255 (1976)	*10*, p. *75*
P. KIRSTEN	(b) **75**, K19 (1976)	*20.3*, p. *283*; 22.8, p. 415
A. V. KIRYUSHIN	(b) **78**, K85 (1976)	*18.1*, p. *227*; *18.2*, p. *236*; 22.8.2, p. 431
U. KISCHKO	(a) **34**, K135 (1976)	*18.2.1*, p. *237*; 22, p. 336
V. P. KISEL	(a) **36**, 297 (1976)	*10.2*, p. *93*; 22.5.2, p. 395
V. F. KISELEV	(a) **37**, K17 (1976)	*13.3*, p. *133*; *19*, p. *254*; 22.6, p. 407
S. KISHIDA	(a) **35**, K57 (1976)	10.2, p. 92; *11*, p. *101*; *20.1*, p. *268*; 22.4.1, p. 380
KISHORE	(a) **35**, 769 (1976)	*10.1*, p. *82*; 21, p. 295
E. N. KISLOVSKII	(a) **33**, 275 (1976)	*1.1*, p. *10*; *10.2*, p. *87*; 22.1.2, p. 351
N. I. KISLUKHA	(b) **75**, 735 (1976)	*13.5*, p. *147*
S. KISS	(a) **33**, K107 (1976)	*1.3*, p. *21*; 21, p. 291; 21.1, p. 304
	(a) **38**, K9 (1976)	*10.1*, p. *85*; 12.1, p. 109; 21, p. 300
A. E. KIV	(a) **36**, 427 (1976)	9, p. 73; 10.2, p. 94; *11*, p. *102*; 14.3, p. 178; 22.1, p. 343; 22.2, p. 361; 22.3, p. 373; 22.4, p. 376
A. KIVERIS	(a) **34**, 377 (1976)	*14.3.4*, p. *196*; *20.3*, p. *281*; 22.4.1, p. 379
	(a) **37**, 321 (1976)	6, p. 61; *13.4*, p. *142*; 14.3.4, p. 199; *20.3*, p. *286*; 22.4.1, p. 381
B. R. KIYAK	(a) **37**, 363 (1976)	*13.4*, p. *143*; *14.3.4*, p. *199*; 20.3, p. 287; 22.4.2, p. 383
H. M. KIZILYALLI	(a) **36**, 499 (1976)	*14.4*, p. *203*; 22.6, p. 406
D. KLÄHN	(b) **77**, 483 (1976)	6, p. 61; *13.5.1*, p. *155*; 22.1.1, p. 348
P. KLEINERT	(b) **73**, 379 (1976)	2, p. 35; *14.3.4*, p. *195*; *20.1*, p. *260*; 22, p. 333
Z. KLEM	(a) **35**, 235 (1976)	*18.2*, p. *232*
V. G. KLEPARSKII	(a) **33**, K117 (1976)	*18.3*, p. *239*; 22.8.2, p. 427
J. G. KLIAVA	(a) **35**, K67 (1976)	*19*, p. *251*; 22.6, p. 405
M. KLIMKIEWICZ	(a) **34**, K107 (1976)	*10.2*, p. *91*; 22.4.3, p. 385
	(a) **35**, K9 (1976)	*10.2*, p. *92*; 22.1.2, p. 354

J. Klimowski	(a) **34**, 697 (1976)	*1.2*, p. *15*; 14.4.1, p. 206; 22.9, p. 435
C. Klingshirn	(b) **77**, 381 (1976)	*13.5.1*, p. *155*; *20.3*, p. *286*; 22.5, p. 390
	(b) **78**, 599 (1976)	*13.5.1*, p. *156*; *20.3*, p. *288*; 22.4, p. 378; 22.4.1, p. 382; 22.4.2, p. 384
K. H. Klohs	(b) **76**, 267 (1976)	*13.51*, p. *153*; *20.3*, p. *284*; 22.4.1, p. 380
G. Kluge	(b) **77**, K63 (1976)	*14.2*, p. *171*
J. Kluge	(b) **77**, 243 (1976)	13.5.1, p. 155; *18*, p. *224*; *20.1*, p. *273*; 22.2.2, p. 365; 22.2.4, p. 371
Th. Klupsch	(b) **73**, 11 (1976)	6, p. 52; *13*, p. *113*; 13.5.4, p. 160; *18.2*, p. *228*
G. G. Knab	(a) **36**, 397 (1976)	*10.2*, p. *94*; 22.5.2, p. 395
B. Knappe	(a) **36**, 617 (1976)	*18.3*, p. *241*; 22.8.2, p. 430
	(a) **38**, K53 (1976)	*18.2.1*, p. *238*; *20.1*, p. *277*; 21.1, p. 314; 21.4, p. 327
K. Kneipp	(a) **36**, K5 (1976)	*6*, p. *58*; 13.5, p. 147; 20.1, p. 271; 20.2, p. 279; 22.8.2, p. 422
I. Knigin	(a) **34**, 391 (1976)	13.4, p. 137; *14.3.4*, p. *196*; 22.1.2, p. 352
H. Knoch	(a) **37**, K135 (1976)	*5*, p. *50*; 22.8.2, p. 431
K. Knorr	(a) **35**, 627 (1976)	*1.2*, p. *17*; 4, p. 46; 22.9, p. 436
J. Kobayashi	(a) **34**, 475 (1976)	*1.1*, p. *11*; *14.4.1*, p. *206*; 22.8.1, p. 421
N. Kobayashi	(b) **73**, K19 (1976)	2, p. 35; *19*, p. *247*; *20.1*, p. *261*; 22, p. 333
S. Kobe	(b) **73**, K65 (1976)	*2*, p. *35*; *18.4*, p. *242*
E. E. Koch	(b) **74**, 485 (1976)	*13.1*, p. *122*; *13.5.1*, p. *151*; 20.1, p. 265; 23, p. 440
K. Sh. Kocharli	(a) **33**, K67 (1976)	*14.3.1*, p. *183*; 14.3.2, p. 187; 22.9, p. 433
	(a) **34**, K77 (1976)	*14.3.1*. p. *183*; 22.9, p. 435
L. A. Kocharyan	(b) **73**, 681 (1976)	*5*, p. *48*
K. D. Kochev	(a) **33**, K137 (1976)	*14.4.1*, p. *206*; *20.1*, p. *263*; 22.8.1, p. 419
H. Köhler	(b) **73**, 95 (1976)	*13.1*, p. *120*; *13.2*, p. *130*; 14.3, p. 173; 22.7, p. 410
	(b) **74**, 591 (1976)	*13.1*, p. *123*; *13.2*, p. *130*; 14.3, p. 175; 22.7, p. 410
	(b) **75**, 127 (1976)	*13.1*, p. *123*; *13.2*, p. *131*; 14.3, p. 176; 22.7, p. 410
	(b) **75**, 441 (1976)	*13.1*, p. *124*; *13.2*, p. *131*; 14.3, p. 177; 22.7, p. 411
	(b) **76**, K15 (1976)	*6*, p. *58*; *20.1*, p. *271*; 22, p. 338
	(b) **78**, 637 (1976)	13.1, p. 129; 13.2, p. 131; 14.3, p. 181; 22.7, p. 412
H.-J. Köhler	(a) **34**, 541 (1976)	*14.2*, p. *169*; 21, p. 293
R. Köhler	(a) **37**, K113 (1976)	*1.4*, p. *28*; 10.2, p. 97; 22.1.2, p. 357
S. Kohn	(b) **75**, 303 (1976)	*13.1*, p. *124*; *20.1*, p. *267*; 22.7, p. 411
S. E. Kohn	(b) **78**, 737 (1976)	4, p. 47; *13.1*, p. *129*; 20.1, p. 278; 22.5, p. 390
P. Koidl	(b) **74**, 477 (1976)	*13.4*, p. *137*; 20.1, p. 265; 20.3, p. 282; 22.2.2, p. 365; 22.4.1, p. 380; 22.4.2, p. 383
A. I. Koifman	(a) **38**, 439 (1976)	*9*, p. *74*; *11*, p. *105*; 14.3.2, p. 189; 22.1.2, p. 358
Th. Kokkinakis	(b) **77**, K49 (1976)	*13.5.2*, p. *157*; 20.1, p. 273; 21.1, p. 311
K. A. Kolesnichenko	(a) **38**, K131 (1976)	*14.3.4*, p. *201*; 22.8.1, p. 425
V. N. Kolesnik	(b) **74**, 471 (1976)	*13.2*, p. *130*; 21, p. 292
J. Koller	(b) **77**, K157 (1976)	*13.1*, p. *128*; 22, p. 340
E. Kolley	(b) **75**, 721 (1976)	*2*, p. *38*; 6, p. 57; *13*, p. *116*; 21, p. 294
W. Kolley	(b) **75**, 721 (1976)	*2*, p. *38*; 6, p. 57; *13*, p. *116*; 21, p. 294
N. P. Kolmakova	(b) **77**, K143 (1976)	*7*, p. *66*; *18.4*, p. *245*
B. T. Kolomiets	(a) **37**, 45 (1976)	2, p. 40; *16*, p. *216*; *20.3*, p. *286*; 22, p. 339
V. V. Kolomoets	(a) **34**, K139 (1976)	*14.3*, p. *176*; 22.1.1, p. 346; 22.1.2, p. 353
A. Kolpakov	(b) **73**, 141 (1976)	*6*, p. *52*; 13.1, p. 120; 13.5.1, p. 149; *20.1*, p. *260*
V. S. Komissarov	(b) **76**, 105 (1976)	*13*, p. *116*; *14.3.4*, p. *197*; 22.2.3, p. 368
L. F. Komolova	(a) **35**, K5 (1976)	*20.3*, p. *283*; 22.4.1, p. 380

Č. Koňák	(a) **34**, K51 (1976)	*1.2*, p. *15*; 22.5, p. 388
M. Kończykowski	(b) **73**, K131 (1976)	12.2, p. 110; *14.3*, p. *174*; 22.2.3, p. 367
	(a) **33**, 421 (1976)	13.4, p. 135; *14.3*, p. *174*; 22.4.3, p. 385
V. Kondawar	(b) **75**, 79 (1976)	*4*, p. *45*; 20, p. 258; 22, p. 336
A. V. Kondrachuk	(a) **34**, 35 (1976)	11, p. 101; *13.3*, p. *132*; *14.3.3*, p. *191*; 22.1.2, p. 353
J. Konetzky	(a) **36**, K77 (1976)	*1.3*, p. *23*; *18.2*, p. *233*; 21.1, p. 310; 21.1.1, p. 320
N. A. Koneva	(a) **37**, 701 (1976)	*1.3*, p. *24*; *10.1*, p. *84*; 21.1, p. 312
E. König	(b) **77**, 657 (1976)	*13.4*, p. *143*; 18.1, p. 227; 19, p. 254
	(b) **78**, 543 (1976)	*13.4*, p. *145*; 18.1, p. 227; 19, p. 255; 20.1, p. 277; 22.5, p. 390
W. König	(b) **78**, K121 (1976)	*6*, p. *64*; *20.1*, p. *278*; 22.5.4, p. 401
Yu. V. Konobeev	(a) **34**, 195 (1976)	*10.1*, p. *80*; 11, p. 100; 21.1, p. 305
P. Konsin	(b) **76**, 487 (1976)	6, p. 58; 13.1, p. 125; *14.4.1*, p. *207*; 22, p. 338
V. Koos	(a) **36**, K47 (1976)	*1.5*, p. *32*; *12.2*, p. *111*; 22.1.2, p. 355
	(a) **37**, 193 (1976)	*1.5*, p. *32*; *12.2*, p. *111*; 22, p. 339
U. Kh. Kopvillem	(b) **73**, K41 (1976)	*6*, p. *53*; 19, p. 247; 22.8, p. 413
S. I. Kornyushin	(a) **34**, K35 (1976)	11, p. 101; *13.3*, p. *132*; *14.3.3*, p. *191*; 22.1.2, p. 353
A. N. Korol	(b) **76**, 413 (1976)	13.1, p. 125; *14.4*, p. *203*; 22.2, p. 361
L. S. Korolyuk	(b) **75**, 299 (1976)	*14.3*, p. *176*; 22.1.1, p. 346
S. L. Korolyuk	(b) **78**, K111 (1976)	*14.3*, p. *182*; 22.1.1, p. 350
B. Korsch	(a) **33**, 461 (1976)	*14.3.2*, p. *187*; *14.3.4*, p. *196*; 16, p. 213; 22.9, p. 433
A. V. Korshunov	(b) **78**, 93 (1976)	13.4, p. 144; 13.5.1, p. 155; *20.1*, p. *275*; 22.9, p. 438
H.-J. Kos	(a) **35**, 121 (1976)	*10.2*, p. *92*; 13.4, p. 138; *20.1*, p. *268*; 22.5.2, p. 394
A. Kosakov	(b) **78**, 681 (1976)	*13.1*, p. *129*; 17, p. 220; 20, p. 259; 22.1, p. 344; 22.1.1, p. 349; 22.1.2, p. 358
N. S. Kosenko	(a) **35**, 735 (1976)	*1.2*, p. *17*; *1.4*, p. *27*
R. Kosicik	(a) **36**, 617 (1976)	*18.3*, p. *241*; 22.8.2, p. 430
J. Kossut	(b) **78**, 537 (1976)	*13.4*, p. *145*; *14.3*, p. *181*; 15, p. 211; 22.4.3, p. 386
K. Kosuge	(a) **35**, 481 (1976)	*18.4*, p. *244*; 22.6, p. 405
M. Kotani	(b) **75**, 621 (1976)	*13.5.1*, p. *153*; 20.3, p. 284; 22.9, p. 436
G. Kote	(b) **74**, K23 (1976)	*18.4*, p. *243*; 21.1, p. 304
L. S. Kothari	(b) **75**, 381 (1976)	*6*, p. *56*
	(b) **78**, 843 (1976)	*6*, p. *56*
M. M. Kotlyarskii	(b) **75**, K123 (1976)	13.5.4, p. 160; *20.1*, p. *269*; 22.8, p. 415
	(b) **76**, 419 (1976)	*13.5.4*, p. *160*; *20.1*, p. *271*; 22.5, p. 389
	(a) **38**, K103 (1976)	*18.3*, p. *241*; 22.8.2, p. 431
E. A. Kotomin	(b) **73**, 81 (1976)	*10.2*, p. *87*; *13.4*, p. *134*; 22.6, p. 402
	(b) **73**, 483 (1976)	*10.2*, p. *89*; *13.4*, p. *135*; 22.6, p. 402
J. Koulmann	(b) **78**, 149 (1976)	2, p. 40; *19*, p. *255*; 22.1.1, p. 349
H. Kousek	(a) **38**, 199 (1976)	*10.1*, p. *85*; *12.1*, p. *109*; 21.1, p. 313
T. P. Kovalenko	(a) **33**, K1 (1976)	*10.2*, p. *88*; 22.5.2, p. 392
	(a) **34**, 445 (1976)	*10.2*, p. *90*; 22.5.2, p. 394
N. M. Kovtun	(a) **36**, 415 (1976)	12.2, p. 111; *18.4*, p. *244*; 22.8.2, p. 429
J. P. Kowall	(a) **38**, 321 (1976)	*9*, p. *74*; 11, p. 104; 22.1.2, p. 358
J. Kowalski	(b) **77**, K1 (1976)	*5*, p. *50*; 22.1.2, p. 356
	(b) **78**, 477 (1976)	*13.1*, p. *129*; 14.3, p. 181; 22.4.2, p. 384
B. Kozarzewski	(b) **74**, 437 (1976)	13.4, p. 137; *14.1*, p. *165*; 21, p. 292
M. G. Kozin	(b) **78**, 103 (1976)	*5*, p. *50*; 18.2, p. 235; 18.4, p. 245; 21.4, p. 327
M. B. Kozintseva	(b) **74**, 199 (1976)	*13.1*, p. *122*; *13.5.4*, p. *160*; 18.3, p. 239; 22.8.2, p. 427

S. N. Kozlov	(a) **37**, K17 (1976)	*13.3*, p. *133*; *19*, p. *254*; 22.6, p. 407
V. K. Kozlov	(b) **78**, 483 (1976)	13.4, p. 145; *20.3*, p. *288*; 22.5.4, p. 401
J. A. Kozubowski	(a) **34**, 711 (1976)	*1.4*, p. *26*; *10.1*, p. *81*; 21.1.1, p. 318
W. Kraak	(b) **74**, K75 (1976)	*13.2*, p. *130*; 14, p. 162; 19, p. 249; 22.7, p. 410
	(b) **77**, K109 (1976)	*13.2*, p. *131*; *14.1*, p. *166*; 19, p. 254; 21.7, p. 332
A. M. van der Kraan	(a) **35**, 137 (1976)	*5*, p. *49*; 18.2, p. 232; 21.1.1, p. 318; 21.4, p. 326
W. D. Kraeft	(b) **73**, 487 (1976)	*13.5*, p. *147*
	(b) **78**, 241 (1976)	*13.5*, p. *148*; *13.5.1*, p. *155*; 22, p. 341
Th. Krajewski	(a) **36**, 253 (1976)	*1.5*, p. *32*; 21.1, p. 309; 21.1.1, p. 319; 22.1.1, p. 348
N. L. Kramarenko	(a) **33**, 773 (1976)	*2*, p. *36*; *20*, p. *257*; 22.9, p. 434
K. Krásenský	(b) **73**, K1 (1976)	*13*, p. *114*; 21, p. 289; 21.1, p. 302; 21.1.1, p. 316; 21.6, p. 329
S. Kraśnicki	(b) **78**, 335 (1976)	*13.5.4*, p. *161*; 18.2, p. 236; 21.1, p. 313
E. P. Krasnoperov	(b) **76**, 357 (1976)	13.1, p. 125; *14.2*, p. *170*; *18.1*, p. *226*; 22.8, p. 416
	(b) **77**, 451 (1976)	*14.2*, p. *171*; 21, p. 298
E. Kratochvílová	(a) **34**, 639 (1976)	*18.3*, p. *240*; 22.8.2, p. 428
H. Krause	(a) **36**, 705 (1976)	*13.4*, p. *141*; *14.3.3*, p. *192*; 22.6, p. 406
E. Krautz	(a) **38**, 237 (1976)	*1.4*, p. *29*; 10.1, p. 85; *17.1*, p. *221*; 21, p. 300
A. F. Kravchenko	(a) **35**, 755 (1976)	*14.3.1*, p. *184*; 22.2.3, p. 368
	(b) **77**, 465 (1976)	*13.5.1*, p. *155*; *20.1*, p. *274*; 22.2.1, p. 364
V. B. Kravchenko	(a) **36**, K1 (1976)	*18.2*, p. *233*; 22.8.2, p. 429
K. Kreher	(a) **34**, K63 (1976)	*20.3*, p. *281*; 22.2.2, p. 365
U. Kreissig	(a) **38**, 139 (1976)	*1.2*, p. *19*; *11*, p. *104*; 22.1.2, p. 357
D. Kremp	(b) **78**, 241 (1976)	*13.5*, p. *148*; *13.5.1*, p. *155*; 22, p. 341
J. Krempaský	(b) **74**, 741 (1976)	*2*, p. *37*; 13.4, p. 137; *14.3*, p. *175*; 15, p. 209; 22, p. 335
E. Kretschmann	(b) **76**, 515 (1976)	*13.3*, p. *133*; *13.5*, p. *147*; 21.6, p. 330
E. W. Kreutz	(a) **34**, 489 (1976)	11, p. 101; *14.3*, p. *175*; 22.1.2, p. 353
M. Kriechbeum	(b) **78**, 653 (1976)	*14.3*, p. *181*; 22.2.3, p. 369
V. G. Krigel	(b) **75**, K133 (1976)	*14.3*, p. *177*; 22.1.3, p. 359
R. Krishnan	(b) **73**, K51 (1976)	13.4, p. 134; *20.1*, p. *261*; 22.8.2, p. 426
	(a) **36**, 39 (1976)	*1.3*, p. *23*; *1.4*, p. *27*; 21, p. 296
N. Kristianpoller	(b) **74**, 245 (1976)	*10.2*, p. *89*; 13.5.1, p. 151; 20.1, p. 263; 20.3, p. 281; 22.5.2, p. 393
M. A. Krivov	(a) **35**, 425 (1976)	11, p. 102; 13.4, p. 139; *14.3*, p. *177*; *20.1*, p. *269*; 22.8, p. 415
H. Krizek	(b) **77**, 725 (1976)	14.1, p. 166; *20.1*, p. *274*; 21.4, p. 327
J. Krizek	(b) **77**, 725 (1976)	14.1, p. 166; *20.1*, p. *274*; 21.4, p. 327
E. Kröger	(b) **76**, 515 (1976)	*13.3*, p. *133*; *13.5*, p. *147*; 21.6, p. 330
R. Kroggel	(a) **37**, K85 (1976)	*10.1*, p. *84*; 21, p. 298; 21.1, p. 311
H. Kronmüller	(b) **74**, 261 (1976)	*10.1*, p. *79*; *14.2*, p. *169*; 21, p. 291
	(a) **34**, 623 (1976)	*10.1*, p. *81*; 18.2, p. 231; 20, p. 258; 21.1, p. 306
	(b) **75**, 647 (1976)	*11*, p. *102*; *18.2*, p. *232*; 21.1, p. 308
	(b) **77**, 581 (1976)	*10.1*, p. *84*; *14.2*, p. *171*; 21, p. 298
	(a) **38**, 261 (1976)	*10.1*, p. *85*; 18.2, p. 236; 21.1.1, p. 321
F. Kroupa	(a) **33**, 555 (1976)	*10.1*, p. *79*; 21.1.1, p. 317
M. Kroupp	(b) **76**, 669 (1976)	*18.2*, p. *233*; 21.1, p. 310; 21.4, p. 327
K. Krug	(b) **78**, K27 (1976)	*13.5.3*, p. *159*; *20.1*, p. *276*; 22.4, p. 377
H. Krüger	(b) **73**, 615 (1976)	*13.2*, p. *130*; 14.3, p. 174; 19, p. 248; 22.7, p. 410
A. E. Kruminš	(a) **33**, K31 (1976)	*14.4.1*, p. *206*; *16*, p. *213*; 22.8.1, p. 419
T. A. Kruze	(a) **36**, 81 (1976)	*1.5*, p. *31*; 2, p. 39; *11*, p. *102*; 22.1.2, p. 355
M. A. Krykin	(b) **76**, 67 (1976)	13.3, p. 132; *14.3.2*, p. *188*; *20.1*, p. *270*
I. V. Krylova	(a) **35**, K119 (1976)	*17*, p. *219*; 21, p. 295
	(a) **37**, K9 (1976)	*17*, p. *220*; 22.6, p. 407

M. Krystek	(a) **37**, 303 (1976)	*13.4*, p. *142*; 17, p. 219; 20.3, p. 286; 22.8.1, p. 424
S. Krzemiński	(b) **74**, K119 (1976)	*18.2*, p. *230*
J. Kubát	(a) **35**, 173 (1976)	*10*, p. *76*
R. Kubiak	(b) **76**, K1 (1976)	*14.2*, p. *170*
L. Kubler	(b) **78**, 149 (1976)	2, p. 40; *19*, p. *225*; 22.1.1, p. 349
A. Kubový	(a) **35**, 391 (1976)	*1.5*, p. *31*; *3*, p. *42*; 14.3.1, p. 183; 22.1.3, p. 359
	(a) **35**, 471 (1976)	*14.3.1*, p. *184*; 22.1.3, p. 359
	(a) **36**, 101 (1976)	*14.3.1*, p. *184*; 22.1.3, p. 359
	(a) **37**, K127 (1976)	*14.3.1*, p. *185*; 22.1.3, p. 360
F. Kuchar	(b) **78**, 643 (1976)	*14.3*, p. *181*; 22.2.3, p. 369
D. Kucharczyk	(a) **37**, 287 (1976)	*1.2*, p. *18*; *8*, p. *70*; 14.4.1, p. 208; 22.8, p. 417
V. I. Kuchnev	(b) **73**, K111 (1976)	*8*, p. *67*; 23, p. 440
A. Yu. Kudzin	(a) **38**, K131 (1976)	*14.3.4*, p. *201*; 22.8.1, p. 425
Š. Kudžmauskas	(a) **37**, 321 (1976)	6, p. 61; *13.4*, p. *142*; 14.3.4, p. 199; *20.3*, p. *286*; 22.4.1, p. 381
R. Kuentzler	(a) **34**, 291 (1976)	*8*, p. *68*; 18, p. 222; 22.8, p. 414
R. Kuhl	(b) **77**, K109 (1976)	*13.2*, p. *131*; *14.1*, p. *166*; 19, p. 254; 21.7, p. 332
A. Kuhn	(a) **36**, 181 (1976)	*1.1*, p. *12*; 22.3.1, p. 375
W. Kühn	(b) **76**, 289 (1976)	12.2, p. 110; *13.1*, p. *125*; *20.1*, p. *270*; 22.1.1, p. 347
A. Kühnel	(b) **77**, 631 (1976)	*1.2*, p. *19*; *14.4.1*, p. *208*
G. Kuka	(b) **76**, 633 (1976)	*13.2*, p. *131*; 14.1, p. 165; 21, p. 297
S. K. Kukharskaya	(a) **33**, K117 (1976)	*18.3*, p. *239*; 22.8.2, p. 427
M. Kulić	(b) **75**, 677 (1976)	10.1, p. 82; *14.2*, p. *170*; 21, p. 294
M. A. Kumakhov	(b) **76**, 119 (1976)	*11*, p. *102*
Narendra Kumar	(b) **77**, K133 (1976)	*20.1*, p. *274*; 22.5.2, p. 397
V. Kumar	(b) **75**, K1 (1976)	*13.4*, p. *138*; 22.8, p. 414; 22.8.1, p. 421
	(b) **76**, 337 (1967)	*18.2*, p. *233*
A. Kumaraswamy	(b) **77**, 215 (1976)	*19*, p. *253*; 22.8.1, p. 423
K. Kumazaki	(a) **33**, 615 (1976)	*4*, p. *45*; 7, p. 65; *12.2*, p. *110*; 22.4, p. 376
	(a) **37**, 579 (1976)	*10.2*, p. *96*; *14.3*, p. *180*; 22.4.2, p. 383
M. Kumeda	(b) **73**, K19 (1976)	2, p. 35; *19*, p. *247*; *20.1*, p. *261*; 22, p. 333
K. D. Kundra	(a) **36**, 517 (1976)	*1.2*, p. *18*; 8, p. 70; 22.3, p. 373
V. Kunigelis	(b) **77**, K113 (1976)	*7*, p. *66*; *14.3.2*, p. *188*; 22.4.1, p. 381
A. B. Kunz	(b) **73**, 193 (1976)	*13.4*, p. *134*; 22.5.1, p. 391
	(b) **75**, 533 (1976)	*13.1*, p. *124*; 20, p. 258; 22.3, p. 373
C. Kunz	(b) **74**, 507 (1976)	*20.1*, p. *265*; 21, p. 292; 21.1, p. 305; 21.1.1, p. 317; 21.4, p. 326
N. I. Kurdiani	(a) **36**, 73 (1976)	13.3, p. 132; *14.3.2*, p. *188*; 20.3, p. 285; 22.2.1, p. 363
J. Kuriata	(b) **74**, 683 (1976)	*19*, p. *250*; 22.5.3, p. 399
	(b) **78**, K55 (1976)	*19*, p. *255*; 22.5, p. 390
J. Kurittu	(b) **74**, 177 (1976)	*6*, p. *54*; 22, p. 335
I. V. Kurnosov	(a) **33**, 773 (1976)	*2*, p. *36*; *20*, p. *257*; 22.9, p. 434
M. Kuroha	(b) **75**, 559 (1976)	*2*, p. *38*; 13, p. 116; 21.1, p. 307; 21.7, p. 332
R. Kuśka	(a) **33**, 313 (1976)	*10.1*, p. *78*; *18.2*, p. *229*; 21.1.1, p. 316
R. N. Kusmin	(a) **38**, K45 (1976)	*5*, p. *51*; 22.1, p. 344
Yu. S. Kusminov	(a) **33**, K137 (1976)	*14.4.1*, p. *206*; *20.1*, p. *263*; 22.8.1, p. 419
E. F. Kustov	(b) **76**, 55 (1976)	13.4, p. 140; *20.1*, p. *270*; *20.3*, p. *284*; 22.8.1, p. 422
R. Kütterer	(a) **38**, 261 (1976)	*10.1*, p. *85*; 18.2, p. 236; 21.1.1, p. 321
H. Kuwabara	(a) **37**, K157 (1976)	13.4, p. 144; *20.3*, p. *287*; 22.3, p. 374
H. Kuwamoto	(b) **73**, 651 (1976)	*13.1*, p. *121*; 17, p. 218; 22.6, p. 403
R. Kužel	(a) **36**, 747 (1976)	*14.3.3*, p. *193*; *14.3.4*, p. *199*; 22.4.3, p. 386
S. A. Kuzmanova	(b) **74**, K173 (1976)	*18*, p. *223*

H. Kuzmany	(a) **34**, 347 (1976)	*7*, p. *65*; *14.3*, p. *174*; 22.4.1, p. 379
V. G. Kvachadze	(b) **77**, K21 (1976)	*6*, p. *60*; *8*, p. *70*; 11, p. 103; 20.1, p. 273; 22.5.2, p. 396
Z. V. Kvinikadze	(a) **38**, K49 (1976)	*8*, p. *71*; *15*, p. *211*; 22.2.4, p. 371
Kwok Leung Yip	(b) **75**, 533 (1976)	*13.1*, p. *124*; 20, p. 258; 22.3, p. 373
R. Lacroix	(a) **35**, K105 (1976)	1.2, p. 17; *19*, p. *252*; 22.8, p. 416
C. Laird	(a) **38**, 315 (1976)	*10.2*, p. *98*; *20*, p. *259*; 22.2.1, p. 364; 22.2.2, p. 365
B. Lalevic	(a) **37**, K177 (1976)	*14.3.1*, p. *185*; *14.3.3*, p. *193*; 22.6, p. 407
Nghi Q. Lam	(a) **35**, K13 (1976)	*9*, p. *73*; *11*, p. *101*; 21.1, p. 307
A. P. Lambros	(b) **76**, 727 (1976)	6, p. 59; *20.1*, p. *272*; 22.7, p. 411
L. G. M. M. Lammers	(a) **35**, K45 (1976)	*18.2*, p. *232*; 21.1.1, p. 318
A. I. Landau	(a) **37**, 709 (1976)	*10.1*, p. *84*; 21.1, p. 312
J. van Landuyt	(a) **33**, 595 (1976)	*1.4*, p. *25*; *10.2*, p. *89*; 22.6, p. 403
	(a) **33**, 723 (1976)	*1.2*, p. *14*; *1.4*, p. *26*; 22.6, p. 403; 22.8.1, p. 419
	(a) **36**, 757 (1976)	*1.1*, p. *12*; *1.2*, p. *18*; 22, p. 339
	(a) **37**, 591 (1976)	*1.2*, p. *19*; 1.4, p. 28; 22, p. 340
	(a) **38**, 185 (1976)	*1.2*, p. *19*; 1.4, p. 29; 22, p. 341
T. G. Langdon	(a) **33**, 375 (1976)	*10.1*, p. *78*; 21, p. 290; 21.1, p. 303
H. Lange	(b) **73**, 633 (1976)	13.1, p. 121; *20.1*, p. *262*; 22.2.4, p. 370
M. Lannoo	(b) **74**, 409 (1976)	*10.1*, p. *79*; *13.4*, p. *136*; 21.6, p. 329
	(b) **74**, K141 (1976)	*13.3*, p. *132*; 21, p. 293
	(a) **35**, 499 (1976)	*1.3*, p. *22*; 10.1, p. 82; 21, p. 295; 21.6, p. 330
D. Lapraz	(a) **36**, 603 (1976)	13.4, p. 141; *20.1*, p. *272*; *20.3*, p. *285*; 22.8.1, p. 423
A. C. McLaren see under letter "M"		
C. Larica	(b) **77**, K11 (1976)	*19*, p. *254*; 22.7, p. 412
P. K. Larsen	(a) **34**, 665 (1976)	*10.2*, p. *91*; 20.1, p. 266; 22.8.2, p. 428
B. M. Latta	(b) **74**, 711 (1976)	*11*, p. *101*; 21, p. 292
A. Lau	(a) **36**, K5 (1976)	*6*, p. *58*; 13.5, p. 147; 20.1, p. 271; 20.2, p. 279; 22.8.1, p. 422
J. Lauckner	(a) **36**, K9 (1976)	*3*, p. *43*; 22.1.3, p. 359
J. W. Laughner	(a) **37**, 235 (1976)	1.4, p. 28; *10.2*, p. *95*; *12.2*, p. *112*; 22.6, p. 407
J. Laugsch	(b) **78**, 211 (1976)	6, p. 62; 13.5.1, p. 155; *20.1*, p. *275*; 22.8.1, p. 424
F. F. Lavrentev	(a) **33**, 289 (1976)	*10.1*, p. *78*; 21.1, p. 303
V. N. Lazukin	(b) **73**, 65 (1976)	2, p. 35; 18, p. 222; *19*, p. *247*; 22.8, p. 413
A. A. Lebedev	(a) **36**, K51 (1976)	*16*, p. *215*; 22.8, p. 416
A. F. Lebedev	(b) **73**, 317 (1976)	*18.4*, p. *242*; *19*, p. *247*; 22.8.2, p. 426
Yu. G. Lebedev	(a) **33**, 325 (1976)	*5*, p. *48*; 18.2, p. 229; 21.1.1, p. 316
E. M. Ledovskaya	(b) **73**, 495 (1976)	6, p. 54; *10.2*, p. *89*; 22.5.2, p. 393
	(b) **76**, 377 (1976)	*13.1*, p. *125*; *13.4*, p. *140*; 22.5.2, p. 395
A. E. Lee	(b) **76**, 661 (1976)	*13.1*, p. *126*; 13.3, p. 133; 20, p. 258; 22.1, p. 344
E. W. Lee	(a) **33**, 483 (1976)	*18.2*, p. *229*; 21.1, p. 303; 21.4, p. 326
	(a) **34**, 383 (1976)	*18.2*, p. *230*; 21.1, p. 305; 21.4, p. 326
P. M. Lee	(b) **76**, 599 (1976)	*13.1*, p. *126*; 14.2, p. 170
S. Leelaprute	(a) **33**, K35 (1976)	*17*, p. *218*; 22.8.1, p. 419
F. Lefaucheux	(a) **35**, 315 (1976)	10.2, p. 92; *20.3*, p. *283*; 22.8.1, p. 421
A. Lefebrve	(a) **34**, 125 (1976)	*1.2*, p. *15*; *1.4*, p. *26*; 22.8.2, p. 427
M. Lefeld-Sosnowska	(a) **35**, K1 (1976)	*1.4*, p. *27*; 10.2, p. 92; 22.1.2, p. 354
	(a) **38**, 139 (1976)	*1.2*, p. *19*; *11*, p. *104*; 22.1.2, p. 357
S. Lefrant	(b) **78**, 779 (1976)	6, p. 64; *10.2*, p. *98*; 13.4, p. 145; 20.1, p. 278; 22.5.2, p. 398
J. M. Leger	(b) **76**, K27 (1976)	*14.3*, p. *178*; 22, p. 338

G. LEHMANN	(a) **34**, 467 (1976)	*10.2*, p. *91*; 20.1, p. 266; 22.6, p. 404; 22.8, p. 414
	(a) **36**, 471 (1976)	*10.2*, p. *94*; 16, p. 215; 20.1, p. 272; 20.3, p. 285; 22.8.1, p. 423
	(a) **37**, K13 (1976)	*3*, p. *43*; 14.3, p. 179; 22, p. 339
	(b) 77, 607 (1976)	*13.4*, p. *142*; 21, p. 298
	(b) 77, K161 (1976)	*13.4*, p. *143*; *14.1*, p. *167*; 21.1, p. 312
	(b) 78, 79 (1976)	13.4, p. 144; *14.3*, p. *180*; *14.3.1*, p. *185*; 22.9, p. 438
	(b) 78, 159 (1976)	8, p. 71; *13.4*, p. *144*; 14.1, p. 167; 14.2, p. 172; 15, p. 211; 21, p. 299
A. YU. LEIDERMAN	(a) **34**, 391 (1976)	13.4, p. 137; *14.3.4*, p. *196*; 22.1.2, p. 352
A. M. LEJUS	(b) **73**, 593 (1976)	*19*, p. *248*; 22.6, p. 402
A. V. LELYAKOW	(b) 77, 59 (1976)	6, p. 60; 13.1, p. 127; *13.5.1*, p. *154*; 20.3, p. 285; 22.5.1, p. 391
M. LEMERCIER	(a) **37**, 521 (1976)	10.2, p. 96; *20*, p. *258*; 22.4.2, p. 383
L. F. LEMMENS	(b) **74**, 45 (1976)	*13*, p. *115*; *14.4*, p. *202*
D. LEMOINE	(a) **34**, 307 (1976)	*14.3*, p. *174*; 22.2.1, p. 362
M. LENGLET	(a) **35**, K141 (1976)	*18.3*, p. *240*; 22.8.2, p. 429
R. LENK	(b) 77, 601 (1976)	*2*, p. *40*; 11, p. 103
G. LEONHARDT	(b) 78, 681 (1976)	*13.1*, p. *129*; 17, p. 220; 20, p. 259; 22.1, p. 344; 22.1.1, p. 349; 22.1.2, p. 358
M. ŁEPEK	(a) **37**, K101 (1976)	*14.3.1*, p. *185*; 22.4.1, p. 381
T. P. LEPÈTRE	(a) **33**, 715 (1976)	*7*, p. *65*; *14.3.4*, p. *196*; 22.2.3, p. 367
B. LEPLEY	(a) **33**, 517 (1976)	*14.3.2*, p. *187*; *16*, p. *213*; 22.4.1, p. 379
M. LEPPIHALME	(a) **33**, 125 (1976)	6, p. 53; 13.5.1, p. 149; *14.3.3*, p. *190*; *16*, p. *213*, 20.1, p. 261; 22.1.2, p. 351
D. LEPSKI	(a) **36**, K17 (1976)	*10.1*, p. *83*; 21.1.1, p. 320
R. LERMER	(a) **33**, 531 (1976)	*18.2*, p. *229*; 21.1, p. 304; 21.1.1, p. 317
V. E. LESNIKOVSKAYA	(a) **34**, 85 (1976)	*2*, p. *36*; *14.3.1*, p. *183*; 14.3.3, p. 190; 22, p. 335
E. LESSMANN	(b) 78, 689 (1976)	*9*, p. *74*; 10.1, p. 86; 21.1, p. 314
D. LESUEUR	(a) **36**, 247 (1976)	2, p. 39; *9*, p. *73*; 21, p. 296
	(a) **36**, 729 (1976)	*11*, p. *103*; 14.1, p. 165; 21.7, p. 332
W. LESZCZYŃSKI	(a) **37**, 669 (1976)	*19*, p. *255*; 21.1. p. 312
KWOK LEUNG YIP see under letter "K"		
W. LEUPERT	(a) **38**, K57 (1976)	*14.3.3*, p. *194*; 22, p. 341
A. P. LEVANYUK	(a) **35**, 605 (1976)	13.4, p. 139; *14.4.1*, p. *207*; *20.1*, p. *269*; 22.8.1, p. 422
R. LEVY	(b) 77, 381 (1976)	*13.5.1*, p. *155*; *20.3*, p. *286*; 22.5, p. 390
P. LEYRAL	(b) **73**, 187 (1976)	*13.1*, p. *120*; *20.1*, p. *260*; 22.2.1, p. 362
L. LI	(a) **33**, 737 (1976)	20.1, p. 263; *20.2*, p. *279*; *20.3*, p. *281*; 22.8, p. 413
S. S. LI	(a) **34**, 255 (1976)	*13.1*, p. *122*; *20.1*, p. *264*; 22.2.4, p. 370
Y. LI	(a) **38**, 171 (1976)	*12.1*, p. *109*; 21, p. 299; 21.3, p. 324
H. LIEBING	(b) **75**, 471 (1976)	*13.5.1*, p. *152*; *20.3*, p. *283*; 22.4.3, p. 385
R. LIEBMANN	(b) **77**, 373 (1976)	*13*, p. *118*
S. LIGENZA	(b) **75**, 315 (1976)	5, p. 49; *13.4*, p. *138*; 18.3, p. 240; 22.8.2, p. 428
	(b) 78, 301 (1976)	5, p. 50; *18.4*, p. *245*; 22.8.2, p. 431
P. LIGOR	(a) **37**, 547 (1976)	*10.2*, p. *96*; *11*, p. *104*; 20.1, p. 275; 22.5.2, p. 397
A. N. LIKHOLET	(a) **33**, K161 (1976)	*14.3.2*, p. *187*; *14.3.3*, p. *190*
	(a) **34**, K71 (1976)	*13.4*, p. *137*; *14.3.3*, p. *191*
Z. LILIENTAL	(a) **34**, K5 (1976)	*10.2*, p. *90*; 22.4.3, p. 385
E. LILLEY	(a) **33**, 229 (1976)	*10.2*, p. *87*; 22.5.2, p. 392
S. K. LILOV	(a) **37**, 143 (1976)	*1.5*, p. *32*; 22.3, p. 374
W. W. LIN	(a) **37**, 617 (1976)	*10.2*, p. *96*; 13.4, p. 143; 20.1, p. 275; 22.5.2, p. 397

R. Lindsey	(b) **75**, 583 (1976)	*5*, p. *49*; *9*, p. *73*; 21.1.1, p. 319
R. Link	(a) **37**, K13 (1976)	*3*, p. *43*; 14.3, p. 179; 22, p. 339
R. Linnebach	(b) **73**, 255 (1976)	13.4, p. 134; *13.5.1*, p. *149*; 20.1, p. 260; *20.3*, p. *280*; 22.2.3, p. 367
H. G. Lipson	(a) **37**, 547 (1976)	*10.2*, p. *96*; *11*, p. *104*; 20.1, p. 275; 22.5.2, p. 397
M. P. Lisitsa	(b) **75**, K51 (1976)	13.5.1, p. 152; *20.1*, p. *267*; 22, p. 336
N. M. Litovcheñko	(a) **33**, K87 (1976)	*13.4*, p. *136*; 16, p. 213; 22.1.2, p. 352
	(a) **35**, K157 (1976)	*13.4*, p. *140*; 16, p. 214; 22.1.2, p. 354
V. G. Litovchenko	(a) **34**, K35 (1976)	11, p. 101; *13.3*, p. *132*; *14.3.3*, p. *191*; 22.1.2, p. 353
	(b) **77**, K71 (1976)	*13.5.1*, p. *155*; *14.2*, p. *171*
M. A. Litovskii	(a) **36**, K145 (1976)	*9*, p. *73*
D. B. Litvin	(b) **76**, 343 (1976)	1.2, p. 17
R. O. Litvinov	(a) **34**, K35 (1976)	11, p. 101; *13.3*, p. *132*; *14.3.3*, p. *191*; 22.1.2, p. 353
V. I. Litvinov	(b) **77**, 71 (1976)	*14*, p. *162*; *18*, p. *224*; 21, p. 297
E. Litwin-Staszewska	(b) **74**, K89 (1976)	*14.3*, p. *175*; 15, p. 209; 22.4.2, p. 383
J. Livage	(b) **73**, 593 (1976)	*19*, p. *248*; 22.6, p. 402
F. P. Livi	(a) **37**, 133 (1976)	*5*, p. *50*; *18.2*, p. *235*; 21.1.1, p. 320
N. N. Lobanov	(a) **38**, K45 (1976)	*5*, p. *51*; 22.1, p. 344
K. Lochner	(b) **76**, 533 (1976)	*16*, p. *215*; 22.9, p. 437
H. Löffler	(a) **33**, 345 (1976)	*1.3*, p. *21*; *10.1*, p. *78*; 14.1, p. 164; 21, p. 290; 21.1, p. 303
	(a) **36**, 127 (1976)	*1.3*, p. *23*; *10.1*, p. *83*; 14.1, p. 165; 21, p. 296; 21.1, p. 309
	(a) **37**, K85 (1976)	*10.1*, p. *84*; 21, p. 298; 21.1, p. 311
A. Loidl	(b) **76**, 581 (1976)	*6*, p. *59*; 7, p. 66; 8, p. 69; *12.2*, p. *111*; 22.5.2, p. 395
M. P. Lokshin	(a) **38**, 131 (1976)	11, p. 104; *14.3.3*, p. *194*; 22.1.2, p. 357
V. M. Loktev	(b) **73**, 415 (1976)	*13.5.1*, p. *149*; *20.1*, p. *261*; 23, p. 440
G. A. Lomakina	(a) **35**, 37 (1976)	3, p. 42; *13.4*, p. *138*; 14.3, p. 176; 20.1, p. 268; 20.3, p. 283; 22.3, p. 372
V. A. Lomonov	(a) **33**, 737 (1976)	20.1, p. 263; *20.2*, p. *279*; *20.3*, p. *281*; 22.8 p. 413
G. Lončar	(a) **35**, 291 (1976)	*1.5*, p. *31*; *3*, p. *42*; 22.6, p. 404
K. Loof	(a) **37**, 467 (1976)	*14.3*, p. *180*; 22.7, p. 412
P. Loose	(b) **73**, 265 (1976)	*13.4*, p. *134*; *13.5.1*, p. *149*; 22.4, p. 376
	(b) **75**, 137 (1976)	*13.5.1*, p. *152*; *20.1*, p. *267*; 20.3, p. 282; 22.4, p. 376
A. Lopez-Otero	(b) **76**, 259 (1976)	13.1, p. 125; *18*, p. *223*; *20.1*, p *270*; 22, p. 337
F. J. López	(b) **78**, K81 (1976)	10.2, p. 98; *19*, p. *256*; 22.5.2, p. 398
J. Lopitaux	(a) **35**, K141 (1976)	*18.3*, p. *240*; 22.8.2, p. 429
B. Lorenz	(b) **77**, K177 (1976)	*14.3*, p. *180*; 22.6, p. 407
K. Lösch	(a) **33**, 473 (1976)	13.1, p. 122; 13.5.1, p. 150; *20.1*, p. *262*; 22.2.2, p. 365
W. Löser	(b) **73**, 179 (1976)	*2*, p. *35*; 13, p. 113; 14, p. 162; *18*, p. *222*
	(b) **74**, K43 (1976)	*8*, p. *67*; *18.4*, p. *243*
D. R. Lovett	(a) **34**, K191 (1976)	*1.6*, p. *34*; 22.3, p. 372
R. N. Lovyagin	(a) **37**, 341 (1976)	*1.5*, p. *32*; 3, p. 43; 22.1.1, p. 348; 22.1.2, p. 356
J. E. Lowther	(b) **77**, 359 (1976)	*2*, p. *39*; 6, p. 60; 13.4, p. 142; 21.4, p. 327
A. F. Lubchenko	(b) **78**, K97 (1976)	*9*, p. *74*; 22, p. 342
V. S. Lubimov	(b) **75**, 91 (1976)	*13.5*, p. *147*; 22, p. 336
	(b) **76**, 505 (1976)	*8*, p. *69*; *13*, p. *117*
	(b) **77**, 441 (1976)	*13*, p. *119*
K. Lubitz	(a) **38**, 261 (1976)	*10.1*, p. *85*; 18.2, p. 236; 21.1.1, p. 321
P. Lucasson	(a) **34**, 513 (1976)	*10.1*, p. *80*; 21, p. 293

K. Lüders	(b) 78, K107 (1976)	*14.2*, p. *172*; *19*, p. *256*; 21, p. 300
T. Łukasiewicz	(a) **37**, K75 (1976)	*20.1*, p. *273*; 22.8, p. 417
K. Łukaszewicz	(a) **37**, 287 (1976)	*1.2*, p. *18*; *8*, p. *70*; 14.4.1, p. 208; 22.8, p. 417
F. Lukeš	(b) 77, 731 (1976)	13.1, p. 127; *14.3.4*, p. *199*; *20.1*, p. *274*; 22.1.1, p. 348
T. Lukes	(b) 78, 563 (1976)	*2*, p. *41*; 13.4, p. 145; *14*, p. *163*
A. I. Lukomskii	(b) 73, K135 (1976)	*13.1*, p. *121*; *20.1*, p. *262*; 22.2.4, p. 370
	(a) **37**, K173 (1976)	*13.1*, p. *128*; 20.1, p. 275; 22.2.4, p. 371
Y. Luspin	(b) 76, 551 (1976)	*1.2*, p. *18*; 7, p. 65; 22.8.1, p. 423
M. J. Luton	(a) **33**, 173 (1976)	*10.1*, p. *78*; 21, p. 289; 21.1, p. 302; 21.3, p. 324
D. L. Lyfar	(b) 76, 183 (1976)	*19*, p. *252*; 22.5.3, p. 400
K. G. Lynn	(b) 77, 277 (1976)	*13*, p. *118*; *14.3*, p. *179*; 22, p. 339
V. S. Lysenko	(a) **38**, 131 (1976)	11, p. 104; *14.3.3*, p. *194*; 22.1.2, p. 357
A. A. Lysykh	(a) **35**, 49 (1976)	*14.3.3*, p. *191*; 22.6, p. 404
S. K. S. Ma	(b) 76, 457 (1976)	6, p. 58; *8*, p. *69*
	(b) 78, 219 (1976)	6, p. 62; *8*, p. *71*
P. J. P. de Maayer	(a) **34**, K143 (1976)	*10.2*, p. *91*; *15*, p. *209*; 22, p. 336
R. M. Mac Farlane	(a) **34**, K13 (1976)	*20.1*, p. *265*; *20.3*, p. *281*; 22.8, p. 414; 22.8.1, p. 420
B. S. Machmudov	(a) **35**, 37 (1976)	3, p. 42; *13.4*, p. *138*; 14.3, p. *176*; 20.1, p. 268; 20.3, p. 283; 22.3, p. 372
M. Maciaszek	(a) **36**, 209 (1976)	*1.1*, p. *12*; 10.2, p. 93; 22.1.2, p. 355
W. Maciejewski	(b) 74, 235 (1976)	*18.2.1*, p. *237*
	(b) 76, K19 (1976)	*18.2.1*, p. *237*
J. D. Mackenzie	(a) **34**, K143 (1976)	*10.2*, p. *91*; *15*, p. *209*; 22, p. 336
W. S. MacKenzie	(a) **33**, 491 (1976)	*1*, p. *9*; *1.4*, p. *25*; 22.6.1, p. 409
A. G. Maddock	(b) 74, 183 (1976)	*5*, p. *48*; 22.5.2, p. 393; 22.5.3, p. 399
S. Mader	(a) **33**, 793 (1976)	*10.2*, p. *89*; 11, p. 100; 22.1.2, p. 352
W. Maenhout-van der Vorst	(a) **37**, 473 (1976)	*10.2*, p. *96*; *14.4*, p. *204*; 22.5.1, p. 391
R. G. Maev	(a) **38**, 77 (1976)	*13.4*, p. *144*; 16, p. 217; 20.1, p. 276; 22.4.1, p. 382
B. Maffeo	(b) 75, 271 (1976)	13.4, p. 138; *19*, p. *250*; 22.5.3, p. 399
	(b) 76, K137 (1976)	*13.4*, p. *141*; *19*, p. *253*; 22.5.3, p. 400
I. N. Magden	(a) **37**, 29 (1976)	10.2, p. 95; *14.3*, p. *179*; 22.1.2, p. 356
	(a) **38**, 45 (1976)	10.2, p. 97; *14.3*, p. *180*; 22.1.2, p. 357
T. J. Magee	(a) **33**, 415 (1976)	*10.2*, p. *88*; *11*, p. *100*; 20.1, p. 261; 22.5.3, p. 399
A. Magerl	(a) **36**, 161 (1976)	*12.1*, p. *108*; 21, p. 296; 21.1, p. 309
H. G. Maguire	(b) 76, 715 (1976)	*13.1*, p. *126*; 17, p. 219; 22.1, p. 344
C. Mai	(a) **36**, K113 (1976)	*10.2*, p. *95*; 22.1.2, p. 356
H. Maier	(b) 78, K65 (1976)	*6*, p. *63*; *13.4*, p. *144*; 20.1, p. 276; 20.3, p. 287; 22.4.1, p. 382
K. Maier	(b) 78, 689 (1976)	*9*, p. *74*; 10.1, p. 86; 21.1, p. 314
R. P. Mainard	(a) **35**, 459 (1976)	8, p. 68; *18.4*, p. *244*; 22.6, p. 405
E. Majchrowska	(b) 77, K91 (1976)	*13.1*, p. *128*; 22.4.4, p. 387
E. F. Makarov	(a) **33**, 783 (1976)	1.2, p. 15; *5*, p. *48*; 22.8.2, p. 427
	(b) 75, 91 (1976)	*13.5*, p. *147*; 22, p. 336
	(b) 76, 505 (1976)	*8*, p. *69*; *13*, p. *117*
	(b) 77, 441 (1976)	*13*, p. *119*
A. Mąkosa	(a) **37**, K57 (1976)	*10.2*, p. *95*; *16*, p. *216*; 19, p. 254; 22.1.2, p. 356
L. A. Makovetskaya	(a) **36**, K141 (1976)	*3*, p. *43*; 22.2.4, p. 371
S. K. Maksimov	(a) **33**, 755 (1976)	*10.2*, p. *89*; 22.2.1, p. 362
A. V. Malakhovskii	(b) 74, K145 (1976)	*18.2*, p. *231*; *20.1*, p. *266*; 22.8.1, p. 421
	(b) 75, K75 (1976)	*18.2*, p. *232*

L. K. Malhotra	(a) **37**, 313 (1976)	*1.2*, p. *18*; 1.5, p. 32; *2*, p. *40*; 22.1.1, p. 348
J. Malinowski	(a) **38**, 161 (1976)	10.2, p. 97; *14.4*, p. *204*; 22.5.1, p. 391
R. Sh. Malkovich	(a) **36**, K145 (1976)	*9*, p. *73*
O. Mallem	(a) **36**, 227 (1976)	13.3, p. 133; *14.3.3*, p. *192*; *16*, p. *215*; 22, p. 338; 22.4.1, p. 381
H. Malter	(b) **74**, 627 (1976)	*13.1*, p. *123*; *20.1*, p. *265*; 22.9, p. 435
	(b) **76**, 559 (1976)	*13.1*, p. *126*; 17, p. 219; 22.9, p. 437
G. Mambriani	(a) **34**, 95 (1976)	*10.2*, p. *90*; 22.1, p. 343; 22.1.1, p. 346; 22.1.2, p. 352
	(a) **34**, 455 (1976)	*10.2*, p. *91*; 22.1, p. 343; 22.1.1, p. 346; 22.1.2, p. 353
S. D. Mamchich	(a) **38**, K5 (1976)	*10.1*, p. *85*; 21, p. 300; 21.1, p. 314
Sh. S. Mamedov	(b) **73**, K69 (1976)	13.5.1, p. 149; *20.1*, p. *261*; 22.8, p. 413
T. N. Mamontova	(a) **37**, 45 (1976)	2, p. 40; *16*, p. *216*; *20.3*, p. *286*; 22, p. 339
L. A. Manakova	(b) **75**, 91 (1976)	*13.5*, p. *147*; 22, p. 336
	(b) **76**, 505 (1976)	*8*, p. *69*; *13*, p. *117*
C. Mande	(b) **75**, 79 (1976)	*4*, p. *45*; 20, p. 258; 22, p. 336
C. Manfredotti	(a) **33**, 563 (1976)	*1.1*, p. *11*; *3*, p. *42*; 22.3.1, p. 375
	(a) **36**, 569 (1976)	*14.3.4*, p. *198*; 22.3, p. 373
	(a) **38**, 685 (1976)	13.4, p. 145; 14.3, p. 182; *16*, p. *217*; 22.3, p. 374
R. Manns	(a) **36**, K93 (1976)	*11*, p. *102*; 21, p. 296; 21.1, p. 310; 21.1.1, p. 320; 22.1.1, p. 348; 22.1.2, p. 355; 22.6, p. 405
	(a) **38**, K155 (1976)	*11*, p. *105*
C. Manohar	(b) **73**, 735 (1976)	*13*, p. *114*; *18*, p. *222*; 21.1, p. 303
C. Manolikas	(a) **33**, K77 (1976)	*1.4*, p. *26*; 22, p. 335
F. Mansfield	(b) **74**, K9 (1976)	*13.4*, p. *136*; *13.5.1*, p. *151*; 13.5.3, p. 159; 20.3, p. 281; 22.4.3, p. 385
A. Mansingh	(a) **38**, K1 (1976)	*2*, p. *41*; *14.3.3*, p. *194*; 22.8.1, p. 424
L. K. Mansur	(a) **37**, 211 (1976)	*10*, p. *76*
S. Mantovani	(a) **35**, 451 (1976)	*13.4*, p. *139*; 14.3, p. 177; 22.1.1, p. 347
	(a) **38**, 109 (1976)	*10.2*, p. *97*; *13.1*, p. *128*; 14.3.4, p. 200; 22.1.1, p. 349
M. Manzel	(a) **34**, 219 (1976)	13.4, p. 137; 14.3.3, p. 190; *14.3.4*, p. *196*; 22.1.2, p. 352
	(a) **37**, 479 (1976)	13.4, p. 143; *14.3.4*, p. *199*; 22.1.2, p. 356
V. G. Manzhelii	(b) **73**, K111 (1976)	*8*, p. *67*; 23, p. 440
L. A. Marasina	(a) **34**, K195 (1976)	*20.3*, p. *282*; 22.2, p. 361
	(a) **38**, 753 (1976)	*20.3*, p. *288*; 22.2, p. 361
M. J. Marcinkowski	(a) **38**, 223 (1976)	*1.4*, p. *29*; 10, p. 76
L. Mărculeşcu	(b) **75**, 375 (1976)	*20.3*, p. *283*; 22.5.2, p. 394
	(b) **77**, 353 (1976)	*10.2*, p. *95*; *20.1*, p. *273*; 22.5.2, p. 396
	(b) **77**, 701 (1976)	*10.2*, p. *95*; 20.1, p. 274; 20.3, p. 287; 22.5.2, p. 397
S. I. Marennikov	(a) **38**, K139 (1976)	*14.4.1*, p. *208*; *16*, p. *217*; 22.8.1, p. 425
J. Marien	(a) **38**, 339 (1976)	*17*, p. *220*; 21, p. 300; 22.4, p. 377
	(a) **38**, 513 (1976)	*17*, p. *220*; 22.4, p. 378
M. M. Marinković	(b) **76**, K85 (1976)	*13.5.4*, p. *161*; *18.2*, p. *234*
K. Marinova	(a) **33**, 195 (1976)	13.5.1, p. 149; *16*, p. *213*; 22.4.2, p. 383; 22.4.4, p. 387
	(b) **75**, 471 (1976)	*13.5.1*, p. *152*; *20.3*, p. *283*; 22.4.3, p. 385
	(b) **77**, K35 (1976)	*13.5.1*, p. *155*; *20.3*, p. *286*; 22.4.3, p. 386
S. V. Marisova	(b) **74**, 453 (1976)	*13.5.1*, p. *151*; *20.3*, p. *282*; 22.9, p. 435
I. A. Markova-Osorgina	(b) **74**, 753 (1976)	10.2, p. 90; 13.5.1, p. 152; *20.3*, p. *282*; 22.5.2, p. 394
T. Markvart	(b) **73**, 689 (1976)	*13.4*, p. *135*; *13.5.1*, p. *150*; 22.9, p. 433
	(b) **74**, 135 (1976)	*13.4*, p. *136*; *13.5.1*, p. *151*; 22.9, p. 434
	(b) **76**, 853 (1976)	*13.4*, p. *136*; *13.5.1*, p. *151*; 22.9, p. 434

Y. Marqueton	(b) **78**, 767 (1976)	*6*, p. *64*; *20.1*, p. *278*; 22.6, p. 408
R. Marquez	(a) **33**, K25 (1976)	*10.2*, p. *88*; 22.6, p. 402
E. Marsch	(b) **78**, K39 (1976)	13, p. 119; *18.2*, p. *236*; *18.4*, p. *245*
A. H. Marshak	(b) **78**, 501 (1976)	13.1, p. 129; *14.1*, p. *167*; *14.3*, p. *181*; 15, p. 211; 21, p. 300; 22, p. 341
R. Martens	(a) **37**, 97 (1976)	*6*, p. *61*; *20.1*, p. *273*; 22.8, p. 417
N. Mårtensson	(b) **75**, 197 (1976)	*13.1*, p. *123*; 17, p. 219; 21, p. 293
C. Marti	(a) **35**, 315 (1976)	10.2, p. 92; *20.3*, p. *283*; 22.8.1, p. 421
J. J. Martin	(a) **37**, 547 (1976)	*10.2*, p. *96*; *11*, p. *104*; 20.1, p. 275; 22.5.2, p. 397
J. W. Martin	(b) **74**, 761 (1976)	6, p. 55; *10*, p. *75*; 12, p. 106
S. Martinuzzi	(a) **34**, K21 (1976)	*14.3.3*, p. *190*; *16*, p. *214*; 22.4.1, p. 379; 22.4.3, p. 385
	(a) **36**, 227 (1976)	13.3, p. 133; *14.3.3*, p. *192*; *16*, p. *215*; 22, p. 338; 22.4.1, p. 381
Cz. Marucha	(a) **37**, K5 (1976)	*8*, p. *70*; 22.2.1, p. 364
E. Maruyama	(b) **73**, K19 (1976)	2, p. 35; *19*, p. *247*; *20.1*, p. *261*; 22, p. 333
M. Maryško	(a) **33**, K133 (1976)	*18.3*, p. *239*; *19*, p. *248*; 22.8.2, p. 427
	(a) **34**, 639 (1976)	*18.3*, p. *240*; 22.8.2, p. 428
	(a) **38**, K163 (1976)	*18.3*, p. *241*; *19*, p. *256*; 22.8.2, p. 432
K. Maschke	(b) **76**, 629 (1976)	*13*, p. *117*; *14.4*, p. *203*; 22.5.2, p. 396; 22.6, p. 406
S. I. Masharov	(b) **73**, 401 (1976)	*6*, p. *53*; *8*, p. *67*; 21, p. 290
P. R. Mason	(a) **36**, 499 (1976)	*14.4*, p. *203*; 22.6, p. 406
T. A. McMath see under letter "Mc"		
R. Mathai	(b) **77**, 215 (1976)	*19*, p. *253*; 22.8.1, p. 423
J. P. Mathieu	(b) **75**, 163 (1976)	*5*, p. *49*; 22.9, p. 436
P. C. Mathur	(a) **36**, 269 (1976)	*14.3.3*, p. *192*; *14.3.4*, p. *198*; 22.2.1, p. 363
M. O. M. de Matos	(b) **75**, 271 (1976)	13.4, p. 138; *19*, p. *250*; 22.5.3, p. 399
T. Matsuda	(a) **36**, 791 (1976)	13.2, p. 131; *14.1*, p. *166*; 21, p. 297; 21.6, p. 331
H. Matsunami	(a) **36**, K129 (1976)	*20.3*, p. *285*; 22.3, p. 373
N. Matsunami	(b) **75**, 483 (1976)	10.2, p. 93; *11*, p. *101*; *20.1*, p. *268*; 22.5.2, p. 395
E. Matsushima	(a) **37**, 579 (1976)	*10.2*, p. *96*; *14.3*, p. *180*; 22.4.2, p. 383
E. Matsuura	(a) **38**, 243 (1976)	13.4, p. 144; *19*, p. *255*; 22.1.2, p. 357
K. Matsuura	(a) **35**, K57 (1976)	10.2, p. 92; *11*, p. *101*; *20.1*, p. *268*; 22.4.1, p. 380
N. Mattern	(a) **38**, 93 (1976)	*1.2*, p. *19*; *3*, p. *44*; 14.3, p. 181; 15, p. 211; 18.1, p. 227; 22.6,1, p. 409
J. W. Matthews	(a) **37**, 499 (1976)	*10.2*, p. *96*; 22.8.2, p. 430
	(a) **38**, 577 (1976)	*10.2*, p. *98*; 22.8.2, p. 431
S. Matthies	(b) **74**, 69 (1976)	*14*, p. *162*; 22, p. 335
	(b) **74**, 531 (1976)	*14*, p. *162*; 22, p. 335
P. Matthys	(a) **35**, K137 (1976)	*3*, p. *42*; *19*, p. *252*; 22.5.2, p. 395
	(b) **77**, 249 (1976)	*19*, p. *254*; 22.5.2, p. 396
	(b) **78**, 553 (1976)	*19*, p. *256*; *20.1*, p. *277*; 22.5.2, p. 398
A. Matulionis	(a) **35**, 43 (1976)	*13*, p. *116*; *14.3.4*, p. *197*; 22.2.1, p. 363; 22.4.3, p. 385
S. N. Matveev	(a) **36**, K1 (1976)	*18.2*, p. *233*; 22.8.2, p. 429
E. E. Matyas	(a) **36**, K141 (1976)	*3*, p. *43*; 22.2.4, p. 371
M. Matyáš	(a) **36**, K137 (1976)	*18.1*, p. *226*; *18.2*, p. *234*; 22.8, p. 417
F. Maury	(a) **34**, 513 (1976)	*10.1*, p. *80*; 21, p. 293
A. A. Mayer	(a) **33**, 737 (1976)	20.1, p. 263; *20.2*, p. *279*; *20.3*, p. *281*; 22.8, p. 413
J. W. Mayer	(a) **33**, 59 (1976)	13.5.1, p. 149; 14.3.4, p. 195; *20.3*, p. *280*; 22.1.2, p. 351
A. Mazel	(a) **35**, 197 (1976)	*1.1*, p. *11*; 22.6, p. 404

T. MAZETS	(a) **35**, K77 (1976)	*2*, p. *38*; *20.1*, p. *268*; 22.8, p. 415
R. MAZUMDAR	(b) **75**, 381 (1976)	*6*, p. *56*
	(b) **78**, 843 (1976)	*6*, p. *56*
J. MAZUR	(b) **76**, K1 (1976)	*14.2*, p. *170*
K. MAZUR	(a) **33**, K155 (1976)	*14.4*, p. *202*; 22.9, p. 434
E. MAZZEGA	(a) **35**, 451 (1976)	*13.4*, p. *139*; 14.3, p. 177; 22.1.1, p. 347
T. C. MCGILL	(a) **33**, 59 (1976)	13.5.1, p. 149; 14.3.4, p. 195; *20.3*, p. *280*; 22.1.2, p. 351
I. T. MCGOVERN	(b) **73**, 307 (1976)	*13.1*, p. *120*; *17*, p. *218*; 20.1, p. 260; 22.3, p. 372
A. C. MCLAREN	(a) **33**, 491 (1976)	*1*, p. *9*; *1.4*, p. *25*; 22.6.1, p. 409
T. A. MCMATH	(a) **38**, 731 (1976)	*20.1*, p. *278*; 22.3, p. 374
M. MEAUDRE	(a) **37**, 633 (1976)	*13.4*, p. *143*; *14.3.4*, p. *199*; 22.4, p. 377
R. MEAUDRE	(a) **37**, 633 (1976)	*13.4*, p. *143*; *14.3.4*, p. *199*; 22.4, p. 377
T. R. MECHTIEV	(b) **73**, K69 (1976)	13.5.1, p. 149; *20.1*, p. *261*; 22.8, p. 413
M. V. MEDVEDEV	(b) **76**, 45 (1976)	*18.1*, p. *226*; 18.2, p. 233; 18.4, p. 244; 21.1, p. 308; 21.1.1, p. 319
AH MEE HOR see under letter "A"		
P. MEHENDRU	(a) **38**, 355 (1976)	*14.3.1*, p. *185*; *14.4*, p. *204*; 22.9, p. 439
P. C. MEHENDRU	(b) **77**, K133 (1976)	*20.1*, p. *274*; 22.5.2, p. 397
	(a) **38**, 355 (1976)	*14.3.1*, p. *185*; *14.4*, p. *204*; 22.9, p. 439
H. MEHRER	(b) **73**, 131 (1976)	*9*, p. *72*; 10, p. 75
	(b) **75**, 361 (1976)	*9*, p. *72*; 10, p. 76
	(b) **78**, 689 (1976)	*9*, p. *74*; 10.1, p. 86; 21.1, p. 314
R. MEHROTRA	(b) **73**, K27 (1976)	*1.6*, p. *34*; *6*, p. *53*; 8, p. 67; 21.1.1, p. 316
V. P. MEIKLYAR	(b) **75**, 731 (1976)	*19*, p. *251*; 22.5.3, p. 399
	(b) **75**, K111 (1976)	*19*, p. *251*; 22.5.3, p. 400
	(b) **77**, K77 (1976)	*19*, p. *251*; 22.5.3, p. 400
B. L. MEILER	(a) **37**, 29 (1976)	10.2, p. 95; *14.3*, p. *179*; 22.1.2, p. 356
H. MEINHOLD	(b) **73**, 105 (1976)	13.1, p. 120; *13.5.1*, p. *149*; *20.1*, p. *260*; 22, p. 333
A. MELANDER	(a) **33**, 255 (1976)	*1.4*, p. *25*; 10.1, p. 78; 21, p. 290; 21.1, p. 303
K. P. MELETOV	(b) **78**, K1 (1976)	*6*, p. *63*; *13.4*, p. *144*; 20.1, p. 276; 20.3, p. 287; 22.9, p. 439
V. G. MELEV	(a) **35**, 425 (1976)	11, p. 102; 13.4, p. 139; *14.3*, p. *177*; *20.1*, p. *269*; 22.8, p. 415
R. J. DIMELFI see under letter "D"		
V. N. MELOV	(a) **36**, 415 (1976)	12.2, p. 111; *18.4*, p. *244*; 22.8.2, p. 429
H. H. MENDE	(b) **75**, K61 (1976)	*19*, p. *251*; 21.1, p. 307
E. A. MENESES	(b) **78**, K35 (1976)	*20.3*, p. *287*; 22.3.1, p. 375
H. MENNIGER	(a) **35**, 639 (1976)	*1.4*, p. *27*; 14.3, p. 177; 22.1.2, p. 354
	(a) **36**, 173 (1976)	*1.4*, p. *27*; *3*, p. *43*; 22.2.4, p. 370
F. MENZINGER	(b) **77**, 309 (1976)	*18.2*, p. *234*; 21.1, p. 311
T. MERCERON	(a) **35**, 681 (1976)	*18.3*, p. *240*; 22.8.2, p. 429
M. MERISALO	(b) **74**, 177 (1976)	*6*, p. *54*; 22, p. 335
A. MERKEL	(a) **37**, K205 (1976)	*18.2*, p. *235*; *19*, p. *255*; 21.1.1, p. 321
C. MERKEL	(b) **77**, 591 (1976)	*13.1*, p. *127*; 13.2, p. 131; 21.2, p. 323
R. MERKER	(a) **33**, K87 (1976)	*13.4*, p. *136*; 16, p. 213; 22.1.2, p. 352
	(a) **35**, K157 (1976)	*13.4*, p. *140*; 16, p. 214; 22.1.2, p. 354
P. MERLE	(a) **35**, 213 (1976)	*10.1*, p. *82*; 12.1, p. 108; 21, p. 294; 21.1, p. 307
P. G. MERLI	(a) **35**, 165 (1976)	*14.3.3*, p. *191*; 22.1.2, p. 353
J. MERLIN	(a) **35**, 213 (1976)	*10.1*, p. *82*; 12.1, p. 108; 21, p. 294; 21.1, p. 307
L. MERTEN	(b) **77**, 221 (1976)	*6*, p. *60*; *14.4.1*, p. *207*
J. MERTSCHING	(b) **77**, 717 (1976)	*13.1*, p. *127*; *18*, p. *224*
J. MESHKAUSKAS	(a) **37**, K39 (1976)	*14.4*, p. *204*; 22, p. 340
U. MESSERSCHMIDT	(a) **34**, 175 (1976)	*10.1*, p. *79*; 21, p. 292
	(a) **38**, 103 (1976)	*10.2*, p. *97*; 22.6, p. 408

R. Metselaar	(a) **34**, 665 (1976)	*10.2*, p. *91*; 20.1, p. 266; 22.8.2, p. 428
J. Metzdorf	(b) **75**, 567 (1976)	13.5.1, p. 153; *18*, p. *223*; *20*, p. *258*; 22.1.1, p. 347
A. Meyer-Laack	(a) **34**, 145 (1976)	*16*, p. *214*; 22.4, p. 376
L. Meyer	(b) **76**, K23 (1976)	*13.1*, p. *125*
	(b) **77**, K79 (1976)	*13.1*, p. *125*
E. Mgbenu	(a) **36**, 217 (1976)	1.5, p. 32; *14.3.1*, p. *184*; 22, p. 338
R. Michael	(a) **36**, K17 (1976)	*10.1*, p. *83*; 21.1.1, p. 320
A. E. Michel	(a) **33**, 793 (1976)	*10.2*, p. *89*; 11, p. 100; 22.1.2, p. 352
H. Michel	(b) **75**, 507 (1976)	*6*, p. *57*; 8, p. 68; *13.4*, p. *139*; 22.5.2, p. 395
M. E. Michel-Beyerle	(a) **37**, 73 (1976)	*14.3.4*, p. *199*; 16, p. 216; 20.3, p. 286; 22.9, p. 437
D. J. Mickish	(b) **73**, 193 (1976)	*13.4*, p. *134*; 22.5.1, p. 391
R. Micnas	(b) **73**, K35 (1976)	*13*, p. *114*; *18.2*, p. *228*
	(b) **76**, 851 (1976)	*18.4*, p. *244*
G. Micocci	(a) **36**, 569 (1976)	*14.3.4*, p. *198*; 22.3, p. 373
A. von Middendorf	(b) **78**, 637 (1976)	*13.1*, p. *129*; *13.2*, p. *131*; 14.3, p. 181; 22.7, p. 412
V. P. Migal	(a) **37**, K21 (1976)	*16*, p. *216*; 22.4.2, p. 383
M. Migitaka	(a) **35**, K115 (1976)	*14.3.3*, p. *192*; 22.1.2, p. 354
	(a) **38**, K41 (1976)	*14.3.3*, p. *194*; 22.1.2, p. 358
D. F. Migley	(b) **73**, K5 (1976)	*15*, p. *209*; 22.7, p. 410
I. D. Mikhailov	(b) **73**, K145 (1976)	*18.4*, p. *242*
M. D. Mikhailov	(a) **35**, 709 (1976)	*9*, p. *73*; 22.9, p. 436
	(a) **36**, 391 (1976)	*9*, p. *73*; 22.9, p. 437
E. Ya. Mikhlin	(a) **35**, K81 (1976)	*10.1*, p. *82*; 21.1.1, p. 318
	(a) **38**, K101 (1976)	*10.1*, p. *82*; 21.1.1, p. 318
A. Milchev	(b) **77**, 571 (1976)	*4*, p. *46*; 13, p. 119; 22.1.1, p. 348
N. V. Mileshkina	(a) **38**, 393 (1976)	*14.3.4*, p. *200*; 22.1.1, p. 349; 22.6, p. 408
A. Miller	(b) **78**, 569 (1976)	*6*, p. *63*; 20.1, p. 277; 22.8, p. 418
K. M. Miller	(b) **78**, 341 (1976)	*10.1*, p. *85*; 12.1, p. 109; 21, p. 299; 21.1, p. 313; 21.6, p. 331
N. D. J. Miller	(a) **37**, 83 (1976)	*18.3*, p. *241*; 22.8.2, p. 430
M. Milleville	(b) **73**, K127 (1976)	*7*, p. *65*; *14.3.4*, p. *195*; 22.1.3, p. 359
B. Mills	(a) **33**, 707 (1976)	*9*, p. *72*; *10.1*, p. *79*; 21.1, p. 304
J. Mimkes	(b) **78**, K137 (1976)	*9*, p. *74*; 10.2, p. 98; 22.1.2, p. 358
Chan Min Shon see under letter "C"		
R. Minakata	(a) **37**, K33 (1976)	*18.2*, p. *235*; 21.1, p. 311; 21.1.1, p. 321
Pham Minh Tan see under letter "P"		
S. S. Minh	(a) **35**, 431 (1976)	*14.4*, p. *203*; *16*, p. *214*; 22.6, p. 405
L. V. Mironova	(a) **38**, 369 (1976)	*1.5*, p. *33*; *14.3.2*, p. *188*; 22.1.2, p. 358; 22.6, p. 408
A. A. Mirtskhulava	(a) **37**, K71 (1976)	*3*, p. *43*; 22.2.4, p. 371
I. A. Mirtskhulava	(a) **37**, K71 (1976)	*3*, p. *43*; 22.2.4, p. 371
A. M. Mishchenko	(a) **38**, 349 (1976)	*16*, p. *217*; *20.1*, p. *277*; 22.1.1, p. 349
V. V. Mishenko	(a) **37**, K37 (1976)	13.4, p. 142; *19*, p. *254*; 22.8, p. 417
S. K. Misra	(b) **75**, 607 (1976)	*19*, p. *251*; 22.5, p. 389
F. P. Missell	(a) **35**, K135 (1976)	*14.2*, p. *170*; 21, p. 295
A. B. Mitin	(b) **77**, 451 (1976)	*14.2*, p. *171*; 21, p. 298
V. V. Mitin	(a) **35**, K65 (1976)	*14.3.1*, p. *184*; 22.1.1, p. 347
	(b) **76**, 407 (1976)	13.1, p. 125; *14.3*, p. *178*; 22.1.2, p. 355
V. Ya. Mitrofanov	(b) **74**, 57 (1976)	13.4, p. 136; *18.2*, p. *230*; 22.1, p. 343
A. Mitsuishi	(b) **73**, 341 (1976)	*6*, p. *53*; *20.1*, p. *260*; 22.5, p. 388
N. Miura	(b) **76**, 787 (1976)	*13.1*, p. *126*; 20.1, p. 272; 22.4.3, p. 386
L. A. Mizandari	(a) **35**, K131 (1976)	*17*, p. *219*; 22.1.1, p. 347
E. Mizera	(a) **34**, K107 (1976)	*10.2*, p. *91*; 22.4.3, p. 385
	(a) **35**, K9 (1976)	*10.2*, p. *92*; 22.1.2, p. 354

J. Mizia	(b) **74**, 461 (1976)	*13*, p. *115*; 18.1, p. 225; 18.2, p. 230; 21.1, p. 305
A. R. Mkrtchyan	(b) **73**, 681 (1976)	*5*, p. *48*
L. K. Mladov	(b) **75**, K51 (1976)	13.5.1, p. 152; *20.1*, p. *267*; 22, p. 336
M. Mocker	(b) **77**, 203 (1976)	*14.3.4*, p. *199*; *16*, p. *215*; 22, p. 339
K. P. Mogilnikov	(a) **38**, K57 (1976)	*14.3.3*, p. *194*; 22, p. 341
F. A. Mohamed	(a) **33**, 375 (1976)	*10.1*, p. *78*; 21, p. 290; 21.1, p. 303
W. Möhling	(a) **37**, K113 (1976)	*1.4*, p. *28*; 10.2, p. 97; 22.1.2, p. 357
E. N. Mokhov	(a) **35**, 37 (1976)	3, p. 42; *13.4*, p. *138*; 14.3, p. 176; 20.1, p. 268; 20.3, p. 283; 22.3, p. 372
L. P. Mokhracheva	(b) **78**, 465 (1976)	*13.1*, p. *129*; 22.3, p. 374
W. Moldenhauer	(a) **34**, 565 (1976)	*1.2*, p. *15*; *14.3*, p. *175*; 15, p. 209; 18.4, p. 243; 22, p. 336
	(a) **35**, K49 (1976)	*1.2*, p. *16*; 15, p. 210; 22.6, p. 404
	(a) **35**, K109 (1976)	*14.3.4*, p. *197*; 22, p. 337
	(a) **37**, 541 (1976)	*14.3*, p. *180*; 15, p. 211; 18.1, p. 227; 22.6, p. 407
	(a) **37**, K143 (1976)	*14.3.3*, p. *193*; 14.3.4, p. 200; 22.6, p. 407
	(a) **38**, 93 (1976)	*1.2*, p. *19*; *3*, p. *44*; 14.3, p. 181; 15, p. 211; 18.1, p. 227; 22.6.1, p. 409
	(a) **38**, K17 (1976)	1.2, p. 20; *15*, p. *211*; 22.6.1, p. 409
	(a) **38**, K23 (1976)	*14.3*, p. *181*; *15*, p. *211*; 22.6, p. 408; 22.6.1, p. 409
H.-J. Möller	(a) **33**, K59 (1976)	*10.2*, p. *88*; 22.1.1, p. 345
R. Molteni	(a) **37**, 399 (1976)	*20.3*, p. *287*; 22.8, p. 417
R. S. Momtaz	(a) **38**, 401 (1976)	*3*, p. *44*; 14.4.1, p. 208; 22.8.1, p. 424
D. A. Moncrieff	(a) **35**, 371 (1976)	*1.4*, p. *27*; *10.2*, p. *92*; 22.5.2, p. 394
G. Mondio	(b) **74**, 317 (1976)	*13.4*, p. *136*; *20.1*, p. *264*; 22.5.2, p. 393
J. Monecke	(b) **74**, K57 (1976)	*13*, p. *115*; *18.4*, p. *243*
	(b) **77**, 527 (1976)	*13.1*, p. *127*; 22.8, p. 417
R. Monnier	(b) **74**, 91 (1976)	6, p. 54; *13.4*, p. *136*; 20.1, p. 263; 22.5.2, p. 393
A. Montaner	(a) **38**, K159 (1976)	*20.1*, p. *278*; 22.9, p. 439
J. Montemayor	(a) **38**, 67 (1976)	*10.2*, p. *97*; 22.5.2, p. 397
K. S. Moodie	(a) **38**, 293 (1976)	*8*, p. *71*; 10.2, p. 98; 22.5.2, p. 398
I. Moore	(b) **73**, K31 (1976)	13, p. 114; *18*, p. *222*; *20*, p. *257*; 22, p. 333
B. L. Mordike	(a) **34**, K59 (1976)	*10.1*, p. *80*; 21, p. 292
	(a) **35**, 303 (1976)	*10.1*, p. *82*; 12.1, p. 108; 21, p. 294
	(a) **36**, K161 (1976)	*12.1*, p. *108*; 21.6, p. 331
R. Moret	(a) **38**, 695 (1976)	*1.2*, p. *20*; 1.3, p. 24; 22, p. 342
R. Morineau	(a) **38**, 559 (1976)	*10.2*, p. *98*; *18.3*, p. *241*; 22.8.2, p. 431
H. Morishima	(a) **33**, 633 (1976)	11, p. 100; *19*, p. *248*; 22.1.2, p. 352
J. W. Moroń	(a) **33**, 313 (1976)	*10.1*, p. *78*; *18.2*, p. *229*; 21.1.1, p. 316
B. V. Morozov	(b) **75**, 423 (1976)	13.1, p. 124; *14.3*, p. *177*; 22.1.2, p. 354
	(a) **35**, 755 (1976)	*14.3.1*, p. *184*; 22.2.3, p. 368
N. P. Morozov	(a) **37**, 57 (1976)	*11*, p. *103*; 22.1.2, p. 356
Yu. N. Morozov	(b) **73**, 455 (1976)	*18.2*, p. *229*; 22.8.2, p. 426
A. J. Morton	(a) **33**, 395 (1976)	1.3, p. 21; *1.4*, p. *25*; 21.1, p. 303
	(a) **35**, 777 (1976)	1.3, p. 23; *1.4*, p. *27*; 21.1, p. 308
H. Morwinski	(a) **34**, K147 (1976)	*1.1*, p. *11*; 22.2.4, p. 370
K. Mosebach	(b) **75**, K89 (1976)	*13.1*, p. *124*; 17, p. 219; 20, p. 258; 22, p. 337
L. Moseley	(b) **78**, 653 (1976)	*2*, p. *41*; 13.4, p. 145; *14*, p. *163*
P. Moser	(a) **36**, 579 (1976)	*10.1*, p. *83*; *18.2*, p. *234*; 21.1.1, p. 320
S. A. Moskalenko	(b) **76**, 497 (1976)	*13.5.1*, p. *154*; *20.3*, p. *285*; 22.4.1, p. 381
	(b) **76**, K51 (1976)	*13.5.1*, p. *153*; 22, p. 338
Y. Motohashi	(a) **37**, 263 (1976)	*10.2*, p. *95*; 12.2, p. 112; 22.8.1, p. 423
C. A. Moyer	(b) **74**, K23 (1976)	*18.4*, p. *243*; 21.1, p. 304
E. Mrosan	(b) **77**, 607 (1976)	*13.4*, p. *142*; 21, p. 298
	(b) **77**, K161 (1976)	*13.4*, p. *143*; *14.1*, p. *167*; 21.1, p. 312

	(b) 78, 159 (1976)	8, p. 71; *13.4*, p. *144*; 14.1, p. 167; 14.2, p. 172; 15, p. 211; 21, p. 299
B. Mrygoń	(b) 77, K57 (1976)	*18.2*, p. *234*
J. Mucha	(a) **37**, K5 (1976)	*8*, p. *70*; 22.2.1, p. 364
H. Mughrabi	(a) **38**, 261 (1976)	*10.1*, p. *85*; 18.2, p. 236; 21.1.1, p. 321
D. Mukherji	(b) 75, K35 (1976)	*13.1*, p. *124*; 22.2.1, p. 363
D. Mullen	(a) **35**, 627 (1976)	*1.2*, p. *17*; 4, p. 46; 22.9, p. 436
D. Müller	(b) 78, 191 (1976)	*14.4.1*, p. *208*; 19, p. 255; 22.9, p. 438
G. Müller	(b) 76, 525 (1976)	*13.4*, p. *141*; 14.3, p. 178; 20.1, p. 271; 22.4, p. 377
H. Müller	(b) 78, 177 (1976)	*9*, p. *74*; *10.1*, p. *85*; 21.1, p. 313; 21.1.1, p. 321
H.-R. Müller	(a) **36**, 617 (1976)	*18.3*, p. *241*; 22.8.2, p. 430
	(a) **38**, K53 (1976)	*18.2.1*, p. *238*; *20.1*, p. *277*; 21.1, p. 314; 21.4, p. 327
K.-H. Müller	(a) **34**, K199 (1976)	*14.3.4*, p. *197*; 22.6, p. 404
	(a) **37**, K201 (1976)	8, p. 71; *14.3.3*, p. *194*; 14.3.4, p. 200; 22, p. 340
P. Müller	(a) **33**, 543 (1976)	*14.3.1*, p. *183*; 22, p. 334
R. H. Müller	(b) 75, 567 (1976)	13.5.1, p. 153; *18*, p. *223*; *20*, p. *258*; 22.1.1, p. 347
W. Müller	(b) 75, 543 (1976)	*13.1*, p. *124*; *14.3.4*, p. *197*; 20.1, p. 268; 20.3, p. 283; 22.2.3, p. 368
U. Müller-Jahreis	(a) **33**, K129 (1976)	*11*, p. *100*; 21.6, p. 329
W. Müller-Lierheim	(a) **35**, K165 (1976)	*1.2*, p. *17*; 14.4.1, p. 207; 20.1, p. 270; 22.8.1, p. 422
	(b) 76, 301 (1976)	*1.2*, p. *17*; 6, p. 58; 20.3, p. 284; 23, p. 440
H. Müllner	(a) **34**, K31 (1976)	*10.1*, p. *80*; 12.1, p. 107; 21.1, p. 305
	(a) **38**, 199 (1976)	*10.1*, p. *85*; *12.1*, p. *109*; 21.1, p. 313
M. Müllner	(b) 76, 581 (1976)	*6*, p. *59*; 7, p. 66; 8, p. 69; *12.2*, p. *111*; 22.5.2, p. 395
G. M. de'Munari	(a) **34**, 95 (1976)	*10.2*, p. *90*; 22.1, p. 343; 22.1.1, p. 346; 22.1.2, p. 352
	(a) **34**, 455 (1976)	*10.2*, p. *91*; 22.1, p. 343; 22.1.1, p. 346; 22.1.2, p. 353
E. Münster	(a) **35**, 521 (1976)	*18.2.1*, p. *237*; 21.1, p. 308
N. V. Murashko	(a) **33**, 783 (1976)	1.2, p. 15; *5*, p. *48*; 22.8.2, p. 427
A. M. Murasik	(b) 75, K79 (1976)	*18.4*, p. *243*; 22, p. 336
K. Murata	(a) **36**, 197 (1976)	*11*, p. *102*; 21, p. 296; 21.1, p. 309; 21.6, p. 330
	(a) **36**, 527 (1976)	*11*, p. *103*; 21.1, p. 310
R. B. Murray	(b) 73, 307 (1976)	*13.1*, p. *120*; *17*, p. *218*; 20.1, p. 260; 22.3, p. 372
R. Murri	(a) **38**, 685 (1976)	13.4, p. 145; 14.3, p. 182; *16*, p. *217*; 22.3, p. 374
S. V. Murthy	(b) 75, K11 (1976)	*19*, p. *251*; 22.8.1, p. 421
	(b) 75, K169 (1976)	*19*, p. *251*; 22.8.1, p. 421
V. R. K. Murthy	(a) **36**, K133 (1976)	*14.4*, p. *203*; 18.3, p. 241; 22.8.2, p. 430
	(a) **38**, 647 (1976)	*14.3*, p. *182*; 15, p. 212; 18.3, p. 241; 22.8.2, p. 431
Y. V. G. S. Murti	(a) **34**, 615 (1976)	*10.2*, p. *91*; 13.4, p. 137; 19, p. 250; 20.1, p. 266; 22.5.2, p. 394
J. P. Muscat	(b) 74, K141 (1976)	*13.3*, p. *132*; 21, p. 293
	(a) **35**, 499 (1976)	*1.3*, p. *22*; 10.1, p. 82; 21, p. 295; 21.6, p. 330
H. E. Müser	(a) **36**, 189 (1976)	*12.2*, p. *111*; *14.4.1*, p. *207*; 22.8, p. 416
P. V. Musienko	(a) **37**, 29 (1976)	10.2, p. 95; *14.3*, p. *179*; 22.1.2, p. 356
V. V. Myakhar	(a) **34**, 437 (1976)	*18.2.1*, p. *237*; 21.1, p. 306; 21.1.1, p. 317
E. N. Myasnikov	(b) 74, 453 (1976)	*13.5.1*, p. *151*; *20.3*, p. *282*; 22.9, p. 435
T. Mydlarz	(b) 76, 357 (1976)	13.1, p. 125; *14.2*, p. *170*; *18.1*, p. *226*; 22.8, p. 416
	(b) 77, 451 (1976)	*14.2*, p. *171*; 21, p. 298

G. Myers	(b) 77, K81 (1976)	*13.1*, p. *128*; 22.1.1, p. 349; 22.1.2, p. 356; 22.2, p. 361; 22.4, p. 377
Sh. I. Nabiev	(a) **36**, 53 (1976)	*14.3.1*, p. *184*; *16*, p. *215*; 22.2.1, p. 363
Yu. V. Naboikin	(a) **33**, 773 (1976)	*2*, p. *36*; *20*, p. *257*; 22.9, p. 434
K. Nachtrieb	(a) **37**, 449 (1976)	*9*, p. *74*; 21.1, p. 312
S. Nadiv	(a) **38**, 177 (1976)	*2*, p. *40*; 11, p. 104; 22, p. 341
B. R. Nag	(b) **73**, K15 (1976)	6, p. 53; *14.3*, p. *173*; 22.2.3, p. 367
	(b) **75**, K35 (1976)	*13.1*, p. *124*; 22.2.1, p. 363
T. Nagarajan	(a) **35**, 379 (1976)	*1.2*, p. *16*; 4, p. 45; 22.5, p. 389
D. Nagchoudhuri	(a) **36**, 139 (1976)	13.3, p. 132; *14.3.3*, p. *192*; 22, p. 338
S. Nagel	(b) **76**, 629 (1976)	*13*, p. *117*; *14.4*, p. *203*; 22.5.2, p. 396; 22.6, p. 406
W. Nägele	(b) **74**, 297 (1976)	8, p. 67; *18.4*, p. *242*; 20.1, p. 264; 22.8.1, p. 420
L. Nagli	(b) **73**, 427 (1976)	*10.2*, p. *88*; 20.3, p. 280; 22.5.2, p. 393
A. T. Nagy	(a) **34**, 705 (1976)	*1.1*, p. *11*; 10.2, p. 91; 22.1.2, p. 353
R. K. Nahar	(a) **36**, 139 (1976)	13.3, p. 132; *14.3.3*, p. *192*; 22, p. 338
P. K. Nair	(a) **38**, K73 (1976)	*14.4*, p. *204*; *16*, p. *217*; 22.4, p. 377
T. Nakajima	(b) 77, K147 (1976)	*14.2*, p. *171*; 21, p. 298
Y. Nakamura	(a) **37**, K33 (1976)	*18.2*, p. *235*; 21.1, p. 311; 21.1.1, p. 321
	(a) **37**, K89 (1976)	*5*, p. *50*; 18.2, p. 235; 21.1, p. 311; 21.1.1, p. 321
Y. O. Nakamura	(b) **75**, 347 (1976)	*13.3*, p. *132*; *13.5.2*, p. *157*; 14.3.2, p. 188
S. Nakashima	(b) **73**, 341 (1976)	*6*, p. *53*; *20.1*, p. *260*; 22.5, p. 388
H. Nakata	(a) **33**, 633 (1976)	11, p. 100; *19*, p. *248*; 22.1.2, p. 352
	(b) **76**, K57 (1976)	*11*, p. *102*
T. Nakayama	(b) **78**, 519 (1976)	*13.5.1*, p. *156*; 20.3, p. 288; 22.9, p. 439
T. Nakazawa	(a) **33**, K35 (1976)	*17*, p. *218*; 22.8.1, p. 419
R. Kh. Nani	(b) **73**, K69 (1976)	13.5.1, p. 149; *20.1*, p. *261*; 22.8, p. 413
	(a) **34**, K115 (1976)	*6*, p. *56*; *20.1*, p. *266*; 22.8, p. 414
H. Nanto	(a) **33**, K125 (1976)	10.2, p. 89; *20.3*, p. *281*; 22.6, p. 403
K. Narahari Rao see under letter "R"		
A. N. Narkulov	(a) **38**, 439 (1976)	*9*, p. *74*; *11*, p. *105*; 14.3.2, p. 189; 22.1.2, p. 358
S. Narita	(b) **73**, 741 (1976)	*13.1*, p. *121*; 18, p. 222; 20.1, p. 262; 22.4.4, p. 387
M. Natarajan	(a) **33**, 427 (1976)	*1.2*, p. *14*; 8, p. 67; 14.4, p. 202; 22.5.4, p. 401
A. N. Nateprov	(a) **35**, K53 (1976)	*13.1*, p. *124*; 14.3, p. 177; 22.3, p. 372
P. Nath	(a) **33**, 333 (1976)	*2*, p. *36*; *14.3.1*, p. *183*; 15, p. 209; 22.1.1, p. 345
	(a) **34**, 405 (1976)	*2*, p. *36*; *20.1*, p. *265*; 22.1.1, p. 346
Rabinder Nath	(a) **33**, K151 (1976)	*16*, p. *213*; 22.5, p. 388
	(a) **37**, 491 (1976)	*16*, p. *216*; 22.4, p. 377
	(a) **38**, K73 (1976)	*14.4*, p. *204*; *16*, p. *217*; 22.4, p. 377
I. Natkaniec	(b) **75**, 105 (1976)	*6*, p. *56*; 20.1, p. 267; 22.9, p. 435
T. Nattermann	(b) **74**, 635 (1976)	*14.2*, p. *169*
	(b) 77, 631 (1976)	*1.2*, p. *19*; *14.4.1*, p. *208*
A. Naurizbaev	(a) **37**, K139 (1976)	*13.4*, p. *143*; 20.1, p. 275; 22.2.3, p. 369
A. N. Nazarov	(a) **38**, 131 (1976)	11, p. 104; *14.3.3*, p. *194*; 22.1.2, p. 357
H. Neddermeyer	(b) **78**, 609 (1976)	*13.1*, p. *129*; *20*, p. *259*; 21.1, p. 314; 21.3, p. 324
D. D. Nedeoglo	(a) **36**, 659 (1976)	*14.3.1*, p. *185*; 22.2.2, p. 365
N. L. Nedzelyak	(a) **38**, 769 (1976)	10.1, p. 86; *14.2*, p. *172*; 21, p. 301
I. Neizvestny	(b) **78**, K19 (1976)	13.1, p. 128; *14.3.4*, p. *200*; *20.1*, p. *276*; 22.1.1, p. 349
I. G. Neizvestnyi	(a) **38**, 349 (1976)	*16*, p. *217*; *20.1*, p. *277*; 22.1.1, p. 349
V. V. Nelaev	(a) **35**, K81 (1976)	*10.1*, p. *82*; 21.1.1, p. 318
	(a) **38**, K101 (1976)	*10.1*, p. *82*; 21.1.1, p. 318
E. Nembach	(a) **34**, 547 (1976)	*18.2*, p. *231*; 21, p. 293
E. Nes	(a) **33**, K5 (1976)	*10.2*, p. *88*; 22.1.2, p. 351

S. NEŠPUREK	(a) **34**, 747 (1976)	13.4, p. 138; *14.3.4*, p. *197*; 22.9, p. 435
H. NEUHÄUSER	(a) **37**, K93 (1976)	*10.1*, p. *84*; 11, p. 103; 21.1, p. 312
H. NEUMANN	(a) **34**, K39 (1976)	*20.1*, p. *265*; 22.2.4, p. 370
	(b) **75**, 493 (1976)	*13.1*, p. *124*; 22.2.4, p. 370
H.-G. NEUMANN	(a) **36**, K47 (1976)	*1.5*, p. *32*; *12.2*, p. *111*; 22.1.2, p. 355
	(a) **37**, 193 (1976)	*1.5*, p. *32*; *12.2*, p. *111*; 22, p. 339
R. A. NEVILLE	(b) **73**, 597 (1976)	*13.5.1*, p. *150*; 20.1, p. 262; 22, p. 334
R. E. NEWNHAM	(a) **37**, 235 (1976)	1.4, p. 28; *10.2*, p. *95*; *12.2*, p. *112*; 22.6, p. 407
NGHI Q. LAM see under letter "L"		
NGO DUONG SINH see under letter "S"		
NGUYEN VAN DONG see under letter "D"		
D. K. NICHUGOVSKII	(a) **33**, 755 (1976)	*10.2*, p. *89*; 22.2.1, p. 362
I. NICOARĂ	(b) **77**, K17 (1976)	*13*, p. *118*; *14*, p. *163*
A. NICULA	(b) **74**, 335 (1976)	*19*, p. *249*; 22.8.1, p. 420
J. NIEBER	(a) **33**, 587 (1976)	*1.1*, p. *11*; 22.1.2, p. 352
	(a) **37**, 529 (1976)	*1.1*, p. *13*; 22.1.2, p. 357
H. P. NIEDERMANN	(b) **78**, 615 (1976)	4, p. 47; *6*, p. *64*; 22, p. 342
R. NIEDERMAYER	(a) **34**, K135 (1976)	*18.2.1*, p. *237*; 22, p. 336
J. NIEWODNICZAŃSKA-ZAWADZKA	(b) **74**, K93 (1976)	*6*, p. *55*; *14.3.3*, p. *191*; 22.4.2, p. 383; 22.4.3, p. 385
G. D. NIGAM	(a) **33**, K101 (1976)	1.3, p. 21; *12.1*, p. *107*; 21.1, p. 304; 21.6, p. 329
Y. NIGARA	(a) **38**, K135 (1976)	*14.4*, p. *205*; 22.5.2, p. 398; 22.5.4, p. 401
G. NIHOUL-BOUTANG	(a) **34**, 331 (1976)	*1.4*, p. *26*; *10.2*, p. *90*; 22.6, p. 403
J. NIKLAS	(b) **73**, K127 (1976)	*7*, p. *65*; *14.3.4*, p. *195*; 22.1.3, p. 359
R. NILSSON	(b) **75**, 197 (1976)	*13.1*, p. *123*; 17, p. 219; 21, p. 293
G. NIMTZ	(a) **38**, K171 (1976)	13.4, p. 146; *20.1*, p. *278*; 22.4.4, p. 387
R. NINK	(a) **35**, 121 (1976)	*10.2*, p. *92*; 13.4, p. 138; *20.1*, p. *268*; 22.5.2, p. 394
T. NIRK	(a) **33**, 75 (1976)	*10.2*, p. *87*; 13.4, p. 135; *14.3*, p. *173*; 22.4.2, p. 383
S. V. NISTOR	(b) **78**, K31 (1976)	*19*, p. *255*; 22.5, p. 390
W. D. NIX	(b) **75**, 573 (1976)	*10*, p. *76*
O. R. NIYAZOVA	(a) **36**, 427 (1976)	9, p. 73; 10.2, p. 94; *11*, p. *102*; 14.3, p. 178; 22.1, p. 343; 22.2, p. 361; 22.3, p. 373; 22.4, p. 376
C. NOGUET	(a) **33**, 155 (1976)	13.4, p. 135; *16*, p. *213*; 22.6, p. 402
E. NORMANTAS	(b) **77**, 667 (1976)	*14.3*, p. *180*; 16, p. 216; 22.1.3, p. 360
P. NOVÁK	(b) **73**, 575 (1976)	*13*, p. *114*; *18.3*, p. *239*; 22.8.2, p. 426
B. V. NOVIKOV	(b) **74**, 81 (1976)	6, p. 54; *13.4*, p. *136*; *13.5.1*, p. *150*; 16, p. 214; 22.1.1, p. 345
P. NOVOTNÝ	(a) **33**, K15 (1976)	*18.2*, p. *229*; 22, p. 334
B. NOWAK	(a) **37**, 669 (1976)	*19*, p. *255*; 21.1, p. 312
J. NOWAK	(b) **74**, 437 (1976)	13.4, p. 137; *14.1*, p. *165*; 21, p. 292
H. NOWOTNY	(b) **77**, 623 (1976)	*14.2*, p. *171*
T. N. NURAKHMETOV	(b) **78**, 529 (1976)	*10.2*, p. *98*; 20.1, p. 277; 22.5.2, p. 398
R. NYHOLM	(b) **75**, 197 (1976)	*13.1*, p. *123*; 17, p. 219; 21, p. 293
G. M. OBERMAIR	(b) **76**, 217 (1976)	*13*, p. *117*
Y. OBI	(a) **36**, K149 (1976)	*1.3*, p. *23*; 14.1, p. 166; 18.2, p. 234; 21.1, p. 310
M. OBIDITSCH	(b) **77**, 677 (1976)	6, p. 61; *14.3*, p. *180*; *20.1*, p. *274*; 22.2.3, p. 368
Y. OCHIAI	(a) **38**, 243 (1976)	13.4, p. 144; *19*, p. *255*; 22.1.2, p. 357
N. ODA	(a) **33**, K35 (1976)	*17*, p. *218*; 22.8.1, p. 419
A. ODAJIMA	(a) **37**, 579 (1976)	*10.2*, p. *96*; *14.3*, p. *180*; 22.4.2, p. 383
G. OELGART	(b) **73**, 615 (1976)	*13.2*, p. *130*; 14.3, p. 174; 19, p. 248; 22.7, p. 410

	(b) 74, K75 (1976)	*13.2*, p. *130*; 14, p. 162; 19, p. 249; 22.7, p. 410
	(b) 75, 189 (1976)	*13.1*, p. *123*; 13.2, p. 131; 19, p. 250; 22.7, p. 411
H. OESTERREICHER	(a) 34, 723 (1976)	*18.2*, p. *231*; *18.4*, p. *243*; 21.1, p. 306; 21.4, p. 326
K. OHMORI	(a) 34, 325 (1976)	1.2, p. 15; *18.3*, p. *240*; 22.8.2, p. 427
S. M. OHR	(a) 38, 553 (1976)	*1.4*, p. *29*; *10.1*, p. *86*; 21, p. 300
P. R. OKAMOTO	(a) 36, 107 (1976)	*1.1*, p. *12*; *1.4*, p. *27*; 21.1.1, p. 319
B. L. OKSENGENDLER	(a) 35, K145 (1976)	*11*, p. *102*; 22.1.2, p. 354
	(a) 38, 439 (1976)	*9*, p. *74*; *11*, p. *105*; 14.3.2, p. 189; 22.1.2, p. 358
A. I. OLEMSKOI	(a) 35, K101 (1976)	*1.3*, p. *23*
N. F. OLIVEIRA, JR.	(a) 33, 573 (1976)	*18.2*, p. *229*; *18.4*, p. *242*; 22, p. 334
A. I. ONIPKO	(b) 73, 699 (1976)	*13.5.1*, p. *150*
A. OPANOWICZ	(a) 33, 195 (1976)	13.5.1, p. 149; *16*, p. *213*; 22.4.2, p. 383; 22.4.4, p. 387
	(b) 75, 471 (1976)	*13.5.1*, p. *152*; *20.3*, p. *283*; 22.4.3, p. 385
	(b) 77, K35 (1976)	*13.5.1*, p. *155*; *20.3*, p. *286*; 22.4.3, p. 386
L. OPILSKA	(a) 35, K183 (1976)	*7*, p. *65*; 22.3, p. 373; 22.3.1, p. 375
A. OPILSKI	(a) 35, K183 (1976)	*7*, p. *65*; 22.3, p. 373; 22.3.1, p. 375
CH. OPITZ	(b) 78, 177 (1976)	*9*, p. *74*; *10.1*, p. *85*; 21.1, p. 313; 21.1.1, p. 321
A. OPPELT	(a) 36, 275 (1976)	18.2, p. 233; *19*, p. *253*; 21.1, p. 309; 21.4, p. 327
	(a) 37, K205 (1976)	*18.2*, p. *235*; *19*, p. *255*; 21.1.1, p. 321
H. OPPERMANN	(a) 37, 541 (1976)	*14.3*, p. *180*; 15, p. 211; 18.1, p. 227; 22.6, p. 407
	(a) 38, 93 (1976)	*1.2*, p. *19*; *3*, p. *44*; 14.3, p. 181; 15, p. 211; 18.1, p. 227; 22.6.1, p. 409
	(a) 38, K23 (1976)	*14.3*, p. *181*; *15*, p. *211*; 22.6, p. 408; 22.6.1, p. 409
V. M. ORERA	(a) 38, 621 (1976)	*10.2*, p. *98*; 20.1, p. 278; 22.5.3, p. 400
A. ORLIUKAS	(a) 37, K39 (1976)	*14.4*, p. *204*; 22, p. 340
W. A. ORTIZ	(a) 35, K135 (1976)	*14.2*, p. *170*; 21, p. 295
D. W. OSBORNE	(a) 34, 291 (1976)	*8*, p. *68*; 18, p. 222; 22.8, p. 414
B. N. OSHCHERIN	(a) 34, K181 (1976)	*1*, p. *9*; *4*, p. *45*; 22, p. 336
	(a) 35, K35 (1976)	*1*, p. *9*; *8*, p. *68*; 22.2, p. 361; 22.4, p. 376; 22.8, p. 415
V. V. OSIKO	(a) 35, 645 (1976)	*13.4*, p. *140*; 20.1, p. 269; 20.3, p. 284; 22.8.2, p. 429
V. V. OSIPOV	(a) 35, 605 (1976)	13.4, p. 139; *14.4.1*, p. *207*; *20.1*, p. *269*; 22.8.1, p. 422
V. OSMININ	(b) 78, 813 (1976)	*10.2*, p. *98*; *20.3*, p. *288*; 22.5.2, p. 398
W. VON DER OSTEN	(b) 75, 613 (1976)	6, p. 57; *10.2*, p. *93*; 20.3, p. 284; 22.5.3, p. 399
E. OSTERTAG	(b) 77, 381 (1976)	*13.5.1*, p. *155*; *20.3*, p. *286*; 22.5, p. 390
V. S. OSTROVSKII	(b) 74, K157 (1976)	*18.4*, p. *243*
M. OSZWAŁDOWSKI	(a) 36, 445 (1976)	13.3, p. 133; *16*, p. *215*; 22.2.3, p. 368
	(a) 37, 675 (1976)	*16*, p. *216*; 22.2.3, p. 369
A. OTTO	(b) 78, 277 (1976)	4, p. 47; *6*, p. *63*; *20.1*, p. *276*; 22, p. 341
H. H. OTTO	(a) 35, K165 (1976)	*1.2*, p. *17*; 14.4.1, p. 207; 20.1, p. 270; 22.8.1, p. 422
M. OVERHAMM	(b) 75, 543 (1976)	*13.1*, p. *124*; *14.3.4*, p. *197*; 20.1, p. 268; 20.3, p. 283; 22.2.3, p. 368
N. OVSYUK	(b) 78, K19 (1976)	13.1, p. 128; *14.3.4*, p. *200*; *20.1*, p. *276*; 22.1.1, p. 349
R. T. OWEN	(a) 34, 255 (1976)	*13.1*, p. *122*; *20.1*, p. *264*; 22.2.4, p. 370
YU. V. OZEROV	(a) 35, 251 (1976)	*13.4*, p. *138*; 16, p. 214; 20.1, p. 268; 20.3, p. 283; 22.4.1, p. 380
	(a) 38, 77 (1976)	*13.4*, p. *144*; 16, p. 217; 20.1, p. 276; 22.4.1, p. 382
D. OŻGO	(a) 34, 697 (1976)	*1.2*, p. *15*; 14.4.1, p. 206; 22.9, p. 435
A. M. OZORIO DE ALMEIDA	(b) 75, 657 (1976)	*1.2*, p. *16*; 14.4.1, p. 207; 22.8.1, p. 421

O. M. Pachev	(b) 75, 91 (1976)	*13.5*, p. *147*; 22, p. 336
A. Paduan Filho	(a) 33, 573 (1976)	*18.2*, p. *229*; *18.4*, p. *242*; 22, p. 334
R. Paetsch	(b) 74, 761 (1976)	6, p. 55; *10*, p. *75*; 12, p. 106
	(b) 75, K105 (1976)	*6*, p. *57*; *10*, p. *76*
V. Paidar	(a) 33, 555 (1976)	*10.1*, p. *79*; 21.1.1, p. 317
K. N. Pak	(b) 74, 673 (1976)	*14.4.1*, p. *206*; 22.8.1, p. 420
P. P. Pal-Val	(a) 38, 383 (1976)	*10.1*, p. *85*; 21.7, p. 332
S. S. Paliwal	(a) 38, 73 (1976)	*10.2*, p. *97*; 13.4, p. 144; *20.3*, p. *287*; 22, p. 341
H. Ö. Pamuk	(a) 37, 695 (1976)	*4*, p. *46*; 12.1, p. 108; 21, p. 299
A. N. Pandey	(b) 78, 399 (1976)	*6*, p. *63*; 22.6, p. 408
R. B. Pandey	(b) 78, 821 (1976)	*18.2*, p. *236*
D. K. Pandya	(a) 34, 405 (1976)	*2*, p. *36*; *20.1*, p. *265*; 22.1.1, p. 346
	(a) 35, 725 (1976)	*2*, p. *38*; *20.1*, p. *269*; 22.1.1, p. 347
	(a) 35, 761 (1976)	2, p. 38; *14.3*, p. *178*; *15*, p. *210*; 20.1, p. 270; 22.7, p. 411
	(a) 36, 89 (1976)	*2*, p. *39*; *14.3.1*, p. *184*; 15, p. 210; 22.1.1, p. 348
A. I. Pankrats	(a) 36, K43 (1976)	*12.2*, p. *111*; *18.4*, p. *244*; 22.8.1, p. 423
M. M. Pant	(b) 73, K27 (1976)	*1.6*, p. *34*; *6*, p. *53*; 8, p. 67; 21.1.1, p. 316
C. Paorici	(a) 34, 651 (1976)	*14.3*, p. *175*; 15, p. 209; 22.8, p. 414
	(a) 36, K33 (1976)	*20.3*, p. *285*; 22.8, p. 416
G. C. Papavassiliou	(b) 77, K49 (1976)	*13.5.2*, p. *157*; 20.1, p. 273; 21.1, p. 311
G. V. Papayan	(b) 78, K59 (1976)	*20.1*, p. *276*; 22.8, p. 418
K. I. Papunashvili	(b) 76, 351 (1976)	*13*, p. *117*; *14.3.4*, p. *197*; 22.1.2, p. 354
V. R. Parameswaran	(a) 33, 639 (1976)	*10.1*, p. *79*; 21.6, p. 329
B. V. Paranjape	(b) 75, 347 (1976)	*13.3*, p. *132*; *13.5.2*, p. *157*; 14.3.2, p. 188
M. Pardavi-Horváth	(b) 78, 353 (1976)	12.1, p. 109; 18.1, p. 227; *18.2*, p. *236*; *18.4*, p. *245*; 21.1.1, p. 321
B. Pardo	(a) 35, 329 (1976)	*1.1*, p. *12*; 22.6, p. 404
D. Parihar	(b) 77, K135 (1976)	*14.1*, p. *167*; 21, p. 298; 21.1, p. 312; 21.2, p. 323
Om Parkash	(a) 36, 669 (1976)	*14*, p. *162*; 15, p. 210; *18*, p. *224*; 22.8.1, p. 423
R. Parker	(a) 33, 189 (1976)	*14.3*, p. *173*; 18.3, p. 239; 22.6, p. 402
V. D. Parkhomenko	(a) 33, K83 (1976)	*18.3*, p. *239*; 22.8.2, p. 427
	(a) 38, 57 (1976)	11, p. 104; *18.3*, p. *241*; 22.8.2, p. 431
	(a) 38, K147 (1976)	11, p. 105; *18.3*, p. *241*; 22.8.2, p. 432
R. Parrot	(b) 76, K5 (1976)	*19*, p. *252*; 22.8.1, p. 422
J. E. Parrott	(a) 34, K43 (1976)	*14.3.3*, p. *191*; 14.3.4, p. 196; 22, p. 335
L. Pasemann	(b) 77, 527 (1976)	*13.1*, p. *127*; 22.8, p. 417
R. Pässler	(b) 76, 647 (1976)	*6*, p. *59*; 13.1, p. 126; *13.4*, p. *141*; 22, p. 339; 22.4.1, p. 381
	(b) 78, 625 (1976)	6, p. 64; 13, p. 119; *13.4*, p. *145*; 22, p. 342; 22.1, p. 344; 22.2, p. 361; 22.4, p. 378
G. Pastori Parravicini	(b) 73, 371 (1976)	*13*, p. *114*; 22, p. 333
K. Pataj	(b) 75, 341 (1976)	12.2, p. 110; *13.5.1*, p. *152*; *20.3*, p. *283*; 22.4.3, p. 385
A. D. Pataraya	(b) 75, K69 (1976)	*18.2*, p. *232*
G. Paternó	(a) 35, K173 (1976)	*14.2*, p. *170*; 21, p. 295; 22.6, p. 405
N. L. Pathak	(a) 38, 355 (1976)	*14.3.1*, p. *185*; *14.4*, p. *204*; 22.9, p. 439
S. A. Patlazhan	(a) 38, 375 (1976)	*12*, p. *106*
V. I. Patoka	(b) 74, 471 (1976)	*13.2*, p. *130*; 21, p. 292
I. B. Patrina	(b) 77, 85 (1976)	*13.4*, p. *142*; *19*, p. *253*; 22.6, p. 406
J. D. Patterson	(b) 74, K115 (1976)	2, p. 37; *8*, p. *68*; *18.2*, p. *230*
P. V. Pavlov	(a) 35, 11, (1976)	11, p. 100; *13.4*, p. *134*; *14.3*, p. *173*; 22.1.1, p. 345; 22.1.2, p. 351
	(a) 36, 81 (1976)	*1.5*, p. *31*; 2, p. 39; *11*, p. *102*; 22.1.2, p. 355
	(a) 37, 57 (1976)	*11*, p. *103*; 22.1.2, p. 356
V. N. Pavlovich	(b) 78, K97 (1976)	*9*, p. *74*; 22, p. 342

L. Pawlak	(b) **75**, K7 (1976)	*19*, p. *251*; 22.8.2, p. 428
A. Pawlikowski	(b) **76**, K123 (1976)	*18.2*, p. *234*
J. M. Pawlikowski	(a) **37**, K183 (1976)	*14.3.4*, p. *200*; 22.4.4, p. 387
H. Peibst	(a) **36**, 173 (1976)	*1.4*, p. *27*; *3*, p. *43*; 22.2.4, p. 370
H. Peisl	(a) **37**, K67 (1976)	*1.2*, p. *18*; 1.3, p. 23; 10.1, p. 84; 21, p. 298
G. P. Peka	(a) **37**, 363 (1976)	*13.4*, p. *143*; *14.3.4*, p. *199*; 20.3, p. 287; 22.4.2, p. 383
G. S. Pekar	(a) **33**, 765 (1976)	10.2, p. 89; *14.3*, p. *174*; 22.4.1, p. 379
	(a) **34**, 397 (1976)	10.2, p. 90; *14.3*, p. *174*; 22.4.1, p. 379
C. Pelletier	(a) **34**, 307 (1976)	*14.3*, p. *174*; 22.2.1, p. 362
C. Pelosi	(a) **36**, K33 (1976)	*20.3*, p. *285*; 22.8, p. 416
J. Pelous	(b) **75**, K101 (1976)	*2*, p. *38*; *6*, p. *57*; 20.1, p. 269; 22.9, p. 436
J. Pelzl	(b) **74**, 193 (1976)	*1.2*, p. *15*; 5, p. 48; 20.1, p. 263; 22.8, p. 414
C. M. Penchina	(b) **74**, 375 (1976)	6, p. 55; *13.5.1*, p. *151*; 14.3.4, p. 196; 20.1, p. 264
J. Peng	(a) **33**, 415 (1976)	*10.2*, p. *88*; *11*, p. *100*; 20.1, p. 261; 22.5.3, p. 399
U. del Pennino	(a) **35**, 451 (1976)	*13.4*, p. *139*; 14.3, p. 177; 22.1.1, p. 347
	(a) **38**, 109 (1976)	*10.2*, p. *97*; *13.1*, p. *128*; 14.3.4, p. 200; 22.1.1, p. 349
G. I. Peresada	(a) **35**, K177 (1976)	*12.2*, p. *110*; 22, p. 337
S. Permogorov	(b) **78**, 389 (1976)	*6*, p. *63*; *13.5*, p. *148*; 20.3, p. 287; 22.4.1, p. 381
G. A. Perov	(a) **37**, 701 (1976)	*1.3*, p. *24*; *10.1*, p. *84*; 21.1, p. 312
F. Perrot	(b) **77**, 517 (1976)	12.2, p. 112; *13.1*, p. *127*; 22, p. 340
P. Petelenz	(b) **73**, 293 (1976)	*13*, p. *113*; 14.4, p. 202; 22.9, p. 433
	(b) **73**, 565 (1976)	6, p. 54; *13*, p. *114*; *13.5.1*, p. *150*; 14.3, p. 174; 22, p. 334
	(b) **74**, 655 (1976)	*6*, p. *55*; *13*, p. *115*; 13.5.1, p. 152; 14.3, p. 175; 22, p. 335
	(b) **78**, 489 (1976)	*13.5.1*, p. *156*; 20.1, p. 277; 22, p. 341
Á. Péter	(a) **33**, K169 (1976)	*10.2*, p. *89*; *14.4*, p. *202*; 22.5.2, p. 393
L. M. Peter	(b) **75**, 621 (1976)	*13.5.1*, p. *153*; 20.3, p. 284; 22.9, p. 436
J. Petersson	(b) **78**, 191 (1976)	*14.4.1*, p. *208*; 19, p. 255; 22.9, p. 438
S. Petersson	(a) **36**, 217 (1976)	1.5, p. 32; *14.3.1*, p. *184*; 22, p. 338
J. Petit	(a) **37**, 257 (1976)	*10.1*, p. *84*; *12.1*, p. *108*; 21.1, p. 311
G. A. Petrakovskii	(a) **36**, K43 (1976)	*12.2*, p. *111*; *18.4*, p. *244*; 22.8.1, p. 423
D. Petrascheck	(a) **36**, 147 (1976)	*1.1*, p. *12*
M. I. Petrashen	(b) **73**, 495 (1976)	6, p. 54; *10.2*, p. *89*; 22.5.2, p. 393
	(b) **76**, 377 (1976)	*13.1*, p. *125*; *13.4*, p. *140*; 22.5.2, p. 395
K. G. Petrashku	(b) **76**, 497 (1976)	*13.5.1*, p. *154*; *20.3*, p. *285*; 22.4.1, p. 381
S. G. Petrosyan	(a) **36**, 53 (1976)	*14.3.1*, p. *184*; *16*, p. *215*; 22.2.1, p. 363
A. I. Petrov	(a) **34**, 73 (1976)	*10.1*, p. *79*; 21, p. 291
V. V. Petrov	(a) **33**, K83 (1976)	*18.3*, p. *239*; 22.8.2, p. 427
	(a) **38**, 57 (1976)	11, p. 104; *18.3*, p. *241*; 22.8.2, p. 431
	(a) **38**, K147 (1976)	11, p. 105; *18.3*, p. *241*; 22.8.2, p. 432
B. Petzel	(a) **37**, K39 (1976)	*14.3.3*, p. *192*; *14.3.4*, p. *198*; 22.2.1, p. 363
K. Peuker	(b) **73**, 141 (1976)	*6*, p. *52*; 13.1, p. 120; 13.5.1, p. 149; *20.1*, p. *260*
	(b) **78**, 711 (1976)	6, p. 64; *14.3.4*, p. *200*; *20.1*, p. *278*; 22, p. 342
J. C. Peuzin	(a) **37**, 119 (1976)	*7*, p. *66*; *14.4.1*, p. *207*; 22.8.1, p. 423
H. Pfeiffer	(b) **74**, 617 (1976)	*18.2*, p. *230*; 19, p. 250
J. C. Pfister	(a) **36**, 679 (1976)	*13.3*, p. *133*; *14.3.3*, p. *192*; 22.1.2, p. 356
Vu Duy Phach see under letter "V"		
P. P. Phakey	(a) **36**, 627 (1976)	*1.4*, p. *28*; 22.1, p. 344
Pham Minh Tan	(a) **37**, 439 (1976)	11, p. 104; *13.4*, p. *143*; *14.3.3*, p. *193*; 22.1.2, p. 356
J. C. Phillips	(b) **77**, 259 (1976)	*1*, p. *9*; *14*, p. *163*; 22.9, p. 437
	(b) **78**, 371 (1976)	*1*, p. *9*; *14*, p. *163*; 22.9, p. 438
	(b) **78**, 471 (1976)	6, p. 63; *14*, p. *163*; 22.9, p. 439

I. G. Pichugin	(a) **34**, K195 (1976)	*20.3*, p. *282*; 22.2, p. 361
	(a) **38**, 753 (1976)	*20.3*, p. *288*; 22.2, p. 361
R. Pickenhain	(b) 77, 571 (1976)	*4*, p. *46*; 13, p. 119; 22.1.1, p. 348
A. Pietraszko	(a) **37**, 287 (1976)	*1.2*, p. *18*; *8*, p. *70*; 14.4.1, p. 208; 22.8, p. 417
G. E. Pike	(b) 75, 289 (1976)	*9*, p. *72*; 14.3, p. 176; 22.1.1, p. 347; 22.1.2, p. 353
A. N. Pikhtin	(a) **38**, 753 (1976)	*20.3*, p. *288*; 22.2, p. 361
M. Pilkuhn	(b) **73**, 255 (1976)	13.4, p. 134; *13.5.1*, p. *149*; 20.1, p. 260; *20.3*, p. *280*; 22.2.3, p. 367
P. K. C. Pillai	(a) **33**, K151 (1976)	*16*, p. *213*; 22.5, p. 388
	(a) **37**, 491 (1976)	*16*, p. *216*; 22.4, p. 377
	(a) **38**, K73 (1976)	*14.4*, p. *204*; *16*, p. *217*; 22.4, p. 377
P. Pinard	(b) **73**, 187 (1976)	*13.1*, p. *120*; *20.1*, p. *260*; 22.2.1, p. 362
	(a) **38**, 85 (1976)	*17*, p. *220*; 21.1, p. 313; 22.1.1, p. 349; 22.5.2, p. 397
A. Pincevičius	(a) **35**, 563 (1976)	*14.3*, p. *177*
A. J. Pindor	(b) **74**, K19 (1976)	*14.2*, p. *169*; 21, p. 291
J. Pinguet	(a) **35**, 431 (1976)	*14.4*, p. *203*; *16*, p. *214*; 22.6, p. 405
L. Pintschovius	(b) 78, 277 (1976)	4, p. 47; *6*, p. *63*; *20.1*, p. *276*; 22, p. 341
R. Piotrzkowski	(b) **73**, K131 (1976)	12.2, p. 110; *14.3*, p. *174*; 22.2.3, p. 367
P. Pipinys	(a) **34**, 377 (1976)	*14.3.4*, p. *196*; *20.3*, p. *281*; 22.4.1, p. 379
	(a) **37**, 321 (1976)	6, p. 61; *13.4*, p. *142*; 14.3.4, p. 199; *20.3*, p. *286*; 22.4.1, p. 381
E. Pippel	(a) **37**, 457 (1976)	1.4, p. 28; *1.5*, p. *32*; *3*, p. *44*; 21, p. 298; 22.6, p. 407
J. Piqueras	(a) **34**, K9 (1976)	*10.1*, p. *80*; 21.1, p. 305
B. Piriou	(b) 75, 117 (1976)	*6*, p. *56*; *20.1*, p. *267*; 22.6, p. 404
S. V. Pirogova	(a) **34**, 319 (1976)	*8*, p. *68*; 10.1, p. 80; 21.1, p. 305
N. Piślewski	(a) **37**, 669 (1976)	*19*, p. *255*; 21.1, p. 312
T. S. Plaskett	(a) **37**, 499 (1976)	*10.2*, p. *96*; 22.8.2, p. 430
	(a) **38**, 577 (1976)	*10.2*, p. *98*; 22.8.2, p. 431
V. Ya. Platkov	(a) **38**, 383 (1976)	*10.1*, p. *85*; 21.7, p. 332
M. N. Pletnev	(b) 78, 57 (1976)	*9*, p. *74*; *10*, p. *76*
Yu. M. Plishkin	(a) **38**, 51 (1976)	*10.1*, p. *85*; 21.1, p. 313
P. Plumelle	(b) **73**, 271 (1976)	*6*, p. *52*; 22.4.3, p. 385
F. Plümer	(a) **33**, K55 (1976)	*18.3*, p. *239*; 22.8.2, p. 426
I. E. Podchinenov	(a) **38**, 51 (1976)	*10.1*, p. *85*; 21.1, p. 313
H.-W. Pohl	(b) 75, 205 (1976)	*14.2*, p. *169*; 21, p. 294
M. Pöhler	(a) **33**, 149 (1976)	*13.1*, p. *121*; *20.1*, p. *261*; 22.1.3, p. 359
B. Pohoryles	(a) **37**, K117 (1976)	*14.3.4*, p. *200*; 22.1.1, p. 349
A. J. Pointon	(a) **35**, 459 (1976)	8, p. 68; *18.4*, p. *244*; 22.6, p. 405
P. Poix	(a) **35**, 153 (1976)	*5*, p. *49*; 18.3, p. 240; 22.8.2, p. 428
E. P. Pokatilov	(b) **73**, 553 (1976)	*20.1*, p. *262*
	(b) **75**, K143 (1976)	*10.2*, p. *93*; *13.5.3*, p. *159*
	(b) 78, 831 (1976)	*14.3.4*, p. *200*; *20.1*, p. *278*
K. I. Pokhodnya	(b) **73**, 335 (1976)	13, p. 113; 17, p. 218; *20*, p. *257*; 22.9, p. 433
S. Pokrzywicki	(b) 75, K7 (1976)	*19*, p. *251*; 22.8.2, p. 428
W. Polanschütz	(a) **38**, 237 (1976)	*1.4*, p. *29*; 10.1, p. 85; *17.1*, p. *221*; 21. p. 300
Yu. Ya. Polishuk	(a) **38**, 131 (1976)	11, p. 104; *14.3.3*, p. *194*; 22.1.2, p. 357
H. Pollak	(b) **74**, K31 (1976)	*5*, p. *49*; *20.1*, p. *264*; 22.8.1, p. 420
E. G. Ponyatovskii	(a) **35**, K177 (1976)	*12.2*, p. *110*; 22, p. 337
G. S. Poo	(b) 78, 278 (1976)	*13.5.4*, p. *161*; *18.2*, p. *235*; 21.1, p. 313
M. Popescu	(a) **35**, 355 (1976)	*2*, p. *37*; *4*, p. *45*; 22.7, p. 411
	(b) 77, 701 (1976)	*10.2*, p. *95*; 20.1, p. 274; 20.3, p. 287; 22.5.2, p. 397
A. Popielewicz	(b) **73**, K47 (1976)	*18.2*, p. *228*
	(b) **74**, 383 (1976)	*18.2*, p. *230*

	(b) **74**, K81 (1976)	*18.1*, p. *225*
	(b) **78**, K115 (1976)	*18.1*, p. *227*; 21.4, p. 327
A. B. Popov	(a) **34**, 761 (1976)	*10*, p. *75*
A. S. Popov	(a) **37**, K53 (1976)	*20.3*, p. *286*; 22.2.2, p. 365
	(a) **38**, K37 (1976)	*14.3.3*, p. *194*; 22.2.2, p. 366; 22.8, p. 418
D. E. Popov	(b) **76**, 779 (1976)	*11*, p. *103*; 13, p. 118; 22.1.2, p. 355
E. A. Popov	(b) **75**, K123 (1976)	13.5.4, p. 160; *20.1*, p. *269*; 22.8, p. 415
L. E. Popov	(a) **37**, 701 (1976)	*1.3*, p. *24*; *10.1*, p. *84*; 21.1, p. 312
S. Popović	(a) **33**, K23 (1976)	*1.1*, p. *10*; 1.2, p. 14; 22.3, p. 372
	(a) **34**, K91 (1976)	*1.1*, p. *11*; 22.1.1, p. 346; 22.8, p. 414
M. Poppi	(a) **35**, K161 (1976)	*18.3*, p. *240*; 22.8.2, p. 429
S. Porowski	(b) **73**, K131 (1976)	12.2, p. 110; *14.3*, p. *174*; 22.2.3, p. 367
	(a) **33**, 421 (1976)	13.4, p. 135; *14.3*, p. *174*; 22.4.3, p. 385
R. Portier	(a) **38**, 629 (1976)	1.2, p. 20; *1.4*, p. *29*
V. S. Potapov	(a) **34**, K163 (1976)	*18.2*, p. *231*
A. V. Potemkin	(a) **33**, 511 (1976)	20.1, p. 263; *20.2*, p. *279*; 20.3, p. 281; 22.8.1, p. 419
A. A. Potseluiko	(a) **34**, 437 (1976)	*18.2.1*, p. *237*; 21.1, p. 306; 21.1.1, p. 317
H.W. Pötzl	(b) **75**, 543 (1976)	*13.1*, p. *124*; *14.3.4*, p. *197*; 20.1, p. 268; 20.3, p. 283; 22.2.3, p. 368
G. M. Pound	(b) **75**, 573 (1976)	*10*, p. *76*
F. Pourarian	(a) **33**, 483 (1976)	*18.2*, p. *229*; 21.1, p. 303; 21.4, p. 326
	(a) **34**, 383 (1976)	*18.2*, p. *230*; 21.1, p. 305; 21.4, p. 326
V. A. Povitskii	(a) **33**, 783 (1976)	1.2, p. 15; *5*, p. *48*; 22.8.2, p. 427
J. Požela	(a) **35**, 43 (1976)	*13*, p. *116*; *14.3.4*, p. *197*; 22.2.1, p. 363; 22.4.3, p. 385
J. Pózniak	(a) **38**, 115 (1976)	*10.2*, p. *97*; 14.4, p. 204; 22.5.2, p. 398
G. Pozzi	(a) **35**, 165 (1976)	*14.3.3*, p. *191*; 22.1.2, p. 353
A. Prasad	(a) **35**, 109 (1976)	*3*, p. *42*; 22.1.2, p. 353
R. Prasad	(a) **38**, 11 (1976)	*1.1*, p. *10*; *1.2*, p. *14*; 10.2, p. 87; 22.3, p. 372; 22.4.1, p. 379; 22.5, p. 388; 22.5.3, p. 399; 22.8, p. 413
D. Preininger	(a) **34**, 227 (1976)	1.3, p. 22; *10.1*, p. *80*; 21, p. 292
M. Prejsa	(a) **35**, 223 (1976)	*1.2*, p. *16*; *2*, p. *37*; 5, p. 49; 21.1.1, p. 318
O. V. Presnyakova	(a) **38**, 447 (1976)	*1.4*, p. *29*; *10.1*, p. *86*; 21.1, p. 314
F. H. Preston	(a) **37**, 427 (1976)	1.1, p. 13; *20.1*, p. *274*; 22.9, p. 438
J. K. Pribram	(b) **74**, 375 (1976)	6, p. 55; *13.5.1*, p. *151*; 14.3.4, p. 196; 20.1, p. 264
A. F. Prikhotko	(b) **73**, 415 (1976)	*13.5.1*, p. *149*; *20.1*, p. *261*; 23, p. 440
R.H. Prince	(b) **78**, 271 (1976)	6, p. 62; *13.5.3*, p. *159*; 22, p. 341
F. Prinz	(a) **37**, 223 (1976)	*10.1*, p. *84*; 12.1, p. 108; 21, p. 297; 21.1, p. 311
V. D. Prochukhan	(a) **36**, K51 (1976)	*16*, p. *215*; 22.8, p. 416
A. Prodan	(a) **35**, 591 (1976)	1.2, p. 16; *1.4*, p. *27*; 22, p. 337
	(a) **36**, K21 (1976)	*1.4*, p. *28*; 10.1, p. 83; 21.1, p. 309
A. V. Prokhorovich	(a) **34**, 777 (1976)	*13.4*, p. *138*; *20.3*, p. *282*; 22.2.1, p. 362
A. I. Proskura	(a) **37**, 363 (1976)	*13.4*, p. *143*; *14.3.4*, p. *199*; 20.3, p. 287; 22.4.2, p. 383
V. Prosser	(b) **75**, K45 (1976)	*16*, p. *214*; *20.1*, p. *267*; 22.8.2, p. 428
	(a) **38**, K53 (1976)	*18.2.1*, p. *238*; *20.1*, p. *277*; 21.1, p. 314; 21.4, p. 327
D. Pudewill	(b) **74**, 485 (1976)	*13.1*, p. *122*; *13.5.1*, p. *151*; 20.1, p. 265; 23, p. 440
M. Pugnet	(a) **35**, 533 (1976)	*13.4*, p. *139*; *14.3.3*, p. *191*; 22.1.2, p. 354
	(a) **36**, 495 (1976)	*13.4*, p. *141*; 14.3.3, p. 192; 22.1.2, p. 355
D. S. Puri	(b) **78**, 113 (1976)	*12.2*, p. *112*; 22.5.2, p. 397
D. I. Pushkarov	(b) **74**, K97 (1976)	*9*, p. *72*; 13.5, p. 147
Kh. I. Pushkarov	(b) **74**, K97 (1976)	*9*, p. *72*; 13.5, p. 147
V. G. Pynko	(a) **34**, 437 (1976)	*18.2.1*, p. *237*; 21.1, p. 306; 21.1.1, p. 317

G. Quelard	(a) **36**, 729 (1976)	*11*, p. *103*; 14.1, p. 165; 21.7, p. 332
A. F. Quenneville	(a) **33**, 265 (1976)	6, p. 53; *11*, p. *100*; 21.1, p. 303
M. Quintard	(a) **37**, 257 (1976)	*10.1*, p. *84*; *12.1*, p. *108*; 21.1, p. 311
A. Quirini	(a) **38**, 685 (1976)	13.4, p. 145; 14.3, p. 182; *16*, p. *217*; 22.3, p. 374
E. Qureshi	(b) 77, 77 (1976)	*6*, p. *60*; *14*, p. *163*
T. V. Quynh	(a) **36**, K85 (1976)	*1.2*, p. *18*; *7*, p. *65*; 22.8, p. 416
J. Rabier	(a) **35**, 259 (1976)	*1.4*, p. *27*; *10.2*, p. *92*; 22.8.2, p. 428
W. A. Rachinger	(b) 74, 785 (1976)	10.1, p. 80; *14.2*, p. *169*; 21, p. 292
M. Rachwalska	(a) **35**, 675 (1976)	*1.2*, p. *17*; 8, p. 69; 22.8, p. 416
B. Rączka	(b) 77, 451 (1976)	*14.2*, p. *171*; 21, p. 298
S. I. Radautsan	(a) **35**, K53 (1976)	*13.1*, p. *124*; 14.3, p. 177; 22.3, p. 372
S. Radhakrishna	(b) 75, K119 (1976)	*19*, p. *251*; 22.5.2, p. 395
	(a) **35**, 715 (1976)	*11*, p. *102*; 19, p. 252; 20.1, p. 269; 22.8.1, p. 422
	(a) **37**, 51 (1976)	13.4, p. 142; *20.3*, p. *286*; 22.5.2, p. 396
N. K. Radhakrishnan	(a) **37**, 719 (1976)	*1*, p. *9*; 10.2, p. 96; 22.8.2, p. 431
S. Radhakrisnan	(a) **38**, K151 (1976)	*14.3.4*, p. *201*; 22.9, p. 439
E. I. Radovanova	(a) **35**, 37 (1976)	3, p. 42; *13.4*, p. *138*; 14.3, p. 176; 20.1, p. 268 20.3, p. 283; 22.3, p. 372
R. K. Radu	(a) **36**, 659 (1976)	*14.3.1*, p. *185*; 22.2.2, p. 365
A. M. Radwan	(b) 76, 605 (1976)	*13.1*, p. *126*; 20, p. 258; 21, p. 297; 22.7, p. 411
	(b) 77, 615 (1976)	*1.3*, p. *23*; 21, p. 298
V. K. Raev	(a) **34**, K163 (1976)	*18.2*, p. *231*
J. Rafałowicz	(a) **37**, K5 (1976)	*8*, p. *70*; 22.2.1, p. 364
M. H. S. Rahim	(a) **34**, 533 (1976)	6, p. 56; *8*, p. *68*; *10.1*, p. *80*; 21.1, p. 306
A. K. Rai	(a) **35**, 719 (1976)	*1*, p. *9*; 22.5, p. 389
B. P. Rai	(a) **36**, 591 (1976)	*14.3.1*, p. *185*; 14.3.4, p. 198; 22.6, p. 406
R. S. Rai	(a) **36**, 699 (1976)	*1.4*, p. *28*; *3*, p. *43*; 22.3, p. 373
	(a) **37**, 139 (1976)	*1.1*, p. *12*; *12.2*, p. *111*; 22.3, p. 373
H. Raidt	(a) **35**, 639 (1976)	*1.4*, p. *27*; 14.3, p. 177; 22.1.2, p. 354
	(a) **36**, 173 (1976)	*1.4*, p. *27*; *3*, p. *43*; 22.2.4, p. 370
T. Rajasekharan	(b) 77, 215 (1976)	*19*, p. *253*; 22.8.1, p. 423
B. Ralph	(a) **33**, 107 (1976)	*1.4*, p. *25*; 21.1.1, p. 316
A. Ramakanth	(b) 77, K123 (1976)	*6*, p. *62*; *19*, p. *255*
P. Ramasamy	(a) **35**, 379 (1976)	*1.2*, p. *16*; 4, p. 45; 22.5, p. 389
O. G. Ramer	(b) 73, 443 (1976)	*18.2.1*, p. *237*; 19, p. 247; 22.8.2, p. 426
H. S. Randhawa	(a) **37**, 313 (1976)	*1.2*, p. *18*; 1.5, p. 32; *2*, p. *40*; 22.1.1, p. 348
N. R. Ranganathan	(b) 74, 667 (1976)	*13.1*, p. *123*; *18*, p. *222*
C. N. R. Rao	(a) **36**, 669 (1976)	*14*, p. *162*; 15, p. 210; *18*, p. *224*; 22.8.1, p. 423
E. V. K. Rao	(a) **33**, 683 (1976)	11, p. 100; *20.1*, p. *263*; 22.2.1, p. 362
K. Narahari Rao	(b) 77, 215 (1976)	*19*, p. *253*; 22.8.1, p. 423
K. V. Rao	(b) 74, K23 (1976)	*18.4*, p. *243*; 21.1, p. 304
K. V. S. Rama Rao	(b) 77, 215 (1976)	*19*, p. *253*; 22.8.1, p. 423
M. K. Rao	(a) **38**, K93 (1976)	*14.3.1*, p. *186*; 22.4.2, p. 384
T. Bhaskar Rao	(a) **35**, 715 (1976)	*11*, p. *102*; 19, p. 252; 20.1, p. 269; 22.8.1, p. 422
T. R. Rao	(b) 75, K31 (1976)	*6*, p. *56*; 20.1, p. 267; 22.8.1, p. 421
L. N. Rashkovich	(a) **38**, 401 (1976)	*3*, p. *44*; 14.4.1, p. 208; 22.8.1, p. 424
W. Rasp	(a) **34**, K203 (1976)	1.6, p. 34, *17*, p. *218*; 22.6, p. 404
E. M. Raspopova	(a) **37**, 45 (1976)	2, p. 40; *16*, p. *216*; *20.3*, p. *286*; 22, p. 339
I. G. Ratishvili	(b) 75, 401 (1976)	*1.2*, p. *16*; *1.3*, p. *22*; 21, p. 294
V. V. Ratnam	(a) **36**, 693 (1976)	*10.2*, p. *94*; 20.3, p. 285; 22.5.3, p. 400
H. Rau	(a) **36**, K55 (1976)	*14.3*, p. *178*; 22.7, p. 411
Ch. J. Raub	(b) 78, K107 (1976)	*14.2*, p. *172*; *19*, p. *256*; 21, p. 300
A. Räuber	(b) 74, 169 (1976)	13.4, p. 136; *19*, p. *248*; 22.8, p. 414
J. Raułuszkiewicz	(b) 74, K93 (1976)	*6*, p. *55*; *14.3.3*, p. *191*; 22.4.2, p. 383; 22.4.3, p. 385

M. Rautenberg	(b) **75**, 553 (1976)	13.5.2, p. 157; *14.1*, p. *165*; *20.1*, p. *268*; 21.7, p. 332
S. Ravelet	(a) **33**, 517 (1976)	*14.3.2*, p. *187*; *16*, p. *213*; 22.4.1, p. 379
R. D. Rawlings	(a) **34**, 117 (1976)	*1.3*, p. *22*; 21.1, p. 304; 21.1.1, p. 317
B. K. Ray	(a) **33**, K101 (1976)	1.3, p. 21; *12.1*, p. *107*; 21.1, p. 304; 21.6, p. 329
J. Ray	(a) **34**, K169 (1976)	*18.1*, p. *225*; *18.2*, p. *231*; 21.1, p. 307; 21.1.1, p. 318
	(a) **35**, 73 (1976)	*5*, p. *49*; 21.1, p. 307; 21.1.1, p. 318
A. Raymond	(b) **76**, 223 (1976)	*13*, p. *117*; *14.3.4*, p. *197*; 22.2.3, p. 368
	(b) **76**, 623 (1976)	*14.3*, p. *178*; 22.2.4, p. 370
	(a) **36**, 133 (1976)	*14.3*, p. *178*; 22.4.1, p. 381
F. Raynal	(b) **76**, K5 (1976)	*19*, p. *252*; 22.8.1, p. 422
H. Rechenberg	(b) **73**, K119 (1976)	*18.2*, p. *229*; 21.1, p. 303; 21.1.1, p. 317
K. V. Reddy	(a) **37**, 687 (1976)	*5*, p. *50*; 18.4, p. 245; 21.1.1, p. 321
A. Rehavy	(b) **74**, 245 (1976)	*10.2*, p. *89*; 13.5.1, p. 151; 20.1, p. 263; 20.3, p. 281; 22.5.2, p. 393
U. Rehberg	(a) **34**, 573 (1976)	14.3, p. 175; *14.4*, p. *202*; 22.9, p. 435
P. Reiche	(a) **33**, 737 (1976)	20.1, p. 263; *20.2*, p. *279*; *20.3*, p. *281*; 22.8, p. 413
W. Reichelt	(a) **37**, 541 (1976)	*14.3*, p. *180*; 15, p. 211; 18.1, p. 227; 22.6, p. 407
	(a) **38**, K23 (1976)	*14.3*, p. *181*; *15*, p. *211*; 22.6, p. 408; 22.6.1, p. 409
B. Reimer	(b) **73**, 709 (1976)	13.1, p. 121; *20.1*, p. *262*; 22.9, p. 433
	(b) **76**, 533 (1976)	*16*, p. *215*; 22.9, p. 437
P. Reineker	(b) **74**, 121 (1976)	*13.5.1*, p. *150*; *19*, p. *248*; 22.9, p. 434
Z. Rek	(a) **34**, 705 (1976)	*1.1*, p. *11*; 10.2, p. 91; 22.1.2, p. 353
A. Reklaitis	(a) **35**, 43 (1976)	*13*, p. *116*; *14.3.4*, p. *197*; 22.2.1, p. 363; 22.4.3, p. 385
D. Renard	(a) **35**, 329 (1976)	*1.1*, p. *12*; 22.6, p. 404
P. Rennert	(b) **77**, 615 (1976)	*1.3*, p. *23*; 21, p. 298
R. Resta	(b) **73**, 371 (1976)	*13*, p. *114*; 22, p. 333
G. Revel	(b) **75**, 433 (1976)	*10.1*, p. *82*; 14.1, p. 165; 21, p. 294; 21.3, p. 324
T. Rewaj	(b) **74**, 683 (1976)	*19*, p. *250*; 22.5.3, p. 399
	(b) **78**, K55 (1976)	*19*, p. *255*; 22.5, p. 390
C. L. Reynolds, Jr	(a) **36**, K109 (1976)	*20.3*, p. *285*; 22.3, p. 373
S. M. Rezende	(b) **73**, 661 (1976)	13, p. 114; 18, p. 222; *20*, p. *257*; 22.8.2, p. 427
B. Rheinländer	(b) **75**, 261 (1976)	13.5.2, p. 157; *18*, p. *223*; *20.1*, p. *267*; 22.2.1, p. 363
B. Riccò	(b) **77**, 287 (1976)	*13*, p. *118*; 21.1, p. 311
D. D. Richardson	(b) **74**, K61 (1976)	10.2, p. 90; *18.2*, p. *230*; *20.1*, p. *264*; 22, p. 335
G. Richter	(a) **38**, K89 (1976)	13.4, p. 145; *16*, p. *217*; 22.2.2, p. 366
J. Richter	(b) **74**, K125 (1976)	*2*, p. *37*; *18.2*, p. *231*
	(b) **77**, K43 (1976)	*2*, p. *39*; *18.2*, p. *234*
R. de Ridder	(a) **33**, 383 (1976)	*1.1*, p. *10*; 1.4, p. 25; 22, p. 334
	(a) **37**, 591 (1976)	*1.2*, p. *19*; 1.4, p. 28; 22, p. 340
	(a) **38**, 185 (1976)	*1.2*, p. *19*; 1.4, p. 29; 22, p. 341
	(a) **38**, 663 (1976)	*1.4*, p. *29*; 10.2, p. 99; 22, p. 342
H. Riedel	(b) **77**, 581 (1976)	*10.1*, p. *84*; *14.2*, p. *171*; 21, p. 298
E. Rieflin	(b) **76**, 547 (1976)	*13.5.1*, p. *154*; 20.1, p. 271; 22.5, p. 389
M. Rigdhal	(a) **35**, 173 (1976)	*10*, p. *76*
M. Riggin	(a) **37**, 51 (1976)	13.4, p. 142; *20.3*, p. *286*; 22.5.2, p. 396
J. Rivas	(a) **37**, 151 (1976)	*18.3*, p. *241*; 22.8.2, p. 430
J.-P. Rivera	(a) **35**, K105 (1976)	1.2, p. 17; *19*, p. *252*; 22.8, p. 416
J. P. Riviére	(a) **38**, 713 (1976)	10.1, p. 86; *11*, p. *105*; 14.1, p. 168; 21.1.1, p. 322
A. Rizzo	(a) **33**, 563 (1976)	*1.1*, p. *11*; *3*, p. *42*; 22.3.1, p. 375

S. Robaszkiewicz	(b) **73**, K35 (1976)	*13*, p. *114*; *18.2*, p. *228*
	(b) **76**, K127 (1976)	*13*, p. *118*; *18*, p. *224*
G. Robbrecht	(a) **37**, 127 (1976)	18.1, p. 226; *18.3*, p. *241*; 18.4, p. 245; 22.8.2, p. 430
G. G. Robbrecht	(a) **34**, 583 (1976)	*1*, p. *9*; 1.2, p. 15; *18.3*, p. *240*; 22.8.2, p. 428
H. Robert	(b) **74**, 217 (1976)	13.5.1, p. 151; *19*, p. *249*; 22.9, p. 434
	(a) **38**, K159 (1976)	*20.1*, p. *278*; 22.9, p. 439
J.-L. Robert	(a) **33**, 715 (1976)	*7*, p. *65*; *14.3.4*, p. *196*; 22.2.3, p. 367
	(b) **76**, 223 (1976)	*13*, p. *117*; *14.3.4*, p. *197*; 22.2.3, p. 368
	(b) **76**, 623 (1976)	*14.3*, p. *178*; 22.2.4, p. 370
M. C. Robert	(a) **35**, 315 (1976)	10.2, p. 92; *20.3*, p. *283*; 22.8.1, p. 421
V. N. E. Robinson	(a) **35**, 371 (1975)	*1.4*, p. *27*; *10.2*, p. *92*; 22.5.2, p. 394
E. M. Rockar	(a) **33**, K91 (1976)	*13.4*, p. *136*; 16, p. 213; 22.1.2, p. 352
R. Rodloff	(a) **37**, K93 (1976)	*10.1*, p. *84*; 11, p. 103; 21.1, p. 312
S. Rogaschewski	(b) **75**, K173 (1976)	*17*, p. *219*; 21.1.1, p. 319
H. Rogass	(a) **36**, K9 (1976)	*3*, p. *43*; 22.1.3, p. 359
J. D. Rogers	(a) **37**, 133 (1976)	*5*, p. *50*; *18.2*, p. *235*; 21.1.1, p. 320
A. L. Roitburd	(a) **35**, 735 (1976)	*1.2*, p. *17*; *1.4*, p. *27*
	(a) **37**, 329 (1976)	1.4, p. 28; *1.5*, p. *32*
A. B. Roitsin	(b) **74**, 323 (1976)	18.1, p. 225; *19*, p. *249*
S. Rolandson	(b) **77**, 693 (1976)	*6*, p. *61*; 12.2, p. 112; 22.5.4, p. 401
S. Rolland	(a) **34**, 307 (1976)	*14.3*, p. *174*; 22.2.1, p. 362
E. P. Romanov	(a) **38**, 433 (1976)	*11*, p. *104*; *14.2*. p. *172*; 21, p. 300
Yu. A. Romanov	(b) **73**, 327 (1976)	13.1, p. 121; *14.3.4*, p. *195*; 22, p. 333
T. Romanovskis	(b) **73**, K13 (1976)	*8*, p. *67*; 22.5.4, p. 401
N. Romeo	(a) **34**, 651 (1976)	*14.3*, p. *175*; 15, p. 209; 22.8, p. 414
	(a) **36**, K33 (1976)	*20.3*, p. *285*; 22.8, p. 416
F. Ronconi	(a) **35**, K161 (1976)	*18.3*, p. *240*; 22.8.2, p. 429
G. Röpke	(b) **74**, K101 (1976)	*14*, p. *162*
	(b) **76**, 695 (1976)	6, p. 59; 8, p. 69; *14.1*, p. *165*
K. Rosan	(a) **38**, 611 (1976)	*10.1*, p. *86*; 14.1, p. 167; 21, p. 301
J. Röseler	(b) **73**, 541 (1976)	13.5.3, p. 159; *14.3.4*, p. *195*; *20.1*, p. *262*; 22, p. 334
P. Rosemann	(a) **36**, 617 (1976)	*18.3*, p. *241*; 22.8.2, p. 430
H. Rosenberger	(a) **35**, K89 (1976)	*19*, p. *251*; 22.9, p. 436
M. Rosenzweig	(b) **75**, 137 (1976)	*13.5.1*, p. *152*; *20.1*, p. *267*; 20.3, p. 282; 22.4, p. 376
M. Rösler	(b) **75**, 633 (1976)	*13.5*, p. *147*; 13.5.2, p. 157; *20.3*, p. *384*; 22.4.1, p. 380
V. I. Roslyakov	(a) **35**, K23 (1976)	*11*, p. *101*; 21, p. 294
P. L. Rossiter	(b) **74**, 785 (1976)	10.1, p. 80; *14.2*, p. *169*; 21, p. 292
K. Rössler	(b) **74**, 193 (1976)	*1.2*, p. *15*; 5, p. 48; 20.1, p. 263; 22.8, p. 414
A. Rost	(a) **33**, 143 (1976)	*14.4.1*, p. *206*; 22.8.1, p. 419
S. J. Rothmann	(a) **35**, K13 (1976)	*9*, p. *73*; *11*, p. *101*; 21.1, p. 307
M. Roulet	(a) **35**, 97 (1976)	*10.2*, p. *92*; 11, p. 101; 22.5.1, p. 391
J. J. Rousseau	(b) **73**, 625 (1976)	*1.2*, p. *14*; *19*, p. *248*; 22.8, p. 413
	(b) **77**, 195 (1976)	*4*, p. *46*; *13.4*, p. *142*; 19, p. 253; 22.8, p. 417
M. Rousseau	(b) **73**, 625 (1976)	*1.2*, p. *14*; *19*, p. *248*; 22.8, p. 413
A. P. Roy	(b) **75**, K31 (1976)	*6*, p. *56*; 20.1, p. 267; 22.8.1, p. 421
D. Roy Chaudhuri see under letter "C"		
L. D. Rozenshtein	(b) **78**, K59 (1976)	*20.1*, p. *276*; 22.8, p. 418
I. Rückmann	(b) **75**, 213 (1976)	*13.5.1*, p. *152*; 20.1, p. 267; 22.4.1, p. 380
Yu. V. Rud	(a) **36**, K51 (1976)	*16*, p. *215*; 22.8, p. 416
V. Rudat	(b) **77**, K63 (1976)	*14.2*, p. *171*
J. Rüdiger	(a) **38**, K45 (1976)	*5*, p. *51*; 22.1, p. 344
A. S. Rudnev	(a) **35**, K23 (1976)	*11*, p. *101*; 21, p. 294
I. M. Rudnik	(a) **37**, 199 (1976)	*14.3*, p. *179*; 22.7, p. 412

P. Rudra	(b) **75**, 145 (1976)	*13*, p. *115*; *18*, p. *223*
K. Ruebenbauer	(b) **74**, K165 (1976)	*5*, p. *49*
R. Ruggeri	(b) **75**, K93 (1976)	2, p. 38; *14.1*, p. *165*; *15*, p. *210*; 21, p. 295
L. Ruggiero	(a) **36**, 569 (1976)	*14.3.4*, p. *198*; 22.3, p. 373
W. Rühle	(b) **73**, 255 (1976)	13.4, p. 134; *13.5.1*, p. *149*; 20.1, p. 260; *20.3*, p. *280*; 22.2.3, p. 367
	(b) **78**, 311 (1976)	*13.1*, p. *128*; 20.3, p. 287; 22.2.1, p. 364
A. L. Ruoff	(a) **38**, 781 (1976)	*12.1*, p. *109*; 21.2, p. 323
P. Runow	(a) **33**, 135 (1976)	1.3, p. 21; *18.1*, p. *225*; *18.2*, p. *229*; 21.1.1, p. 316
W. Ruppel	(a) **33**, 195 (1976)	13.5.1, p. 149; *16*, p. *213*; 22.4.2, p. 383; 22.4.4, p. 387
	(b) **75**, 471 (1976)	*13.5.1*, p. *152*; *20.3*, p. *283*; 22.4.3, p. 385
R. Ruppin	(b) **74**, 647 (1976)	*13.5.2*, p. *157*; *20.1*, p. *265*; 21, p. 292
A. P. Rusakov	(b) **75**, 73 (1976)	*4*, p. *45*; 12.2, p. 110; 22.1, p. 343; 22.2, p. 361; 22.4, p. 376; 22.5, p. 389
L. Rusiniak	(b) **78**, 265 (1976)	13.1, p. 128; *19*, p. *255*; 20.1, p. 276; 22.2.3, p. 369
E. Ruske	(a) **35**, 477 (1976)	*1.5*, p. *31*; 2, p. 38; *20.1*, p. *269*; 22, p. 337
	(a) **35**, K113 (1976)	2, p. 38; *20.1*, p. *270*; 22, p. 337
R. Ruskov	(a) **37**, 295 (1976)	*5*, p. *50*; 18.4, p. 245; 22.6, p. 407
M. Russo	(a) **35**, K173 (1976)	*14.2*, p. *170*; 21, p. 295; 22.6, p. 405
K. A. Rustamov	(b) **76**, 437 (1976)	*6*, p. *58*; *13.4*, p. *140*; 22.8.2, p. 429
G. Ruszczynski	(b) **76**, 427 (1976)	13.4, p. 140; *20.1*, p. *271*; 22.4.1, p. 380
G. Ruymbeek	(b) **78**, K89 (1976)	*2*, p. *41*; 18, p. 224; 20, p. 259; 22, p. 342
A. V. Ruzhnikov	(b) **74**, 689 (1976)	13.1, p. 123; *14.4.1*, p. *206*; *20.1*, p. *266*; 22.8.1, p. 420
A. L. Rvachev	(a) **37**, K21 (1976)	*16*, p. *216*; 22.4.2, p. 383
S. M. Ryabchenko	(b) **76**, 183 (1976)	*19*, p. *252*; 22.5.3, p. 400
P. V. Ryabko	(a) **35**, 575 (1976)	*10.1*, p. *82*; 12.1, p. 108
K. P. Ryaboshapka	(a) **35**, 575 (1976)	*10.1*, p. *82*; 12.1, p. 108
C. Rybka	(b) **76**, K35 (1976)	*10.2*, p. *93*; 22.9, p. 437
M. A. Rzaev	(a) **34**, K7 (1976)	*13.4*, p. *137*; *14.3.3*, p. *190*; 22.2.3, p. 367
H. Rzewuski	(a) **38**, 139 (1976)	*1.2*, p. *19*; *11*, p. *104*; 22.1.2, p. 357
V. A. Saakyan	(a) **38**, K49 (1976)	*8*, p. *71*; *15*, p. *211*; 22.2.4, p. 371
F. Sacchetti	(b) **77**, 309 (1976)	*18.2*, p. *234*; 21.1, p. 311
A. V. Sachenko	(a) **38**, 131 (1976)	11, p. 104; *14.3.3*, p. *194*; 22.1.2, p. 357
A. Sadoc	(b) **78**, 779 (1976)	6, p. 64; *10.2*, p. *98*; 13.4, p. 145; 20.1, p. 278; 22.5.2, p. 398
S. A. Sadreddinov	(a) **33**, K67 (1976)	*14.3.1*, p. *183*; 14.3.2, p. 187; 22.9, p. 433
	(a) **34**, K77 (1976)	*14.3.1*, p. *183*; 22.9, p. 435
V. A. Safonov	(a) **38**, 45 (1976)	10.2, p. 97; *14.3*, p. *180*; 22.1.2, p. 357
Y. Sagara	(a) **33**, 691 (1976)	1.5, p. 31; *2*, p. *36*; 22.7, p. 410
H. Saha	(a) **37**, 625 (1976)	13.3, p. 133; *14.3.3*, p. *193*; *16*, p. *216*; 22.1.2, p. 357; 22.2.1, p. 364; 22.4.3, p. 386
V. C. Sahni	(b) **75**, K31 (1976)	*6*, p. *56*; 20.1, p. 267; 22.8.1, p. 421
V. Saile	(b) **74**, 485 (1976)	*13.1*, p. *122*; *13.5.1*, p. *151*; 20.1, p. 265; 23, p. 440
R. W. Sailer	(a) **36**, 189 (1976)	*12.2*, p. *111*; *14.4.1*, p. *207*; 22.8, p. 416
F. Saint-Yves	(a) **35**, 533 (1976)	*13.4*, p. *139*; *14.3.3*, p. *191*; 22.1.2, p. 354
S. Saito	(a) **34**, 671 (1976)	*18.3*, p. *240*; *19*, p. *250*; 22.8.2, p. 428
G. Saitta	(b) **74**, 317 (1976)	*13.4*, p. *136*; *20.1*, p. *264*; 22.5.2, p. 393
A. P. Sakalas	(a) **35**, 571 (1976)	*13.4*, p. *139*; *14.3*, p. *177*; 22.8, p. 415
S. G. Sakharova	(b) **74**, 471 (1976)	*13.2*, p. *130*; 21, p. 292
V. M. Sakosarenko	(a) **34**, 541 (1976)	*14.2*, p. *169*; 21, p. 293

O. V. SAKSAGANSKII	(a) **36**, 73 (1976)	13.3, p. 132; *14.3.2*, p. *188*; 20.3, p. 285; 22.2.1, p. 363
L. G. SAKVARELIDZE	(a) **37**, K71 (1976)	*3*, p. *43*; 22.2.4, p. 371
E. Y. SALAEV	(b) **73**, K69 (1976)	13.5.1, p. 149; *20.1*, p. *261*; 22.8, p. 413
E. YU. SALAEV	(a) **34**, K115 (1976)	*6*, p. *56*; *20.1*, p. *266*; 22.8, p. 414
N. M. SALANSKII	(a) **36**, 779 (1976)	18.4, p. 245; *19*, p. *253*; 22.8.1, p. 423
R. SALCHOW	(b) 77, 373 (1976)	*13*, p. *118*
M. SALEEM	(a) **38**, K27 (1976)	*20.3*, p. *288*; 22.4, p. 377
N. SALEH	(a) **37**, 39 (1976)	14.3.3, p. 193; *16*, p. *216*; 22.2.3, p. 368
J. SALIK	(a) **38**, 177 (1976)	*2*, p. *40*; 11, p. 104; 22, p. 341
E. SALJE	(a) **33**, K165 (1976)	*7*, p. *65*; *20.1*, p. *263*; 22.8, p. 413
	(a) **37**, K187 (1976)	13.1, p. 128; *14.3*, p. *180*; *20.1*, p. *275*; 22.6, p. 408
A. N. SALUGIN	(a) **33**, 783 (1976)	1.2, p. 15; *5*, p. *48*; 22.8.2, p. 427
R. J. SALZBRENNER	(a) **36**, 117 (1976)	*10.1*, p. *83*; 21.1.1, p. 319
V. V. SAMARTSEV	(b) **76**, 55 (1976)	13.4, p. 140; *20.1*, p. *270*; *20.3*, p. *284*; 22.8.1, p. 422
J. R. SAMBLES	(b) **75**, 155 (1976)	*19*, p. *250*; 21, p. 293
K. SAMINADAYAR	(a) **36**, 679 (1976)	*13.3*, p. *133*; *14.3.3*, p. *192*; 22.1.2, p. 356
A. SAMOĆ	(a) **36**, 735 (1976)	*13.4*, p. *142*; 14.3.4, p. 198; 14.4, p. 203
	(a) **37**, 271 (1976)	*13.4*, p. *142*; 14.3.4, p. 199; 14.4, p. 204; 22.9, p. 438
M. SAMOĆ	(a) **36**, 735 (1976)	*13.4*, p. *142*; 14.3.4, p. 198; 14.4, p. 203
	(a) **37**, 271 (1976)	*13.4*, p. *142*; 14.3.4, p. 199; 14.4, p. 204; 22.9, p. 438
A. G. SAMOILOVICH	(b) **75** 299 (1976)	*14.3*, p. *176*; 22.1.1, p. 346
A. A. SAMOKHVALOV	(b) **73**, 455 (1976)	*18.2*, p. *229*; 22.8.2, p. 426
G. SAMUEL	(a) **34**, 615 (1976)	*10.2*, p. *91*; 13.4, p. 137; 19, p. 250; 20.1, p. 266; 22.5.2, p. 394
YU. M. SANDLER	(b) **74**, K85 (1976)	6, p. 55; *10*, p. *75*
N. K. SANYAL	(a) **35**, 387 (1976)	*2*, p. *37*; *18*, p. *223*; 22.9, p. 436
G. V. SAPARIN	(a) **35**, K5 (1976)	*20.3*, p. *283*; 22.4.1, p. 380
M. N. SAPOZHNIKOV	(b) **75**, 11 (1976)	*6*, p. *52*; 13.4, p. 134; 20.1, p. 260; *20.3*, p. *280*; 22.9 p. 433
R. S. SARASWAT	(b) **78**, K73 (1976)	*19*, p. *256*; 22.8.1, p. 424
R. M. SARDARLY	(a) **34**, K115 (1976)	*6*, p. *56*; *20.1*, p. *266*; 22.8, p. 414
C. M. SARGENT	(a) **33**, 281 (1976)	*1.3*, p. *21*; *12.1*, p. *107*; 21.1, p. 303
K. K. SARKAR	(b) **75**, 101 (1976)	*6*, p. *56*; 22.6, p. 404
S. K. SARKAR	(b) **77**, 329 (1976)	*4*, p. *46*; 22.5.2, p. 396
S. E. SARKISOV	(a) **33**, 737 (1976)	20.1, p. 263; *20.2*, p. *279*; *20.3*, p. *281*; 22.8, p. 413
	(a) **34**, K109 (1976)	*20.2*, p. *279*; *20.3*, p. *282*; 22.8.2, p. 428
V. KH. SARKISOV	(a) **36**, K25 (1976)	*14.3.4*, p. *198*; 22.8.1, p. 422
	(a) **37**, K111 (1976)	*14.3.4*, p. *198*; 22.8.1, p. 422
J. SÁRKÖZI	(a) **36**, K105 (1976)	*10.2*, p. *94*; 14.4, p. 203; 22.5.2, p. 396
G. SARRABAYROUSE	(a) **37**, K105 (1976)	14.3.2, p. 188; *14.3.3*, p. *193*
T. SASAKI	(a) **34**, 339 (1976)	*17*, p. *218*; *19*, p. *249*; 22.8.1, p. 420
D. H. SASTRY	(a) **33**, 173 (1976)	*10.1*, p. *78*; 21, p. 289; 21.1, p. 302; 21.3, p. 324
M. SATHE	(a) **36**, 453 (1976)	*13*, p. *118*; *14.3.4*, p. *198*; 22, p. 339
R. SATI	(b) **73**, 353 (1976)	*10.2*, p. *87*; 13.4, p. 134; 22.5.2, p. 392
H. SATO	(b) **73**, 723 (1976)	13.2, p. 130; *14.1*, p. *164*; 21, p. 290
T. SATOW	(a) **33**, 691 (1976)	1.5, p. 31; *2*, p. *36*; 22.7, p. 410
G. A. SAUNDERS	(a) **33**, 367 (1976)	*12.2*, p. *110*; 22.1, p. 343; 22.2, p. 361; 22.4, p. 376
N. I. SAVATEEV	(a) **37**, K165 (1976)	*1.2*, p. *19*; 5, p. 50; 21, p. 299
M. M. SAVIN	(a) **35**, 747 (1976)	*10.1*, p. *82*; 12.1, p. 108; 21.1, p. 308; 21.3, p. 324
	(a) **38**, 761 (1976)	*10.1*, p. *86*; 21, p. 301; 21.1, p. 315; 21.3, p. 325; 21.6, p. 331

E. M. Savitskii	(a) **34**, 319 (1976)	*8*, p. *68*; 10.1, p. 80; 21.1, p. 305
	(a) **37**, K165 (1976)	*1.2*, p. *19*; 5, p. 50; 21, p. 299
A. M. Savvinov	(b) **73**, 317 (1976)	*18.4*, p. *242*; *19*, p. *247*; 22.8.2, p. 426
B. Sawicka	(b) **77**, K1 (1976)	*5*, p. *50*; 22.1.2, p. 356
J. Sawicki	(b) **77**, K1 (1976)	*5*, p. *50*; 22.1.2, p. 356
G. Sberveglieri	(a) **34**, 651 (1976)	*14.3*, p. *175*; 15, p. 209; 22.8, p. 414
	(a) **36**, K33 (1976)	*20.3*, p. *285*; 22.8, p. 416
C. M. Scala	(b) **73**, K115 (1976)	*10.2*, p. *89*; 19, p. 248; 22.6, p. 403
	(b) **75**, K167 (1976)	*19*, p. *251*; 22.8, p. 415
P. J. Scanlon	(b) **74**, 711 (1976)	*11*, p. *101*; 21, p. 292
H. E. Schaefer	(b) **75**, 173 (1976)	10.1, p. 81; *11*, p. *101*; 14.1, p. 165; *18.2*, p. *231*; 21.1, p. 307
	(b) **75**, 647 (1976)	*11*, p. *102*; *18.2*, p. *232*; 21.1, p. 308
	(b) **78**, 139 (1976)	*11*, p. *104*; *18.2*, p. *235*; 21.1, p. 313
L. M. Schafe	(b) **74**, 785 (1976)	10.1, p. 80; *14.2*, p. *169*; 21, p. 292
M. Schäfer	(b) **76**, K63 (1976)	*4*, p. *46*; *20.1*, p. *271*; 22.2, p. 361
C. Schaffrin	(a) **35**, 79 (1976)	*9*, p. *72*; 14.4, p. 202; 22.8.1, p. 421
F. W. Schapink	(a) **33**, K121 (1976)	*1.4*, p. *26*; *10*, p. *75*
	(a) **38**, K107 (1976)	*1.4*, p. *30*; 21, p. 301
A. Scharmann	(a) **34**, 297 (1976)	1.6, p. 34; 13.3, p. 218; *17*, p. *218*; 22.6, p. 403
H. Schaumburg	(a) **34**, K173 (1976)	*10.2*, p. *92*; 13.4, p. 138; *20.1*, p. *266*; 22.1.1, p. 346
D. Scheerlinck	(b) **73**, 223 (1976)	*18.1*, p. *225*; *18.4*, p. *242*; 22.8.2, p. 426
L. Schenk	(a) **35**, 511 (1976)	*14.3.4*, p. *197*; 22.2.3, p. 368
C. Scherer	(b) **75**, 465 (1976)	*8*, p. *68*; *18.1*, p. *226*
G. Schiffmacher	(a) **33**, K9 (1976)	*10.2*, p. *88*; 22.6, p. 402
P. Schiller	(a) **38**, 569 (1976)	*10.1*, p. *86*; 21, p. 301
W. Schiller	(b) **74**, K101 (1976)	*14*, p. *162*
A. Schlachetzki	(a) **37**, K151 (1976)	*14.3.3*, p. *193*; 22.2. 1, p. 364
W. Schlenk	(a) **33**, 217 (1976)	*17*, p. *218*; 21, p. 290
M. Schlüter	(b) **75**, 303 (1976)	*13.1*, p. *124*; *20.1*, p. *267*; 22.7, p. 411
	(b) **77**, 295 (1976)	*13.1*, p. *127*; 22.5.1, p. 391
	(b) **78**, 737 (1976)	4, p. 47; *13.1*, p. *129*; 20.1, p. 278; 22.5, p. 390
D. Schmid	(b) **74**, 609 (1976)	13.5.1, p. 152; *19*, p. *250*; 22.8.1, p. 420
Ph. Schmid	(b) **73**, 587 (1976)	*6*, p. *54*; 20.1, p. 262; 22.5, p. 388
B. Schmidt	(b) **76**, 307 (1976)	*8*, p. *69*; 13.1, p. 125; 14.2, p. 170; 21, p. 296
D. Schmidt	(a) **36**, 617 (1976)	*18.3*, p. *241*; 22.8.2, p. 430
G. Schmidt	(a) **33**, 143 (1976)	*14.4.1*, p. *206*; 22.8.1, p. 419
	(b) **76**, K47 (1976)	1.2, p. 17; *12.2*, p. *111*; *14.4*, p. *203*; 22.8.1, p. 422
	(a) **36**, K85 (1976)	*1.2*, p. *18*; *7*, p. *65*; 22.8, p. 416
H. H. Schmidt	(b) **73**, 87 (1986)	*1.1*, p. *10*
W. Schmidt	(a) **34**, K147 (1976)	*1.1*, p. *11*; 22.2.4, p. 370
W. D. Schmidt	(a) **36**, K153 (1976)	*14.2*, p. *170*; 14.3.3, p. 193; 22.6, p. 406
R. Schnakig	(b) **77**, 657 (1976)	*13.4*, p. *143*; 18.1, p. 227; 19, p. 254
	(b) **78**, 543 (1976)	*13.4*, p. *145*; 18.1, p. 227; 19, p. 255; 20.1, p. 277; 22.5, p. 390
E. Schneider	(b) **74**, 261 (1976)	*10.1*, p. *79*; *14.2*, p. *169*; 21, p. 291
G. Schneider	(b) **74**, K75 (1976)	*13.2*, p. *130*; 14, p. 162; 19, p. 249; 22.7, p. 410
H. A. Schneider	(a) **36**, K165 (1976)	*6*, p. *60*; *8*, p. *70*; 22.1.2, p. 356
J. Schneider	(b) **73**, 533 (1976)	*13.4*, p. *135*; 19, p. 248; 20.1, p. 261; 20.3, p. 280; 22.8.2, p. 426
	(b) **74**, 169 (1976)	13.4, p. 136; *19*, p. *248*; 22.8, p. 414
	(a) **36**, K59 (1976)	*1.3*, p. *23*; *18.2*, p. *233*; 21.1, p. 309; 21.1.1, p. 320

	(a) **36**, K77 (1976)	*1.3*, p. *23*; *18.2*, p. *233*; 21.1, p. 310; 21.1.1, p. 320
	(b) 77, K103 (1976)	*14*, p. *163*
V. E. SCHNEIDER	(a) **35**, 61 (1976)	*1.2*, p. *16*; 22.5, p. 389; 22.5.4, p. 401
T. SCHOBER	(a) **33**, 673 (1976)	*1.2*, p. *14*; *1.4*, p. *25*; 10.1, p. 79; 21, p. 291
G. SCHOECK	(a) **38**, 539 (1976)	*10*, p. *77*
D. W. SCHOENFELD	(a) **34**, 255 (1976)	*13.1*, p. *122*; *20.1*, p. *264*; 22.2.4, p. 370
H. W. SCHÖPGENS	(a) **33**, 135 (1976)	1.3, p. 21; *18.1*, p. *225*; *18.2*, p. *229*; 21.1.1, p. 316
M. SCHOTT	(b) **77**, K119 (1976)	*14.3*, p. *180*; 22.9, p. 438
K. D. SCHOTTE	(a) **34**, 593 (1976)	*14.1*, p. *165*; 21, p. 293
J. SCHREIBER	(b) **74**, K125 (1976)	*2*, p. *37*; *18.2*, p. *231*
	(b) **78**, 199 (1976)	*2*, p. *40*; *13*, p. *119*
	(b) **78**, 577 (1976)	*2*, p. *41*; *13*, p. *119*; 22, p. 341
G. SCHROEDER	(a) **34** 623 (1976)	*10.1*, p. *81*; 18.2, p. 231; 20, p. 258; 21.1, p. 306
M. SCHUBERT	(a) **33**, 149 (1976)	*13.1*, p. *121*; *20.1*, p. *261*; 22.1.3, p. 359
E. SCHUBERTH	(b) **76**, 301 (1976)	*1.2*, p. *17*; 6, p. 58; 20.3, p. 284; 23, p. 440
W. SCHÜLE	(b) **78**, 689 (1976)	*9*, p. *74*; 10.1, p. 86; 21.1, p. 314
E. SCHULLER	(b) **76**, 759 (1976)	*13.5*, p. *147*; *20.1*, p. *272*; 22.6, p. 406
	(b) **78**, 123 (1976)	*6*, p. *62*; 20.1, p. 275; 22.6, p. 408
D. SCHULTZE	(a) **33**, 737 (1976)	20.1, p. 263; *20.2*, p. *279*; *20.3*, p. *281*; 22.8, p. 413
W. SCHULZ	(b) **78**, 183 (1976)	*1.6*, p. *34*; 22.5.2, p. 397
K.-R. SCHULZE	(b) **75**, 493 (1976)	*13.1*, p. *124*; 22.2.4, p. 370
F. SCHUSZTER	(a) **38**, K123 (1976)	*10.1*, p. *86*; *14.1*, p. *168*; 21, p. 301; 21.1, p. 315
C. SCHWAB	(b) **76**, 363 (1976)	*19*, p. *252*; 22.8, p. 416
E. SCHWAB	(b) **76**, 547 (1976)	*13.5.1*, p. *154*; 20.1, p. 271; 22.5, p. 389
P. SCHWELLINGER	(a) **36**, 335 (1976)	*18.2.1*, p. *237*; 21.1.1, p. 320
N. SCHWENTNER	(b) **74**, 485 (1976)	*13.1*, p. *122*; *13.5.1*, p. *151*; 20.1, p. 265; 23, p. 440
D. B. SCOTT	(a) **34**, 263 (1976)	*14.3.2*, p. *187*; *14.3.3*, p. *190*; 14.3.4, p. 196; 22.1.2, p. 352
C. H. SEAGER	(b) **75**, 289 (1976)	*9*, p. *72*; 14.3, p. 176; 22.1.1, p. 347; 22.1.2, p. 353
E. A. SECCO	(a) **33**, 427 (1976)	*1.2*, p. *14*; 8, p. 67; 14.4, p. 202; 22.5.4, p. 401
A. SEEGER	(b) **73**, 131 (1976)	*9*, p. *72*; 10, p. 75
	(b) **75**, 361 (1976)	*9*, p. *72*; 10, p. 76
	(a) **36**, 579 (1976)	*10.1*, p. *83*; *18.2*, p. *234*; 21.1.1, p. 320
M. S. SEEHRA	(b) **74**, 145 (1976)	*19*, p. *248*; 22.5, p. 388
V. SEETHARAMAN	(a) **36**, 39 (1976)	*1.3*, p. *23*; *1.4*, p. *27*; 21, p. 296
Y. SÉGUI	(a) **33**, 247 (1976)	*14.3.1*, p. *183*; 14.3.2, p. 187; 22.9, p. 433
H. K. SEHGAL	(a) **37**, 313 (1976)	*1.2*, p. *18*; 1.5, p. 32; *2*, p. *40*; 22.1.1, p. 348
T. SEIDENKRANZ	(a) **33**, 205 (1976)	*12.1*, p. *107*; 21.1, p. 302; 21.3, p. 324
V. N. SELEZNEV	(a) **36**, 779 (1976)	18.4, p. 245; *19*, p. *253*; 22.8.1, p. 423
G. YA. SELYUTIN	(b) **75**, 233 (1976)	*5*, p. *49*; 18, p. 223
G. K. SEMIN	(a) **33**, K63 (1976)	*19*, p. *247*; 22.9, p. 433
	(a) **33**, K111 (1976)	*19*, p. *248*; 22.9, p. 434
S. SENGUPTA	(b) **77**, 329 (1976)	*4*, p. *46*; 22.5.2, p. 396
M. SENONER	(b) **75**, 213 (1976)	*13.5.1*, p. *152*; 20.1, p. 267; 22.4.1, p. 380
W. SENSKE	(b) **75**, 553 (1976)	13.5.2, p. 157; *14.1*, p. *165*; *20.1*, p. *268*; 21.7, p. 332
V. V. SERDYUK	(a) **33**, K145 (1976)	13.4, p. 136; *16*, p. *213*; 22.4.2, p. 383
V. M. SERGEEVA	(a) **37**, 199 (1976)	*14.3*, p. *179*; 22.7, p. 412
R. SERNEELS	(b) **74**, 359 (1976)	*1.1*, p. *11*
J. SERUGHETTI	(a) **33**, 405 (1976)	*10.2*, p. *88*; 22.8.1, p. 419
	(a) **36**, 637 (1976)	*10.2*, p. *94*; 22.8.1, p. 423
	(a) **37**, 263 (1976)	*10.2*, p. *95*; 12.2, p. 112; 22.8.1, p. 423

S. M. Shahinyan	(b) 77, K171 (1976)	*13*, p. *119*; *20.1*, p. *274*; 22, p. 340
S. N. Shakhmaev	(a) **34**, K187 (1976)	*10.1*, p. *81*; 21, p. 293; 21.6, p. 330
S. I. Shakhovtsov	(a) **34**, K35 (1976)	11, p. 101; *13.3*, p. *132*; *14.3.3*, p. *191*; 22.1.2, p. 353
U. A. Shakirov	(a) **37**, 681 (1976)	9, p. 74; *11*, p. *104*; *14.3.2*, p. *188*; 22.1.2, p. 357
A. N. Shalygin	(b) 78, K85 (1976)	*18.1*, p. *227*; *18.2*, p. *236*; 22.8.2, p. 431
V. F. Shamray	(b) 77, K129 (1976)	*14.2*, p. *171*; 22.8, p. 417
L. I. Shanskii	(b) 73, 415 (1976)	*13.5.1*, p. *149*; *20.1*, p. *261*; 23, p. 440
Yu. P. Sharkeev	(a) **37**, 701 (1976)	*1.3*, p. *24*; *10.1*, p. *84*; 21.1, p. 312
K. C. Sharma	(b) 76, 337 (1976)	*18.2*, p. *233*
	(b) 76, K75 (1976)	*2*, p. *39*; *13*, p. *118*; 21.2, p. 323
	(b) 78, K93 (1976)	*2*, p. *41*; 22.5, p. 390
N. D. Sharma	(b) 78, K125 (1976)	*5*, p. *51*; 6, p. 64; 21.1, p. 314
R. V. Sharma	(b) 76, K75 (1976)	*2*, p. *39*; *13*, p. *118*; 21.2, p. 323
	(b) 78, K93 (1976)	*2*, p. *41*; 22.5, p. 390
Vijender Sharma	(a) **36**, 269 (1976)	*14.3.3*, p. *192*; *14.3.4*, p. *198*; 22.2.1, p. 363
Y. K. Sharma	(a) **34**, K87 (1976)	*14.3.3*, p. *191*
G. R. Sharp	(b) 75, 607 (1976)	*19*, p. *251*; 22.5, p. 389
G. Sharp-Dent	(b) 75, 155 (1976)	*19*, p. *250*; 21, p. 293
M. I. Sheboul	(b) 73, 165 (1976)	*13.5.1*, p. *149*; 22.5, p. 388
	(b) 73, 475 (1976)	13.1, p. 121; *13.5.1*, p. *150*; 20.1, p. 261; 20.3, p. 280; 22.5, p. 388
	(b) 74, 523 (1976)	*13.5.1*, p. *151*; 20.3, p. 282; 22.5, p. 388
	(b) 76, K89 (1976)	*13.5.1*, p. *154*; 22.4.1, p. 381
S. S. Sheinin	(b) 73, 431 (1976)	*1.1*, p. *10*; 21, p. 290; 21.6, p. 329
	(a) **38**, 151 (1976)	*1.4*, p. *28*; 21.1, p. 313
	(a) **38**, 675 (1976)	*1.4*, p. *29*; 21.1, p. 314
M. K. Sheinkman	(a) **33**, 765 (1976)	10.2, p. 89; *14.3*, p. *174*; 22.4.1, p. 379
	(a) **34**, 397 (1976)	10.2, p. 90; *14.3*, p. *174*; 22.4.1, p. 379
D. I. Sheka	(b) 76, 413 (1976)	13.1, p. 125; *14.4*, p. *203*; 22.2, p. 361
E. F. Sheka	(b) 75, 105 (1976)	*6*, p. *56*; 20.1, p. 267; 22.9, p. 435
	(b) 78, 325 (1976)	*6*, p. *63*; *20.1*, p. *276*; 22.9, p. 438
	(b) 78, K1 (1976)	*6*, p. *63*; *13.4*, p. *144*; 20.1, p. 276; 20.3, p. 287; 22.9, p. 439
I. Yu. Shelestina	(b) 73, K145 (1976)	*18.4*, p. *242*
Y. R. Shen	(b) 75, 303 (1976)	*13.1*, p. *124*; *20.1*, p. *267*; 22.7, p. 411
T. D. Shermergor	(a) **38**, 375 (1976)	*12*, p. *106*
Shi-yu Wu	(b) 74, 349 (1976)	*6*, p. *55*; *10*, p. *75*
	(b) 77, K5 (1976)	*6*, p. *60*; 10, p. 76
M. Shiga	(a) **37**, K33 (1976)	*18.2*, p. *235*; 21.1, p. 311; 21.1.1, p. 321
	(a) **37**, K89 (1976)	*5*, p. *50*; 18.2, p. 235; 21.1, p. 311; 21.1.1, p. 321
A. Ya. Shik	(a) **36**, 53 (1976)	*14.3.1*, p. *184*; *16*, p. *215*; 22.2.1, p. 363
T. Shimizu	(b) 73, K19 (1976)	2, p. 35; *19*, p. *247*; *20.1*, p. *261*; 22, p. 333
	(b) 74, K39 (1976)	*2*, p. *36*; *13*, p. *115*; 22.3, p. 372
B. K. Shishediev	(a) **35**, 709 (1976)	*9*, p. *73*; 22.9, p. 436
T. Shishido	(a) **38**, 453 (1976)	*1.1*, p. *13*; 1.4, p. 29; 22.1.1, p. 350
E. V. Shitova	(a) **36**, 81 (1976)	*1.5*, p. *31*; 2, p. 39; *11*, p. *102*; 22.1.2, p. 355
B. I. Shklovskii	(b) 76, 475 (1976)	*14.1*, p. *165*; *14.4*, p. *203*
O. I. Shklyarevskii	(a) **35**, 49 (1976)	*14.3.3*, p. *191*; 22.6, p. 404
G. M. Shmelev	(b) 76, 811 (1976)	*14.3*, p. *179*
S. Z. Shmurak	(b) 74, 753 (1976)	10.2, p. 90; 13.5.1, p. 152; *20.3*, p. *282*; 22.5.2, p. 394
K. A. Shoaib	(a) **35**, 181 (1976)	*1.2*, p. *16*; 22.6, p. 404
Chan Min Shon see under letter "C"		
V. S. Shpinel	(b) 78, 103 (1976)	*5*, p. *50*; 18.2, p. 235; 18.4, p. 245; 21.4, p. 327
K. N. Shrivastava	(b) 78, 749 (1976)	*6*, p. *64*; *18.2*, p. *236*; 22.5.4, p. 401

E. L. SHTRUM	(a) **33**, 765 (1976)	10.2, p. 89; *14.3*, p. *174*; 22.4.1, p. 379
	(a) **34**, 397 (1976)	10.2, p. 90; *14.3*, p. *174*; 22.4.1, p. 379
M. M. SHUKLA	(b) **76**, 371 (1976)	*6*, p. *58*; 21.6, p. 330
	(b) **77**, 433 (1976)	*6*, p. *61*; 21, p. 298; 21.1, p. 312; 21.5, p. 328; 21.6, p. 331
YU. M. SHULGA	(b) **73**, 335 (1976)	13, p. 113; 17, p. 218; *20*, p. *257*; 22.9, p. 433
S. G. SHULMAN	(b) **73**, K137 (1976)	13.1, p. 122; *20.1*, p. *262*; 21.4, p. 326
	(b) **75**, K153 (1976)	*20.1*, p. *269*; 21, p. 295
L. A. SHUVALOV	(a) **36**, K85 (1976)	*1.2*, p. *18*; *7*, p. *65*; 22.8, p. 416
YU. M. SHVARTS	(a) **35**, K65 (1976)	*14.3.1*, p. *184*; 22.1.1, p. 347
I. SHY-YIH WANG	(b) **77**, 295 (1976)	*13.1*, p. *127*; 22.5.1, p. 391
S. K. SIDOROV	(a) **33**, K83 (1976)	*18.3*, p. *239*; 22.8.2, p. 427
	(b) **76**, K113 (1976)	*18.2*, p. *234*; 18.4, p. 244; 21.1, p. 310; 21.1.1, p. 320
	(a) **38**, 57 (1976)	11, p. 104; *18.3*, p. *241*; 22.8.2, p. 431
	(a) **38**, 433 (1976)	*11*, p. *104*; *14.2*, p. *172*; 21, p. 300
	(a) **38**, K147 (1976)	11, p. 105; *18.3*, p. *241*; 22.8.2, p. 432
J. HÖNER ZU SIEDERDISSEN see under letter "H"		
E. SIEGEL	(b) **77**, 229 (1976)	18.2.1, p. 238; *19*, p. *254*; 21.1, p. 310; 21.1.1, p. 320
V. SIEGEL	(a) **34**, K203 (1976)	1.6, p. 34; *17*, p. *218*; 22.6, p. 404
W. SIEGEL	(a) **35**, 269 (1976)	6, p. 56; 13.4, p. 139; *14.3*, p. *176*; 22.8, p. 415
	(a) **36**, 491 (1976)	*6*, p. *59*; *14.3*, p. *179*; 22.8, p. 417
K. E. SIEKIERSKA	(b) **74**, 183 (1976)	*5*, p. *48*; 22.5.2, p. 393; 22.5.3, p. 399
D. SIEMERS	(a) **34**, 547 (1976)	*18.2*, p. *231*; 21, p. 293
K. SIERAŃSKI	(b) **77**, K91 (1976)	*13.1*, p. *128*; 22.4.4, p. 387
J. D. SIEVERT	(a) **37**, 205 (1976)	*18.2*, p. *235*; 21.1.1, p. 320
E. SIGMUND	(b) **76**, 325 (1976)	6, p. 58; 13.4, p. 140; *20.1*, p. *270*
V. I. SILANTEV	(b) **74**, 471 (1976)	*13.2*, p. *130*; 21, p. 292
E. A. SILINSH	(a) **34**, 747 (1976)	13.4, p. 138; *14.3.4*, p. *197*; 22.9, p. 435
	(b) **75**, 247 (1976)	*13.3*, p. *132*; 14.3.3, p. 191; 16, p. 214; 22.9, p. 436
A. SIMON	(b) **74**, 297 (1976)	8, p. 67; *18.4*, p. *242*; 20.1, p. 264; 22.8.1, p. 420
J. P. SIMON	(a) **38**, 207 (1976)	10.1, p. 85; *13.4*, p. *144*; 21, p. 300; 21.2, p. 323
M. I. SIMONOVA	(b) **73**, 455 (1976)	*18.2*, p. *229*; 22.8.2, p. 426
Z. ŠIMŠA	(a) **34**, 639 (1976)	*18.3*, p. *240*; 22.8.2, p. 428
J. ŠIMŠOVÁ	(a) **33**, K133 (1976)	*18.3*, p. *239*; *19*, p. *248*; 22.8.2, p. 427
	(a) **34**, 639 (1976)	*18.3*, p. *240*; 22.8.2, p. 428
	(a) **38**, K163 (1976)	*18.3*, p. *241*; *19*, p. *256*; 22.8.2, p. 432
G. SINGH	(a) **36**, 699 (1976)	*1.4*, p. *28*; *3*, p. *43*; 22.3, p. 373
	(a) **37**, 139 (1976)	*1.1*, p. *12*; *12.2*, p. *111*; 22.3, p. 373
K. SINGH	(a) **36**, 591 (1976)	*14.3.1*, p. *185*; 14.3.4, p. 198; 22.6, p. 406
K. P. SINGH	(b) **78**, 677 (1976)	*6*, p. *64*; 20.1, p. 277; 22.5.1, p. 391
NAWAB SINGH	(b) **76**, K39 (1976)	*10.2*, p. *93*; *20.1*, p. *271*; 22.5.2, p. 395
R. K. SINGH	(b) **78**, 677 (1976)	*6*, p. *64*; 20.1, p. 277; 22.5.1, p. 391
R. N. SINGH	(b) **78**, K45 (1976)	2, p. 40; *14.1*, p. *167*; 21.2, p. 323
SATBIR SINGH	(a) **38**, 355 (1976)	*14.3.1*, p. *185*; *14.4*, p. *204*; 22.9, p. 439
S. R. SINGH	(a) **36**, 699 (1976)	*1.4*, p. *28*; *3*, p. *43*; 22.3, p. 373
	(a) **37**, 139 (1976)	*1.1*, p. *12*; *12.2*, p. *111*; 22.3, p. 373
V. P. SINGH	(b) **78**, 381 (1976)	*6*, p. *63*; 8, p. 71; 12.1, p. 109; 21.2, p. 323
R. M. SINGRU	(b) **74**, K161 (1976)	*13.1*, p. *123*; 22.1.1, p. 346
NGO DUONG SINH	(a) **37**, 533 (1976)	*13.3*, p. *133*; 14.3.2, p. 188; 22.1.2, p. 357
A. P. B. SINHA	(b) **73**, 651 (1976)	*13.1*, p. *121*; 17, p. 218; 22.6, p. 403
O. P. SINHA	(a) **38**, 73 (1976)	*10.2*, p. *97*; 13.4, p. 144; *20.3*, p. *287*; 22, p. 341
R. SINON	(a) **33**, 661 (1976)	*14.3.2*, p. *187*; 14.3.3, p. 190; 22, p. 334
	(a) **36**, 235 (1976)	*14.3.2*, p. *188*; 14.3.3, p. 192; 22.1.2, p. 355

	(a) **38**, 123 (1976)	*14.3.2*, p. *188*; 14.3.3, p. 194; 22.1.2, p. 357
E. P. Sinyavskii	(b) 77, K39 (1976)	*18*, p. *224*; *20.1*, p. *273*
M. Sinyukov	(b) 78, K19 (1976)	13.1, p. 128; *14.3.4*, p. *200*; *20.1*, p. *276*; 22.1.1, p. 349
M. P. Sinyukov	(a) **38**, 349 (1976)	*16*, p. *217*; *20.1*, p. *277*; 22.1.1, p. 349
D. I. Sirota	(b) 77, K143 (1976)	*7*, p. *66*; *18.4*, p. *245*
N. N. Sirota	(b) 73, K135 (1976)	*13.1*, p. *121*; *20.1*, p. *262*; 22.2.4, p. 370
E. I. Sirotinin	(a) **35**, K23 (1976)	*11*, p. *101*; 21, p. 294
J. Sitek	(a) **35**, 223 (1976)	*1.2*, p. *16*; *2*, p. *37*; 5, p. 49; 21.1.1, p. 318
T. N. Sitenko	(a) **36**, K157 (1976)	13.3, p. 133; *14.3.2*, p. *188*; 22.2.1, p. 363
J. De Sitter	(b) 73, 527 (1976)	*5*, p. *48*; *18.3*, p. *239*; 22.8.2, p. 426
S. Sivaraman	(a) **34**, 615 (1976)	*10.2*, p. *91*; 13.4, p. 137; 19, p. 250; 20.1, p. 266; 22.5.2, p. 394
J. Sivardière	(b) 76, 77 (1976)	*13*, p. *116*; *18.2*, p. *233*
J. Škácha	(a) **38**, K143 (1976)	*2*, p. *41*; *14.3.1*, p. *186*; 22, p. 342
E. Skaistys	(b) 75, K23 (1976)	*13.5.1*, p. *152*; *20.3*, p. *283*; 22.4.1, p. 380
L. Skála	(b) 74, K177 (1976)	*13*, p. *115*; 21, p. 293
M. Skibowski	(b) 74, 485 (1976)	*13.1*, p. *122*; *13.5.1*, p. *151*; 20.1, p. 265; 23, p. 440
E. M. Skok	(b) 75, 423 (1976)	13.1, p. 124; *14.3*, p. *177*; 22.1.2, p. 354
	(a) **35**, 755 (1976)	*14.3.1*, p. *184*; 22.2.3, p. 368
T. Skośkiewicz	(a) **37**, K163 (1976)	*14.2*, p. *172*; 21, p. 299
M. J. Škrinjar	(b) 74, 103 (1976)	*6*, p. *54*; *13.5.1*, p. *150*
A. Słebarski	(a) **36**, K171 (1976)	*8*, p. *70*; 18.2, p. 234; 21.4, p. 327
M. Šljukić	(a) **34**, K91 (1976)	*1.1*, p. *11*; 22.1.1, p. 346; 22.8, p. 414
I. Sh. Slobodetskii	(b) 73, K9 (1976)	*1.1*, p. *10*
T. Słupkowski	(a) **34**, 657 (1976)	*14.3*, p. *175*; 22.9, p. 435
V. Šmíd	(b) 74, K169 (1976)	*2*, p. *37*; *14.3.4*, p. *196*
I. A. Smirnov	(b) 73, K137 (1976)	13.1, p. 122; *20.1*, p. *262*; 21.4, p. 326
	(b) 75, K153 (1976)	*20.1*, p. *269*; 21, p. 295
	(b) 76, K27 (1976)	*14.3*, p. *178*; 22, p. 338
L. I. Smirnov	(a) **33**, 67 (1976)	*1.3*, p. *21*; *9*, p. *72*; 21, p. 289; 21.6, p. 329
G. F. Smirnova	(b) 73, K135 (1976)	*13.1*, p. *121*; *20.1*, p. *262*; 22.2.4, p. 370
	(a) **37**, K173 (1976)	*13.1*, p. *128*; 20.1, p. 275; 22.2.4, p. 371
T. A. Smirnova	(a) **37**, K37 (1976)	13.4, p. 142; *19*, p. *254*; 22.8, p. 417
T. V. Smirnova	(b) 77, K97 (1976)	*14.1*, p. *166*; *14.3.1*, p. *185*; 22.6, p. 407
A. W. Smith	(a) **34**, 117 (1976)	*1.3*, p. *22*; 21.1, p. 304; 21.1.1, p. 317
B. E. Smith	(a) **38**, K167 (1976)	*1.4*, p. *30*; 12.1, p. 109; 21, p. 301
J. H. Smith	(a) **33**, K43 (1976)	14.1, p. 164; *18.2*, p. *229*; 21.1, p. 303
E. M. Smokotin	(a) **34**, K47 (1976)	*12.2*, p. *110*; 22.8.2, p. 427
V. Šnejdar	(a) **34**, 505 (1976)	*13.3*, p. *132*; *14.3.3*, p. *191*; 22, p. 335
M. Šob	(b) 73, K1 (1976)	*13*, p. *114*; 21, p. 289; 21.1, p. 302; 21.1.1, p. 316; 21.6, p. 329
J. Sobhanadri	(a) **36**, K133 (1976)	*14.4*, p. *203*; 18.3, p. 241; 22.8.2, p. 430
	(a) **38**, 647 (1976)	*14.3*, p. *182*; 15, p. 212; 18.3, p. 241; 22.8.2, p. 431
A. A. Sobol	(a) **35**, 645 (1976)	*13.4*, p. *140*; 20.1, p. 269; 20.3, p. 284; 22.8.2, p. 429
H. Sodolski	(a) **34**, 657 (1976)	*14.3*, p. *175*; 22.9, p. 435
I. A. Sokolik	(b) 77, 265 (1976)	13.5.1, p. 155; *14.3.1*, p. *185*; *16*, p. *215*; 20.3, p. 286; 22.9, p. 437
F. F. Sokolov	(b) 75, 669 (1976)	13.4, p. 139; *20.1*, p. *268*; 22, p. 337
	(b) 76, K131 (1976)	*6*, p. *59*; *13.5.1*, p. *154*
N. S. Sokolov	(b) 74, 81 (1976)	6, p. 54; *13.4*, p. *136*; *13.5.1*, p. *150*; 16, p. 214; 22.1.1, p. 345
O. B. Sokolov	(b) 74, 35 (1976)	*13*, p. *115*; 21.1, p. 304

V. J. Sokolov	(a) **35**, 37 (1976)	3, p. 42; *13.4*, p. *138*; 14.3, p. 176; 20.1, p. 268; 20.3, p. 283; 22.3, p. 372
Zh. D. Sokolovskaya	(a) **35**, K177 (1976)	*12.2*, p. *110*; 22, p. 337
A. Soliman	(a) **37**, 39 (1976)	14.3.3, p. 193; *16*, p. *216*; 22.2.3, p. 368
A. V. Solomonov	(a) **38**, 753 (1976)	*20.3*, p. *288*; 22.2, p. 361
B. V. Solovev	(b) 77, 505 (1976)	*4*, p. *46*; 6, p. 61; 18.1, p. 226; 19, p. 254; 22.9, p. 438
V. S. Solovev	(a) **33**, 53 (1976)	*14.2*, p. *169*; 21, p. 289
T. Soma	(b) 74, K67 (1976)	1.1, p. 11; *12.1*, p. *107*; 21, p. 291
	(b) 76, 753 (1976)	*6*, p. *59*; 12.2, p. 111; 13.1, p. 126; 22.1, p. 344; 22.1.1, p. 348; 22.1.2, p. 355
	(b) 78, 229 (1976)	*1.1*, p. *13*; 1.2, p. 19; 22.1.2, p. 357
K. Somogyi	(a) **35**, 659 (1976)	13.4, p. 140; *14.3*, p. *178*; 22.1.2, p. 354; 22.2.2, p. 365
	(a) **37**, 653 (1976)	*14.3*, p. *180*; 22.2.2, p. 365
R. S. Sorbello	(b) 77, 141 (1976)	*6*, p. *60*; 7, p. 66; 14.1, p. 166
V. Soriano	(b) 77, 739 (1976)	*13.4*, p. *143*; *19*, p. *254*; 22.4, p. 377
Z. Šourek	(a) **35**, 55 (1976)	*10.2*, p. *92*; 22.1.2, p. 353
	(a) **36**, 509 (1976)	*10.2*, p. *94*; 22.1.2, p. 355
M. de Souza	(b) 77, 685 (1976)	1.2, p. 19; *6*, p. *61*; 10.2, p. 95; *20.1*, p. *274*; 22, p. 340
H. N. Spector	(a) **38**, K61 (1976)	*7*, p. *66*
B. Speit	(a) **36**, 471 (1976)	*10.2*, p. *94*; 16, p. 215; 20.1, p. 272; 20.3, p. 285; 22.8.1, p. 423
F. Spiegelberg	(b) 77, 233 (1976)	*6*, p. *60*; *13.5.1*, p. *154*; 20.1, p. 272; 22.4.1, p. 381
P. Spindler	(a) **37**, 449 (1976)	*9*, p. *74*; 21.1, p. 312
G. Splitt	(a) **37**, K131 (1976)	*7*, p. *66*; *14.2*, p. *171*; 21, p. 299
F. Sprenger	(a) **36**, 483 (1976)	13.4, p. 141; *14.3*, p. *179*; *16*, p. *215*; 22.5, p. 389
G. P. Srivastava	(b) 77, 131 (1976)	*6*, p. *60*; *8*, p. *70*
O. N. Srivastava	(a) **35**, 719 (1976)	*1*, p. *9*; 22.5, p. 389
P. L. Srivastava	(b) 78, K45 (1976)	2, p. 40; *14.1*, p. *167*; 21.2, p. 323
R. S. Srivastava	(a) **36**, 591 (1976)	*14.3.1*, p. *185*; 14.3.4, p. 198; 22.6, p. 406
V. K. Srivastava	(a) **38**, K77 (1976)	*14.4*, p. *204*; 22.9, p. 439
J. L. Staehli	(b) 75, 451 (1976)	*13.5.1*, p. *152*; 20.3, p. 283, 22.1.1, p. 347
A. Stancanelli	(b) 75, K93 (1976)	2, p. 38; *14.1*, p. *165*; *15*, p. *210*; 21, p. 295
J. Stanek	(b) 77, K1 (1976)	*5*, p. *50*; 22.1.2, p. 356
N. Starbov	(a) **38**, 161 (1976)	10.2, p. 97; *14.4*, p. *204*; 22.5.1, p. 391
B. P. Starikov	(a) **35**, 645 (1976)	*13.4*, p. *140*; 20.1, p. 269; 20.3, p. 284; 22.8.2, p. 429
V. I. Startsev	(a) **33**, 289 (1976)	*10.1*, p. *78*; 21.1, p. 303
	(a) **38**, 383 (1976)	*10.1*, p. *85*; 21.7, p. 332
G. Staupendahl	(a) **33**, 149 (1976)	*13.1*, p. *121*; *20.1*, p. *261*; 22.1.3, p. 359
W.-H. Steeb	(b) 74, K71 (1976)	13, p. 115; *18.2*, p. *230*; *18.4*, p. *243*
	(b) 78, K39 (1976)	13, p. 119; *18.2*, p. *236*; *18.4*, p. *245*
R. Stegmann	(b) 73, 615 (1976)	*13.2*, p. *130*; 14.3, p. 174; 19, p. 248; 22.7, p. 410
E. Steinbeiss	(b) 73, K95 (1976)	*13*, p. *114*; *20*, p. *257*; 22.6, p. 403
E. Steiner	(b) 73, 131 (1976)	*9*, p. *72*; 10, p. 75
	(b) 75, 361 (1976)	*9*, p. *72*; 10, p. 76
Yu. N. Stelmakhov	(a) **38**, K103 (1976)	*18.3*, p. *241*; 22.8.2, p. 431
S. I. Stenin	(a) **33**, 435 (1976)	*1.5*, p. *31*; 2, p. 36; 10.2, p. 88; 22.1.1, p. 345; 22.6, p. 402
J. Šternberk	(a) **33**, K15 (1976)	*18.2*, p. *229*; 22, p. 334
A. D. Stewart	(a) **36**, K119 (1976)	*10.1*, p. *83*; 11, p. 103; 12.1, p. 108; 21.1, p. 310
G. A. Stewart	(a) **38**, 533 (1976)	*19*, p. *256*; 21.1.1, p. 321
A. Steyerl	(a) **33**, 531 (1976)	*18.2*, p. *229*; 21.1, p. 304; 21.1.1, p. 317
J. Stoemenos	(a) **34**, 243 (1976)	*1.3*, p. *22*; 10.2, p. 90; 22.7, p. 410

S. D. Stojanović	(b) **74**, 103 (1976)	*6*, p. *54*; *13.5.1*, p. *150*
H. J. Stolz	(b) **78**, 277 (1977)	4, p. 47; *6*, p. *63*; *20.1*, p. *276*; 22, p. 341
M. Stordeur	(b) **78**, K103 (1976)	*15*, p. *212*; 22.7, p. 412
R. Strehlow	(b) **73**, 541 (1976)	13.5.3, p. 159; *14.3.4*, p. *195*; *20.1*, p. *262*; 22, p. 334
A. M. Strokova	(a) **35**, 747 (1976)	*10.1*, p. *82*; 12.1, p. 108; 21.1, p. 308; 21.3, p. 324
K. de Strooper	(a) **37**, 127 (1976)	18.1, p. 226; *18.3*, p. *241*; 18.4, p. 245; 22.8.2, p. 430
E. E. Strumban	(a) **35**, 251 (1976)	*13.4*, p. *138*; 16, p. 214; 20.1, p. 268; 20.3, p. 283; 22.4.1, p. 380
	(a) **38**, 77 (1976)	*13.4*, p. *144*; 16, p. 217; 20.1, p. 276; 22.4.1, p. 382
B. M. Strunin	(a) **34**, 761 (1976)	*10*, p. *75*
	(a) **35**, 551 (1976)	*10*, p. *76*
J. E. Strutt	(a) **33**, 229 (1976)	*10.2*, p. *87*; 22.5.2, p. 392
G. Stühmer	(b) **76**, 547 (1976)	*13.5.1*, p. *154*; 20.1, p. 271; 22.5, p. 389
	(a) **37**, 105 (1976)	*1.2*, p. *18*; *20*, p. *258*; 22.5.4, p. 401
M. Suezawa	(a) **36**, 263 (1976)	*10*, p. *76*
K. Sugawara	(b) **78**, K7 (1976)	*14.1*, p. *167*; *18.4*, p. *245*; 21, p. 299
K. Suk	(a) **34**, 639 (1976)	*18.3*, p. *240*; 22.8.2, p. 428
	(a) **38**, K163 (1976)	*18.3*, p. *241*; *19*, p. *256*; 22.8.2, p. 432
Z. Sułek	(b) **74**, K5 (1976)	*19*, p. *249*; 22.8.1, p. 420
Yu. Suleymanov	(a) **34**, K195 (1976)	*20.3*, p. *282*; 22.2, p. 361
M. Süli	(a) **38**, 321 (1976)	*9*, p. *74*; 11, p. 104; 22.1.2, p. 358
K. Sumino	(a) **36**, 263 (1976)	*10*, p. *76*
R. Suryanarayanan	(b) **73**, K137 (1976)	13.1, p. 122; *20.1*, p. *262*; 21.4, p. 326
	(b) **75**, K153 (1976)	*20.1*, p. *269*; 21, p. 295
T. Suski	(a) **35**, K165 (1976)	*1.2*, p. *17*; 14.4.1, p. 207; 20.1, p. 270; 22.8.1, p. 422
H. Süssmann	(a) **36**, K55 (1976)	*14.3*, p. *178*; 22.7, p. 411
	(a) **37**, 467 (1976)	*14.3*, p. *180*; 22.7, p. 412
M. Suszynska	(a) **34**, 165 (1976)	*10.2*, p. *90*; 22.5.2, p. 393
H. H. Sutherland	(b) **77**, K53 (1976)	13.4, p. 142; *13.5.1*, p. *155*; 22.4, p. 377
A. Suzuki	(b) **74**, 579 (1976)	*13.5.1*, p. *151*; *20.3*, p. *282*; 23, p. 440
	(a) **36**, K129 (1976)	*20.3*, p. *285*; 22.3, p. 373
K. Suzuki	(b) **76**, 787 (1976)	*13.1*, p. *126*; 20.1, p. 272; 22.4.3, p. 386
	(b) **78**, K77 (1976)	*8*, p. *71*; 22.1.1, p. 349
	(a) **38**, K119 (1976)	*10.2*, p. *99*; 22.4, p. 378
S. Suzuki	(a) **33**, 789 (1976)	14.4.1, p. 206; *20.1*, p. *263*; 22.8, p. 413
Y. Suzuki	(a) **33**, 691 (1976)	1.5, p. 31; *2*, p. *36*; 22.7, p. 410
M. S. Svirskii	(b) **77**, 9 (1976)	*13*, p. *113*; 14.1, p. 164; 14.2, p. 169; 18.2, p. 228; 18.4, p. 242; 21, p. 289
	(b) **77**, 403 (1976)	*13*, p. *113*; 14.1, p. 164; 14.2, p. 169; 18.2, p. 228; 18.4, p. 242; 21, p. 289
V. I. Svitov	(a) **35**, K119 (1976)	*17*, p. *219*; 21, p. 295
	(a) **37**, K9 (1976)	*17*, p. *220*; 22.6, p. 407
M. L. Swanson	(a) **33**, 265 (1976)	6, p. 53; *11*, p. *100*; 21.1, p. 303
J. Świątek	(a) **38**, 285 (1976)	*13.4*, p. *144*; *14.3.1*, p. *185*; 22.9, p. 439
J. Sworakowski	(a) **36**, 735 (1976)	*13.4*, p. *142*; 14.3.4, p. 198; 14.4, p. 203
	(a) **37**, 271 (1976)	*13.4*, p. *142*; 14.3.4, p. 199; 14.4, p. 204; 22.9, p. 438
V. F. Synorov	(a) **33**, K161 (1976)	*14.3.2*, p. *187*; *14.3.3*, p. *190*
	(a) **34**, K71 (1976)	*13.4*, p. *137*; *14.3.3*, p. *191*
B. I. Sysoev	(a) **33**, K161 (1976)	*14.3.2*, p. *187*; *14.3.3*, p. *190*
	(a) **34**, K71 (1976)	*13.4*, p. *137*; *14.3.3*, p. *191*
A. W. Szafrański	(a) **37**, K163 (1976)	*14.2*, p. *172*; 21, p. 299

J. Szatkowski	(b) 77, K91 (1976)	*13.1*, p. *128*; 22.4.4, p. 387
E. Szczepaniak	(b) 74, K81 (1976)	*18.1*, p. *225*
L. von Szentpaly	(b) **76**, 817 (1976)	*13.5.1*, p. *154*; 20.1, p. 272; 22.9, p. 437
T. Szörényi	(a) **37**, K169 (1976)	2, p. 40; *19*, p. *255*; 22.8.1, p. 424
Z. Szücs	(b) **78**, K107 (1976)	*14.2*, p. *172*; *19*, p. *256*; 21, p. 300
J. Szumowski	(b) **74**, 683 (1976)	*19*, p. *250*; 22.5.3, p. 399
W. Szymańska	(b) **74**, K89 (1976)	*14.3*, p. *175*; 15, p. 209; 22.4.2, p. 383
J. Szymański	(b) **76**, 439 (1976)	6, p. 58; 13.1, p. 125; *20.1*, p. *271*; 22, p. 338
	(b) 77, 347 (1976)	*6*, p. *60*; 22, p. 339
H. Szymczak	(b) 74, 225 (1976)	*6*, p. *55*; *19*, p. *249*; 22.4.1, p. 379; 22.6, p. 403
A. Szytuła	(a) **35**, K69 (1976)	*18.2*, p. *232*; *18.4*, p. *244*; 22.8, p. 415
	(a) **38**, 721 (1976)	*18.4*, p. *246*; 22.8, p. 418
K. Tabuchi	(b) **73**, 341 (1976)	*6*, p. *53*; *20.1*, p. *260*; 22.5, p. 388
Yu. M. Tairov	(a) **37**, 143 (1976)	*1.5*, p. *32*; 22.3, p. 374
N. Tajabor	(a) **34**, K191 (1976)	*1.6*, p. *34*; 22.3, p. 372
S. Takács	(a) **33**, 85 (1976)	*14.2*, p. *169*; 21, p. 289
	(a) **37**, 279 (1976)	10.1, p. 84; *14.2*, p. *171*
M. Takagi	(a) **33**, 789 (1976)	14.4.1, p. 206; *20.1*, p. *263*; 22.8, p. 413
S. Takahashi	(a) **33**, K141 (1976)	*18.2*, p. *230*; 21.1, p. 304; 21.1.1, p. 317
K. Takanaka	(b) **75**, 279 (1976)	*13*, p. *116*; *18*, p. *223*
S. Takemoto	(a) **34**, 671 (1976)	*18.3*, p. *240*; *19*, p. *250*; 22.8.2, p. 428
N. Takeuchi	(a) **33**, K125 (1976)	10.2, p. 89; *20.3*, p. *281*; 22.6, p. 403
S. Takeuchi	(b) **75**, 559 (1976)	*2*, p. *38*; 13, p. 116; 21.1, p. 307; 21.7, p. 332
	(a) **38**, K119 (1976)	*10.2*, p. *99*; 22.4, p. 378
A. Tamura	(b) **73**, 391 (1976)	*11*, p. *100*; 13, p. 114; 21, p. 290; 22.5.2, p. 393
Pham Minh Tan see under letter "P"		
M. Tanaka	(a) **38**, 453 (1976)	*1.1*, p. *13*; 1.4, p. 29; 22.1.1, p. 350
S. Tanaka	(b) **76**, 787 (1976)	*13.1*, p. *126*; 20.1, p. 272; 22.4.3, p. 386
T. Tanaka	(a) **36**, K129 (1976)	*20.3*, p. *285*; 22.3, p. 373
R. P. Tandon	(a) **38**, K1 (1976)	*2*, p. *41*; *14.3.3*, p. *194*; 22.8.1, p. 424
A. Tanga	(b) 77, 341 (1976)	*10.2*, p. *95*; 20.1, p. 273; *20.3*, p. *286*; 22.5.2, p. 396
S. Tanigawa	(b) **73**, 517 (1976)	1.3, p. 21; *10.1*, p. *78*; 21, p. 290
Ch. Tantzscher	(b) **78**, 79 (1976)	13.4, p. 144; *14.3*, p. *180*; 14.3.1, p. 185; 22.9, p. 438
S. J. Tao	(a) **35**, K17 (1976)	*10.1*, p. *82*; 21.1.1, p. 318
M. Tapiero	(a) **33**, 155 (1976)	13.4, p. 135; *16*, p. *213*; 22.6, p. 402
L. V. Tarakanova	(b) **75**, K143 (1976)	*10.2*, p. *93*; *13.5.3*, p. *159*
L. Tarricone	(a) **36**, K33 (1976)	*20.3*, p. *285*; 22.8, p. 416
M. Tasson	(a) **37**, 119 (1976)	*7*, p. *66*; *14.4.1*, p. *207*; 22.8.1, p. 423
L. Taurel	(b) 78, 779 (1976)	6, p. 64; *10.2*, p. *98*; 13.4, p. 145; 20.1, p. 278; 22.5.1, p. 398
A. Tausend	(b) 75, K149 (1976)	*6*, p. *57*; *20.1*, p. *269*; 22, p. 337
M. Taut	(b) **73**, 151 (1976)	*6*, p. *52*; *13.1*, p. *120*; 13.2, p. 130; 21.3, p. 324
	(b) **76**, 605 (1976)	*13.1*, p. *126*; 20, p. 258; 21, p. 297; 22.7, p. 411
	(b) 77, 165 (1976)	*13.1*, p. *127*; *14.1*, p. *166*; 21, p. 297
	(b) 77, 543 (1976)	*13.1*, p. *127*; 13.5.2, p. 157; *14.1*, p. *166*; 14.4, p. 204; 20.1, p. 274; 21.3, p. 324
	(b) 78, 169 (1976)	*13.1*, p. *128*; 13.5.2, p. 158; *14.1*, p. *167*; 14.4, p. 204; 20.1, p. 275; 21.3, p. 324
K. N. R. Taylor	(b) 77, 725 (1976)	14.1, p. 166; *20.1*, p. *274*; 21.4, p. 327
H. Teichler	(b) 75, 205 (1976)	*14.2*, p. *169*; 21, p. 294
L. Tejedor	(a) **34**, K17 (1976)	*1.2*, p. *15*; 21.1.1, p. 317
D. E. Temkin	(a) **34**, 413 (1976)	*1.4*, p. *26*; 10, p. 75
G. van Tendeloo	(a) **33**, 383 (1976)	*1.1*, p. *10*; 1.4, p. 25; 22, p. 334

	(a) **33**, 723 (1976)	*1.2*, p. *14*; *1.4*, p. *26*; 22.6, p. 403; 22.8.1, p. 419
	(a) **36**, 757 (1976)	*1.1*, p. *12*; *1.2*, p. *18*; 22, p. 339
	(a) **37**, 591 (1976)	*1.2*, p. *19*; 1.4, p. 28; 22, p. 340
	(a) **38**, 185 (1976)	*1.2*, p. *19*; 1.4, p. 29; 22, p. 341
	(a) **38**, 663 (1976)	*1.4*, p. *29*; 10.2, p. 99; 22, p. 342
A. Tepore	(a) **36**, 569 (1976)	*14.3.4*, p. *198*; 22.3, p. 373
S. Teratani	(a) **35**, K57 (1976)	10.2, p. 92; *11*, p. *101*; *20.1*, p. *268*; 22.4.1, p. 380
A. S. Terekhov	(b) **77**, 465 (1976)	*13.5.1*, p. *155*; *20.1*, p. *274*; 22.2.1, p. 364
S. F. Terekhova	(b) **75**, K51 (1976)	13.5.1, p. 152; *20.1*, p. *267*; 22, p. 336
E. I. Terukov	(a) **37**, 541 (1976)	*14.3*, p. *180*; 15, p. 211; 18.1, p. 227; 22.6, p. 407
	(a) **38**, K23 (1976)	*14.3*, p. *181*; *15*, p. *211*; 22.6, p. 408; 22.6.1, p. 409
B. Tesche	(a) **33**, 461 (1976)	*14.3.2*, p. *187*; *14.3.4*, p. *196*; 16, p. 213; 22.9, p. 433
S. I. Teslenko	(a) **34**, K195 (1976)	*20.3*, p. *282*; 22.2, p. 361
D. I. Tetelbaum	(a) **35**, 11 (1976)	11, p. 100; *13.4*, p. *134*; *14.3*, p. *173*; 22.1.1, p. 345; 22.1.2, p. 351
	(a) **36**, 81 (1976)	*1.5*, p. *31*; 2, p. 39; *11*, p. *102*; 22.1.2, p. 355
	(a) **37**, 57 (1976)	*11*, p. *103*; 22.1.2, p. 356
W. Tews	(b) **73**, K95 (1976)	*13*, p. *114*; *20*, p. *257*; 22.6, p. 403
A. Thanailakis	(b) **76**, 727 (1976)	6, p. 59; *20.1*, p. *272*; 22.7, p. 411
J. Thery	(a) **38**, 595 (1976)	*1.4*, p. *29*; 22.8.2, p. 431
J. Thiemann	(b) **73**, 155 (1976)	10.2, p. 87; *19*, p. *247*; 22.5.2, p. 392
A. R. Thölén	(a) **38**, 739 (1976)	*1.4*, p. *30*; 10.2, p. 99; 22, p. 342
H. Thomas	(a) **35**, 697 (1976)	*14.3.3*, p. *192*; *20.3*, p. *284*; 22.3, p. 373
	(a) **36**, 365 (1976)	*14.3.3*, p. *192*; *20.3*, p. *285*; 22.3, p. 373
J. M. Thomas	(a) **37**, 271 (1976)	*13.4*, p. *142*; 14.3.4, p. 199; 14.4, p. 204; 22.9, p. 438
P. Thomas	(b) **76**, 439 (1976)	1.5, p. 31; *14.3.1*, p. *184*; 22.1.1, p. 347
R. K. Thorpe	(a) **33**, K97 (1976)	*5*, p. *48*; 22.1, p. 343
G. Thummes	(b) **75**, K61 (1976)	*19*, p. *251*; 21.1, p. 307
B. Thuss	(a) **35**, K49 (1976)	*1.2*, p. *16*; 15, p. 210; 22.6, p. 404
	(a) **38**, K13 (1976)	*1.2*, p. *19*; *18.1*, p. *227*; 22.6.1, p. 409
M. Tieng	(a) **34**, K43 (1976)	*14.3.3*, p. *191*; 14.3.4, p. 196; 22, p. 335
W. Tiersch	(b) **74**, K43 (1976)	*8*, p. *67*; *18.4*, p. *243*
J. W. J. Timans	(b) **73**, 283 (1976)	*6*, p. *53*; *20.1*, p. *260*; 22.5.3, p. 399
S. F. Timashev	(b) **76**, 67 (1976)	13.3, p. 132; *14.3.2*, p. *188*; *20.1*, p. *270*
Yu. P. Timofeev	(b) **78**, 483 (1976)	13.4, p. 145; *20.3*, p. *288*; 22.5.4, p. 401
M. I. Timoshechkin	(a) **35**, 645 (1976)	*13.4*. p. *140*; 20.1, p. 269; 20.3, p. 284; 22.8.2, p. 429
C. J. Tinsley	(a) **33**, 189 (1976)	*14.3*, p. *173*; 18.3, p. 239; 22.6, p. 402
R. S. Tiwari	(a) **35**, 719 (1976)	*1*, p. *9*; 22.5, p. 389
E. N. Tkalenko	(a) **37**, 199 (1976)	*14.3*, p. *179*; 22.7, p. 412
Yu. A. Tkhorik	(a) **35**, K65 (1976)	*14.3.1*, p. *184*; 22.1.1, p. 347
S. M. Todorov	(b) **75**, K51 (1976)	13.5.1, p. 152; *20.1*, p. *267*; 22, p. 336
J. Todorović	(a) **35**, K69 (1976)	*18.2*, p. *232*; *18.4*, p. *244*; 22.8, p. 415
	(a) **38**, 721 (1976)	*18.4*, p. *246*; 22.8, p. 418
K.-D. Toepfer	(a) **35**, 131 (1976)	1.2, p. 16; *14.4.1*, p. *207*; 22.9, p. 436
R. Tognato	(a) **33**, 579 (1976)	*10.1*, p. *79*; 12.1, p. 107; 21, p. 291
K. R. Tokhirov	(a) **35**, K145 (1976)	*11*, p. *102*; 22.1.2, p. 354
A. M. Tolkachev	(b) **73**, K111 (1976)	*8*, p. *67*; 23, p. 440
G. Tolkiehn	(b) **73**, K99 (1976)	*13.5.1*, p. *150*; *20.3*, p. *280*; 23, p. 440
W. Tolksdorf	(b) **73**, 533 (1976)	*13.4*, p. *135*; 19, p. 248; 20.1, p. 261; 20.3, p. 280; 22.8.2, p. 426
E. I. Tolpygo	(b) **76**, 407 (1976)	13.1, p. 125; *14.3*, p. *178*; 22.1.2, p. 355

P. M. Tomchuk	(b) **76**, 385 (1976)	*13*, p. *117*; *14.3.4*, p. *197*; 22, p. 338
	(b) **78**, 843 (1976)	*13*, p. *117*; *14.3.4*, p. *197*; 22, p. 338
Z. Tomkowicz	(a) **38**, 721 (1976)	*18.4*, p. *246*; 22.8, p. 418
T. Tomov	(a) **37**, 295 (1976)	*5*, p. *50*; 18.4, p. 245; 22.6, p. 407
J. Topolsky	(a) **35**, K17 (1976)	*10.1*, p. *82*; 21.1.1, p. 318
B. S. Tošić	(b) **74**, 103 (1976)	*6*, p. *54*; *13.5.1*, p. *150*
	(b) **76**, K85 (1976)	*13.5.4*, p. *161*; *18.2*, p. *234*
A. Tóth	(a) **33**, K47 (1976)	*10.2*, p. *88*; *14.4*, p. *202*; 22.5.2, p. 392
J. Toušková	(a) **36**, 747 (1976)	*14.3.3*, p. *193*; *14.3.4*, p. *199*; 22.4.3, p. 386
P. A. Tove	(a) **36**, 217 (1976)	1.5, p. 32; *14.3.1*, p. *184*; 22, p. 338
Tran Huu Danh	(a) **33**, 295 (1976)	*2*, p. *35*; *14.3*, p. *173*; 15, p. 209; 22.8, p. 413
V. Travnikov	(b) **78**, 389 (1976)	*6*, p. *63*; *13.5*, p. *148*; 20.3, p. 287; 22.4.1, p. 381
M. M. Tribel	(b) **77**, 265 (1976)	13.5.1, p. 155; *14.3.1*, p. *185*; *16*, p. *215*; 20.3, p. 286; 22.9, p. 437
M. J. Tricker	(a) **33**, K97 (1976)	*5*, p. *48*; 22.1, p. 343
A. Z. Trifonov	(a) **35**, 709 (1976)	*9*, p. *73*; 22.9, p. 436
	(a) **36**, 391 (1976)	*9*, p. *73*; 22.9, p. 437
E. P. Trifonova	(a) **38**, K37 (1976)	*14.3.3*, p. *194*; 22.2.2, p. 366; 22.8, p. 418
R. S. Tripathi	(b) **78**, 563 (1976)	*2*, p. *41*; 13.4, p. 145; *14*, p. *163*
O. A. Troitskii	(a) **35**, K151 (1976)	*3*, p. *42*; 21, p. 295; 21.1, p. 308
V. A. Troyan	(a) **35**, 779 (1976)	*1.2*, p. *17*; *1.5*, p. *31*; 11, p. 102; 21, p. 295; 21.1, p. 308; 21.1.1, p. 319; 22.1.1, p. 347
E. M. Trukhanov	(a) **33**, 435 (1976)	*1.5*, p. *31*; 2, p. 36; 10.2, p. 88; 22.1.1, p. 345; 22.6, p. 402
J. S. Tsakadze	(a) **38**, 769 (1976)	10.1, p. 86; *14.2*, p. *172*; 21, p. 301
G. G. Tsebulya	(b) **75**, K51 (1976)	13.5.1, p. 152; *20.1*, p. *267*; 22, p. 336
I. A. Tsenov	(a) **35**, 709 (1976)	*9*, p. *73*; 22.9, p. 436
	(a) **36**, 391 (1976)	*9*, p. *73*; 22.9, p. 437
A. A. Tsertsvadze	(a) **35**, K131 (1976)	*17*, p. *219*; 22.1.1, p. 347
M. A. Tserunian	(a) **35**, K181 (1976)	*20*, p. *258*
S. V. Tsivinskii	(a) **33**, 301 (1976)	*10.2*, p. *88*; 22.1.1, p. 345; 22.1.2, p. 351; 22.2.3, p. 367; 22.5.2, p. 392
	(a) **36**, 11 (1976)	*10.1*, p. *78*; *10.2*, p. *87*; 21, p. 289; 21.1, p. 302; 21.1.1, p. 316; 21.6, p. 329; 22.1.1, p. 345; 22.2.3, p. 367; 22.5.2, p. 392
V. A. Tskhai	(b) **78**, 465 (1976)	*13.1*, p. *129*; 22.3, p. 374
K. Tsuno	(a) **35**, 437 (1976)	*18.2*, p. *232*; 21.1, p. 308; 21.1.1, p. 319
V. A. Tsurin	(a) **33**, 325 (1976)	*5*, p. *48*; 18.2, p. 229; 21.1.1, p. 316
I. Tsurumi	(a) **35**, K57 (1976)	10.2, p. 92; *11*, p. *101*; *20.1*, p. *268*; 22.4.1, p. 380
N. Tsuya	(a) **34**, 325 (1976)	1.2, p. 15; *18.3*, p. *240*; 22.8.2, p. 427
V. F. Tsvetkov	(a) **37**, 143 (1976)	*1.5*, p. *32*; 22.3, p. 374
A. F. Tulinov	(a) **35**, K23 (1976)	*11*, p. *101*; 21, p. 294
S. S. Tumanyan	(b) **73**, 681 (1976)	*5*, p. *48*
T. Tuomi	(a) **33**, 125 (1976)	6, p. 53; 13.5.1, p. 149; *14.3.3*, p. *190*; *16*, p. *213*; 20.1, p. 261; 22.1.2, p. 351
	(a) **36**, 461 (1976)	*10.2*, p. *94*; 22.1.3, p. 360
E. A. Turov	(b) **74**, 35 (1976)	*13*, p. *115*; 21.1, p. 304
V. V. Turovtsev	(b) **78**, 103 (1976)	*5*, p. *50*; 18.2, p. 235; 18.4, p. 245; 21.4, p. 327
I. I. Tychina	(a) **35**, 571 (1976)	*13.4*, p. *139*; 14.3, p. 177; 22.8, p. 415
E. Tyssen	(a) **38**, K171 (1976)	13.4, p. 146; *20.1*, p. *278*; 22.4.4, p. 387
P. Tzanetakis	(b) **75**, 433 (1976)	*10.1*, p. *82*; 14.1, p. 165; 21, p. 294; 21.3, p. 324
S. Uchida	(b) **76**, 787 (1976)	*13.1*, p. *126*; 20.1, p. 272; 22.4.3, p. 386
H. Ueba	(b) **75**, 501 (1976)	6, p. 57; *13.5.1*, p. *153*; 13.5.2, p. 157; *20.1*, p. *268*; 22, p. 336

O. Uemura	(a) **33**, 691 (1976)	1.5, p. 31; *2*, p. *36*; 22.7, p. 410
Y. Uesu	(a) **34**, 475 (1976)	*1.1*, p. *11*; *14.4.1*, p. *206*; 22.8.1, p. 421
K.-D. Ufert	(a) **34**, K83 (1976)	*8*, p. *68*; *12.2*, p. *110*; 14.3.1, p. 183; 22.6, p. 404
	(a) **36**, K71 (1976)	*8*, p. *69*; *14.3.3*, p. *192*; 14.3.4, p. 198; 22.6, p. 405
A. I. Ugulava	(b) **76**, K97 (1976)	*19*, p. *253*
C. Uher	(b) **73**, 379 (1976)	6, p. 53; 13.1, p. 121; *15*, p. *209*; 21.7, p. 332
	(b) **75**, K137 (1976)	6, p. 57; *15*, p. *210*; 21.7, p. 332
N. Uhle	(b) **74**, K15 (1976)	13.5.3, p. 159; *20.1*, p. *264*; 22.6, p. 403
	(b) **78**, K27 (1976)	*13.5.3*, p. *159*; *20.1*, p. *276*; 22.4, p. 377
Ch. Uihlein	(b) **76**, 613 (1976)	*13.5.1*, p. *154*; *20.1*, p. *271*; 22.5.1, p. 391
V. A. Ulanov	(b) **75**, 731 (1976)	*19*, p. *251*; 22.5.3, p. 399
J. Ullrich	(a) **34**, K31 (1976)	*10.1*, p. *80*; 12.1, p. 107; 21.1, p. 305
B. Ulrici	(b) **74**, 601 (1976)	*6*, p. *55*; *20.1*, p. *265*; 22.2.2, p. 365
L. M. Umarov	(b) **78**, 325 (1976)	*6*, p. *63*; *20.1*, p. *276*; 22.9, p. 438
T. Umenai	(b) **78**, 229 (1976)	*1.1*, p. *13*; 1.2, p. 19; 22.1.2, p. 357
M. Umeno	(a) **37**, 561 (1976)	*1.1*, p. *13*; 22.1.1, p. 349
	(a) **38**, 701 (1976)	*1.1*, p. *13*; 22.1.1, p. 350
	(b) **76**, 207 (1976)	*2*, p. *38*
	(b) **76**, 841 (1976)	*2*, p. *39*; 22.1.3, p. 359
K. Unger	(b) **75**, 493 (1976)	*13.1*, p. *124*; 22.2.4, p. 370
	(b) **78**, 681 (1976)	*13.1*, p. *129*; 17, p. 220; 20, p. 259; 22.1, p. 344; 22.1.1, p. 349; 22.1.2, p. 358
S. Unterricker	(b) **73**, K89 (1976)	*5*, p. *48*; 18.2, p. 229; 21, p. 291; 21.1.1, p. 317
K. S. Upadhyaya	(b) **78**, 399 (1976)	*6*, p. *63*; 22.6, p. 408
G. C. Upreti	(b) **78**, K73 (1976)	*19*, p. *256*; 22.8.1, p. 424
M. I. Urbakh	(b) **76**, 93 (1976)	*13*, p. *116*; *13.1*, p. *124*; 21.2, p. 323
A. A. Urusovskaya	(a) **36**, 397 (1976)	*10.2*, p. *94*; 22.5.2, p. 395
R. G. Usmanov	(b) **76**, 55 (1976)	13.4, p. 140; *20.1*, p. *270*; *20.3*, p. *284*; 22.8.1, p. 422
M. M. Usmanova	(a) **35**, 37 (1976)	3, p. 42; *13.4*, p. *138*; 14.3, p. 176; 20.1, p. 268; 20.3, p. 283; 22.3, p. 372

R. Vacher	(b) **75**, K101 (1976)	*2*, p. *38*; *6*, p. *57*; 20.1, p. 269; 22.9, p. 436
R. Vaglio	(a) **35**, K173 (1976)	*14.2*, p. *170*; 21, p. 295; 22.6, p. 405
J. K. Vaid	(a) **38**, K1 (1976)	*2*, p. *41*; *14.3.3*, p. *194*; 22.8.1, p. 424
J. Vaitkus	(a) **34**, 769 (1976)	*14.3*, p. *176*; *16*, p. *214*; 20.3, p. 282; 22.1.1, p. 346
V. G. Vaks	(a) **35**, 61 (1976)	*1.2*, p. *16*; 22.5, p. 389; 22.5.4, p. 401
L. M. Valiev	(a) **35**, K85 (1976)	8, p. 68; 14, p. 162; 15, p. 210; *18.1*, p. *226*; *18.4*, p. *244*; 22.8, p. 415
R. Z. Valiev	(a) **35**, 403 (1976)	*10.1*, p. *82*; 21, p. 294; 21.1, p. 307
L. Valkūnas	(b) **75**, K23 (1976)	*13.5.1*, p. *152*; *20.3*, p. *283*; 22.4.1, p. 380
M. Valli	(b) **77**, 341 (1976)	*10.2*, p. *95*; 20.1, p. 273; *20.3*, p. *286*; 22.5.2, p. 396
E. R. Vance	(b) **77**, K85 (1976)	*10.2*, p. *96*; 20.1, p. 274; 22.6, p. 407
R. E. Vandenberghe	(a) **34**, 583 (1976)	*1*, p. *9*; 1.2, p. 15; *18.3*, p. *240*; 22.8.2, p. 428
M. Vandevyver	(b) **73**, 271 (1976)	*6*, p. *52*; 22.4.3, p. 385
M. Vaněček	(a) **33**, K31 (1976)	*14.4.1*, p. *206*; *16*, p. *213*; 22.8.1, p. 419
	(b) **75**, K45 (1976)	*16*, p. *214*; *20.1*, p. *267*; 22.8.2, p. 428
M. Vanhaelst	(a) **35**, K137 (1976)	*3*, p. *42*; *19*, p. *252*; 22.5.2, p. 395
	(b) **77**, 249 (1976)	*19*, p. *254*; 22.5.2, p. 396
	(b) **78**, 553 (1976)	*19*, p. *256*; *20.1*, p. *277*; 22.5.2, p. 398
T. G. Vardosanidze	(b) **76**, K97 (1976)	*19*, p. *253*
R. A. Varin	(a) **34**, 711 (1976)	*1.4*, p. *26*; *10.1*, p. *81*; 21.1.1, p. 318
J. Varvas	(a) **33**, 75 (1976)	*10.2*, p. *87*; 13.4, p. 135; *14.3*, p. *173*; 22.4.2, p. 383

L. VASANELLI	(a) **38**, 685 (1976)	13.4, p. 145; 14.3, p. 182; *16*, p. *217*; 22.3, p. 374
P. VAŠEK	(b) **77**, K67 (1976)	*10.1*, p. *83*; *14.1*, p. *166*; 21.6, p. 331
A. S. VASILKOVSKII	(a) **33**, 275 (1976)	*1.1*, p. *10*; *10.2*, p. *87*; 22.1.2, p. 351
P. G. VASSILEV	(a) **37**, K25 (1976)	14.1, p. 166; *15*, p. *210*; *20.1*, p. *273*; 21.1, p. 311
V. A. VASSILEV	(a) **37**, 45 (1976)	2, p. 40; *16*, p. *216*; *20.3*, p. *286*; 22, p. 339
K. I. VASU	(a) **35**, 769 (1976)	*10.1*, p. *82*; 21, p. 295
V. S. VAVILOV	(a) **36**, 427 (1976)	9, p. 73; 10.2, p. 94; *11*, p. *102*; 14.3, p. 178; 22.1, p. 343; 22.2, p. 361; 22.3, p. 373; 22.4, p. 376
I. VECCHI	(a) **35**, 165 (1976)	*14.3.3*, p. *191*; 22.1.2, p. 353
A. VAN VEEN	(a) **34**, 277 (1976)	*10.1*, p. *80*; 21, p. 292
	(a) **37**, 371 (1976)	*10.1*, p. *84*; *11*, p. *103*; 21, p. 298
H. VEESER	(a) **37**, 161 (1976)	*7*, p. *66*; 12.2, p. 111; *14.4.1*, p. *208*; 20.1, p. 273; 22.8.1, p. 423
N. B. VELCHEV	(b) **75**, 423 (1976)	13.1, p. 124; *14.3*, p. *177*; 22.1.2, p. 354
J. N. J. VAN DER VELDEN	(a) **35**, 137 (1976)	*5*, p. *49*; 18.2, p. 232; 21.1.1, p. 318; 21.4, p. 326
I. VELTRUSKÝ	(b) **73**, 575 (1976)	*13*, p. *114*; *18.3*, p. *239*; 22.8.2, p. 426
S. VENGRIS	(a) **35**, 339 (1976)	*14.3.2*, p. *188*; *16*, p. *214*; 22.1.3, p. 359
J. VENNIK	(b) **76**, 701 (1976)	*6*, p. *59*; *20.1*, p. *272*; 22.6, p. 406
K. A. VERKHOVSKAYA	(a) **33**, K137 (1976)	*14.4.1*, p. *206*; *20.1*, p. *263*; 22.8.1, p. 419
M. P. VERMA	(b) **78**, 113 (1976)	*12.2*, p. *112*; 22.5.2, p. 397
G. VERMIGLIO	(b) **74**, 317 (1976)	*13.4*, p. *136*; *20.1*, p. *264*; 22.5.2, p. 393
V. D. VERNER	(a) **33**, 755 (1976)	*10.2*, p. *89*; 22.2.1, p. 362
P. VEYSSIÈRE	(a) **35**, 259 (1976)	*1.4*, p. *27*; *10.2*, p. *92*; 22.8.2, p. 428
P. J. VICCARO	(a) **37**, 133 (1976)	*5*, p. *50*; *18.2*, p. *235*; 21.1.1, p. 320
J. VÍCENS	(a) **33**, 497 (1976)	*1.2*, p. *14*; *1.4*, p. *25*; 22, p. 334
YU. A. VIDADI	(a) **33**, K67 (1976)	*14.3.1*, p. *183*; 14.3.2, p. 187; 22.9, p. 433
	(a) **34**, K77 (1976)	*14.3.2*, p. *183*; 22.9, p. 435
O. VIGIL	(a) **36**, 33 (1976)	*13.4*, p. *141*; 16, p. 215; *20.1*, p. *271*; 22.8, p. 416
I. VILFAN	(b) **78**, K131 (1976)	*6*, p. *64*; 13.5.3, p. 159; *14.3*, p. *182*; 22.9, p. 439
M. VILLA	(b) **75**, 179 (1976)	*19*, p. *250*; 22.8.1, p. 421
L. I. VINOKUROVA	(b) **78**, 353 (1976)	12.1, p. 109; 18.1, p. 227; *18.2*, p. *236*; *18.4*, p. *245*; 21.1.1, p. 321
J. VIŠČAKAS	(a) **35**, 563 (1976)	*14.3*, p. *177*
Š. VIŠŇOVSKÝ	(a) **38**, K53 (1976)	*18.2.1*, p. *238*; *20.1*, p. *277*; 21.1, p. 314; 21.4, p. 327
R. VISOCEKAS	(a) **35**, 315 (1976)	10.2, p. 92; *20.3*, p. *283*; 22.8.1, p. 421
A. KASI VISWANATH	(b) **75**, K119 (1976)	*19*, p. *251*; 22.5.2, p. 395
D. VIVIEN	(b) **73**, 593 (1976)	*19*, p. *248*; 22.6, p. 402
S. V. VLACHOS	(b) **76**, 727 (1976)	6, p. 59; *20.1*, p. *272*; 22.7, p. 411
V. I. VLADIMIROV	(a) **34**, 73 (1976)	*10.1*, p. *79*; 21, p. 291
A. V. VLASOV	(b) **78**, 353 (1976)	12.1, p. 109; 18.1, p. 227; *18.2*, p. *236*; *18.4*, p. *245*; 21.1.1, p. 321
R. M. VLASOVA	(b) **78**, K59 (1976)	*20.1*, p. *276*; 22.8, p. 418
K. M. VAN VLIET	(b) **78**, 501 (1976)	13.1, p. 129; *14.1*, p. *167*; *14.3*, p. *181*; 15, p. 211; 21, p. 300; 22, p. 341
	(b) **78**, 667 (1976)	*14.3*, p. *181*; *15*, p. *211*; 21, p. 300; 22, p. 342
M. VODĂ	(b) **78**, K31 (1976)	*19*, p. *255*; 22.5, p. 390
YU. A. VODAKOV	(a) **35**, 37 (1976)	3, p. 42; *13.4*, p. *138*; 14.3, p. 176; 20.1, p. 268; 20.3, p. 283; 22.3, p. 372
E. VOGT	(a) **34**, 11 (1976)	1.3, p. 21; 18.1, p. 225; *18.2*, p. *228*; *18.4*, p. *242*; 21, p. 289; 21.1, p. 302; 21.6, p. 329
C. VOIGT	(a) **37**, 205 (1976)	*18.2*, p. *235*; 21.1.1, p. 320
F. VOIGT	(b) **73**, 11 (1976)	6, p. 52; *13*, p. *113*; 13.5.4, p. 160; *18.2*, p. *228*

G. Voigt	(a) **35**, 639 (1976)	*1.4*, p. *27*; 14.3, p. 177; 22.1.2, p. 354
	(a) **36**, 173 (1976)	*1.4*, p. *27*; *3*, p. *43*; 22.2.4, p. 370
J. Voigt	(b) **75**, 213 (1976)	*13.5.1*, p. *152*; 20.1, p. 267; 22.4.1, p. 380
	(b) **76**, 313 (1976)	13.5, p. 147; *13.5.1*, p. *153*; *20.1*. p. *270*; 20.3, p. 284; 22, p. 337
	(b) **77**, 233 (1976)	*6*, p. *60*; *13.5.1*, p. *154*; 20.1, p. 272; 22.4.1, p. 381
B. Voigtsberger	(a) **35**, K89 (1976)	*19*, p. *251*; 22.9, p. 436
J. P. Voitchovsky	(b) **75**, 303 (1976)	*13.1*, p. *124*; *20.1*, p. *267*; 22.7, p. 411
G. Vojta	(b) **76**, 695 (1976)	6, p. 59; 8, p. 69; *14.1*, p. *165*
U. Voland	(a) **36**, K165 (1976)	*6*, p. *60*; *8*, p. *70*; 22.1.2, p. 356
T. R. Volk	(a) **33** K137 (1976)	*14.4.1*, p. *206*; *20.1*, p. *263*; 22.8.1, p. 419
G. Völkel	(b) **77**, K31 (1976)	*19*, p. *254*; 22.4.1, p. 381
V. G. Vologin	(a) **33**, K83 (1976)	*18.3*, p. *239*; 22.8.2, p. 427
	(a) **38**, 57 (1976)	11, p. 104; *18.3*, p. *241*; 22.8.2, p. 431
	(a) **38**, K147 (1976)	11, p. 105; *18.3*, p. *241*; 22.8.2, p. 432
E. B. Volzhan	(b) **75**, K69 (1976)	*18.2*, p. *232*
S. V. Vonsovskii	(b) **77**, 9 (1976)	*13*, p. *113*; 14.1, p. 164; 14.2, p. 169; 18.2, p. 228; 18.4, p. 242; 21, p. 289
	(b) **77**, 403 (1976)	*13*, p. *113*; 14.1, p. 164; 14.2, p. 169; 18.2, p. 228; 18.4, p. 242; 21, p. 289
E. van der Voort	(a) **36**, 375 (1976)	*10.2*, p. *94*; 22.6, p. 405
S. A. Vorobev	(b) **76**, 779 (1976)	*11*, p. *103*; 13, p. 118; 22.1.2, p. 355
Yu. K. Voronko	(a) **35**, 645 (1976)	*13.4*, p. *140*; 20.1, p. 269; 20.3, p. 284; 22.8.2, p. 429
Yu. V. Voronov	(b) **78**, 483 (1976)	13.4, p. 145; *20.3*, p. *288*; 22.5.4, p. 401
W. Maenhout-van der Vorst see under letter "M"		
K. Vos	(b) **76**, 769 (1976)	13.1, p. 126; 13.5.1, p. 154; *20.1*, p. *272*; 22.5, p. 389
V. V. Voskoboynikov	(a) **34**, 85 (1976)	*2*, p. *36*; *14.3.1*, p. *183*; 14.3.3, p. 190; 22, p. 335
R. Voszka	(a) **36**, K105 (1976)	*10.2*, p. *94*; 14.4, p. 203; 22.5.2, p. 396
V. I. Vovnenko	(a) **34**, 777 (1976)	*13.4*, p. *138*; *20.3*, p. *282*; 22.2.1, p. 362
S. Vratislav	(a) **37**, K47 (1976)	*1.1*, p. *12*; 22.8.2, p. 430
K. J. de Vries	(a) **35**, 281 (1976)	*10.2*, p. *92*; *14.4*, p. *203*; 22.8, p. 415
Vu Duy Phach	(b) **77**, 381 (1976)	*13.5.1*, p. *155*; *20.3*, p. *286*; 22.5, p. 390
B. E. Vugmeister	(b) **76**, 161 (1976)	*18.1*, p. *226*; *19*, p. *252*
A. Ya. Vul	(a) **36**, 53 (1976)	*14.3.1*, p. *184*; *16*, p. *215*; 22.2.1, p. 363
V. I. Vybornov	(b) **76**, K51 (1976)	*13.5.1*, p. *153*; 22, p. 338
V. B. Vykhodets	(a) **34**, 787 (1976)	*9*, p. *72*; 10.1, p. 81; 21, p. 293; 21.1, p. 306
D. N. Vylegzhanin	(a) **38**, 409 (1976)	*20.2*, p. *279*; *20.3*, p. *288*; 22.8.2, p. 431

A. Wachniewŝki	(b) **76**, K73 (1976)	*18.2*, p. *233*; *18.4*, p. *244*
	(b) **77**, 319 (1976)	*18.2.1*, p. *238*
P. Wachter	(b) **74**, 151 (1976)	*18.4*, p. *242*; *20.1*, p. *263*; 22.9, p. 434
R. H. Wade	(a) **37**, 247 (1976)	*1.5*, p. *32*; *2*, p. *40*; 22.1.1, p. 348
M. Wagner	(b) **75**, 507 (1976)	*6*, p. *57*; 8, p. 68; *13.4*, p. *139*; 22.5.2, p. 395
	(b) **76**, 325 (1976)	6, p. 58; 13.4, p. 140; *20.1*, p. *270*
	(b) **78**, 615 (1976)	4, p. 47; *6*, p. *64*; 22, p. 342
N. Wagner	(b) **74**, 249 (1976)	*13.1*, p. *122*; 17, p. 218; 22.7, p. 410
	(b) **75**, K157 (1976)	*1.1*, p. *12*; *13.1*, p. *124*; 22.7, p. 411
W. E. Wahnsiedler	(b) **73**, 651 (1976)	*13.1*, p. *121*; 17, p. 218; 22.6, p. 403
F. V. Wald	(a) **38**, 253 (1976)	*3*, p. *44*; 10.2, p. 97; 14.3, p. 181; 20.1, p. 277; 22.4.3, p. 386
P. R. Wallace	(b) **74**, 23 (1976)	13.1, p. 122; *14.3*, p. *174*; 22.2.1, p. 362
P. W. Wallace	(a) **38**, 321 (1976)	*9*, p. *74*; 11, p. 104; 22.1.2, p. 358

F. Walz	(a) **37**,	151 (1976)	*18.3*, p. *241*; 22.8.2, p. 430
W. Wanarski	(a) **34**,	697 (1976)	*1.2*, p. *15*; 14.4.1, p. 206; 22.9, p. 435
J. Shy-Yih Wang see under letter "S"			
B. M. Wanklyn	(a) **36**,	307 (1976)	*3*, p. *43*; *19*, p. *253*; 22.8.1, p. 422
G. H. Wannier	(b) **76**,	217 (1976)	*13*, p. *117*
W. Wardzyński	(b) **75**,	341 (1976)	12.2, p. 110; *13.5.1*, p. *152*; *20.3*, p. *283*; 22.4.3, p. 385
T. Warminski	(a) **34**,	K119 (1976)	*10.2*, p. *91*; 22.1.2, p. 353
	(a) **35**,	K9 (1976)	*10.2*, p. *92*; 22.1.2, p. 354
A. V. R. Warrier	(b) **73**,	K123 (1976)	*20.1*, p. *262*; 22.5.3, p. 399
T. Wasiutyński	(b) **76**,	175 (1976)	4, p. 46; *6*, p. *57*; 12.2, p. 110; 22.9, p. 437
Y. Watanabe	(a) **34**,	671 (1976)	*18.3*, p. *240*; *19*, p. *250*; 22.8.2, p. 428
F. E. Wawner	(a) **36**,	K109 (1976)	*20.3*, p. *285*; 22.3, p. 373
A. Weber	(a) **34**,	519 (1976)	*10.2*, p. *91*; 22.8.2, p. 428
F. Weber	(a) **33**,	651 (1976)	*13.3*, p. *132*; *14.3.2*, p. *187*; 14.3.4, p. 196; 22.1.1, p. 345
	(a) **34**,	107 (1976)	14.3.2, p. 187; *14.3.4*, p. *196*; 22.1.1, p. 346
J. Weber	(b) **78**,	699 (1976)	6, p. 64; *13.5.1*, p. *156*; *20.3*, p. *288*; 22.5.1, p. 391
J. Webster	(b) **76**,	661 (1976)	*13.1*, p. *126*; 13.3, p. 133; 20, p. 258; 22.1, p. 344
R. Wedell	(b) **76**,	119 (1976)	*11*, p. *102*
W. Wegener	(b) **78**,	585 (1976)	*13.1*, p. *129*; 22.1, p. 344
R. C. Weger	(b) **74**,	K115 (1976)	2, p. 37; *8*, p. *68*; *18.2*, p. *230*
G. Weill	(b) **76**,	K27 (1976)	*14.3*, p. *178*; 22, p. 338
G. Weiser	(b) **73**,	105 (1976)	13.1, p. 120; *13.5.1*, p. *149*; *20.1*, p. *260*; 22, p. 333
	(b) **73**,	709 (1976)	13.1, p. 121; *20.1*, p. *262*; 22.9, p. 433
	(b) **76**,	817 (1976)	*13.5.1*, p. *154*; 20.1, p. 272; 22.9, p. 437
D. O. Welch	(b) **77**,	277 (1976)	*13*, p. *118*; *14.3*, p. *179*; 22, p. 339
W. Weller	(b) **76**,	K23 (1976)	*13.1*, p. *125*
	(b) **77**,	K79 (1976)	*13.1*, p. *125*
G. Wende	(a) **36**,	557 (1976)	*18.3*, p. *240*; *19*, p. *253*; 22.8.2, p. 430
	(a) **37**,	505 (1976)	*3*, p. *44*; 22.8.2, p. 430
H. Wendel	(b) **78**,	277 (1976)	4, p. 47; *6*, p. *63*; *20.1*, p. *276*; 22, p. 341
M. Wendt	(a) **35**,	347 (1976)	*3*, p. *42*; 22.8.2, p. 429
	(a) **36**,	253 (1976)	*1.5*, p. *32*; 21.1, p. 309; 21.1.1, p. 319; 22.1.1, p. 348
S. Wendt	(b) **77**,	631 (1976)	*1.2*, p. *19*; *14.4.1*, p. *208*
H. Wenzl	(a) **33**,	673 (1976)	*1.2*, p. *14*; *1.4*, p. *25*; 10.1, p. 79; 21, p. 291
W. Werncke	(a) **36**,	K5 (1976)	*6*, p. *58*; 13.5, p. 147; 20.1, p. 271; 20.2, p. 279; 22.8.1, p. 422
H. G. Wessjohann	(b) **77**,	535 (1976)	*13.5*, p. *147*; 14.4, p. 204; 22.1, p. 344
R. M. Westervelt	(b) **74**,	727 (1976)	*13.5.1*, p. *152*; 22.1.1, p. 346; 22.1.2, p. 353
	(b) **76**,	31 (1976)	*13.5.1*, p. *153*; 22.1.1, p. 347; 22.1.2, p. 354
K. H. Westmacott	(a) **35**,	185 (1976)	*10.1*, p. *81*; 21, p. 294; 21.3, p. 324
F. W. de Wette	(b) **76**,	457 (1976)	6, p. 58; *8*, p. *69*
	(b) **78**,	219 (1976)	6, p. 62; *8*, p. *71*
P. W. Whippey	(a) **37**,	51 (1976)	13.4, p. 142; *20.3*, p. *286*; 22.5.2, p. 396
A. M. White	(b) **74**,	K9 (1976)	*13.4*, p. *136*; *13.5.1*, p. *151*; 13.5.3, p. 159; 20.3, p. 281; 22.4.3, p. 385
W. B. Whitten	(b) **74**,	401 (1976)	*13.5.1*, p. *151*; 20.3, p. 281; 22.9, p. 434
R. W. Whitworth	(a) **38**,	299 (1976)	*10.2*, p. *98*; 22.5.2, p. 398
N. J. Wiegman	(a) **35**,	K45 (1976)	*18.2*, p. *232*; 21.1.1, p. 318
F. van de Wiele	(a) **33**,	661 (1976)	*14.3.2*, p. *187*; 14.3.3, p. 190; 22, p. 334
	(a) **36**,	235 (1976)	*14.3.2*, p. *188*; 14.3.3, p. 192; 22.1.2, p. 355
	(a) **38**,	123 (1976)	*14.3.2*, p. *188*; 14.3.3, p. 194; 22.1.2, p. 357
B. Wielke	(a) **33**,	241 (1976)	*10.1*, p. *78*; 21, p. 290; 21.1, p. 302
	(a) **38**,	539 (1976)	*10*, p. *77*

E. Wiener-Avnear	(b) **75**, 717 (1976)	*13.5*, p. *147*; 20.1, p. 269; 22.5, p. 389
H. D. Wiesinger	(a) **36**, 61 (1976)	*1.3*, p. *23*; *14.2*, p. *170*; 21, p. 296
H. Wiesner	(a) **37**, K59 (1976)	*1.3*, p. *23*; *18.2*, p. *233*; 21.1, p. 309; 21.1.1, p. 320
H. W. de Wijn	(b) **76**, 11 (1976)	5, p. 48; *13*, p. *113*; *18*, p. *222*; 19, p. 247; 21, p. 289; 21.1, p. 302; 21.1.1, p. 316; 21.4, p. 326
A. F. Williams	(b) **74**, 183 (1976)	*5*, p. *48*; 22.5.2, p. 393; 22.5.3, p. 399
J. O. Williams	(a) **37**, 271 (1976)	*13.4*, p. *142*; 14.3.3, p. 199; 14.4, p. 204; 22.9, p. 438
P. Williams	(a) **34**, K97 (1976)	*10.1*, p. *81*; 21.1, p. 306
R. H. Williams	(b) **73**, 307 (1976)	*13.1*, p. *120*; *17*, p. *218*; 20.1, p. 260; 22.3, p. 372
W. S. Williams	(b) **75**, 533 (1976)	*13.1*, p. *124*; 20, p. 258; 22.3, p. 373
F. Willig	(a) **33**, 461 (1976)	*14.3.2*, p. *187*; *14.3.4*, p. *196*; 16, p. 213; 22.9, p. 433
F. Willmann	(a) **34**, K173 (1976)	*10.2*, p. *92*; 13.4, p. 138; *20.1*, p. *266*; 22.1.1, p. 346
C. H. Wilts	(b) **73**, 443 (1976)	*18.2.1*, p. *237*; 19, p. 247; 22.8.2, p. 426
J. Windscheif	(b) **74**, K15 (1976)	13.5.3, p. 159; *20.1*, p. *264*; 22.6, p. 403
J. Winter	(b) **74**, 193 (1976)	*1.2*, p. *15*; 5, p. 48; 20.1, p. 263; 22.8, p. 414
H. Wipf	(a) **33**, 525 (1976)	8, p. 67; *9*, p. *72*; 14.1, p. 164; 21.1, p. 304
	(a) **38**, 611 (1976)	*10.1*, p. *86*; 14.1, p. 167; 21, p. 301
R. Wirths	(a) **33**, 135 (1976)	1.3, p. 21; *18.1*, p. *225*; *18.2*, p. *229*; 21.1.1, p. 316
D. Wobig	(b) **75**, K149 1(976)	*6*, p. *57*; *20.1*, p. *269*; 22, p. 337
M. Wöhlecke	(b) **73**, 265 (1976)	*13.4*, p. *134*; *13.5.1*, p. *149*; 22.4, p. 376
	(b) **75**, 137 (1976)	*13.5.1*, p. *152*; *20.1*, p. *267*; 20.3, p. 282; 22.4, p. 376
L. Wojtczak	(b) **77**, K57 (1976)	*18.2* p. *234*
E. Wolf	(a) **38**, 93 (1976)	*1.2*, p. *19*; *3*, p. *44*; 14.3, p. 181; 15, p. 211; 18.1, p. 227; 22.6.1, p. 409
G. Wolf	(a) **36**, K99 (1976)	*8*, p. *69*; 14.2, p. 170; 21, p. 296
	(a) **36**, K125 (1976)	*8*, p. *70*; 14.2, p. 170; 21, p. 297
	(a) **37**, 179 (1976)	*8*, p. *70*; 13.1, p. 127; 21, p. 297
	(a) **37**, 485 (1976)	*8*, p. *70*; 14.2, p. 171; 21, p. 299
M. Wolf	(a) **34**, K199 (1976)	*14.3.4*, p. *197*; 22.6, p. 404
	(a) **37**, K201 (1976)	8, p. 71; *14.3.3*, p. *194*; 14.3.4, p. 200; 22, p. 340
W. G. Wolfer	(a) **37**, 211 (1976)	*10*, p. *76*
E. Wolfgang	(a) **34**, 553 (1976)	*1.2*, p. *15*; *1.3*, p. *22*; 22.7, p. 410
	(a) **35**, 89 (1976)	*1.2*, p. *16*; 1.3, p. 22; *18.2.1*, p. *237*; 22.7, p. 411,
C. Wölk	(b) **73**, 255 (1976)	13.4, p. 134; *13.5.1*, p. *149*; 20.1, p. 260; *20.3*, p. *280*; 22.2.3, p. 367
J. Woltersdorf	(a) **34**, 183 (1976)	*1.4*, p. *26*; *1.5*, p. *31*; 21.6, p. 330
	(a) **37**, 457 (1976)	1.4, p. 28; *1.5*, p. *32*; *3*, p. *44*; 21, p. 298; 22.6, p. 407
B. Y. Woo	(a) **34**, 263 (1976)	*14.3.2*, p. *187*; *14.3.3*, p. *190*; 14.3.4, p. 196; 22.1.2, p. 352
T. Wosiński	(a) **37**, K57 (1976)	*10.2*, p. *95*; *16*, p. *216*; 19, p. 254; 22.1.2, p. 356
Shi-Yu Wu see under letter "S"		
W. Wüchner	(b) **74**, 297 (1976)	8, p. 67; *18.4*, p. *242*; 20.1, p. 264; 22.8.1, p. 420
H.-J. Wünsche	(b) **73**, 245 (1976)	13.1, p. 120; *14.3.3*, p. *190*; 14.3.4, p. 195; 22.2.1, p. 362
	(b) **76**, 639 (1976)	*13.5.1*, p. *154*; *20.1*, p. *272*; 22, p. 339
M. Wurlitzer	(a) **33**, K55 (1976)	*18.3*, p. *239*; 22.8.2, p. 426
W. Wycisk	(a) **37**, 183 (1976)	*10.1*, p. *83*; 14.1, p. 166; 21.1, p. 311
K. I. Wysokiński	(b) **77**, K139 (1976)	*14.1*, p. *167*

Yu. V. Yablokov	(b) **78**, K69 (1976)	1.2, p. 20; *6*, p. *64*; *13.4*, p. *145*; 19, p. 256; 22.8, p. 418
A. Ya. Yablonskii	(a) **35**, 645 (1976)	*13.4*, p. *140*; 20.1, p. 269; 20.3, p. 284; 22.8.2, p. 429
V. Yakhot	(b) **76**, 451 (1976)	*6*, p. *58*; 23, p. 440
A. A. Yakovenko	(a) **36**, K51 (1976)	*16*, p. *215*; 22.8, p. 416
S. Yamada	(a) **37**, K157 (1976)	13.4, p. 144; *20.3*, p. *287*; 22.3, p. 374
M. Yamamoto	(a) **33**, 625 (1976)	1.3, p. 21; *18.1*, p. *225*; *18.2*, p. *229*; 21.1, p. 304
R. Yamamoto	(b) **73**, K103 (1976)	*10.1*, p. *79*; 21, p. 291
	(b) **74**, 159 (1976)	*10.1*, p. *79*; 21, p. 291
	(b) **75**, 279 (1976)	*13*, p. *116*; *18*, p. *223*
T. Yamamoto	(a) **35**, 437 (1976)	*18.2*, p. *232*; 21.1, p. 308; 21.1.1, p. 319
	(a) **37**, 607 (1976)	*18.2*, p. *235*; 21.1, p. 312
	(a) **38**, 361 (1976)	*18.2*, p. *236*; 21.1.1, p. 321
K. Yamanaka	(a) **37**, K157 (1976)	13.4, p. 144; *20.3*, p. *287*; 22.3, p. 374
E. Yang	(b) **78** 803 (1976)	*13*, p. *119*
I. K. Yanson	(a) **33**, 53 (1976)	*14.2*, p. *169*; 21, p. 289
	(a) **35**, 49 (1976)	*14.3.3*, p. *191*; 22.6, p. 404
H. Yatom	(b) **74**, 647 (1976)	*13.5.2*, p. *157*; *20.1*, p. *265*; 21, p. 292
J. H. Yee	(b) **77**, K81 (1976)	*13.1*, p. *128*; 22.1.1, p. 349; 22.1.2, p. 356; 22.2, p. 361; 22.4, p. 377
Kwok Leung Yip see under letter "K"		
Y. Yokoyama	(b) **75**, 483 (1976)	10.2, p. 93; *11*, p. *101*; *20.1*, p. *268*; 22.5.2, p. 395
K. Yonemitsu	(b) **73**, 723 (1976)	13.2, p. 130; *14.1*, p. *164*; 21, p. 290
	(a) **36**, 791 (1976)	13.2, p. 131; *14.1*, p. *166*; 21, p. 297; 21.6, p. 331
M. H. Yoo	(b) **77**, 181 (1976)	*9*, p. *74*; *10.1*, p. *83*; 21, p. 297; 21.1, p. 310
A. Yoshihara	(a) **34**, 369 (1976)	*14.4.1*, p. *206*; 22.8.1, p. 420
D. M. Yudin	(b) **77**, K27 (1976)	*2*, p. *39*; 11, p. 103; *19*, p. *254*; 22.8.1, p. 423
G. F. Yuldashev	(a) **35**, 37 (1976)	3, p. 42; *13.4*, p. *138*; 14.3, p. 176; 20.1, p. 268; 20.3, p. 283; 22.3, p. 372
M. S. Yunusov	(a) **35**, K145 (1976)	*11*, p. *102*; 22.1.2, p. 354
	(a) **37**, 681 (1976)	9, p. 74; *11*, p. *104*; *14.3.2*, p. *188*; 22.1.2, p. 357
	(a) **38**, 439 (1976)	*9*, p. *74*; *11*, p. *105*; 14.3.2, p. 189; 22.1.2, p. 358

Yu. V. Zabara	(a) **38**, K131 (1976)	*14.3.4*, p. *201*; 22.8.1, p. 425
H. Zabel	(a) **37**, K67 (1976)	*1.2*, p. *18*; 1.3, p. 23; 10.1, p. 84; 21, p. 298
W. Zacharko	(b) **76**, K1 (1976)	*14.2*, p. *170*
M. A. Zaikovskaya	(a) **35**, K145 (1976)	*11*, p. *102*; 22.1.2, p. 354
V. I. Zaitsev	(a) **38**, 447 (1976)	*1.4*, p. *29*; *10.1*, p. *86*; 21.1, p. 314
J. Zajíček	(a) **37**, K47 (1976)	*1.1*, p. *12*; 22.8.2, p. 430
A. P. Zakharchuk	(b) **75**, K51 (1976)	13.5.1, p. 152; *20.1*, p. *267*; 22, p. 336
M. I. Zakharova	(a) **37**, K121 (1976)	*1.2*, p. *19*; *1.3*, p. *24*; 21.1, p. 313
T. Zaleski	(a) **35**, K69 (1976)	*18.2*, p. *232*; *18.4*, p. *244*; 22.8, p. 415
A. V. Zalesskii	(b) **73**, 317 (1976)	*18.4*, p. *242*; *19*, p. *247*; 22.8.2, p. 426
V. M. Zaletin	(b) **77**, 465 (1976)	*13.5.1*, p. *155*; *20.1*, p. *274*; 22.2.1, p. 364
M. M. Zaripov	(b) **75**, 731 (1976)	*19*, p. *251*; 22.5.3, p. 399
	(b) **78**, K69 (1976)	1.2, p. 20; *6*, p. *64*; *13.4*, p. *145*; 19, p. 256; 22.8, p. 418
V. S. Zasimov	(a) **38**, K45 (1976)	*5*, p. *51*; 22.1, p. 344
M. Závětová	(a) **35**, K77 (1976)	*2*, p. *38*; *20.1*, p. *268*; 22.8, p. 415
S. Zawisz	(b) **77**, 367 (1976)	*13.5.2*, p. *157*; *18.2*, p. *234*
S. Zazubovich	(b) **78**, 813 (1976)	*10.2*, p. *98*; *20.3*, p. *288*; 22.5.2, p. 398
Z. Zboiński	(b) **74**, 561 (1976)	*14.3*, p. *175*; 22.9, p. 435
G. G. Zdorovtseva	(a) **35**, 779 (1976)	*1.2*, p. *17*; *1.5*, p. *31*; 11, p. 102; 21, p. 295; 21.1, p. 308; 21.1.1, p. 319; 22.1.1, p. 347

J. J. Żebrowski	(a) **37**, 407 (1976)	*18.2.1*, p. *238*
V. P. Zenchenko	(b) 77, K39 (1976)	*18*, p. *224*; *20.1*, p. *273*
V. P. Zentsov	(b) 75, K111 (1976)	*19*, p. *251*; 22.5.3, p. 400
	(b) 77, K77 (1976)	*19*, p. *251*; 22.5.3, p. 400
G. Zerweck	(b) 78, K15 (1976)	*14.2*, p. *172*
I. S. Zheludev	(b) 73, 317 (1976)	*18.4*, p. *242*; *19*, p. *247*; 22.8.2, p. 426
A. I. Zhernovoy	(b) 74, K79 (1976)	*19*, p. *249*
V. A. Zhikharev	(b) 74, K35 (1976)	*19*, p. *249*
E. A. Zhilinskaya	(b) 73, 65 (1976)	2, p. 35; 18, p. 222; *19*, p. *247*; 22.8, p. 413
R. A. Zhitnikov	(b) 76, K109 (1976)	*19*, p. *253*; 22.5.2, p. 396
P. G. Zhumatii	(b) 75, 61 (1976)	*14.3*, p. *176*; *15*, p. *209*; 22, p. 336
A. M. Ziatdinov	(b) 78, K69 (1976)	1.2, p. 20; *6*, p. *64*; *13.4*, p. *145*; 19, p. 256; 22.8, p. 418
A. Zięba	(a) **35**, K69 (1976)	*18.2*, p. *232*; *18.4*, p. *244*; 22.8, p. 415
	(a) **38**, 721 (1976)	*18.4*, p. *246*; 22.8, p. 418
E. Ziegler	(a) **35**, 269 (1976)	6, p. 56; 13.4, p. 139; *14.3*, p. *176*; 22.8, p. 415
	(a) **36**, 491 (1976)	*6*, p. *59*; *14.3*, p. *179*; 22.8, p. 417
K. Ziegler	(a) **37**, K97 (1976)	*14.3*, p. *179*; *15*, p. *211*; 22, p. 340
	(a) **37**, K147 (1976)	13.4, p. 143; *16*, p. *216*; 22, p. 340
	(b) 78, K23 (1976)	13.1, p. 128; *20.1*, p. *276*; 22, p. 341
M. Ziegler	(b) 77, 153 (1976)	*6*, p. *60*; 20.1, p. 272; 22.6, p. 406
J. P. Zielinger	(a) **33**, 155 (1976)	13.4, p. 135; *16*, p. *213*; 22.6, p. 402
E. Zielińska-Rohozińska	(a) **35**, K1 (1976)	*1.4*, p. *27*; 10.2, p. 92; 22.1.2, p. 354
J. Zieliński	(b) 74, 437 (1976)	13.4, p. 137; *14.1*, p. *165*; 21, p. 292
	(b) 74, 779 (1976)	*13*, p. *115*; *18.2*, p. *230*; 21.1, p. 305
	(b) 78, 319 (1976)	*13.1*, p. *128*; *18.2*, p. *236*; 21.1, p. 313
R. Zieliński	(a) **34**, 657 (1976)	*14.3*, p. *175*; 22.9, p. 435
P. Ziesche	(b) 78, 169 (1976)	*13.1*, p. *128*; 13.5.2, p. 158; *14.1*, p. *167*; 14.4, p. 204; 20.1, p. 275; 21.3, p. 324
V. E. Zilbervarg	(b) 74, 199 (1976)	*13.1*, p. *122*; *13.5.4*, p. *160*; 18.3, p. 239; 22.8.2, p. 427
G. Zimmerer	(b) 73, K99 (1976)	*13.5.1*, p. *150*; *20.3*, p. *280*; 23, p. 440
	(b) 74, 579 (1976)	*13.5.1*, p. *151*; *20.3*, p. *282*; 23, p. 440
M. Zimmermann	(a) **37**, 179 (1976)	*8*, p. *70*; 13.1, p. 127; 21, p. 297
	(a) **37**, 485 (1976)	*8*, p. *70*; 14.2, p. 171; 21, p. 299
R. Zimmermann	(b) 75, 633 (1976)	*13.5*, p. *147*; 13.5.2, p. 157; *20.3*, p. *284*; 22.4.1, p. 380
	(b) 76, 191 (1976)	*13.5.1*, p. *153*; 13.5.2, p. 157; 22, p. 337
V. I. Zinenko	(b) 78, 721 (1976)	*1.2*, p. *20*; *14.4*, p. *205*; 22.9, p. 439
J. Zirke	(b) 75, K149 (1976)	*6*, p. *57*; *20.1*, p. *269*; 22, p. 337
J. Zmija	(a) **37**, K75 (1976)	*20.1*, p. *273*; 22.8, p. 417
M. Zoaeter	(a) **38**, 505 (1976)	*14.3.4*, p. *201*; 22, p. 342
V. V. Zolotarev	(b) 76, K31 (1976)	*13.5.4*, p. *160*; 18.2, p. 233
E. I. Zorin	(a) **35**, 11 (1976)	11, p. 100; *13.4*, p. *134*; *14.3*, p. *173*; 22.1.1, p. 345; 22.1.2, p. 351
	(a) **36**, 81 (1976)	*1.5*, p. *31*; 2, p. 39; *11*, p. *102*; 22.1.2, p. 355
	(a) **37**, 57 (1976)	*11*, p. *103*; 22.1.2, p. 356
E. Zsoldos	(a) **34**, 519 (1976)	*10.2*, p. *91*; 22.8.2, p. 428
W. Żurek	(b) 74, K165 (1976)	*5*, p. *49*
M. Zvára	(b) 75, K45 (1976)	*16*, p. *214*; *20.1*, p. *267*; 22.8.2, p. 428
A. K. Zvezdin	(b) 73, 317 (1976)	*18.4*, p. *242*; *19*, p. *247*; 22.8.2, p. 426
D. Zwingel	(b) 77, 171 (1976)	*19*, p. *253*; 22.6, p. 406

19. Magnetic Resonance

20. Optical Properties
    20.1 Spectra. Optical Constants (X-Ray Spectra See 20)
    20.2 Lasers
    20.3 Luminescence (X-Ray Spectra See 20)

21. Metals. Alloys
    21.1 Transition Metals (Group Sc to Zn) and Their Alloys
        21.1.1 Fe and Fe Alloys
    21.2 Alkali Metals
    21.3 Alkali-Earth Metals
    21.4 Rare-Earth Metals
    21.5 Actinides
    21.6 Noble Metals
    21.7 Semi-Metals (Semiconducting Alloys See 22.7)

22. Semiconductors and Ionic Crystals
    22.1 Elements
        22.1.1 Germanium
        22.1.2 Silicon
        22.1.3 Group VI Elements
    22.2 III–V Compounds
        22.2.1 Arsenides
        22.2.2 Phosphides
        22.2.3 Antimonides
        22.2.4 Mixed Crystals
    22.3 IV–IV, II–V, and III–VI Compounds
        22.3.1 Mixed Crystals
    22.4 II–VI Compounds
        22.4.1 Sulphides
        22.4.2 Selenides
        22.4.3 Tellurides
        22.4.4 Mixed Crystals
    22.5 Halides
        22.5.1 Silver Halides
        22.5.2 Alkali Halides
        22.5.3 Alkali-Earth Halides
        22.5.4 Mixed Crystals
    22.6 Simple Oxides (CdO, ZnO See 22.4)
        22.6.1 Mixed Crystals
    22.7 Semiconducting Intermetallic Compounds
    22.8 Tertiary and Higher Compounds
        22.8.1 Salts of Oxy-Acids (Carbonates, Phosphates, Silicates Including Mica etc.)
        22.8.2 Spinels, Garnets, Ferrites
    22.9 Organic Semiconductors

23. Solidified Gases

For comments on the use of the classification scheme see p. 6.

Manuscripts and letters for physica status solidi (b) — basic research and physica status solidi (a) — applied research

should be sent to

DDR-102 Berlin, Neue Schönhauser Str. 20

or

Internationaler Buchversand GmbH, 1 Berlin 19, Eschenallee 21.

The object of "physica status solidi (b)" and "physica status solidi (a)" is to publish rapidly important and topical results concerning fundamental research in the field of solid state physics and aspects of application and experimental techniques of fundamental solid state research, respectively. The issues of "physica status solidi (b)" and "physica status solidi (a)" will be published on the 1st and the 16th day of each month, respectively. Either series will contain the following categories:

1. Review articles written by arrangement with the editors. They will be remunerated.
2. Original papers selected from those which have not been submitted for publication elsewhere, after being reviewed. (The latest date for acceptance of contributions is the 8th and the 23rd day of each month for "physica status soldi (b)" and "physica status solidi (a)", respectively, 50 days before the delivery of the journal.)
3. Short notes published in the off-set part of the journal. They are, if necessary after being reviewed, selected by the editorial staff. (The latest date for acceptance of manuscripts for this part is the 12th and the 27th day of each month for "physica status solidi (b)" and "physica status solidi (a)", respectively, 16 days before the delivery of the journal.)
4. Pre-printed titles of papers to be published in the next issue of both series.

Editorial Information

Manuscripts should be submitted preferentially in English. Besides, papers in German, French, and Russian are admitted. The final decision on the incorporation of the papers in the series (b) or (a) is taken by the editorial staff after consultation of the referee. Authors are asked to observe the following rules:

1. Manuscripts have to be written as concisely as possible. As most, 20 and 4 pages (28 lines per page; abstracts, tables, references, figures, and captions of figures included, where 3 figures are counted as about one page) are admitted for original papers and short notes, respectively. All manuscripts should be type-written on one side only. Two copies are required. Badly arranged manuscripts are returned.
2. The author's name, initials, and institution should be given.
3. One abstract of about ten lines in English and one in German, French, or Russian are required for original papers.
4. Because of lack of space conclusions (if any) must not have the character of a summary (i. e. substantially a repetition of the contents of the abstract).
5. Formulae should not be written too small and with pencil. Separate lines for formulae are desirable. Indistinct writing will lead to a return of the manuscript.
6. Drawings should be enclosed on separate sheets of paper, not exceeding the maximum and minimum sizes of 20×30 and 10×15 cm^2, respectively. Photographs should be submitted, on white paper, only for review articles and original papers. Duplicates are required.
7. Captions for the figures should be submitted on separate sheets of paper, and special instructions with regard to the size of illustrations should be given on these.
8. References should be numbered in order of appearance and should be given, on a separate sheet, in the form used in physica status solidi.
9. Units, symbols, etc. should be used in accordance with the recommendations given by the IUPAP.
10. Each paper should contain a proposal for the subject classification according to the classification scheme appearing on the second and third cover of each issue.

The authors will be sent galley proofs of review articles and original papers. Proof reading of short notes and pre-printed titles will be done by the editorial staff.

For each original paper, and short note, a total of 75 reprints is available free of charge. The number of reprints free of charge per review article is 25. Requests for additional reprints (against payment) should be given together with the submitted manuscript. **The deadline for orders of additional reprints is the 15th and the 30th day of each month** for "physica status solidi (b)" and "physica status solidi (a)", respectively, two weeks before the delivery of the journal.

Bei Fragen zur Produktsicherheit wenden Sie sich bitte an:
If you have any questions regarding product safety,
please contact:

Walter de Gruyter GmbH
Genthiner Straße 13
10785 Berlin
productsafety@degruyterbrill.com